The Notodontidae of the Indonesian Archipelago (Lepidoptera)

The Notodontidae of the Indonesian Archipelago (Lepidoptera)

Volume 1

By

Alexander Schintlmeister

TUTA SUB AEGIDE PALLAS · 1683 ·

BRILL

LEIDEN | BOSTON

Cover illustrations: (front) *Tarsolepis malayana* Nakamura, 1976. Borneo, Sabah, Crocker Range, 10. ix. 2007. (Photo: Kenichiro Nakao); (back) *Omichlis rufofasciata* Bethune-Baker, 1904. West Papua, Keerom district, Tekai camp near Usku, 25. ii. 2007. (Photo: Michael Thirnbeck).

Library of Congress Cataloging-in-Publication Data

Names: Schintlmeister, Alexander, author.
Title: The notodontidae of the Indonesian archipelago (lepidoptera) / by Alexander Schintlmeister.
Description: Leiden ; Boston : Brill, [2020] | Includes bibliographical references and index.
Identifiers: LCCN 2020006066 (print) | LCCN 2020006067 (ebook) | ISBN 9789004354241 (volume 1 ; hardback) | ISBN 9789004354258 (volume 1 ; ebook)
Subjects: LCSH: Lepidoptera--Indonesia. | Lepidoptera--Malay Archipelago. | Malay Archipelago.
Classification: LCC QL556.I5 S35 2020 (print) | LCC QL556.I5 (ebook) | DDC 595.7809598--dc23
LC record available at https://lccn.loc.gov/2020006066
LC ebook record available at https://lccn.loc.gov/2020006067

Typeface for the Latin, Greek, and Cyrillic scripts: "Brill". See and download: brill.com/brill-typeface.

ISBN 978-90-04-35424-1 (hardback)
ISBN 978-90-04-35425-8 (e-book)

This book is printed on acid-free paper and produced in a sustainable manner.

Contents

Introduction

What is known of Indonesian moths is currently limited. Even so, the literature is voluminous but restricted often to merely brief descriptions of new species without illustrations. There are a few older publications, by Van Eecke, 1930 (de Heterocera van Sumatra) or Gaede in Seitz, 1930 (Die Großschmetterlinge der Erde), providing colored illustrations. These treat only small fractions of Indonesian moths, and the presented identifications and synonyms are often proven erroneous.

More recent publications, such as Barlow, 1982 (An Introduction to the Moths of South East Asia), Holloway, 1983 (Notodontidae in the series Moths of Borneo), or several publications of the Heterocera Sumatrana Society e.V., e.g. Schintlmeister, 1994 greatly expanded our knowledge of the fauna of Sundaland, i.e., the Malay Peninsula, Borneo, and Sumatra. The Notodontidae of the Philippines (Schintlmeister & Lourens, 2010), Vietnam (Schintlmeister, 1997) and Thailand (Schintlmeister & Pinratana, 2007) have already been published. A taxonomic-faunistic study of the Notodontidae of the Indonesian archipelago, in particular its eastern parts, is missing. Such a work is important for a more complete view of Oriental Notodontidae and provides base data for further phylogenetic or biogeographic studies.

The purpose of this book is to enable the user to identify most of the Oriental Notodontidae.

Summary

This is the first volume of a two-volume work on the Notodontidae or prominent moths of the Indonesian archipelago. The study area covers also the territories of Malaysia, Brunei and Papua New Guinea. The present volume deals with 298 species of the subfamilies Dudusinae, Scranciinae, Ceirinae, Cerurinae and Dicranurinae. All of them are illustrated in both sexes as adults in color and in their genitalia. A distribution map (by dots) and bionomic data to each species are provided. In several taxonomic notes 64 species and 22 subspecies are described as new to science and five new genera are errected. From syntypes series 20 lectotypes are designated to stabilize the nomenclature. 35 new synonymies are recognized and nine new combinations are introduced. The status of 20 taxa has been changed in the present work.

Acknowledgements

The author is grateful for the loan of specimens, permission to dissect their genitalia, provision of genitalia slides for scanning, and permission to illustrate images of specimens and genitalia taken from their holdings to: Carnegie Museum of Natural History, Pittsburgh (John Rawlins), Florida Museum of Natural History, Gainesville (Dr. A. Sourakov, Dr. J. Heppner), Museum Witt, Munich (Dr. h.c. Thomas Witt†, Dr. Wolfgang Speidel, Harry Sulak), National Museum of Natural History, Washington D.C. (Dr. Scott Miller, Dr. Don Harvey, Dr. Michel G. Pogue, Dr. P. Goldstein), Naturhistorisches Museum, Vienna (Dr. Martin Lödl, Dr. Sabine Gaal), the Natural History Museum UK, London (Dr. Geoff Martin, Dr. Martin Honey, Dr. Alberto Zilli), Naturalis Biodiversity Center Leiden (Rob de Voss), Oxford University Museum of Natural History (Dr. James Hogan), Staatliche Museum für Naturkunde Karlsruhe (Dr. Robert Trusch), Staatliches Museum für Naturkunde Stuttgart (Dr. Rajaei Hossein, Daniel Bartsch), Zentrum für Biodokumentation Entomologie, Reden (Andreas Werno), Zoologische Staatssammlung, Munich (Dr. Axel Hausmann, Ulf Buchsbaum), Museum für Naturkunde – Leibniz-Institut für Evolutions- und Biodiversitätsforschung an der Humboldt-Universität zu Berlin (Dr. Wolfram Mey).

Tibor Csövarie, Budapest and Jan-Peter Rudloff, Roßleben, mounted and scanned many genitalia, which were used for this publication.

The author is also much obliged for fruitful discussion, information, technical support, material of Indonesian Notodontidae to Dr. Jeremy D. Holloway, Siegfried Ihle, Filderstadt, Stanley Jakl, Prague and Jakarta, Dr. Wolfram Mey, ZMHU, Berlin, Dr. Scott E. Miller, US National Museum, Washington D.C., Dr. Wolfgang Nässig, Forschungsinstitut und Naturmuseum Senckenberg, Frankfurt/Main, Danny Nielsson, Kalvehave, Michal Rindos, Institute of Entomology, Ceske Budejovice, Dr. Wolfgang Speidel, ZSM and MWM, Munich, and Rob de Voss, NBC, Leiden.

Rob de Voss provided a gazetteer of W Papua with geographical coordinates and altitudes (database), which was very helpful for tracing collecting sites.

Special thanks go to William Clapp, Dresden, for linguistic corrections.

The Family Notodontidae

The family Notodontidae, or "Prominent Moths", have an almost worldwide distribution, being absent from the Arctic regions and New Zealand, and comprise some 4500 described species in about 550 different genera (Schintlmeister, 2013). Schintlmeister, 2008 classified the Palaearctic Notodontidae into 10 subfamilies, all of which occur in Indonesia. This classification is followed and supplemented here.

The Notodontidae are part of the superfamily Noctuoidea, which is the largest superfamily of the Lepidoptera and includes more than 80,000 described species. Noctuoidea are characterized by the presence of a metathoracic tympanal organ in the adults and the presence of two MD setae on the larval metathorax (Kristensen 1998: 355). Antennae are in most males bipectinated with long rami, but a part toward the antenna tip remains filiform. Female antennae are bipectinated with shorter rami than in the males or filiform with ciliae. The forewing pattern is noctuoid: basal, postmedial and marginal fasciae as well as a discal spot and reniform stignum are present in most species. Hindwings are usually uniformly brown with a medial shadow. Most groups display costal and anal markings. Male genitalia are characterized by the presence of a pair of socii and often by a gnathal process. The saccular area of the valvae is membranous, but in some genera strongly sclerotized and well developed. The phallus (= aedeagus) often displays modifications at the tip, such as spines and hooks. The everted endophallus (= vesica) bears often moveable stellate spicules, which are sometimes seen in the corpus bursae after copulating. A carina and other sclerotized structures are regularly present in the endophallus. If the endophallus is fully everted, its shape is of taxonomic value, in particular in cases of closely related species. 8th abdominal segments are in Notodontidae in most genera modified, bearing lobes, spines and projections and also specifically sclerotizations. Female genitalia have two pairs of apophyses. Some groups display a large, strongly sclerotized postvaginal plate of specific shape. Many species bear a signum on corpus bursae as seen also in other families.

Notodontidae larvae are in most species oligophagous. Groups of more closely related species feed usually in the same plant families. Species of the subfamily Ceirinae Matsumura, 1929 are restricted to monocotyledons, mainly palms and bamboos. Caterpillars of Cerurinae Butler, 1881 and Pygaerinae Duponchel, 1845 have host plants in Salicaceae and Flacourtiaceae. Most Dicranurinae Duponchel, 1845 feed on Fagaceae, but occasionally Rosceae, Juglandaceae, Ulmaceae, Styracaceae and other plant families are among their host plants. For most Oriental notodontids, however, the larvae and their host plants are still unknown. Only a very few species in the Notodontidae are in Indonesia known for their economic importance in the defoliating of fruit trees, e.g. *Tarsolepis* for the Rambutan tree (Nephelium lappaceum, Sapindaceae).

Methodological Notes

The study area covers the whole territory of Indonesia and also includes the territories of Malaysia, Brunei and Papua New Guinea.

The arrangement of the species is as follows:

The number of each species corresponds with the numbers in the checklist and on the plates. The author and the year of the description are cited according to current knowledge. Square brackets are not used to indicate differences between the true year of publication and those imprinted in the title of the original publication.

The original combination of the description of a taxon is given in brackets. The next row listed the status of the primary type (holotype, lectotype) and its type locality, because taxonomy is basing on name-bearing types.

If a taxon was not described from a holotype by monotypy or by original designation, lectotypes have been designated to stabilize the nomenclature. Species group taxa are listed in their original spellings regardless of the gender of the current generic combination. The type locality is given in the spelling of the original description. The country (actual state) is added in square brackets [ ] if not given in the original description. The type depository, where a primary type is actually stored, or where it was last deposited is given. In most cases the author was able to verify the type depository by personal examination.

In the following row references are given, if a species was treated also in the monograph of the Palearctic Notodontidae (Schintlmeister, 2008) which provides further information on taxonomy, bionomics and distribution. The second reference mentions on the *World Catalogue of Insects, Notodontidae & Oenosandridae* (Schintlmeister, 2013), which gives information on nomenclatural issues of a species, e. g. method of designation of the primary types, but includes also lists with full references and type localities of the synonyms.

Taxonomic notes deal with the designation of lectotypes and taxonomical or nomenclatural changes, such as new synonymies or status changes. Subspecific context, sometimes with descriptions of new subspecies, is also discussed in this section.

The **Diagnosis** is not a full description of the species. Together with the illustrations of imagines and their genitalia it should enable the reader to recognize an insect at the species level only. This section includes short notes on individual and geographical variation.

Bionomics lists the frequency and habitats of occurrence of the adults. Usually altitudes of occurrence for the moths are given. Where habitat preferences, host plants of larvae, or other details relating to the life history are known, they are mentioned.

The **Distribution** section is accompanied by a map, which illustrates the entire distribution of the species in the study area. This would include also distribution dots in Australia or Thailand. The distribution dots refer predominately to specimens examined by the author. Literature references are used very critically and in cases where the risk of misidentification of species seems to be high, or in other doubtful cases, such references are not used for mapping. It should be noted, that an empty area does not automatically mean that the species is absent from the region.

The color plates are derived from single photographs, which were photographed according to the method of Eckweiler, 2001 with a neutral grey background (40% black). It is the intention of the author to illustrate a male and a female of each taxon at species and subspecies level. The number of the species refers to the number in the checklist and the systematic section. The origin of the illustrated specimen is given only approximately, e.g. country or island, as in the legends of the genitalia plates. All illustrated imagines are 110% of the natural size. Genitalia figures may differ in magnifications between various genera, but in such cases a scale bar is provided on the plates.

Material and Its Collectors

Material has been studied from the following sources:

Among the oldest known Indonesian notodontids, which are still preserved, there are Javan insects collected by Thomas **Horsfield** (1773–1859) for the East Indian Company. His results were published by Horsfield & Moore, 1858 and 1860 as a two-volume catalogue. The collection of the East Indian Company is now in the holdings of The Natural History Museum, London (NHMUK).

Antwerp Edgar **Pratt** [1852-1920] travelled together with his son Henry in 1901 for the first time in British New Guinea in the Owen Stanley Range (Mt. Kebea, Dinawa, Aru River and other localities) for Lord Lionel Walter Rothschild. In 1905 and 1907 A. E. Pratt collected with two further of his sons, Felix and Charles, in Dutch New Guinea (Fak Fak, Arfak Mts.); in 1910 and 1911 the Pratt's visited in the Cyclops Mountains in northern New Guinea and spent six months on the island of Biak. From 1914 Pratt collected exclusively for James John Joicey (Hill Museum; the collections were given 1929 to NHMUK) in W Papua: Arfak Mts., Wandammen Mts. and in 1915 in Waigeu. In 1919 the three Pratt brothers, Charles, Felix and Joseph set off on new collecting adventures. They visited the island of Ceram (Mt. Manusela) in 1919 and 1920 and then moved on to New Guinea (Weyland Mts., Menoo River and Mt. Kunupi). 1921 the Pratt brothers collected in South-west Sumatra and in 1922 in Buru. The Pratts are doubtless the most successful insect collectors in Indonesia in regard to quantity and quality.

Albert Steward **Meek** (1871–1943) collected for Lord Rothschild from 1894 until 1923 in New Guinea (Owen Stanley Range, Mt. Goliath) and many surrounding islands (Vulcan Isl., Goodenough Isl., Rook Island, New Britain, New Hannover, Admirality Isl., Dampier Isl.)

From 1924 and 1925 Meek sent his brother-in-law, Albert Frederic **Eichhorn** (?–1931), on expeditions to New Britain, New Hannover and the Solomon Isls.

A. Eichhorn and his brother George spent also five months in the remote Hydrographer Mts. in British New Guinea. All this material was sent to Lord Rothschild at Tring and is now in the holdings of NHMUK.

Evelyn Lucy **Cheesman** (1882–1969) explored on four solo expeditions between 1933 and 1939 on Mt. Kokoda (S Papua New Guinea), Cyclops Mts. (W Papua), Yapen Isl. and Waigeu Isl. and collected huge amounts of material. From Yapen Island in the course of four months she collected 52,000 insects for the British Museum, now NHMUK.

Jorinus Pieter Adrianus **Kalis** (1899–1949) was a very successfull Dutch professional collector in Indonesia. He explored from 1931 Eastern Java for E. van Delden. From 1932–1938 he collected Lepidoptera in long series in Java, Bali, Lombok, Sulawesi and the Salayer Isl. mainly for Lord Rothschild.

Maria Ernestine **Walsh** (1881–1973) was a resident in Java from 1911 and made entomological and botanical

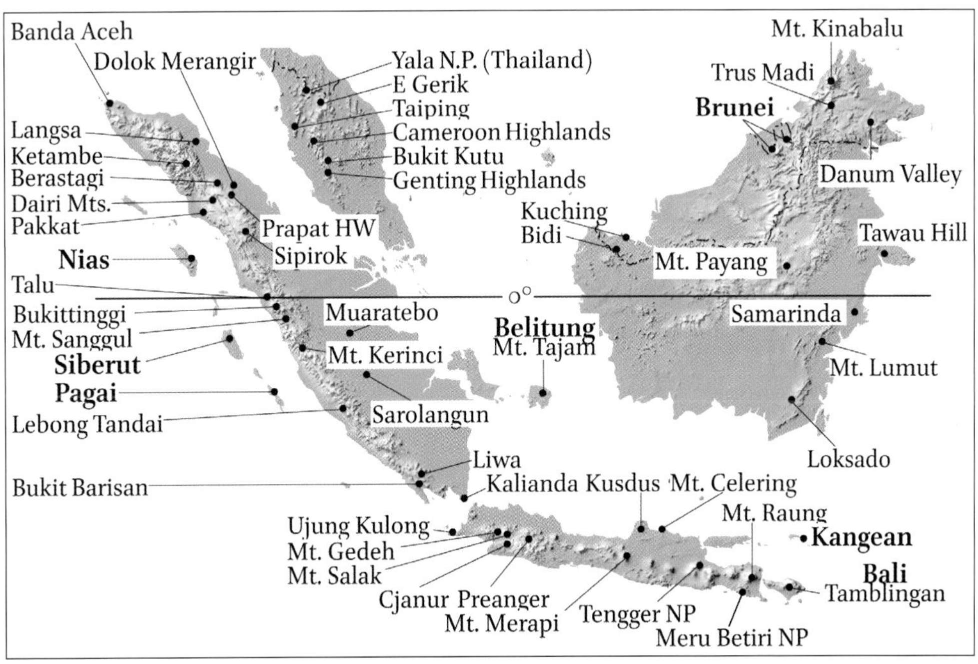

FIGURE 1
Important localities in the Malayan Peninsula, Borneo, Sumatra, and Java. Island names are printed in bold.

collections in Java, Borneo, and S Sumatra for NHMUK and NBC.

From within the holdings of NHMUK important collections used for this publication, are also the entire material collected and used by Jeremy D. **Holloway**, for the series *Moths of Borneo*, in particular from Mt. Kinabalu and Brunei). Holloway collected valuable material for „Project Wallace" of the Royal Entomological Society of London 1985 in N. Sulawesi (Dumogo-Bone National Park) and for „Operation Raleigh" 1987 in Seram (Mt. Manusuela).

Henry **Barlow** collected and observed moths in the Genting Highlands in Malaysia permanently over the past 40 years. The author visited him in 1981 and studied his collection. Barlow has transferred parts of his collection to NHMUK.

In 2010, three Dutch natural history collections were merged to form the Naturalis Biodiversity Center Leiden (NBC): the Nationaal Natuurhistorisch Museum (formerly Rijksmuseum van Natuurlijke Historie, RMNH), the Zoölogisch Museum Amsterdam and the Wageningen University collections. With the latter the important collection curated by Walter Karl Johann **Roepke** (1882–1961) came to NBC. It includes material collected by Mrs. M. E. Walsh, J. P. A. Kalis, and W. Roepke mainly in Java, Sumatra and Sulawesi.

Johan Marie Alexander van **Groenendael** (1896-1980) came as a physician in 1931 to Java. He collected by himself butterflies and moths in Java, Bali Sumatra and Flores, but he bought Indonesian and Australian moths in huge numbers from J. P. A. Kalis, W. Stüber and others. The van Groenendael collection contains about 800.000 Lepidoptera, but most of it is still packed in envelopes and not yet pinned and spread.

Wilhelm Carl Julius **Stüber** (1877–1942) went in 1917 to Hollandia (= Jayapura) and there collected chiefly dragonflies for M. A. Lieftinck, who described from this material more than 100 species as new to science. Later he worked until 1939 for J. M. A. v. Groenendael and sent him (and others) numerous moths. The Groenendael collection was donated to the Zoölogisch Museum Amsterdam and was later transferred to NBC.

Lambertus Johannes **Toxopeus** (1894–1951) was resident in Java. He collected in Sumatra, Java, Sulawesi and Buru. He made particularly important collections as a participant in the third Archbold Expedition to the central Mountains of Dutch New Guinea 1938–1939. The valuable material collected by Toxopeus is concentrated at least in NBC, but some parts should be in the Bogor Zoology Museum, was inaccessible to the author during a visit in 1998.

NBC also houses material, which the author examined, from W Papua, collected beginning in 2006 by the Papua Insects Foundation (Rob de **Voss** and others) during several expeditions. Rob de Voss also made contact with the late Bruder Henk van **Mastrigt** (1946–2015), who was from 1974 resident in Jayapura. Van Mastrigt sent the author a complete photo set of notodontids of van Mastrigt's rich collection, of which some are used here for the plates. The collection of van Mastrigt is now stored in the Koleksi Serangga Papua at the Universitas Cenderawasih in Waena, near Jayapura.

The Museum für Naturkunde – Leibniz-Institut für Evolutions- und Biodiversitätsforschung an der Humboldt-Universität zu Berlin (ZMHU) holds Otto **Staudinger**'s „exotic Lepidoptera" collection of, which was purchased in 1907. In this collection the specimens from Indonesia are not numerous, but notable contributions of material from

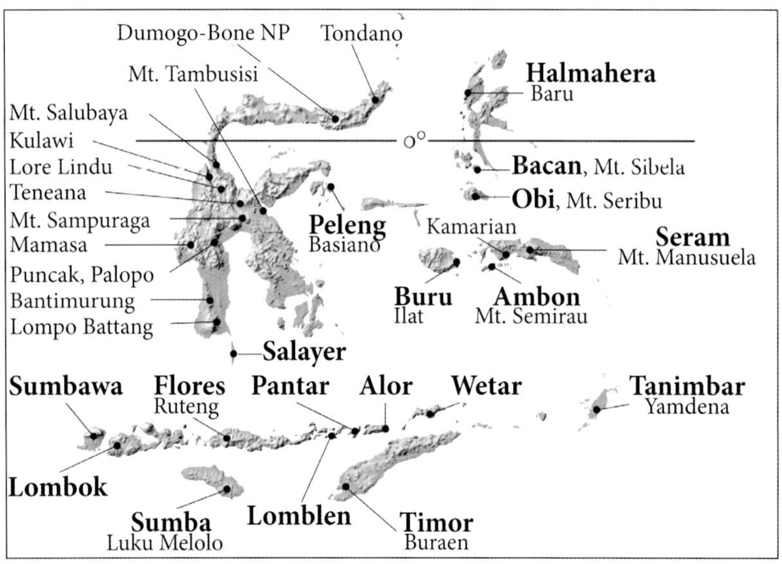

FIGURE 2
Important localities in Sulawesi, the Lesser Sundas and the Moluccas. Island names are printed in bold.

Sumatra, Borneo, Java and the Key Isls. by many different collectors, which were engaged by O. Staudinger, such as Bernhard Hagen (1853-1919), Heinrich Kühn(1860-1906), A. F. Eichhorn, Paul Staudinger(1859–1933, son of Otto Staudinger). Important are also samplings by S. G. **Bürgers** of whose life nothing is known, during the Kaiserin-Augusta-Fluß-Expedition 1912-1913 on the river Sepik in German New Guinea (Kaiser-Wilhelmsland). Wolfram **Mey**, the lepidoptera-curator of ZMHU, collected recently on the Malayan Peninsula (2001), Borneo (2005, 2015) and Java (2015, 2016). Several open problems could be solved with this material.

The Zoologische Staatssammlung, Munich (ZSM) houses currently the Thomas Witt foundation's Oriental Notodontidae, which include apart from doublets of the coll. A. Schintlmeister, numerous specimens collected by local collectors (via Ron Brechlin) from Sulawesi, the lesser Sundas, Seram and the collection of Karl **Martini**, Ingolstadt from Borneo. Of particular importance among the holdings of ZSM are the Sumatran Notodontidae, collected by Eduard Wilhelm **Diehl** (1917–2003). Diehl was a German surgeon who came in 1961 to Sumatra, where he remained until his death. He explored restlessly the insect fauna of Sumatra and sent huge numbers of the insects he collected, to various museums and scientists. He also invited many entomologists to Sumatra, including the author, providing support in the form of cars and generators for their expeditions. Many species of several insect orders have been described from his material.

The collection of the author, A. Schintlmeister in Dresden, is probably the largest collection (n > 60.000 specimens) of Indonesian Notodontidae. He collected from 1979 until 1998 in some parts of Indonesia (Sumatra, Nias, Java, Bali, Sulawesi, Buru, Ambon, Seram) and received collections from several colleagues such as Ron Brechlin, Pasewalk, Karel Cerny, Innsbruck, Karl Martini, Ingolstadt, Stefan Naumann, Berlin, Wolfgang A. Nässig, Frankfurt/Main, Kenichiro Nakao, Tokyo, Ulrich and Laela Paukstadt, Wilhelmshaven, Hermann Schnitzler, Frechen, Victor Sinjaev, Moscow, Rikio Sato, Niigata, Michael Schaarschmidt, Leipzig; the late Eduard W. Diehl, and the late Lutz W. Kobes. Particularly important among the Indonesian Notodontidae in the collection of Schintlmeister is material collected by Stanley Jakl, Jakarta and Prague who undertook between 1998 until 2013 many long-term expeditions to S Kalimantan (Loksado), W Sumatra, Siberut, Belitung, Java, Bali, Kangean, Sumbawa, Sumba, Alor, Pantar, Loblen, Timor, Obi, Bacan, Buru, Tanimbar, and W Papua (Arfak Mts.). He visited these locations usually for several times and his collections from these areas for the most part representative. Doublets of this rich material were given to the Museum Witt, Munich and are now in ZSM, Munich.

Further Material has been studied from the following sources:

Carnegie Museum of Natural History, Pittsburgh (contains material collected by W. Stüber in W Papua and recent material collected during the Project Wallace in Sulawesi).

Florida Museum of Natural History, Gainesville (contains material collected by J. Heppner in Sumatra and in Sulawesi).

Museum Wiesbaden, Naturhistorische Sammlungen (contains the collection of Arnold Pagenstecher, 1837–1913).

National Museum of Natural History, Washington D.C. (particularly important are the samplings by Scott E. Miller 1983 and Vitor O. Becker 1992 in PNG).

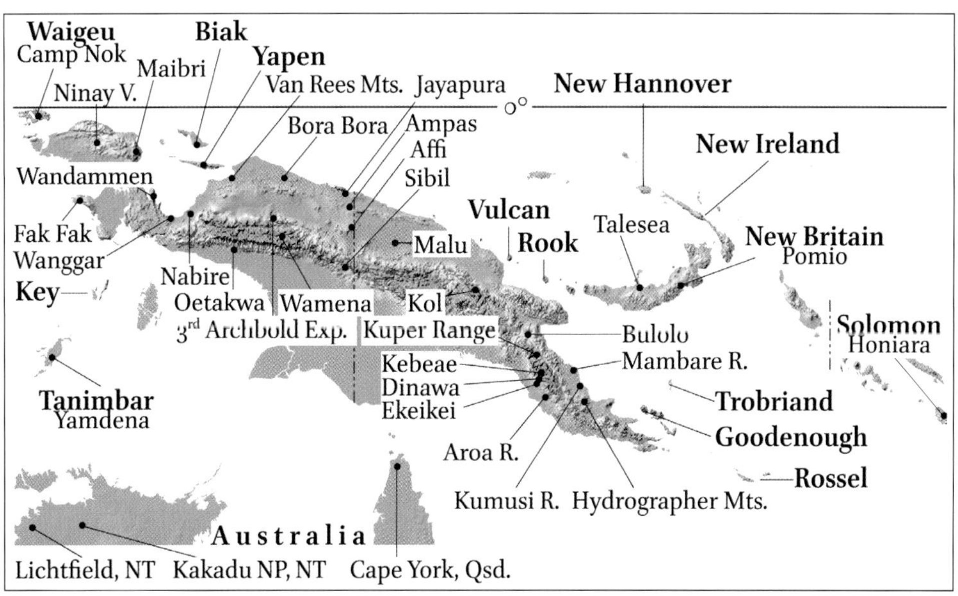

FIGURE 3
Important localities in New Guinea and adjacent islands. Island names are printed in bold.

Naturhistorisches Museum, Vienna (including samplings of Popp in Malacca 1912).

Oxford University Museum of Natural History.

Private collection of Siegfried Ihle, Filderstadt.

Royal Belgian Institute of Natural Sciences, Brussels (coll. E. van Delden).

Staatliche Museum für Naturkunde Karlsruhe holds the collection of R. Bender, 1905–1993, of Sumatran Notodontidae, which includes much and valuable material collected by E. W. Diehl.

Staatliches Museum für Naturkunde Stuttgart.

Zentrum für Biodokumentation Entomologie, Reden (containing material collected by E. W. Diehl and others in Sumatra).

Morphological Terms

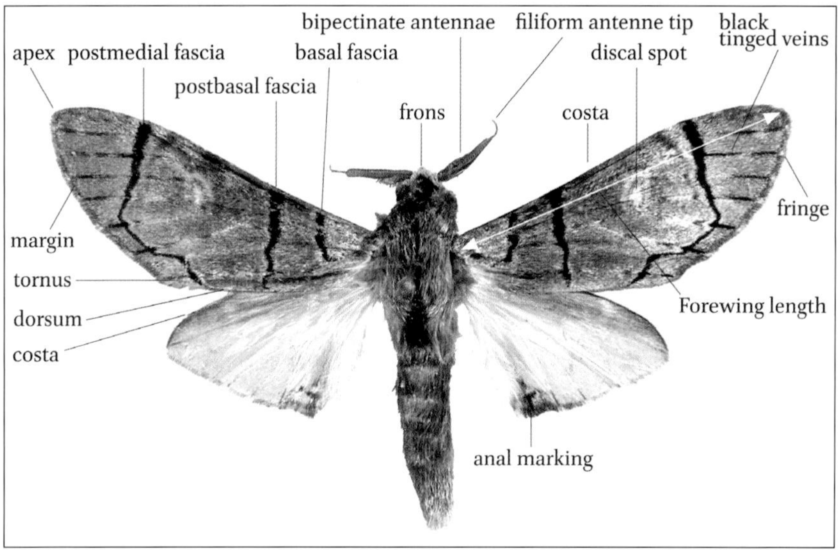

FIGURE 4
Wing pattern of *Teleclita flavisticta* Gaede, 1930.

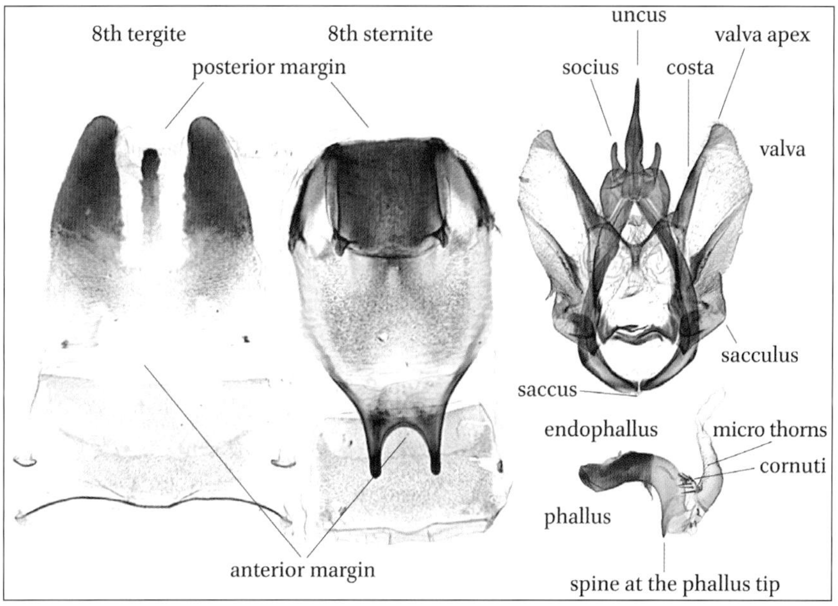

FIGURE 5
Male genitalia of *Ambadra macula* Kobayashi.

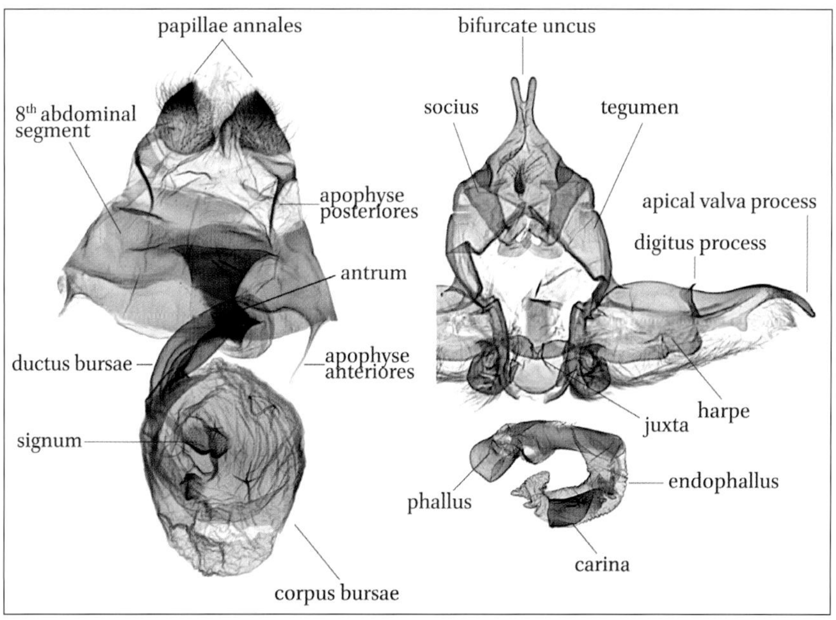

FIGURE 6
Female and male genitalia of *Cascera latifasciata* Gaede.

Abbreviations

The following abbreviations are used:

PNG	–	Papua New Guinea (state)
W Papua	–	Indonesian territory of New Guinea Island (formerly Irian Jaya)
Malaya	–	The Malayan Peninsula (W Malaysia)
New Guinea	–	The Island of New Guinea in its entirety
NT	–	Northern Territory (Australia)
Qsd	–	Queensland (Australia)
Borneo	–	The Island of Borneo in its entirety
Kalimantan	–	Indonesian part of Borneo Island.
Sundaland	–	Borneo, the Malayan Peninsula and Sumatra in its entirety
Indochina	–	Laos, Vietnam, Thailand, Cambodia and eastern part of Myanmar

ANIC	–	Australian National Insect Collection, Canberra
NHMUK	–	The Natural History Museum, London
MWM	–	Museum Witt, Munich
NBC	–	Netherlands Centre for Biodiversity Naturalis, Leiden
NHM	–	Naturhistorisches Museum, Vienna
USNM	–	National Museum of Natural History, Washington DC
FLNM	–	Florida Museum of Natural History, Gainesville
ZMHU	–	Museum für Naturkunde – Leibniz-Institut für Evolutions- und Biodiversitätsforschung an der Humboldt-Universität zu Berlin
ZSM	–	Zoologische Staatssammlung, Munich

HT	–	Holotype
PT	–	Paratype
ST	–	Syntype
LT	–	Lectotype
PLT	–	Paralectotype
GU	–	Genitalia dissection

S	–	South of the Greenwich Observatory
E	–	East of the Greenwich Observatory

An abbreviation before the genital number indicates the institution or collection where the slide is storaged.

BM	–	Notodontidae genitalia slides in NHMUK, London
MWM	–	Genitalia slides in MWM, Munich
MV	–	Genitalia slides in NHM, Vienna
ZSM	–	Genitalia slides „sp." in ZSM, Munich
GU	–	Genitalia slides in coll. A. Schintlmeister, Dresden.

Check-list

Dudusinae

Dudusa Walker, 1865
1. ***synopla*** Swinhoe, 1907
   = *Dudusa sphingiformis rufobrunnea* Mell, 1922
   = *Dudusa fumosa* Matsumura, 1925
   = *Dudusopsis horishana* Matsumura, 1929
2. ***roepkei*** **spec. nov.**
3. ***celebensis*** Roepke, 1944
4. ***vethi*** *vethi* (Snellen, 1892)
   vethi ***borneensis*** Roepke, 1944
   vethi ***javana*** Roepke, 1944
   vethi ***ferruginea*** **ssp. nov.**

Tarsolepis Butler, 1872
Subgenus ***Tarsolepis*** Butler, 1872
5. ***remicauda*** *remicauda* Butler, 1872
6. ***punicea*** Kobayashi, 2011
7. ***adultus*** **spec. nov.**
8. ***sommeri*** (Hübner, 1824)
   = *Tarsolepis javana* Swinhoe, 1907
9. ***dinawensis dinawensis*** Bethune-Baker, 1904
   dinawensis ***extera*** **ssp. nov.**

Subgenus ***Tarsolepisoides*** Nakamura, 1976
10. ***malayana*** Nakamura, 1976

Subgenus ***Stigmatophorina*** Mell, 1922
11. ***sericeus*** Rothschild, 1917

Gangarides Moore, 1866
12. ***rosea*** (Walker, 1865)
13. ***splendidus*** Schintlmeister, 1994
14. ***wallacei*** Kobayashi & Kishida, 2008
15. ***brunneus*** Schintlmeister, 1994
16. ***vittipalpis*** (Walker, 1869)
   = *Gangarides irregularis* Schintlmeister, 1994
17. ***vardena*** Swinhoe, 1892
   = *Gangarides punctata* Mell, 1922
   = *Gangarides corticis* Van Eecke, 1929

Diehleana **gen. nov.**
18. ***gigantea*** (Druce, 1909) **comb. nov.**
   = *Euhampsonia magnifica* Rothschild, 1917
19. ***gilesi*** (Holloway, 1983) **comb. et stat. nov.**

Euhampsonia Dyar, 1897
20. ***roepkei*** *roepkei* Holloway, 1983
   roepkei ***sumatrana*** **ssp. nov.**
   roepkei ***javana*** **ssp. nov.**

Scranciinae

Ortholomia Felder, 1861
   = *Besida* Walker, 1865
   = *Osica* Walker, 1865
   = *Parathemerastis* Bethune-Baker, 1916
21. ***moluccana*** Felder, 1861
   = *Osica turneri* Bethune-Baker, 1904
   = *Osica funerea* Bethune-Baker, 1904
   = *Parathemerastis turneri melanistis* Bethune-Baker, 1916 **syn. nov.**
22. ***gilolo*** **spec. nov.**
23. ***subtilis*** **spec. nov.**
24. ***xylinata*** (Walker, 1865)
   = *Osica turneri* form *albiplaga* Gaede, 1930
   = *Besida vinvalva* Schaus, 1928
25. ***besida*** (Schintlmeister, 1994)
26. *osica* ***sulawesiensis*** Kobayashi, 2012

Pseudogargetta Bethune-Baker, 1904
   = *Blakeia* Kiriakoff, 1967
27. ***diversa*** Bethune-Baker, 1904
28. ***orbiculata*** Kobayashi, 2012
29. ***marmorata*** (Kiriakoff, 1967)

Gargettiana Gaede, 1930
   = *Parosica* Kiriakoff, 1967
30. ***punctatissima*** *punctatissima* (Bethune-Baker, 1916)
   = *Gargetta melanosticta* Joicey & Talbot, 1917
   = *Parosica nomo* Kiriakoff, 1967 **syn. nov.**
   punctatissima ***tritus*** **ssp. nov.**

Polychoa Turner, 1907
   = *Parameces* Turner, 1936
31. ***aeruginosa*** Schintlmeister, 2013 **stat. nov.**
32. ***bakami*** **spec. nov.**
33. ***haliga*** **spec. nov.**
34. ***cheesmanae*** **spec. nov.**
35. ***flavibasis*** (Rothschild, 1917)
36. ***rafinor*** **spec. nov.**
37. ***bipuncta*** (Rothschild, 1917)
38. ***metallica*** Kiriakoff, 1967

Archigargetta Kiriakoff, 1967
   = *Roepkeella* Kiriakoff, 1968
39. ***amydra*** (Turner, 1903)
   = *Themerastis amalopa* Turner, 1905
   = *Gargetta albostigmata* Rothschild, 1917

 | DOI:10.1163/9789004354258_002

= *Pseudogargetta funebris* Gaede, 1930
= *Archigargetta cyclopea* Kiriakoff, 1967
= *Archigargetta diakonoffi* Kiriakoff, 1967
40. ***viridigrisea*** (Dudgeon, 1898)
= *Stauropus clothus* Swinhoe, 1899
= *Pseudogargetta fuscicollis* Gaede, 1930
= *Roepkeella tornalis* Kiriakoff, 1974

Lasioceros Bethune-Baker, 1904
= *Litoscelis* Turner, 1944
41. ***aroa*** *aroa* Bethune-Baker, 1904
= *Euprora tanyphylla* Turner, 1929
42. ***dentilinea*** Joicey & Talbot, 1917
43. ***asudara*** **spec. nov.**
44. ***impurata*** **spec. nov.**
45. ***sola*** Kobayashi, 2011

Noctuola **gen. nov.**
46. ***diehlongata*** (Schintlmeister, 2002)

Phycidopsis Hampson, 1893
= *Sentana* Gaede, 1930
47. ***sundana*** Schintlmeister, 1993 **stat. nov.**
48. ***segregarius*** **spec. nov.**
49. ***sulawesiana*** Schintlmeister, 1993

Stictogargetta Kiriakoff, 1968
50. ***albimacula*** (Hampson, 1893)
51. ***lithosidia*** (Hampson, 1895)

Gargetta Walker, 1865
52. ***costigeroides*** Holloway, 1983
53. ***costigerana*** **spec. nov.**
54. ***hampsoni*** *hampsoni* Schintlmeister, 1981
55. ***divisa*** *divisa* Gaede, 1930
56. ***tompua*** Kiriakoff, 1970
57. ***acarodes*** (Turner, 1903)
58. ***brevisocius*** **spec. nov.**
59. ***curvaria*** *curvaria* Hampson, 1893
= *Stictogargetta umbrina* Kiriakoff, 1974 **syn. nov.**
curvaria ***palua*** Kiriakoff, 1970
60. ***alora*** **spec. nov.**

Thacona Walker, 1864
= *Porsica* Walker, 1866
= *Scrancia* Holland, 1893
61. ***costivitta*** Walker, 1865
= *Gargetta circumducta* Gaede, 1930
62. ***rufocostata*** (Gaede, 1930)
63. ***benderi*** (Schintlmeister, 1981)
64. ***sidaonta*** (Kiriakoff, 1970)
65. *ingens* ***inopinata*** (Holloway, 1983)
66. ***albescens*** (Gaede, 1930)
67. ***triplicepunctata*** (Gaede, 1930)
68. ***punctifascia*** (Hampson, 1897)
69. ***fallax*** (Holloway & Bender, 1985)

Hijracona Holloway, 2011
70. ***punctifascioides*** Holoway, 2011

Ceirinae

Ramesa Walker, 1855
Subgenus ***Ramesa*** Walker, 1855
= *Leucaniochroa* Felder & Felder, 1874
71. ***tosta*** Walker, 1855
= *Ramesa luridivitta* Hampson, 1893
Subgenus ***Togarishachia*** Matsumura, 1925
= *Poncetia* Kiriakoff, 1962
72. ***albistriga*** (Moore, 1879)
= *Pydna kanshireiensis* Wileman, 1914
= *Togarishachia argenteopicta* Matsumura, 1925
= *Stauropus sphingoïdes* van Eecke, 1929
= *Pydna aperta* de Joannis, 1929

Brykia Gaede, 1930
= *Pseudoturnaca* Kiriakoff, 1967
73. ***horsfieldi*** (Moore, 1860)
= *Allata albonotata* Walker, 1862
= *Brykia ferruginea* Nakamura, 1976
74. ***samarinda*** (Kiriakoff, 1967)

Hyperaeschra Butler, 1880
= *Epiphalerina* Gaede, 1934
= *Epiphalera* Gaede, 1930
75. ***pallida*** Butler, 1880
= *Phalera ochropis* Hampson, 1910

Changea Schintlmeister & Fang, 2001
74. ***siamensis*** Schintlmeister, 2007

Turnaca Walker, 1865
= *Rebita* Kiriakoff, 1962
= *Rhabdoturnaca* Kiriakoff, 1967
77. ***subcarnea*** Warren, 1903
78. ***carnea*** **spec. nov.**

Tensha Matsumura, 1925
= *Hypambadra* Kiriakoff, 1974
79. ***delineivena*** (Swinhoe, 1894)

= *Hybambadra speculigera* Kiriakoff, 1974 **syn. nov.**
79. ***postobscura*** Holloway, 1983
80. ***pizzicata*** **spec. nov.**

Canatensha **gen. nov.**
82. ***mediofascia*** (Rothschild, 1917) **comb. nov.**
= *Turnaca thiaucourti* Holloway & Bender, 1985

Ambadra Moore, 1883
= *Corma* Moore, 1881 (homonym)
= *Suriga* Kiriakoff, 1962
= *Atornoptera* Kiriakoff, 1974
83. ***rafflesi*** (Moore, 1860)
= *Ceira pallida* Heylaerts, 1891 **syn. nov.**
= *Pydna bryantia* Schaus, 1928.
= *Turnaca rafflesi borneensis* Roepke, 1943 **syn. nov.**
= *Turnaca rufescens* Gaede, 1930 **syn. nov.**
84. ***canescens*** Schintlmeister & Lourens, 2010
= *Ambadra arta* Kobayashi, 2012 **syn. nov.**
= *Ambadra gracilis* Kobayashi, 2011
= *Ambadra angustifolia* Schintlmeister, 2013 **syn. nov.**
85. ***miscera*** **spec. nov.**
86. ***celebensis*** (Roepke, 1943)
87. ***macula*** Kobayashi, 2011
88. ***parva*** Kobayashi, 2011
89. ***tondanoa*** Kobayashi, 2011
90. ***vittata*** Kobayashi, 2011
91. ***gracilis*** Holloway, 2011
92. ***batara*** **spec. nov.**
93. ***discocellularis*** (Kiriakoff, 1974)
94. ***similis*** (Dierl, 1984)
95. ***melanostriga*** (Schintlmeister, 1994) **spec. rev.**
96. ***owadai*** Kobayashi, 2011
97. ***mentawai*** **spec. nov.**
98. ***montana*** Holloway, 1982

Ambadrana **gen. nov.**
99. ***punctata*** (Holloway, 1983) **comb. nov.**
= *Turnaca altipunctata* Holloway, 1983
100. ***barlowi*** (Schintlmeister, 1994) **comb. nov.**
101. ***taruma*** **spec. nov.**
102. ***spinifera*** (Holloway, 1983) **comb. nov.**
103. ***furcata*** (Dierl, 1984) **comb. nov.**

Pydnella Roepke, 1943
104. ***rosacea*** (Hampson, 1896)
= *Pydnella monticola* Roepke, 1943
= *Pydna rosacea nesiotes* Kiriakoff, 1960

Niganda Moore, 1879
= *Stenadonta* Hampson, 1895
= *Eutornopera* Hampson, 1895
105. ***strigifascia*** Moore, 1879
= *Niganda strigifascia insularis* Kiriakoff, 1960 **syn. nov.**
= *Bireta* (*Niganda*) *strigifascia coelestis* Kiriakoff, 1962 **syn. nov.**
106. ***pallens*** **spec. nov.**
107. ***ancipitalis*** Schintlmeister, 2007

Postniganda **gen. nov.**
108. ***pudens*** (Schintlmeister, 2007) **comb. nov.**

Norraca Moore, 1881
Oraura Kiriakoff, 1962
109. ***longipennis*** Moore, 1881
= *Norraca uncinata* Semper, 1902
= *Norraca ordgara* Schaus, 1928

Saliocleta Walker, 1862
Subgenus *Armiana* Walker, 1862
110. ***lativitta*** Walker, 1862
= *Spatalia* ***notodontina*** van Eecke, 1929
Subgenus *Saliocleta* Walker, 1862
= *Ceira* Walker, 1865
= *Oglana* Kiriakoff, 1968
111. ***nonagrioides*** Walker, 1862
= *Bireta* (*Norraca*) ***sabulosa*** Kiriakoff, 1962
112. ***sordida*** (Roepke, 1943)
113. *odrana* ***celebica*** (Kiriakoff, 1970)
114. ***ediehli*** Schintlmeister, 1994
115. ***commutatis*** Schintlmeister & Lourens, 2010
116. ***widagdoi*** Schintlmeister, 1994
117. ***beshda*** **spec. nov.**
118. ***malayana*** (Schintlmeister, 1994)
Subgenus *Biraia* Kiriakoff, 1962
119. ***armata*** (Kiriakoff, 1974)
120. ***nannion*** Kiriakoff, 1974
121. ***straminea*** (Roepke, 1943)

Chalepa Kiriakoff, 1959
= *Rosiora* Kiriakff, 1962
122. ***pendleburyi*** **spec. nov.**
123. ***takaloa*** **spec. nov.**

Periergos Kiriakoff, 1959
= *Pydna* Walker, 1856 (homonym)
= *Hunyada* Kiriakoff, 1962
= *Loudonta* Kiriakoff, 1962
= *Mismia* Kiriakoff, 1962

= *Eupydna* Fletcher, 1980
124. ***postrubra*** (Swinhoe, 1903)
= *Pydna testacea* ***indonesiae*** Kiriakoff, 1962
= *Pydna* ***backanensis*** Lemee & Tams, 1950
125. ***hunyada*** (Swinhoe, 1903)
= *Pydna marconia* Schaus, 1928
= *Pydna callista* West, 1932
= *Rosiora* (*Hunyada*) *venosa* Kiriakoff, 1962
126. ***spinosa*** (Holloway, 1983)

Cerurinae

Neocerura Matsumura, 1929
127. ***liturata*** (Walker, 1855)
= *Cerura damodara* Moore, 1866
= *Dicranura argentea* Felder & Felder, 1874
= *Cerura arikana* Matsumura, 1927
= *Cerura liturata baibarana* Matsumura, 1927
= *Furcula hapala* West, 1932
128. ***longinquus*** **spec. nov.**

Kamalia Koçak & Kemal, 2006
= *Paracerura* Schintlmeister, 2002 (homonym)
129. ***malaysiana*** *malaysiana* (Holloway, 1982)
malaysiana ***niasica*** (Schintlmeister, 1994)
malaysiana ***nigresco*** **ssp. nov.**
malaysiana ***jakli*** (Schintlmeister, 2002)
130. ***minahassae*** (Holloway, 1982)
131. ***fulva*** **spec. nov.**
132. ***timorensis*** (Kiriakoff, 1970)
133. ***multipunctata*** (Bethune-Baker, 1904)
134. ***solomonica*** **spec. nov.**
135. ***amoa*** (Holloway, 1979

Dicranurinae

Liparopsis Hampson, 1893
136. *postalbida* ***brevispinus*** **ssp. nov.**
137. ***barlowi*** Holloway, 1985
138. ***sundana*** Holloway, 1983
139. ***dierli*** Bender & Bender, 1984

Pantanopsis Kiriakoff, 1974
140. ***diehli*** Kiriakoff, 1974
141. ***celebensis*** (Gaede, 1930)

Harpyia Ochsenheimer, 1810
= *Hoplitis* Hübner, 1819
= *Hybocampa* Lederer, 1853
= *Damata* Walker, 1855
= *Damatoides* Matsumura, 1927
= *Macrohoplitis* Roepke, 1943
142. ***kalisi*** (Roepke, 1943) **stat. rev.**
143. ***kobesi*** **spec. nov.**
144. ***koeppeli*** **spec. nov.**

Teleclita Turner, 1903
= *Pseudoteleclita* Kiriakoff, 1968
145. *strigata* ***cathana*** (Schaus, 1928) **stat. nov.**
= *Teleclita* ***sundana*** Holloway, 1983
strigata ***mediocana*** **ssp. nov.**
strigata ***multifascia*** **ssp. nov.**
strigata ***insignifica*** (Rothschild, 1917) **stat. nov.**
strigata ***cinnamomea*** (Rothschild, 1917) **stat. nov.**
146. ***rasilis*** **spec. nov.**
147. ***flavisticta*** Gaede, 1930

Pheressaces Turner, 1903, **stat. rev.**
148. ***siegfriedihlei*** **spec. nov.**

Destolmia Walker, 1855
149. ***lineata*** Walker, 1855

Stauroplitis Gaede, 1930.
= *Grangula* Kiriakoff, 1967
= *Briachisia* Kiriakoff, 1968
= *Antichadisra* Kiriakoff, 1974
150. ***annulata*** *annulata* Gaede, 1930
= *Antichadisra dentata* Kiriakoff, 1974

Oxoia Kiriakoff, 1967
= *Panteleclita* Kiriakoff, 1974
151. ***smaragdiplena*** (Walker, 1862)
= *Heterocampa antonia* Druce, 1901
= *Somera oxoia* Swinhoe, 1904
152. ***viridipicta*** (Kiriakoff, 1974)
153. ***irrorativiridis*** *irrorativiridis* (Bethune-Baker, 1904)
irrorativiridis ***triangularis*** (Gaede, 1930) **stat. nov.**

Rodneya Kiriakoff, 1967
154. ***caudata*** Kiriakoff, 1967

Stauropus Germar, 1812
Subgenus ***Stauropus*** Germar, 1812
= *Terasion* Hübner, 1819
= *Neostauropus* Kiriakoff, 1967
= *Mesostauropus* Nakamura, 1974
155. ***major*** van Eecke, 1929
= *Stauropus albimacula* Gaede, 1930

156. *alternus* ***brunnea*** (Schintlmeister, 1981)
alternus ***niasicus*** (Schintlmeister, 1981)
alternus ***hofmanae*** Schintlmeister, 2003
alternus ***sulawesianus*** Schintlmeister, 2003
alternus ***melastomatis*** Felder, 1861
157. ***pusilla*** **spec. nov.**
Subgenus ***Chlorostauropus*** Kiriakoff, 1968
158. ***viridissimus*** *viridissimus* Bethune-Baker, 1904
viridissimus ***habrochlora*** Turner, 1922 **stat. nov.**
viridissimus ***amboynica*** Oberthür, 1911
viridissimus ***naumanni*** Schintlmeister, 2003
viridissimus ***boeroe*** **ssp. nov.**
159. ***talasea*** **spec. nov.**
160. ***vanmastrigti*** **spec. nov.**
161. ***caerulea*** **spec. nov.**
162. ***affinis*** Rothschild, 1917
163. ***pratti*** Bethune-Baker, 1904
164. ***schaarschmidti*** Schintlmeister, 2003
165. ***evanescens*** Gaede, 1930
166. ***pelz*** **spec. nov.**
167. ***loki*** Schintlmeister, 2003
Subgenus ***Benbowia*** Kiriakoff, 1967
168. *virescens* ***obscuratus*** Van Eecke, 1929
= *Benbowia dudgeoni* Kiriakoff, 1967
169. ***elisabethae*** *elisabethae* Dierl, 1981
elisabethae ***kiriakoffi*** Holloway, 1983 **stat. nov.**

Kakostauropus Gaede, 1930
170. ***glaucoviridis*** *glaucoviridis* (Rothschild, 1917)
glaucoviridis ***seramica*** **ssp. nov.**

Syntypistis Turner, 1907.
= *Quadricalcarifera* Strand, 1916
= *Egonocia* Marumo, 1920
= *Stauropodopsis* Roepke, 1944
= *Taiwa* Kiriakoff, 1967
= *Vaneeckia* Kiriakoff, 1968
171. ***kebeae*** (Bethune-Baker, 1904)
= *Stauropus dubiosus* Bethune-Baker, 1904, **syn. nov.**
= *Syntypistis chloropasta* Turner, 1907 **syn. nov.**
= *Stauropus mediobrunnea* Bethune-Baker, 1916
= *Stauropus leucocraspedus* Joicey & Talbot, 1917
= *Stauropus melanogramma* Joicey & Talbot, 1917
= *Syntypistis opaca* Turner, 1922
= *Desmaeocraera rufotegula* Gaede, 1930
= *Desmeocraera frugilegus* form *aeruginosus* Gaede, 1930 **syn. nov.**
= *Quadricalcarifera ceramensis* Kiriakoff, 1967 **syn. nov.**
172. ***dasychirinus*** (Roepke, 1944)
= *Quadricalcarifera paranga* Kiriakoff, 1970
173. ***grisescens*** (Roepke, 1944)
= *Quadricalcarifera ardjuna* Kiriakoff, 1967
174. ***frugilegus*** (Rothschild, 1917)
= *Quadricalcarifera eichhorni* Kiriakoff, 1970, **syn. nov.**
175. ***tisada*** **spec. nov.**
176. ***mapia*** **spec. nov.**
177. *charistera* ***celebensis*** (Roepke, 1944)
charistera ***fraseriana*** (Kiriakoff, 1967)
= *Quadricalcarifera ferrea* Kiriakoff, 1967
= *Quadricalcarifera rhypara* Kiriakoff, 1970
= *Quadricalcarifera bambusicola* Kiriakoff, 1974
178. ***triplagosus*** (Rothschild, 1917)
= *Quadricalcarifera nitidula* Kiriakoff, 1967 **syn. nov.**
= *Quadricalcarifera testacea* Kiriakoff, 1967 **syn. nov.**
= *Quadricalcarifera didyma* Kiriakoff, 1967 **syn. nov.**
= *Quadricalcarifera trivia* Kiriakoff, 1967 **syn. nov.**
179. ***seramica*** Kobayashi, 2011
180. ***trisospylus*** (Joicey & Talbot, 1917)
= *Desmeocraera trisopylus* form *roseus* Gaede, 1930
181. ***famelica*** (Kiriakoff, 1967)
182. ***quadrivittata*** (Kiriakoff, 1967
= *Quadricalcarifera nana* (Kiriakoff, 1967) **syn. nov.**
183. ***histrionica*** (Kiriakoff, 1967)
184. ***dinawa*** (Bethune-Baker, 1904)
185. ***bella*** (Bethune-Baker, 1904)
= *Stauropus rufescens* Rothschild, 1917
= *Quadricalcarifera murina* Kiriakoff, 1967
186. ***mediogriseus*** *mediogriseus* (Gaede, 1930)
mediogriseus ***inviridis*** **ssp. nov.**
mediogriseus ***binaia*** **ssp. nov.**
187. ***leucaspis*** **spec. nov.**
188. ***papuana*** **spec. nov.**
189. ***mixta*** (Bethune-Baker, 1916)
= *Stauropus leucophaeus* Rothschild, 1917 **syn. nov.**
190. ***chloriolus*** (Joicey & Talbot, 1917)
= *Stauropus viridinitens* Rothschild, 1917 **syn. nov**
191. ***viridigriseus*** (Rothschild, 1917)
= *Syntypistis sciera* Turner, 1922
192. ***alboviridis*** (Rothschild, 1917)
193. ***flavicollis*** (Rothschild, 1917)

= *Quadricalcarifera poecilochroa* Kiriakoff, 1967 **syn. nov**
194. ***nitidus*** (Rothschild, 1917)
195. ***gemmea*** **spec. nov.**
196. ***purpurascens*** (Rothschild, 1917)
197. ***triguttata*** (Kiriakoff, 1967)
= *Quadricalcarifera transcaerulea* Holloway, 1976
198. ***variegata*** (Kiriakoff, 1974)
199. ***uskwara*** (Kiriakoff, 1970)
200. ***miyanoi*** Kobayashi, 2011
201. ***cognata*** **spec. nov.**
202. ***sula*** Kobayashi, 2011
203. ***paukstadtorum*** **spec. nov.**
204. ***umbrosa*** (Matsumura, 1927)
= *Desmeocraera perdicula* Bryk, 1950
= *Desmeocraera okurai* Okano, 1960
= *Quadrticalcarifera unicolor* Kiriakoff, 1974
= *Quadricalcarifera hasegawai* Nakamura, 1976
205. ***garuda*** (Schintlmeister, 1994)
206. ***kiriakoffi*** Schintlmeister & Lourens, 2010
= *Quadricalcarifera alboviridis* Kiriakoff, 1970 (Homonym)
207. *palladina* ***viridimargo*** (Kiriakoff, 1967)
palladina ***sasak*** **ssp. nov.**
208. ***comatus*** *comatus* (Leech, 1898)
= *Quadricalcarifera viridimaculata* Matsumura, 1922
= *Quadricalcarifera bioculata* Kiriakoff, 1967
= *Quadricalcarifera fasciata tanakai* Nakamura, 1976
= *Quadricalcarifera fasciata malayana* Nakamura, 1976
comatus ***majapahit*** **ssp. nov.**
209. ***trioculata*** (Holloway, 1976)
210. ***viridipicta*** (Wileman, 1910)
= *Stauropus chlorotricha* Hampson, 1912
= *Quadricalcarifera marginalis* Matsumura, 1925
= *Quadricalcarifera kusukusuana* Matsumura, 1929
= *Stauropus mixtus* van Eecke, 1929
= *Stauropus permixta* Kiriakoff, 1968
= *Quadricalcarifera medioviridis* Kiriakoff, 1964
= *Quadricalcarifera doloka* Kiriakoff, 1967
= *Quadricalcarifera eusebia* Kiriakoff, 1974
211. ***stanleyi*** **spec. nov.**
212. ***pallidifascia*** *pallidifascia* (Hampson, 1893)
= *Quadricalcarifera centro-brunnea* Matsumura, 1927
= *Quadricalcarifera concentrica* Matsumura, 1927
= *Stauropus ovalis* van Eecke, 1929
pallidifascia ***germana*** (Rothschild, 1917)
213. ***vaneeckei*** **spec. nov.**
214. *nigribasalis* ***tropica*** (Kiriakoff, 1974)
= *Quadricalcarifera notoprocta* Yang, 1995
215. ***alleni*** (Holloway, 1983)
= *Syntypistis aspera* Kobayashi & Kishida, 2004 **syn. nov.**

Parasinga Kiriakoff, 1967
216. ***lichenina*** *lichenina* (Butler, 1880)
= *Desmeocraera triangulum* Gaede, 1930
= *Stauropus grisescens* Roepke, 1951 (homonym)
= *Quadricalcarifera roepkei* Nakamura, 1960
= *Parasinga bicolor* Kiriakoff, 1977
217. ***muliercula*** **spec. nov.**
218. ***cinerascens*** Kiriakoff, 1974
= *Parasinga subapicalis* Kiriakoff, 1974
= *Parasinga tuhana* Holloway, 1976
= *Quadricalcarifera taeniata* Kiriakoff, 1977
219. ***harmani*** Holloway, 1983
220. ***insufficiens*** (Gaede, 1930)
= *Quadricalcarifera pallidicollis* Kiriakoff, 1967

Cascera Walker, 1865: 460.
221. ***marginata*** Rothschild, 1917
222. ***inconcisa*** Swinhoe, 1892
= *Cascera flavovirens* Rothschild, 1917 **syn. nov.**
223. ***bella*** Bethune-Baker, 1904
= *Cascera irrorata irrorata* Rothschild, 1917 **syn. nov.**
= *Cascera irrorata pallida* Rothschild, 1917 **syn. nov.**
= *Cascera variegata* Rothschild, 1917 **syn. nov.**
224. ***latifasciata*** Gaede, 1930
225. ***albiscripta*** Rothschild, 1917
= *Cascera albiscripta* ***virens*** Rothschild, 1917 **syn. nov.**
226. ***olivacea*** Rothschild, 1917
= *Cascera olivacea flavolavata* Rothschild, 1917 **syn. nov.**
227. ***perscripta*** Rothschild, 1917
228. ***violetta*** Gaede, 1930
229. ***callima*** Bethune-Baker, 1916
= *Cascera archboldiana* Kiriakoff, 1967 **syn. nov.**

Omichlis Hampson, 1895
230. ***dimorpha*** Kiriakoff, 1974
= *Omichlis dimorpha pallida* Kiriakoff, 1974
231. ***bifurcata*** **spec. nov.**
232. ***polymorpha*** **spec. nov.**
233. ***dubia*** **spec. nov.**
234. ***griseola*** Bethune-Baker, 1904

235. ***renata*** Gaede, 1930
236. ***pratti*** *pratti* Bethune-Baker, 1904
 = *Omichlis parallelistriga* Rothschild, 1915 **syn. nov.**
 pratti ***falcigera*** **ssp. nov.**
237. ***erythra*** Bethune-Baker, 1916
238. ***paragriseola*** **spec. nov.**
239. ***opacitas*** **spec. nov.**
240. ***rudloffi*** **spec. nov.**
241. ***disiuncta*** **spec. nov.**
242. ***priskae*** **spec. nov.**
243. ***lineata*** Bethune-Baker, 1908
244. ***plagiosa*** Joicey & Talbot, 1917
245. ***trilineata*** Bethune-Baker, 1908
246. ***dinawa*** Bethune-Baker, 1904
247. ***ochromacula*** **spec. nov.**
248. ***baliema*** **spec. nov.**
249. ***calyptis*** (Swinhoe, 1892)
 = *Omichlis hampsoni* Bethune-Baker, 1904 **syn. nov.**
250. ***sero*** **spec. nov.**
251. ***ochracea*** Bethune-Baker, 1904
 = *Omichlis ochracea* f. *immaculata* Gaede, 1930
 = *Omichlis macrospila* Kiriakoff, 1970 **syn. nov.**
252. ***leucosticta*** Joicey & Talbot, 1917
253. ***transversa*** Bethune-Baker, 1908
254. ***strigata*** Bethune-Baker, 1908
255. ***geminata*** Gaede, 1930
256. ***buergersi*** **spec. nov.**
257. ***albidilinea*** Gaede, 1930
258. ***stueberi*** **spec. nov.**
259. ***superne*** **spec. nov.**
260. ***plagata*** Bethune-Baker, 1908
261. ***rufofasciata*** Bethune-Baker, 1904
262. ***pseudolibatrix*** Rothschild, 1917
263. ***mediofasciata*** Rothschild, 1917

Omestia Bethune-Baker, 1908
264. ***bella*** Bethune-Baker, 1908
 = *Omichlis plusiotis* Rothschild, 1917 **syn. nov.**

Netria Walker, 1855
265. ***viridescens*** *viridescens* Walker, 1855
266. ***torajae*** Schintlmeister, 2006
267. ***longisacci*** Schintlmeister, 2006
268. ***insularis*** Schintlmeister, 2006
269. *arisemna* ***maiorvirens*** **ssp. nov.**
 arisemna ***andreasmuelleri*** Schintlmeister, 2006 **stat. nov.**
 arisemna ***tarasovi*** Schintlmeister, 2006 **stat.nov.**
 arisemna ***bipartita*** Schintlmeister, 2006 **stat.nov.**
270. ***jakli*** Schintlmeister, 2006
271. ***carentis*** Schintlmeister, 2006
272. ***multispinae*** *multispinae* Schintlmeister, 2006

Cerasana Walker, 1862
 = *Billetia* de Joannis, 1900
273. ***anceps*** *anceps* Walker, 1862
 anceps ***ursulae*** Schintlmeister, 2004
274. ***alleni*** Holloway, 1983
275. ***pagenstecheri*** *pagenstecheri* Schintlmeister & Lourens, 2010

Somera Walker, 1855
276. *viridifusca* ***sumatrana*** Schintlmeister, 1993
 viridifusca ***celebica*** Schintlmeister, 1993
 viridifusca ***javana*** **ssp. nov.**
277. ***viriviri*** Kobayashi, 2012
278. ***virens*** *virens* Dierl, 1976
 = *Somera similis* Nakamura, 1976
 = *Somera haxairei* Orhant, 1995
279. ***brillians*** Gaede, 1930

Psegmaphora Gaede, 1930
 = *Tolmiana* Kiriakoff, 1974
280. ***tripunctata*** Gaede, 1930
 = *Tolmiana cyanosticta* Kiriakoff, 1974

Shachia Matsumura, 1919
281. ***vernalis*** *vernalis* (Gaede, 1930)
 vernalis ***infuscata*** (Gaede, 1930)

Pheraspis Turner, 1903
282. ***calida*** **spec. nov.**

Notodontella Roepke, 1943
 = *Podocryptula* Bryk, 1949
283. ***ferrifusa*** (Dudgeon, 1898)
 = *Notodontella nieuwenhuisi* Roepke, 1943
 = *Podocryptula nana* Bryk, 1949
284. ***floresiensis*** **spec. nov.**

Chloroceramis Kiriakoff, 1968 **stat.rev.**
 = *Maguila* Kiriakoff, 1968
285. ***viridinota*** (Hampson, 1896)
 = *Fentonia maguila* Schaus, 1928
286. ***albuala*** **spec. nov.**

Medanella Kiriakoff, 1974
287. ***subterminalis*** Kiriakoff, 1974

Antiphalera Gaede, 1930
= *Grangulina* Kiriakoff, 1974
288. ***sumatrana*** (Kiriakoff, 1974)

Formofentonia Matsumura, 1925
289. *orbifer* ***cremea*** **ssp. nov.**
orbifer ***madena*** (Schaus, 1928)
orbifer ***sampuraga*** **ssp. nov.**

Betashachia Matsumura, 1925
= *Mesaeschra* Kiriakoff, 1964
= *Pseudofentonia* Kiriakoff, 1964
= *Apistaeschra* Kiriakoff, 1964
= *Antifentonia* Kiriakoff, 1967
290. *angustipennis* ***tropica*** Kiriakoff, 1974

Fentonia Butler, 1881.
= *Urocampa* Staudinger, 1892
= *Neoshachia* Matsumura, 1925
= *Hoplitina* Gade, 1930
= *Subwilemanus* Kiriakoff, 1964
291. ***sumatrana*** Kiriakoff, 1974
292. ***helena*** Kiriakoff, 1974
293. ***hollowayi*** Schintlmeister, 2007
294. ***talboti*** *talboti* (Gaede, 1930)
= *Stauropus pulverulentus* Talbot, 1926 (Homonym)
= *Fentonia elongata* Holloway, 1976
talboti ***ingridae*** Holloway & Bender, 1985
295. ***bipunctus*** (Rothschild, 1917)

Neopheosia Matsumura, 1920
= *Parafentonia* Kiriakoff, 1964
= *Hemifentonia* Kirkiakoff, 1967
296. ***fasciata*** *fasciata* (Moore, 1888)
= *Neopheosia fasciata formosana* Okano, 1959
= *Neopheosia fasciata japonica* Okano, 1955
fasciata ***buruana*** **ssp. nov.**
297. ***albiplaga*** Gaede, 1930

Fusadonta Matsumura, 1920
= *Pheosilla* Kiriakoff, 1964
298. ***albipuncta*** (Gaede, 1930)

To be continued in vol. II

Systematic Account

Dudusinae

Dudusa Walker, 1865: 446.
Type species: *Dudusa nobilis* Walker, 1865
Schintlmeister 2008: 28; Schintlmeister 2013: 132.

Diagnosis. Large adults. Antennae are brown to black and bipectinate except for the last 25–30% to the tip, which is filiform. Female has shorter rami. Forewing is ochrous to yellow – brown with fuscous brown fascia from costa toward the tornus. Abdomen is black with yellowish brown markings characterized by yellow or pale brown hairs of taxonomic value. Apical tufts have spatulate scales. The forewing displays well developed postbasal and postemedial fascia. Marginal and submarginal fasciae of all wings are marked as rows of black U-shaped patches. A fuscous brown band runs from the costa to the tornus and there is a prominent white costal patch in the postbasal area. Discal spot of inconspicous brown; below the discal spot most species have a silver patch. Hindwings are fuscous-brown with pale blackish and yellowish-brown anal markings. Sexual dimorphism limited.

Uncus is broad and deeply bilobed, a pair of socii are long. Saccus is large and triangular. Valvae display an apical flap. Sacculus bears a small process toward the apical part. Phallus is slender, either slightly curved or straight and long. It's shape has diagnostic value at species level. Female genitalia are characterized by a very long ductus bursae and a small corpus bursae without signum.

Larvae of *Dudusa* display a prominent yellow and black or brown pattern and have several long spines. Caterpillar's hostplants belong to the Aceraceae and Sapindaceae.

This oriental genus is represented by ten known species, which are found in the Himalayas, S India, Myanmar, China, Korea, Japan, Indochina, the Philippines, and Indonesia west of Buru. The highest diversity of species is found in Indochina.

1. ***Dudusa synopla*** Swinhoe, 1907: 205.
Lectotype: ♀, [India], Khasia Hills – NHMUK, London **by present designation.**
Schintlmeister 2008: 30; Schintlmeister 2013: 133.

Taxonomic note. *Dudusa synopla* was described of "three examples" which are accord with the original description all from "Kasia Hills". The author traced in NHMUK one ♀ of the syntype series. To stabilize the nomenclature this ♀ with the following labels is hereby designated as lectotype: "Khasia Hills | ♀", red ringed, circular label "Type", "*Dudusa* | *synopla* ♀ | Swinhoe type", yellow label "Photo done by A. Schintlmeister #4973". A lectotype label was added.

Diagnosis. The largest species of *Dudusa.* Forewings orange brown. *Dudusa synopla* is recognized best by the continous pale brown dorsal line on the abdomen. Intensity and width of the brown dorsal line on the abdomen may vary individually and appear sometimes reduced and broken. The brown fascia from the costa toward the tornus are broadened at the dorsum.

Male genitalia are as described under genus diagnosis. Phallus is slender and rather straight. A spine on the phallus distinguishes *synopla* from other species. Female genitalia have a broad 8th abdominal segment. Apophyses posteriores are slightly shorter than in *vethi.* Ductus bursae is clearly longer and corpus bursae larger than in other congeners. No signum.

Bionomics. Adults are infrequent in Indonesia, appearing from the lowlands (10 m) up to an altitude of 1900 m.

Larva was reared in mainland Asia on *Sapindus mukurosii* and *Schleichera trijuga* (Sapindaceae) (Sugi, 1987: 282 and Holloway, 1983: 20).

Distribution. *Dudusa synopla* is widely distributed in NE India, Myanmar, Vietnam, Thailand, Laos, Cambodia, S China, Taiwan and the Malayan Peninsula, Sumatra, Nias, Borneo. 2 ♂♂ are from Central Buru (Identity confirmed by genitalia dissection BM 2252), far away from the hitherto known distribution area.

2. ***Dudusa roepkei* spec. nov.** (figs. 2a–c).
Holotype: ♂, Indonesia, Java, Gunung Gedeh, Pangrango National Park, 6°47'S, 106°54'E, 1250 m, iv.–vii.1996 – in coll. A. Schintlmeister, Dresden.

Paratypes (20 ♂♂, 1 ♀):
Java: 4 ♂♂, Gunung Gedeh, Pangrango National Park, 6°47'S, 106°54'E, 1250 m, iv.–vii. 1996 (GU 59-23); 1 ♂, E Java, Slopes of Mt. Argopuro, 1000-1200 m, iv.–vii. 1994, leg. J. Pasternak; 1 ♂, Djoenggo-Ardjoeno [= Gn. Arjuno], 1500 m, ix. 1937(Kalis); 1 ♂, Manggis Laswang, 1935 (Walsh); 4 ♂♂, 1 ♀, East Java, Nongkodjadjar, 4000 ft. i. 1934 (A.M.R. Wegner); 1 ♂, dto., ii. 1934; 2 ♂♂, dto., iii.

 | DOI:10.1163/9789004354258_002

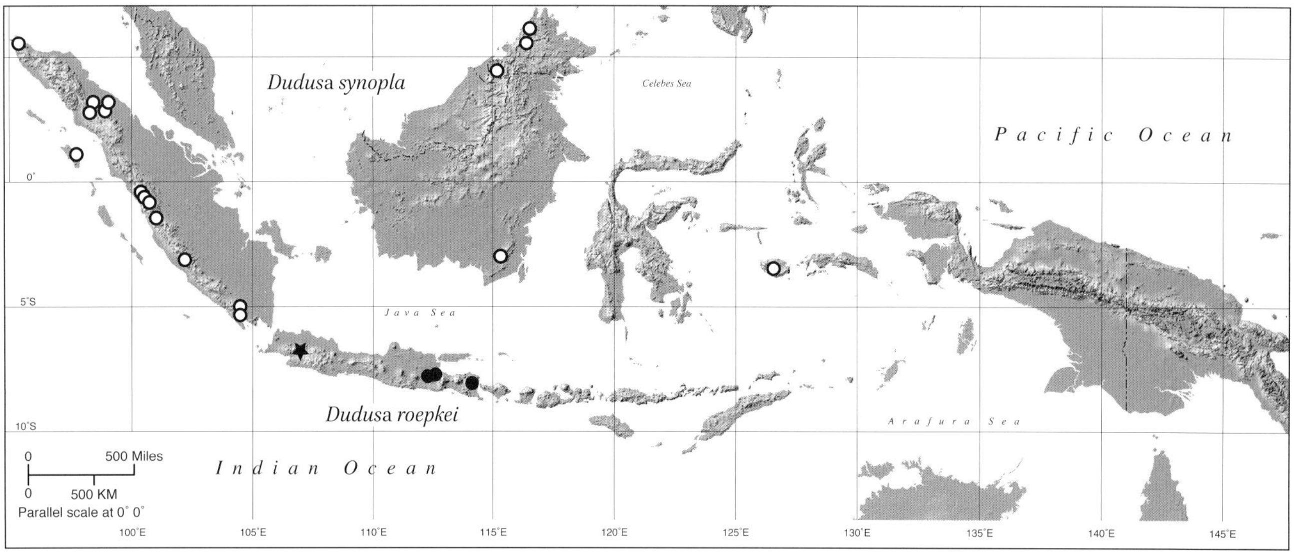

FIGURE 7 Distribution of *Dudusa synopla* and *Dudusa roepkei*.

1934; 1 ♂, dto., iv. 1934; 2 ♂♂, dto., x. 1934; 1 ♂, dto., xi. 1934; 1 ♂, dto., xii. 1934; 1 ♂, East Java, Treffen, 3000', v. 1932 (J.P.A. Kalis).

Etymology: Named after Walter Karl Johann Roepke (b. 18. ix. 1882, Hohensalza – d. 7. ii. 1961, Wageningen) who published among others also an important paper on Indonesian *Dudusa* (Roepke, 1944).

Diagnosis. Forewing length ♂♂ 39–41 mm, ♀ 52 mm and much smaller than *D. synopla*. Forewing length of 12 ♂♂ from Sumatran and 8 ♂♂ from Borneo specimens are on average 49 mm. *D. roepkei* resembles by its coloration and wing pattern *synopla*. The orange-brown forewing color is darker, and the silver patch on the forewing are smaller than in *synopla*. The only female of *roepkei* (fig. 2c) have paler colored forewings as in *synopla*.

Male genitalia resemble *synopla*, but the valva is shorter and narrower. Phallus show great differences, which lead to the classification as a distinct species and not as a subspecies. The spine on the phallus, which is characteristically for *synopla*, is absent in *roepkei*. The tip of the everted endophallus is strongly curved in *roepkei*, but rather straight in *synopla*. The posterior margin of the 8th sternite is less rounded and straight, while in *synopla* triangular notched.

Bionomics. The adults were observed infrequently between 1000–1500 m above sea level.

Distribution. Restricted to Java.

3. ***Dudusa celebensis*** Roepke, 1944: 79; fig. 2 (*Dudusa nobilis celebensis*).

Lectotype: ♀, [Indonesia, Sulawesi, Minahassa tenggara], Gurupahi – NBC, Leiden.

Schintlmeister 2013: 132.

Taxonomic note. *Dudusa nobilis celebensis* was described from 1 ♂, "N Cel., Tondano"; 1 ♀, "Gurupahi" and 3 ♂♂, "Todjambu near Paloppo" without designation of a holotype. The author traced the type series in NBC, Leiden. To stabilize the nomenclature, hereby the ♀ with the following labels is designatd as lectotype: handwritten "Goeroepalu | N. Celebes | 20. 7. 1917 | W. Kaudern"; yellow label "Photo done by A. Schintlmeister #8227". A lectotype label will be added. The lectotype is illustrated here as fig. 3d; the left forewing is damaged. In the original description Gurupahi was given as type locality. The remaining syntypes becomes paralectotypes.

Diagnosis. *Dudusa celebensis* is readily recognized by yellow rings between the black abdominal segments. It is the smallest Indonesian *Dudusa* species. The forewing pattern resembles much of *synopla*, but the silver forewing spot is smaller and of circular shape. The brown band from the costa is not broadened toward the tornus.

Male genitalia are distinguishable by the pointed valva shape. Sacculus bears a small spine as illustrated and the phallus is characteristically zig-zag shaped.

In the Philippines the sister species of *celebensis*, *Dudusa minor* Schintlmeister, 1993 occur in several subspecies. They are separable by male genitalia, which resemble rather *D. vethi* (Schintlmeister & Lourens, 2010: 42).

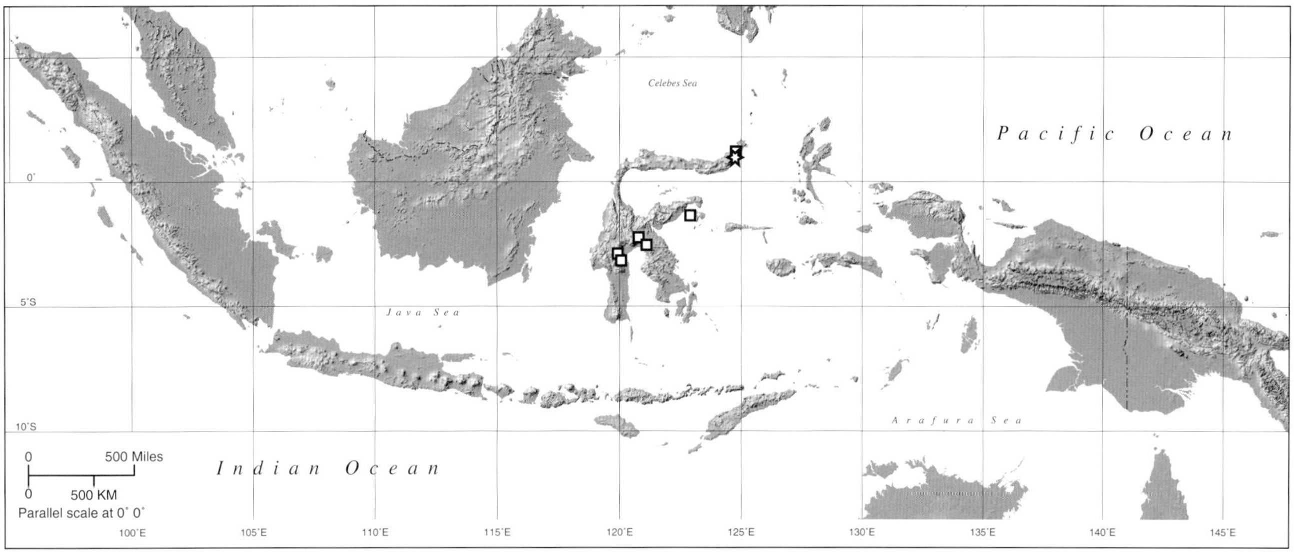

FIGURE 8 Distribution of *Dudusa celebensis*.

Bionomics. Adults occur in Sulawesi commonly at medial altitudes between 700–1300 m, but was also found in lowland forestries.

Distribution. *Dudusa celebensis* is restricted to Sulawesi and Peleng Island.

4. ***Dudusa vethi*** (Snellen, 1892): 40, pl. 4: 1, 2 (*Crinodes vethi*).
   Holotype: ♂, [Indonesia], Sumatra, Soeroelangoen [= Sarolangun] – NBC, Leiden.
   Schintlmeister 2013: 133.

Taxonomic notes. *Dudusa nobilis javana* Roepke, 1944: 79, fig. 1, was described of an unspecified number of ♂♂, ♀♀, Java, Buitenzorg [= Bogor]; Perbawattee; Djunggo-Ardjuno; Nonkodjadjar; Idjen Plains, 1.000–1.500 m – NBC, Leiden.

To stabilize the nomenclature, hereby a ♂ with the following labels is designated as lectotype: "Java | 25. ii. '31 M 2.5.0 | Buitenzorg ... [not readable] | B...] leg. F. Dupont "yellow label "Photo done by A. Schintlmeister #8234". A lectotype label will be added. The remaining 2 ♂♂, 1 ♀ from Buitenzorg traced by the author in NBC becomes paralectotypes.

The lectotype of *javana* is conspecific with *Dudusa vethi,* but represent a distinct subspecies. Distinguished by shorter forewing length and general paler appearance. Male genitalia have a longer sacculus process and thinner, slightly more curved phallus than other subspecies of *vethi.*

Further subspecies of *D. vethi* are:

Dudusa vethi borneensis Roepke, 1944: 79, fig. 3 (*Dudusa nobilis borneensis*).
   Holotype: ♂, [Indonesia], SE Born[eo], Samarinda – NBC, Leiden.

This subspecies appear large. The number of yellowish hairs on the black abdomen is reduced. Male genitalia differ from other subspecies of *vethi* by the rather triangular shaped valva, the reduced size of the sacculus process and the long and slender phallus, which is clearly curved.

Dudusa vethi ferruginea **ssp. nov.** (figs. 4e–f, 4i–j)
   Holotype: ♂, Indonesia, Bali, Bedugul, 1200 m, 8°17'S, 115°09'E, xi. 2005, leg. St. Jakl – in coll. A. Schintlmeister, Dresden.

Paratypes (12 ♂♂, 4 ♀♀):
Bali: 1 ♂, 2 ♀♀, Bedugul, 1200 m, 8°17'S, 115°09'E, xi. 2005 leg. St. Jakl; 2 ♂♂, Bedugul District, Tamblingan Nat. P., 1300 m, 8°14'S, 115°08'E, i–ii.2004 leg. S. Jakl; 3 ♂♂, 1 ♀, Bedugul, 1300 m, 8°16'S, 115°09'E, ii. 2005 leg. St. Jakl; 2 ♂♂, 1 ♀, Tamblingan Lakes, 1300 m, 8°16'S, 115°09'E, vii. 2004, leg. St. Jakl; 4 ♂♂, W. Bali Reserve, Titap, near Papaan 550 m, 8°25'N, 114°58'E, 29.-30.1998, leg. Jakl, Schintlmeister & Cervenka (GU 59-22).

Diagnosis of ssp. *ferruginea*. Forewing lengths ♂♂ 32–34 mm, ♀♀ 45–46 mm. Forewing length compared to ssp. *javana* is 2–3 mm shorter. Ssp. *ferruginea* is distinguished by the fuscous rusty brown color and less contrasting pattern

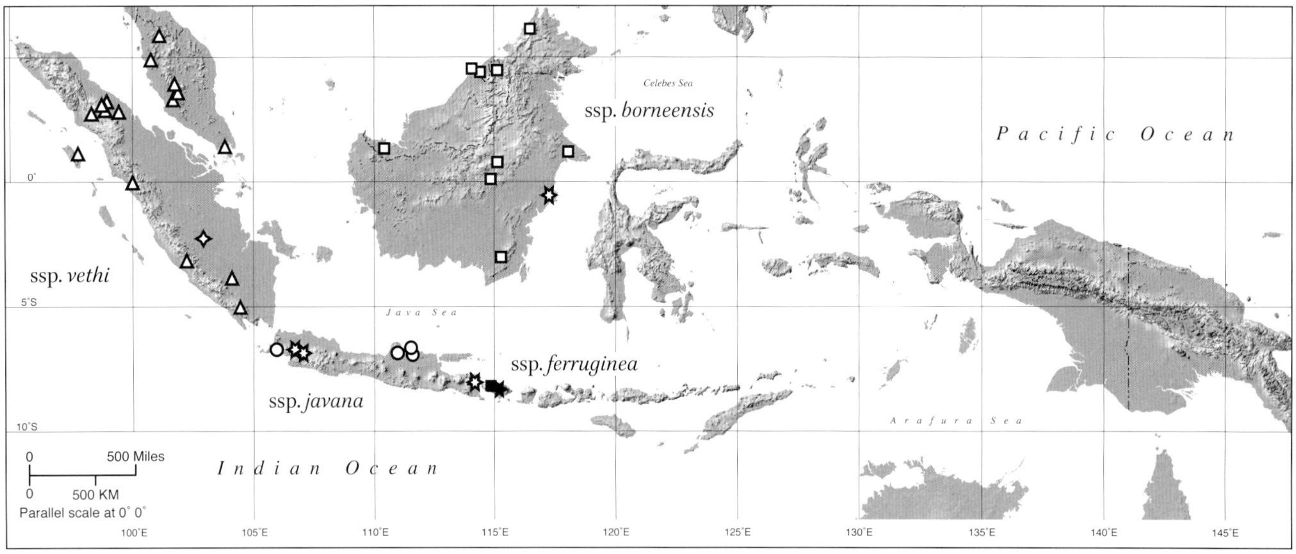

FIGURE 9 Distribution of *Dudusa vethi*.

of the forewings. Silver spot on the forewing reduced in size. The illustrated female (fig. 4l) represents a particularly fuscous individual; most specimens of the type series do match in external appearance to the illustrated holotype. Sexual dimorphism is limited.

Male genitalia are virtually identical to those of ssp. *javana*. The rather straight phallus is a diagnostic character. In ssp. *vethi* and ssp. *borneensis* the phallus is curved.

Diagnosis. *Dudusa vethi* is separable from its congeners by the reduced pale dorsal line of the abdomen, marked with pale yellowish-brown flocci. The weakest development of the dorsal line appears in ssp. *borneensis*. The ssp. *javana* and also ssp. *ferruginea* show the dorsal line stronger. Ssp. *ferruginea* display fuscous colored forewings, while ssp. *borneensis* is the palest subspecies. Forewing pattern as described in the genus diagnosis. The brown band from costa to the forewing tornus is not enbroadened at the margin. The silver spots are smaller than in other congeners. Sexual dimorphism is limited.

Male genitalia of *D. vethi* are well distinguished by a larger saccular process, which is situated near the valva apex.

Bionomics. The common moth prefer in Indonesia the lowlands up to 500 m above sea level, but a few specimens were observed at altitudes up to 1600 m.

Distribution. *Dudusa vethi vethi* occur in S Thailand, Malaya, Sumatra and Nias. Ssp. *javana* is restricted to Java, whereas ssp. *ferruginea* is endemic to Bali. *D. vethi borneensis* flies in Borneo and Palawan.

Tarsolepis Butler, 1872: 125.

Type species: *Tarsolepis remicauda* Butler, 1872.

Schintlmeister 2008: 31; Schintlmeister 2013: 450.

Diagnosis. Medium and large notodontids. Forewings are violet-brown. Antennae of males are bipectinated; the last 40–60% toward the tip is filiform. Antennae of females are filiform. Prominent external characters are the large triangular silver spots on the forewing and the well defined submarginal fascia. Costa of forewings pale brown. Abdomen has an apical tuft of spatulate scales. Underside of the abdomen bears near to the thorax a pair of brushes, which is in most species contrasting red.

Male genitalia resembles *Dudusa* and have a prominent bifurcate uncus and a pair of rather straight and slender socii. Valvae bear a harpe, which is not present in *Dudusa*. Phallus is long and straight. Everted endophallus bear a serrated plate. Female genitalia are characterized by a very long ductus bursae and a large appendix bursae. Corpus bursae is small. No signum.

Due to different structures in antennae, the silver markings on the forewings and differences of male genitalia (uncus and valvae) the species are classified in four subgenera.

Larvae feed on various species of *Acer* (Aceraceae), *Sapindus* and *Nephelium* (both in Sapindaceae).

Currently 16 species are known in the genus. *Tarsolepis* species are distributed in India, Indochina, China, Korea, Japan, Taiwan, the Philippines through the islands of Indonesia to New Guinea and New Britain. The highest diversity of species occur in Indochina.

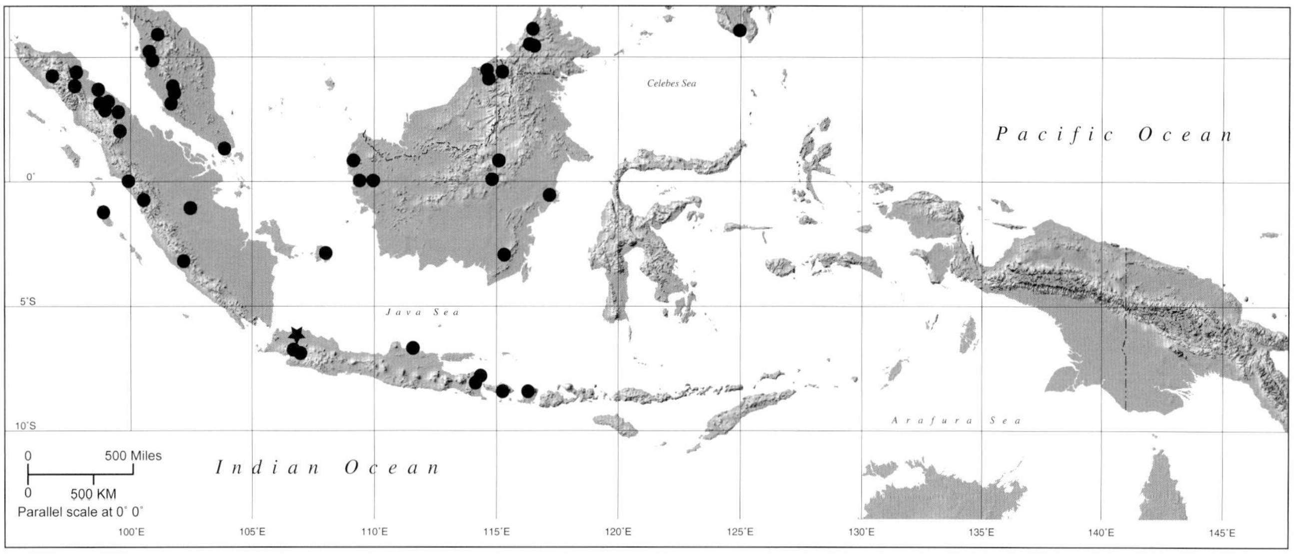

FIGURE 10 Distribution of *Tarsolepis remicauda.*

Subgenus *Tarsolepis* Butler, 1872

5. ***Tarsolepis remicauda*** (Butler, 1872): 125; pl. 8.
   Holotype: ♂, [Indonesia], Java, Batavia – NHMUK, London.
   Schintlmeister 2008: 31, Schintlmeister 2013: 451.

Taxonomic notes. *Tarsolepis remicauda* occur in several subspecies:

Tarsolepis remicauda captura Schintlmeister, 1997: 51; pl. 2: 1, 2; fig. GU1.; holotype: ♂, N Vietnam, Tam Dao, 60 km NW Hanoi, 21°24'N, 105°20'E, 900 m – coll. A. Schintlmeister, Dresden differs from ssp. *remicauda* by its darker hindwings; male genitalia have a larger saccular processes of valvae and shorter phallus.

Tarsolepis remicauda fuscata Schintlmeister & Lourans, 2010: 46; figs. 74-76a; holotype: ♂, Philippinen, Palawan, Mt. Salakot res., 9°51'N, 118°38'E, 800 m – Naturhistorisches Museum, Vienna have the most fuscous external appeareance of all populations and the adults are larger than ssp. *remicauda*.

Diagnosis. The largest *Tarsolepis* in Indonesia. Male antennae are bipectinated over more than 60% of their length; the remainder is filiform. The other Indonesian congeners have the filiform portion between 55–60% of the antenna. Female antennae are filiform. A prominent external character are the two large triangular silver spots of the forewing, which touch the submarginal fascia.

Male genitalia are similar to *T. sommeri*, but the cucullus is broader and its tip rounded. 8th tergite is characteristically sclerotized; the notch of the 8th sternite at the posterior margin is taller in comparison to *sommeri.* Female genitalia are well distinguished from the other congeners by a very long ductus bursae and long and narrow corpus bursae. No signum.

Bionomics. Adults occur commonly in Sundaland up to altitudes of 1800 m.

Distribution. *T. remicauda remicauda* is distributed throughout Philippines (except Palawan), the Malayan Peninsula, Borneo, Sumatra, Siberut, Belitung, Java, Bali and Lombok. Ssp. *captura* occur in Indochina and S. China; ssp. *fuscata* is restricted to Palawan.

6. ***Tarsolepis punicea*** Kobayashi, 2011: 171; figs. 1-4, 7, 7a–i, 10, 10a.
   Holotype: ♂, [Indonesia], Sulawesi, Bantimurung, Gunung Makaroewa, 800 m – National Science Museum, Tokyo.
   Schintlmeister 2013: 451.

Diagnosis. Similar to *T. remicauda,* but forewing length slightly (1–3 mm) shorter and less reddish tinged wing coloration. Antennae of males are strongly bipectinated over 50% of its length; the remainder is filiform. Female antennae are filiform. The silver spot on the forewing does not touch the submarginal fascia. Sexual dimorphism is limited.

Male genitalia of *punicea* resemble *remicauda* but differ significantly by the absence of a saccular triangular

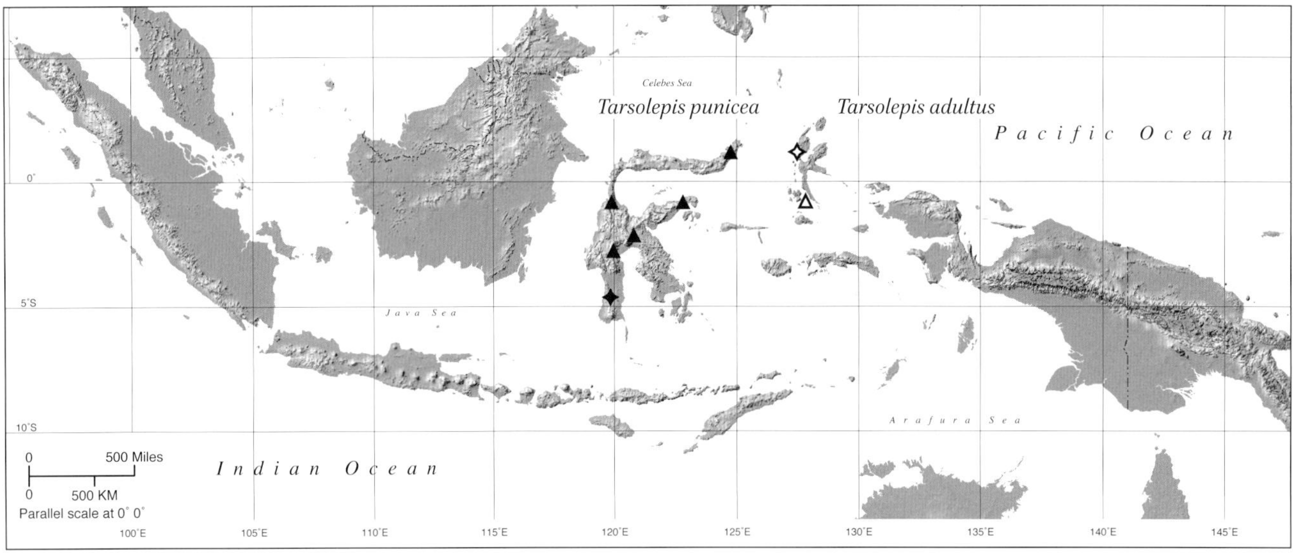

FIGURE 11 Distribution of *Tarsolepis punicea* and *adultus*.

protrusion. Digitus process of the harpe is shorter than *remicauda* and of triangular shape. Phallus is large and strongly curved at the tip. The 8th sternite deeply notched as in *T. adultus*. Female genitalia are chracaterized by the shapes of the ostium, a rather small appendix bursae and a very long corpus bursae. No signum.

The species is known from a few specimens only (n = 10) to the author, which show limited individual variability.

Bionomics. The moth was caught infrequently at altitudes between 800 m and 1300 m.

Distribution. Endemic to Sulawesi.

7. ***Tarsolepis adultus*** spec. nov. (figs. 7a–c)
Holotype: ♂, Indonesia, N Moluccas, Halmahera, 15 km SW Baru, Mt. Talagaranu, 1°21'N, 127°32'E, 600 m, 22.–31. i. 1996 leg. Sinjaev & Tarasov – in coll. A. Schintlmeister, Dresden.

Paratypes (4 ♂♂):
Halmahera: 2 ♂♂, Straße Baru-Basale, Gn. Talagarama, 600 m, 2.–7. iii. 1997, leg. S. Naumann; 1 ♂ Halmahera, vi. 1998 (MWM 1.753).
Bacan: 1 ♂, Bacan, 6 km N Labuna, 4. vi. 1998, (GU 95-01).

Diagnosis. Forewing length ♂♂ 34.5–37 mm. The holotype spans 37 mm. The species resembles *T. punicea* much, but is slightly larger; forewing length *punicea* ♂♂ 32–34 mm. The general appearance is paler and the silver triangle in the medial area of the forewings is significantly broader in *adultus*. It does not touch the grey submarginal area, but is closer to this field than in *punicea*. The prominent contrasting fuscous discal spot on the hindwings underside show a extension toward the base. This line is absent in *punicea*. Both species display a red tuft ventrally at the base of the abdomen as do most *Tarsolepis* males.

Male genitalia are similar to those of *punicea*. A saccular triangular protrusion as seen in *remicauda* is absent in both species. The following characteristics separate *adultus* from *punicea*: arms of the divided uncus are thicker and shorter; socii are shorter, tapered and more compact than in *punicea*. Cucullus is knob-shaped. Phallus is curved and bear a small and strong sclerotized nodule. 8th tergite of both species are differently sclerotized as illustrated, and the margins of the notch of 8th sternite run parallel, not triangular as in *punicea*.

Bionomics. The new species occur infrequently and was caught together with *T. dinawensis extera* in rainforests at 600 m above sea level.

Distribution. Known from Halmahera and Bacan.

8. ***Tarsolepis sommeri*** (Hübner, 1824): pl. 197: 1, 2.
Neotype: ♂, [Indonesia], East Java – NHMUK, London.
Schintlmeister 2013: 451.

Diagnosis. Species of the *sommeri*-group (*T. sommeri*, *splendida* Schintlmeister, 1993 from the Philipines, *elephantorum* Bänziger, 1988 from Indochina and *dinawensis*)

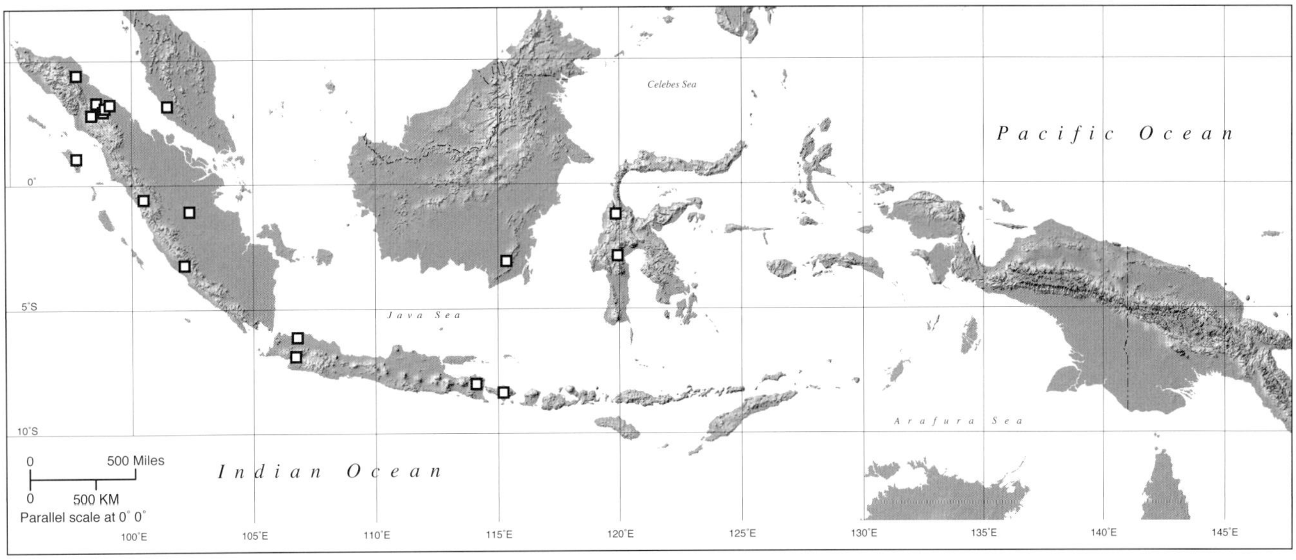

FIGURE 12 Distribution of *Tarsolepis sommeri*.

have the tuft on the abdomen's underside usually not red but brownish colored. *Tarsolepis sommeri* is the smallest Indonesian *Tarsolepis*. Forewing length ♂♂ are 20–31 mm and so distinguishable from the similar species of the *remicauda* group. The filiform portion of male antenna is about 45% of its total length. Forewing color is light brown. Apart from the forewing length there is a considerable individual variation in *sommeri* in wing coloration and in shape of the silver triangles. Hindwings vary from white to fuscous brown. Sexual dimorphism is limited.

Male genitalia are recognizable by a pair of sacculus processes. Thickness of the long uncus arms is subject of individual variation. Phallus is straight and longer than *dinawensis*. Endophallus bear an area of many strongly sclerotized small thorns at the carina. 8th tergite is specifically sclerotized. Female genitalia are characterized by the rather straight margin of the ostium. Ductus bursae very long and slender; appendix bursae is small. Corpus bursae is ellpsoid and narrow, without a signum.

Bionomics. *T. sommeri* occur infrequently up to 1300 m, but most specimens were observed below 500 m.

Distribution. Distributed in Borneo, Malaya, Sumatra, Nias, Siberut, Java, Bali and Sulawesi.

9. ***Tarsolepis dinawensis*** Bethune-Baker, 1904: 373.
Lectotype: ♂, [Papua New Guinea], Dinawa [4.000 ft.] – NHMUK, London **by present designation**.
Schintlmeister 2013: 451.

Taxonomic notes. *Tarsolepis dinawensis* was described from a pair from Dinawa. Both syntype specimens the author traced in NHMUK. To stabilize the nomenclature, hereby a ♂ with the following labels is designated as lectotype: "Dinawa, | B. New Guinea | 4000ft. | Aug. 1902 | Coll. A. E. Pratt" green circular label "Type", blue label Notodontidae genitalia slide No. 1195♂", yellow label "Photo done by A. Schintlmeister #4962. A lectotype label will be added. The remaining ♀ with same collecting label and also with green circular label "Type" become a paralectotype.

There is a good series of *dinawensis* from Halmahera, which represents a distinct subspecies, described hereunder.

Tarsolepis dinawensis extera **ssp. nov.** (figs. 9a, 9e, 9g–h)
Holotype: ♂, Indonesia, Halmahera, Gunung Saululu, 12. ii. 1998, leg. A. Sumbala – coll. A. Schintlmeister, Dresden.

Paratypes (13 ♂♂, 9 ♀♀):
Halmahera: 3 ♂♂, Gunung Ngosiwa, 5. ii. 1999, leg. S. Naumann; 10 ♂♂, 9 ♀♀, Gunung Gagagut, 1.–5. ii. 1999 (GU 71-40, GU 73-41, GU 73-42, GU 73-45) leg. S. Naumann; 1 ♂, Gunung Saululu, 12. ii. 1998, leg. A. Sumbala.

Diagnosis of ssp. *extera*. Forewing lengths ♂♂ 28.0–30.5 mm, ♀♀ 32–35 mm. The holotype spans 30.5 mm. General appearance of the new ssp. is darker than ssp. *dinawensis*. Forewing costa is ash-grey in ssp. *extera*, but pale reddish-brown in ssp. *dinawensis*. Forewings of ssp. *extera* lack all

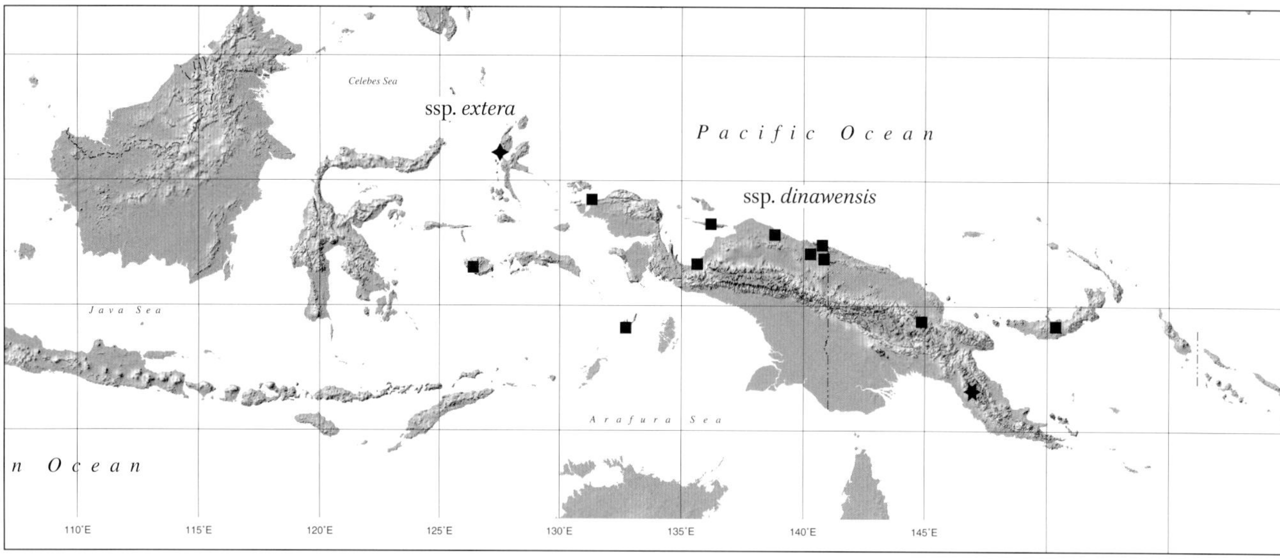

FIGURE 13 Distribution of *Tarsolepis dinawensis*.

reddish-brown pattern and the silver triangle in the medial area is longer.

Male genitalia differ from ssp *dinawensis* in the longer arms of the divided uncus and in a shorter phallus. Valvae shape, shape of saccular processes and the triangular protrusion on the saccus is subject to individual variation and is less useful for discrimination. Female genitalia somewhat resemble *sommeri,* but the antrum is concave shaped at posterior margin.

Diagnosis. *Tarsolepis dinawensis* is the eastern vicariant of *T. sommeri.* It differs from *sommeri* by its darker general appearance, particularly the less contrasting greyish tinged forewing costa. The upper silver triangle of the forewings does not touch the pale brownish submarginal fascia.

Male genitalia differ from *sommeri* by broader saccular valva process. Everted endophallus bears many small thornes, which differ in shape from *sommeri.* 8th sternite is less deeply notched and 8th tergite differs in its sclerotizations from *sommeri.* The female genitalia of ssp. *extera* are characterized by the large and concave postvaginal plate. Apophyses posteriores are, together with *T. malayana,* the shortest within the genus. Antrum is rather short. Ductus bursae is long and very slender.

Bionomics. Adults occur infrequently in New Guinea, but commonly in Halmahera up to 1300 m. Most specimens has been found up to 800 m above sea level.

Distribution. *T. dinawensis dinawensis* is occurs in Buru, Key Islands, New Guinea and New Britain. The ssp. *extera* is restricted to Halmahera.

Subgenus *Tarsolepis* Nakamura, 1976: 35.
Type species: *Tarsolepis malayana* Nakamura, 1976.

Female antennae are bipectinated with a filiform tip, similar to those of the males. Male genitalia have an undivided uncus. Valvae display a costal process.

10. ***Tarsolepis malayana*** Nakamura, 1976: 35; figs. 1, 2, 21a, b.
Holotype: ♂, West Malaysia, Fraser's Hill – National Institute of Agricultural Science, Tokyo.
Schintlmeister 2008: 33, Schintlmeister 2013: 451.

Diagnosis. This large species is readily recognized by the brown-violet color of their forewings. The upper silver triangle touches the paler submarginal area of forewing.

Male genitalia have an unbilobed uncus. Valvae display a prominent basal process on the costa. Female genitalia have a relatively large 8th abdominal segment. Antrum is short and ductus bursae is significantly shorter than in the subgenus *Tarsolepis*.

Bionomics. This beautiful species flies commonly and occur up to 1450 m above sea level, but lower altitudes up to 800 m are preferred.

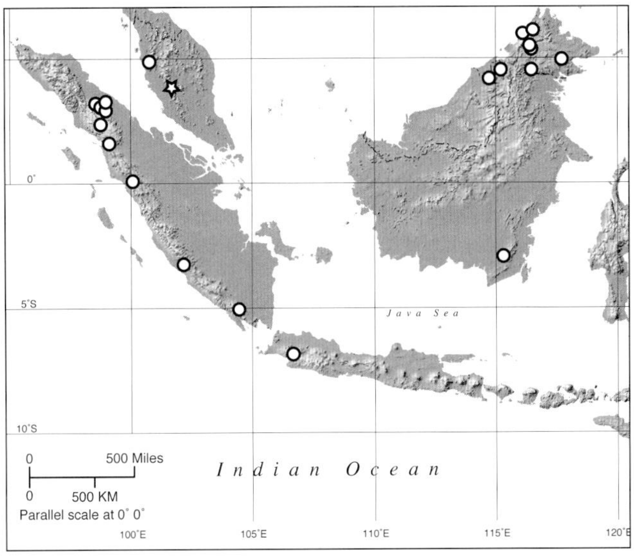

FIGURE 14 Distribution of *Tarsolepis malayana*.

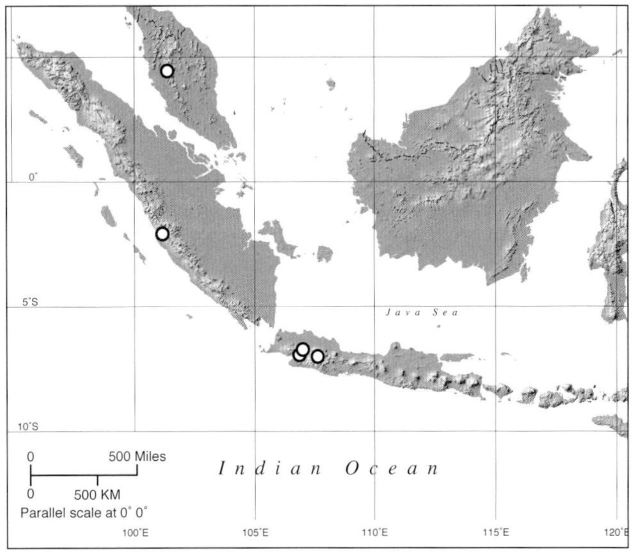

FIGURE 15 Distribution of *Tarsolepis sericeus*.

Distribution. Distributed in Myanmar, Thailand, Laos, Vietnam, southwestern China (Yunnan), the Malayan Peninsula, Borneo, Sumatra, Java.

Subgenus *Stigmatophorina* Mell, 1922: 122.

Type species: *Stigmatophorina* Mell, 1922.

Antennae bipectinated almost until to the tip in both sexes. The last 5–10 % until the tip are filiform. Silver spots on the forewings are reduced. Male genitalia are as in the subgenus *Tarsolepis*.

11. ***Tarsolepis sericeus*** Rothschild, 1917: 252; pl. 5: 16.
Holotype: ♂, [Indonesia], Java – NHMUK, London.
Schintlmeister 2008: 34, Schintlmeister 2013: 450.

Taxonomic note. Schintlmeister & Pinratana, 2007: 37 placed *Stigmatophorina hammamelis* Mell, 1922, described from China, Guangdong as a subjective synonym of *sericeus*, because of external similarities of the Thailand populations to specimens from Java and Sumatra. When genitalia of a Sumatran male were dissected for this book it appeared, that they are very different from those from Thailand and SE China in the length of saccular process, the saccus shape, phallus length and the shape of 8th sternite (fig. 11A). Thus there cannot be any doubt that *hammamelis* represent a distinct species (**stat. rev.**).

Diagnosis. This large and fuscous species is unmistakable by its small silver discal spot. The submarginal area of forewings is of same fuscous brown color as in other parts of the wing. Hindwings are of a paler brown with an indistinct fuscous discal spot. The female is unknown, but is presumable similar to the illustrated female (fig. 11B) of *T. hammamelis*.

Male genitalia have a long bifurcate uncus and very short socii. Valvae are characterized by their long sacculus process, but there is no second saccular protrusion on the sacculus as in other *Tarsolepis*.

Bionomics. The species is rare and only a very few, mostly old males (n < 10) are known. Adults were observed in Sumatra and West Malaysia between 350 m and 1300 m above sea level.

Distribution. Known on the Malayan Peninsula, Sumatra and Java.

Gangarides Moore, 1866: 821.
Type species: *Gangarides dharma* Moore, 1866.
Schintlmeister 2013: 174.

Diagnosis. Medium sized to large moths of red or reddish-brown colored wings and body. The genus is characterized by broad shaped forewings, often displaying a wavy margin. Antennae in both sexes are strong plumous and bipectinated; the last 10–15% toward the tip are filiform. Forewings have a white or black discal spot and four transverse fasciae. Postmedial fascia of forewings ends on the dorsum usually in a white spot. Sexual dimorphism is limited.

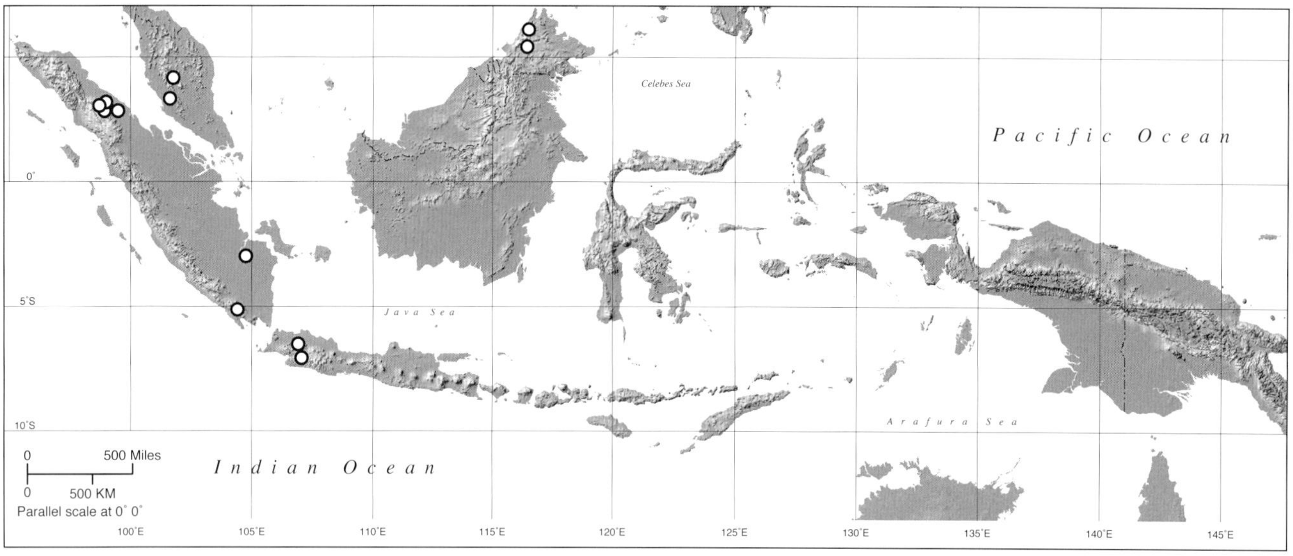

FIGURE 16 Distribution of *Gangarides rosea.*

Male genitalia have a reduced uncus and socii are absent. Valvae are of an unique rounded shape with prominently developed costal and saccular processes. Juxta bear a long y-shaped process. Phallus is long, slender and curved; at the tip are sometimes scale like structures. Carina of the everted endophallus is strongly sclerotized. 8th sternite is concave at the posterior margin and in its shape of taxonomic value. The 8th tergit often concave at posterior margin, but a few species display a convex rounded margin. The shape of the 8th abdominal segments is variable. Female genitalia have an enlarged antevaginal plate, a rather small antrum and a short ductus bursae. Corpus bursae bear a small reniform or circular signum.

Adults of *Gangarides* rest head down like imagines of the closely related genus *Euhampsonia.*

The genus is currently comprised of nine species, which are distributed in the Himalaya, China, Korea, Myanmar, entire Indochina, the Philippines, Sundaland, Java, Sulawesi and the Moluccas. The highest diversity of species is found in Sumatra and in Indochina.

12. ***Gangarides rosea*** (Walker, 1865): 513 (*Apona rosea*).
Holotype: ♂, [India], Hindostan – Type not located in NHMUK.
Schintlmeister 2008: 42, Schintlmeister 2013: 175.

Diagnosis. *G. rosea, brunneus, splendius* and *wallacei* are similar to each other in their external apparance. The margins of the forewings are less serrated than in *brunneus* or *vittipalpis* and not straight as in *splendidus. G. rosea* from Sundaland has the lightest brownish forewing of these four species, and the forewing is paler than the hindwing. From Java there are 2 ♂♂ (genitalia confirmed) with fuscous reddish-brown forewings, and a less wavy margin. A white tornal spot on forewing is small and sometimes inconspicuous. A white surrounded black discal spot is small and less contrasting. The postmedial area often contains several indistinct blackish patches. Hindwings are more deeply red colored than in *brunneus.* Sexual dimorphism is limited.

Male genitalia resembles *splendidus* and *wallacei* by a large costal valva process. The saccular process of *rosea* is thicker compared to that of *splendidus.* Phallus is short and thicker than in the other congeners. 8th sternite is characteristically emarginated with two acute bilateral peaks. Shape of the emargination and the peaks are subject to individual variation as illustrated. 8th abdominal segment of the female genitalia exhibits a serrated marigin and reduced ductus bursae. Antevaginal plate in female genitalia is rectangular shaped with a pair of relatively small flaps. Ostium has a pair of elongated lobes. Corpus bursae bears a circular signum.

Bionomics. *G. rosea* appears in Sundaland infrequently and prefers lower altitudes up to 500 m, but in Himalaya it was observed up to 1700 m above sea level.

Distribution. Distributed in northern India, Nepal, the Andaman Isls., Myanmar, Indochina, southwestern China (Yunnan), the Malayan Peninsula, Borneo, Palawan, Sumatra and Java.

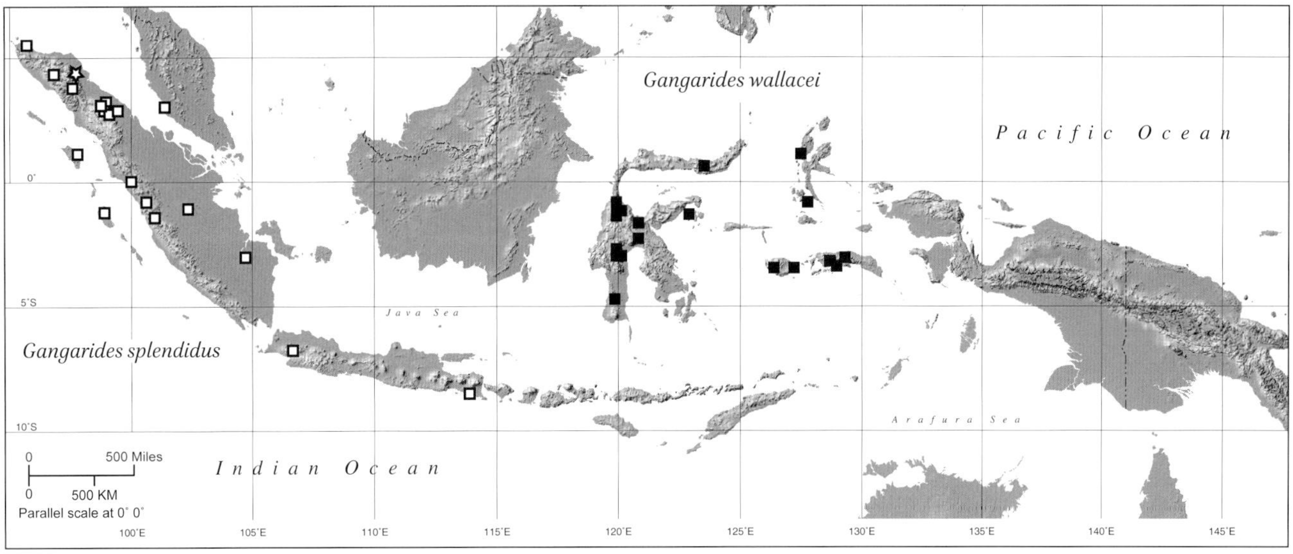

FIGURE 17 Distribution of *Gangarides splendidus* and *wallacei*.

13. ***Gangarides splendidus*** Schintlmeister, 1994: 217; pl. 1: 3, 4; GU-pl. 1: 3.

Holotype: ♂, Indonesia, Sumatra, Aceh, 20 km NW Langsa, 4°32'N, 97°45'E, 80 m – coll. A. Schintlmeister, Dresden.

Schintlmeister 2013: 175.

Taxonomic note. A series (n = 8 ♂♂) from Siberut Isl. is on average smaller (forewing lengths 31–34 mm) than the Sumatran population (35–39 mm, n = 27 ♂♂), but markings and male genitalia are seemingly, apart from individual variability, not different.

Diagnosis. *G. splendidus* is the most reddish-tinged species among the Indonesian *Gangarides*. Margin of the forewings is wavy and straight in shape. Forewings vary from rusty to fuscous reddish-brown; the white tornal spot is relatively large. The black discal spot on forewings is surrounded in white and contrasting. Postmedial area contains a row of black and pale reddish-brown patches. The brown fasciae of forewings are well developed. Female resembles the male but has more intensely red-tinged wings.

Male genitalia are similar to *rosea* and *wallacei*. They are distinguished by the slender saccular processes of the valva and the shape of the tegumen, which display a pair of long protrusions near the uncus. The posterior margin of the 8th sternite is variable but at least not distinguishable from *wallacei*. Female genitalia have a straight and serrated margin of the 8th abdominal segment. Antevaginal plate has a pair of large semicircular flaps. Ostium have no lobes as in *wallacei* or *rosea*. Corpus bursae bears a smaller signum.

Bionomics. *G. splendidus* is in Sumatra the most common *Gangarides* and occur up to 1700 m, but most specimens were observed at lower altitudes up to 500 m.

Distribution. Distributed in Thailand, the Malayan

Peninsula, Sumatra, Nias, Siberut and Java.

14. ***Gangarides wallacei*** Kobayashi & Kishida, 2008: 184; figs. 1, 16, 16a–c

Holotype: ♂, [Indonesia, Maluku], Seram [Isl.] – National Science Museum, Tokyo.

Schintlmeister 2013: 176.

Diagnosis. *G. wallacei* is the sister species of *rosea* and *splendidus*. It is best distinguished from them by the black punctate discal spot of the forewing, which is not encircled by white. Shape of the forewing margin is similar to *splendidus*. Postmedial area of forewing often contains a row of large and light brownish patches. Sexual dimorphism is limited; females are for the most part more red-tinged.

Male genitalia are similar to those of *splendidus* and differ by their shorter tegumen protrusions near the uncus, the taller uncus and the relatively long saccular processes. Phallus is kinky at the tip. 8th sternite is as in *splendidus* and individual width variation is similar as well. Female genitalia are characterized by a sinous

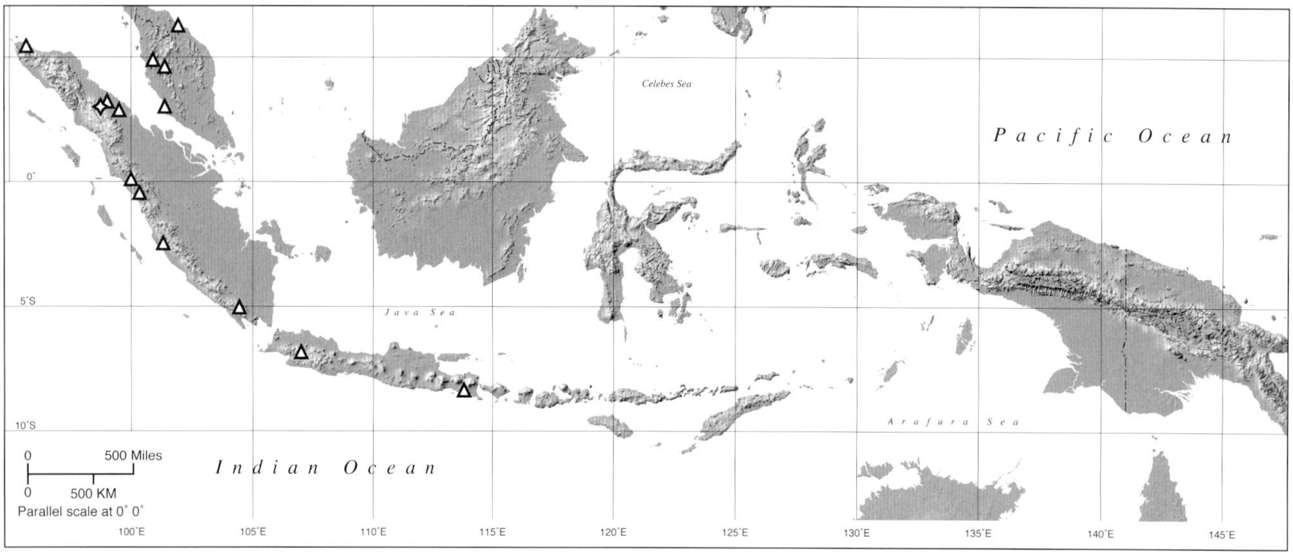

FIGURE 18 Distribution of *Gangarides brunneus*.

margin of the 8th abdominal segment. The antevaginal plate is large with two lobes. Ductus bursae is short; corpus bursae bears a circular signum.

Bionomics. *G. wallacei* is a common species and occurs up to altitudes of 1500 m, but most specimens were taken at lower altitudes up to 700 m.

Distribution. Distributed in Sulawesi, Peleng, Bacan, Halmahera, Buru and Seram.

15. ***Gangarides brunneus*** Schintlmeister, 1994: 217; pl. 1: 5, 6; GU-pl. 1: 4.
Holotype: ♂, Indonesia, North Sumatra, Sindar Raya 2, 15 km SW Sindar Raya, 3°10'N, 98°56'E, 500 m – coll. A. Schintlmeister, Dresden.
Schintlmeister 2013: 175.

Diagnosis. The species is readily recognized by its small size, the less contrasting wing coloration and the strong serrated, slightly falcate margin of the forewings. Postmedial area of the forewing contain a row of prominent blackish patches. The black discal spot on the forewings is very small and in some specimens not visible. The tornal spot is whitish, less contrasting and inconspicuous. Sexual dimorphism is limited, but forewing margin of the female is less sinous.

Male genitalia are distinguished by the short saccular processes of valvae and by a short tegumen process near the saccus. Phallus is less curved than that of other congeners, with a small projection in the middle part. 8th sternite is triangularly notched at the posterior margin. 8th tergit is less modified and not bilobed at the margin. Female genitalia have a broad 8th abdominal segement, having a sinous margin. Ostium is large with a pair of tapered lobes. Ductus bursae is compact and short, but longer than in *G. rosea, splendidus* or *wallacei*. Corpus bursae bears a reniform signum.

Bionomics. The moth occurs in Sumatra infrequently and rare in the Malayan Peninsula. They appear at lower altitudes up to 500 m. One male was taken in Sumatra, Gunung Sanggul, at 1700 m above sea level.

Distribution. Distributed in southern Myanmar, southern Thailand, the Malayan Peninsula, Sumatra and Java.

16. ***Gangarides vittipalpis*** (Walker, 1869): 90 (*Lonomia vittipalpis*).
Holotype: ♀, [N. India], near Benares [= Varanasi] – NHMUK, London.
Schintlmeister 2008: 42, Schintlmeister 2013: 175.

Diagnosis. *G. vittipalpis* is distinguished by several indistinct brown fasciae in the postmedial area of the forewing. Forewing color varies from pale reddish-brown, as in *G. brunneus,* to intensly red-tinged. The black discal spot is not encircled by white; the white tornal spot is prominent and constrastingly developed.

Male genitalia are very different from their congeners and readily recognized by the large subscaphium and a pair of tapered tegumen processes. Valvae bear no processes. Phallus is relatively short and curved. 8th sternite is widely bilobed with tapered lobes at the posterior

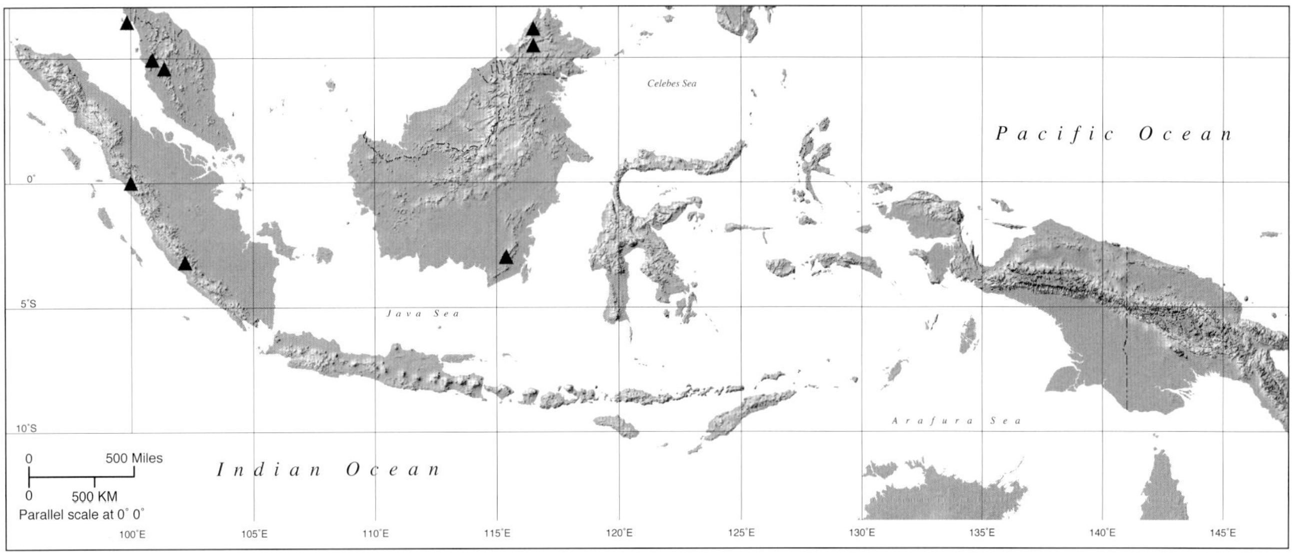

FIGURE 19 Distribution of *Gangarides vittipalpis*.

margin. The lobes are somewhat variable as illustrated. 8th tergite is convex rounded at posterior margin and less modified. Female genitalia have a serrated margin in the 8th abdominal segment. Postvaginal plate is very large with a pair of rounded lobes. Ductus bursae is of the same length as in *G. brunneus*, but slenderer. Corpus bursae has a small circular signum.

Bionomics. The species is common and widespread in continental Asia. In Sumatra and Borneo it is a rare species, reported only from a few localities (1100 and 1500 m above sea level).

Distribution. Distributed in northern India, Nepal, Myanmar, southwestern China (Yunnan), Indochina, the Malayan Peninsula, Borneo and Sumatra.

17. ***Gangarides vardena*** Swinhoe, 1892: 270; pl. 7: 3.
Holotype: ♂, Borneo – Oxford University Museum of Natural History.
Schintlmeister 2008: 42, Schintlmeister 2013: 175.

Taxonomic note. This species have clearly different valvae in relation to the genitalia of other Indonesian *Gangarides*, but in fact they are a development of a serrated sacculus of some continental species in the genus (*G. rufinus* Schintlmeister, 1998, *G. dharma* Moore, 1866).

Diagnosis. Easy to distinguish from its congeners by the small size and the black patch near the apex of the forewings. *G. vardena* is, however a variable species, where the color of the wings varies (in any combinations) from pale yellowish-brown to splendid fuscous red or red-brown. Sometimes the black apex patch of the forewings is weakly developed or absent. Hindwings vary from pale brown to red. Most specimens have only the anal and basal region reddish. Females tend to have a less wavy forewing margin and show more reddish-tinged hindwings.

Male genitalia are very well distinguished by their strongly sclerotized and serrated valvae which bear no processes. Other features such as Y-shaped juxta process and the slender and curved phallus are as in most *Gangarides*. 8th sternite is well distinguished by a pair of tapered processes at the posterior margin. 8th tergite is slighty bilobed at the posterior margin. Female have a large postvaginal plate without projections. Antrum is large; ductus bursae is long and slender. It is the longest within the genus. Corpus bursae bears a strongly sclerotized knob-shaped signum.

Bionomics. The species is widespread and common. It occurs from the lowland (150 m) up to 1700 m above sea level.

Distribution. Occuring in Myanmar, Thailand, Laos, Vietnam, the Malayan Peninsula, Borneo and Sumatra.

***Diehleana* gen. nov.**
Type species: *Nadata gigantea* Druce, 1909 **by present designation.**

Taxonomic note. *Nadata gigantea* was transferred by Gaede, 1934: 9 to *Euhampsonia* Dyar, 1897 and by Kobayashi & Kishida, 2008: 181 to *Gangarides*. But it is also

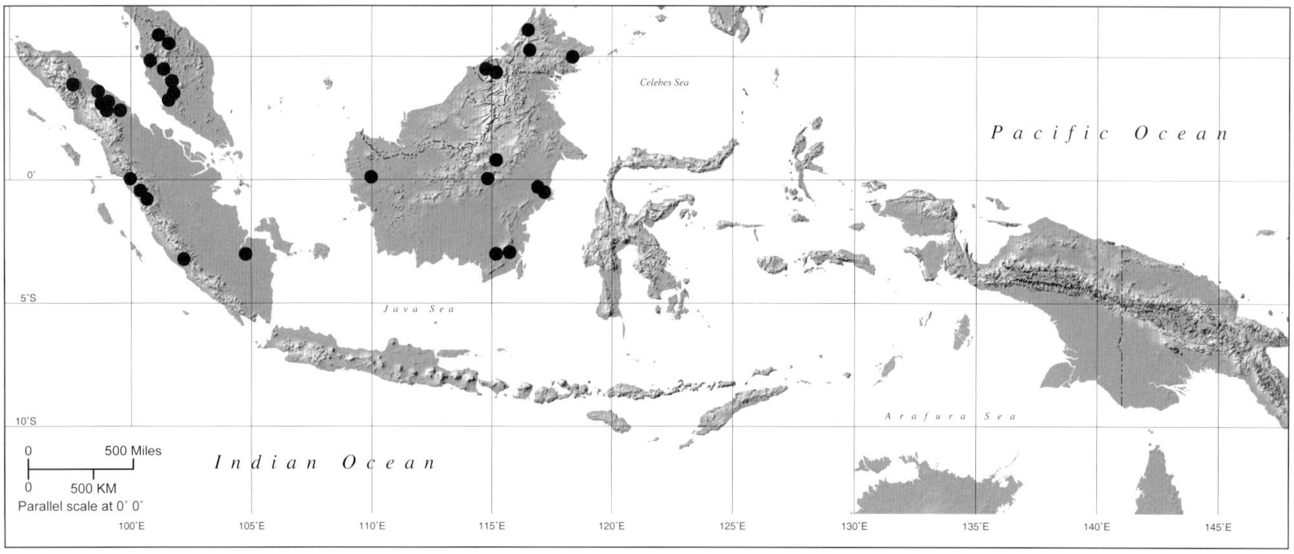

FIGURE 20 Distribution of *Gangarides vardena.*

misplaced in *Gangarides* based on the structural differences in the adults including the morphology of their genitalia.

Etymology. The genus is named for the late Dr. Eduard W. Diehl (b. 12. ii. 1917 in Saarbrücken d. 19. viii. 2003 in Medan, Sumatra), a great explorer of Sumatran Lepidoptera. He guided and instructed the author in collecting under tropical conditions during his first expedition to Sumatra and Nias Island in 1979.

Diagnosis. The new genus include currently two large (forewing length ♂♂ 45–60 mm, the few females at the author's disposal being much larger) species. Antennae in both sexes are strongly plumous and bipectinated as in *Gangarides.* The last 10–15% at the tip are filiform. Adults share externally several features with *Euhampsonia* such as: a strongly serrated forewing margin with an acute apex and the venation as well. It differs from *Euhampsonia* by a punctuate fine black forewing discal spot and several brown fasciae. *Gangarides,* which is also closely related, have a less elongated forewing shape. Hindwings are orange to reddish-brown. Sexual dimorphism is limited. The female of *Diehleana* is much larger than the male.

Male genitalia have, contrary to *Gangarides,* a small but well developed uncus. Socii are absent. *Euhampsonia* have a large and long often bifurcate uncus and a pair of large socii. The tegumen of *Diehleana* is compact and of circular shape. Near the uncus there is a pair of tapered tegumen processes. Juxta bear a y-shaped weakly sclerotized process, similar to *Gangarides.* Saccus is not pronounced. Valva is short, with two tapered processes at the apex and sacculus. The harpe produces a unique fan-shaped strongly sclerotized, serrated structure (digitus projection). Phallus is rather straight; the endophallus bears a thorny microstructure at the tip. There are no cornuti. 8th sternite has centrally at the posterior margin a pair of tapered processes. Female genitalia of *D. gigantea* have large papillae anales and large and compact apophyses. 8th abdominal segment is broad. The postvaginal plate is in comparison to *Gangarides* very small. Ductus bursae is long and slender. The circular corpus bursae has no signum.

The genus is currently comprised of two species, which are distributed in southern Thailand, Sundaland and – disjunct – New Guinea.

18. ***Diehleana gigantea*** (Druce, 1909): 347 (*Nadata gigantea*) **comb. nov.**

Holotype: ♀, [Indonesia, W Papua], Dutch New Guinea, Fak Fak, 1.700 ft. – NHMUK, London.

Schintlmeister 2013: 175.

Diagnosis. Unmistakable by its size and the strong serrated margin of the forewings. It differ from *D. gilesi* by its greater forewing length and the presence of a white patch near the forewing tornus. Individual variability is limited. The female resembles the male.

Bionomics. The species is common and was mostly observed in lowland forests up to 500 m. But there are also a few records from 750 m up to 1700 m.

Distribution. Distributed in W Papua. Not yet found in Papua New Guinea.

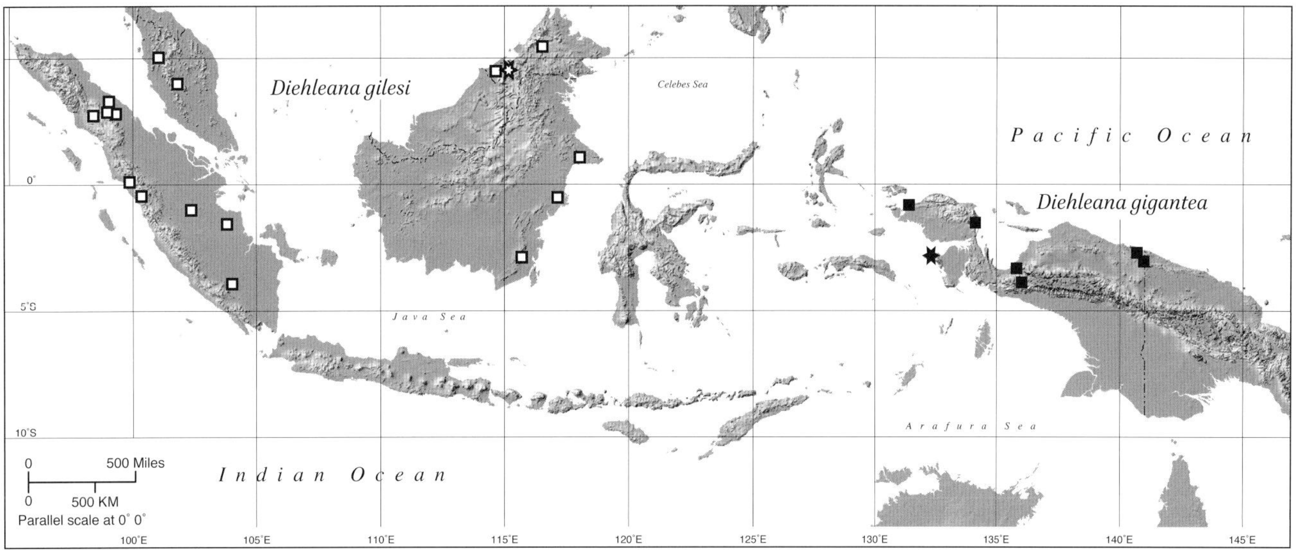

FIGURE 21 Distribution of *Diehleana gigantea* and *gilesi*.

19. ***Diehleana gilesi*** (Holloway, 1983): 33; pl. 2: 10; fig. 25. (*Euhampsonia gigantea gilesi*) **stat. nov.**
Holotype: ♂, Brunei, Ulu Temburong, 300 m – NHMUK, London.
Schintlmeister 2013: 175.

Taxonomic note. The taxon was described as a subspecies of *gigantea*. But it is raised here to specific level as a genuine species, because of some structural differences of the male genitalia; see below.

Diagnosis. A large species, which is readily recognized by its size and the serrated forewing margin. It differs from *D. gigantea* by its smaller wingspan and the absence of a white tornal patch. The female is unknown.

Male genitalia of both species differ as follows. The uncus of *gigantea* is bifid, in *gilesi* not. The juxta of *gilesi* is bifurcate, in *gigantea* Y-shaped. The digitus projection of the valva is smaller in *gilesi*. At the saccus there is a prominent process. The upper valva arm is reduced in size in *gigantea*, but long and pointed in *gilesi*. Phallus of *gigantea* is longer than in *gilesi*. 8th sternite displays in *gigantea a* central triangularly notched area. In *gilesi* it is not splayed but the processes are curved. 8th sternite of *gilesi* is gently rounded at the caudal margin, and broadly notched.

Bionomics. The species is rare but seemingly widely distributed. Most specimens were taken in lowland forests up to 500 m. But there are is also a record from Sumatra at 1700 m altitude.

Adults of *D. gilesi* rest head down like the imagines of the closely related genus *Euhampsonia*.

Distribution. Distributed in southeastern Myanmar, southern Thailand, the Malayan Peninsula, Borneo and Sumatra.

Euhampsonia Dyar, 1897: 16.
Type species: *Trabala niveiceps* Walker, 1865.
Schintlmeister 2013: 154.

Large moths. Males have long, the females short bipectinate brown antennae. The pectinations reach nearly to the tip. The margin of the forewings is of a prominent, serrated shape. The yellowish pale discal spot on the forewings is characteristic for the genus.

A large bifurcate uncus, a pair of long socii and a costal valva process characterize the male genitalia. Distinguished also by the shape of the 8th sternite, which is deeply bilobed. Females have long apophyses posteriores and a large 8th abdominal segment. Corpus bursae is ovoid and bear a reniform signum.

The resting position of the adults is head down as in *Gangarides*. Hostplants of the larvae of the palaearctic species are *Quercus* spp. (Fagaceae).

Euhampsonia is represented by eight species, distributed from the Himalaya, Myanmar, Mainland China, the Far East, Japan, Taiwan, Indochina and Sundaland. The greatest diversity of the species is found in Indochina.

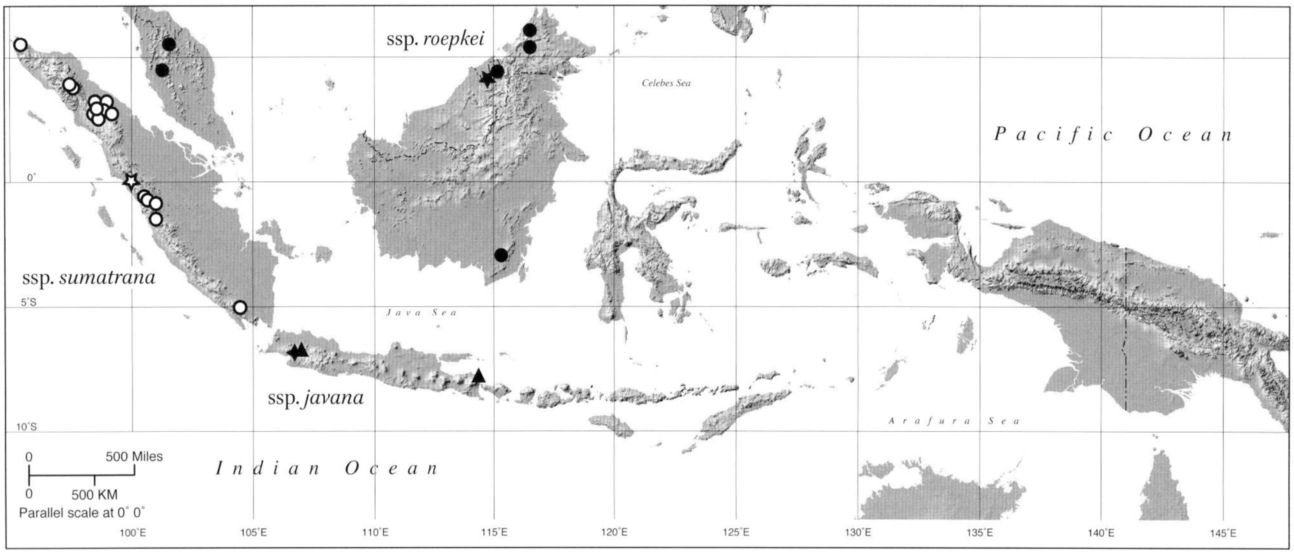

FIGURE 22 Distribution of *Euhampsonia roepkei*.

20. ***Euhampsonia roepkei*** Holloway, 1983: 32; pl. 2: 9; fig. 31.

Holotype: ♂, [Malaysia], Borneo, Sarawak, Gunong Mulu Nat. Park. Site 5, Camp 4, Mulu 1.780 m – NHMUK, London.

Schintlmeister 2013: 154.

Taxonomic notes. *Euhampsonia roepkei* varies geographically in respect of the male genitalia. The following subspecies are recognized:

Euhampsonia roepkei roepkei Holloway, 1983

Diagnosis of ssp. *roepkei*. Uncus deeply notched; phallus tapered and strong curved. 8th sternite is notched with a small processes on each side of the posterior margin.

A dissected male from the Malayan Peninsula does not differ from the Bornean material.

Euhampsonia roepkei sumatrana **ssp. nov.** (figs. 20c, 20i).

Holotype: ♂, Indonesia, North Sumatra, Brastagi, 1600 m, Nebelwald, 3°01'N, 98°31'E, 21. viii. 1979, leg. E. Diehl & A. Schintlmeister – in coll. A. Schintlmeister, Dresden.

Paratypes (50 ♂♂, 9 ♀♀):

Sumatra: 2 ♂♂, Aceh, Ketambe, 1. xii. 1980 (GU 23-61); 1 ♂, Gunung Leuser N.P., Ketambe, 35 km N Kutacane, 400 m, 3°40'N, 97°40'E, leg. Malicky; 1 ♂, Aceh, Ketambe, Alas valley, 400 m, 28. xii. 1980; 1 ♂, Dairi Mts., 1600 m, 10. x. 1980; 1 ♂, Dairi Mts., 15 km W Sidikalang, 850 m; 1 ♂, Paritohan, Porsea, 90 m, 13. ix. 1980; 1 ♂, Dolok Merangir, 180 m, x. 1970–i. 1971, leg. Dr. E. W. Diehl; 1 ♂, dto. ix. 1975; 2 ♂♂, dto. 30.x.–9. xii. 1967; 2 ♂♂, dto., 18. vi. 1967; 3 ♂♂, Prapat, Holzweg 2, 30 km S Pematang Siantar, 1050 m, 2°48'N, 98°55'E, 30. viii.–28. ix. 1991, leg. Graul & Schintlmeister; 1 ♂, dto., 31. iii. 1993, leg. R. Brechlin; 1 ♂, dto., 16. iv. 1985, leg. Dr. Diehl; 1 ♀, dto., 3. xii. 1972; 1 ♂, dto. 10. viii. 1980; 2 ♀♀, dto., 18. x. 1982, leg. Dr. Diehl (GU42-32); 1 ♀, dto., 31. i. 1984, leg. W. Nässig & Diehl; 1 ♂, Brastagi, 1600 m, 3°01'N, 98°31'E, 21. viii. 1979, leg. E. Diehl & A. Schintlmeister; 1 ♂, Berastagi, 1500 m, viii. 1985; 1 ♂, dto., 1800 m, 6. x. 1972 leg. Diehl; 1 ♂, Tele, 1600 m, 1974; 2 ♂♂, Tobasee, Samosir Isl., Partunghoan, 1600 m, 2°37'N, 98°44'E, 20. ix. 1991; 1 , Aek Tarum, 7. vi. 1978, leg. Dr. E.W. Diehl (GU 180); 1 ♂, 30 km SW Solok, Gunung Talang, TV-Station, 1600 m, 1°00'S, 100°38'E, 7.–9. x. 1981, leg. Schintlmeister & Widagdo; 11 ♂♂, 1 ♀, ibid, 1500 m, vi. 2005 leg. Jakl; 3 ♂♂, 1 ♀, Lake Data, Mt. Talang, 1550 m, 1°01'S, 100°42'E, xii. 2006, leg. S. Jakl; 4 ♂♂, Mt. Sanggul, 1700 m, 0°30'S, 100°30'E, xii. 2004, leg. S. Jakl; 3 ♂♂, 2 ♀♀, Mt. Kerinci, iii. 2005, leg. Jakl; 1 ♀, Bukit Subang, 22 km E Padang, 1200 m, 0°40S, 100°12'E, 18.–20. x. 1981, leg. Schintlmeister; 3 ♂♂, 28 km E Liwa, 1000 m, 5°02'S, 104°18'E, 31. x. 1981, leg. A. Schintlmeister (GU 43-31).

Diagnosis for ssp. *sumatrana*. Forewing lengths ♂♂ 41–53 mm, ♀♀ 52–60 mm. The subspecies is not externally different by appearance from Bornean populations.

Male genitalia differ from ssp. *roepkei* by their thicker and less deeply notched uncus tip, more compact socii and a shorter phallus. 8th sternite is somewhat variable, but the short processes on the side on their posterior margins, which are characteristic for ssp. *roepkei,* are absent.

Female genitalia have the postvaginal plate more deeply notched, but only one female of ssp. *sumatrana* has been dissected.

Euhampsonia roepkei palawana **ssp. nov.** (fig. 20e).

Holotype: ♂, Philippines, Palawan, Mt. Mantalingahan, abandoned village, Kibyawon in prim rainforest, 950 m, 8°46'59.0"N, 117°42'06.7" E, 26.–30. v. 2001, leg. A. & S. Zwick – in coll. A. Schintlmeister, Dresden.

Paratypes (5 ♂♂, 3 ♀♀):

Palawan: 2 ♂♂, 3 ♀♀, Mt. Mantalingahan, 600–800 m, 2.–12. viii. 2000, leg. D. Mohagan (GU 73-32, GU 95-100); 2 ♂♂, Mt. Mantalingahan, house of Ening, 1100 m, 8°47'16.1"N, 117°42'03.3" E, 21.–22. v. 2001, leg. A. & S. Zwick; 1 ♂, Mt. Mantalingahan, abandoned village, Kibyawon, 950 m, 8°46'59.0"N, 117°42'06.7" E, 16.–19. v. 2001, leg. A. & S. Zwick.

Diagnosis for ssp. *palawana*. Forewing lengths ♂♂ 42–44 mm; ♀♀ 50 and 51 mm. Forewing length is on average 2 mm shorter than those of ssp. *roepkei* and ssp. *sumatrana*. Forewing pattern ise not different to ssp. *roepkei*.

Male genitalia have have a longer and narrower uncus, which is less deeply notched in comparison to ssp. *roepkei* or ssp. *sumatrana*. The phallus is the shortest of all known subspecies of *roepkei*. 8th sternite lacks the spines at the posterior margin.

Euhampsonia roepkei javana **ssp. nov.** (figs. 20g–h, 20k).

Holotype: ♂, Indonesia, W. Java, Pagrango N. P., Mt. Kencana, 1450 m, 15. iii. 2007 – in coll. A. Schintlmeister, Dresden.

Paratypes (15 ♂♂):

Java: 4 ♂♂, Pagrango N. P., Mt. Kencana, 1450 m, 15. iii. 2007; 4 ♂♂, Pagrango 1625 m, 30 km SE Bogor, 6°30' S, 107°10'E, 16.-20. ii. 1996, leg. V. Sinaev & E. Afonoin; 6 ♂♂, Mt. Gedeh, Pagrango N. P., 6°47' S, 107° E, 1250 m, iv.–vi. 1996 (GU 42-30, GU 95-99); 1 ♂, East Java, Mt. Baluran, 600–1000 m, iv. 1996.

Diagnosis for ssp. *javana*. Forewing lengths ♂♂ 39–42 mm. It is significantly smaller than ssp. *roepkei* (forewing lengths 46–48 mm). The forewing pattern is as in ssp. *roepkei*. The female is unknown.

Uncus of male genitalia is shorter, narrower and less deeply notched at the tip. The valva costa is curved in the apical part. Phallus is rather straight in the middle part and less curved as in comparison to other subspecies of *roepkei*. 8th sternite is sinous at the posterior margin and without spines.

Diagnosis. The species in Sundaland is rather unmistakable by size and shape of the serrated forewings with its acute apex and the prominent yellowish discal patch. Forewing color, in particular the greenish tone, is subject to individual variation. Sexual dimorphism is limited; the females are much larger than the males. The group of allopatrically sister species, *E. serratifera* Sugi, 1994 (Indochina and China), *E. niveiceps* (Walker, 1865) (Himalaya) and *formosana* (Matsumura, 1925) from Taiwan are externally very similar but *roepkei* is by far the largest species in the group.

Male genitalia are distinct by the bifurcate uncus, the costal valva process and the shape of the curved phallus. 8th sternite is bilobed. Female has a large and postvaginal plate notched at its posterior. Antrum is very broad. Ductus bursae is short. The circular corpus bursae bears a reniform signum.

Bionomics. The moth is common and appears between 180 m and 1700 m above sea level. Adults rests with its head down as do the closely related species of *Gangarides* and *Diehleana* head down.

Distribution. The ssp. *roepkei* is distributed in Borneo, the Malayan Peninsula and S Thailand. Subspecies *sumatrana* is restricted to Sumatra and ssp. *javana*. Endemic to Java.

Scranciinae

Ortholomia Felder, 1861: 39.
Type species: *Ortholomia moluccana* Felder, 1861.
Schintlmeister 2008: 47; Schintlmeister 2013: 307.

Middle-sized moths of elongated forewing shape. Tornus of forewings sharply angled. Antennae of males are bipectinate with short rami; those of females are filiform. Forewing color is fuscous brown to black. An indistinct whitish tornal spot near the dorsum of the forewing, often accompanied by deep black spot, is a characteristic pattern element of the genus *Ortholomia*. Toward the base of the forewing a prominent black dorsal streak is visible. The species of the genus are very similar in their external appearance; however, they are well distinguished by their male genitalia.

Genitalia of *Ortholomia* are characterized by the modified shape of the sacculus. Uncus is long and slender, sometimes with knob-shaped tip. 8th sternites and tergites are not significantly modified, but the second abdominal segment bears a pair of larger cteniophores, which is probably a characteristic feature of *Ortholomia*. Female genitalia have long apophyses posteriores and relatively short apophyses anteriores. Postvaginal plate bears a wing-shaped strongly sclerotized structure. The ductus bursae is compact and rather short; the corpus bursae being large and of circular shape. A diagnostic feature of the genus is the large circular signum in the corpus bursae which bears prominent transverse sclerotizations of specific value.

Ortholomia is distributed from S China, Indochina, Sundaland, the Philippines, and Indonesia up to Australia and the Solomon islands. 11 species are currently known, with highest diversity found in Sumatra and Mindanao. Most of the species occur allopatrically.

21. ***Ortholomia moluccana*** Felder, 1861: 40.
Holotype: ♂, [Indonesia], Amboina – NHMUK, London.
Schintlmeister 2013: 307.

Taxonomic notes. *Osica glauca* Walker, 1865, described from Australia, was placed by Gaede, 1934: 145 as a synonym of *O. moluccana*. The populations from Australia are not separable by external appearance from Papuan specimens due to the width variations in external appearance, but they are different in respect of several important characters: socii are slenderer, sacculus of the valvae have a shorter spine on the left side, clasper is larger and more rounded. The phallus is not pointed at the tip. Australian populations has been described as *Osica* Walker, 1865, and where placed in synonymy by Gaede, 1934: 145. *O. glauca* is reinstated as a distinct species (**stat. rev.**).

Parathemerastis turneri var. *melanistis* Bethune-Baker, 1916: 383 was described from a single ♀ from Mount Kebeae, Papua New Guinea. It was raised to specific level by Gaede, 1930: 683. Examination of the holotype in NHMUK, including the genitalia slide BM 58, showed, that it is conspecific as also *O. turneri* Bethune-Baker, 1904, with *Ortholomia moluccana* and thus belongs to the most common form, with blackish forewings and a contrasting white rectangular discal patch (fig. 21g). The taxon *melanistis* is placed hereby as a junior synonym of *moluccana* (**syn. nov.**).

Diagnosis. *O. moluccana* is the largest species within the genus; forewing length about 4 mm longer than in the largest other congeners. Only a few individuals of *moluccana* are comparable in size to *O. xylinata*. The forewing color of *moluccana* varies from black, grey, fuscous brown up to pale yellowish brown. The prominent white discal patch of quadrangular shape is a diagnostic character for identification, if present. If it is replaced by a brown or black spot the visibility is limited to fuscous forewings. Sexual dimorphism is limited.

Male genitalia are distinct within the group (apart from *glauca*) by the tapered uncus. The bulbous shaped sacculus, slightly asymmetrically developed on both valvae, bears a long and tapered process. The shape of the phallus is conical toward the tip. The second abdominal segment bears a pair of large cteniophores. Female genitalia are as described under genus, but the wing shaped structures of postvaginal plate are rather small. Signum is strongly sclerotized and show X-shaped transverse bands, which separates *moluccana* from the other congeners.

Bionomics. The species flies most commonly in the lowlands of W Papua up to an altitude of 500 m. A few specimens have been caught from 1100 m up to 1850 m above sea level. In the adjacent islands, the species was found sparsely and only at lower altitudes.

Distribution. Known in New Guinea in its entirety, Biak (Schouten Isls.), Tobriand Isl., Vulcan Isl., Goodenough

 | DOI:10.1163/9789004354258_004

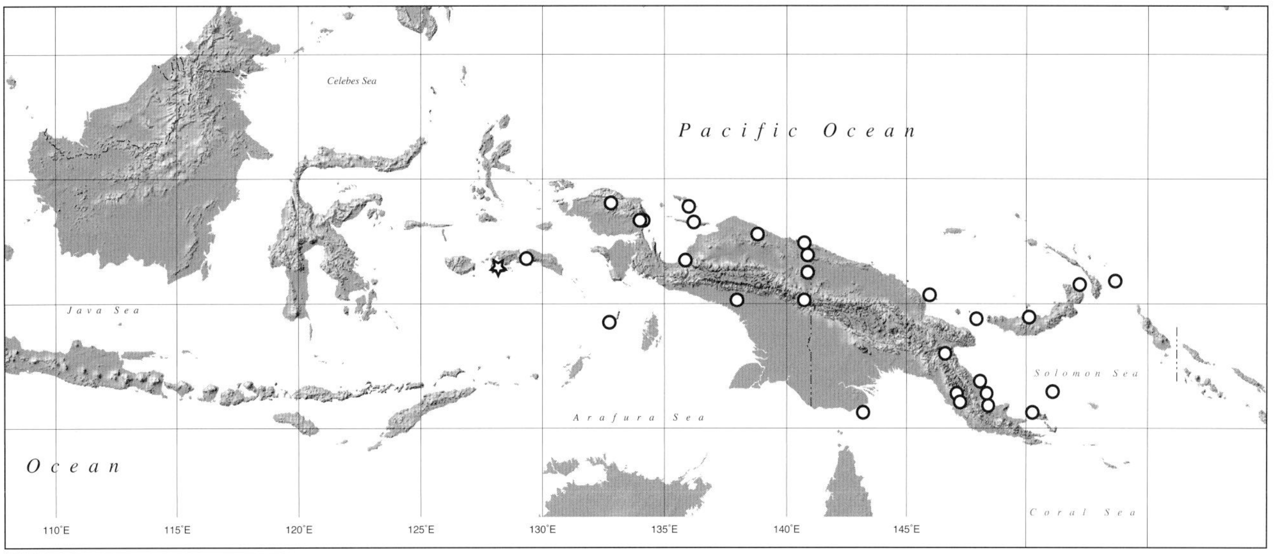

FIGURE 23 Distribution of *Ortholomia moluccana*.

Isl., New Britain, the Solomons (genitalia not checked), Key Islands and Ambon.

22. *Ortholomia gilolo* spec. nov. (figs. 22a–c).

Holotype: ♂, Indonesia, Halmahera, 15 km SW Baru, Mt. Talagaranu, 1°21'N, 127°32'E, 150 m, 22.-31.vi.1996, leg. V. Sinjaev & E. Tarasov (GU 79-06) – in coll. A. Schintlmeister, Dresden.

Paratypes (1 ♀):

Halmahera: 1 ♀, 23 km SW of Tobelo, Tunuo Camp, 1°32'40"N, 127°23'50"E, 19.-22.ix.1995, leg. de Jong & Ansari.

Etymology: Named after the former name of Halmahera Island: Gilolo.

Diagnosis. Forewing length ♂ 20.5 mm, ♀ 25 mm. This species is similar in external appearance, including wing shape, to *O. moluccana*, but clearly (4 mm to 6 mm) smaller in forewing length. A black tornal spot is present on forewings in both sexes, accompanied in the male by an indistinct and less visible whitish patch. Sexual dimorphism is – apart from the forewing length, limited.

Male genitalia are different to *moluccana* by rather rectangular, untapered uncus shape. Uncus is slightly bilobed at the tip; socii are uncurved and broader compared to *molucana* or *glauca*. The bulbous sacculus of valva resembles *moluccana,* but the process is broad and rounded. Tegumen bears at valva base a tapered process, which is not seen in *moluccana*. Shape of the valva is broader and more rounded at the apex than *moluccana*. Phallus is straight, broad and short. 8th sternite is gently bilobed, but the rounded 8th tergite bears a small central projection at the posterior margin. The female genitalia have not yet been dissected.

Bionomics. The only known two specimens were taken at medium altitudes approximately 600 m above seal level.

Distribution. Restricted to Halmahera.

23. *Ortholomia subtilis* spec. nov. (figs. 23a–i).

Holotype: ♂, Indonesia, Nusa Tenggara Timur, Alor Isl., Moru, 8°16'S, 125°05'E, 500 m, 22. iii.–3. iv. 2006, leg. St. Jakl – in coll. A. Schintlmeister, Dresden.

Paratypes (15 ♂♂, 16 ♀♀):

Tanimbar: 2 ♂♂, 2 ♀♀, Yamdema Isl., 20 km NE Saumlaki, Loulun village, 7°47'S, 131°18'E, 150 m, 5.–28. xii. 2006, leg. St. Jakl (GU 96-19);

Alor: 6 ♂♂, 2 ♀♀ Moru, 8°16'S, 125°05'E, 500 m, 1.–8. iii. 2006, leg. St. Jakl (GU 92-83, (GU92-85); 5 ♂♂, 8 ♀♀ dto. but 22. iii.–3. iv. 2006 (GU41-61a, 43-01a, GU 92-84);

Pantar: 4 ♀♀, Tanah Labang, 8°19'S, 124°16'E, 350 m, 9.–21. iii. 2006 (GU 92-86);

W Timor: 1 ♂, 35 km SE Kupang, Buraen, 10°15'S, 123°49'E, 350 m, 26. i.–9. ii. 2006;

Lomblen: 1 ♂, Lamalera, 8°33'S, 123°25'E, 450 m, 8.–14. ix. 2006 (GU 19-26a);

Sumba: 1 ♂, Luku Melob, 500 m, 10°S, 120°10'E v. 2006 (GU95-20); 1 ♀, Padira Tana, Hutan Logolalang, 22. xii. 2002, leg. A. Saito.

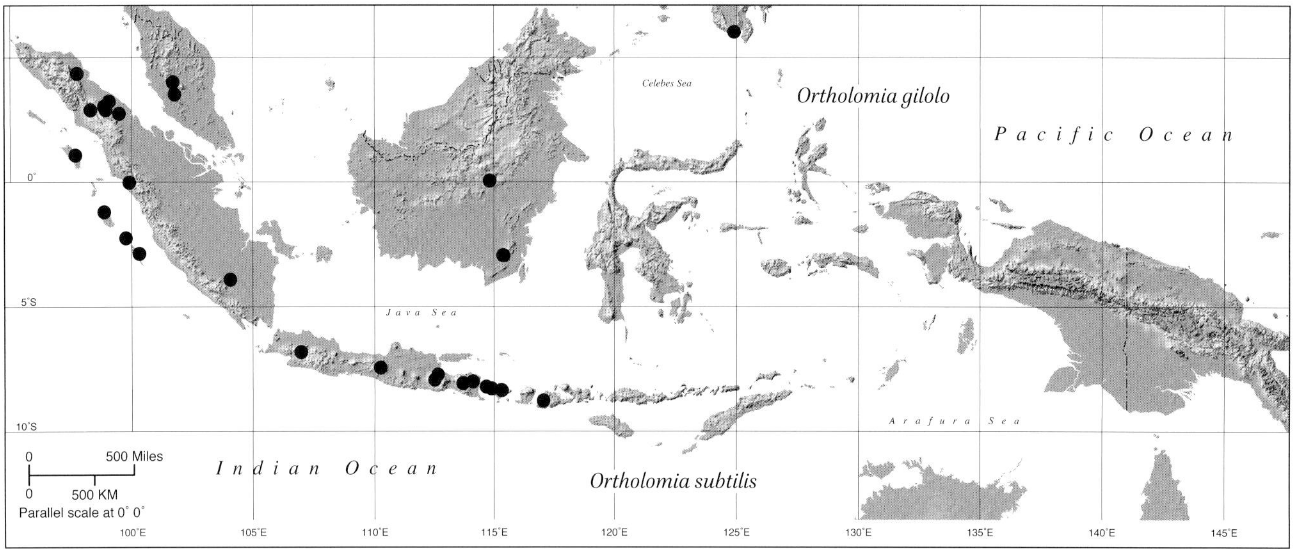

FIGURE 24 Distribution of *Ortholomia gilolo* and *subtilis*.

Taxonomic note: Between the populations from different islands there are different shapes in the sacculus of valvae and differences in the two dissected female genitalia as well, which indicates that several subspecies could occur in various islands. Due to the limited material at the author's disposal it is currently not possible to separate these populations in a taxonomically proper way.

Diagnosis. Forewing length ♂♂ 20.0–22.5 mm, ♀♀ 21–26 mm. The smallest species in the genus is on average 6 mm smaller than *moluccana*. Forewing color may vary from pale yellowish brown and ash-grey to blackish-grey. Discal spot is pale grey with an ash-grey core. The black tornal spot of the forewing is visible, but not contrasting. Some fasciae marked as rows of black dots, but the medial fascia is partly developed as a blackish band. Hind wings are dark brown.

Male genitalia are similar to those of *O. gilolo*. Uncus is of quadrangular shape, slightly bilobed. Socii are robust and of similar shape as in *gilolo*, but thicker than in *moluccana*. Sacculus is bulbous as in *moluccana* and *gilolo*, but there are no processes. Tegumen processes at the valva base are absent. At the base of the sacculus there is a rounded and thicker appendix, which is not seen in the other congeners. Valvae shape is as in *gilolo*. Phallus resembles *gilolo* in shape, but is significantly longer. 8th abdominal segment is similar to *gilolo*. The 8th tergite lacks the small central projection, which is characteristically for *gilolo*. Female genitalia have strongly sclerotized papillae anales and long apophyses posteriores. Postvaginal plate bears a large, wing shaped and strongly sclerotized structure, which varies in its shape between the dissected females from Alor and Pantar as illustrated. Ostium is large and of a diagnostic rectangular shape. Between the dissected females from Alor and Pantar it varies in width. The circular, large signum in the corpus bursae differs from all other congeners by the curved shape of the strongly sclerotized transverses. From the center of signum a further strongly sclerotized band is to be seen.

Bionomics. The species appears infrequentely at altitudes lower than 500 m.

Distribution. Occurring on various Islands of the Nusa Tenggara (Sumba, Lomblen, Pantar, Alor, Timor) and Tanimbar.

24. ***Ortholomia xylinata*** (Walker, 1865): 456 (*Besida xylinata*).

Syntypes: 3 specimens (♂♂, ♀♀) [Indonesia], Java – NHMUK, London.

Schintlmeister 2008: 47, Schintlmeister 2013: 308.

Taxonomic note. The taxonomy and synonymy of this difficult complex is described in detail in Schintlmeister & Lourens, 2010: 62. Individual variation is presented there as well, including the illustration of 12 imagines and 12 genitalia figures.

Diagnosis. This inconspicuous species displays pale to fuscous brownish or blackish forewings with brown pattern. A white tornal spot on the forewing is present in most adults. As shown in Schintlmeister & Lourens, 2010, it is not possible to separate *O. xylinata* and *osica* in the

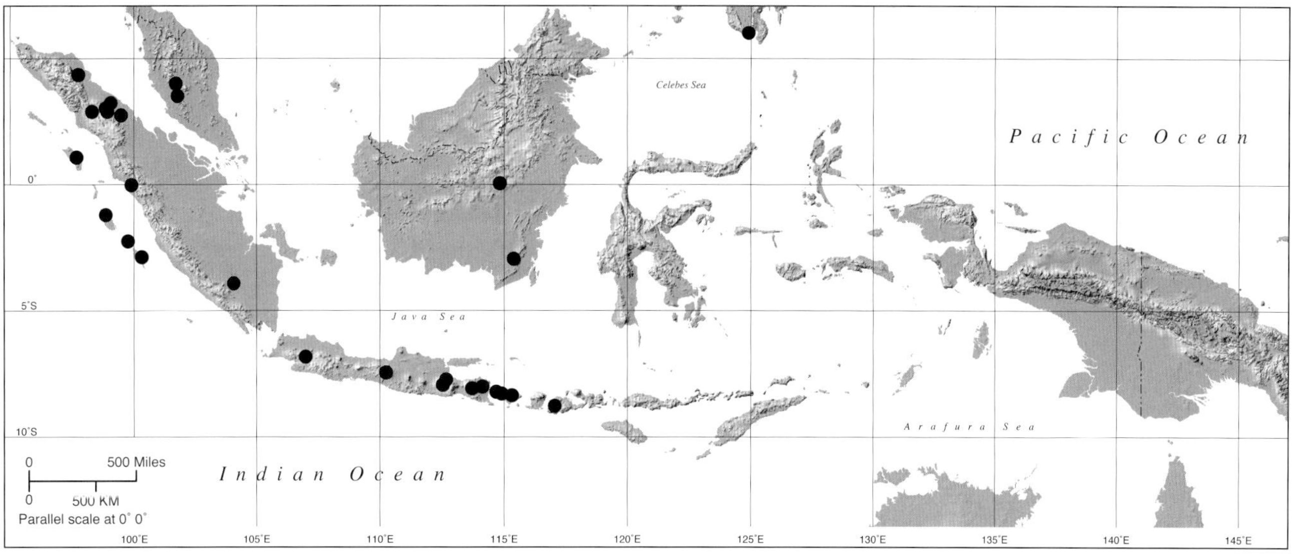

FIGURE 25 Distribution of *Ortholomia xylinata*.

Philippines according to external appearance.

Male genitalia are well characterized by the strongly sclerotized sacculus, which ends in a tapered process. Shape of the strongly sclerotized sacculus is subject of a wide individual variation, but no other species in the genus have a comparable and diagnostic sacculus structure. Female genitalia show the transverse structures in the signum in two unconnected streaks. The other dissected species of the genus have the center of the signum strongly sclerotized.

Bionomics. The species occurs locally common to a wider range of habitats and also in urban areas as well. Most adults were observed at lower altitudes up to 500 m, but single specimens were taken up to 1250 m above sea level.

Distribution. Distributed in southwestern China (Yunnan), Thailand, Laos, Vietnam, Malaya, the entire Philippines, Sumatra, Nias, Mentawei Islands (Sipura and Pagai), Kalimantan, Java, Bali, and Sumbawa.

25. ***Ortholomia besida*** (Schintlmeister, 1994): 219; pl. 4: 13; GU-pl. 3: 1 (*Besida besida*).
 Holotype: ♂, Indonesia, West Sumatra, Mt. Talamau, Westseite, 12 km S. Talu, 0°08'N, 99°59'E, 500 m – coll. A. Schintlmeister, Dresden.
 Schintlmeister 2013: 307.

Diagnosis. *O. besida* displays pale brownish forewings with black pattern. As far as is known, blackish colored specimens do not occur (n = 16 ♂♂, 2 ♀♀ examined). A white tornal spot is present on the forewings. *Ortholomia besida* is separable from *xylinata* by the narrower shape of its forewing. Forewing lengths in both species are same.

Male genitalia have aprominently serrated margin of the large and strongly sclerotized sacculus. Female genitalia are distinguished from other congeners by the large and modified antevaginal plate and the Y-shaped, strongly sclerotized band in the signum.

Bionomics. The species may occur commonly in clusters. Lower altitudes of up to 500 m are preferred. However, one male was collected in Sumatra, Berastagi at an altitude of 1400 m.

Distribution. Known from the Malayan Peninsula from one male sampled near the border to Thailand (confirmed by genitalia dissection) – and Sumatra.

26. ***Ortholomia osica*** Schintlmeister & Lourens, 2010: 66; figs. 104, 105, 107, 108.
 Holotype: ♂, The Philippines, E Luzon, Aurora, S. Madre, Dapalan riv. S of Dilalongan, 16°02.707"N, 121°42.673'E, 50 m – NHM, Vienna.
 Schintlmeister 2013: 307.

Taxonomic note. In Sulawesi a distinct subspecies, *Ortholomia osica sulawesiensis* Kobayashi, 2012: 7, figs. 4, 5, 9a–e; holotype: ♂, Indonesia, Sulawesi, Puncak Palopo, 900m – National Museum of Nature and Science, Tokyo occurs. Ssp. *sulawesiensis* differ from nominotypical subspecies by the shape of the uncus tip and the saccular spine of the valve.

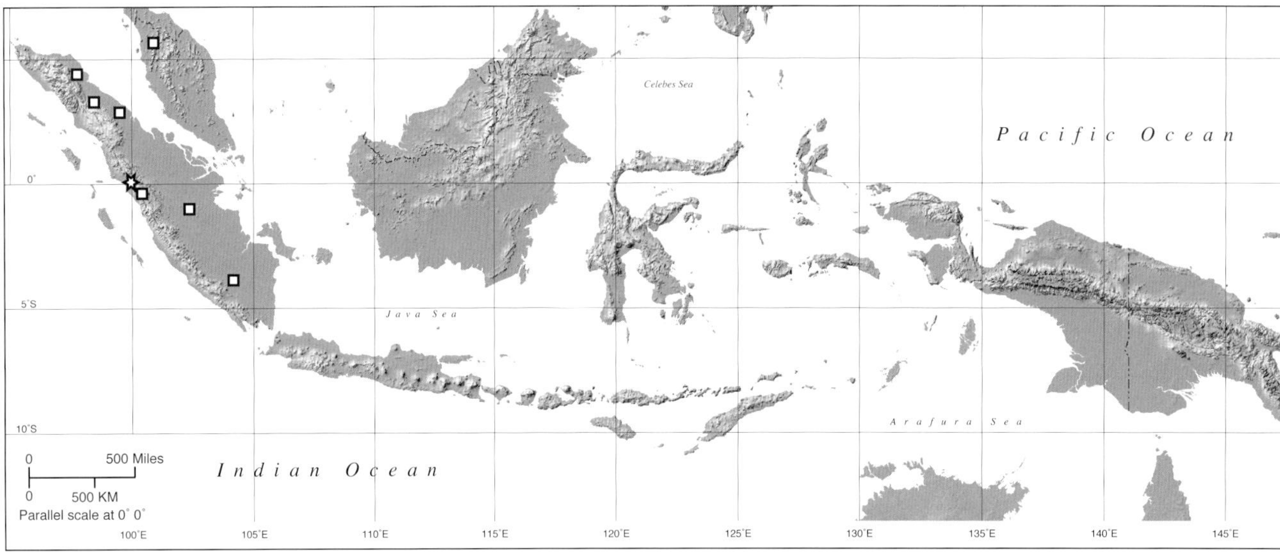

FIGURE 26 Distribution of *Ortholomia besida.*

Diagnosis. A variable species occuring in brownish and blackish forms. Width variation is presented in Schintlmeister & Lourens, 2010. Due to overlapping variation it is not separable by external features from *O. xylinata*.

Male genitalia of *osica* have a large and strongly sclerotized sacculus, which displays a smooth margin and a spine. Sacculus shape is somewhat variable as shown in Schintlmeister & Lourens, 2010: 64, but the ssp. *sulawesiensis* has a significant expanded sacculus. Phallus is arched, which is unique in *Ortholomia*. Female genitalia resemble *besida*. Postvaginal plate is less modified as in *besida* or *xylinata*. Signum is most similar to *xylinata,* but smaller. Sclerotized transverse structures are as in *xylinata*.

Bionomics. *O. osica sulawesiensis* occur in Sulawesi infrequently in rainforests between 1000–1200 m above seal level. 1 ♂ was collected in Bantimurung at an altitude of 200 m.

Distribution. *Ortholomia osica osica* is widespread in the Philippines on all major islands. In Sulawesi and Peleng the ssp. *sulawesiensis* occur.

Pseudogargetta Bethune-Baker, 1904: 373.

Type species: *Pseudogargetta diversa* Bethune-Baker, 1904.

Schintlmeister 2008: 47; Schintlmeister 2013: 356.

Medium sized moths with bipectinate antennae, of which the last 20% towards the tip are filiform in both sexes. The inconspicous imagines are mostly brown or blackish colored with deep black pattern. All known species of *Pseudogargetta* are very similar in external appearance. A pale grey discal patch on the forewings is indistinct and often even less visible or absent. The medial area bear two black lines in its center. In the basal area a black dorsal shadow. A black tornal patch occur in *diversa* and *orbiculata*.

Male genitalia have a robust and large strongly curved uncus. Socii are absent. Valva have an apical process and a second costal process near the apex. Phallus is long, slender and curved. Female genitalia of *Pseudogargetta* have not been dissected yet.

The genus comprises three species and is distributed in SE China, the Malayan Peninsula, Borneo, Sulawesi, Halmahera, Obi, Ambon and New Guinea, including some adjecent islands.

27. ***Pseudogargetta diversa*** Bethune-Baker, 1904: 373; pl. 5: 4.

Lectotype: ♂, [Papua New Guinea], Dinawa – NHMUK, London **by present designation.**

Schintlmeister 2013: 356.

Taxonomic note. *Pseudogargetta diversa* was described of a pair from Dinawa and Ekeikei. Both syntype specimens were traced by the author in NHMUK. To stabilize the nomenclature, hereby the ♂ with the following labels is designated as lectotype: "Dinawa, | B. New Guinea | 4000ft. | Mai–July. 1902 | Coll. A. E. Pratt" green circular label "Type", blue label "Notodontidae genitalia prep. No. 34 ♂, yellow label "Photo done by A. Schintlmeister #4820. A lectotype label was added. The male genitalia of the lectotype are

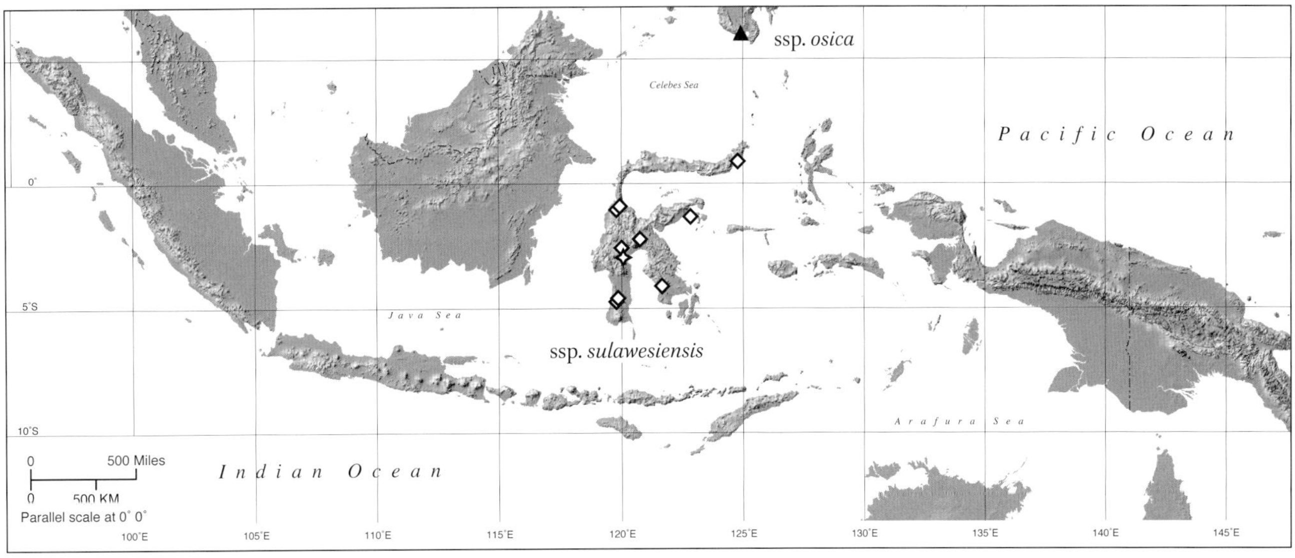

FIGURE 27 Distribution of *Ortholomia osica*.

illustrated here in fig. 27a. The remaining ♀ from Ekeikei, 1500 ft. March-April 1903 and also with a green circular label "Type" becomes a paralectotype.

Diagnosis. Forewing pattern is as described in the genus diagnosis above. Forewing colors vary from pale yellowish-brown, grey to black. Sometimes an ellipsoid pale greyish discal patch occurs. *P. diversa* is generally larger than *orbicularia* or *marmorata,* but from W Papua the author examined also a rather small male as well. Sexual dimorphism is limited.

Male genitalia are distinguished from other congeners by the large uncus, which bears a small process at the tip. Juxta has a large, stronger sclerotized process, similar to *marmorata*. The costal valva process of *diversa* is small and positioned more basally. Phallus is thin and very long, much longer than in both the other species. 8th sternite concave at posterior margin. 8th tergite is slightly bilobed.

Bionomics. *Pseudogargetta diversa* is rather rare in collections (n < 30). Most adults were observed at lower altitudes up to 500 m above seal level, but the lectotype was collected at 1220 m elevation.

Distribution. Known from W Papua, Papua New Guinea, Yapen Isl. Ferguson Isl.

28. ***Pseudogargetta orbiculata*** Kobayashi, 2012:6; figs. 3, 8a–c.

Holotype: ♂, Indonesia, Bantimurung, Ujung Padang, 200m – National Museum of Nature and Science, Tokyo.

Taxonomic note: There are only four specimens at the authors disposal, each from a different island. The dissected males differ somewhat in their genitalia and might represent different subspecies.

Diagnosis. Very similar to *P. diversa*. Forewing pattern consists of a basal fascia, a medial area and a postmedial fascia, which are marked as deep black dots. Discal spot is indistinctly developed as a pale circle. Fringe of all wings is checkered whitish and fuscous-brown. Hindwings are fuscous-brown.

Male genitalia are characterized by a very long and curved, tapered uncus, which has no processes at the tip as seen in *diversa.* Valvae are not modified but bear at the apex a pair of processes of different length. Juxta is specifically shaped and sclerotized. It is similar to *diversa,* but shorter than in *P. marmorata.* Phallus is curved, slender, and shorter than in *diversa.* 8th tergite is specifically sclerotized and bilobed at posterior margin. 8th sternite is concave at posterior margin; *diversa* displays contrary a convex margin. 8th abominal segments on specimens from Ambon, Halmahera and Obi are slightly different in shape and sclerotizations as illustrated. Sexual dimorphism is limited, but the only one known female is much larger than the males.

Bionomics. Adults are rare (n = 9) and was collected between 200 m and 1200 m above sea level.

Distribution. Known from Sulawesi, Banggai Isl., Halmahera, Obi and Ambon.

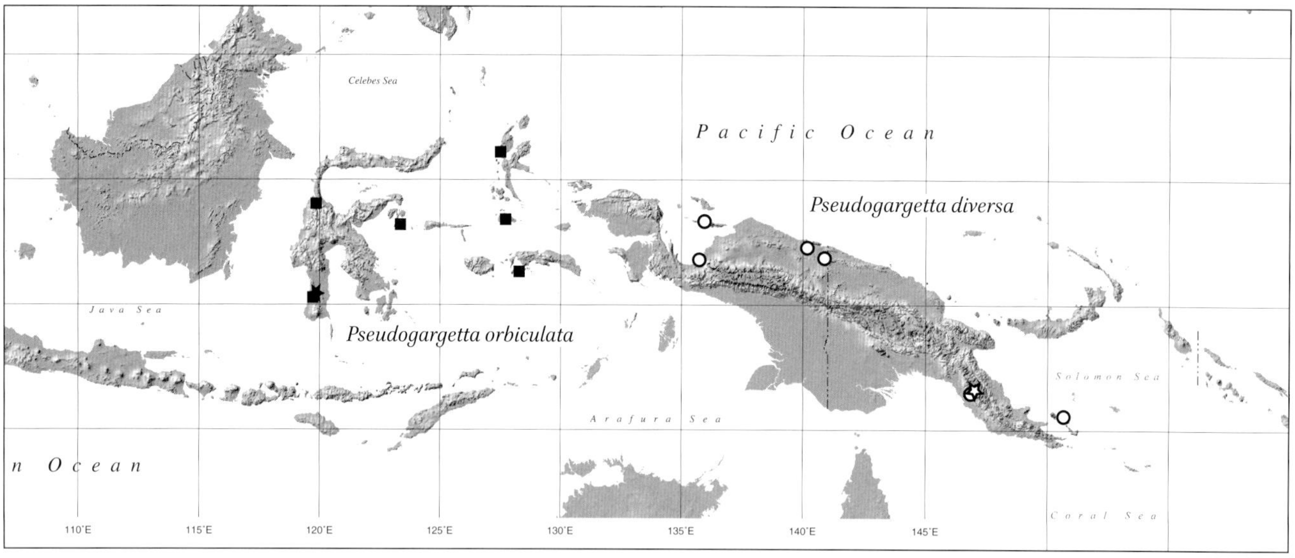

FIGURE 28 Distribution of *Pseudogargetta diversa* and *orbiculata*.

29. ***Pseudogargetta marmorata*** (Kiriakoff, 1967): 41; fig. 4 (*Blakeia marmorata*).
Holotype: ♂, [Indonesia], S. E. Borneo, Samarinda – NHMUK, London. Schintlmeister 2013: 356.

Diagnosis. A characteristic feature and helpful for identification is the prominent black patch between the basal and medial area of the forewings. Forewing color may vary individually from cream, ash-grey, fuscous rusty-brown or black. Sometimes the costal area contrasts whitish colored on a fuscous background.

Male genitalia are readiy separable from those of their congeners by the bifurcate uncus. Juxta is long and specifically sclerotized, similar to *diversa* but very different from *orbiculata*. Near the apex two processes arise, of which the costal one is very long and slender. Phallus is straight and slender but becomes broad toward in the vicinity of the base.

Bionomics. *P. marmorata* occur infrequently and mostly at lower altitudes up to 500 m. A few specimens were taken at 1200 m above seal level.

Distribution. Distributed in SE China (Guangxi), Vietnam, Thailand, the Malayan Peninsula and Borneo.

Gargettiana Gaede, 1930: 617.
Type species: *Gargetta melanosticta* Joicey & Talbot, 1917 [= *Gargetta punctatissima* Bethune-Baker, 1916].
Schintlmeister 2013: 178.

Middle-sized moths with filiform antennae in both sexes. Forewings are brown to black with many blackish dots. Several forewing fasciae are marked as rows of blackish dots. The circular discal spot is often inconspicuously developed. It is accompanied by a prominent, contrasting squarish fuscous field below. Sexual dimorphism is limited.

Male genitalia are characterized by a long and slender uncus. Socii are half the length of the uncus. From a bulbous sacculus structure arises many long setae. Valva apex bears also some long setae. Phallus is short and thick. 8th abdominal segment is less modified. The second abdominal segment bears a pair of large cteniophores. Female genitalia are still unknown.

This monotypic genus occurs in all of New Guinea, the Moluccas (Halmahera, Bacan, Ambon, Seram, Buru), and Australia.

30. ***Gargettiana punctatissima*** Bethune-Baker, 1916: 382 (*Gargetta punctatissima*).
Lectotype: ♂, [Papua New Guinea], British New Guinea, Mount Kebea, 3600 ft. – NHMUK, London **by present designation**.
Schintlmeister 2013: 178.

Taxonomic notes. *Gargetta punctatissima* was described of a pair from Mount Kebeae and Babooni, 3000 ft. The author traced the male syntype in NHMUK. To stabilize the nomenclature, herewith the ♂ with the following labels is designated as lectotype: "Mount Kebea, | B. New Guinea | 3,600-ft. | July. 1903 | A. E. Pratt" blue label

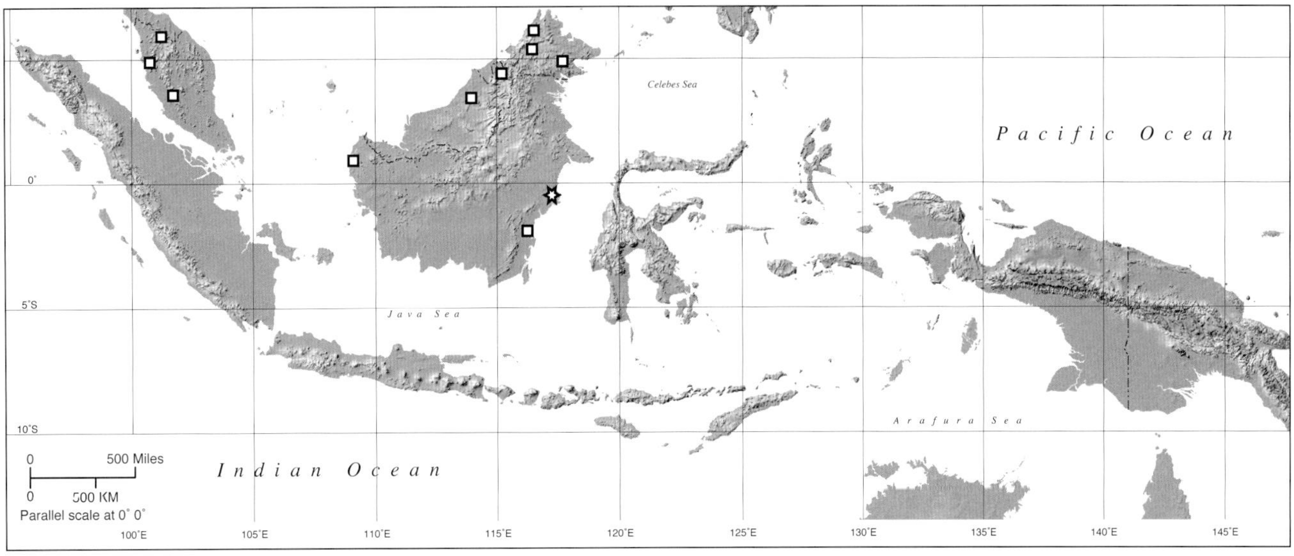

FIGURE 29 Distribution of *Pseudogargetta marmorata*.

"Notodontidae genitalia prep. No. 27" yellow label "Photo done by A. Schintlmeister #4842. A lectotype label was added.

Parosica nomo Kiriakoff, 1967 38; fig. 1 (holotype: ♂, West New Guinea, Mt. Nomo, south of Mt. Bougainville – NHMUK, London) was described in an own genus *Parosica* Kiriakoff, 1967. Kiriakoff 1968 placed *Parosica* as a subjective synonym of *Gargettiana*. Examination of the holotype of *nomo* showed, that it is an individual fuscous form, occuring as such not rarely among *punctatissima* in W Papua. Male genitalia of *nomo* (the probable holotype) as illustrated by Kiriakoff, 1967 does not differ virtually from other material dissected of *punctatissima*. *P. nomo* sinks therefore as a junior synonym of *punctatissima* (**syn. nov.**).

A single specimen from Buru differs in its male genitalia from dissected populations from W Papua, Halmahera and Bacan. It is described here as a subspecies of *punctatissima*, but might represent a distinct species, should more material from Buru becomes available for examinations to verify the differences:

Gargettiana punctatissima tritus **ssp.nov.** (figs. 30c, 30h)
Holotype: ♂, Indonesia, SE Buru Isl., Ilat, Mt. Remajah, 250 m, 3.–20. i. 2013 leg. St. Jakl (GU 93-58) – coll. A. Schintlmeister, Dresden.

No paratypes.

Diagnosis of ssp. *tritus*. Forewing length ♂ 21 mm. The holotype is somewhat worn and thus paler than specimens from W Papua or Halmahera. Forewing pattern of ssp. *tritus* is seemingly not different from ssp. *punctatissima*, except for the fuscous patch below thediscal spot, which is more indistinct. Female is unknown.

Male genitalia differ through their slenderer and shorter uncus. The bulbous sacculus bear a triangular projection, which is not tapered at the tip as in ssp. *punctatissima*, but rounded. Phallus of ssp. *tritus* is significantly shorter than ssp. *punctatissima*. 8th abdominal segments of *tritus* differ slightly in their sclerotization from ssp. *punctatissima* as illustrated.

Diagnosis. Best to recognized by the many fuscous dots on the forewings and the squarish patch in the medial area. Hindwing uniform brown with yellowish brown marginal fascia. Sometimes fuscous blackish brown colored individuals (f. *nomo*) occur in both sexes. Sexual dimorphism is limited.

Male genitalia are distinguished by their large uncus, a bulbous sacculus that resembles to several *Ortholomia* species, and a short phallus with two spines at the tip. Female genitalia are yet unknown.

Bionomics. The species occur commonly and mostly at lower altitudes up to 500 m, but one specimen was collected in W Papua at 1700 m.

Distribution. *G. punctatissima punctatissima* is distributed in Australia (Qsd., NT), New Guinea in its entirety, Yapen isl., Halmahera, Bacan, Seram. *G. punctatissima tritus* occurs in Buru.

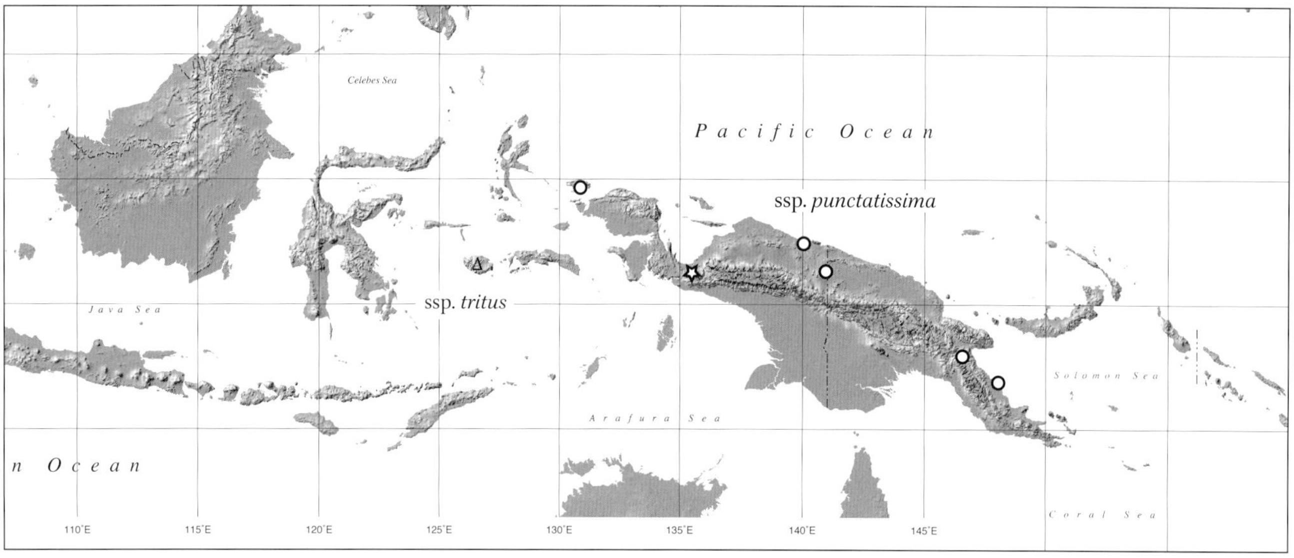

FIGURE 30 Distribution of *Gargettiana punctatissima*.

Polychoa Turner, 1907: 681.
Type species: *Poychoa styphlopis* Turner, 1907.
Schintlmeister 2013: 348.

Middle sized notodontids with bipectinated antennae in both sexes, but the females have shorter rami. Forewing shape is rather rectangular, e.g. the costa and dorsum running almost parallel. Forewing pattern consists of contrasting, metallic green scales on a fuscous brown to black background. Most species display in the medial area of forewing two black dots, surrounded by whitish or pale grey. Postmedial fascia is marked as a row of contrasting white dots. Hindwings are fuscous brown, mostly with a paler yellowish-tinged base. Sexual dimorphism is limited, but due to the rareness of most species and their large range of individual variability it is sometimes difficult to recognize corresponding females and males.

Male genitalia have a deeply divided uncus and a pair of long socii. Uncus arms are shorter than socii, often half their length. Valvae have a large, strongly sclerotized and bulbous sacculus with a characteristic striulate structure. Apical part of the valva is elongated, but not pointed at apex. Phallus is short and curved. Everted endophallus is specifical shaped and of taxonomic value. 8th sternite is twice bilobed on posterior and anterior margins. 8th abdominal segments are specifically sclerotized. Female genitalia have two pairs of long to medium-sized and slender apophyses. 8th abdominal segment is rather slender; ductus bursae short and broad. Corpus bursae is large, bearing a large ellipsoid signum, which is weakly sclerotized.

Polychoa is poorly known and the number of known Papuan specimens to be found in collections is limited (n < 40). The genus is currently comprised of nine species, of which six are distributed in New Guinea. In the Moluccas (Halmahera, Seram) and Australia (Queensland) further species occur.

31. *Polychoa aeruginosa* Schintlmeister, 2013: 348, (*Polychoa styphlopis aeruginosa* Gaede), **stat. nov.**
Holotype: ♀, [Indonesia, W Papua], Neu Guinea, Weyland Mts., Mt. Kunupi, 2000 m – NHMUK, London.
Schintlmeister 2013: 348.

Taxonomic note. Gaede, 1930: 633 described in Seitz *Polychoa styphlopis* form *aeruginosa* from "Neu Guinea" as type locality. He mentioned the enormous variability and placed *Gargetta albostigmata* Rothschild, 1917 and also *Gargetta flavibasis* Rothschild, 1917 as well, all described from New Guine, as further "forms" to *styphlopis* and wrote "alle von Neu-Guinea". So, it is clear, that he did not treat *aeruginosa* as a geographic form (subspecies) but as an aberration, e.g. an infrasubspecific entity. In Lepidopterorum Catalogus (Gaede, 1934: 44) he listed these three taxa as "ab". Schintlmeister, 2013: 348 used first the combination "*Polychoa styphlopis aeruginosa*" and made *aeruginosa* nomenclaturally available.

Polychoa styphlopis Turner, 1907: 681 (holotype: ♂, N Queensland, Kuranda – ANIC) is a species restricted to Australia. Distinguished from *aeruginosa* by its external appearance, broader and more rounded forewing shape, and in its male genitalia, short and broad uncus arms, longer valva and in particular having a contrasting shape of the much longer phallus. The taxon *aeruginosa* is raised therefore to specific level (**stat. nov.**).

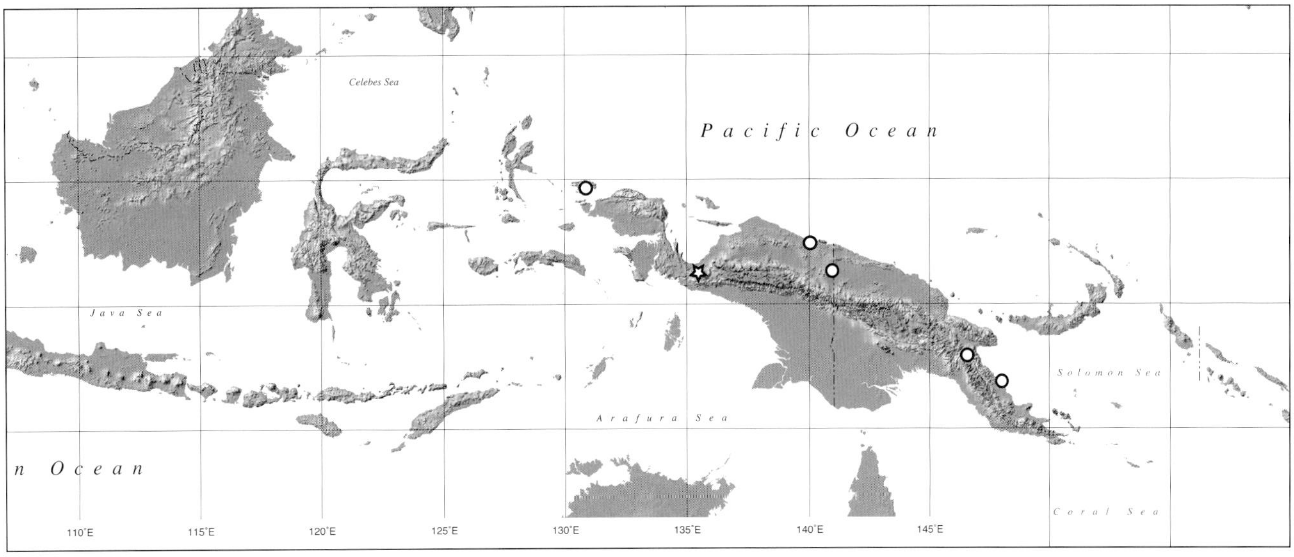

FIGURE 31 Distribution of *Polychoa aeruginosa*.

Diagnosis. *Polychoa aeruginosa* occur often and in both sexes with a diagnostically cyan apical patch on the forewing, but there are also specimens with less prominently developed apical patches. Forewings are fuscous brown mixed sparsely with metallic green scales. Postmedial and marginal fasciae are marked as rows of contrasting white dots. Three blackish spots in the medial area of forewing are surrounded in white. Hindwings are brown with a paler brown base. Sexual dimorphism is limited. For separation from the following two species, see below.

Male genitalia are remarkably small in size. Uncus divided and the slender arms are short, only 1/3 the length of the socii. Valvae are broader in the terminal part than other congeners. The bulbous sacculus part is rather small. Phallus is predominated by its strongly curved shape. 8th sternite is not bilobed at the posterior margin. Female genitalia have not yet been dissected.

Bionomics. The moth was collected infrequently between 800 m and 2000 m above sea level.

Distribution. Distributed in the whole of New Guinea. A single male comes from Waigeu isl. (genitalia not dissected).

32. ***Polychoa bakami*** **spec. nov.** (figs. 33a–b).

Holotype: ♂, Indonesia, N Moluccas, Halmahera, 15 km SW Baru, Mt. Talagaranu, 1°21'N, 127°32'E, 600 m, leg. Sinjaev & Tarasov – in coll. A. Schintlmeister, Dresden.

Paratype (1 ♂):

Halmahera: 1 ♂, 15 km SW Baru, Mt. Talagaranu, 1°21'N, 127°32'E, 600 m leg. Sinjaev & Tarasov (GU 96-41).

Diagnosis. Forewing lengths ♂♂ 19–20 mm. The species is similar to *P. aeruginosa* and *haliga* but is generally 2–3 mm smaller in forewing length. Forewings are fuscous brown mixed with metallic green scales. In the medial area are three blackish dots, of which the upper one is surrounded in a cream. Postmedial fascia is marked as a row of white dots, but the marginal fasciae of white dots, as they occur in *aeruginosa* and *haliga* are absent. The submarginal area is paler than the remainig forewings. The female is unknown.

Male genitalia are characterized by slender uncus arms, which are approximately half the length of the socii. In *haliga* the uncus arm is much broader and 75% of the socii length. Valvae are clearly longer than in *aeruginosa*; they extend over the socii. Sacculus is found with very large circular bulbous projection. It is together with *flavibasis* the largest within the genus. The slender apical part of the valva is longer than in *aeruginosa*. Valva apex is slightly pointed, in *haliga* more rounded. Phallus resembles that of *haliga* in shape: slightly curved and thicker than in *styphlopis* or *aeruginosa*. It bears at the tip a more strongly sclerotized rounded process. The everted endophallus show a sclerotized microstructure (very small thornes), which are occur seemingly exclusively in *bakami*. 8th sternite is slightly bilobed at the posterior margin; 8th tergite rather narrow and rounded.

Bionomics. The new species was caught at 600 m above sea level.

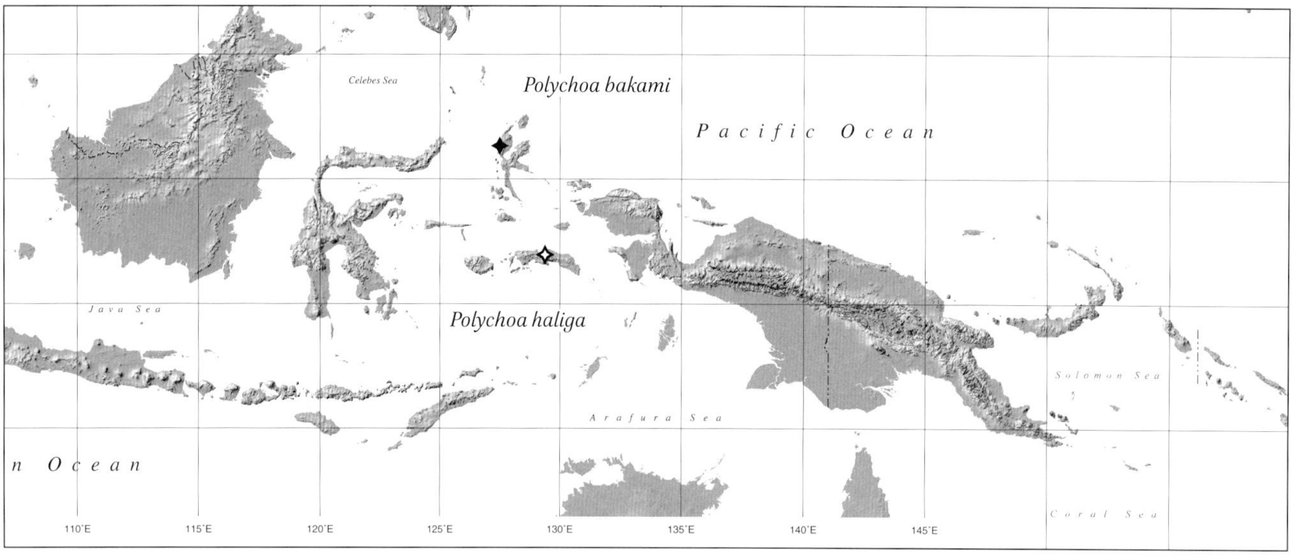

FIGURE 32 Distribution of *Polychoa bakami* and *haliga*.

Distribution. Restricted to Halmahera.

33. ***Polychoa haliga*** **spec. nov.** (figs. 33a–33c).
Holotype: ♂, Indonesia, Seram, Gunong Kobipoto, summit ridge, Site 7: 2, 1470 m, 3°04'S, 129°30'E, viii.–ix. 1987, Operation Raleigh, leg. J. D. Holloway, D. T. Jones et. al. – NHMUK, London.

Paratypes (1 ♂, 1 ♀):
Seram: 1 ♂, 1 ♀, Gunong Kobipoto, summit ridge, Site 7: 1, 1470 m, 3°04'S, 129°30'E, viii.–ix. 1987, Operation Raleigh, leg. J. D. Holloway, D. T. Jones et. al. (BM 10313988).

Diagnosis. Antennae in both sexes black bipectinated until the tip. Forewing lengths ♂♂ 20.5–21.0 mm; ♀ 20.5 mm. *P. haliga* resembles *P. aeruginosa* and also *bakami*. It is the largest of these three species and about 2–3 mm larger than *aeruginosa*. Forewings are fuscous brown, mixed with metallic green scales. There are more of these green scales than in *aeruginosa* or *bakami*. The medial area of forewing display three black dots, which are surrounded with cream or white. Postmedial and marginal fasciae are marked as rows of small but contrasting white dots. Sexual dimorphism is limited.

Male genitalia resemble *P. bakami*. Uncus is large and deeply divided. It is broader than in *aeruginosa* or *bakami* and uncus arms are approximately 75% of the socii length (in *bakami* 40%). Socii are large, tip kinked. Tegumen robust developed and broader than in *bakami*. Valva is long and narrow with rounded apex, less tapered as in *bakami* or *aeruginosa*. Sacculus have a bulbous projection, which is smaller as in *bakami*. Phallus is slightly curved and longer and thicker than that of *styphlopis* or *aeruginosa*. It differs from *bakami* by the larger and tapered process at the tip. Endophallus has no sclerotized microstructures (thornes) as *bakami* does. 8th sternite is wider and more deeply bilobed at the posterior margin than *bakami*. 8th tergite is broader than that of *bakami*. Female genitalia have not yet been dissected.

Bionomics. Adults was observed in three specimens at 1470 m above sea level in montane forest.

Distribution. Endemic to Seram, Mt. Manusuela range.

34. ***Polychoa cheesmanae*** **spec. nov.** (figs. 34a–d).
Holotype: ♂, Indonesia, W Papua, Yapen Isl., near Ambaidiru, 1000 m, 1°46.626' S, 136°134.34' E, 10.–13. vi. 2010, leg. G. Bretschneider (genitalia slide Ihle 03-2017) in coll. S. Ihle, Filderstadt.

Paratypes (1 ♂, 2 ♀♀):
Yapen: 1 ♀, Distr. Serui, 6 km from Montembu, 6.–9. ix. 2013, leg. G. Bretschneider (genitalia slide Ihle 04-2017).
W Papua: 1 ♀, Waris distr., Ampas, 30. viii. 1938, leg. W. Stüber; 1 ♂, Sarmi Distr., Foja
Mts., Furu Camp, Dabra, 600 m.

Etymology. Named for Lucy Evelyn Cheesman (b. 8. x. 1882 Westwood Kent – d. 15. iv. 1969 London). She explored on four solo expeditions between 1933 and 1939 at Mt. Kokoda (S Papua New Guinea), Cyclops Mts. (W Papua), Yapen isl. and Waigeu isl. and collected a huge amount of material. From Yapen Island during 4 months she

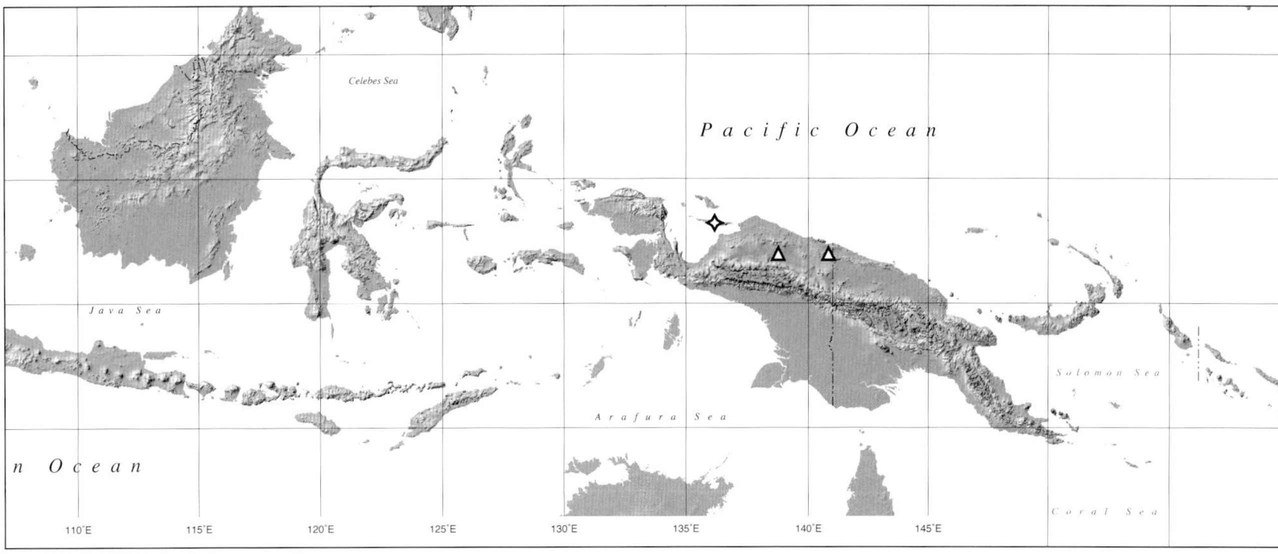

FIGURE 33 Distribution of *Polychoa cheesmanae.*

collected 52,000 insects for the British Museum (Natural History), now NHMUK. Many dots of Papuan notodontids showing in distribution maps base on material collected by her. She collected, among others also the holotype of *Polychoa metallica.*

Diagnosis. Forewing lengths ♂♂ 23 mm; ♀ 20.5 -22.0 mm. Antennae are in both sexes brown bipectinated until the tip. Rami are shorter than in *P. aeruginosa, bakami* or *haliga.* Forewings are fuscous brown mixed with a few metallic green scales. A rectangular postbasal and medial area bear many green scales near the costa. A large white costal blotch near the forewing apex is a diagnostic feature and distinguish *cheesmanae* from other congeners. Postmedial fascia marked as a row of white dots. Hindwing brown with paler base. A female from Yapen Isl. (fig. 34e) display the white costal blotch weakly developed, but a further female from W Papua have the blotch as the male have.

Male genitalia has a small and bifurcate uncus. Uncus length is < 50% of socii length. Socii are straight and tapered. Tegumen as in *bakami.* Valva long and straight, not curved as in *haliga.* The bulbous projection of sacculus is smaller as in *bakami.* Phallus is slightly curved and shorter than in *haliga.* Phallus tip have a triangular projection, which is diagnostic in its shape. Everted endophallus is larger than in other congeners and bears sclerotized microstructures as in *bakami.* 8th sternite is bilobed at the posterior margin. 8th tergite is of rectangular shape. Posterior margin is straight and not rounded as all other known *Polychoa* have. Female genitalia are relatively large. Antrum is rectangular shaped and longer as in *bipuncta.* Ductus bursae is short. Corpus bursae is circular shaped and large. Signum is of rectangular shape and not pointed at the tips as in *bipuncta.*

Bionomics. *P. cheesmanae* was observed between 600–1000 m above sea level.

Distribution. Known from W Papua and Yapen Isl.

35. ***Polychoa flavibasis*** (Rothschild, 1917): 256; pl. 5: 14 (*Gargetta flavibasis*).
Holotype: ♂, [Indonesia, W Papua], Dutch New Guinea, Snow Mts., near Oetakwa River up to 3500 ft. – NHMUK, London.
Schintlmeister 2013: 348.

Diagnosis. Forewings are blackish-brown, mixed metallic green, in particularly along the costa and basal area. There are three black dots in the medial area surrounded pale brown. Occasionally a pale yellowish-brown marked postbasal and a medial fascia occur. Postmedial and marginal fasciae marked by fine white dots. Distinguished from other congeners by yellowish basal and medial area of hindwing. Postmedial area of hindwing fuscous brown. Female is sexual dimorphic, displaying green, not blackish-brown, forewings with fuscous brown postbasal and postmedial bands. Hindwings are less contrastingly yellowish colored.

Male genitalia have a short and broad bifurcate uncus, about 30% of the length of the socii. Valva is as long as in *haliga,* but the bulbous sacculus projection is rather more ellipsoid than circular and larger than in *haliga.* Phallus is

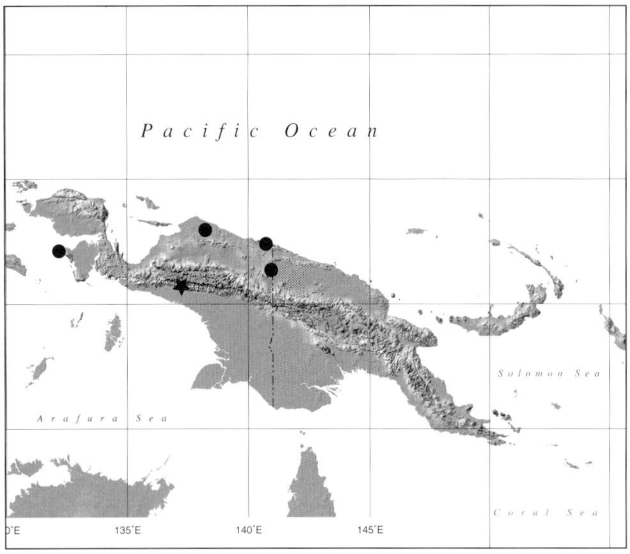

FIGURE 34 Distribution of *Polychoa flavibasis.*

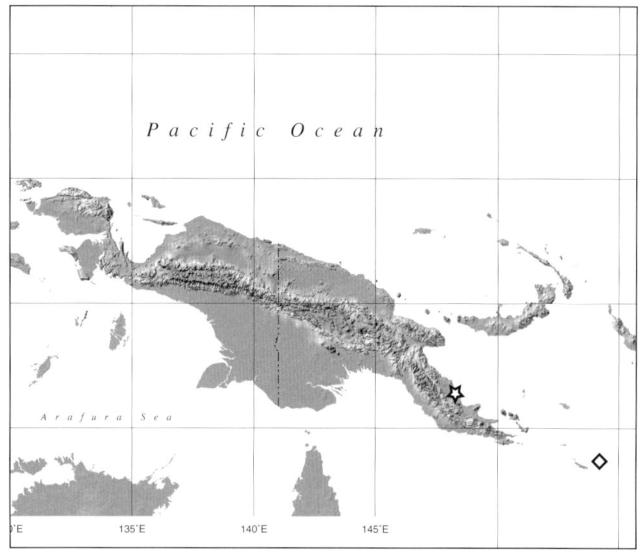

FIGURE 35 Distribution of *Polychoa rafinor.*

of similar shape as in *haliga*, but the projection a the tip is of circular shape and not as a process. 8th sternite is slightly bilobed at the posterior margin. 8th tergite have a central projection at posterior margin. Anterior margins of both 8th abdominal segments are bilobed.

Bionomics. The moth occur infrequently between 600–1050 m above sea level.

Distribution. Known only from W Papua.

36. *Polychoa rafinor* spec. nov. (figs. 36a–e).

Holotype: ♂, Papua New Guinea, Kumusi River, low elev., May 1907, leg. A. S. Meek (BM 10313986) – NHMUK, London.

Paratypes (2 ♂♂, 1 ♀):

PNG: 1 ♂, 1 ♀, Kumusi River, low elev., v. 1907; dto. viii.–ix. 1907, leg. A.S. Meek (BM 128).

Rossel Isl.: 1 ♂, Mt. Rossel, 2100 ft., xii. 1915 (W.F. Eichhorn).

Diagnosis. Forewing lengths ♂♂ 19–21 mm; ♀ 20 mm. Antennae brown bipectinate until the tip in both sexes. Forewing shape is narrow, apex almost rectangularly angled. Forewings are metallic green. There are three contrasting rectangular blackish brown fields along the costa, which distinguishing *rafinor* from its other congeners. Postmedial fascia marked as several indistinct white dots. Submarginal fascia is fuscous brown and somewhat indistinct. Hindwing is chocolate brown; the basal area is not paler colored. The only female collected has the medial area black, but perhaps this is not an expression of sexual dimorphism but simply an individual variant.

Male genitalia have a short bifurcate uncus. Uncus length is of 35% of the length of the socii. Tegumen is compact as in *haliga*. Valvae are long, somewhat curved with a rounded apex. Sacculus have a bulbous projection of medium size. Phallus is slightly curved. At the tip there is a smaller triangular projection. This projection is the smallest of all known *Polychoa*. Female genitalia are distinguished by the constricted basal part of ductus bursae. Corpus bursae is rather long than circular and bears an ellipsoid signum, which is among the smallest of *Polychoa* females.

Bionomics. Only four specimens are currently known, which has been taken at lower elevations up to 640 m.

Distribution. Known from New Guinea and Rossel Isl.

37. *Polychoa bipuncta* (Rothschild, 1917): 256; pl. 4: 13 (*Gargetta bipuncta*).

Holotype: ♀, [Papua New Guinea], North-east British New Guinea, Mambare River, Biagi, 5000 ft. – NHMUK, London.

Schintlmeister 2013: 348.

Taxonomic note. *Polychoa bipuncta* was described after a female and so far only females are known, which matches well to the holotype in their genitalia. The presumable male (fig. 37c) bases on an image that was sent by the late H. van Mastrigt to the author. It resembles by the pair of white dots on forewing to the heavy demaged holotype of

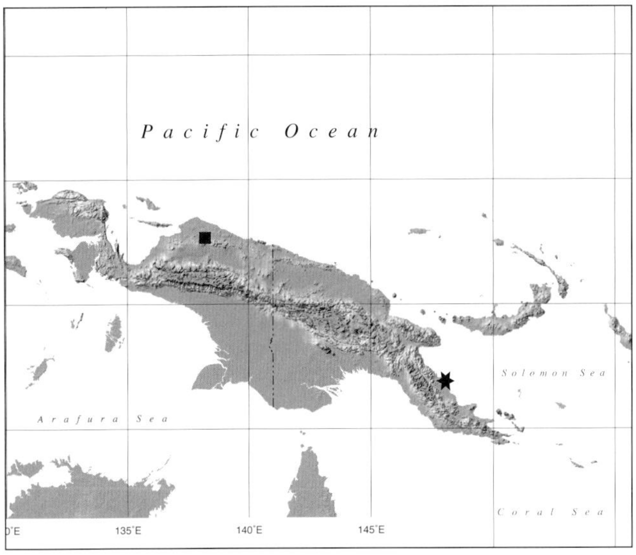

FIGURE 36 Distribution of *Polychoa bipuncta*.

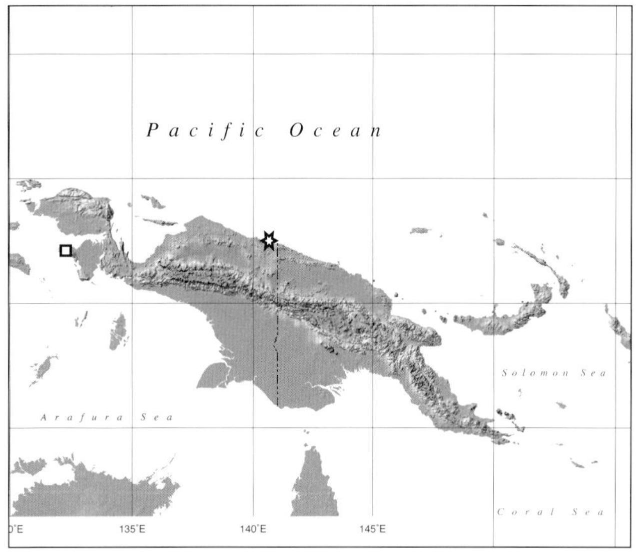

FIGURE 37 Distribution of *Polychoa metallica*.

bipuncta. The identity, however, to that species remains uncertain and also genitalia are not yet dissected.

Diagnosis. Antennae are brown and bipectinated in both sexes until the tip. Forewings are elongated and narrow, which are rather pointed at the apex in the females. Forewings display many metallic green scales, but they are less shiny than in *rafinor*. Broad fuscous brown fasciae occurring in basal, postbasal, medial and, more indistinct, in the submarginal area. In the medial area there are three black dots, surrounded pale brown. Some specimens, including the holotype, display two further contrasting white dots. Hindwings are uniform brown as in *rafinor*.

Female genitalia are distinct by a short antrum, and a very short ductus bursae. Corpus bursae bear a large ellipsoid signum, which is larger than in *rafinor*.

Bionomics. A few females (n = 4) were taken at 1500 m above sea level. The illustrated male was collected at 2200 m.

Distribution. The females are exclusively known from the type locality in Papua New Guinea. The attributed male was collected in W Papua (Foja Mts.).

38. ***Polychoa metallica*** Kiriakoff, 1967: 39; fig. 3.
Holotype: ♂, [Indonesia, W Papua], New Guinea, Cyclops Mts., 2.000 ft. – NHMUK, London.
Schintlmeister 2013: 348.

Diagnosis. Forewings are brownish-black with fields of contrasting metallic green. makes the species unmistakable. Metallic shining green fields are situated along costa and dorsum of forwing. In medial area there is an indistinct small black discal dot, surrounded by cream. Postmedial fascia marked as a row of whitish dots. Hindwings are uniform fuscous brown. Female have similar forewing markings as the male but forewing shape is narrower.

Male genitalia have not been examined by the author. For fig. 38a the illustration from the original description by Kiriakoff, 1967 was used. Uncus relativ long, valva curved with rounded apex. Sacculus projection large. Female genitalia have not yet been dissected.

Bionomics. The holotype was taken at 600 m. Only three specimen (n = 3) are known currently.

Distribution. Known from W Papua.

Archigargetta Kiriakoff, 1967: 38.
Type species: *Archigargetta cyclopea* Kiriakoff, 1967 [= *Cascera amydra* Turner, 1903].
Schintlmeister 2013: 44.

Diagnosis. Medium sized moths. Antennae of males bipectinate; last 20–25% toward the tip are filiform. The female's antennae are filiform. Forewings are fuscous violet brown to reddish brown, sometimes mixed with metallic shining green scales. A diagnostic feature of *Archigargetta* is the prominent contrasting black dorsal patch near the tornus on forewing. Male has a fine deep black postbasal fascia, while the sexual dimorphic females display a

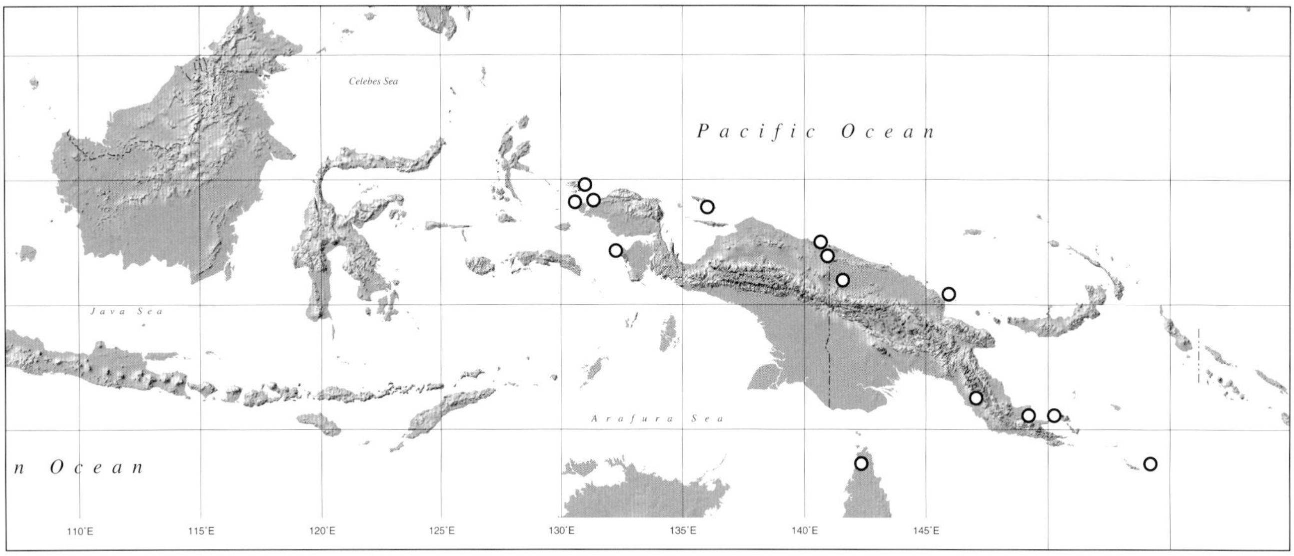

FIGURE 38 Distribution of *Archigargetta amydra*.

contrasting blackish-brown and curved band. Hindwings are chocolate brown without markings in both sexes.

Male genitalia have a pair of small and slender socii. Valvae are elongated with a tapered apex. The abdomen bears in second abdominal segment pair of large cteniophores (fig. 39b).

Archigargetta underwent a taxonomical revision by Holloway 1983: 29, who recognized at least two species only, which schema is followed here. But probably there are more species in the *viridigrisea* complex as variation in uncus and socii from various places indicates.

Caterpillar of an *Archigargetta* from S India was reared by Bell on wild guava, *Careya arbores* (Lecythidaceae) (Holloway, 1983: 31).

The genus is distributed from the Himalaya, India in its entirety, Sri Lanka, the Andaman Isls., Indochina, Sundaland, entire Indonesia including New Guinaea and Australia.

39. ***Archigargetta amydra*** (Turner, 1903): 74 (*Cascera amydra*).

Holotype: ♀, Australia, N. Q[ueensland], Townsville – ANIC, Canberra.

Schintlmeister 2013: 44.

Taxonomic note. Holloway, 1983: 30 noticed, that genitalia of holotypes of *Archigargetta cyclopea* Kiriakoff, 1967 and *Archigargetta diakonoffi* Kiriakoff, 1967, described as from the Waigeu islands and Biak respectively, differ slightly from mainland specimens. Unfortunately there is not enough material available, to check evidence and significance of these differences. Male genitalia of several dissected Australian specimens do not differ from those from New Guinea.

Diagnosis. Almost indistinguishable by its external appearance from its western sister species *A. viridigrisea*; however it seems that most males of *viridigrisea* have slightly longer and narrower forewings. Individual variability of forewing pattern is large. There are individ uals with almost uniform blackish or grey forewings. But there are also forms with contrasting white marked discal spots. Most males, but not all, displaying a small cream apical dot on forewings. Female is sexual dimorphic with much broader shaped forewings, where the fuscous basal area is separated from the paler medial area.

Male genitalia are distinguished from *viridigrisea* by short uncus and large basal valva process, which is absent in *viridigrisea*. Phallus is long and much narrower than *viriodogrisea*. It display at the tip a ring of several strongly sclerotized small spines. 8th abdominal segments are less modified, but the 8th sternite is more deeply notched at the posterior margin than is *viridigrisea*. Female genitalia have not yet been dissected.

Bionomics. Adults are rare in New Guinea. The few known specimens ($n < 15$) were taken mostly at low elevations of up to 350 m; one male was collected at 800 m above sea level.

Distribution. Distributed in Australia (N Queensland), New Guinea and many of the adjacent islands: Rossel Islands, Goodenough Islands, New Britain, Kar Kar Island, Biak, Batanta Isl., Waigeu.

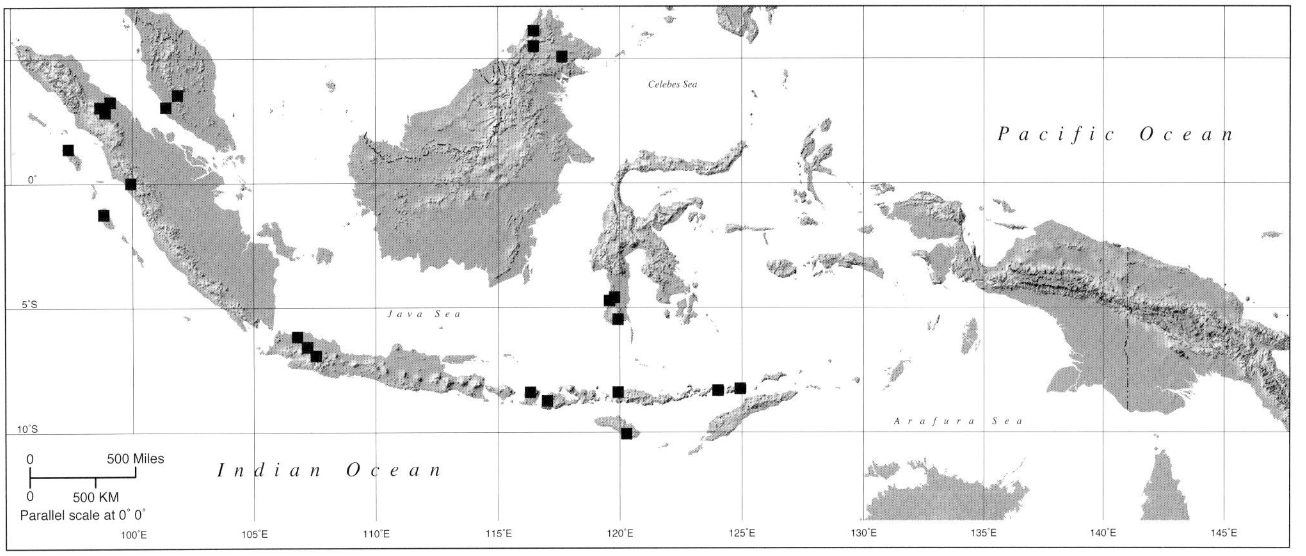

FIGURE 39 Distribution of *Archigargetta viridigrisea*.

40. ***Archigargetta viridigrisea*** (Dudgeon, 1898): 626; pl. 1: 3 (*Gargetta viridigrisea*).
Lectotype: ♀, [India, Sikkim], Sikhim, 1.800 feet – NHMUK, London.
Schintlmeister 2013: 44.

Taxonomic note. Dissection of several males from various islands showed remarkable variability in shapes of uncus, socii, and particularly in the phallus shape. Due to the rareness of imagines mostly single males from various smaller islands were available for dissection. Thus eventually morphological differences cannot be confirmed by further dissections. It is therefore currently not possible to evaluate the taxonomic value of the various forms. The taxa *Stauropus clothus* Swinhoe, 1899: 110 (type locality: [India, Karnataka], Karwar), *Pseudogargetta fuscicollis* Gaede, 1930: 618 (type locality: „Malaya, Borneo, Java, Celebes") and *Roepkeella tornalis* Kiriakoff, 1974: 381; pl. 1: 5; fig. 6 (type locality: NO-Sumara, Dolok Merangir, 180 m) are currently placed in synonymy. These names are available for subspecies or species names of the *viridigrisea* complex.

Diagnosis. Almost indistinguishable by its external appearance from its eastern sister species *A. amydra,* but it seems that most males of *viridigrisea* have slightly longer and narrower forewings. Individual variability of forewing pattern is large and comparable to *amydra*. There are males with almost uniform blackish forewing, forms with contrasting white discal spot, and a paler brownish medial area. Sometimes greenish scales are mixed in the fuscous ground color of the forewing. Such forms have not been observed yet in *amydra*. Female is sexual dimorphic and display a fuscous basal area, which is separated from medial area.

Due to the scarcity of known females, the range of individual variation is not properly known, but seemingly individual variation of females is contrary to the males, limited.

Male genitalia are distinguished readily from *amydra* by the long and tapered uncus and a pair of filigreed socii. Shape and length of uncus and socii vary considerably as illustrated. Phallus is short and thick, variable in its shape. Sometimes there is a prominent, strong sclerotized process in the middle part of the phallus. 8th abdominal segments are less modified, but the 8th sternit is less notched at posterior margin than *amydra*. Sclerotization of 8th tergite is not homogenous as in *amydra* but display a specific pattern. Female genitalia have not yet been dissected.

Bionomics. *A. viridigrisea* occur infrequently at lower and medium altitudes up to 700 m. One male was taken in N Sumatra at 1050 m and another male on Mt. Kinabalu at 1930 m above sea level.

Distribution. *Archigargetta viridigriseus* occur in the whole of India, Sri Lanka, the Andamans, Myanmar, Indochina, the Malayan Peninsula, Sumatra, Nias, Siberut, Borneo, Palawan, Java, Lombok, Flores, Sumbawa, Sumba, Alor, Pantar and Sulawesi.

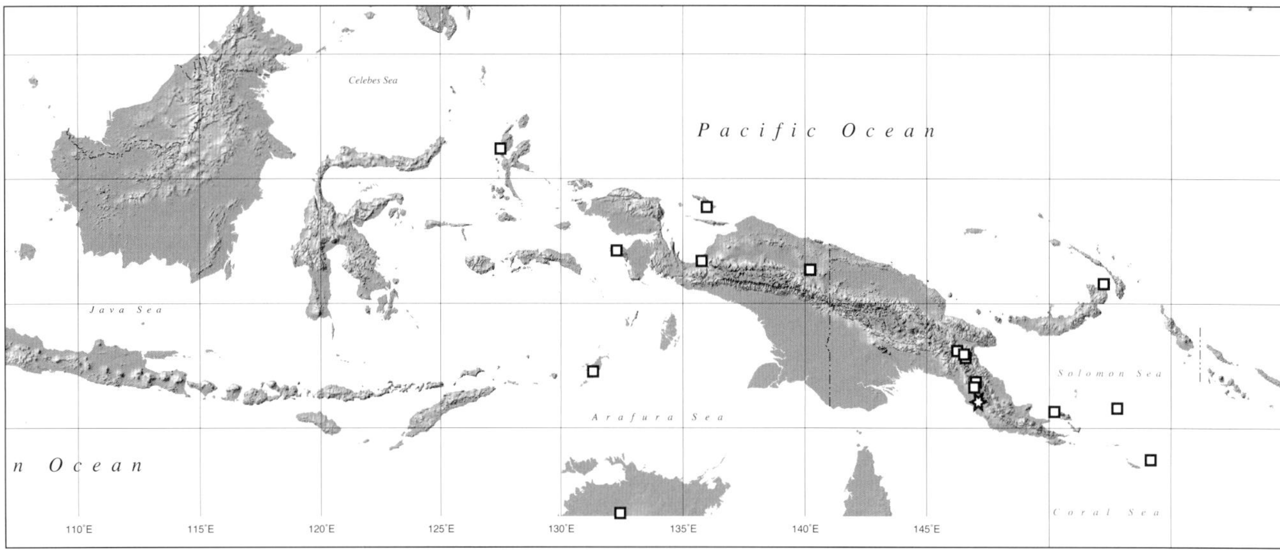

FIGURE 40 Distribution of *Lasioceros aroa*

Lasioceros Bethune-Baker, 1904: 380.
Type species: *Lasioceros aroa* Bethune-Baker, 1904.
Schintlmeister 2013: 232.

Diagnosis. Smaller moths, which are distinguished by elongated and narrow shape of the forewings. Antennae are very long and filiform in both sexes, except males of *L. aroa*. Characteristic pattern features on forewing are a blackish reniform stigma and a pair of white dorsal patches. On forewing costa, near the apex, three black dots occur. Hindwings are pale brown with a slightly darker postmedial area. Males have abdomen a black anal tuft, which is absent in females. Female resembles to their corresponding males, but wings are usually slightly darker.

Male genitalia resembles to some extent Noctuidae. Uncus is slender, tapered and often curved; socii are absent. Valvae are of prominent rounded shape with a stronger sclerotized process along the costa. Uncus and valva bear tufts of long and dense setae. Juxta is well sclerotized, broad and notched. Tegumen bears several structures, such as small knobs or other projections. Saccus is reduced in size, notched or bilobed. 8th abdominal segments are characteristically sclerotized and shaped, often with projections at the anterior margin. Female genitalia have very small 8th abdominal segment. Both pairs of apophyses are long. Postvaginal plate comes with a tapered lobe at the side. Antrum is strongly sclerotized but rather small. Ductus bursae is extremely long; corpus bursae is disc-shaped, with a large ellipsoid signum.

Lasioceros is distributed as six species from Fiji, New Guinea (including many adjacent islands), New Caledonia, Australia (Queensland, Northern Territory), the Moluccas, East Nusa Tenggara, Sulawesi, and the Philippines.

41. *Lasioceros aroa* Bethune-Baker, 1904: 381.
Holotype: ♂, [Papua New Guinea], Upper Aroa River – NHMUK, London.
Schintlmeister 2013: 232.

Taxonomic note. Males from Halmahera and New Britain are larger, with 20 mm forewing length, as compared to New Guinean populations. The smallest specimens are seen in Australia (16 mm forewing length) and in Tanimbar (18 mm). However, no significant genital differences among these populations were found. *Lasioceros aroa vitiensis* Robinson, 1975 from Fiji differ by their significantly broader shaped forewings in both sexes and might represent a distinct species (genitalia not dissected).

Diagnosis. Males of *aroa* are unmistakable by very long tufty black rami in the middle part of male antenna. The last 25% toward the antenna tip is filiform. Forewing color varies from white and cream to pale brown. Forewing costa is usually whitish. On the costa near apex there are three contrasting black dots. There is a discal streak in contrasting black, often accompanied by a second black or fuscous brown patch. Sometimes the basal and medial area of the forewings are filled in black, but mostly only a few black dots are in both area. White dorsal patches are situated in the tornus and the medial area of forewing. Female is sexually dimorphic having a fuscous brown forewing color.

Male genitalia have a slender, short, and curved uncus.

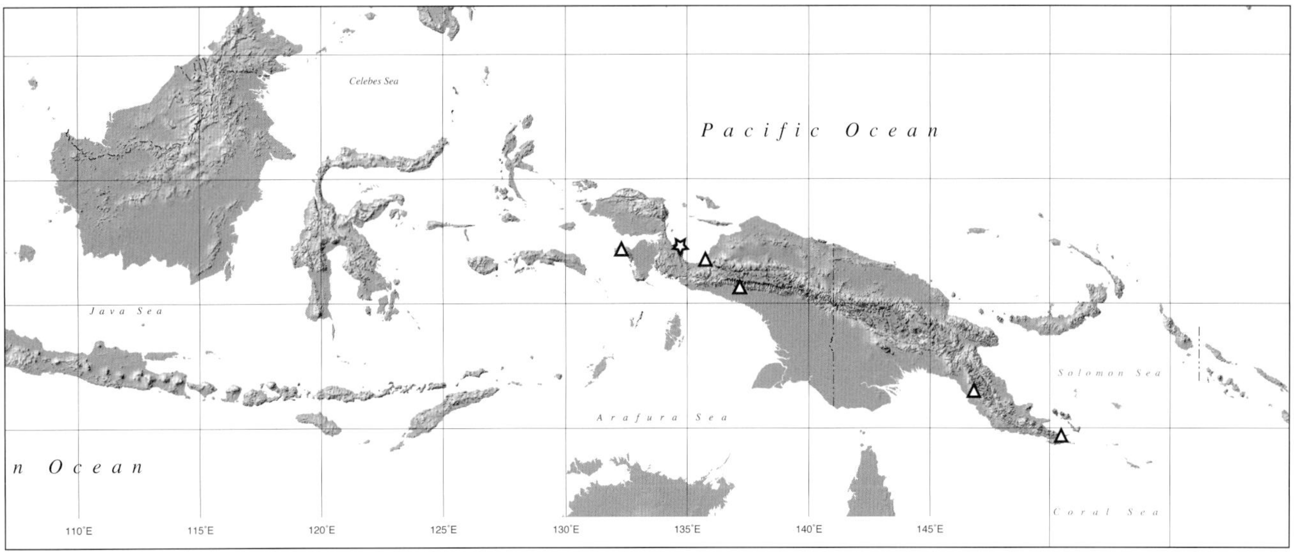

FIGURE 41 Distribution of *Lasioceros dentilinea*

Distinguished from other congeners by the shape of the modified tegumen which bears many projections. Sacculus of valva developed as a large projection. Costal process of the rounded valavae is shorter than in other *Lasioceros*. Phallus is slightly curved with a diagnostically hook at the tip. One dissected male has a short phallus with an abberrant shape of the tip (fig. 41b). 8th abdominal segments are rather slender and unmistakeably by their sclerotization and their shapes of the anterior margin. 8th tergite have a central projection at the anterior margin. Female genitalia are distinguished by the shape of the short antrum and the large ellipsoid signum.

Bionomics. *L. aroa* occurs infrequently at lower altitudes up to 300 m, but a few specimens were taken in PNG up to 1700 m above sea level.

Distribution. Distributed in New Guinea in its entirety, Rossel Islands, Goodenough Islands, Woodlark, New Britain, Biak, Halmahera, Tanimbar, Australia (Queensland, Northern Territory) and New Caledonia. In Fiji the ssp. *vitiensis* Robinson, 1975 occurs.

42. ***Lasioceros dentilinea*** Joicey & Talbot, 1917: 63; pl. 2: 6.
Holotype: ♂, [Papua New Guinea], Wandammen Mtns. 3.000-4.000 feet. – NHMUK, London.
Schintlmeister 2013: 232.

Diagnosis. Forewing color vary individually from light brownish-yellow to fuscous chocolate-brown. Costal area mixed in most specimens with whitish scales. In the postmedial area on the costa there is a larger, indistinct blackish patch, which separates *dentilinea* from the following species. Hindwings are pale brown with slightly more fuscous postmedial area. Female resembles male, but have no black anal tuft of the abdomen.
Male genitalia are larger than *L. aroa* and bear as the following species, a pair of rounded additional flaps arising from the tegumen. The function of these flaps is unknown. Uncus is short, tapered, and slightly curved. Valva is rounded in its shape and bear many dense setae. Phallus is straight with a characteristic "fading" sclerotization intensity at the tip. Everted endophallus displays a pair of antler-like projections. 8th abdominal segments are specifically shaped at the anterior margins and also specifically in their sclerotizations. Female genitalia have not yet been dissected.

Bionomics. *L. dentilinea* occur infrequently and only a few ($n < 10$) specimens are known, all collected at altitudes between 450–1100 m.

Distribution. Distributed in New Guinea.

43. ***Lasioceros asudara*** **spec. nov.** (figs. 43a–d).
Holotype: ♂, Papua New Guinea, Biagi, Mambare River, 5000 ft., ii. 1906, leg. A. S. Meek – NHMUK, London.

Paratypes (19 ♂♂, 1 ♀):
PNG: 3 ♂♂, 1 ♀, Hydrographer Mts., 2500 ft., ii. 1918, leg. Eichhorn Bros.; 1 ♂, dto. v. 1918 (BM 10313977); 1 ♂, Mt. Kebeae, 3600 ft., vi. 1903, A. E. Pratt (BM 10313978); 5 ♂♂, Kumusi River, vii. 1907, leg. A.S. Meek; 1 ♂, Upper Aroa River, ii. 1903 (A.S. Meek); 1 ♂, Haldana

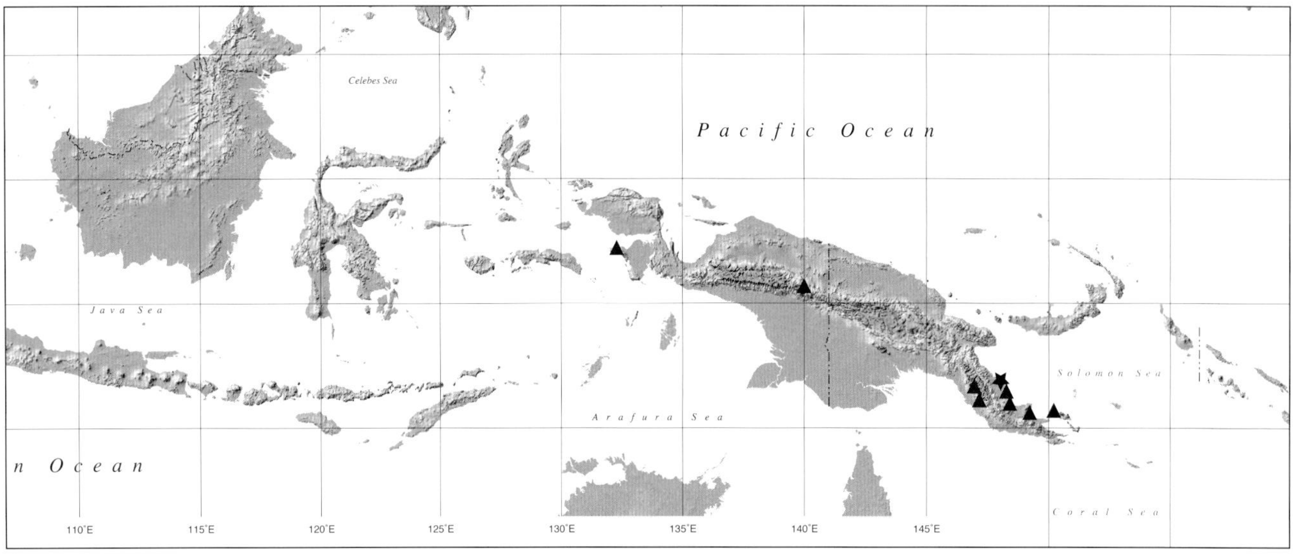

FIGURE 42 Distribution of *Lasioceros asudara*.

Collingwood Bay, iv. 1907 (A. S. Meek); 1 ♂, Dinawa, 4000', Coll. A. E. Pratt;
Goodenough Isl.: 3 ♂♂, 2500–4000 ft., v. 1913, (A. S. Meek);
W Papua: 1 ♂, Fak Fak, 1700 ft., Nov. 1907, J. Pratt; 1 ♂, dto. xii. 1907; 1 ♂, Jayawijaya Mts., Kec Abenaho, Pass Valley, 1950 m, 3°51'S, 139°05'E, 11.–17. ii. 2005, leg. R. de Vos.

Diagnosis. Forewing lengths ♂♂ 18–19 mm; ♀ 20 mm. Antennae of males and females are filiform. This is similar to *L. dentilinea,* but the shape of the forewing of *asudara* in both sexes is narrower. Forewing color is chocolate-brown. The upper half of the forewing along costa is mixed in contrasting white. This area varies individually in its extension. Discal streak is black, sharper than in *dentilinea* and accompanied by a blackish patch toward the base. On costa there is a fine black streak instead of a larger blotch as in *dentilinea*. Marginal fascia of forewing is fuscous brown with small contrasting white dots. Postmedial fascia is marked as a row of fine black dots. Hindwings are uniform fuscous brown, darker than in *dentilinea* and there is no paler basal area. Female is like the male in this respect, but tip of the abdomen has no blackish tuft as in the male.

Male genitalia differ from *dentilinea* by their longer uncus, which is slightly curved and tapered. Tegumen bears in the upper part a pair of small projections. Valva is of rounded shape and with many dense setae as in *dentilinea*. At base of valva a process arises, which is absent in *dentilinea*. Costal process of valva less compact and weaker sclerotized than in *dentilinea*. Sacculus of valve is smaller and bears fewer projections or other structures. As in *dentilinea* a pair of rounded flaps occur. Phallus is narrow and slightly curved. The phallus tip is stronger sclerotized, narrower and tapered in comparison to *dentilinea*. Everted endophallus resembles *dentilinea* in shape and with its antler like projections. 8th sternite differ in the semicircular and less widely notched anterior margin from *dentilinea*. 8th tergite shows the more strongly sclerotized portion anteriorally concave, which is in *dentilinea* straight. Anterior margin of 8th tergite less curved than in *dentilinea*. Female genitalia have not yet been dissected.

Bionomics. The species is much more common than *dentilinea*. Adults were taken at medium altitudes between 500 m and 1200 m. One male was collected at 1950 m above sea level.

Distribution. Distributed in Papua New Guinea, including Goodenough Isl.

44. *Lasioceros impurata* spec. nov. (figs. 44a–d).
Holotype: ♂, Indonesia, W Timor, Buraen, 35 km SE Kupang, 10°15'S, 123°49'E, 350 m, 26. i.–9. ii. 2006, leg. St. Jakl (GU 93-15) – in coll. A. Schintlmeister, Dresden.

Paratypes (2 ♀♀):
Alor: 1 ♀, Moru, 500 m, 22. iii.–3. iv. 2006 leg. St. Jakl.
Pantar: 1 ♀, Tanah Labang, 350 m, 8°19'S, 124°16'E, 9.-21. iii. 2006, (GU 93-11).

Diagnosis. Forewing lengths ♂ 16 mm, ♀♀ 19 mm. Male is similar from its external appearance to *L. sola*, but smaller. Forewing color of male is light brown. Costal area near the apex whitish tinged. Discal streak is fine and black. Near

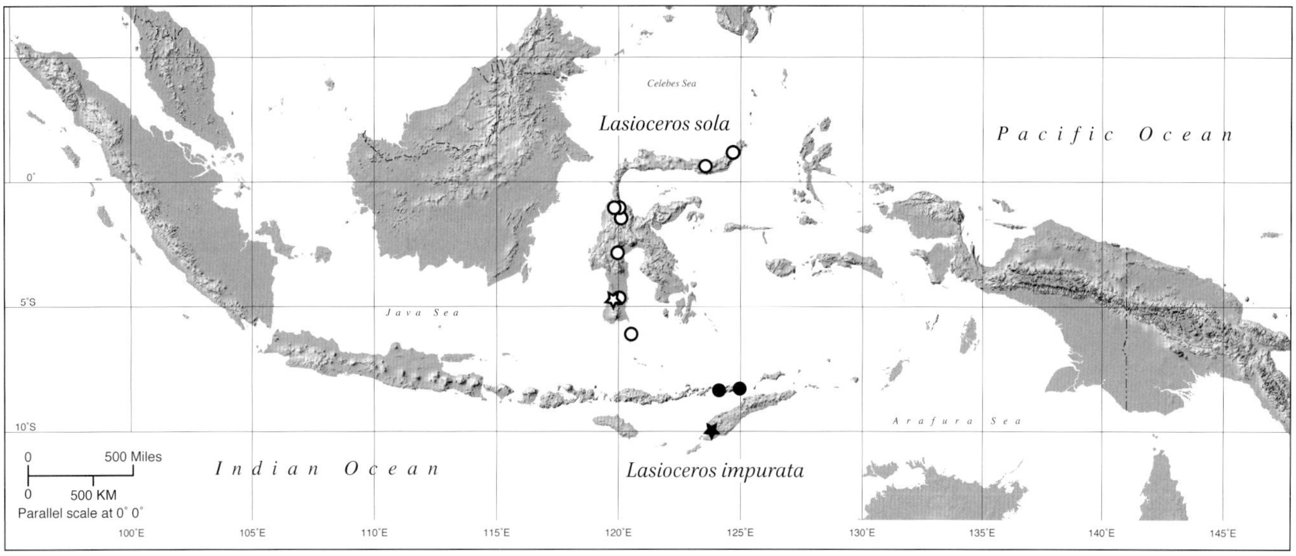

FIGURE 43 Distribution of *Lasioceros impurata* and *sola*.

the costa three black costal spots are visible. Veins in submarginal area of forewing are black tinged. A white tornal spot is present, which is mostly absent in *sola*. Hindwings are in both sexes uniformly brown. In *sola* the basal and medial area of hindwing are light brown. Female is larger and forewings are darker brown. Postmedial area of forewing displays a more intensively white pattern and the white tornal patch is therefore more contrasting.

Male genitalia are similar to those of *L. euteles* (West, 1932) from the Philippines and *L. sola*, but are distinguished by shapes of uncus and phallus. Uncus of *impurata* is straight, short and tapered, slightly thicker than in other similar *Lasioceros* species. Valva costa is of roundish shape and costal process stronger sclerotized than in other *Lasioceros*. Sacculus is large and triangular shaped; saccular projection is shorter than in *sola*. Phallus is straight, with a diagnostic broad tip. A hook, which is characteristically of *euteles* and *sola* are only rudimentary developed. 8th sternite displays at anterior margin a pair of strongly sclerotized lobes, which are at a broader distance to each other than in *sola*. 8th tergite is slightly concave at the posterior margin (in *sola* straight) and the more strongly sclerotized area conves is bordered toward the anterior margin. Female genitalia display rather long apophyses. Antrum is bilobed on all margins. Ductus bursae is very long and slender. Corpus bursae bearing a very large ellipsoid signum. Both strongly sclerotized streaks in the middle of the signum are situated close to each orther, but more widely apart in *euteles*.

Bionomics. This rare species (n = 3) was collected at 350 and 500 m above sea level.

Distribution. Found in the East Nusa Tenggara: Timor, Alor and Pantar.

45. ***Lasioceros sola*** Kobayashi, 2011: 307; figs. 15; 38a–d.
Holotype: ♂, Indonesia, Sulawesi Bantimurung, Gunung Makaroewa, 800 m – National Science Museum, Tokyo.
Schintlmeister 2013: 232.

Taxonomic note. A single male from Seleyer Isl. (figs. 45a, 45d) is smaller in size and is recognized by the presence of an indistinct whitish tornal spot on forewing. Male genitalia are also smaller than in the Sulawesian material and differ slightly in the shape of phallus and, in particular, by the shape and sclerotizations of the 8th abdominal segment. Probably representing a distinct subspecies, but provisionally placed in *sola*.

Diagnosis. Similar in external appearance to *L. euteles* from the Philippine and the previous species, *L. impurata*. Males are distinguished in that they have no white tornal spot, which the both other species have. *L. sola* is larger than *impurata*. Hindwings are light brown with darker brown postmedial area. Female is unknown.

Male genitalia are characterized by rather short and curved uncus. Sacculus is folded almost ellipsoid, not triangular as in *impuratra*. The clasper is longer than in *impurata* and the knob at the tip is smaller. Phallus shows a prominent hook at the tip, which is the feature to separating the species best from its congeners. The phallus surface is in its middle section serrated, but smooth in *impurata*.

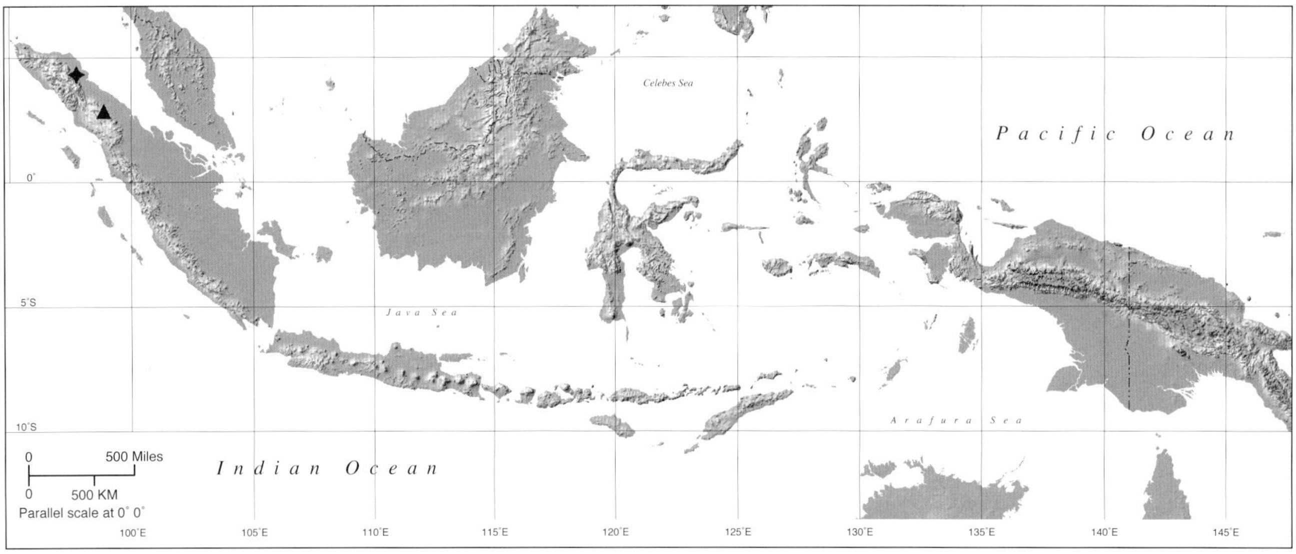

FIGURE 44 Distribution of *Noctuola diehlongata.*

Bionomics. Adults occuring infrequently and has been taken at medium altiudes between 700 m and 1700 m above sea level.

Distribution. Distributed in Sulawesi and Seleyer Isl.

Noctuola gen. nov.

Type species: *Porsica diehlongata* Schintlmeister, 2002.

Diagnosis. A smaller notodontid which resembles somewhat to a noctuid moth by external apperance. It is well characterized by its very narrow and elongated forewing, with a pointed apex in both sexes. The last part of forewing costa is rounded toward the apex. Venation of wings is as in *Phycidopsis.* Antennae are long in both sexes. Male has bipectinate antennae with short rami; the remaining 40% is filiform toward the tip. Antennae of female are filiform. Both, males and females have very long legs with a pair of long spures on hindlegs. Forewing of male uniformly dark brown. Discal streak is black, surrounded by a circular pale brown patch. Female is sexual dimorphics have forewing shape broader, like females of *P. albovittata.* Brown discal streak is as in male, but the surroundings areas are cream colored and of ellipsoid shape. Costal area is pale brown and the remaining part of the forewing fuscous brown. Hindwings are in both sexes pale brown with slightly darker postmedial areas.

Male genitalia resembles those of *Lasioceros* rather than *Phycidopsis.* Uncus is reduced in length and of triangular shape; socii are absent. Tegumen is rather small. Saccus is more strongly sclerotized and relatively large. Juxta is U-shaped sclerotized. Valva is of an unique angled shape with large rhomboid shaped apical part. Sacculus of valva is large and strongly sclerotized. Digitus proccess is asymmetric on both valvae. Valva costa bears a large and strongly sclerotized projection at the base. Phallus is straight and short with a tapered process at the tip. Endophallus is without modifications. 8th sternite displays a diagnostically X-shaped strongly sclerotized structure. 8th tergite is conical with stronger sclerotization at posterior margin. Female genitalia have small papillae annales and a large 8th abdominal segment. Apophyses posteriores are long and slender; apophyses anteriores are reduced in size. Postvaginal plate is large. Antrum is broad, becoming narrow toward the ductus burse. Ductus bursae broad and short. Corpus bursae is long and of ellipsoid shape. It bears a small, ellipsoid signum.

The genus is monotypic and known from Vietnam and Sumatra.

46. *Noctuola diehlongata* (Schintlmeister, 2002): 165; figs. 1–4 (*Porsica diehlongata*) **comb. nov.**

Holotype: ♂, Indonesia, Sumatra, South Aceh, Langsa I, 4°32'N, 97°45'E, 80 m – coll. A. Schintlmeister, Dresden.

Schintlmeister 2013: 457.

Diagnosis. See under the generic diagnosis above. Males are unmistakable by the unique forewing shape.

Bionomics. Only a few specimens (n = 6) are known, which were taken at 80 m and 1350 m above sea level.

Distribution. Known from Vietnam and Sumatra.

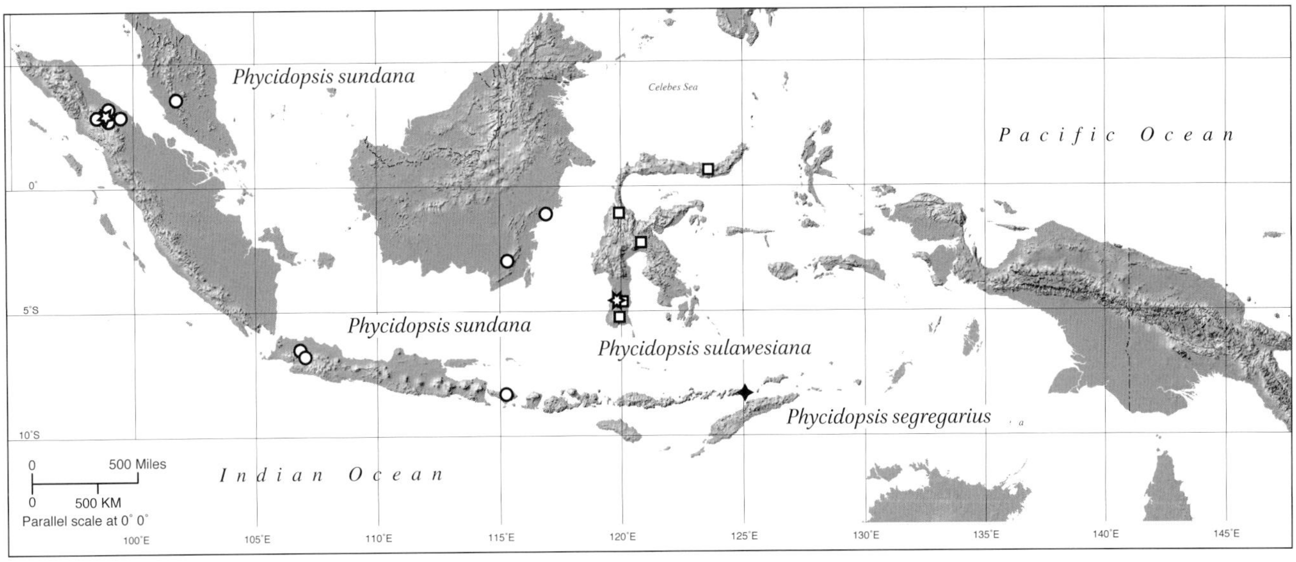

FIGURE 45 Distribution of *Phycidopsis sundana*, *sulawesiana* and *segregarius*.

Phycidopsis Hampson, 1893: 380.

Type species: *Phycidopsis albovittata* Hampson, 1893.

Schintlmeister 2013: 345.

Smaller moths, which are distinguished by an elongated and narrow forewing shape. Antennae are very long, and males have pectination over 3/4 of their lengths, whereas those of the females are entirely filiform. The elongated and narrow wingshape resembles *Noctuola, Porsica* and *Gargetta.* Costal area of forewing has contrasting whites and includes a narrow area of contrasting fuscous brown color. discal spot on forewing is indistinctly marked as a fuscous brown streak, surrounded by white scales. Medial area is bounded by two blackish fasciae, which are often indistinctly developed. Marginal fascia on all wings are fuscous brown with small white dots. Sexual dimorphism is limited.

Male genitalia resembles *Porsica* and *Gargetta.* Uncus is bilobed, small and short. A asymmetric gnathos is present. Valvae are elongated with a large digitus process, which is different shaped on both valva arms. Valva apex is slightly falcate. Phallus is straight and displays minor differences on specific level. 8th sternite specifically sclerotized. They bear at anterior margin a pair of lobes. Female genitalia have relatively small papillae annales but a long 8th abdominal segment. Postvaginal plate has a pair of lobes. Antrum is large and strong sclerotized; ductus bursae is short. Corpus bursae is large and of ellipsoid in shape. It bears a strongly sclerotized signum.

The genus is comprised of four species distributed in the whole of India, Sri Lanka, southern China, Taiwan, Indochina, the Philippines, Sundaland, Sulawesi and Alor.

47. ***Phycidopsis sundana*** Schintlmeister, 1993: 118: pl. 11: 1 **stat. nov.** (*Phycidopsis albovittata sundana*).

Holotype: ♂, Indonesien, Nord-Sumatra, Prapat, Holzweg 3, 1.350 m – coll. A. Schintlmeister, Dresden.

Schintlmeister 2008: 52; Schintlmeister 2013: 345.

Taxonomic note. *Phycidopsis sundana* was originally placed as a subspecies of *Phycidopsis albovittata* Hampson, 1893, described from Ceylon, Pundaloya. Comparison of male genitalia with material from N India and SW China shows, that *sundana* differ in the shape of the gnathos, the narrow shape of the valvae and shape of phallus much from continental Asian populations. *P. sundana* is therefore raised to species level (**stat. nov.**).

Diagnosis. All known species of *Phycidopsis* are very similar by external appearance (see genus diagnosis). A useful character in discriminating is length and shape of the fuscous brown costal strip on forewings. This field is in *sundana* longer than in all its other congeners.

Uncus of *sundana* has knob-shaped arms reduced in length, but among the three dissected males from Sumatra specimens also occur with longer uncus arms, comparable to *sulawesiana* (fig. 47a). Left arm of the asymmetric gnathos is narrow. All other known species' left arms are broader. Right arm is of circular shape. Juxta is unique in its shape, with two projections. Valva comes with a large sacculus. Digitus process is large and of circular shape. Phallus of the illustrated male (fig. 45a) is very long; those of other specimens are shorter, comparable to other species of the genus. 8th sternite is of conical shape. Projections at the anterior margin are shorter than in *segregarius*. 8th

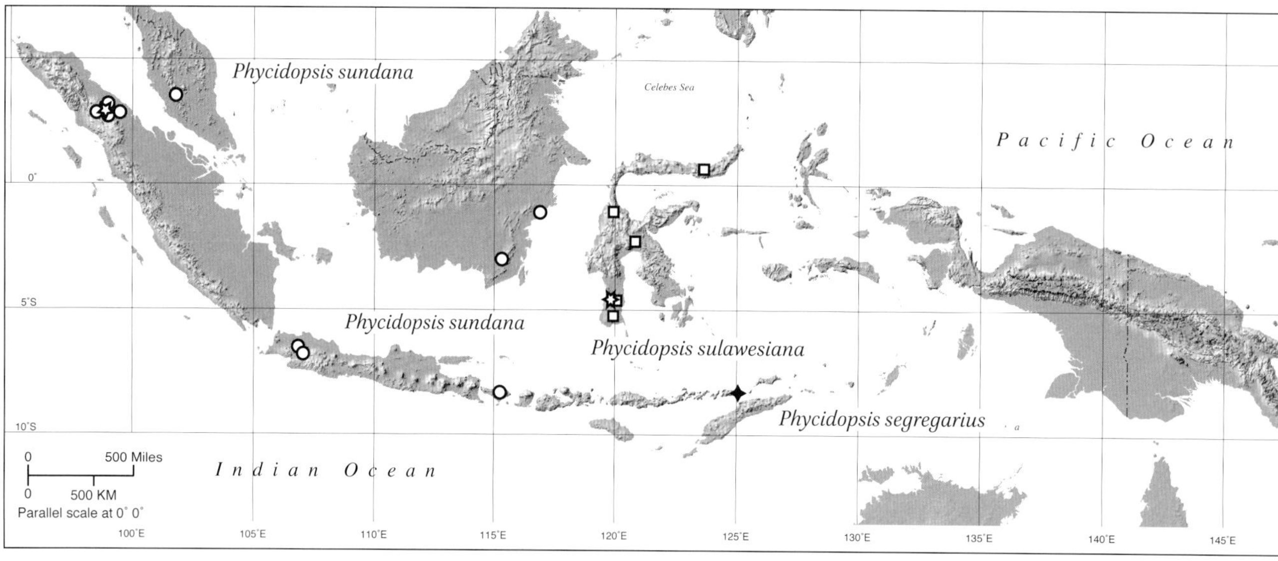

FIGURE 45 Distribution of *Phycidopsis sundana, sulawesiana* and *segregarius*.

tergite is diagnostically notched; other congeners exhibit straight posterior margins. Female genitalia are characterized by large lobes at the postvaginal plate. Antrum is large and broader than in *sulawesiana*. Signum of the long corpus bursae is hear-shaped and small.

Bionomics. Adults occur infrequently between 300–1600 m above sea level. Most specimens were observed at altitudes of about 1000 m.

Distribution. The Malayan Peninsula, Borneo, Sumatra, Java, Bali.

48. ***Phycidopsis segregarius* spec. nov.** (figs. 48a–d).
Holotype: ♂, Indonesia, Alor Isl., Nusa Tenggara Timur, Moru, 500 m, 8° 16'S 125°05'E, 1.–8. iii. 2006 leg. St. Jakl (GU 96-30) – coll. A. Schintlmeister, Dresden.

Paratype (1 ♂):
Alor: 1 ♂, Moru, 500 m, 8° 16'S 125°05'E, 1.–8. iii. 2006 leg. St. Jakl (GU 93-10).

Diagnosis. Forewing lengths ♂♂ 15.0–15.5 mm, 2 mm shorter than in the forewing than in the externally similar *P. sundana* or *sulawesiana*. In the series of *sundana* (n = 33 ♂♂) there are two small males only, which span 15 mm and 16 mm. The fuscous brown costal strip near the apex of the narrow forewings is in *segregarius* 4 mm long; in *sundana* 7–8 mm and in *sulawesiana* 6–7 mm. Whitish forewing pattern is diffusely distributed. Discal spot is not visible. The female is unknown.

Male genitalia differ from *sundana* and *sulawesiana* by small und subtle uncus knobs, which are slightly asymmetrically, and by the more robust asymmetrically shaped gnathoi. Valva has a very large sacculus and digitus process is circula,r with a fine and sharp spine. Phallus is straight, somewhat variable in the length and in shape of its base. Phallus is thicker than in *sundana* or *sulewesiana*. 8th sternite is rather squarish than conical as in *sundana*. At the anterior margin there is a pair of relatively long lobes, longer than in *sundana* or *sulawesiana*. 8th sternite is distinguished by its specifically sclerotization. 8th tergite is rather straight at the posterior margin.

Bionomics. Only two males are known, both were taken at 500 m.

Distribution. So far known only from Alor Island.

49. ***Phycidopsis sulawesiana*** Schintlmeister, 1993: 119: pl. 11: 2.
Holotype: ♂, Indonesia, Sulawesi Selatan, Bantimurung, 40 km NEE Ujung Padang, 4°56' S 119°39'E, 200 m – coll. A. Schintlmeister, Dresden.
Schintlmeister 2013: 345.

Diagnosis. Very similar in external appearance to other congeners. The fuscous brown costalstrip of the forewing is longer than in *seggregarius*, but slightly shorter than in *sundana*.

Male genitalia are distinct by shape of the asymmetrical gnathoi. Right arm is pointed at the tip, left arm straight and not curved as seen in other *Phycidopsis*. Sacculus of

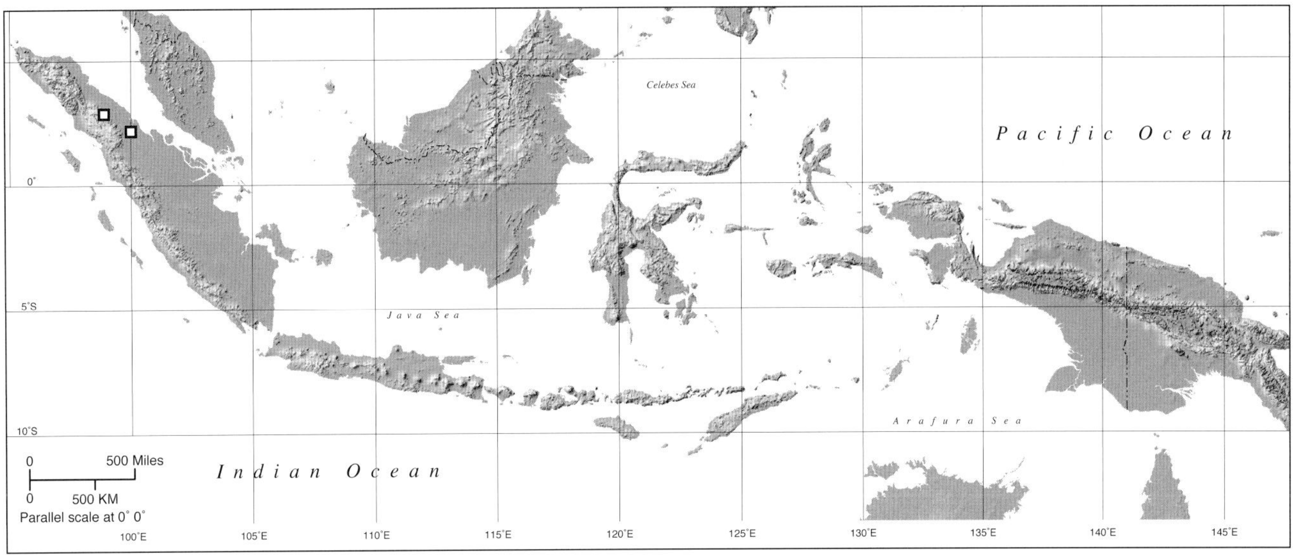

FIGURE 46 Distribution of *Stictogargetta albimacula*.

valva is narrower than in other species. Digitus process is large and of specific shape. Phallus is slenderer than in *seggregarius* or *sundana*. 8th abdominal segments are not modified in their sclerotizations and there are also no projections at the anterior margin of 8th sternite. Female genitalia resembles those of *sundana*, but lobes of the antevaginal plate are smaller, and the antrum is distinctivly narrower. Corpus bursae is rather rectangular in shape. Signum is rectangular and much longer than in *sundana*.

Bionomics. Adults occurs infrequently at altitudes between 200 and 1400 m above sea level.

Distribution. Endemic to Sulawesi.

Stictogargetta Kiriakoff, 1968: 27.
Type species: *Phycidopsis albimacula* Hampson, 1893.
Schintlmeister 2013: 431.

Medium sized moths. Antennae are very long and bipectinated. Last 30% toward the tip is filiform. Females have shorter rami than males. Forewings are brown with right-angled tornus. Discal streak is transverse and fuscous brown. Postmedial, submarginal and marginal fasciae are marked as rows of fine white dots. Hindwing brown with fine white dots on the marginal fascia. Sexual dimorphism is limited.

Male genitalia have a reduced uncus; socii are absent. Saccus is small. Valvae are strongly sclerotized and robust. Sacculus of valva has a large apical process. Phallus is straight, medium sized. 8th abdominal segments are specifically sclerotized. 8th ternite bears a pair of lobes at the anterior margin.

Stictogargetta is represented by two species, distributed from the Himalaya, the whole of Indochina, the Malayan Peninsula, Borneo and Sumatra.

50. ***Stictogargetta albimacula*** (Hampson, 1893): 136 (*Gargetta albimacula*).
Holotype: ♂, [India], Sikhim – NHMUK, London.
Schintlmeister 2013: 431.

Diagnosis. Readily recognized by the forewing shape and a prominent white patch on the dorsum. Discal streak is indistinct. Marginal fascia of all wings contains rows of small but contrasting white dots.

Male genitalia are rather unmistakable by the large and compact tegumen. Valvae are of ellipsoid shape and pointed apically. Sacculus process is long and pointed. Phallus is bulbous in the middle part. 8th tergite characteristically sclerotized and bilobed at the anterior margin. The only known female was not dissected.

Bionomics. Only a few specimens of this moth are known ($n < 10$). Adults were observed in Sumatra at 1100 m above sea level, but in Thailand one specimen has been taken at 2100 m.

Distribution. Known from northeastern India (Sikkim, Assam), Thailand and Sumatra.

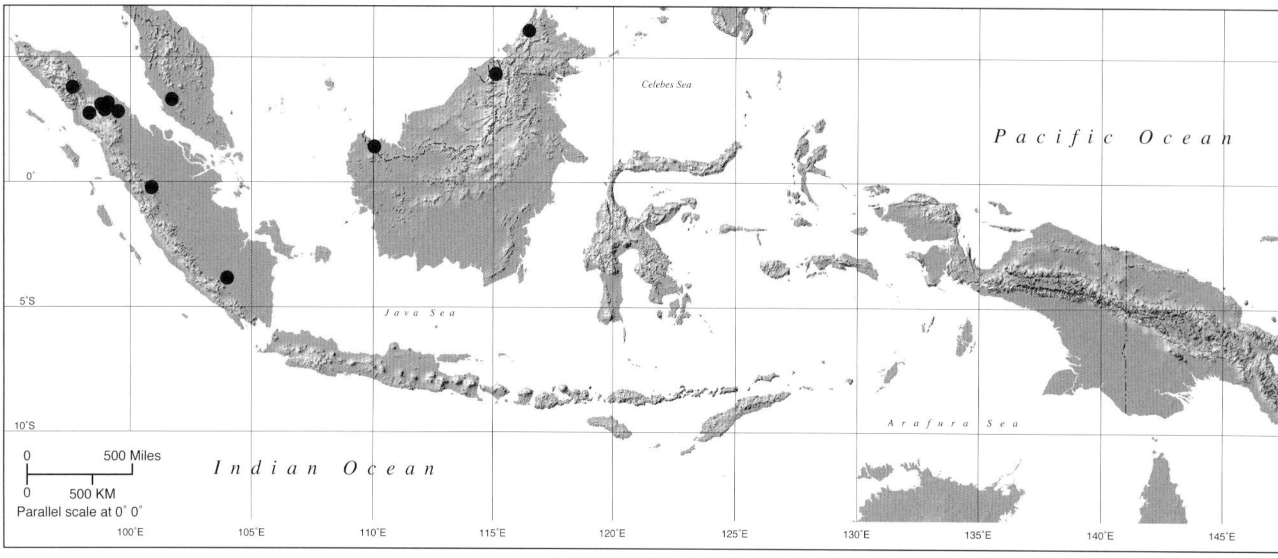

FIGURE 47 Distribution of *Stictogargetta lithosidia.*

51. ***Stictogargetta lithosidia*** (Hampson, 1895): 136 (*Gargetta lithosidia*).
Holotype: ♂, [India], Sikhim – NHMUK, London.
Schintlmeister 2013: 431.

Diagnosis. Forewings are brown with a contrasting black streak surrounded by a pale brown area. Postmedial fascia consist of a fine blackish zig-zag line; submarginal fascia is straight and of prominent contrasting cream colors. Hindwing pale brown with an indistinct lighter brown medial fascia. The only known female does not differ from the male in its external appearance.

Male genitalia have a short squarish uncus. Valva ending apically in a long process. Sacculus process is longer than in *S. albimacula*. 8th sternite is deeply notched; 8th tergite is diagnostically shaped with a pair of lobes at the anterior margin. Female genitalia have not yet been dissected.

Bionomics. *S. lithosidia* flies in Sumatra commonly at altitudes between 150–1100 m, but occur infrequently in Borneo.

Distribution. Distributed in northeastern India (Sikkim), the Malayan Peninsula, Sumatra and Borneo.

Gargetta Walker, 1865: 455.
Type species: *Gargetta costigera* Walker, 1865.
Schintlmeister 2013: 177.

Medium sized to smaller moths. Antennae are in both sexes bipectinated with long rami in both sexes. The last 15–25% toward the tip are filiform. Forewings are elongated and narrow, almost uniform brown, but the discal spot is developed as a deep black transverse streak. Most species have a separated basal area. Veins are often black tinged; marginal fascia marked fuscous brown. Near the forewing apex a fuscous patch occur. Sexual dimorphism is limited.

Male genitalia have a widely emarginate uncus, and a pair of robust socii. Sacculus bears one or two processes of taxonomic value. Phallus is rather short and straight, less modified. 8th abdominal segments are similarly sclerotized in all *Gargetta*. 8th sternite concave; 8th tergite is gently convex at posterior margin. Female genitalia have a broad but short 8th abdominal segment. Papillae annales are lage; apophyses medium sized and slender. Antrum is large, ductus bursae is rather short but broad. Signum is present.

The genus is at present known to be comprised of ten species, distributed in the Himalaya, India, the Andamans, Myanmar, China, Indochina, Sundaland up to the Moluccas and Australia. The greatest species diversity occurs in Indochina.

52. ***Gargetta costigeroides*** Holloway, 1983: 21; pl. 1: 6; fig. 14.
Holotype: ♂, Brunei, Ulu Temburing, Base Camp, 350 m – NHMUK, London.
Schintlmeister 2013: 177.

Taxonomic note. Two dissected females from Sumatra (figs. 52c, 52d) differ greatly in its genitalia (shapes of postvaginal plate and antrum), which cannot be interpreted or

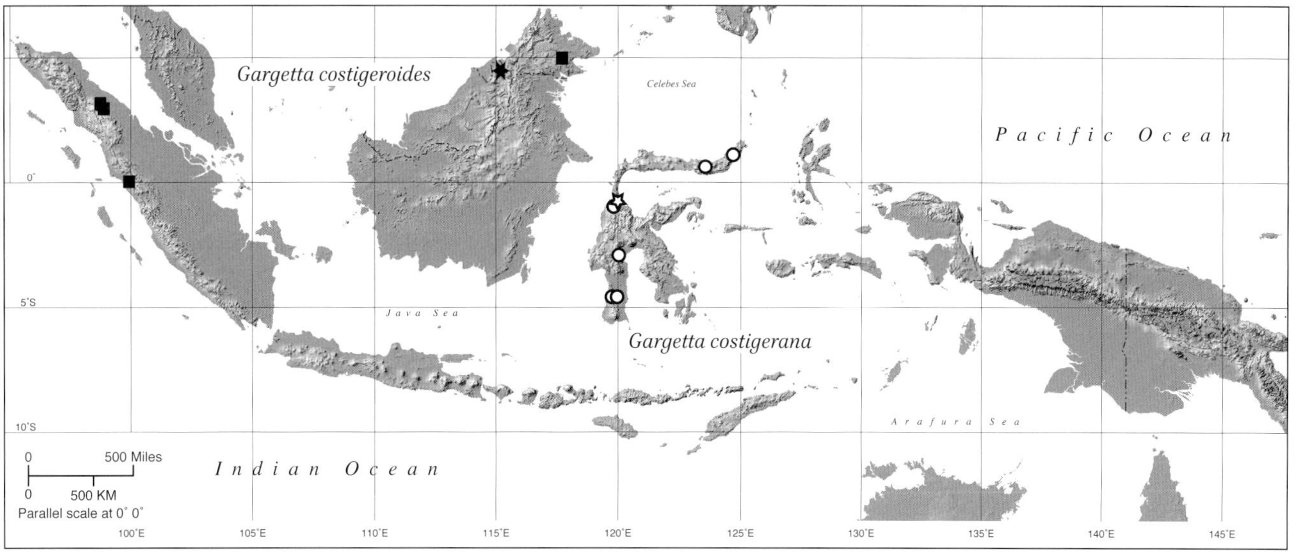

FIGURE 48 Distribution of *Gargetta costigeroides* and *costigerana*.

confirmed due to lack of further females. Also male genitalia of Bornean and Sumatran are somewhat different (shapes of socii, digitus process, 8th abdominal segments), but these differences are on lower level.

Diagnosis. *Gargetta costigeroides* is very similar to *G. costigera* Walker, 1865, which occurs in India and Indochina. It is best to distinguished from the latter by broader shaped forewings in both sexes with a more rounded outer margin. All wings are chocolate brown with reduced markings. A black discal streak on the forewing is prominent and short, but a male from Sumatra displays a weakly developed streak. *Gargetta costigerana* differ from *costigeroides* having a longer and finer discal streak. Sexual dimorphism is limited, but female have broader shaped wing.

Male genitalia have a widely emarginate uncus, and a pair of compact socii. They differ from *costigera* that the uncus arms are rather ellipsoid. In *costigerana* uncus arm is slender and tapered. Sacculus process is large. Phallus is curved, longer and slenderer than *costigera* and *costigerana*. Female genitalia are as described under the genus diagnosis. They differ from other congeners by the shape of the postvaginal plate, with a pair of slender lobes and the antrum shape. Signum is long and broader than in other *Gargetta*.

Bionomics. Only a few specimens (n = 5) are known; they were collected in lowland forests at altitudes between 200–350 m.

Distribution. Known from Borneo and Sumatra.

53. ***Gargetta costigerana* spec. nov.** (figs. 53a–f).
Holotype: ♂, Indonesia, Central Sulawesi, Gunung Salubaja, 1000 m, 0° 45'S 120°00'E, 9.–10. ii. 1995 leg. V. Sinjaev & A. Tarasov – coll. A. Schintlmeister, Dresden.

Paratypes (75 ♂♂, 17 ♀♀):
Sulawesi: 3 ♂♂, 1 ♀ Bantimurung, 40 km NEE Ujung Padang, 4°56' S 119°39'E, 200 m, 2.–9. v. 1984, leg. A. Schintlmeister (GU 07-28, GU 96-35; GU 96-36); 1 ♂, Tjamba near Maros, ii. 1938 (J. P. A. Kalis); 2 ♂♂, Pangean near Maros, 2,000 ft., iii. 1938, (J. P. A. Kalis) 1 ♂, Todjamboe, 14. vii. 1936 leg. Toxopeus; 1 ♂, Gunung Makaroewa, Bantimurung NP, 800 m, 8. iv. 2010, leg. Akihiko Miyano; 1 ♂, Puncak, Palopo, 900–1300 m, x. 1997 (MWM 8.111); 1 ♂, Puncak, Palopo, 1300 m, viii. 1995; 1 ♂, Gunung Ambang Reserve, Road Kotambagu-Motoling, 850 m, 15.–18. i. 2001 leg. M. Schaarschmidt (MWM 14.971); 60 ♂♂, 15 ♀♀, Paloe [= Palu], G. Rangkoenau, 900', Nov. 1936, leg. J.P.A. Kalis; 2 ♂♂, 1 ♀, Dumoga Bone N.P., 540 m, 12.–13. ii. 1985, leg. J. D. Holloway (BM 1470); 1 ♂, Dumoga- Bone Natl. Park, 16. – 20. x. 1985, 220 m, leg. J. B. Heppner (FLMNH-MGCL specimen 289691); 1 ♂, dto. 21.–25. x. 1985 (FLMNH-MGCL specimen 289691).

Diagnosis. Forewing lengths ♂♂ 20.0–25.5 mm (average 21 mm); ♀♀ 23–24 mm. Similar to *G. costigeroides*. Forwings are fuscous brown with a paler yellowish- brown transverse area from the base to the margin. Some specimens including the holotype display this pale brown area in contrast to the fuscous ground color. Discal streak is deep black, fine and longer than in *costigeroides*. Marginal fascia of all wings are black and interrupted by whitish

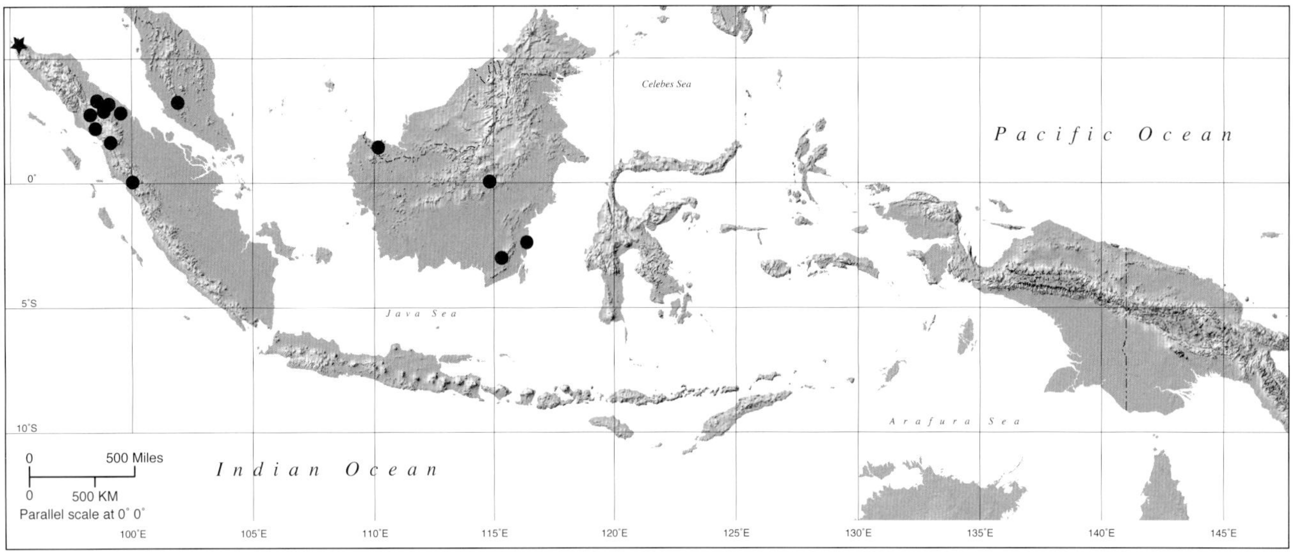

FIGURE 49 Distribution of *Gargetta hampsoni*.

dots, where the veins cross. Veins in submarginal area in most specimens are blackish tinged. Fuscous brown postbasal, medial and postmedial fasciae at costa are well visible and indistinct in the remainig forewing areas. Hindwings are of uniform, fuscous brown. Medial fascia is indistinctly developed as a slightly more fuscous shadow developed. Sexual dimorphism is limited.

Male genitalia resembles those of *costigeroides*, but differ in longer and tapered uncus arms. Socii are short and slender, as in *costigera*. Valvae are broader than in *costigera* or *costigeroides*. Sacculus is large and the apical process is longer and slenderer than in *costigera*, but shorter than *costigeroides*. Phallus is slightly curved, shorter and thicker than *costigeroides*. 8th abdominal segments are characteristically sclerotized as in *G. costigera*, the 8th tergite gently bilobed at posterior margin.

Bionomics. The species occur infrequently in the lowlands (200 m) but appears more commonly at medium altitudes between 800 m and 1300 m.

Distribution. Endemic to Sulawesi.

54. ***Gargetta hampsoni*** Schintlmeister, 1981: 286; figs. 5a–c.
Holotype: ♂, [Indonesia], Sumatra, Aceh, Banda Aceh – coll. A. Schintlmeister, Dresden.
Schintlmeister 2013: 178.

Diagnosis. Narrow forewings are rather uniform chocolate brown. The external appearance of the forewings suggests a striolate pattern, by the fuscous tinged veins. The marginal fascia with whitish dots are as in *costigerana*. The female resembles the male, apart from the broader shaped forewings.

Male genitalia have a rectangular shaped uncus arm, but the tip is pointed as in *costigera*. Socii are small and not prominently developed. Phallus straight and narrower toward the tip. The 8th tergite triangular deeply notched.

Bionomics. *G. hampsoni* occur in Sumatra commonly up to 1350 m, but seemingly rare in Malaya and Borneo.

Distribution. The ssp. *hampsoni* is distributed in Sumatra, Malaysia, Borneo and Palawan. The remaining parts of the Philippines (Mindanao, Negros, Mindoro) are populated by the ssp. *occulta* Schintlmeister & Lourens, 2010.

55. ***Gargetta divisa*** Gaede, 1930: 615; pl. 81: d.
Holotype: ♂, [Indonesia, Java, Malang] – ZMHU, Berlin.
Schintlmeister 2008: 50; Schintlmeister 2013: 177.

Taxonomic notes. The true type locality of *Gargetta divisa* is Java and not "Burma, Malakka" as stated in the original description and as it was shown by Schintlmeister & Lourens 2010: 78.

From the Philippines, Marinduque, *Gargetta divisa orienta* Schintlmeister & Lourens, 2010 was described, which is in fact a junior synonym to *Gargetta luzonica* Semper, 1896, (**syn. nov.**) Schintlmeister & Lourens 2010: 80 associated *luzonica* errorneously with *curvaria*, but since the type of *luzonica* was recently located and examined by the author at Senckenberg Forschungsinstitut und

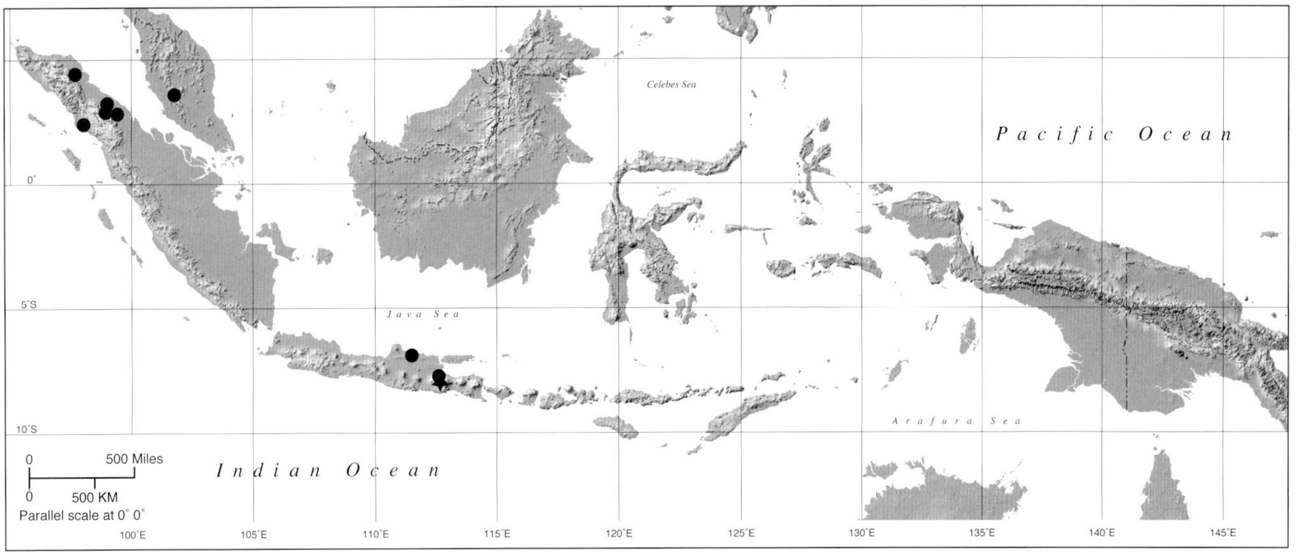

FIGURE 50 Distribution of *Gargetta divisa*

Naturmuseum, Frankfurt/Main it became clear, that it is conspecific with *divisa*.

The traced syntypes – a pair – were labeled as lectotype and paralectotype, but the author was unable to find a publication for designation of a lectotype. To stabilize the nomenclatur the ♂, with the following labels is hereby designated as lectotype: "Manila 30.6.28"; red framed handwritten label (in Semper's hand) "*Gargetta luzonica* type Semper"; "69"; pink label "Lecto-typus"; yellow label "Photo done by A. Schintlmeister #8939". The remaining female also from Manila, which bear a pink label "Paralecto-typus" becomes a paralectotype.

Diagnosis. A variable species. The examined series from Java (n = 8) from different years are short in forewing length, but such small specimens occur also in Sumatra among larger specimens. Discal spot of forewing is brown, surrounded by pale brown color. Forewings in ssp. *dives* often uniformly brown colored. Specimens with clearly marked postbasal and postmedial fasciae on forewing (fig. 55d) represent only individual forms. *G. tompua* differs from *divisa* by a contrasting blackish brown postbasal area and a fuscous tornal patch. Postbasal and postmedian fasciae are clearly marked.
Male genitalia are characterized by the shape of the tapered and diverging uncus lobes, which have differing lengths. Socii are rather short but thicker than in the other similar species. Sacculus bears a slender process. Phallus is straight and longer than in *acarodes* or *bervisocius*. 8th abdominal segments in shape and sclerotization are like *acarodes* and *brevisocius*; 8th tergite is notched at the posterior margin. Female genitalia are distinguished by the shape of antrum, which has a pair of rounded lobes at posterior margin. Ductus bursae is short and strongly sclerotized. Corpus bursae very long. It bears a long rectangularly shaped signum.

Bionomics. *G. divisa* is a common species, which is often associated with cultivated and urban habitats. It occurs up to 1150 m, but lower altitudes up to 500 m are seemingly preferred. Hostplant of the caterpillar is *Bridelia tomentosa* in Euphorbiaceae (Roepke, 1943: 3).

Distribution. *Gargetta divisa divisa* is distributed from northern India, Indochina, the Malaysian Peninsula, Sumatra and Java. The ssp. *luzonica* occur in the Philippines (Luzon, Marinduque, Mindoro, Panay).

56. ***Gargetta tompua*** (Kiriakoff, 1970: 108, fig. 7.
Holotype: ♂, [Indonesia, Sulawesi], SW Celebes, Tjamba [= Camba] near Maros, 1500 ft. – NHMUK, London.
Schintlmeister 2013: 177.

Taxonomic note. *Gargetta tompua, G. acarodes* and *G. brevisocius* are closely related and might represent a circle of allopatric subspecies. They are treated here as different species because of great differences in shapes of uncus and phallus.

Diagnosis. Similar to *G. acarodes* and *brevisocius*. *G. tompua* is best recognized by contrasting fuscous brown postbasal area of the forewing, which is bordered by two white/blackish fasciae. Postmedial fascia is cream-colored

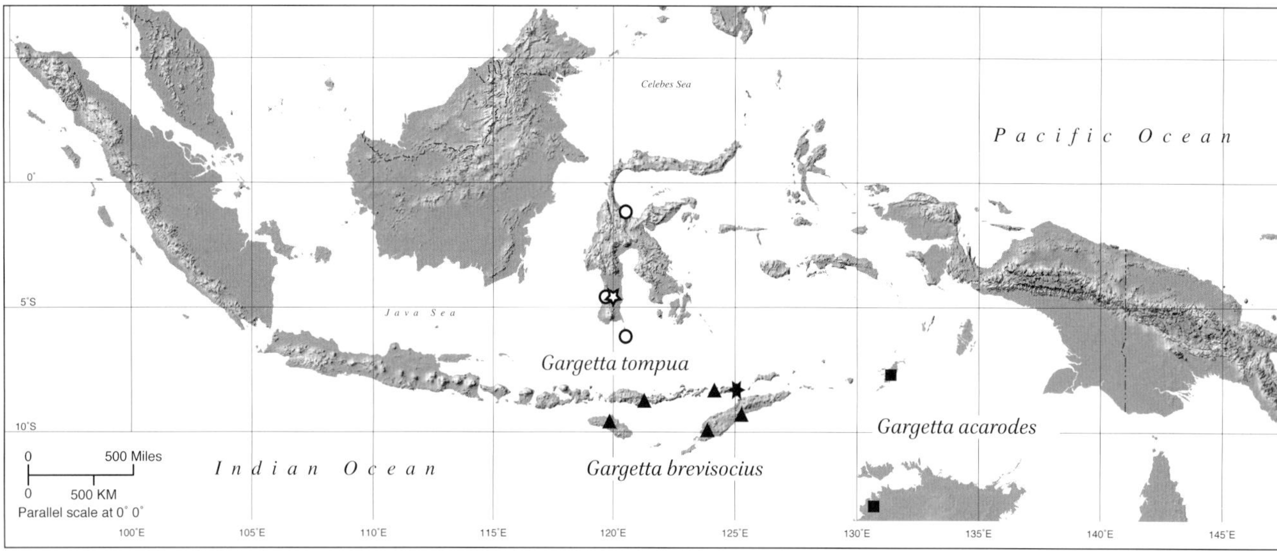

FIGURE 51 Distribution of *Gargetta tompua, acarodes* and *brevisocius.*

with a row of black dots. The tornal spot is distinct blackish brown.

Male genitalia are similar to *acarodes* and *brevisocius* Uncus is widely notched and the lobes less rounded and narrower than *acarodes.* A second dissected male from Seleyer Isl. (BM 10313981) resembles in uncus shape more the illustrated holotype (fig. 56b). Socii are significantly longer and narrower than *brevisocius.* Valvae are similar to *acarodes* but clearly thicker than *brevisocius.* Sacculus is large and its process somewhat variable in its shape as illustrated. Phallus is longer and slenderer than *acarodes* or *brevisococius.* 8th abdominal segment is as in *acarodes,* but sclerotization of 8th sternite is different. Female genitalia have not as yet been dissected.

Bionomics. The species occurs infrequently at altitudes of around 500 m.

Distribution. Known from Sulawesi and Seleyer Isl.

57. ***Gargetta acarodes*** (Turner, 1903): 71 (*Lasioceros acarodes*) **comb. nov.**
Holotype: ♂, Australia, N. Q[ueensland], Townsville – ANIC, Canberra. ww
Schintlmeister 2013: 232.

Taxonomic note. *Lasioceros acarodes* is here combined with *Gargetta* because of similarities in external appearance with *G. divisa* and chiefly by male genitalia, which differ in many aspects (notched uncus, present socii, shape of valvae) from *Lasioceros,* but matching *Gargetta.*

Diagnosis. There is only one rubbed male from Tanimbar Isl. The diagnosis was made from a homogenous series (n = 18) from Australia.

Similar to *divisa* and *brevisocius,* but wing color is less reddish and more fuscous. Postbasal area of forewing is more fuscous, but less contrasting than in *G. tompua.* A fuscous tornal patch on the forewing is present. Sometimes forewing pattern appears very indistinct. Sexual dimorphism is limited.

The species differ in male genitalia from *divisa* and *brevisocius* by its large and rounded uncus lobes and long socii. Valvae bear a small sacculus process like *divisa.* Phallus is shorter than in *divisa* and *brevisocius.* 8th abdominal segments are simlar to those of *divisa* and *brevisocius.* 8th sternite is broader and 8th tergite less deep notched than *divisa.* Female genitalia resemble by the shape of antrum with its pointed lobes rather to *brevisocius* than *divisa.* The unique shape of the signum is a diagnostic feature.

Bionomics. *G. acarodes* is known from a single male from Indonesia taken at 150 m above sea level, but flies quite commonly in Australian lowlands, below 200 m altitude.

Distribution. Distributed in northern Australia (Queensland, Northern Territory). A single male is known from Tanimbar, Yamdena Isl.

58. ***Gargetta brevisocius* spec. nov.** (figs. 58a–h).
Holotype: ♂, Indonesia, Nusa Tenggara Timur, Alor Isl., Moru, 8°16'S, 125°05'E, 500 m, 22. iii.–3. iv. 2006, leg. St. Jakl – in coll. A. Schintlmeister, Dresden.

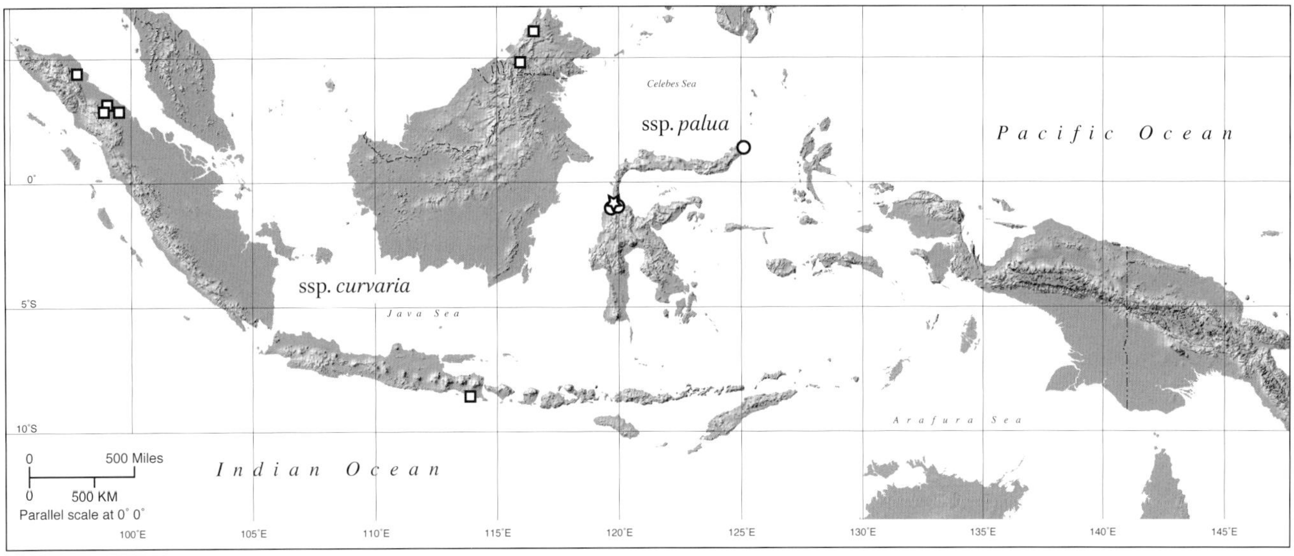

FIGURE 52 Distribution of *Gargetta curvaria.*

Paratypes (59 ♂♂, 11 ♀♀):

Alor: 9 ♂♂, Moru, 8°16'S, 125°05'E, 500 m, 1.–8. iii. 2006, leg. St. Jakl (GU 91-05a, GU 96-56); 7 ♂♂, 3 ♀♀, dto. but 22. iii.–3. iv. 2006 (GU 93-05);

Pantar: 10 ♂♂, 3 ♀♀, Tanah Labang, 8°19'S, 124°16'E, 350 m, 9.–21. iii. 2006 (GU 54-80a);

W Timor: 20 ♂♂, 4 ♀♀, 35 km SE Kupang, Buraen, 10°15'S, 123°49'E, 350 m, 26. i.–9. ii. 2006 (GU 54-89a, GU 96-57) leg. St. Jakl; 9 ♂, ibid, 20.–27. ii. 2006 (GU 51-77a);

E Timor: 1 ♂, Portug. Timor, Suai, 1.–7. xii. 1912 (E. Wahr);

Flores: 1 ♂, 1 ♀, Pautai Selatan, 100 m, 5. xii. 1953 (JMA van Groenendael);

Sumba: 1 ♂, near Lewapaku, at km 57 on Waingapu to Waikabubak rd. 500 m, 9.– 12. xii. 1985, leg. J. D. Weintraub (USMN 127.523).

Taxonomic note. *Gargetta brevisocius* is a variable species as illustrated: forewing length, paleness of forewing, development of the fasciae and also shape of uncus and phallus in male genitalia as well are subjects to variation. Populations of different islands seem slightly different in external appearance and in their male genitalia. However, due to wider individual variability and also possibly to seasonal variation – the material was collected only during a few days of a single year – a taxonomic splitting into subspecies is currently not possible.

Diagnosis. Forewing length ♂♂ 15–20 mm (average 19 mm, about 1 mm larger than *acarodes*); ♀♀ 18–22 mm. Similar to *G. acarodes* in external appearance, but color of the adult is more reddish tinged and rami of the antennae are significant longer. Most specimens have a fuscous filled area between basal and medial area of forewing. There is a fuscous apical patch near costa and a blackish tornal patch. Veins in the submarginal area are contrastingly blackish tinged, more prominently than in *acarodes.* Sometimes whitish patterns along the fasciae occure, which is not seen in *acarodes.* Hindwings are paler than *acarodes* and the fuscous submarginal area is more contrasting. Sexual dimorphism is limited; females have antennae with shorter rami and slightly broader forewings.

Male genitalia are distinguished by a widely notched uncus, where the lobes are of rather triangular shape and not elliptically rounded as in *acarodes*. Socii are significantly (60%) shorter than in *acarodes.* Valvae are club-shaped; they bear a small sacculus process as in *acarodes* or *dives*. Phallus is longer than in *acarodes* and less modified. 8th abdominal segments are as in *acarodes*, but 8th tergite is more deeply notched at the posterior margin. Female genitalia have large papillae anales. Apophyses are medium sized and slender as is characteristic of *Gargetta.* Shape of antrum with pointed lobes resembles rather more *acarodes* than *dives*. It is, however, longer than *acarodes*. Ductus bursae is short and broad. Corpus bursae is long ellipsoid. Signum is larger than *acarodes* and in its notched shape is characteristically.

Bionomics. *G. brevosocius* flies commonly at lower altitudes up to 500 m.

Distribution. Distributed in Sumba, Pantar, Alor and Timor island.

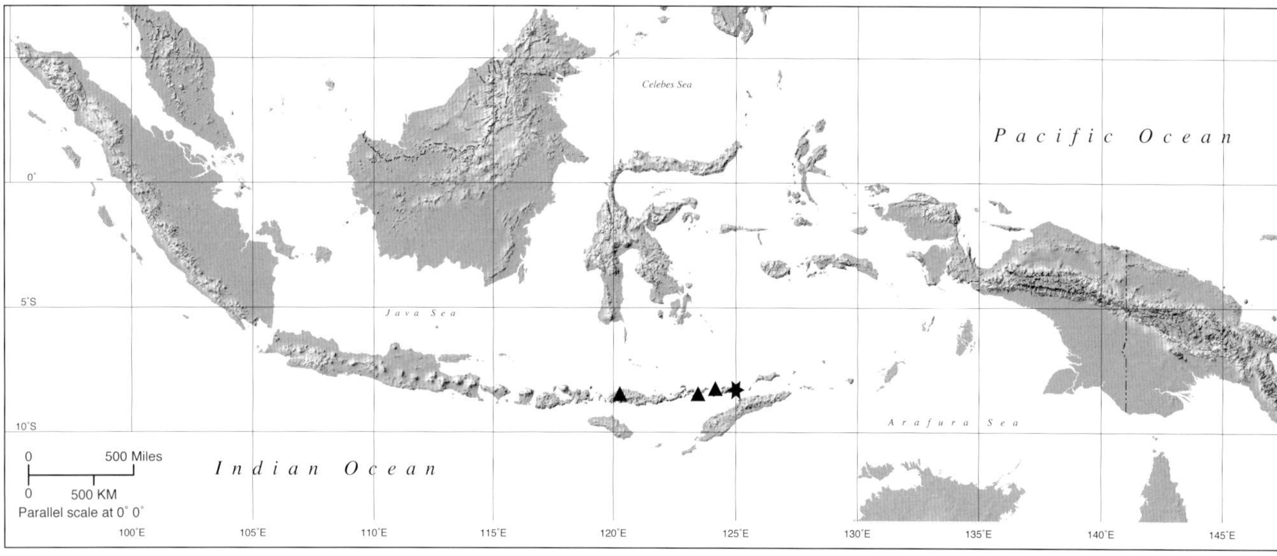

FIGURE 53 Distribution of *Gargetta alora.*

59. ***Gargetta curvaria*** Hampson, 1893: 136.
Syntypes: unspecified number of ♂♂, [India], Sikhim; [Myanmar], Momeit – NHMUK, London.
Schintlmeister 2008: 51; Schintlmeister 2013: 177.

Taxonomic notes. *G. curvaria* is polytypic. In Sulawesi occur *Gargetta curvaria palua* Kiriakoff, 1970: 106, fig. 6 (*Porsica palua*). Holotype: ♂, W Celebes, Paloe [= Palu] district, Mt. Tompoe – NHMUK, London. Distinguished from other subspecies of *curvaria* by minor differences in shape of the digitus process and the phallus tip. Due to the reareness of Sulawesian material in collections only one male was dissected and these differences should be confirmed by further dissecrtions.

As mentioned before, the lectotype of *Gargetta luzonica* is conspecific with *G. divisa* and not with *G. curvaria.* Populations from the Philippines are, however, morphologically distinct (shape of uncus and sacculus process of the valva – see Schintlmeister & Lourens 2010: 81) from those of Sundaland and India. They were described as *Gargetta dyspines* West, 1932 from Luzon and are reinstalled here as a subspecies: *Gargetta curvaria dyspines* **stat. nov.**

From Sumatra *Stictogargetta umbrina* Kiriakoff, 1974: 380; pl. 1: 4; fig. 5 was described. Holotype: ♂, NO-Sumara, Dolok Merangir, 180 m – ZSM, München. It was subsequently downgraded by Schintlmeister & Lourens, 2010: 80 as a subspecies of *curvaria.* Examination of further males from Sumatra and comparison with north Indian material showed, that apart from a slightly individual variation, there are no significant differences between Sundanian and Indian populations. Hence *umbrina* sinks to *curvaria* **syn. nov.**

Diagnosis. Readily recognized by narrow forewing shape and a prominent blackish dorsal patch near the medial area. Discal spot is marked as a black dot, surrounded by pale yellowish brown. Basal area of most males is pale-yellowish brown colored, contrasting with the fuscous medial area. Female sexual dimorphics with ash grey forewings lack the black dorsal spot.

Male genitalia have a characteristically, notched uncus. Socii are absent. Valvae bear a pair of digitus processes on each side, where the smaller process differs slightly in its length from the right process of the left valva. Phallus exhibits a characteristic strongly sclerotized spine, which appear in some individuals curved like a hook. 8th abdominal segments are diagnostically sclerotized but resembles those of the next species, *alora* (see below). Female genitalia have a short ductus bursae and a large, circular corpus bursae. The prominent large signum is reniform with a prickly structure.

Bionomics. The moth occurs locally and infrequently mostly at lower altitudes up to 500 m above sea level, but a few males were caught in Sumatra at 1050 m and in Luzon at 1600 m above sea level.

Distribution. *Gargetta curvaria curvaria* occurs in northeastern India, southern China, Myanmar, Indochina, Borneo, Sumatra, Java and Flores. Ssp. *palua* flies in Sulawesi and ssp. *dyspines* is restricted to the Philippines.

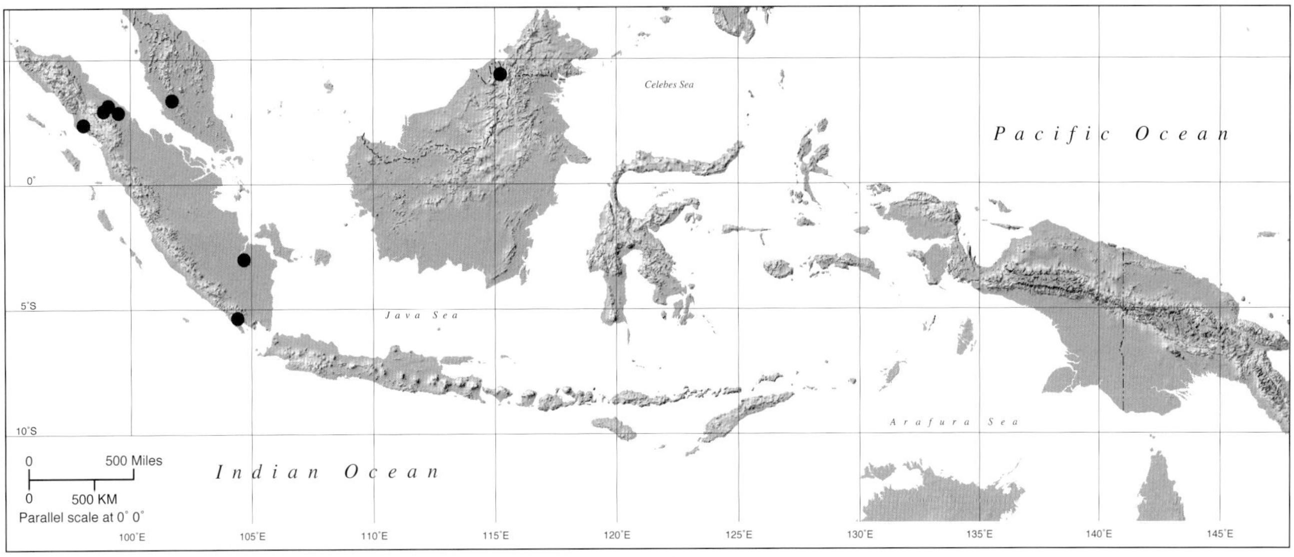

FIGURE 54 Distribution of *Thacona costivitta*.

60. ***Gargetta alora*** **spec. nov.** (figs. 60a–g).

Holotype: ♂, Indonesia, Nusa Tenggara Timur, Alor Isl., Moru, 8°16'S, 125°05'E, 500 m, 1.–8. iii. 2006, leg. St. Jakl – in coll. A. Schintlmeister, Dresden.

Paratypes (8 ♂♂, 1 ♀):

Alor: 5 ♂♂, Moru, 8°16'S, 125°05'E, 500 m, 1.–8. iii. 2006, leg. St. Jakl (GU 95-58);

Pantar: 2 ♂♂, Tanah Labang, 8°19'S, 124°16'E, 350 m, 9.–21. iii. 2006, leg. St. Jakl (GU 95-60);

Lomblen: 1 ♂, Lamalera, 450 m, 8°33'S, 123°25'E, 8.–14. xi. 2006, leg. St. Jakl.

Flores: 1 ♀, Mt. Ramaka (E), 3 km S Mano, 18 km SE Ruteng, Primärwald, 1270 m, 17.–21. v. 1996, leg. Dr. R. Brechlin (MWM 14.979).

Diagnosis. Forewing lengths ♂♂ 15 .0- 16.5 mm, ♀, 20 mm. The new species resembles to *G. curvaria*, but *alora* is paler and rather cream-colored. Forewings are elongated as in *curvaria,* but forewing length is 1 - 1.5 mm shorter. Antennae are bipectinated in both sexes with short rami. Last 10 % approaching the tip are filiform. Forewing pattern resembles *curvaria*, but the blackish dorsal spot is circular, not ellipsoid. Veins of the forewings are less blackish tinged in comparison to *curvaria*. The fuscous spots on forewing costa are, due to the paler ground color of *alora* more contrasting. Hindwings are pale brown with an indistinct fuscous discal spot. The only worn female at the author's disposal resembles the male but lacks the fuscous dorsal spot on forewing.

Male genitalia differ from *curvaria* by a squarish uncus, which is less widely notched than in *curvaria*. Valva processes are as in *curvaria,* but the left smaller process is significantly longer. There is some variability in length and shape of this process as illustrated. Phallus differs greatly in its shape from *curvaria:* it is much thicker, and at the tip there is a prominent buckle. As in *curvaria* the endophallus bears a fine but strongly sclerotized spine. Female genitalia are similar to those of *curvaria* with a robust 8th abdominal segment. Papillae annales are as in *curvaria,* but the apophyses posteriores are shorter. Ductus bursae is thicker and longer than in *curvaria*. The prominent signum is more deeply notched.

Bionomics. Imagines was collected rarely and mostly at lower altitudes up to 500 m above sea level, but the female from Flores was taken at 1270 m.

Distribution. The species is restricted to the Eastern Nusa Tenggara: Alor, Pantar, Lomblen and Flores.

Thacona Walker, 1865: 169.

Type species: *Thacona costivitta* Walker, 1865.

Schintlmeister 2013: 456.

Medium sized adults. Males have bipectinated antennae with short rami; the last 30% to the tip are filiform. Females' antennae are filiform. Forewings are narrow and elongated. Forewing color is whitish to pale brown. Forewing pattern consists of a well-developed, pale surrounded discal spot, a black apical patch at the costa, and a marginal fascia, marked as a row of black dots. Often a

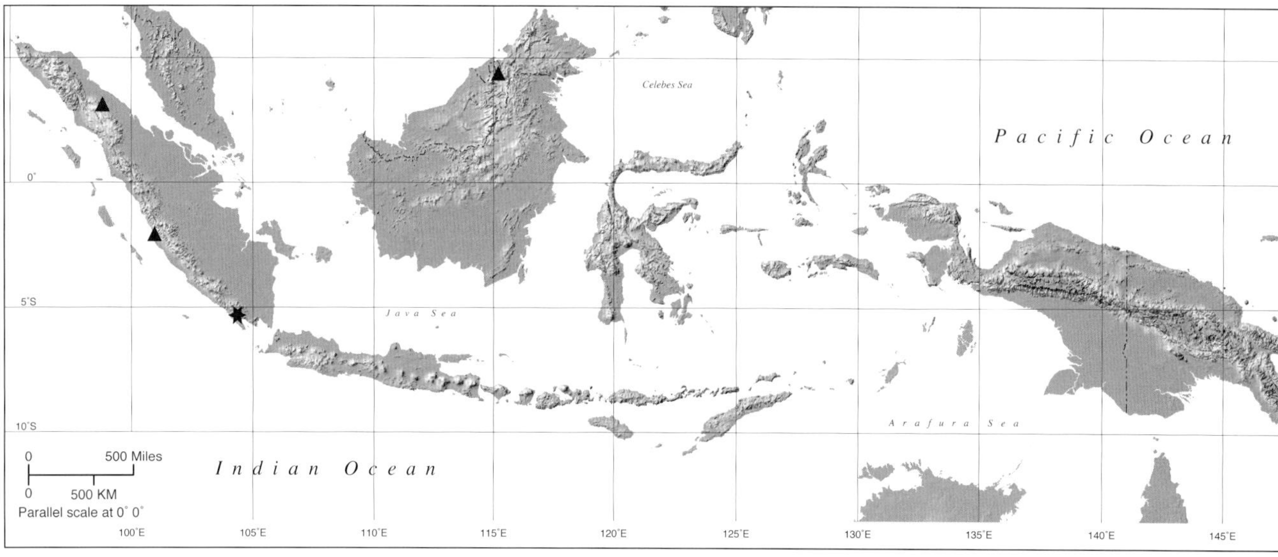

FIGURE 55 Distribution of *Thacona rufocostata.*

fuscous dorsal spot near the base and a fuscous tornal spot occur. Hindwings are pale brown with a fuscous brown submarginal area. Sexual dimorphism is limited.

Male genitalia have a relatively small and slender uncus, except in *T. benderi*. Socii are absent. Valvae are elongated and costa is often strongly sclerotized. Harpes are asymmetric on both valva arms. They are bifurcate, sometimes trifurcate. Phallus is straight and rather short. 8th abdominal segments are bilobed at the anterior margin but generally less modified. Female genitalia have a rather small 8th abdominal segment. Apophyses are compact and short. Antrum large, ductus bursae is long and thick. Corpus bursae is pyriform with a strongly sclerotized reniform signum.

Thacona comprises 15 oriental species, distributed in the Himalaya, S China, Indochina, Sundaland and Sulawesi with greatest diversity in Sundaland. As Schintlmeister & Witt (2015: 20) showed, the genus *Thacona* occurs also in Africa, but it is currently not known how many of the nominal 58 African species formerly listed in *Scrancia* are really congeneric with *Thacona* or to be placed in other genera.

61. *Thacona costivitta* Walker, 1865: 169.
Holotype: ♂, [Malaysia], Sarawak – NHMUK, London.
Schintlmeister 2013: 456.

Diagnosis. The species is unmistakable in the Oriental region, but in Africa some externally similar species occurs (Witt & Schintlmeister 2015: 215). Forewing displays a broad brownish-red costal area and a whitish surrounded grey discal streak. The lower part of the forewing is greyish-white with sparsely blackish pattern. Marginal fascia is marked as rows of contrasting black dots. Hindwings are greyish-brown with indistinct paler medial fascia.

Male genitalia have a small, conically shaped uncus. Harpe is on the left side bifurcate, but on right valva the lower harpe process is reduced in length. Phallus is arched and less modified. 8th abdominal segments are not specifically sclerotized as in most other oriental *Thacona*. 8th sternite is notched at the anterior margin and both lobes are rather more rectangularly shaped than rounded as they are in most other congeners. Female genitalia are charatcerized by their large antrum. Ductus bursae is short; corpus bursae is large and ellipsoid. It bears a prominent reniform signum.

Bionomics. *Thacona costivitta* occur infrequent and mostly at lower altitudes of up to 500 m above sea level.

Distribution. Distributed in Sumatra, the Malayan Peninsula and Borneo.

62. *Thacona rufocostata* (Gaede, 1930): 616; pl. 81: d (*Gargetta rufocostata*).
Holotype: ♂, [Indonesia], Sumatra, Barisan-Gebirge, 800 m – NHMUK, London.
Schintlmeister 2013: 457.

Diagnosis. An unmistakable species characterized by the very narrow, elongated forewings and a prominent, rusty-red colored costal area. The remaining part of the forewing is pale grey with many blackish dots. Hindwings are

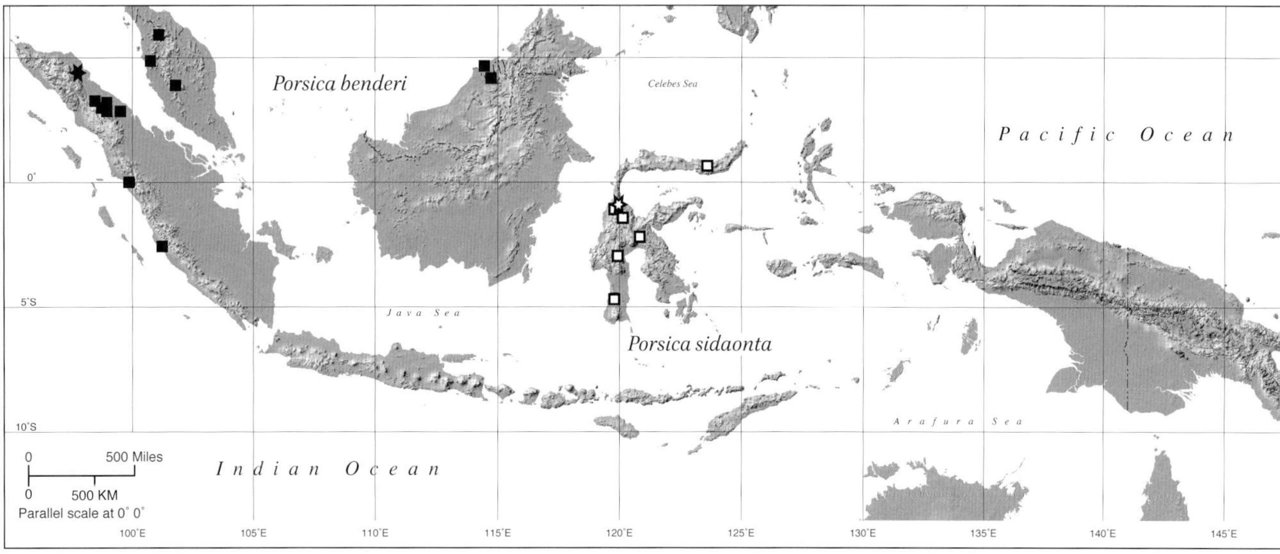

FIGURE 56 Distribution of *Thacona benderi* and *sidaonta*

pale brown with slightly more fuscous postmedial ares. The female is unknown.

Male genitalia resembles somewhat to *costivitta*, but the uncus is slenderer and not of conical shape. Valva is pointed at the apex. Harpe is bifurcate, where the smaller, lower arm is asymmetrically shaped on both valvae. Phallus is rather straight and relatively long.

Bionomics. This is a rare species, which has been taken only a few times (n < 10) at lower altitudes up to 500 m above sea level.

Distribution. Known from Sumatra and Borneo.

63. ***Thacona benderi*** (Schintlmeister, 1981): 291; figs. 7a–b (*Porsica benderi*).
Holotype: ♂, [Indonesia], Sumatra, Aceh, 20 km NW Langsa, 97°45‘ E, 4°32‘ N – coll. A. Schintlmeister, Dresden.
Schintlmeister 2013: 456.

Diagnosis. Forewing shape is very narrow. Closely related to *P. benderoides* Schintlmeister, 1993 and *P. intermediata* Schintlmeister, 1993, both described from the Philippines. *P. sidaonta* from Sulawesi is a further species belonging to this lineage. *P. benderi* is, however, readily distinguished readily by their more contrasting black markings on cream forewing. There is a pair of large costal spots in the postbasal and apical areas and a corresponding pair of black patches on the dorsum of forewing. Marginal fascia marked a a row of large black patches. Discal streaks of light grey are somewhat indistinct and surrounded by a contrasting larger light brownish area. An accompanying larger black patch at the postmedial fascia is a characteristic feature of *benderi*. Hindwings are light brown with broad darker brown postmedial area. Female resembles the male, but forewing shape is slightly broader.

Male genitalia have a squarish uncus which is unusual for *Thacona*. Harpe of valvae is trifurcate, which is a diagnostic feature; all other congeners have bifurcate harpes. Female genitalia have not yet been dissected.

Bionomics. Adults of *T. benderi* are infrequent and prefer seemingly lower altitudes of up to 500 m. Only a few specimens have been taken at 1000 m.

Distribution. Restricted to Sumatra, the Malayan Peninsula and Borneo.

64. ***Thacona sidaonta*** (Kiriakoff, 1970): 105; figs. 9 (*Porsica sidaonta*).
Holotype: ♂, [Indonesia, Sulawesi], W Celebes, Paloe District [= Palu], Sidaonta, 4500 ft. – NHMUK, London.
Schintlmeister 2013: 457.

Diagnosis. External appearance resemble those of *T. benderi*, but forewings are broader. Most specimens of *sidaonta* display less contrasting black patterns and forewing color is cream and not white as in *benderi*. Development and intensity of blackish pattern on forewings is subject of individual variation. Female resembles male, but wings are more broadly shaped.

Male genitalia have a long and narrow a uncus. Harpe of valvae is bifurcate and broadly notched. The lower

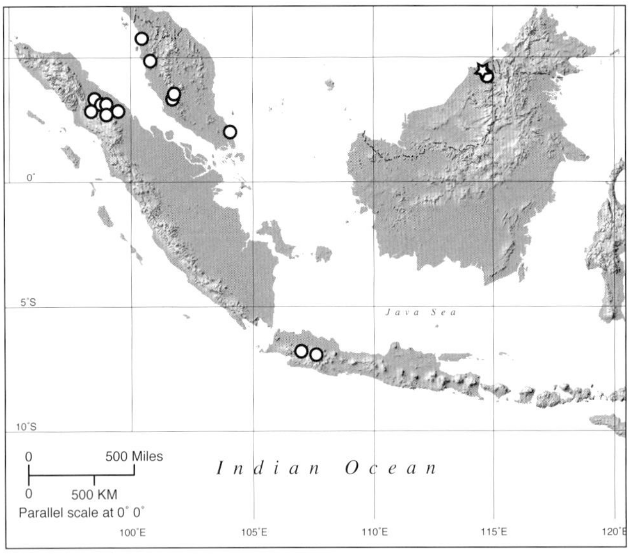

FIGURE 57 Distribution of *Thacona ingens inopinata*.

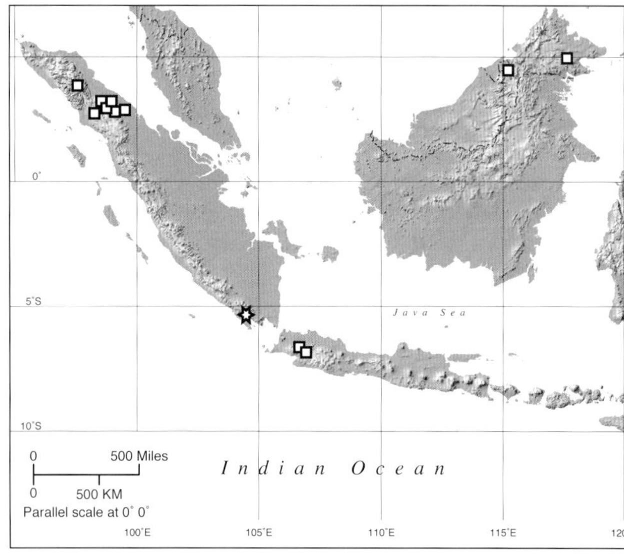

FIGURE 58 Distribution of *Thacona albescens*.

process has a sharp pointed tip. Juxta have, contrary to other congeners, a pair of long lobes. Phallus is straight, buckled before the tip. Female genitalia are distinguished by a small antrum, a long ductus bursae and a large pyriform corpus bursae. Signum is heart shaped.

Bionomics. Adults are common and occur at medium altitudes between 800 m and 1400 m.

Distribution. Endemic to Sulawesi.

65. ***Thacona ingens*** (Walker, 1866): 1823 (*Porsica ingens*).
Holotype: ♂, Silhet – NHMUK, London.
Schintlmeister 2008: 51; Schintlmeister 2013: 457.

Taxonomic note. In Sudaland the ssp. *inopinata* (Holloway, 1983): 24; pl. 1: 9; fig. 10, holotype ♂, Brunei, Labi, Rempayoh primary forest, 75 m – NHMUK, London occur. It differs from ssp. *ingens* by shorter rami in the bipectinated antennae of the males, richer developed blackish pattern, and in male genitalia by a slightly different shape of the harpe (upper projections is shorter than the lower arm). Uncus shape varies individually and is not useful in distinguishing the two subspecies.

Diagnosis. Forewing color is is olive-greenish fawn and wing shape is narrow, and elongated. *T. ingens* is distinguished by the postbasal and postmedial fasciae, which are white and well visible. A discal streak in a whitish area of light brown is in most specimens not visible. Sexual dimorphism is limited.

Male genitalia are relatively small. Uncus being broad and of triangular shaped. Valvae have a broad more strongly sclerotized costa and a large sacculus. Harpe is broad and bifurcate; the upper projection is of equal length or shorter than the lower process. Phallus is straight. Shape and sclerotizations of the 8th abdominal segments are as is common in *Thacona*. Female genitalia have not yet been dissected.

Bionomics. Adults occuring infrequently at medium altitudes between 500 m and 1150 m. Caterpillar's hostplant of ssp. *ingens* in India was reported as *Antidesma* spec. (Phyllanthaceae) (Gardner 1943).

Distribution. *Thacona ingens inopinata* occur in Sumatra, the Malayan Peninsula, Borneo and Java. Ssp. *ingens* is distributed in northeastern India, Myanmar, Thailand, Vietnam and southern China.

66. ***Thacona albescens*** (Gaede, 1930): 616 (*Gargetta albescens*).
Holotype: ♂, [Indonesia], Sumatra, Barisan-Gebirge, 800 m – NHMUK, London.
Schintlmeister 2013: 456.

Diagnosis. A variable species. Most specimens display white basal and postbasal areas on their forewings and contrasting blackish dorsal patches. Some individuals have this area filled with fuscous brown. Characteristic markings are a circular, white, discal patch with a fine fuscous streak inside and a white patch near the forewing apex. The apical patch is fuscous brown and small.

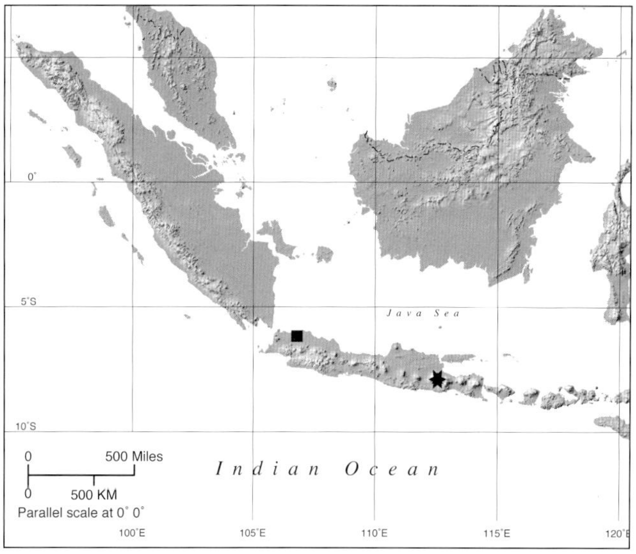

FIGURE 59 Distribution of *Thacona triplicepunctata*.

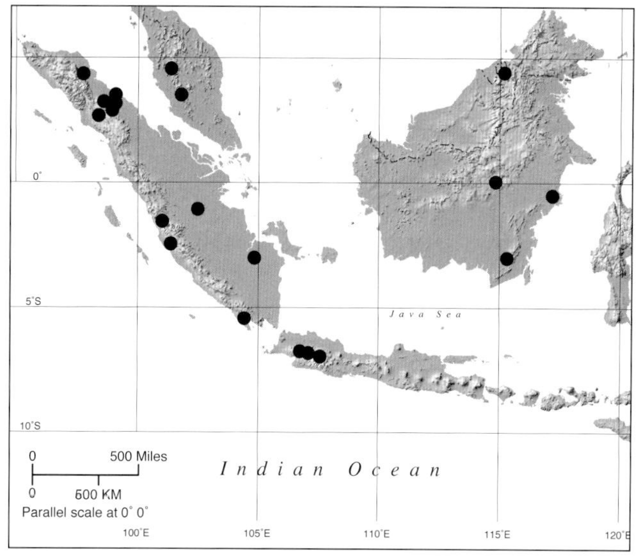

FIGURE 60 Distribution of *Thacona punctifascia*.

Marginal fascia is marked as a row of fuscous brown patches. Hindwings are brown with darker brown postmedial area. Females resembles the males and display similar range of individual variation.

Male genitalia of *albescens* are similar to *T. sidaonta,* but uncus is shorter and juxta lobes are significantly shorter and more strongly sclerotized. Clasper of valva is bifurcate and displays a third, reduced central lobe. Juxta bears two short but strongly sclerotized lobes. Phallus is straight, not buckled as in *sidaonta.* Female genitalia have a broad antrum of medium length. Ductus bursae is long and broad. Corpus bursae is relatively small and of circular shape. It bears a reniform shaped signum.

Bionomics. *Thacona albescens* is the most common *Thacona* and occur in a range from 200 m up to 1500 m above sea level.

Distribution. *Thacona albescens* is distributed in Borneo, Sumatra and Java.

67. ***Thacona triplicepunctata*** (Gaede, 1930): 616; pl. 80: g (*Gargetta triplicepunctata*).
Holotype: ♂, [Indonesia], Java [Malang] – NHMUK, London.
Schintlmeister 2013: 457.

Taxonomic note. *Thacona triplicepunctata* is known from a few old specimens (n = 4) without exact data. They show remarkeable resemblance to the S Indian *Thacona ferreopicta* (Hampson, 1900). Unfortunately it was not possible to examine genitalia of both species. On other hand a small series of *T. albescens* from Java differ clearly externally from *triplicepunctata* and genitalia of a dissected Javan male of *albescens* do not differ from Sumatran males. The status of *triplicepunctata* and its relation to *ferreopicta* remains therefore somewhat doubtful.

Diagnosis. *Thacona triplicepunctata* is distinguished by the yellowish-brown forewing color and the absence of visible white pattern elements. The discal streak is brown, surrounded by a paler brown. Medial and marginal fasciae are marked as rows of fuscous brown spots. All known specimens are remarkable small. Hindwings are light brown with darker brown postmedial area. Sexual dimorphism is limited and the only female matches the holotype exactly.

Genitalia of both sexes are still unknown.

Bionomics. *Thacona triplicepunctata* is known from a few old (>100 years) specimens only.

Distribution. Endemic to Java.

68. ***Thacona punctifascia*** (Hampson, 1900): 40 (*Gargetta punctifascia*).
Holotype: ♂, [NW India], Khásis – NHMUK, London.
Schintlmeister 2013: 457.

Diagnosis. Forewings and patagia are uniform yellowish-brown. Medial, postmedial, submarginal and marginal fasciae on forewing marked as rows of black dots. A discal streak is present as a fine blackish stripe. Occasionally a

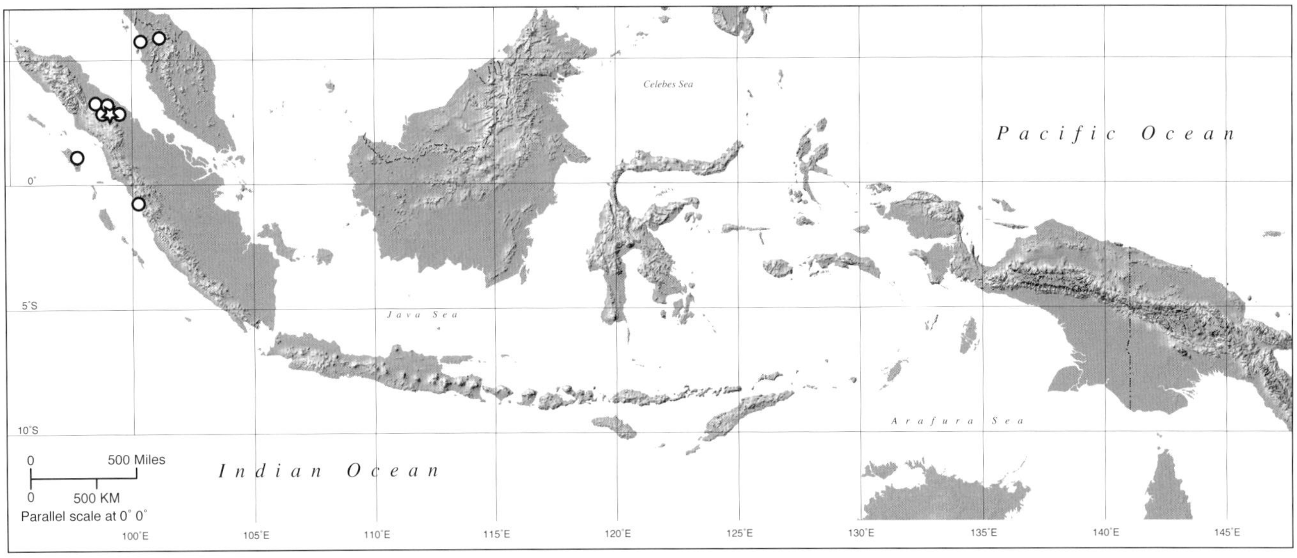

FIGURE 61 Distribution of *Thacona fallax*.

fuscus dorsal patch occurs. Range of individual variation width of forewing pattern overlapping in *punctifascia* and *fallax*. *Thacona punctifascia* might be separated by hindwing coloration, with are in *punctifascia* light brown and displying a darker brown postmedial area. In *fallax* hindwings are a uniform brown. For separation from *Hirjacona punctifascioides* see there. Sexual dimorphism in *punctifascia* is limited; females have broader forewings.

Male genitalia have a rather slender uncus and a single processed clasper. Both parts are variable in its shape. Phallus is buckled. *T. fallax* resembles in its genitalia much to *punctifascia*, but is distinguished by shapes of the bifurcate harpe and a straight, unbuckled phallus. Female genitalia have a very large antrum, a short ductus bursae and a large corpus bursae with characteristically shaped signum. Readily separable from *fallax* by the larger antrum and the smaller signum.

Bionomics. Adults occur commonly and flies from 200 m up to 1500 m. Holloway (1983: 24) reported *Aporusa chinensis* (Phyllanthaceae) as the hostplant of the caterpillar in Hongkong.

Distribution. *Thacona punctifascia* is distributed in northeastern India, Myanmar, Indochina, southern China, including Hainan, the Malayan Peninsula, Borneo, the entire Philippines, Sumatra and Java.

69. ***Thacona fallax*** (Bender & Holloway, 1985): 28, pl. 13: 5, 6; GA: 23 (*Porsica fallax*).

Holotype: ♂, [Indonesia], NO-Sumatra, Aek Tarum, 250 km südl. Dolok Merangir – Staatliches Museum für Naturkunde Karlsruhe.

Schintlmeister 2013: 457.

Diagnosis. External appearance of *fallax* is very similar to *punctifascia*. Rami of bipectinated antennae are shorter and forewing shape is slightly narrower than in *punctifascia*. Occasionally a fuscous dorsal patch occur, which is sharper and more contrasting than in *punctifascia*. The remaining forewing pattern is as in *punctifascia* (see there). Range of indivdual variability is in *fallax* seemingly wide as confirmed by several genitalia dissections. Some individuals, in particular females, are not distinguishable from *punctifascia* merely by forewing shape and their markings. Most specimens of *fallax* are distinguished by a fuscous reddish brown patagia, narrower shaped forewings a contrasting fuscous reddish brown patch on the dorsum of the forewing. Best feature differentiating both species from another are the uniformly brown hindwings of *fallax*, whereas in *punctifascia* the basal and medial part are lighter brown.

Male genitalia resemble *punctifascia*, but the uncus is swollen at the base, and harpe is bifurcate with the lower projection angled. Phallus is not buckled, but straight. Female genitalia readily distinguished from *punctifascia* by their smaller antrum and a longer ductus bursae. Corpus bursae is not circular, but rather rectangular and long. Signum is, as in *punctifascia*, shaped like a butterfly.

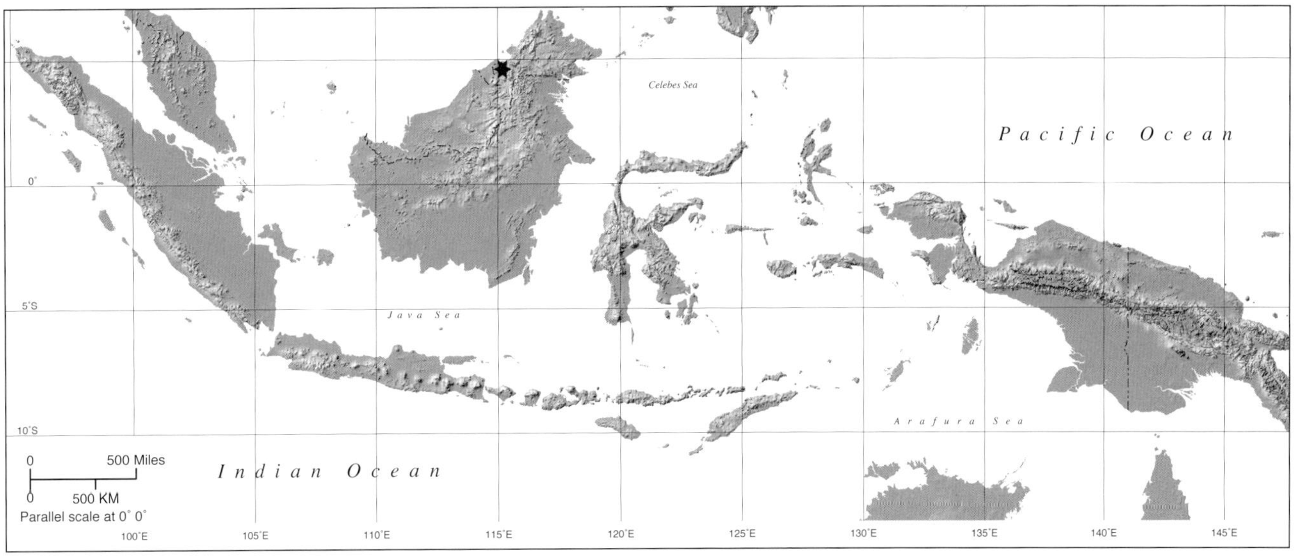

FIGURE 62 Distribution of *Hijracona punctifascioides.*

Bionomics. *Thacona fallax* occur in Sumatra seemingly more common than *punctifascia*. The species prefer medium altitudes between 1000–1500 m. A few specimens were taken in the lowlands at 200–500 m.

Distribution. *Thacona fallax* is probably widely distributed. The following mentioned occurrences have been confirmed to date by genitalia dissections: northeastern India (Meghalaya), Vietnam, Thailand, the Malayan Peninsula, Sumatra and Nias.

Hijracona Holloway, 2011: 349.

Type species: *Hijracona punctifascioides* Holloway, 2011.

Schintlmeister 2013: 216.

Diagnosis. External appearance of male as in *Thacona fallax* or *punctifascia*, but forewing shape is narrower and apex less rounded. The female is unknown.

Male genitalia are very different from *Thacona*. Uncus is large, T-shaped at the tip. Socii are absent. Tegumen is large and robust. Valva is of paddle-like shape, bearing a large flap arising from the sacculus. Phallus is long and slender, slightly arched. 8th sternite has a pair of lobes at the posterior margin and central projection in the central part. 8th tergite has at the anterior margin a pair of large rectangularly shaped projections. The 7th sternite bears a large central process at the anterior margin, which is unique in Notodontidae.

The genus is monotypic and known from Borneo.

70. ***Hijracona punctifascioides*** Holloway, 2011: 350; pl. 7: 93.

Holotype: ♂, Brunei, Muzium, Ulu Temburong, 300 m – NHMUK, London.

Schintlmeister 2013: 216.

Diagnosis. Similar to *Thacona fallax and punctrifascia*, but forewings slightly narrower and apex less rounded. The only known specimen, the holotype male, is rubbed, but it seems, that the marginal fascia contains no blackish dots. Hindwings are light brown with darker postmedial area as in *punctifascia*. The female is unknown.

Male genitalia are unmistakable, see above under genus diagnosis.

Bionomics. The holotype was taken in a lowland forest at 300 m.

Distribution. Known only from Brunei.

Ceirinae

Note. *Biula lignilutea* Walker, 1857: 714 was described with wrong type locality "Java". The holotype female in NHMUK, London, bear a label "N India". The species, which belongs to the Ceirinae, does not occur in Indonesia and is therefore not treated here.

Ramesa Walker, 1855: 981 (key), 1016.
Type species: *Ramesa tosta* Walker, 1855.
Schintlmeister 2008: 53; Schintlmeister 2013: 371.

Taxonomic note. There are two groups of species, which differ by external appearance as well as in their male genitalia. However, they are closely related. *Ramesa* sensu stricto comprises two species, *R. tosta* and *R. docilis* (Walker, 1857). For the remaining 9 species *Togarishachia* Matsumura, 1925 can be applied as subgenus name due to morphological differences.

Diagnosis. Most of *Ramesa* species have similarity to some cutworm moths (Noctuidae, Hadeninae Tribus Mythmnini). Antennae of male bipectinated, those of females filiform. Forewing color is fuscous violet brown. Basic pattern of forewings consist in subgenus *Togarishachia* of a white streak in the submarginal area and a blackish discal patch, accompanied by a second black patch. In subgenus *Ramesa* the white streak is absent and discal patch rather small and not accompanied by a second black patch. In *Ramesa* sexual dimorphism is limited, while in subgenus *Togarishachia* sexualdimorphic females occuring, which are usually much paler than their corresponding males. Male genitalia of *Ramesa* s. str. have a small, triangular shaped uncus, which is in subgenus *Togarishachia* bifurcate (exception: *R. albistriga*). Socii are much larger than uncus. Valvae of *Togarishachia* display a saccular projection at the base of diagnostic value, which is absent in subgenus *Ramesa*. Phallus long and arched. 8th sternite at posterior margin rounded, but at the anterior margin there is a pair of projections. 8th tergite notched. Female genitalia have long apophyses, a large, strong sclerotized antrum, short ductus bursae. Corpus bursae circular with small signum in subgenus *Ramesa* but in subgenus *Togarishachia* the signum is larger and forms bended line.

Ramesa species became famous as tear drinkers of Mammalia, including humans (Bänziger 1988, 1989, 1992). Larvae feeding on Monocotyledoneae such as bamboo, grasses and probably also palms as the most members of this subfamily do. *Ramesa* comprises eleven species. In Thailand it represents a taxonomically difficult complex of hitherto eight known species.

Ramesa species are distributed in Pakistan, India, Sri Lanka, Indochina, southern China, Japan, Taiwan, The Malayan Peninsula, Sumatra and Java.

71. ***Ramesa tosta*** Walker, 1855: 1017.
Holotype: ♀, North India – NHMUK, London.
Schintlmeister 2008: 53; Schintlmeister 2013: 372.

Diagnosis. A smaller species within the genus, which is best to recognize by the prominent marginal fascia, marked as rows of black and white dots. Postmedial fascia marked as a row of fine blackish dots. Forewings dark reddish-brown with paler area, runing parallel to dorsum. Hindwings are whitish brown. Sexual dimorphism is limited; hindwings are more fuscous than in males.

Male genitalia have a small triangular shaped uncus and a pair of large and thick Sociii. Valvae narrower and less rounded than in *T. albistriga*. Phallus long, slender and arched. 8th sternite displays only short projections at anterior margin. 8th tergite wider notched at posterior margin than other congeners. Female genitalia are characterized by a central projection on the vaginal plate. Antrum large, ductus bursae very short. Corpus bursae circular. It bears a small, strongly sclerotized wing-shaped signum.

Bionomics. *R. tosta* is known in Indonesia only from a pair labelled "Surabaya", which is more than 100 years old. Adults are generally rare at light and occur in mainland asia up to 700 m, but preferring altitudes below 300 m. Caterpillar was reared in Japan on rice plants, *Oryza* in Poaceae (Sugi, 1987: 285).

Distribution. Distributed in northeastern India, Sri Lanka, Myanmar, China, Japan, Taiwan, Indochina. There is a distribution gap between Thailand and Java. Probably introduced in Java with rice plants.

Subgenus ***Togarishachia*** Matsumura, 1925: 399.
Type species: *Togarishachia argenteopicta* Matsumura, 1925 [= *Niganda albistriga* Moore, 1879].

 | DOI:10.1163/9789004354258_005

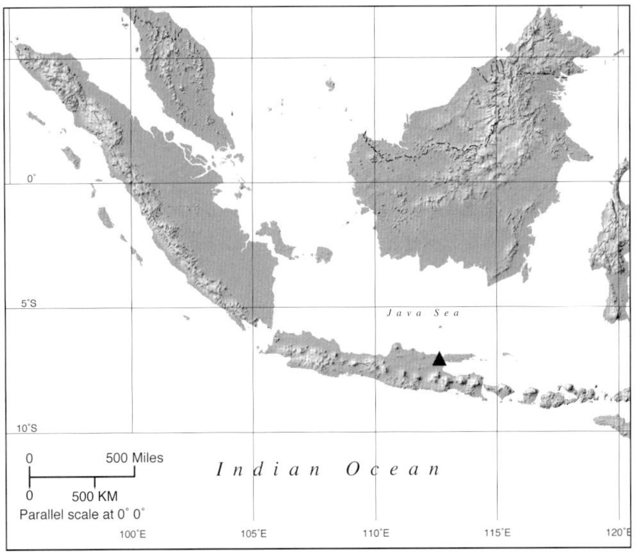

FIGURE 63 Distribution of *Ramesa tosta*

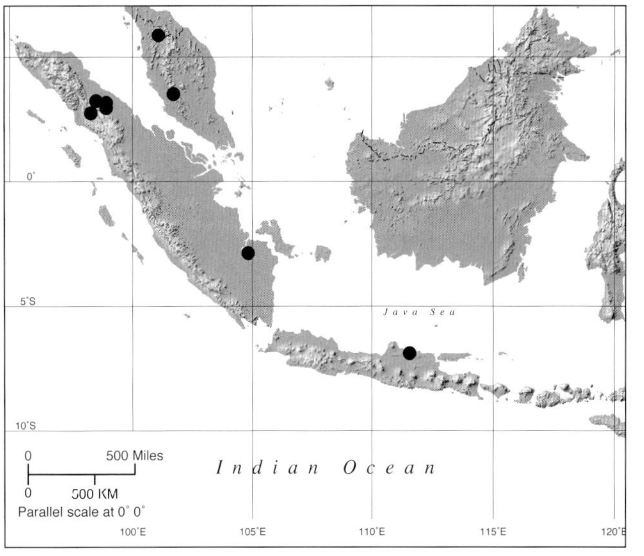

FIGURE 64 Distribution of *Ramesa albistriga*

72. ***Ramesa albistriga*** (Moore, 1879): 64 (*Niganda albistriga*).
Holotype: ♂, [India], Darjiling [= Darjeeling] – ZMHU, Berlin. Schintlmeister 2008: 53; Schintlmeister 2013: 371.

Diagnosis. Patagia fuscous brown. Males have reddish to yellowish-brown colored forewing. Basal, postbasal and postmedial fasciae are marked a black dots. Discal patch is contrasting black. Medial fascia developed as a fuscous shadow. A characteristic feature is in both sexes a contrasting white streak in submarginal area. Hindwings are fuscous brown with pale brown fringe. Sexual dimorphism is limited. Female having paler forewings with reduced black pattern.

Male genitalia have a small and triangular shaped uncus. Socii long and robust, slightly curved. Valvae broad and rounded; costa bears a smaller projection, which vary individually in size. Sacculus projection is large, variable in its shape. Phallus large, but shorter and thicker than *R. tosta*. Endophallus bears a strongly sclerotized carina of triangular shape; diverticulum large. 8th sternite displays in the center a pair of strongly sclerotized fields with a central window. Anterior margin have centrally a pair of projections. 8th tergite deeply notched at posterior margin. Female genitalia have a deeply notched antevaginal plate. Antrum large, ductus bursae reduced and corpus bursae disc-shaped with a large signum.

Bionomics. The most widespread and common *Ramesa*. Adults are on the wings from sea-level up to 2,200 m, often also in urban areas.

Distribution. *Ramesa albistriga* is distributed in Nepal, northeastern India, Myanmar, China, Taiwan, Indochina, the Malayan Peninsula, Sumatra and Java.

Brykia Gaede, 1930: 644.
Type species: *Allata albonotata* Walker, 1862 [= *Eumeta horsfieldi* Moore, 1860].
Schintlmeister 2008: 55; Schintlmeister 2013: 66.

Diagnosis. Antennae are black and in both sexes bipectinated; the last 25–40% until the tip are filiform. Forewing have a medial tooth like tuft of scales. Near the forewing apex there are three white costal patches. Discal spot is black. A characteristic feature is a triangular white spot in the submarginal area. Forewing veins are tinged brown or blackish in basal and postmedial area. Submarginal fascia marked as a row of black dots. Sexual dimorphic females show reduced forewing pattern and much broader forewings.

Uncus long; socii robust and long. Valvae narrow, sacculus large and bear a process. Phallus short and slightly arched. 8th sternite notched at anterior margin and characteristically sclerotized. 8th tergite notched at posterior margin. Female genitalia have a large 8th abdominal segment and a strongly sclerotized deeply notched antevaginal plate. Ductus bursae short, corpus bursa circular.

There are three described species in *Brykia* ranging from southeastern China, Indochina, Sundaland and Java.

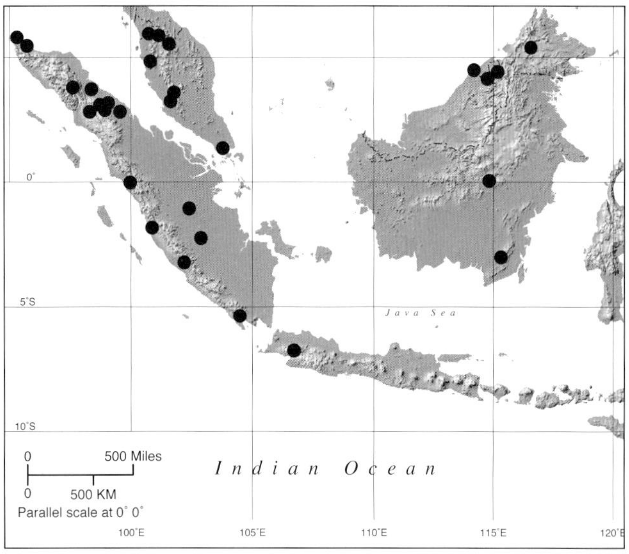

FIGURE 65 Distribution of *Brykia horsfieldi*.

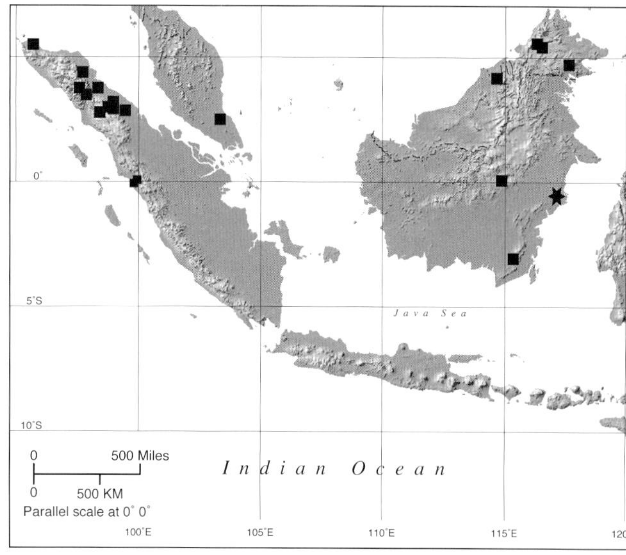

FIGURE 66 Distribution of *Brykia samarinda*.

73. ***Brykia horsfieldi*** (Moore, 1860): 430 (*Eumeta horsfieldi*).
Holotype: ♂, [Indonesia], Java – NHMUK, London.
Schintlmeister 2008: 56; Schintlmeister 2013: 66.

Diagnosis. Variable species. There are pale yellowish-brown individuals as well as almost uniform deep blackish-violet colored moths. Also wingspan is variable. Best character for identification is a white patch near the margin of the forewings which is accompanied by a deep black streak toward the base. Veins are often black tinged. Some specimens display a small black discal spot on forewing.

Male genitalia have arched valves and a large, serrated sacculus with a short process. Phallus bear at the tip a hook and a serrated process. 8th sternite slightly bilobed at posterior margin and bearing two smaller lobes. The 8th tergite notched on both sides. Female genitalia show a slightly notched antevaginal plate. Ductus bursae is longer and slenderer than in *samarinda*. Corpus bursae large and circular with a circular signum.

Bionomics. *B. horsfieldi* is a common moth and occur from sea level up to 1600 m. Most specimens were taken below 1000 m, often in urban places but also in primary rain forests.

Distribution. Distributed in Myanmar, Thailand, Cambodia, Malaya, Borneo, Palawan, Sumatra, Pulau We, and Java.

74. ***Brykia samarinda*** (Kiriakoff, 1967): 42; fig. 5 (*Pseudoturnaca samarinda*).
Holotype: ♂, [Indonesia, Kalimantan], S. E. Borneo, Samarinda – NHMUK, London.
Schintlmeister 2013: 66.

Diagnosis. Forewing color vary from fawn, fuscous red, brown to grey. Often a fuscous medial fascia with an indistinct fuscous discal patch occur. Submarginal area of forewing bear a small white spot, which is hard to see in pale specimens. A black shadow near forewing tornus is a diagnostic character. Veins are whitish tinged. Sexual dimorphism is limited, but females occuring mostly brown.

Male genitalia have triangular shaped socii. Valvae broad with a large sacculus processes. They vary individually and are often asymmetrically shaped on each valva. Phallus arched. Endophallus bear two long and sharp spines. 8th sternite notched and specifically sclerotized. Female genitalia have a deep notched antevaginal plate. Ductus bursae reduced in length; corpus bursae disc-shaped and small. No signum.

Bionomics. Adults are occuring infrequently at altitudes between 50–1150 m.

Distribution. Distributed in Borneo and Sumatra. A single male is known from the Malayan Peninsula, Endau Rompin N. P.

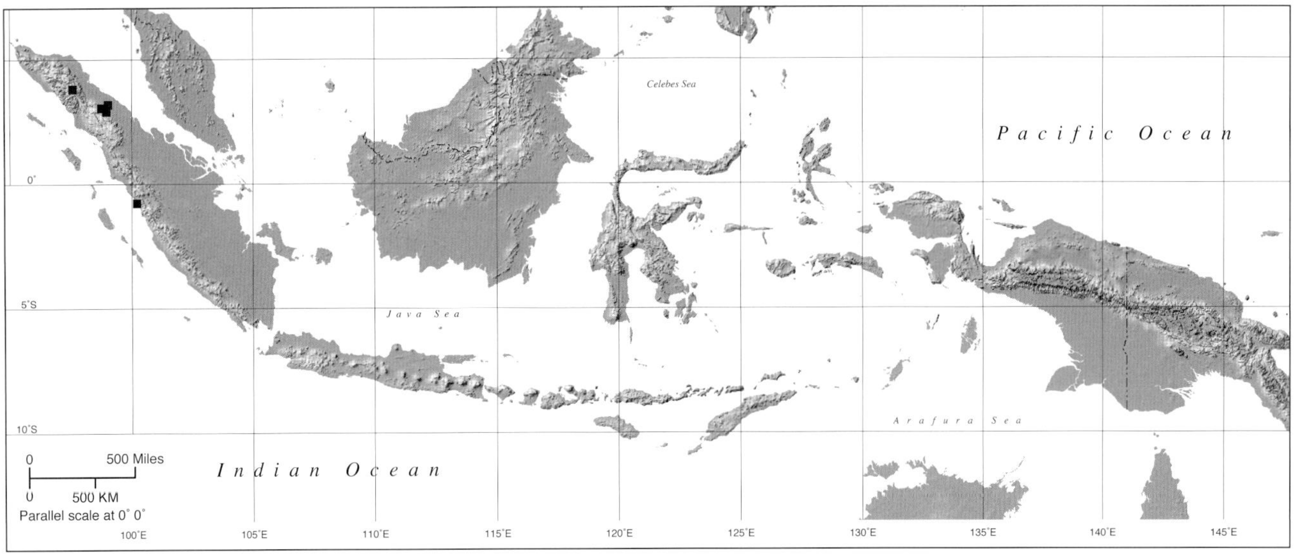

FIGURE 67 Distribution of *Hyperaeschra pallida*.

Hyperaeschra Butler, 1880: 65.

Type species: *Hyperaeschra pallida* Butler, 1880.

Schintlmeister 2008: 56; Schintlmeister 2013: 222.

Diagnosis. Antennae bipectinated in both sexes; last 15 % to the tip are filiform.Forewings brownish-yellow, with a large whitish discal patch, cored with a small greyish dot. A prominent black streak extend from medial area to margin accompanied by a small white marginal patch. There is a further black basal streak on the forewings. Sexual dimorphism is limited.

Male genitalia have a trapezoid uncus and a pair of large Sociii, which bear each two further processes. Valvae are reduced in size and simple. Tegumen robust. Phallus slender and bended. It bears a mall spine at the tip. 8th tergite characteristically cross-shaped sclerotized. 8th sternite have a pair of central processes and a central projection at the posterior margin. 1st abdominal segment bears centrally a pair of cteniophores. Female genitalia are in comparison to male genitalia small. Both pairs of apophyses are long. Ductus bursae long and slender. Corpus bursae small and rather squarish shaped. It displays a small, streak-shaped signum.

Hyperaeschra is a monotypic genus, distributed in the Himalaya, southern China, Indochina, Sri Lanka, Sumatra, and the Philippines.

75. ***Hyperaeschra pallida*** Butler, 1880: 65.

Holotype: ♂, [NE India], Darjiling [= Darjeeling] – NHMUK, London.

Schintlmeister 2008: 56; Schintlmeister 2013: 222.

Diagnosis. See under the genus diagnosis above.

Bionomics. *Hyperaeschra pallida* is a common species that occur in Sumatra from the lowlands up to 1050 m, in Sikkim up to 2000 m.

Distribution. See above under genus.

Changea Schintlmeister & Fang, 2001: 35.

Type species: *Changea yangguifei* Schintlmeister & Fang, 2001.

Schintlmeister 2008: 57; Schintlmeister 2013: 78.

Diagnosis. Antennae are black or grey and bipectinated until the tip with long rami in both sexes. Forewings are brown or yellowish-brown. Forewing pattern is often reduced; but some individuals display postbasal and postmedial fasciae, marked as indistinct fuscous brown patches. Dorsum often darker than the remaining forewing. Sexual dimorphism is limited.

Male genitalia are characterized by a broad and rather rectangular shaped, bifurcate uncus and in particular by the large and notched Sociii. Valvae strongly sclerotized, arched, and narrow. Tegumen robust. Phallus short and straight; sometimes it bears a process. 8th sternite notched at posterior and anterior margins. 8th tergite slightly bilobed at the posterior margin. Female genitalia have a large 8th abdominal segment. Antevaginal plate large and notched. Ductus bursae slender. Corpus bursae small and circular. No signum.

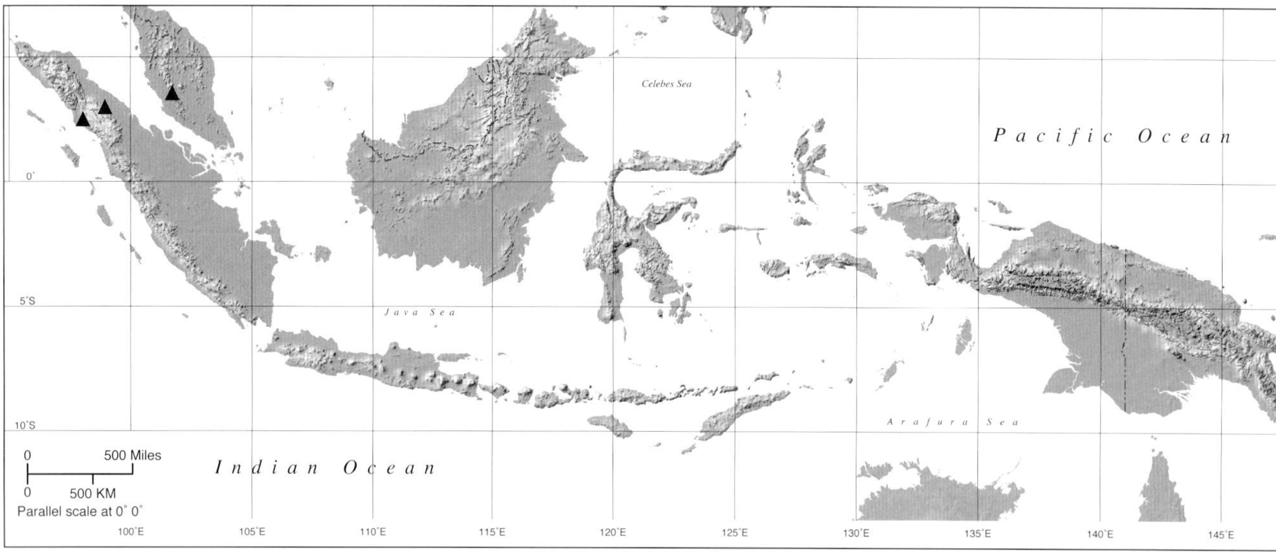

FIGURE 68 Distribution of *Changea siamensis*.

The genus comprises three species, which are distributed in northeastern India, southern China, Indochina, the Malayan Peninsula and Sumatra.

76. *Changea siamensis* Schintlmeister, 2007: 61; pl. 8: 43.
Holotype: ♂, Thailand, Khao Yai Nat. Park, 800 m – NHMUK, London.
Schintlmeister 2013: 78.

Diagnosis. This inconspicous, brownish-yellow colored species has black and tufted antennae, which is the best feature to identify it readily. Along the forewing dorsum there is a blackish shadow. Some individuals from Thailand, fig. 70e, show a postbasal and submarginal fasciae on forewings, which are marked as a rows of fuscous patches. Sexual dimorphism is limited.

Male genitalia display a deeply notched rectangular shaped uncus. Socii are large and have projections are at the tip, which vary individually in its shape. Valvae are rectangular, bended and rather slender. Phallus bear a large process arising from its central part. 8th abdominal segments are specifically sclerotized and notched at posterior margin. Female genitalia have very long apophyses posteriores, a notched antevaginal plate and a strongly sclerotized ostium of unique shape. Ductus bursae slender and short; corpus bursae disc-shaped and relatively small.

Bionomics. A rare species, which is known from Sumatra and the Malayan Peninsula from a few (n = 11), specimens taken between 600–1050 m. In Thailand a male was collected at 2300 m above sea level.

Distribution. Thailand, Cambodia, the Malayan Peninsula and Sumatra.

Turnaca Walker, 1865: 454.
Type species: *Turnaca acuta* Walker, 1865.
Schintlmeister 2008: 58; Schintlmeister 2013: 465.

Diagnosis. Exernal appearance of imagines resembles much to *Tensha, Canatensha* and *Ambadra,* which are doubtless closely related genera. Antennae are brown and bipectinated in both sexes; the last 25–35% to the tip are filiform. Forewings are elongated and greyish or yellowish-brown. Forewing shape is pointed at the apex. From base to margin a brown transversal streak is running along vein M2. Postmedial and marginal fasciae marked as blackish dots, but sometimes reduced in visibility. A characteristic feature are two whitish triangular shaped areas in postmedial sector above and below the transversal streak. Hindwings vary from brown to white without any markings. Sexual dimorphism is limited, but females regularly much larger than males.

Male genitalia have a long, slender and bifurcate uncus, slightly notched in *subcarnea.* Socii are long, slender and curved. Sacculus of valva is large and bears several projections. Phallus short and arched. 8th sternite less modified, but slightly notched at posterior margin. There are no projections at the abdominal segments as in other closely related genera. Female genitalia of *T. subcarnea* have long apophyses posteriores, reduced apophyses anteriores, and a large 8th abdominal segment. Antevaginal plate is large,

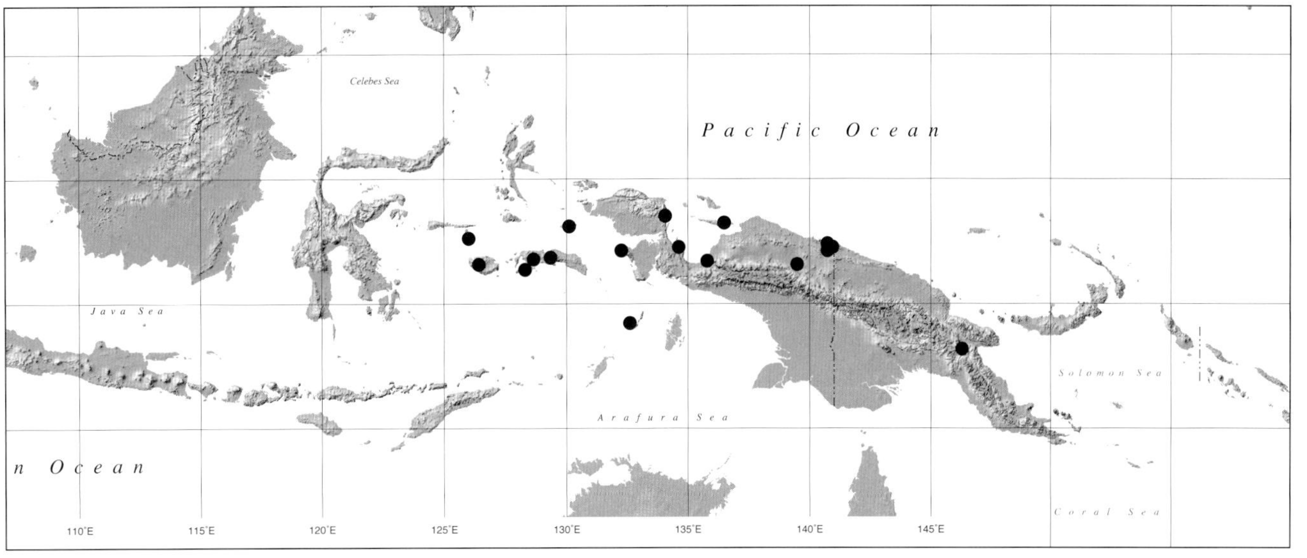

FIGURE 69 Distribution of *Turnaca subcarnea*.

of disc-shaped shape and deeply notched. Ductus bursae is short and narrow. Surface of corpus bursae is corrugated and display no signum.

The genus comprises five described species. Distributed in entire India, S China and, disjunct, in New Guinea and the Moluccas.

77. ***Turnaca subcarnea*** Warren, 1903: 120.
Holotype: ♀, [Papua New Guinea] British New Guinea – NHMUK, London.
Schintlmeister 2013: 466.

Diagnosis. A variable species. Wingspan and forewing shape are subject of individual variation as illustrated. Forewing and hindwing coloration vary from brown, orange brown or pale yellowish brown. Markings on forewing are often reduced and also uniform specimens without blackish markings occur. Most specimen display a brown transversal streak from base to margin. Sexual dimorphism is limited. The females have a slightly acuter apex of the forewings. Also female are variable in size and coloration

Contrary to width variation by external appearance male genitalia remarkable homogenous, also in populations of remote islands like Sanana. Uncus long, slender and of rectangular shape; slightly notched at the tip. Socii long, arched and pointed at tip. Shape of valvae squarish. Sacculus large with a pair of projections at the tip. Costa of valva slightly variable in shape. Phallus short and curved. The everted endophallus bear a pair of large and sharp needles, which makes the male genitalia rather unmistakable. 8th abdominal segments less modified of circular shape. Female genitalia are as described under the genus diagnosis.

Bionomics. Adults are common in urban areas, but much more rare in primary forests. They were collected mostly at lower altitudes up to 550 m. In Papua New Guinea a few specimens were taken at 1700 m above sea level.

Distribution. Distributed in entire New Guinea and in the islands of Yapen, Misool, Sanana, Buru, Ambon, Seram, Key.

78. ***Turnaca carnea* spec. nov.** (figs. 78a, 78b).
Holotype: ♂, Indonesia, W Papua, 17 km of Bora Bora, Gautier/Foja Mountains, 1200 m, 2°28.259' S, 138°51.288'E, 7.–8. vi. 1016, leg. G. Bretschneider (Ihle09-2017) – in coll. S. Ihle, Filderstadt.

Paratypes (2 ♂♂):
W Papua: 1 ♂, Dutch New Guinea, Humboldt Bay dist., Wembi, 4. viii. 1937, leg. W. Stüber; 1 ♂, Hollandia [= Jayapura], 8. vii. 1937.

Diagnosis. Forewing length ♂♂ 19.5–23.0 mm. The holotype spans 23 mm. Antennae brown, bipectinated, with long rami. Last 30% toward the tip filiform.

Forewing pure grey not tinged brown as in *subcarnea*. Forewing pattern similar to *subcarnea*. Postmedial fascia marked a row of black dots. Along the dorsum there are a few blackish dots. From base to margin a fine, blackish transversal line runs. Hindwing brown with contrasting fuscous chocolate-brown anal region. The latter feature is

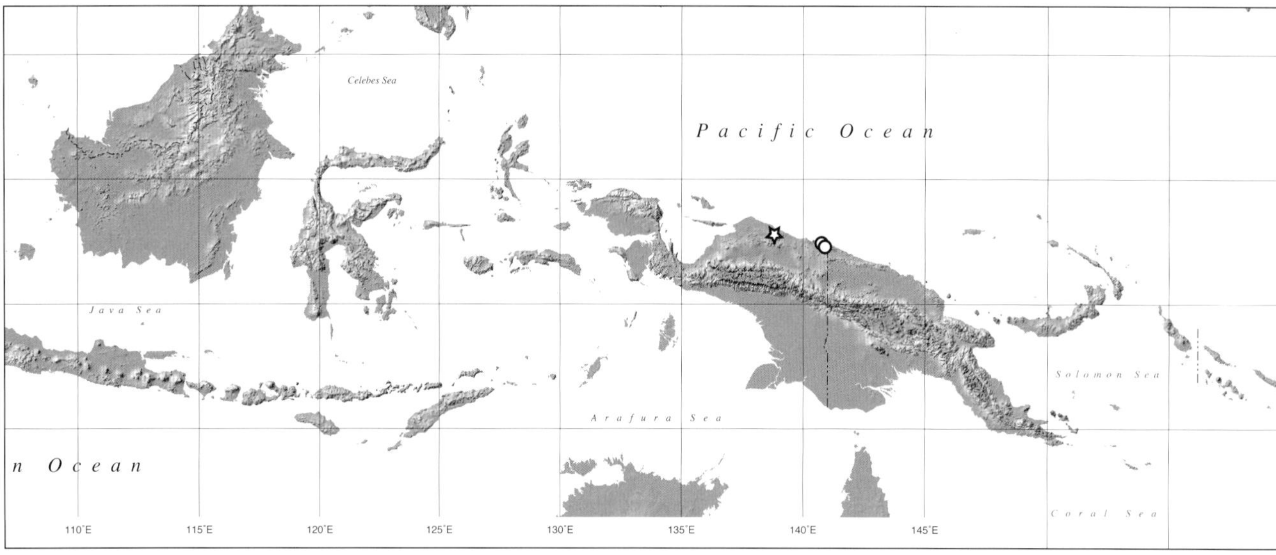

FIGURE 70 Distribution of *Turnaca carnea*.

a good character to separate *carnea* from *subcarnea*. Individual variability is limited. Female is unknown.

Male genitalia are distinct by the deeply bifurcate and long uncus. A pair of slender and long socii measure 70% of uncus length. In *subcarnea* socii are approximately of same length as uncus. Valvae of squarisch shape. Sacculus well developed and strongly sclerotisized, but smaller than in *subcarnea*. It bears a long and tapered spine with pointed tip. Phallus arched with pointed tip. Everted endophallus bears no curnuti or spines, but a few (< 10) strongly sclerotized microthornes. 8th sternite of rhomboid shape, posterior margin convexely with a small notch and concave anterior margin. 8th tergite less modified.

Bionomics. Three males are hithertho known, which were collected at lower altitudes (< 200 m) and 1200 m above sea level. In Humboldt Bay, Wembi both species, *subcarnea* and *carnea* were observed together.

Distribution. Known from W Papua.

Tensha Matsumura, 1925: 392.

Type species: *Tensha striatella* Matsumura, 1925.

Schintlmeister 2008: 59; Schintlmeister 2013: 455.

Diagnosis. Antennae brown, bipectinate in both sexes; the last 40–60% toward the tip are filiform. External appearance of adults are as described before under *Turnaca*. A characteristic feature of *Tensha* males is a whitish discal patch in medial area, accompanied by an indistinct light brown area between dorsum. Females are sexualdimorphic, lacking both patches and show the whitish triangles in postmedial area of forewing more contrastingly.

Uncus is diverse in various species and bear a prominent gnathos. Uncus base is mostly broad and shoulder like. Socii are usually short, but in some species (e.g. *delineivena*) they appears large. Valvae are small, but the pleurit is large. Valva costa is buckled. Phallus thick, short and arched. Near the tip there is a buckle and tip is ending in a pointed process. Everted endophallus bears sclerotized microstructures of fine thorns in several rows. 8th sternite has a pair of strongly sclerotized and pointed lobes, which occur in reduced size also in some *Ambadra* species. Anterior margin is slightly bilobed. 8th tergite less modified. Female genitalia have medium sized and robust apophyses. Antrum small, ductus bursae medium sized, corpus bursae is large and bear a small reniform signum.

Tensha comprises seven described species, which are distributed in the Himalayas, Myanmar, Indochina, S China, Taiwan, Malaya and Sumatra.

79. ***Tensha delineivena*** (Swinhoe, 1894): 159 (*Turnaca delineivena*).

Syntypes: 2 ♀♀, [NE India, Meghalaya] Khasia Hills, Cherra Punji – NHMUK, London.

Schintlmeister 2008: 59; Schintlmeister 2013: 455.

Taxonomic note. *Hypambadra speculigera* Kiriakoff, 1974: 386; pl. 1: 6; fig. 9; holotype: ♂, Nord-Sumatra, 30 km SW Siantar, Holzweg 2, 1050 – ZSM, Munich, was described in a own genus. Schintlmeister & Pinratana 2007: 62 recognized the conspecifity of Sumatran, Thailand and Indian

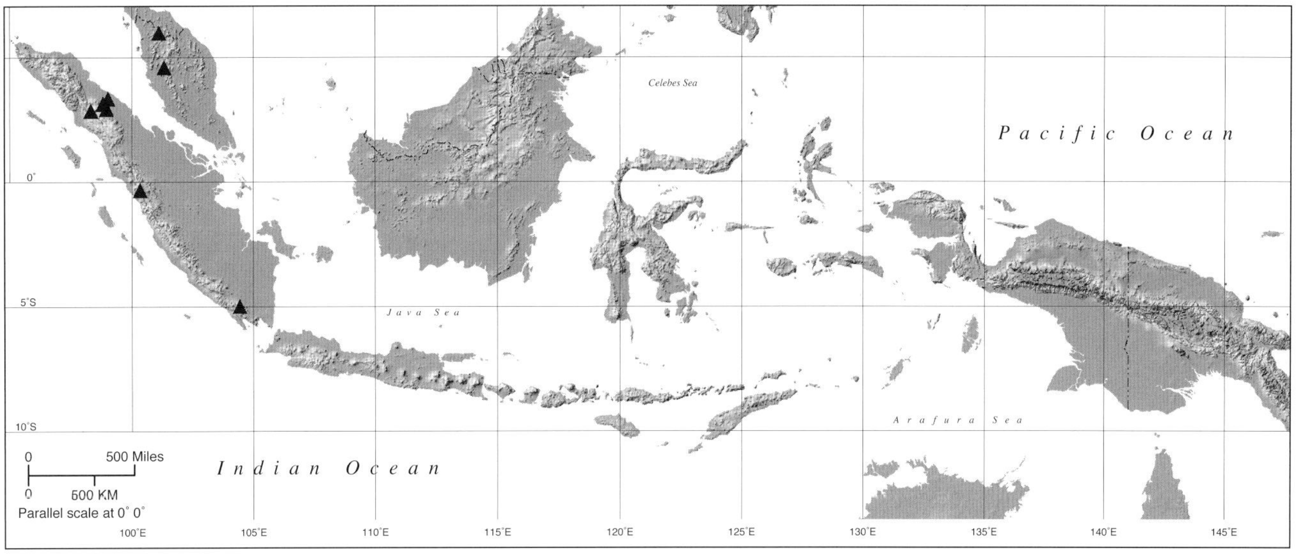

FIGURE 71 Distribution of *Tensha delineivena*.

delineivena–populations but kept *speculigera* as a Sumatran subspecies of *delineivena*. Now, more (n = 6) male genitalia were dissected and examined. Apart from individual variability, no significant taxonomic differences were evaluated. Thus *speculigera* is placed here in synonymy to *delineivena* (**syn. nov.**).

Diagnosis. This is the largest *Tensha*. Distinguished from *postobscura* and *pizzicata* by an indistinct, black tornal patch of forewing. Forewing color vary from reddish-brown to yellowish-brown. Discal streak whitish, often indidistinct; in the medial area a large paler brownish blotch occur. Medial vein fine black, surrounded greenish-yellow. Hindwings of same color as forewings. Female displays an acute apex of forewing and the margin is roughly serrated, rather then rounded as in *postobscura* and *pizzicata*.

Male genitalia are readily to distinguish by shape of the tapered – not bifurcate – uncus and the large gnathos. Shape and thickness of socii vary individually. Phallus arched with small buckle. Antevaginal plate of female genitalia have two lobes, which are pointed at the tip. Ductus bursae thick; corpus bursae disc-shaped and bear a reniform signum.

Bionomics. Imagines occure commonly at altitudes between 140–1700 m; in Darjeeling up to 1900 m. Most Sumatran specimens were observed in rainforests at altitudes of 1000 m.

Distribution. Widely distributed from the Himalaya, southeastern China, Indochina, Malaya and Sumatra.

80. ***Tensha postobscura*** Holloway, 1983: 42; pl. 3: 14, 15; fig. 38.

Holotype: ♂, [Malaysia], Borneo, Sarawak, Gunong Mulu Nat. Park, Camp 4, Mulu, 1.790 m – NHMUK, London.

Schintlmeister 2013: 456.

Diagnosis. Very similar to *T. deineivena*, but smaller. Forewing colored reddish brown. From white discalpatch a whitish shadow runs to margin. For separation from sister species *T. pizzicata* see there. Female resembles male but the apex of forewings acute, not rounded. Margin of forewings rounded. Two examined females are very different in size and markings (figs. 80b, 80d).

Male genitalia are unmistakable by their unique modified U–shaped uncus, which bear a pointed process on each arm. Socii rectangularly shaped. Valvae relatively small and rounded. Phallus short with large buckle in its central part. The 8th sternite display a pair of central sclerotized processes as in its other congeners. Female genitalia have not yet been dissected.

Bionomics. *T. postobscura* occur frequently at altitudes between 800 and 1790 m.

Distribution. Endemic to the Island of Borneo.

81. ***Tensha pizzicata*** **spec. nov.** (figs. 81a–d).

Holotype: ♂, Indonesia, Sumatra, Sitahoan, 2. viii. 1981, leg. Dr. E. Diehl – in coll. A. Schintlmeister, Dresden.

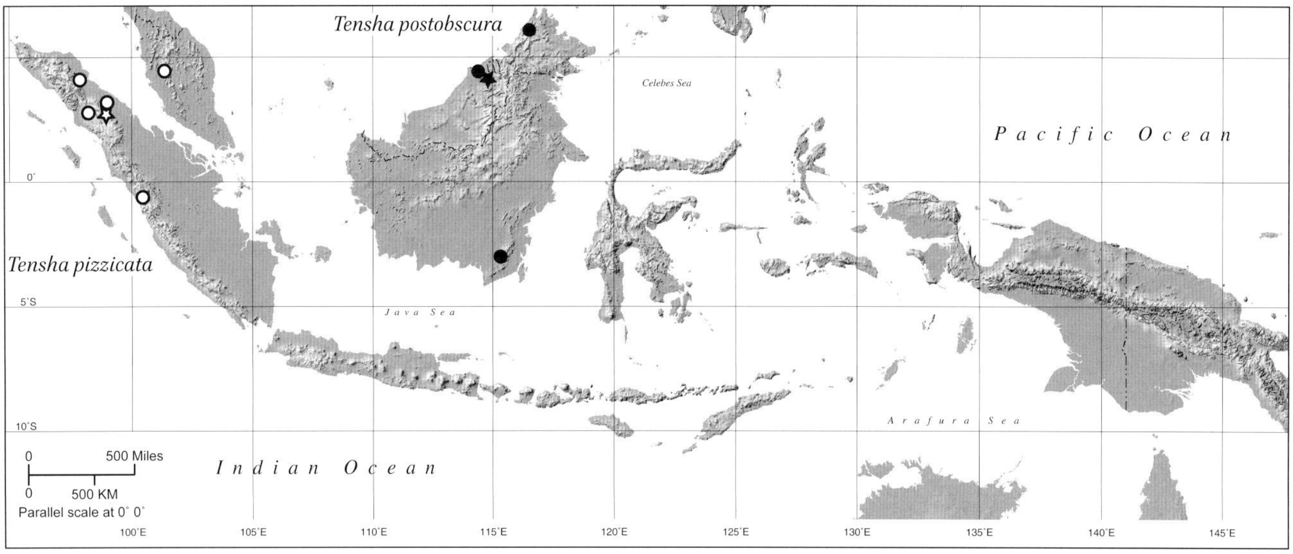

FIGURE 72 Distribution of *Tensha postobscura* and *pizzicata*.

Paratypes (10 ♂♂, 2 ♀♀,):

Sumatra: 1 ♂, Ost-Aceh, Gajo Hochland, 28.–31. viii. 1978, leg. Dr. E. W. Diehl (GU 19-85); 1 ♀, Prapat, Holzweg 2, 2. ix. 1985, leg. Dr. E. Diehl (GU 96-78); 1 ♂, Siantar, Holzweg 3, 20. iv. 1969, legit. E. W. Diehl; 2 ♂♂, Straße Siantar–Prapat, 14 km vor P., 1150 m, 25.–28. xii. 1975, leg. Dr. E. W. Diehl; 1 ♂, Dolok Merangir, 1.–29. x. 1967, leg. Dr. E. Diehl (MWM 29.109); 1 ♂, Dolok Merangir, 180 m, 1973, leg. E. Diehl; 1 ♂, Dairi Mts., 1500 m, 8. ii. 1970, leg. Dr. E. Diehl; 1 ♂, dto. 8. ix. 1978; 1 ♀, Lumban Julu, vic. Parapat, 6. iii. 1987; 1 ♂, Sumatera Barat, Mt. Sanggul, 1250 m, 2°44'S, 100°23'E, 1700 m, v. 2004, leg. S. Jakl.

Further material (no paratype): 1 ♂, W. Malaysia, Cameroon Highland, South of Ringlet, 1300 m, 19. xi. 2009.

Diagnosis. Forewing length ♂♂ 20.0 mm, ♀, 22.5 mm. This is the Sumatran vicariant of and closely related to *T. postobscura*. External appearance is similar as in *postobscura*, but forewings are 1 mm shorter. Wing coloration is less reddish tinged. Forewing displays a small whitish discal-patch, and a small cream colored blotch below the discal patch, which is smaller than in *delineivena*. From the discal streak a fine whitish band running toward margin. Veins are blackish tinged. Hindwings are of a more darker brown than in *postobscura*. Sexual dimorphic female resembles by whitish stripes in postmedial area rather *delineivena*, but forewing length is 8 - 9 mm shorter and outer margin more rounded.

Male genitalia resembles *postobscura*, but uncus tip is V–shaped with pointed arms. A small central spin in the notch is a diagnostic feature of *pizzicata*. Gnathos is shorter than in *postobscura* and displays a further small projection. Socii triangular shaped and smaller than *postobscura*. Valvae elongate with acute apex. Phallus less arched and longer than *postobscura*. Phallus tip bears hook-like projection. Everted endophallus displays many strongly sclerotized micro thorns. 8th sternite bears centrally a pair of sclerotized processes which are longer than in *postobscura*. Anterior margin is sligtly bilobed. 8th tergite have a straight, not rounded, posterior margin. Female genitalia have a circular antevaginal plate, a medium sized antrum and a slender ductus bursae. Corpus bursae is disc-shaped with a striolate surface. Signum is reniform and small.

Bionomics. The few known (n = 12) specimens were taken between 180–1500 m, most specimens were observed about 1000 m above sea level.

Distribution. Known from north Sumatra. One male, identified from photo only and not in the paratype series, comes from the Malayan Peninsula.

Canatensha gen. nov.

Type species: *Tensha nigrodorsalis* Schintlmeister, 1997.

Diagnosis. Medium sized adults of grey forewing color. Male antennae bipectinated; the last 30–40% to the tip are filiform. Female antennae are contrary to its related genera like *Tensha* or *Ambadra* complete filiform. External appearance resembles much *Tensha* but a white discal patch is absent. Forewings are elongate. Postmedial

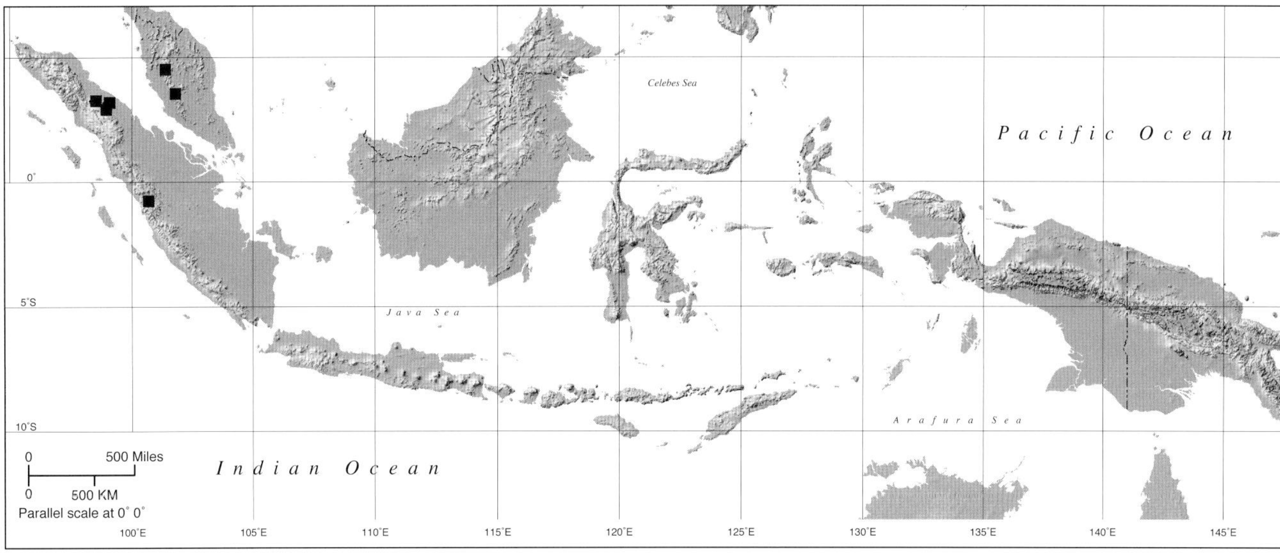

FIGURE 73 Distribution of *Canatensha mediofascia*.

fascia is marked as a row of well developed black dots. Postbasal fascia occasionally visible. Fringe of forewings checkered light grey and blackish. Near forewing apex a smaller blackish indistinct streak occure. Dorsum of forewing is darkened. Hindwings are chocolate-brown. Females display a prominently broad fuscous brown longitudinal fascia from base to forewing margin.

Male genitalia have a tapered, triangularly shaped uncus and a pair of long, curved and pointed Socii. Gnathos is absent. Valvae resemble those of *Tensha*. Anellar membrane is very large and bear dense setae, which is an unique generic feature. Phallus curved and less modified. 8th abdominal segments bilobed at anterior margin. 8th sternite deeply notched at posterior margin. Female genitalia have medium sized apophyses. Ductus bursae slender and long. Appendix bursae is disc-shaped and large. Corpus bursae circular shaped with striolate structure on its surface. Signum inconspicous and disc-shaped.

Canatensha is represented by two allopatric species only. Distributed in NE India, Myanmar, Indochina, the Malayan Peninsula and Sumatra.

81. ***Canatensha mediofascia*** (Rothschild, 1917): 257; pl. 7: 14 (*Turnaca mediofascia*).
Syntypes: 2 ♀♀, [India, Meghalaya], Assam, Khasia Hills – NHMUK, London.
Schintlmeister 2013: 456.

Diagnosis. See also under the genus description above. *C. mediofascia* is separable from *C. nigridorsalis* by lighter grey forewing color and weaker developed fuscous pattern. Female display the longitudinal streak on forewing weaker developed.

Male genitalia are as described above. They differ from *nigridorsalis* by longer socii and a costal projection of valvae. Shapes of socii and uncus are subject of individual variation. Sometimes the notch at the uncus tip is very small. 8th sternite well distinguished from *nigridorsalis* by a deep and very wide notch.

Female genitalia of *mediofascia* differ from *nigridorsalis* by larger antrum.

Bionomics. *C. mediofascia* occur in Sumatra and in Malaya uncommonly between 1050 and 1500 m, in Thailand up to 2000 m above sea level.

Distribution. Occuring in northeastern India, Myanmar, Thailand, the Malaya and Sumatra.

Ambadra Moore, 1883: 16.
Type species: *Eumeta rafflesi* Moore, 1860.
Schintlmeister 2008: 60, Schintlmeister 2013: 27.

Diagnosis. Medium sized imagines. Antennae of in both sexes bipectinated; distal 20–30% toward the tip is filiform. Male have long rami, those of female are short. Forewings mostly yellowish or brownish colored; only a few species appear grey or chocolate-brown. Forewing shape elongate. Forewing veins are tinged black or brown. There is a longitutinal black fascia running along vein M2 from the base to the margin. Submarginal and occasionally also postmedial fascia are marked as rows of black dots.

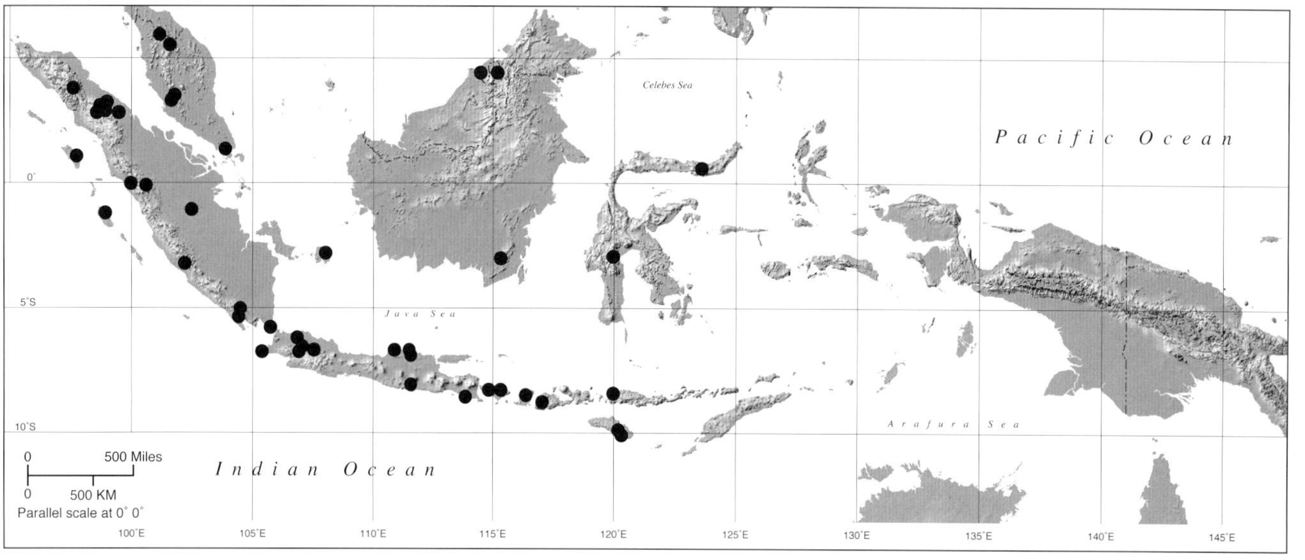

FIGURE 74 Distribution of *Ambadra rafflesi.*

Females are sexual dimorphic and display rather greyish forewings with an acute apex. Their pattern is characterized by fuscous and light stripes from the base to the margin.

Male genitalia are homogenous within the genus. Uncus is long and slender and broad at base. Socii slender and well developed. Valvae weakly sclerotized, broad, and of rhomboid shape. Phallus arched with a tapered tip. Everted endophallus bear rows of sclerotized small thorns, sharp and long needles and occasionally also spines, see fig. 81f, which are of taxonomic value. 3rd abdominal segment bears a pair of cteniophores (fig. 81g). 8th sternite anteriorly bilobed. It has centrally subapical projections and centrally in most species a weakly sclerotized "window". Female genitalia have medium sized apophyses. Antevaginal plate is of specifically shape and deeply notched. Ductus bursae narrow. Corpus bursae is large, of circular or ellipsoid shape with a crinkly structure at the surface. No signum.

Caterpillars are feeding oligophagous on various grasses (Poaceae) and palme trees (Arecaceae). *Ambadra* is a taxonomic difficult complex of many species (n > 32), of which about 25 are currently described yet. They are distributed from the Himalaya, S India, Sri Lanka, Indochina, Sundaland, the Lesser Sunda Islands, Sulawesi and the Philippines. The genus is particularly rich in species in Sulawesi.

82. *Ambadra rafflesi* (Moore, 1860): 430 (*Eumeta rafflesi*).
Holotype: ♂, [Indonesia], Java – NHMUK, London.
Schintlmeister 2008: 60, Schintlmeister 2013: 29.

Taxonomic notes. *Ambadra rafflesi* is variable in external appearance as well as in its male genitalia. It was not possible to correlate various forms to geographical areas, but it is likely, that environment conditions such as humidity might have an influance to external appearance including the size of the adults.

To secure the taxonomic identity of the species n = 32 male genitalia from all parts of the distribution area were examined.

Turnaca rufescens Gaede, 1930: 619; holotype: ♂, Sumatra, Barisan-Gebirge, 800 m – NHMUK, London, is recognized as an individual form of *Ambadra rafflesi* and sinks therefore as a junior synonym (**syn. nov.**) This distinct form depicted in fig. 82k, occure rarely in Sumatra, Siberut, Malaya and Thailand among unmottled specimens of *rafflesi.* Male genitalia (n = 3 dissected males) are not separable from those of unmottled *rafflesi* individuals.

Turnaca rafflesi borneensis Roepke, 1943: 102; holotype: ♂, S. E. Borneo – NBC, Leiden, and *Ceira pallida* Heylaerts, 1891: CCCCXVI; holotype: ♂, Java, Buitenzorg [= Bogor] – NBC, Leiden, depicted in, fig. 82i, are further junior synonyms of *A. rafflesi* (**syn. nov.**), as they fall in the width variation of *rafflesi.*

Diagnosis. Coloration of forewings vary from light grey, grey, pale brown, to yellowish-brown. Sometimes the pattern, e.g. submarginal fascia, blackish apical streak, post basal fascia, is well developed, other individuals are almost uniform. The form *rufescens* display brown mottled pattern on yellowish-brown ground. Discal spot is marked as an indistinct whitish patch, but often not visible. Forewing length and shape of the forewings are subjects of

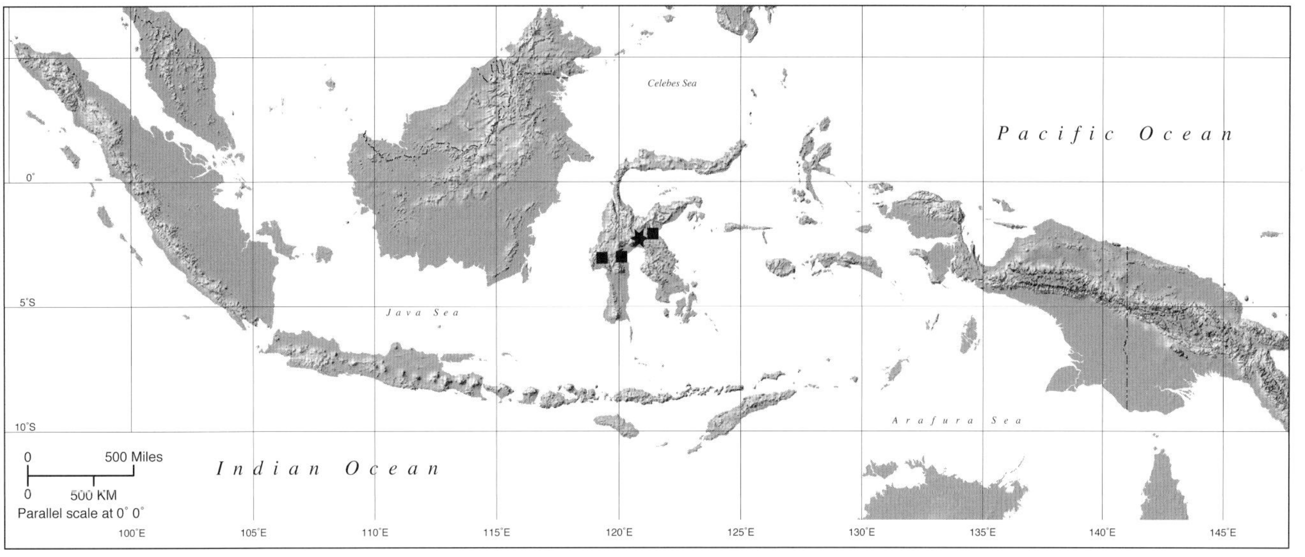

FIGURE 75 Distribution of *Ambadra canescens*.

individual variation as illustrated. Best external character to recognize *A. rafflesi* are a fuscous shadow in the tornus of forewing and fuscous reddish-brown hindwings. Female is sexual dimorphic, displaying a greyish to greyish brown wing color with a prominent longitudinal fascia from the base to the margin.

Male genitalia vary individually by shape of uncus tip, which might be knob-shaped or notched in different stages as illustrated. Socii are also variable in shape. They are about 50–58% of the uncus length. Valvae are variable in shape of margin, but apex is tapered and less rounded than in other congeners. A characteristic character is the strongly curved phallus, which ends in a strongly sclerotized spine. Everted endophallus displays a characteristric area of needle-like cornuti and four rows of specifically micros thornes, see fig. 82f. 8th sternite is deeply bilobed at anterior margin. The central, stronger sclerotized part bears two projections of variable shape. The "window" in the central part is heart shaped. Female genitalia are characterized by a large, deeply notched antevaginal plate. Ductus bursae is slender. Corpus bursae is disc-shaped and bear no signum.

Bionomics. The most common *Ambadra* in Sumatra and Java, which occur more commonly in urban areas such as plantations, villages, open landscapes. In rainforests adults are rarely seen. Moths were observed in lowlands up to 500 m. Only a few specimens were taken up to 1050 m. Hostplant of the caterpillars includes also the introduced African oil palme (*Elaeis guineensis* – Arecaceae) (Holloway, 1983: 40), but they feed very probably also other on Monocotyledones such as bamboo or other palms. *Ambadra rafflesi* was probably introduced with oil palmes to Sulawesi and other islands.

Distribution. Found in Vietnam, Thailand, Cambodia, the Malayan Peninsula, Borneo, Palawan, Sumatra, Nias, Siberut, Belitung, Java, Bali, Lombok, Sumba, Sumbawa, Flores and Sulawesi.

84. ***Ambadra canescens*** Schintlmeister & Lourens, 2010: 93; figs. 153, 154, 155 (*Ambadra sibena canescens*).
Holotype: ♂, Indonesia, Sulawesi, Mt. Sampuraga, 2°10‘S, 120°45‘E, 1400 m – coll. A. Schintlmeister, Dresden.
Schintlmeister 2013: 29.

Taxonomic notes. *A. canescens* was originally described as a subspecies of *Ambadra sibena* Schintlmeister, 1993 from Mindanao, and was raised to species level by Kobayashi, 2011: 306. The type series of *canescens* comprises of 48 ♂♂ and 2 ♀♀. Since Kobayashi, 2011 pointed out, that many externally very similar *Ambadra* species occur sympatrically in Sulawesi, it became clear, that the type series of *canescens* includes two different species.

Thus Kobayashi 2011: 303, figs. 14, 33a–e described *Ambadra gracilis;* holotype: ♂, Indonesia, Sulawesi, Mt. Tambusisi – National Science Museum, Tokyo in november 2011, which is a junior homonym of *Ambadra gracilis* Holloway, 2011, The latter paper by Holloway was published in july 2011. Schintlmeister 2013: 27 poposed *angustifolia* as a replacement name (published 1st august 2013), but two month earlier Kobayashi proposed *arta* as replacement name, which is therefore nomenclatural an available

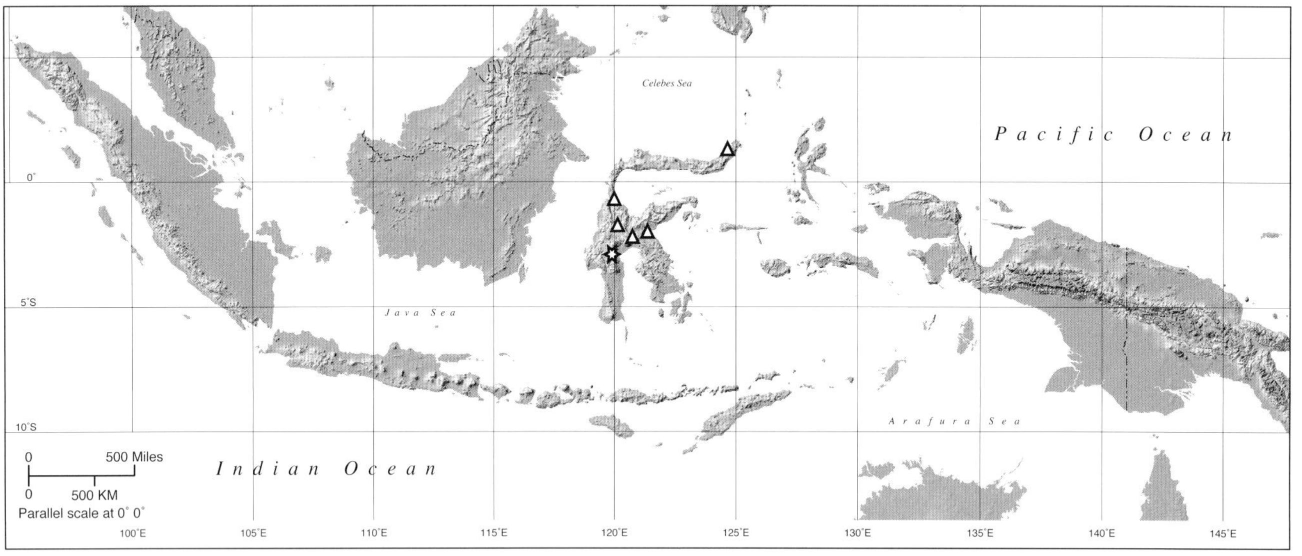

FIGURE 76 Distribution of *Ambadra miscera*.

name for that species and *angustifolia* is an unnecessary replacement name (**syn. nov.**). Comparison of holotype of *canescens* with color illustrations of holotype of *arta* given in the original description showed, that both taxa are conspecific and *arta* must sink as a junior synonym of *canescens* (**syn. nov.**). The remaining second species is described below as *Ambadra miscera.*

Diagnosis. *Ambadra canescens* differ from the other similar congeners in Sulawesi by the light greyish brown appearance and the narrow shaped forewings. The wings are not yellow tinged. Hindwings are white, with light reddish-brown anal region. *A. canescens* have the lightest hindwings of all known congeners in Sulawesi. Range of individual variability includes forewing shape and intensity of the blackish forewing pattern. Female is unknown.

Male genitalia have a long and slender uncus, which is bulged halfaway and pointed at the tip. Socii are longer than *miscera* and *celebensis*, and measures about 50% of the uncus length. Phallus strongly curved, long, with tapered tip. Phallus length vary individually. Everted endophallus bear several spines and microthornes, which are weakly sclerotized. 8th sternite of specific shape as illustrated. Its central projections are large. The "window" is at anterior margin bordered by a strongly sclerotized triangular area. Lobes at anterior margin of 8th sternite are shorter than in *celebensis* and *miscera*.

Bionomics. The moth was observed commonly but locally, between 700–2000 m above sea level.

Distribution. Endemic to Sulawesi.

85. ***Ambadra miscera* spec. nov.** (figs. 85a–g).

Holotype: ♂, Indonesia, Sulawesi, Puncak, Palopo, 1000 m, 2°55'S, 120°05'E, 25.–31. i. 1995, leg. Sinjaev & Tarasov – coll. A. Schintlmeister, Dresden.

Paratypes (28 ♂♂, 2 ♀♀):

Sulawesi: 4 ♂♂, Puncak, Palopo, 1000 m, 2°55'S, 120°05'E, 25.–31. i. 1995, leg. Sinjaev & Tarasov (GU 54-63); 1 ♂, dto., 13.–14. ii. 1995; 1 ♂, 1100 m, 25. ix. 1995, leg. S. Naumann: 2 ♂♂, Puncak, Palopo, 3°S, 120°09'E. xii. 1998 (GU 88-52); 1 ♂, dto., 900 m–1300 m, iii. 1997 (GU 54-74); 5 ♂♂, Mt. Sampuraga, 2°10'S, 120°45'E, 1400 m, 1.–6. ii. 1995, leg. Sinjaev & Tarasov (GU 41-41, GU 54-60); 5 ♂♂, 2 ♀♀, dto., 11.–12. ii. 1995 (GU 41-35); 1 ♂, Mt. Salubaya, 1000 m, 0°45'S, 120°E, 9.–10. ii. 1995 leg. Sinjaev & Tarasov (GU 54-64); 3 ♂♂, Tambusisi, x. 1997; 3 ♂♂, Namo, Straße Palopo-Gimpu, 650 m, 21.–22. 1995, leg. S. Naumann; 1 ♂, Dumogo-Bone N.P., Headquarter, 350 m, 29. iii. 2000, leg. S. Naumann; 1 ♂, Maros region, Camba area, 7-800 m, vi. 2017, leg. Benny de Groof.

Forewing lengths ♂♂ 19–22 mm, ♀♀ 25 mm. Antennae are in both sexes bipectinate; last 33% toward the tip is filiform. Similar to *A. canescens*, but forewing shape is broader. Forewings are light greyish-brown. Forewing pattern mostly indistinct. Postmedial fascia marked as a row of fuscous brown patches. Along the dorsum a fuscous brown shadow occur. Veins occasionally blackish tinged. Hindwings are reddish-brown, darker than in *canescens*. Female is sexual dimorphic having three fuscous brown longitudinal stripes from base to margin of forewing. Apex of forewing falcate.

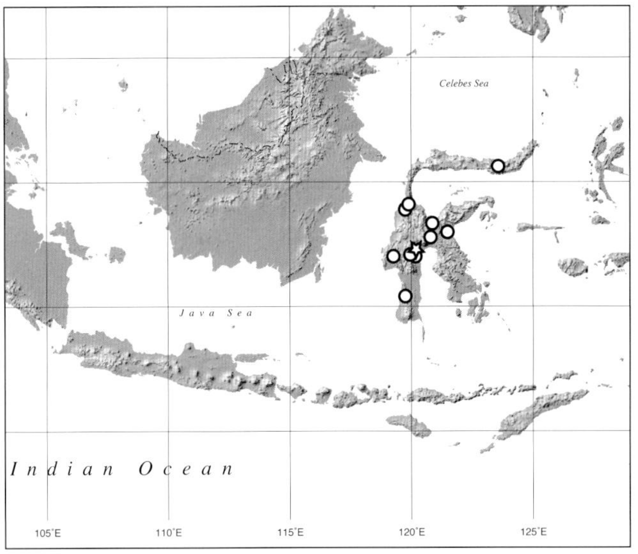

FIGURE 77 Distribution of *Ambadra celebensis.*

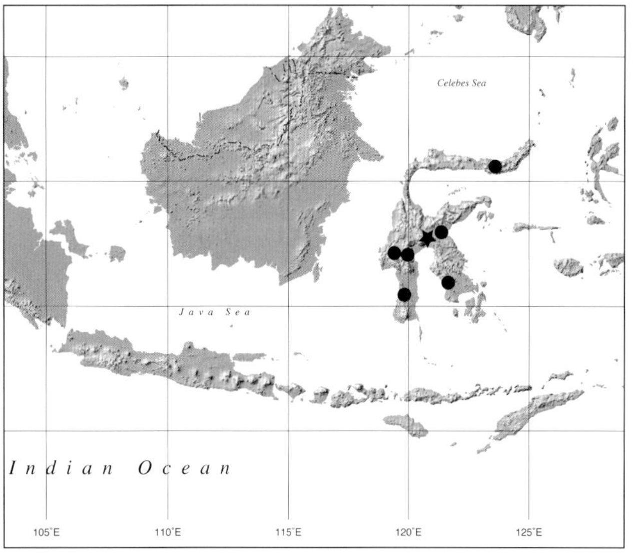

FIGURE 78 Distribution of *Ambadra macula.*

Male genitalia resembles *canescens.* Uncus long and slender; socii are shorter and slenderer than in *canescens,* about 30% of uncus length. Phallus shaped as *canescens,* but the tip ends in a sharp tapered and curved hook. Central projections of 8th sternite are reduced, much smaller than in *canescens.* The central part lacks the "window" as seen in *canescens.* At anterior margin a pair of long projections arise, wich are often of different length. Female genitalia are as described in genus diagnosis; they have a large and triangular notched antevaginal plate. Appendix bursae is large and disc-shaped. Corpus bursae circular shaped with a crinkly structure at surface. No signum.

Bionomics. A common species which occur at altitudes between 350–1400 m.

Distribution. Endemic to Sulawesi.

86. ***Ambadra celebensis*** (Roepke, 1943): 102; fig. 14a, b (*Turnaca celebensis*).
Holotype: ♂, [Indonesia, Sulawesi], Centr. Celebes, Todjambu, 1.500 m – NBC, Leiden.
Schintlmeister 2013: 27.

Diagnosis. *A. celebensis* is the largest *Ambadra* in Sulawesi and distinguished by the darkest hindwings. Forewings are broad shaped and display a rich black pattern. Submarginal and marginal fasciae marked as rows of blackish dots. Postmedial fascia developed as a blackish shadow. which is a good character for separation from other similar species. There is a similarity to an individual form of *A. macula* with weakly developd black pattern, but the latter is 2 mm smaller in forewing length. Individual variation of *celebensis* is less wide than in *macula*; sometimes a brownish shadow on forewing dorsum occur. Hindwings of *celebensis* are darker reddish-brown than *macula.* A single male, depicted in fig. 84d, resembles externally much to *macula,* but identity was confirmed by genitalia dissection (GU 96-87). The only examined female resembles the male, but blackish pattern on forewing is reduced.

Male genitalia very similar to *macula.* Uncus is long and slender, bulged in the basal part. Socii differ from *macula,* being shorter – < 30% of uncus length – slenderer and curved. Phallus short, not rectangularly angled like *macula,* with tapered and strong sclerotized tip. 8th sternite anteriorly bilobed with shorter processes than *miscera.* Central projections are smaller than in *macula.* Differences in sclerotizations of 8th abdominal segments between *celebensis* and *macula* are subtle and indicated in figs. 84a and 85b. Female genitalia have not yet been dissected.

Bionomics. Adults occur frequently between 900–2000 m.

Distribution. Endemic to Sulawesi.

87. ***Ambadra macula*** Kobayashi, 2011: 288; figs. 2-4, 21a–e, 22a–e.
Holotype: ♂, Indonesia, Sulawesi, Sampuraga – National Museum of Nature and Science, Tokyo.
Schintlmeister 2013: 28.

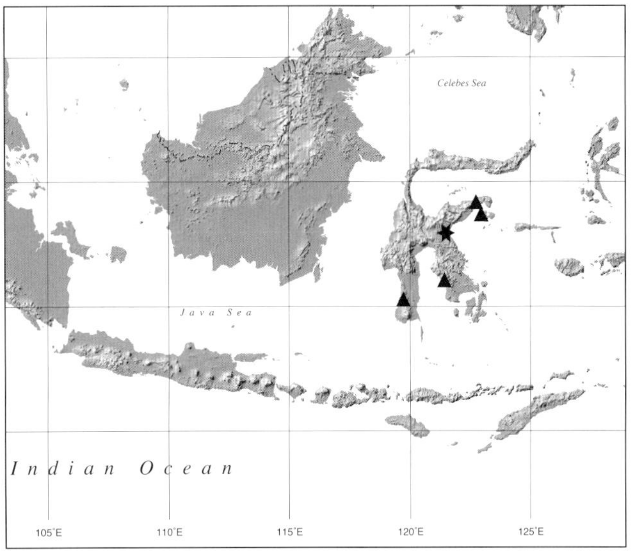

FIGURE 79 Distribution of *Ambadra parva*.

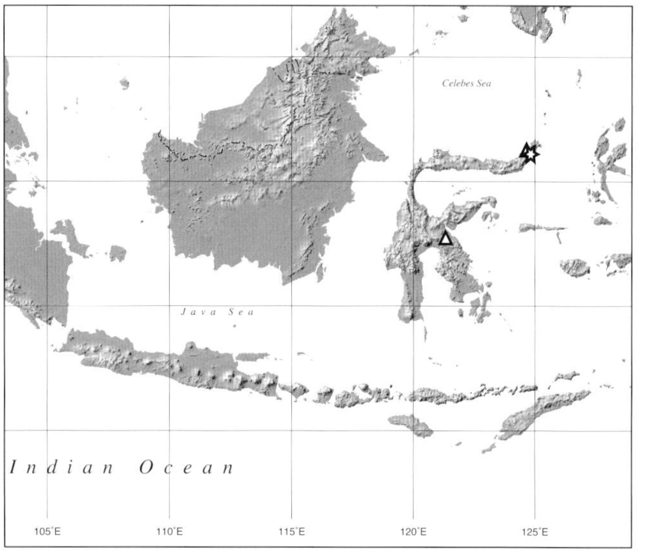

FIGURE 80 Distribution of *Ambadra tondanoa*.

Diagnosis. A variable species. Most common and most easy to identify are specimens with two black contrasting patches on the straw colored forewings. The form with a black tranversal line from the base to the margin of the forewings and a contrasting black rectangular stigma is also readily to identify; but such specimens occur more rarely. Apart from this there is a morphe, that resemble very much to *A. celebensis*, but all wings are more yellow tinged. Female is unknown.

Male genitalia are similar to *celebensis*, but having shorter – 30% of the length of the uncus – socii and a thicker uncus at base. Both, uncus and Socii, vary individually in its shapes. Phallus is arched and shorter than *celebensis*. Everted endophallus bear four rows of microthornes, while *celebensis* have only two. Differences in sclerotizations of 8th abdominal segments between *celebensis* and *macula* are subtle and indicated in figs. 84a and 85b.

Bionomics. *A. macula* is the most common *Ambadra* in Sulawesi. It was observed from 200 m up to 1400 m above sea level.

Distribution. Endemic to Sulawesi.

88. ***Ambadra parva*** Kobayashi, 2011: 294; figs. 9, 25a–c.
Holotype: ♂, Indonesia, Sulawesi, Mt. Tambusisi – National Museum of Nature and Science, Tokyo.
Schintlmeister 2013: 28.

Diagnosis. The smallest *Ambadra* species in Sulawesi. Also distinguished by a radial longitudinal fascia on forewing, which runs from base to authordian area. Dorsum in basal and medial area fuscous reddish-brown. In the medial area there are several blackish spots and postmedial fascia marked as a row of black dots. Apart from this form also specimens with reduced blackish pattern occur, but the black medial patch remains visible. Hindwing light yellowish-brown. Female is unknown.

Male genitalia have a very long and slender uncus and relatively short Socii, about 25% of uncus length. Juxta wave-shaped and well sclerotized. Phallus rectangular angled and thick. Endophallus bear a variable number of larger, strongly sclerotized spines grouped in a ring. 8th sternite anteriorly slightly bilobed. The projections in the central part are remarkable similar to those of *A. canescens*.

Bionomics. *A. parva* is rare. Including the type series, only n = 8 specimens are known to the author, which were taken between 150–600 m above sea level.

Distribution. Distributed in Sulawesi and Peleng island.

89. ***Ambadra tondanoa*** Kobayashi, 2011: 292; figs. 8, 24a–d.
Holotype: ♂, Indonesia, N-Sulawesi, Tondano, Mt. Makaweiben, 1.000 m – National Museum of Nature and Science, Tokyo.
Schintlmeister 2013: 29.

Diagnosis. Wingshape and markings resemble *A. celebensis*, but forewing length is about 5 mm shorter. Forewing pattern resembles *A. parva*, but there is no black patch in

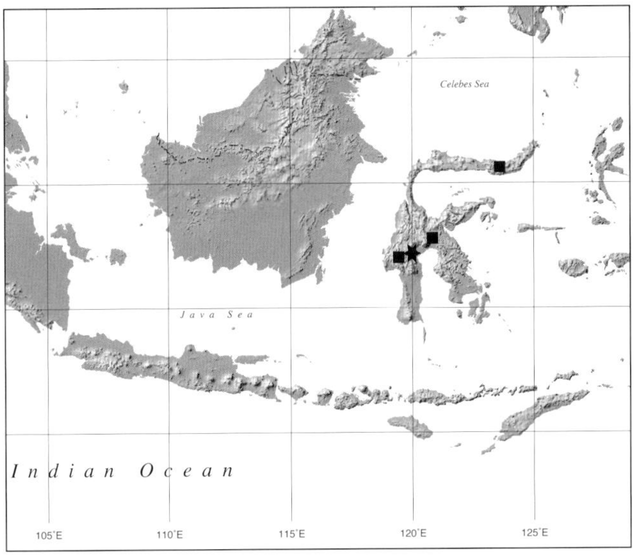

FIGURE 81 Distribution of *Ambadra vittata*.

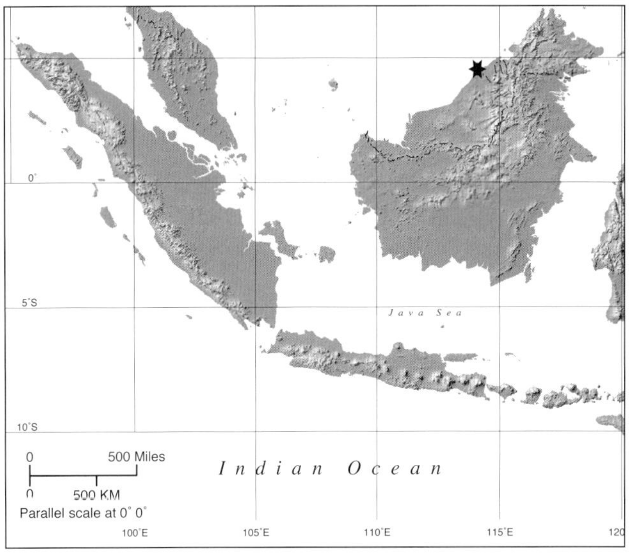

FIGURE 82 Distribution of *Ambadra gracilis*.

medial area. A radial longitutinal fascia on forewing runs from base to authordian area. Postmedial fascia marked as a row of black dots. Along the dorsum there is a brown shadow as in *parva*. Specimens with weakly developed fuscous pattern occur as an individual form. Hindwings are pale reddish-brown. A female tentatively attributed to *parva* have almost no markings on forewing, but show the longitutinal line and – also weakly developed – a postmedial fascia.

Male genitalia differ from *celebensis* by less sharp and triangularly uncus tip. Socii short, 25% of uncus length. Phallus strongly curved, almost semicircular. Phallus tip is ending in a curved and tapered spine. Everted endophallus bear fields with needle-like spines. 8th sternite deeply bilobed at the convexly anterior margin. The "window", e.g. the weakly sclerotized central part, is small. Central projections are reduced. Female genitalia have not yet been dissected.

Bionomics. The few known specimens (n = 4) were taken between 500–1000 m above sea level.

Distribution. Endemic to Sulawesi.

90. ***Ambadra vittata*** Kobayashi, 2011: 289; figs. 5, 6, 23a–d.
Holotype: ♂, Indonesia, Sulawesi, Puncak Dingin, 1.700 m – National Museum of Nature and Science, Tokyo.
Schintlmeister 2013: 29.

Taxonomic note. Three dissected male genitalia differ in uncus shape from those of the holotype as illustrated in fig. 88. Other features of genitalia match the description of Kobayashi and there is no reason to assume the existence of a further similar species.

Diagnosis. *A. vittata* resembles by its external appearance rather to *macula*, but forewing length is 3 mm shorter. Postbasal, medial, postmedial and marginal fasciae are marked as rows of black dots. Along the dorsum a fuscous brown shadow occur. Submarginal fascia marked as a row of black dots. The black longitudinal streak from base to margin is sometimes very strongly developed, but there are also individuals lacking this streak. Hindwings are pale reddish-brown as in *macula*. Female resembles the male but forewing apex is acute.

Male genitalia have uncus bulged in the middle part. Socii rather long, about 40% of uncus length. The sacculus of the valves display two stronger sclerotized oval protuberances, which are not seen in other *Ambadra* species. Phallus slender and slightly curved. 8th sternite displays a pair of small central processes. 8th tergite convex shaped at the posterior margin.

Bionomics. A rare species (n = 8 specimens known), that was observed between 1000–2000 m.

Distribution. Endemic to Sulawesi.

91. ***Ambadra gracilis*** Holloway, 2011: 353; pl. 7: 5; fig. 95.
Holotype: ♂, W. Malaysia, Sarawak, Baram dist. – NHMUK, London.
Schintlmeister 2013: 28.

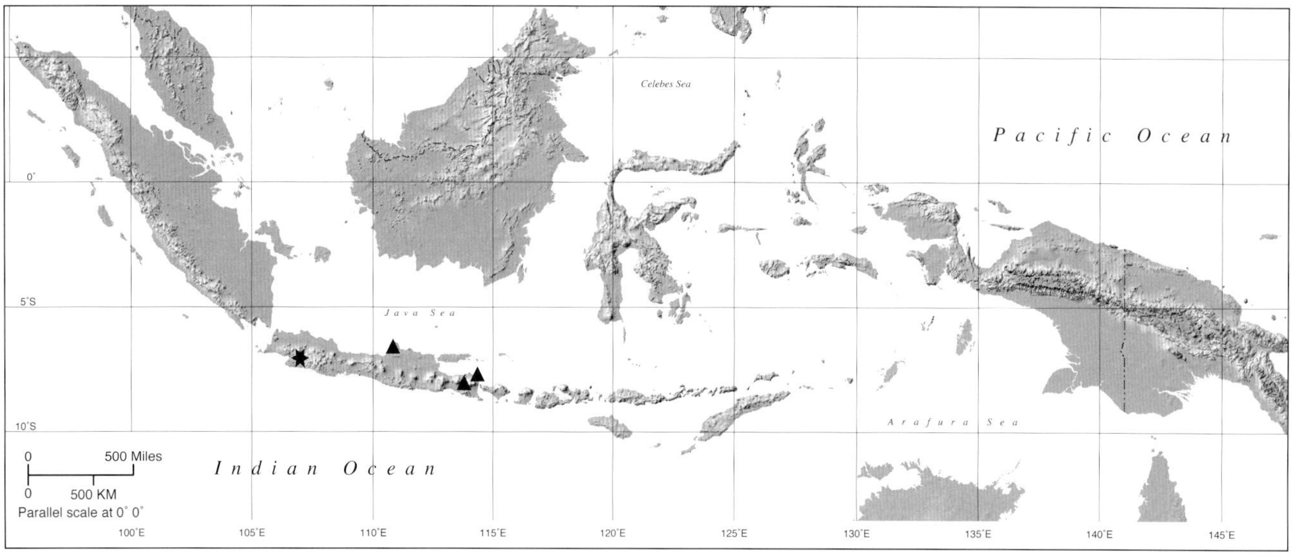

FIGURE 83 Distribution of *Ambadra batara.*

Diagnosis. Forewing length only 15 mm– the smallest known *Ambadra*. Forewings are narrow. Dorsal area brownish shadowed. Forewing pattern resembles *A. parva*. Medial fascia consists from several unregularly brown patches. Postmedial and submarginal fasciae developed as rows of fuscous brown dots. Hindwings are light yellow. Female is unknown.

Male genitalia have a short uncus. Socii are of same length as uncus. Tegumen and in particular saccus are large. Valvae are rather small. Phallus curved and display a bifid tip. 8th sternite have a large "window", and a pair of small central processes. Anterior margin of 8th sternite bears a pair of long processes. 8th tergite is notched at its posterior margin.

Bionomics. The holotype specimen of *A. gracilis* comes from the coast and was taken at 50 m above sea level.

Distribution. Known only from the type locality in Borneo.

92. ***Ambadra batara*** **spec. nov.** (figs. 92a–c).

Holotype: ♂, Indonesia, W Java, Cianur reg., Sukanegara, 7°06' S, 107°07' E, 1050 m, 18.–21. iii. 2007, leg. St. Jakl (GU 88-53) – in coll. A. Schintlmeister.

Paratypes (8 ♂♂):

Java: 3 ♂♂, Cianur reg., Sukanegara, 7°06' S, 107°07' E, 1050 m, 18.–21. iii.2007, leg. St. Jakl; 3 ♂♂, Mt. Muria, 15 km N Kusdas, 1055 m, 6°35'S, 110°53'E, 21.–22. i. 1998, leg. Jakl, Schintlmeister & Cervenka (GU 78-71, GU 54-68); 2 ♂♂, Sempolan, Mt. Ranung, 410 m, 8°10'S, 113°40'E, 28. i. 1998 leg. Jakl, Schintlmeister & Cervenka; 1 ♀, Baluran NP, 60 m, 7°41'S, 111°30'E, 31. i. 1998, leg. Jakl, Schintlmeister & Cervenka (GU 97-02).

Diagnosis. Forewing length ♂♂ 16–18 mm; ♀ 17 mm. A small species, which resembles to small individuals of *A. rafflesi.* Antennae are in both sexes orange-brown, bipectinated, except last filiform part of 20% toward the tip. Forewings narrow and light yellowish-brown colored. Forewing pattern often weakly developed. The illustrated holotype display well developed pattern, which consist of basal, postbasal, postmedial, submarginal and marginal fasciae, marked as rows of fuscous brown patches. Two patches in the postbasal area are larger and also in weakly marked specimens visible. A longitutinal fascia runs along vein M1 from the base to the margin of forewing. This line is often indistinct and sometimes not visible. Hindwings are reddish-brown. There is also a resemblance to *Ambadrana taruma,* which occur sympatrically, but the latter species displays a prominent blackish patch in medial area instead of a postbasal fascia. Sexual dimorphic female of *batara* lacks the postbasal fascia, but have a black longitudinal band from base to margin. Forewing shape of females have margin rather straight and apex acute.

Male genitalia have a slender and long uncus, which is slightly longer than Socii. Uncus tip not pointed, but squarish. Valvae shape and in particular the bulbous, large sacculus are unmistakable. Near the valva base a large costal projection of rounded shape occur. Phallus less arched ending in a sharp pointed spine. 8th abdominal segment resembles most to those of *A. modesta.* 8th sternite lacks a

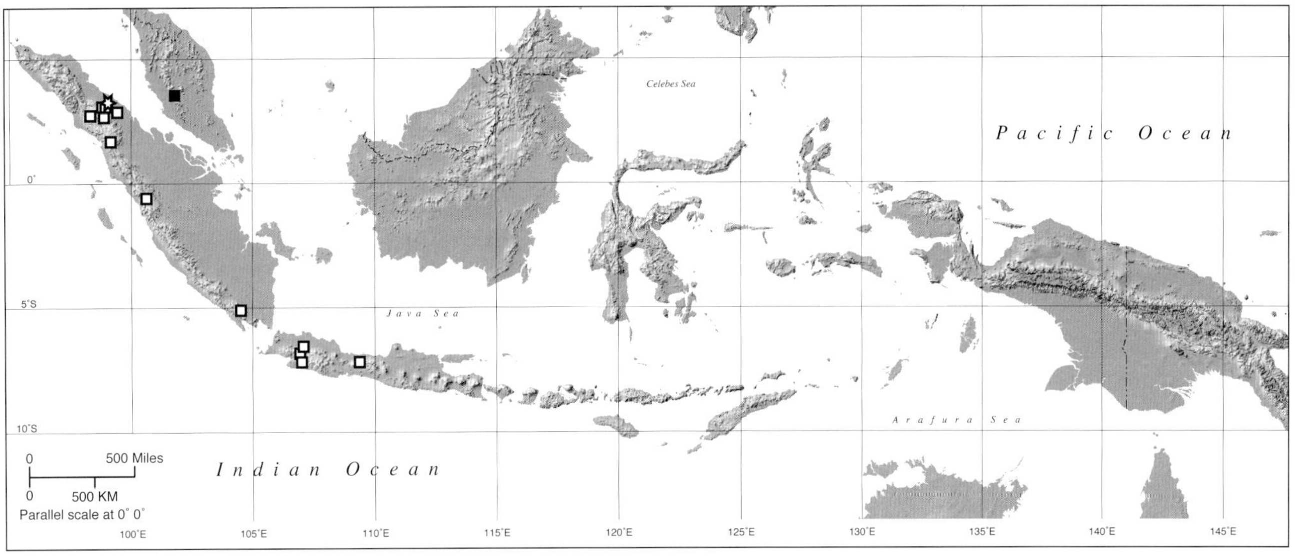

FIGURE 84 Distribution of *Ambadra discocellularis*.

"window"; central processes are short and anterior margin is slightly bilobed. Female genitalia resembles somewhat to *rafflesi*, but antevaginal plate is much smaller and only slightly notched. Corpus bursae bear a prominent signum, which ends in a pair of spines.

Bionomics. Type series was taken between 60 and 1055 m above sea level.

Distribution. Endemic to Java.

93. ***Ambadra discocellularis*** (Kiriakoff, 1974): 374; pl. 1: 1; fig. 1 (*Atornoptera discocellularis*).
Holotype: ♂, [Indonesia], Sumatra, Dolok Merangir – ZSM, Munich.
Schintlmeister 2013: 27.

Taxonomic note. Holloway, 1982: 207, argued, that *A. discocellularis* might be synonymous with *A. stigmatica* Gaede, 1930, described from S. India. But *A. stigmatica* is externally similar to *A. similis,* while *A modesta* Schintlmeister, 2007, described from China, Hainan also occuring in Indochina and NW India is externally more similar to *discocellularis*. A single male from Malaya matching in its genitalia well to Sumatran populations, but forewings are fawn as in *A. similis.*

Diagnosis. *Ambadra discocellularis* is distinguished by fuscous chocolate-brown color of all wings and resembles externally only to *modesta*. But the latter species have 2–3 mm shorter forewings. Forewing pattern of *discocellularis* consists of a small indistinct whitish discal spot, a submarginal fascia, marked as row of blackish dots and a fuscous brown dorsal shadow. The paleness of wings vary slightly individually as illustrated. Sexual dimorphic female have in submarginal area of forewing two whitish stripes.

Male genitalia have a curved uncus and a pair of long and slender socii which are sidewards directed. Uncus vary individually: sometimes tip is tapered, other specimens have a knob-like tip and also shape of bended uncus is variable. Phallus have a tapered tip, is curved, narrow. It is longer and more strongly curved than in *modesta.* Everted endophallus bears a variable number of microthorns in three rows, a single neddle-like long spine at the tip and in the central part a group of two to four long spines. 8th sternite bears a pair of small central lobes. Posterior rounded and anterior margin is bilobed. Female genitalia less modified. Corpus bursae is ovoid and large. It displays a striolate surface structure. No signum.

Bionomics. Moths occur frequently between 300–1700 m above sea level, most commonly above 1000 m.

Distribution. Distributed in Sumatra and Java; a single male comes from Malaya.

94. ***Ambadra similis*** (Dierl, 1984): 18; figs. 1, 2. (*Turnaca similis*) **stat. rev.**
Holotype: ♂, [Indonesia], Sumatra, [Pematang] Siantar, Holzweg 2, 1.050 m – ZSM, Munich.
Schintlmeister 2013: 29.

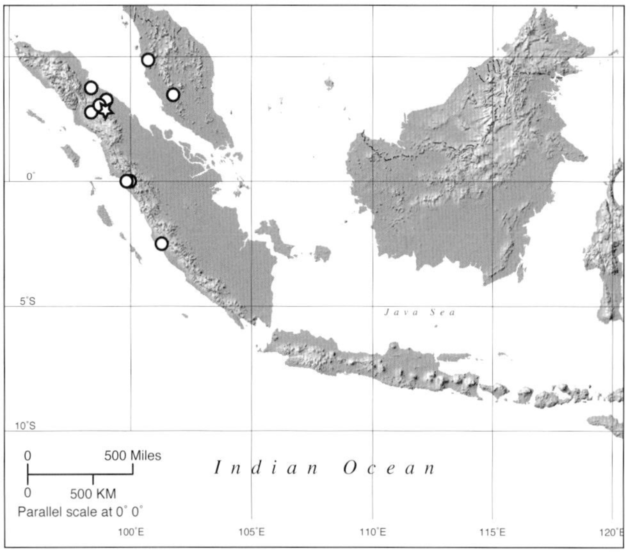

FIGURE 85 Distribution of *Ambadra similis*.

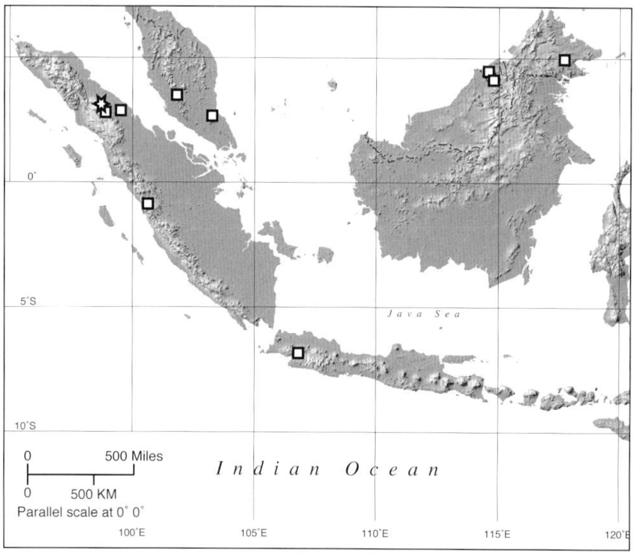

FIGURE 86 Distribution of *Ambadra melanostriga*.

Taxonomic notes. *A. similis* was placed by Schintlmeister, 1994: 222 as a junior synonym of *A. stigmatica* Gaede, 1930: 618; pl. 81: f; holotype: ♂, S. India, Karwar – NHMUK, London. The latter species differ however greatly in its male genitalia by short and thick phallus and a pair of long and thick Socii. Thus *A. similis* is reinstalled here as a genuine species (**stat. rev.**).

The *Ambadra suriga* (Schaus, 1928)–complex is taxonomically difficult. Range of individual variability in external appearance is wide and females are unknown.

Male genitalia are seemingly distinguished only by subtle differences in shapes of uncus tip, phallus tip and structures of the everted endophallus. Ratio of socii length to uncus length is also useful for separation of species. Such minor differences in morphology of male genitalia would usually indicate separation on subspecific level, but most of these species occuring sympatrically.

All illustrated imagines of the group were dissected in their genitalia to confirm correct identification.

Diagnosis. Forewing color yellow to fawn. A blackish tinged line runs along the vein M1 from base to forewing margin, accompanied from a further blackish line above from medial area to apex. Discal patch is indistinct and weakly whitish marked. A fuscous patch in medial area around the longitudinal fascia is weaker than in *A. melanostriga, owadai,* and *suriga.* Hindwings vary from yellowish-brown to reddish-brown, which separates *similis* from *mentawai.* The latter have darker hindwings. Female is unknown.

Male genitalia have a tapered uncus with pointed uncus tip. Ratio uncus length to socii length is 35%. Shape of valvae vary according to preparation and position in the slide. Sacculus is edged. Juxta is weavy, but of variable shape. Phallus is thick and strongly bended, about 100°. Phallus tip is like an arrow, see fig. 94b. Everted endophallus displays five rows of microthorns and a central field with many needle-like spines. 8th sternite bears a pair of small central lobes. Posterior margin rounded and anterior margin bilobed. 8th tergite is less modified.

Bionomics. *A. similis* occurs commonly in Sumatra and infrequent in Malaya between 350–1100 m above sea level. Most specimens were taken at 1000 m.

Distribution. Restricted to the Malayan Peninsula and Sumatra.

95. ***Ambadra melanostriga*** (Schintlmeister, 1994): 221; pl. 3: 10; pl. 3: 8. (*Turnaca melanostriga*), **spec. rev.**
Holotype: ♂, [Indonesia], N Sumatra, Sindar Raya 2, 15 km SW Sindar Raya, 3°10‘N, 98°56‘E – coll. A. Schintlmeister, Dresden.
Schintlmeister 2013: 28.

Taxonomic notes. Schintlmeister & Pinratana, 2007: 66 placed *melanostriga* as a junior synonym of *A. suriga* (Schaus, 1928) because of similarities in its male gentalia. But since Kobayashi 2011 showed, that in Sundaland several different species occur, it is clear, that *melanostriga* is a distinct species (**spec. rev.**), separable by phallus shape. Kobayashi, however, also illustrated in his paper a specimen of *melanostriga* but wrongly attributed it as *A. rafflesi.*

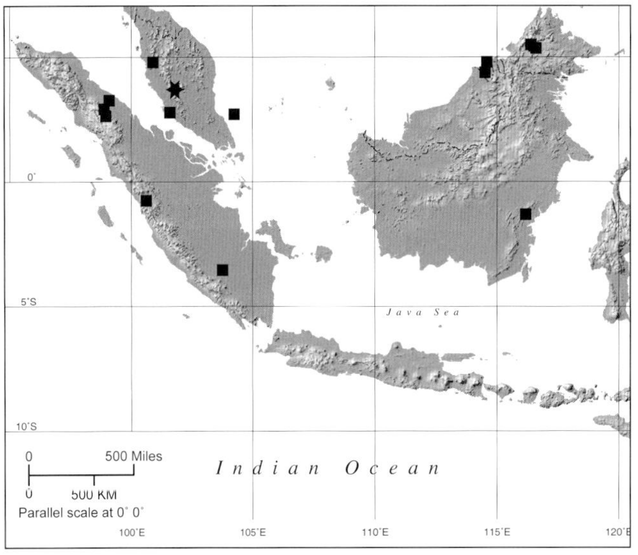

FIGURE 87 Distribution of *Ambadra owadai*.

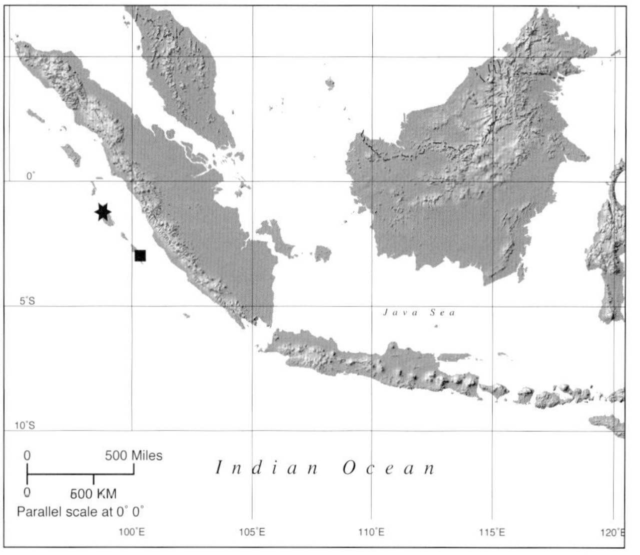

FIGURE 88 Distribution of *Ambadra mentawai*.

Diagnosis. The species occur in two externally different forms. The nominotypical form *melanostriga* is unmistakable by the prominent black longitudinal streaks on forewing, which resembles somewhat to a melanic individual form of Sulawesian *A. macula*. But apart from this form also specimens occur, which are almost indistinguishable by external appearance from *A. owadai*. Both species have the fuscous spot in the medial area larger developed than *similis* or *mentawai*. Remainig forewing pattern as described in *similis* (see above). Female is unknown.

Male genitalia are characterized by a fan shaped uncus tip, which is sometimes slightly notched. Socii length is about 40% of uncus length. Phallus tip differ from *owadai* beeing shorter than in *similis* and sharp spine, not bifurcate. Everted endophallus bear a rectangular, long field of nedle-shaped cornuti.

Bionomics. The nominotypical blackish form occur seemingly rare. Only n = 10 specimens are known. The species occur between 50–1600 m above sea level.

Distribution. Known from Borneo, Sumatra and Java.

96. ***Ambadra owadai*** Kobayashi, 2011: 296; figs. 10, 27a–g.
Holotype: ♂, Malaysia, Pahang, Raub Bukit Fraser, 1300 m – National Science Museum, Tokyo.
Schintlmeister 2013: 28.

Diagnosis. Very similar to *A. melanostriga*, but blackish medial patch of forewing more strongly developed. Due to range of individual variability sometimes individuals with weaker pattern occur, as depicted in fig. 96e, that can be separated only by genitalia dissection. Pattern elements on forewing as described in *A. similis*. Female is unknown.

Male genitalia are as in *melanostriga*. Uncus tip fan shaped and notched; socii about 50% of uncus length. Phallus distinguished by a bifurcate tip and an area of needle like spines of circular shape. 8th abdominal segments are as in *melanostriga*.

Bionomics. *A. owadai* occur commonly between 50 m–1400 m.

Distribution. Known from Borneo, Sumatra, the Malayan Peninsula, including Pulau Tioman and S Thailand.

97. ***Ambadra mentawai*** **spec. nov.** (figs. 97 a, 97b).
Holotype: ♂, Indonesia, Mentawai Islands, N Siberut Isl., 1°05' S, 98°46' E, 300 m, ii. 2005, leg. St. Jakl – in coll. A. Schintlmeister.

Paratypes (3 ♂♂):
Sumatra, Mentawai Isls: 1 ♂, N Siberut Isl., 1°05' S, 98°46' E, 300 m, ii. 2005, leg. St. Jakl (GU 84-52); 1 ♂, Siberut Isl., Mt. Malancan, 17.–20. i. 2004 leg. St. Jakl; 1 ♂, N Pangai Isl., Sikakaop, 26. ii.–7. iii. 1992 leg. Dr. E. W. Diehl (GU 41-50).

Diagnosis. Forewing length ♂♂ 17–19 mm. Forewings fawn. Submarginal fascia of forewing marked as a row of fuscous brown dots. Holotype specimen displays several brown patches in the postbasal area of forewing. Paratypes does not show them. Longitudinal streak along the radial vein weakly developed and fuscous patch in medial

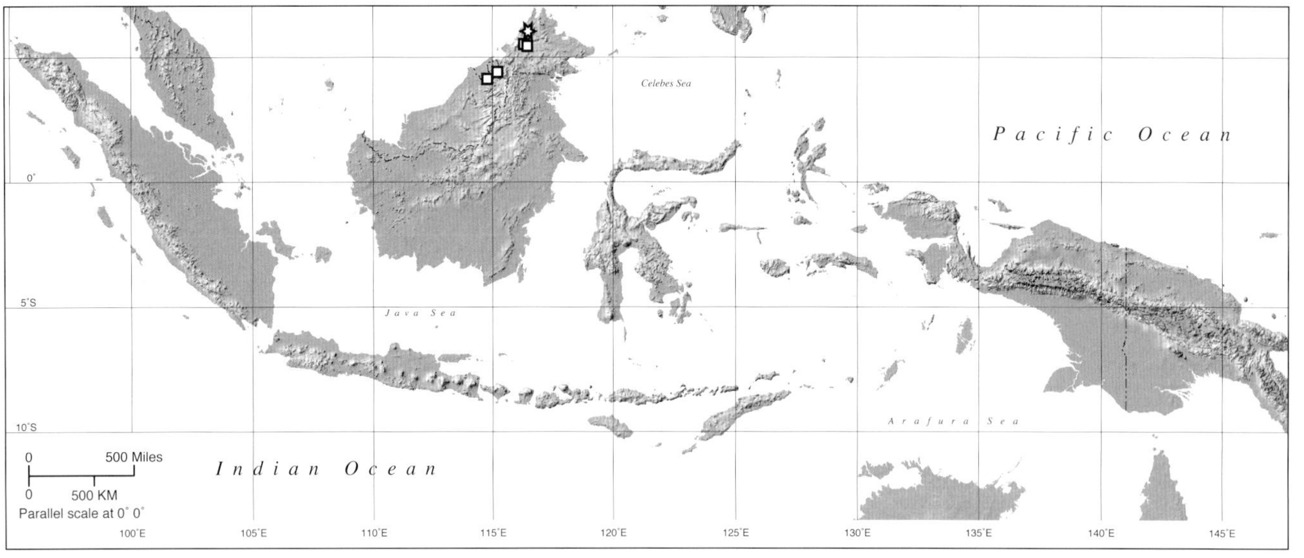

FIGURE 89 Distribution of *Ambadra montana*.

area absent. External appearance resembles *A. similis*, but forewings are broader und more round shaped. Hingwings are dark reddish-brown, which separates *mentawai* from *similis*. Female is unknown

Male genitalia are distinguished from *melanostriga* and *owadai* by a tapered uncus. Uncus is shorter than in *similis*. Ratio uncus length to length of socii is 35%. Juxta of *mentawai* is deeper bilobed than in *similis*. Phallus thicker and shorter than in *similis*, approximately rectangularly arched. Everted endophallus bears a field of needle-shaped cornuti. 8th sternite short and of a squarish shape rather than oval as in *similis* or *owadai*. Posterior margin slightly convex and anterior margin bilobed.

Bionomics. Type series (n = 4) has been collected in different years and localities at lower altitudes near the coast, up to 50 m above sea level.

Distribution. Restricted to the Sumatran Mentawai island group: Siberut and Pagai Islands.

98. *Ambadra montana* (Holloway, 1984): 207.
Holotype: ♂, [Malaysia, Borneo], Sabah, Mt. Kinabalu, Park H.Q., 1.620 m – NHMUK, London.
Schintlmeister 2013: 28.

Diagnosis. Large species. Distinguished by violet-brown tinged basal and postbasal area on forewing. Also violet tinged on costa near the apex. A whitish discal spot is well contrasting on fuscous yellowish-brown forewing. Veins are black tinged On the radial vein a fine black longitudinal line runs from base to margin. Hindwings are reddish or orange-brown. Female is unknown.

Male genitalia have a long uncus, which is curved upwards and backwards over its distal half. Also specimens with rather straight uncus occur as depicted in fig. 96a. Uncus is longer than in any other *Ambadra* and the ratio uncus length to socii length is 27%. Valvae are as in *similis*. Juxta is characteristically trapezoid shaped and strongly sclerotized. Phallus stout and arched. Phallus tip is and differ by its blunted shape from other congeners.

Bionomics. *A. montana* occur frequently in the montane zone above 1100 m up to 1620 m.

Distribution. Known hitherto only from Sabah/Borneo. Holloway 1982: 207 mentioned on a pair without abdomen from the Malayan Peninsula, which may probably belong to *montana*, but currently there is no evidence for an occurence outside of Borneo.

***Ambadrana* gen. nov.**
Type species: *Turnaca barlowi* Schintlmeister, 1994.

Taxnomic note. The genus *Ambadra* consists of numerous species. A group of five closely related Sundanian species differ structural in its male genitalia. Already in the past some authors (e.g. Holloway, 1982, Holloway, 1983, Schintlmeister 1994) considered these differences and placing this group in *Turnaca*. But *Turnaca*, as showed before, is a distinct genus and is not closely related to this species group. The new genus *Ambadrana* errected below,

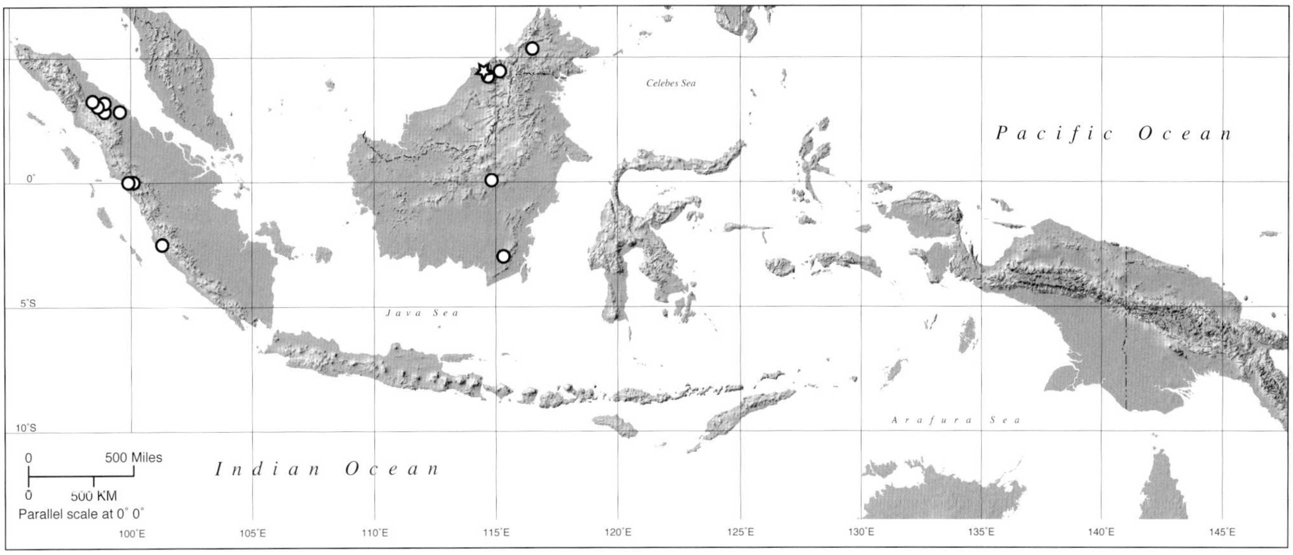

FIGURE 90 Distribution of *Ambadrana punctata*.

could also treat as a subgenus of *Ambadra*, because morphological differences are limited. But then consequentely many other taxa on generic level within the large subfamily Ceirinae, must reclassified as subgenera from consistence reasons.

Diagnosis. Small sized notodontids of fawn or yellowish color. Antennae of males bipectinated; the last 5-10% to the tip are naked. Female antennae, as far as females known, are filiform. External appearance are as in *Ambadra*. Characteristric pattern elements of forewing includes a fuscous longitudinal fascia along the radial vein and a blackish patch in the mdian area below this fascia. Submarginal fascia of the forewing marked as a row – of brown dots.

Male genitalia are different to it closest relatives, the genus *Ambadra*. They are much smaller, only of approximately half size of *Ambadra*. Uncus short and of triangular shape with broad base. Socii are robust, not pointed at the tip and of same length as uncus. Tegumen massive developed; saccus reduced. Valvae small, ovoid shaped and less modified. Phallus short and thick, slightly curved. Endophallus bears a field of cornuti. 8th sternite are contrary to *Ambadra* bilobed at posterior margin and bear often prominently developed lobes. Central lobes, which are characteristic for *Ambadra*, are absent. 8th tergit slightly bilobed at posterior margin. Female genitalia have relatively long apophyses. Antevaginal plate enlarged and displays prominently processes. Ductus bursae of medium size and strongly sclerotized. Corpus bursae disc-shaped with striolate structure on its surface. Signum very large and of circular shape.

Ambadrana comprises five Sundanian species, with highest diversity in Sumatra. *Turnaca fortunatorum* Schintlmeister, 1993, described from Palawan, must also placed in *Ambadrana*, **comb. nov.**

99. ***Ambadrana punctata*** (Holloway, 1983): 43; pl. 4: 1; fig. 40 (*Turnaca punctata*).
Holotype: ♂, Brunei, Labi, 30–60 m – NHMUK, London.
Schintlmeister 2013: 28.

Taxonomic note. Due to individual variability not all individuals of *A. punctata*, *barlowi* and *furcata* are separable with certainty according to their external appearance. Most examined specimens has been identified as *punctata*, which is seemingly the most common species in the complex, confirmed by 17 genitalia dissections.

Diagnosis. Black longitudinal transverse line from base to margin of forewing well developed. Most specimens display a submarginal fascia on forewings, marked as a row of brown dots. Some individuals show a second postmedial fascia. A few specimens from Sumatra have forewings brownish tinged. Individual forms with a larger blackish patch in the medial area (25 of 87 examined males = 29%) of forewing occuring quite often. All these forms were dissected by their male genitalia and their conspecifity with *punctata* was confirmed. Hindwings are light brownish. The only examined female resembles the male, but forewing broader and pattern lacks the black medial patch.

Male genitalia are readily to separate from its congeners by shape of the narrow and tapered uncus in its upper part. Socii are large and variable in its shape as

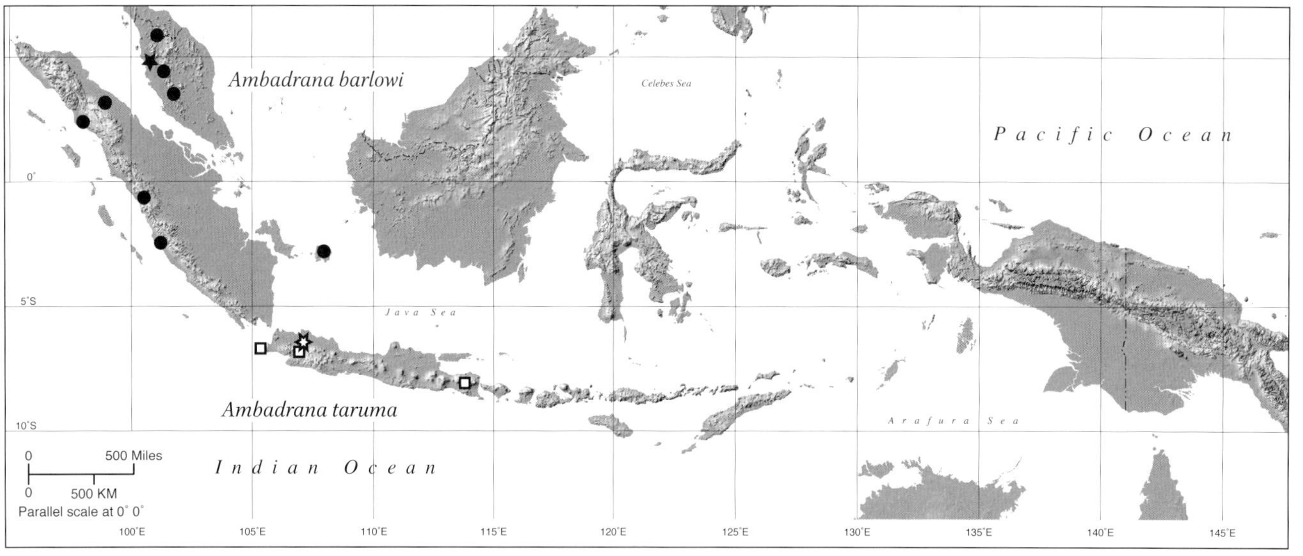

FIGURE 91 Distribution of *Ambadrana barlowi* and *taruma*

illustrated. Valvae vary individually in size and shape. Phallus stout with two areas of strongly sclerotizedzed microthorns at the everted endophallus. 8th sternite specifically shaped at the anterior margin with a central cleft between two lobes. 8th tergite slightly bilobed as in *barlowi*. Female genitalia have a large antevaginal plate, but without processes as seen in other congeners. Ductus bursae broad and strongly sclerotized; corpus bursae bears a strongly sclerotized small signum.

Bionomics. *A. punctata* occur commonly mostly at altitudes between 1000–1800 m. A few specimens were taken in the lowlands between 0–350 m.

Distribution. Known from Borneo and Sumatra.

100. ***Ambadrana barlowi*** (Schintlmeister, 1994): 220; pl. 3: 3, 6; GU-pl. 3: 4 (*Turnaca barlowi*).
Holotype: ♂, W Malaysia, Taiping – coll. A. Schintlmeister, Dresden.
Schintlmeister 2013: 27.

Diagnosis. Very similar to *A. punctata* in external appearance, but forewing shape slightly broader and black longitudinal band weaker developed. Postmedial fascia marked as a row of brownish dots. The inconspicous species is variable in size and also forewing color vary from light yellowish to yellowish-brown. Female resembles males, but forewings are broader and forewing pattern lacks the black medial patch on forewing.

Male genitalia are characterized by short and broad triangular shaped uncus with a rounded tip. Valvae are slightly larger in comparison to *punctata*. Phallus stout with a characteristically bump in the middle part. Everted endophallus bears two areas with microthorns. 8th sternite specifically triangular notched. A single male (fig. 98b) differ in uncus shape and notch of the 8th sternite. But these differences are expression of individual variation and does not indicate specifically differences. Female genitalia are similar to *S. taruma*. Antevaginal plate have a pair of flaps, from where large and bifurcate processes arise. Corpus bursae bears a disc-shaped signum with a V-shaped and strongly slecrotised band.

Bionomics. *A. barlowi* occur infrequently. Adults were collected mostly below 500 m; but one male was taken in Sumatra at 1200 m above sea level.

Distribution. Distributed in S Thailand, the Malayan Peninsula, Sumatra and Belitung Island.

101. ***Ambadrana taruma*** **spec. nov.** (figs. 101a–d).
Holotype: ♂, Indonesia, W Java, Cianur reg., Sukanegara, 7°06' S, 107°07' E, 1050 m, 18.–21. iii. 2007, leg. St. Jakl – in coll. A. Schintlmeister.

Paratypes (6 ♂♂, 1 ♀):
Java: 4 ♂♂, 1 ♀, Cianur reg., Sukanegara, 7°06' S, 107°07' E, 1050 m, 18.–21. iii. 2007, leg. St. Jakl (GU 84-52, GU 96-93); 1 ♂, Sempolan, Mt. Ranung, 8°10'S, 113°40'E, 410 m,

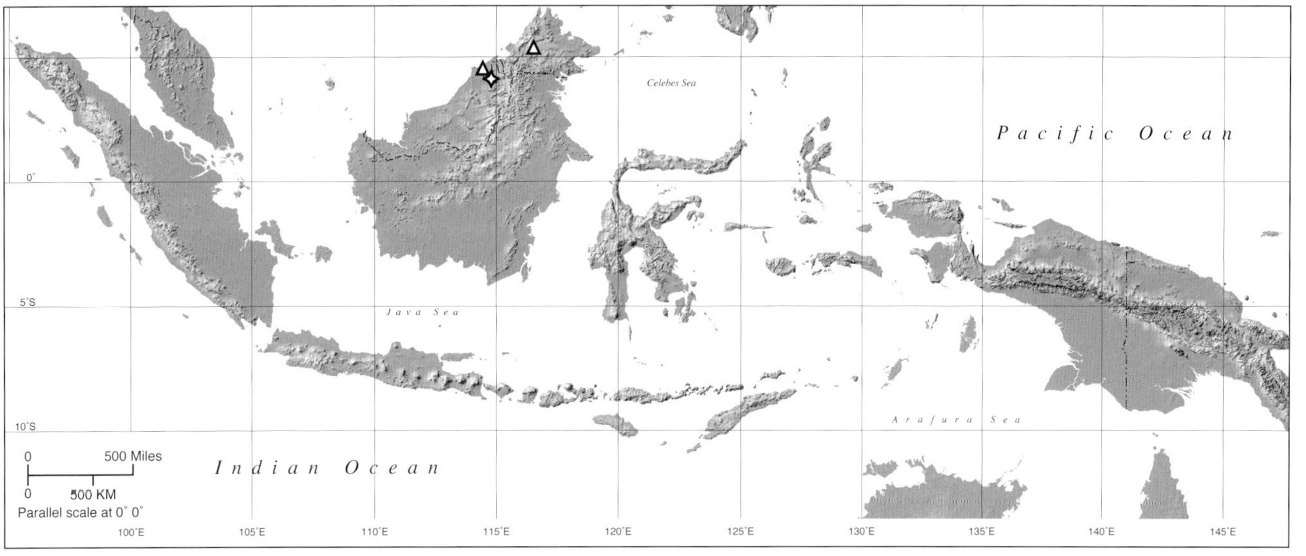

FIGURE 92 Distribution of *Ambadra spinifera*.

28. i. 1998, leg. Jakl, Schintlmeister & Cervenka; 1 ♂, Gedeh, Pangrango N.P., Gunung Kencana, 350 m, 6°37'S, 106°06'E, 15.–18. iii. 2007 (GU 93-47); 1 ♂, Ujung Kulon Nat. Park, 100 m, 6°50'S, 105°35'E leg. St. Jakl (GU 53-20).

Etymology. Named after the ancient Tarumanagara kingdom in W Java in fourth century.

Taxonomic note. This is the vicariant sister species of *A. barlowi.* It is treat here as a distinct species and not as a subspecies because of great differences in phallus and shapes of the postvaginal plates of the female genitalia.

Diagnosis. Forewing length ♂♂ 17–19 mm, a single male from Ujung Kulon spans only 15 mm and is 1–2 mm larger than other congeners. Forewing length ♀ 21 mm. Forewings fawn, slightly tinged brownish. Forewing pattern is as in *barlowi: a* prominent blackish patch in the medial area and a postmedial fascia of brown dots are present. *Ambadra batara,* an external similar species from Java, lacks this patch. Submarginal and postmedial fasciae marked as rows of fuscous brown dots. Along the dorsum of forewing there is a fuscous brown shadow. Hindwings reddish-brown, darker than in *barlowi.* A worn female resembles the male, but lacks the black patch in the medial area of forewing.

Male genitalia resembles *barlowi,* but uncus is longer and not triangular shaped. Uncus is narrower and uncus tip not pointed in comparison to *punctata.* Valvae ovale; sacculus more expanded toward the valva apex than in *barlowi.* Juxta specifically shaped and sclerotized, similar to *barlowi.* Phallus differ have a constriction in the middle part. Phallus tip displays a strongly sclerotized micro-thorns as seen in *A. punctata* or *barlowi.* Everted endophallus have a strongly sclerotized carina. 8th sternite is triangular notched at posterior margin. 8th tergite slightly bilobed at anterior margin and specifically sclerotized as illustrated. Female genitalia have both pairs of apophyses long and display an enlarged antevaginal plate. A pair of large and pointed processes arises from there, resembling to *S. barlowi.* But the latter species have clearly bifurcate processes, which is not the case in *taruma.* Ductus bursae strongly sclerotized. Corpus bursae disc-shaped with a large ellipsoid signum, divided by a straight and strong sclerotized streak. Corpus bursae and signum are much larger than in *barlowi.*

Bionomics. Adults occuring infrequently at elevations between 100 -1000 m.

Distribution. Endemic to Java.

102. ***Ambadrana spinifera*** (Holloway, 1983): 44; pl. 4: 3; fig. 41 (*Turnaca spinifera*).

Holotype: ♂, [Malaysia], Borneo, Sarawak, Gunong Mulu Nat. Park, Site 21, W Melinau Gorge, 130 m – NHMUK, London.

Schintlmeister 2013: 29.

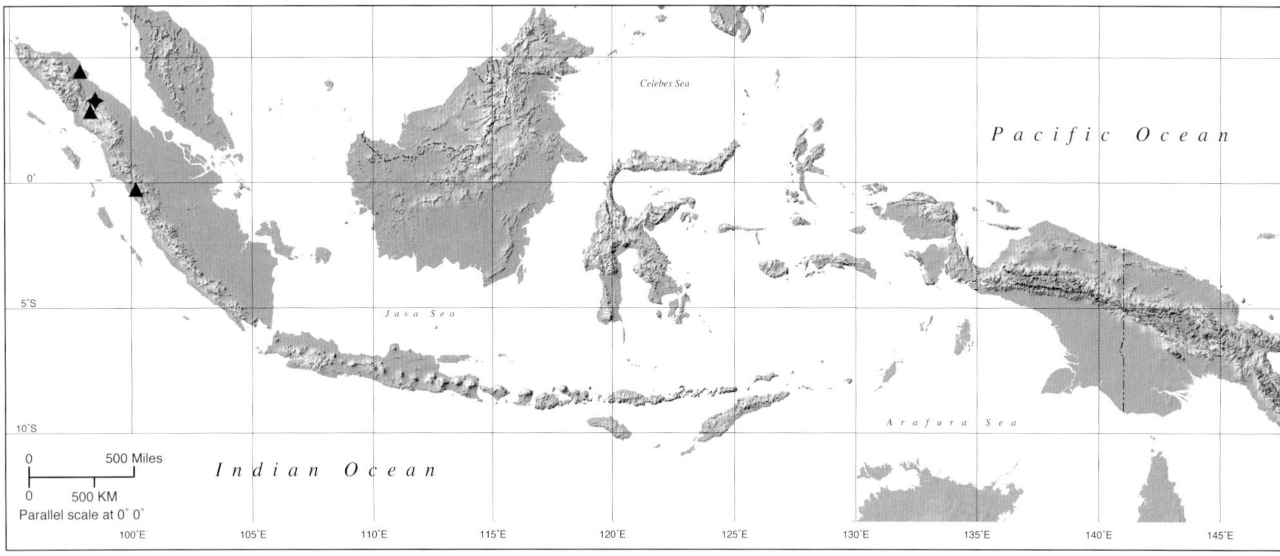

FIGURE 93 Distribution of *Ambadrana furcata*.

Diagnosis. Externally similar to *A. punctata* but smaller. Forewings display two prominent blackish longitudinal fasciae, which are also occasionally occuring in *punctata*, but less prominent. Dorsum of forewing is accompanied by a fuscous brown shadow. Hindwings are reddish-brown, darker than in *punctata*. Female is unknown.

Male genitalia are distinguished by axe-head shaped socii and the presence of a pair of gnathoi. Valvae are small; sacculus bears a large spine. Phallus rather straight and narrow. Everted endophallus bears four rows of microthornes. 8th sternite specifical shaped with a pair of large projections at posterior margin.

Bionomics. Known from three males, taken at altitudes of 20 m, 130 m and 1050 m.

Distribution. Endemic to Borneo.

103. ***Ambadrana furcata*** (Dierl, 1984): 19; figs. 3, 4 (*Turnaca furcata*).

Holotype: ♂, [Indonesia], Sumatra, Brastagi, 1.000 m – ZSM, Munich.

Schintlmeister 2013: 28.

Diagnosis. Externally very similar to *A. punctata*. The few known specimens, confirmed by genitalia dissection, are variable in forewing pattern and paleness and each specimen appear different. All specimens show a well developed brown dorsal shadow on forewing. Female is unknown.

Male genitalia unmistakable by a triangular notched, broad uncus. Socii robust; they bear a secondary smaller process. Phallus is narrow and slightly arched with a large spine at the tip. 8th sternite resembles somewhat to *spinifera* but the processes are much slenderer and curved.

Bionomics. Only five specimens of *A. furcata* are known, which has been taken in montane areas between 1000 m -1600 m.

Distribution. Endemic to Sumatra.

Pydnella Roepke, 1943: 93.

Type species: *Pydnella monticola* Roepke, 1943 [= *Pydna rosacea* Hampson, 1896].

Schintlmeister 2008: 62, Schintlmeister 2013: 368.

Diagnosis. Antennae of male bipectinated with short rami; female antennae filiform. Forewing narrow of rounded wingshape. Color of adults vary from brownish to pink. Female slightly sexual dimorphic with falcate forewings.

Male genitalia have a bifurcate long and slender uncus. Socii bifurcata with long processes. Valva have a small costal process at base; sacculus strong sclerotisized with a sharp and long spine. Margin of valva concave. Juxta prominent sclerotized with a pair of central lobes. Phallus axe shaped at base. Everted endophallus have a small and

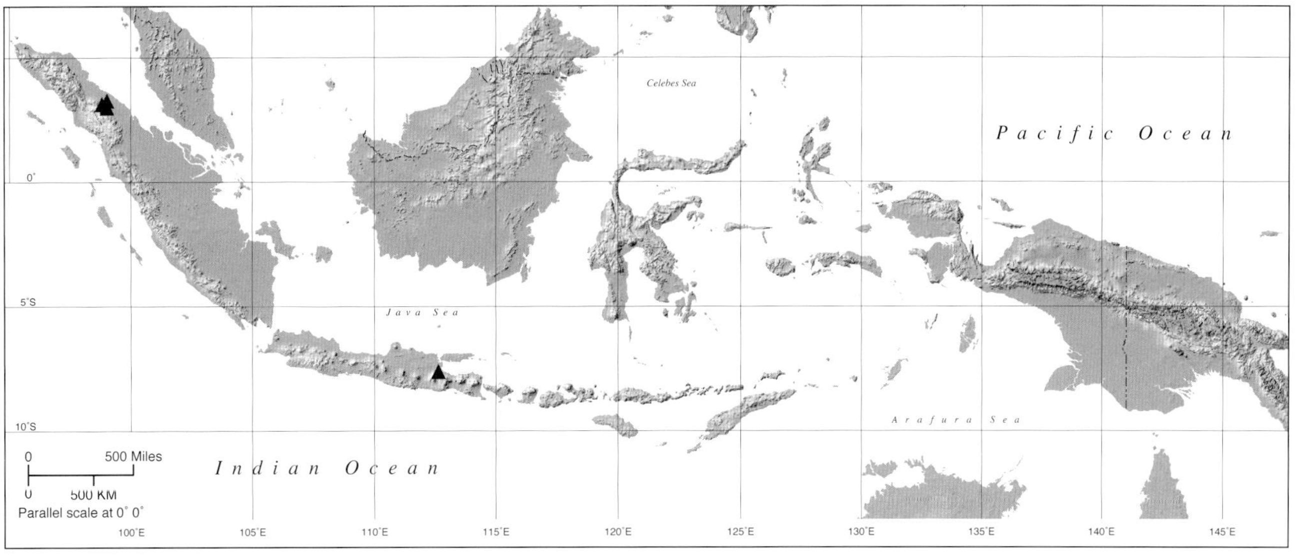

FIGURE 94 Distribution of *Pydnella rosacea*.

strong sclerotized carina. 8th sternite with a pair of central processes at posterior margin. Female genitalia have a large 8th abdominal segment. Antevaginal plate large and serrated at margin. Apophyses, massive but not very long. Ductus bursae short; corpus bursae circular with weakly sclerotized signum.

The genus is monotypic. For distribution see below.

104. ***Pydnella rosacea*** (Hampson, 1896): 458 (*Pydna rosacea*).

Syntypes: 2 ♂♂, [India], Khasis; [Myanmar], Pegu – NHMUK, London.

Schintlmeister 2008: 62, Schintlmeister 2013: 368.

Diagnosis. Males are readily to recognize by wingshape, pinkish brown colored forewings and a prominent black costal spot. Discal spot whitish but less prominent. Female sexual dimorphic lacks the black costalspot and have a falcate forewing apex.

Male and female genitalia are as described under genus diagnosis.

Bionomics. *Pydnella rosacea* occur in Sumatra and Java not rare but rather locally in montane forests at altitudes between 1000 m–1500 m.

Distribution. Distributed in NE India, Myanmar, SW China (Yunnan), Myanmar, Indochina, Sumatra and Java.

Niganda Moore, 1879: 63.

Type species: *Niganda strigifascia* Moore, 1879.

Schintlmeister 2008: 60, Schintlmeister 2013: 282.

Diagnosis. Medium sized to smaller moths with elongated and narrow forewing and long abdomen. Antennae of males bipectinated with very short rami; in females antennae are filiform. Forewing yellowish, olive or brownish. Most species display a contrasting paler costal strip. Black discal streak on forewing usually prominent marked, but in a few species of the *radialis* group weakly developed.

Male genitalia have a large and broad bifurcate uncus. Socii large; juxta prominent and specifically sclerotized. A diagnostic feature of *Niganda* are the paddle shaped sclerite arising laterally from the vinculum. Everted endophallus bear several larger cornuti. Female genitalia have a large and strong sclerotized 8th sternite. Ductus bursae broad; corpus bursae ovale. Most species (except *N. strigifascia*) display a prominent large signum.

Niganda is distributed in entire India, southern parts of China (Yunnan-Zheijang), Indochina, the Malayan Peninsula, Sumatra and Java. The genus is very diverse in species in Indochina. Currently 11 species are described, but several further species awaiting their descriptions in Indochina.

105. ***Niganda strigifascia*** Moore, 1879: 63.

Syntypes: 1 ♂, 1 ♀, [NE India], Darjiling [= Darjeeling] – ZMHU, Berlin.

Schintlmeister 2008: 61; Schintlmeister 2013: 282.

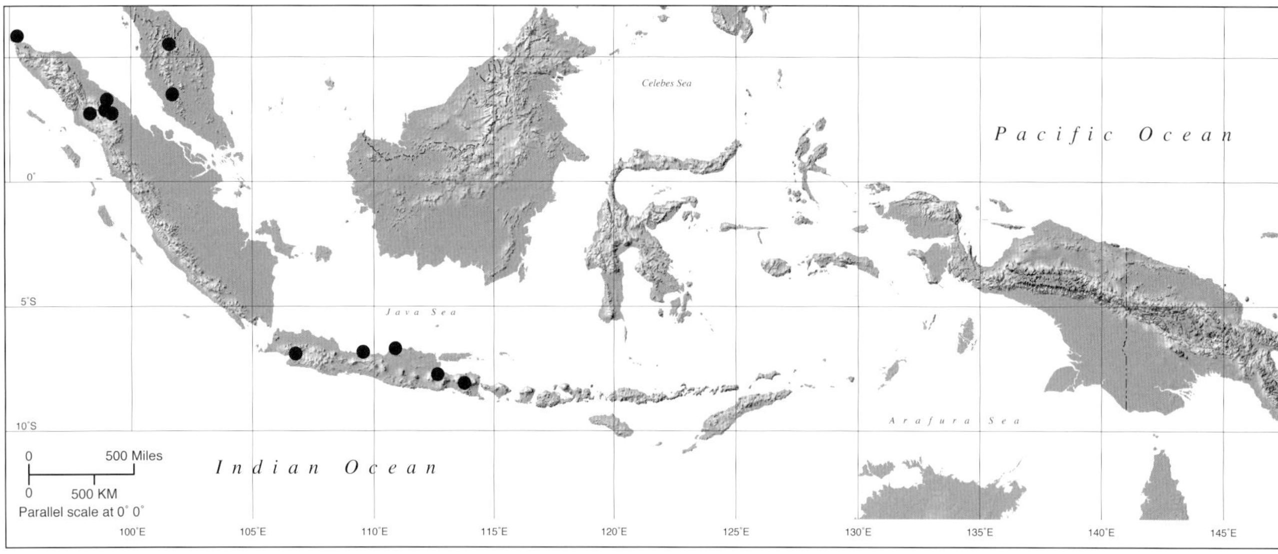

FIGURE 95 Distribution of *Niganda strigifascia.*

Taxonomic note. Kiriakoff described two subspecies of *N. strigifascia:*

Niganda strigifascia insularis Kiriakoff, 1960: 6; fig. 2; holotype: ♂, [Indonesia], Java est, Poenten Ardjoeno, 4.000‘ ZMHU, Berlin – and *Bireta* (*Niganda*) *strigifascia coelestis* Kiriakoff, 1962: 185; holotype: ♂, Chine, Chekiang [= Zhejiang], Mokanshan – ZFMK, Bonn.

Males and females of *strigifascia* vary much in the paleness of the wings and also in their size. But this variation is correlated with climatic influences rather than geographical subspecification as material from several years from one locality shows. Male genitalia (shape of uncus) are also subject of individual variation. From these reasons it is currently not possible to separate subspecies of *N. strigifascia* and both taxa mentioned above are placed here as junior synonyms of *N. strigifascia* (**syn. nov.**).

Diagnosis. *Niganda strigifascia* is the largest species in the genus. Males are externally to identify by characteristically shaped whitish discal spot. Forewing display a submarginal and marginal fasciae, marked as rows of black dots. Sexual dimorphic female lacks often the white discal spot. A prominent silver colored area at margin of forewing apex is a diagnostic character. Males and females vary individually in the paleness of the wings and also in their size.

Male genitalia vary in the shape of the bifurcate uncus as well as in the shape of the large Socii. A characteristic feature is a strong sclerotized and serrated bulge at the base of the valve. This projection arrising from vinculum is rather less large in comparison to other congeners. Phallus short and stout with serrated tip. 8th abdominal segments inconspicous, but 8th sternite bilobed with specifical sclerotization. Female genitalia have a large ductus bursae and large ovale corpus bursae. No signum.

Bionomics. Adults occuring in Sundaland commonly in urbanized lowlands as well as in montainous forests up to 1100 m; in Thailand up to 1700 m above sea level.

Distribution. *N. strigifascia* occur from northeastern India, Myanmar, China, Indochina, the Malayan Peninsula, Sumatra, including Pulau We/Aceh, and Java.

106. *Niganda pallens* spec. nov. (figs. 106a–e).

Holotype: ♂, Malaysia, West Pahang, Genting Tea Estate, 2000’ 3°22’N, 101°47’E, 20. iii. 1977, leg. H. Barlow (GU 16-12)– in coll. A. Schintlmeister.

Paratypes (7 ♂♂, 1 ♀):

Malaya: 1 ♀, West Pahang, Genting Tea Estate, 2000’ 3°22’N, 101°47’E, 20. iii. 1977, leg. H. Barlow; 1 ♂, dto., 12. iv. 1983; 1 ♂, dto., 14. i. 1978; 1 ♂, dto., 24. iv. 1978; 1 ♂, dto., 22. xii. 1983 (BM 1247); 1 ♂, Prov. Perak, 25 km E Gerik, 500 m, 13.–17. ix. 2001, leg. H. Deumer & E. Görgner; 1 ♂, Pahang, Teranum, 400 m, 12. viii. 2001, leg. W. Mey (ZMHU 08-2017); **Thailand:** 1 ♂, Chumphon Prov., 5 km S Phato, 155 m, 9°46’N, 98°47’E, 6. vii. 2008, leg. Langer, Löffler & Naumann.

Diagnosis. Forewing length ♂♂ 19–20 mm, a single male from Thailand spans only 16 mm. Forewing length of ♀ 23 mm. Forewings pale olive colored. Basal and postmedial

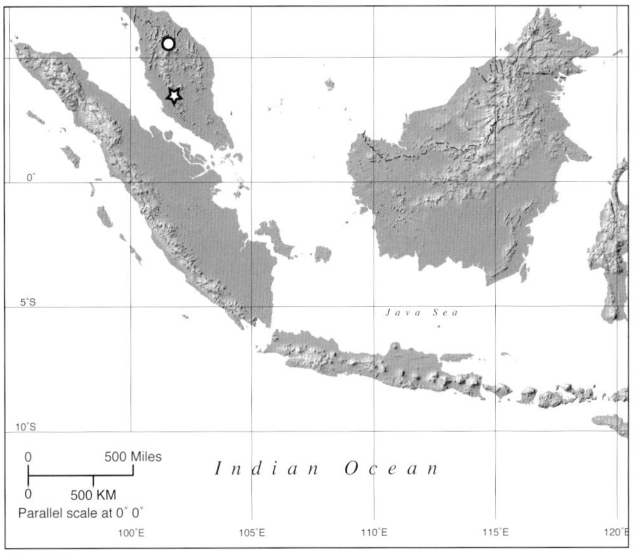

FIGURE 96 Distribution of *Niganda pallens*

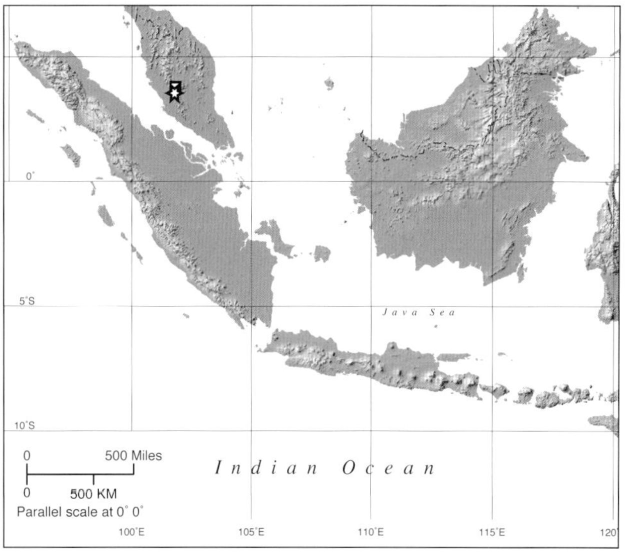

FIGURE 97 Distribution of *Niganda ancipitalis*.

fasciae developed as fine blackish lines. Postbasal fascia indistinct developed as a fuscous shadow. Discal streak weak and less visible. Fringe white-olive checkered. Hindwing pale brown, olive tinged. Sexual dimorphic Female has yellowish brown forewing. Marginal area blackish filled. Hindwing dark brown. *N. pallens* is the sister species of *Niganda cyttarosticta* (Hampson, 1895), which was described of Bhutan. It differ from the latter by shorter forewing length and absence of brown fields along the costa. Discal streak in *cyttarosticta* is promient large and contrasting black.

Male genitalia of *pallens* have a large and bifurcate uncus, where the left uncus lobe is smaller then the right one. *N. cyttarosticta* have uncus slightly notched. Socii are slender and straight, whereas *cyttarosticta* have much thicker Socii, which are knob-shaped at tip. Valva of *pallens* simple, sacculus bulbous enlarged. Tegumen have near saccus a pair of projections as characteristic within the genus. Phallus long and straight. Tip have a bulbous projection, which is much larger developed than in *cyttarosticta*. A dissected paratype of *pallens* (fig. 103b) have an additional long and pointed spine. 8th sternite concave rounded posterior margin with a pair of lobes at sides, which are absent in *cyttarosticta*. 8th tergite longer and narrower in *pallens* than in *cyttarosticta*.

Bionomics. Only a few specimens (n < 10) are hitherto known in collections, but the species occur not rare in Genting Highlands, West Malaysia. Imagines were taken from 150 m–600 m above sea level.

Distribution. Distributed on the Malayan Peninsula including south Thailand.

107. ***Niganda ancipitalis*** Schintlmeister, 2007: 73; pl. 10; fig. 57.
Holotype: ♂, W Malaysia, Pahang, Genting tea Estate, 600 m – coll. A. Schintlmeister, Dresden.
Schintlmeister 2013: 282.

Taxonomic note. The *Niganda radialis* (Gaede, 1930) species group is a taxonomic difficult complex, which is diverse in Indochina. In West Malaysia seemingly only one species occur.

Diagnosis. Most members of this species group are very similar by external appearance: forewing shape rectangular; costal area contrasting cream; Postmedial fascia blackish. Separable from other similar species of *radialis* group by presence of a submarginal row of black dots. *N. ancipitalis* vary individually in forewing length. Female is sexualdimorphic and resembles to other females of the *radialis*-group, but appear paler in external appearance. A characteristic feature is as in males presence of a postmedial fascia marked as a row of fine blackish dots. Other blackish forewing pattern is reduced.

Male genitalia are readily to recognize by shape of the bifurcated uncus tip and shape of the ellipsoid juxta. Paddle shaped sclerite arising from vinculum appear larger than in other congeners. Phallus slender, everted endophallus bear four prominent spines. Female genitalia have not yet been dissected.

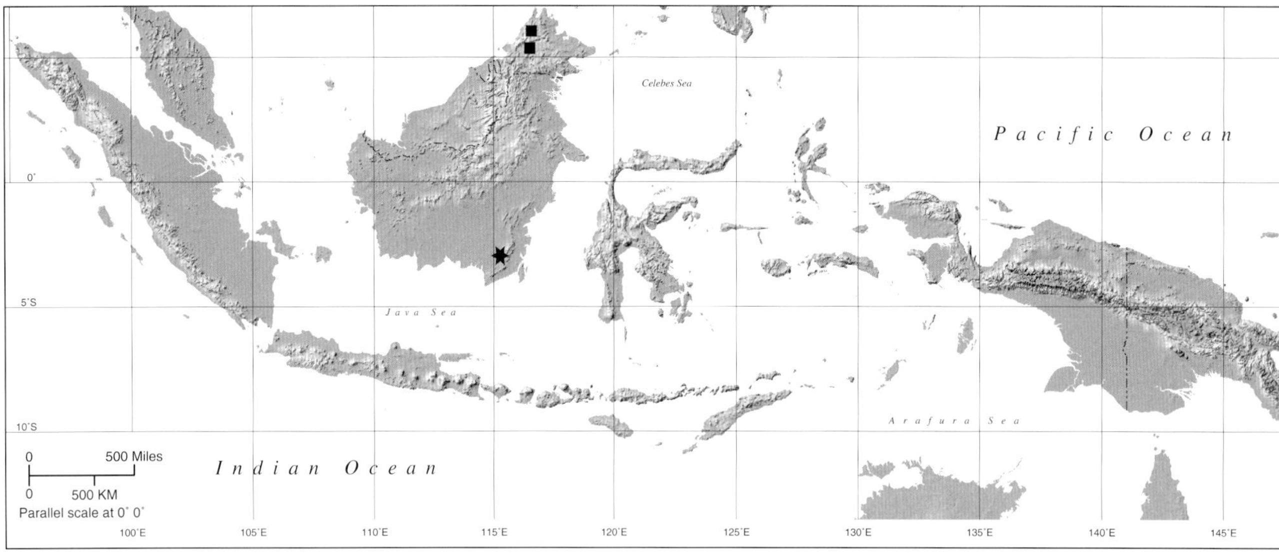

FIGURE 98 Distribution of *Postniganda pudens*.

Bionomics. The type series of *N. ancipitalis* (n = 4) was collected by the author during one night in a secondary forest at 600 m; a further old male was taken in Tras, near the Fraser's Hill.

Distribution. Known only from the Malayan Peninsula.

Postniganda **gen. nov.**

Type species: *Niganda pudens* Schintlmeister, 2007.

Diagnosis. Antennae in both sexes bipectinated with very short rami. Color of all wings and the body brown to blackish brown. Forewing shape resembles *Niganda:* narrow, elongated and rectangular shaped. Pattern consist from a basal, medial fascia and submarginal fasciae, marked as rows of deep black dots. Submarginal fascia is accompanied in apical area by two contrasting white dots, which is a diagnostic character for identifying the species. Discal spot marked as two deep black streaks, which are also in more fuscous specimens visible. Sexual dimorphism is limited.

Male genitalia are structural very different from *Niganda*, where the species was placed originally. Uncus deeply triangular notched. Both uncus lobes are triangular. Socii are long, slender and curved with a pointed tip. Socii are longer than uncus. Valvae are elongate and bear at apex a large tapered process. Phallus short, arched with a hook at the tip. 8th abdominal segments are bilobed on both margins. Posterior margin are in both segments stronger sclerotized. Female genitalia have robust and medium sized apophyses posteriores and shorter apophyses anteriores. 8th abdominal segment. Ductus bursae large of circular shape long and slender. Corpus bursae large and of circular shape. Signum medium sized and reniform, shaped; strong sclerotized.

The genus is monotypic and known currently only from Borneo and Thailand.

108. ***Postniganda pudens*** Schintlmeister, 2007: 70; pl. 10; fig. 51, 52.

Holotype: ♂, Indonesia, Kalimantan Selatan, 30 km E Kandagan, 15 km NE of Loksado, 2°52' S, 115° 38' E, 800 m – coll. A. Schintlmeister, Dresden.

Schintlmeister 2013: 283.

Diagnosis. See above under the genus diagnosis.

Bionomics. The species is rare and is known from a few specimens (n < 10) from 350m–1150 m altitude.

Distribution. Reported from Borneo and Thailand.

Norraca Moore, 1881: 340.

Type species: *Norraca longipennis* Moore, 1881.

Schintlmeister 2013: 283.

Diagnosis. Medium sized adults of orange or yellowish-brown color. Antennae of male bipectinated with short rami; those of femalea are filiform. Forewing shape elongated, narrow and rounded at apex. Discal streak of males

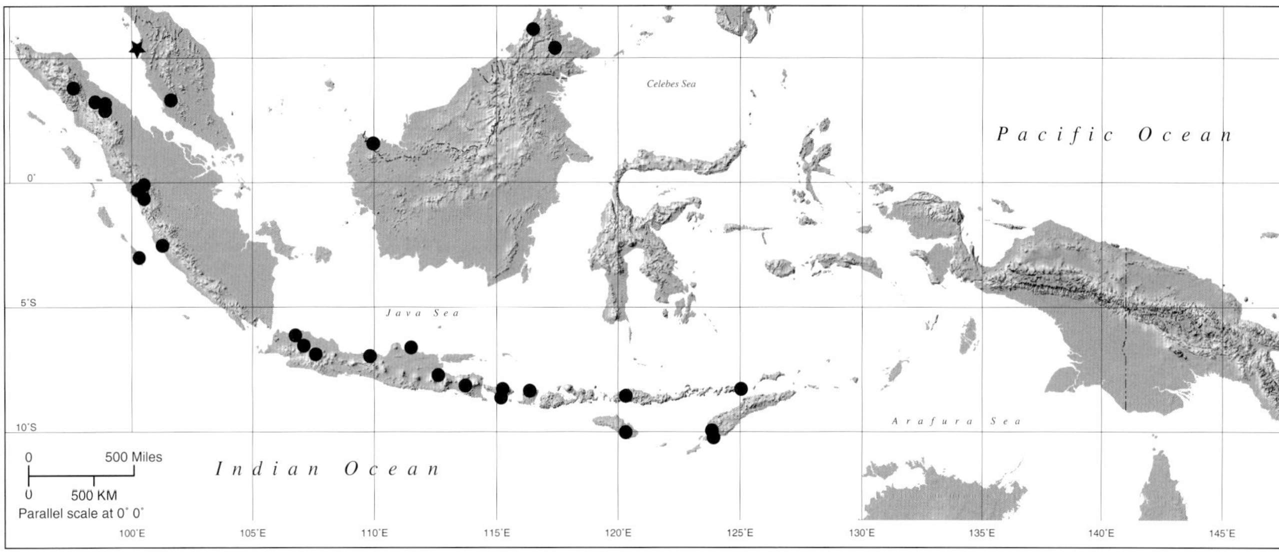

FIGURE 99 Distribution of *Norraca longipennis*.

consists from three white portions, which is a characteristic genus feature. Postmedial and submarginal fasciae indistinct but present. Hindwing of same color as forewing. Sexualdimorphic females lacks the discal markings and forewing is falcate at apex.

Male genitalia have a large, triangular shaped uncus; uncus tip knob or fanshaped. A pair of gnathoi appear small and inconspicous. Socii large with several edges and processes. Valva elongate displaying a costal projection near base. A triangular shaped sclerite arising laterally from the vinculum as in *Niganda*. Phallus long with bifurcate tip. Everted endophallus bear a stronger sclerotized carina and occasionally several spines. 8th sternite have a pair of sidewards directed projections. Central part, in particular the less sclerotized "window" resembles *Saliocleta*. 8th tergite bilobed at posterior margin and and characteristically sclerotized. Female genitalia have a large 8th abdominal segment. Apophyses posteriores long; apophyses anteriores very short. Appendix bursae large and of circular shape. Ductus bursae rather massive and short; corpus bursae large and circular without signum.

Norraca comprise three species. Distributed in entire India, China, Indochina, Borneo, the Philippines and from Malaya, Sumatra, Java up to Timor.

109. ***Norraca longipennis*** Moore, 1881: 340.

Holotype: ♀, [W Malaysia], Penang – NHMUK, London.

Schintlmeister 2008: 88; Schintlmeister 2013: 284.

Diagnosis. Best to recognize by the cream discal markings of male and the acute forewing shape. Forewing and hindwing color vary from orange brown to light yellowish brown and also forewing length is subject of individual variation. In some males there is no pattern on the forewings visible. It seems, that adults collected above 1000 m altitude are larger and darker than specimens from the lowlands. Female is sexual dimrophic. It lacks the discal markings and have a falcate apex of forewing. The acute and convexe shaped hindwings are rather unique in the Notodontidae.

Male genitalia are best to recognize by the large triangular shaped uncus with fan shaped tip and the lobe at the costa of the valves. Phallus have a process, which is bifurcate at the tip. Endophallus bear several asterisk-shaped cornuti. There is some variation in shape of uncus tip, costal valva lobe and phallus shape, but no geographical correlation was found. Female genitalia are best to identify by the prominent appendix bursae. Ductus bursae rather massive and short; corpus bursae large and circular. No signum.

Bionomics. Adults occur more commonly in urbanized habitats or in secondary shrub vegetation in lowlands up to 300 m. But moth flies also in mountainous rainforests up to 1700 m above sea level.

Distribution. Widely distributed: entire India, Sri Lanka, Andaman Islands, China including Hainan, Indochina, Penang and Borneo Not known yet from the Malayan Peninsula. In Indonesia: Sumatra, Mentawei Isls., Java, Bali, Lombok, Flores, Sumba, Alor, Timor.

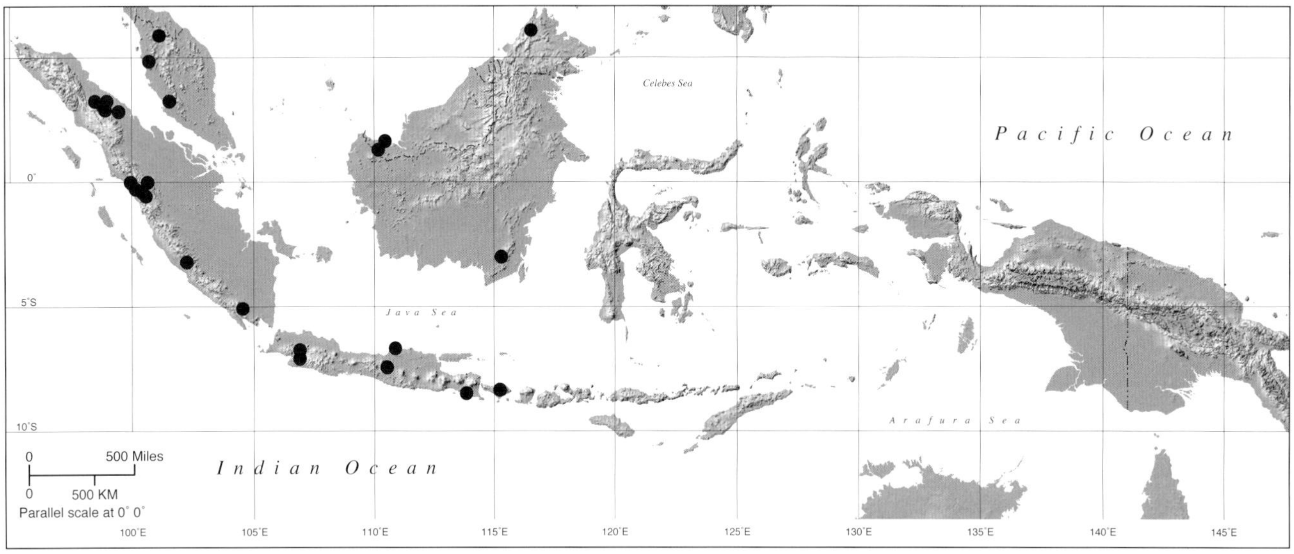

FIGURE 100 Distribution of *Saliocleta lativitta.*

Saliocleta Walker, 1862: 124.

Type species: *Saliocleta nonagrioides* Walker, 1862.

Schintlmeister 2013: 386.

Diagnosis. Medium sized moth of yellow brownish color of wings and body. Antennae of males are bipectinated with short rami; female antennae are filiform. Forewing are rounded at apex, tornus is in most species angled. Forewing pattern indistinct and consists of a transversal fascia from base to apex and a fuscous spot below the whitish discal spot. Hindwings vary from fuscous brown to light yellowish. Females are sexualdimorphic having broader forewings with pointed apex. Hindwings of females are in most species paler than in corresponding males.

Male genitalia are homogenous within the genus and differences between the species are often subtle. Characteristic features in male genitalia are a robust, but not very long uncus with rounded tip. In subgenus *Norracana* Kiriakoff, 1962 uncus appear bifurcate. Socii are rather broad, sometimes armed with further processes. Juxta prominent and specifically sclerotized and of taxonomic value. From sacculus of valva a brush of long setae arise Phallus usually straight and slender. Everted endophallus bear cornuti. The 8th abdominal segments resembles somewhat to *Norraca.* The 8th sternite is rounded at its anterior margin and display a pair of subapical processes. The middle part toward the marigin is weakly sclerotized and appear like a window, which is smaller than in *Norraca*. The female genitalia have a large 8th abdominal segment, and a long ductus bursae. the corpus bursae have a signum.

The genus is currently comprized of 47 plus some ten undescribed species, which are grouped into six subgenera. Three of them (*Armiana* Walker, 1862, *Biraia* Kiriakoff, 1962 and *Saliocleta*) occuring in Indonesia. The center of diversity is Indochina and China. The genus occur in entire India, China, Indochina, Borneo, the Philippines and Indonesia up to Sulawesi.

Subgenus *Armiana* Walker, 1862: 142.

Type species: *Armiana* Walker, 1862

Characterized by medial tooth like tuft of scales ("prominent-tooth") on forewing a large uncus, elongated valvae and a long phallus with long spines on everted endophallus.

110. *Saliocleta lativitta Walker,* 1862: 142 (*Armiana lativitta*).

Holotype: ♂, [Malaysia], Borneo, Sarawak – Oxford University Museum of Natural History.

Schintlmeister 2013: 387.

Diagnosis. Rather unmistakable by a prominent-tooth on forewing. Discal spot on forewing marked as silver streak. Postbasal, medial and submarginal fasciae whitish. Medial and postmedial area in most cases brown filled, but sometimes yellow colored. Hindwing chocolate brown.

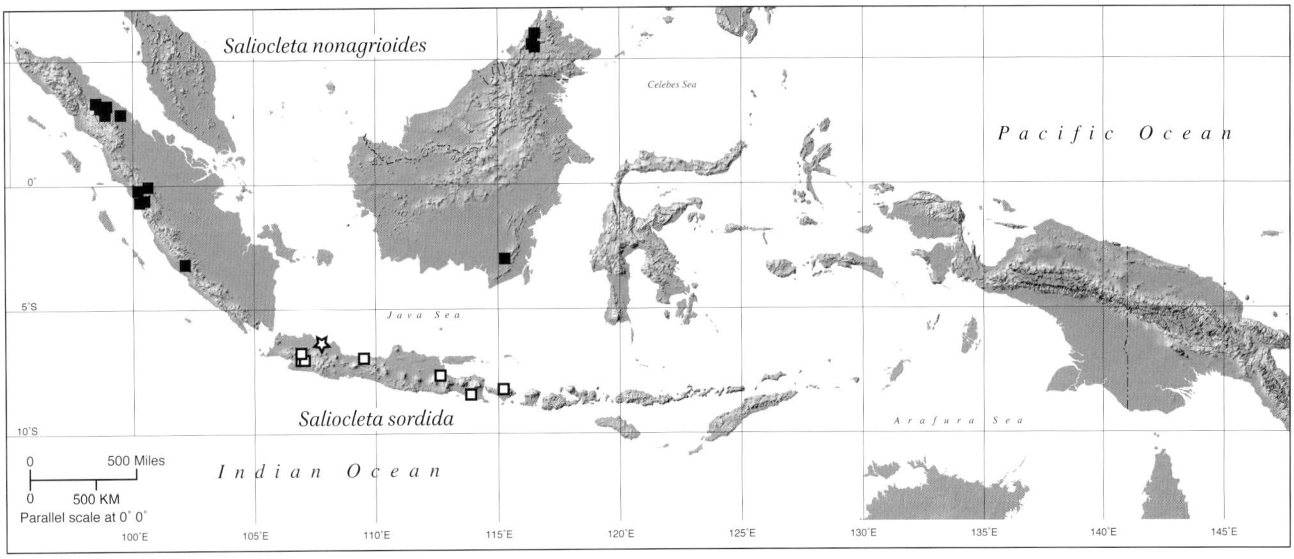

FIGURE 101 Distribution of *Saliocleta nonagrioides* and *sordida*.

Range of individual variability limited. Sexual dimorphism is limited; females are larger and have broader shaped forewings.

Male genitalia have a broad and large uncus and a pair of relatively small Socii. Juxta semicircular sclerotized. Valva long with rounded apex. Phallus, straight, narrow and long. Everted endophallus bear three diagnostically and strong slerotized large needle like spines. 8th abdominal segments long and diagnostically sclerotized. 8th sternite convex at posterior margin, but concave notche at anterior margin. 8th tergite notched at posterior margin. Female genitalia have long apo-physes posteriores an short but robust apophyses anteriores. Ductus bursae strong sclerotized and long. Corpus bursae circular shaped; it bear a reniform signum.

Bionomics. The species occur commonly in primary and secondary forests, less common in urbanized areas. Most specimens were taken between 1000–1300 m above sea level, but in Thailand the species is seemingly restricted to lower altitudes below 500 m.

Distribution. Distributed in Thailand, the Malayan Peninsula, Borneo, Sumatra, Java and Bali.

Subgenus *Saliocleta* Walker, 1862

111. ***Saliocleta nonagrioides*** Walker, 1862: 124.
Holotype: ♂, [Malaysia], Borneo, Sarawak – Oxford University Museum of Natural History.
Schintlmeister 2013: 390.

Taxonomic note. A variable species; it is currently not clear if *nonagrioides* consists from a complex of species. There is no evidence, that geographical subspecies occuring. Holotype of *nonagrioides* would match according to hindwing pattern figs. 109m–o with corresponding male genitalia fig. 109f. On other hand majority (> 90%) of examined specimens have uniform chocolate brown hindwing and fig. 111k would match exact by external apperance the holotype of *Bireta* (*Norraca*) *sabulosa* Kiriakoff, 1962: 199; fig. 53 (Holotype: ♂, Sumatra occid., Lebong Tandai – NHMUK, London). Male genitalia of *sabulosa*-holotype is illustrated in fig. 111c. The latter taxon was placed in synonymy to *nonagrioides* by Schintlmeister & Lourens, 2010: 105.

Diagnosis. Best to separate from its other congeners in Sundaland by its fuscous chocolate-brown hindwings and rounded, not angled tornus of forewing. The elongated forewings vary from yellow, orange to yellowish brown. Forewings display a larger but indistinct fuscous brown apical streak. Postmedial and submarginal fasciae are marked as rows of fine black dots. Occasionally a basal and a postmedial fascia, also consisting as rows of black dots occure. In a few cases males display a blackish blotch in the medial area of the forewings, but many specimens display this blotch weakly developed. Hindwing vary from uniform fuscous chocolate brown to lighter orange brown with contrasting fuscous chocolate brown anal area. Female is sexual dimorphic having apex of the forewing pointed and hindwing lighter.

Male genitalia are recognizeable by shape of the bulbous and pointed tip of phallus and shape of the ellipsoid

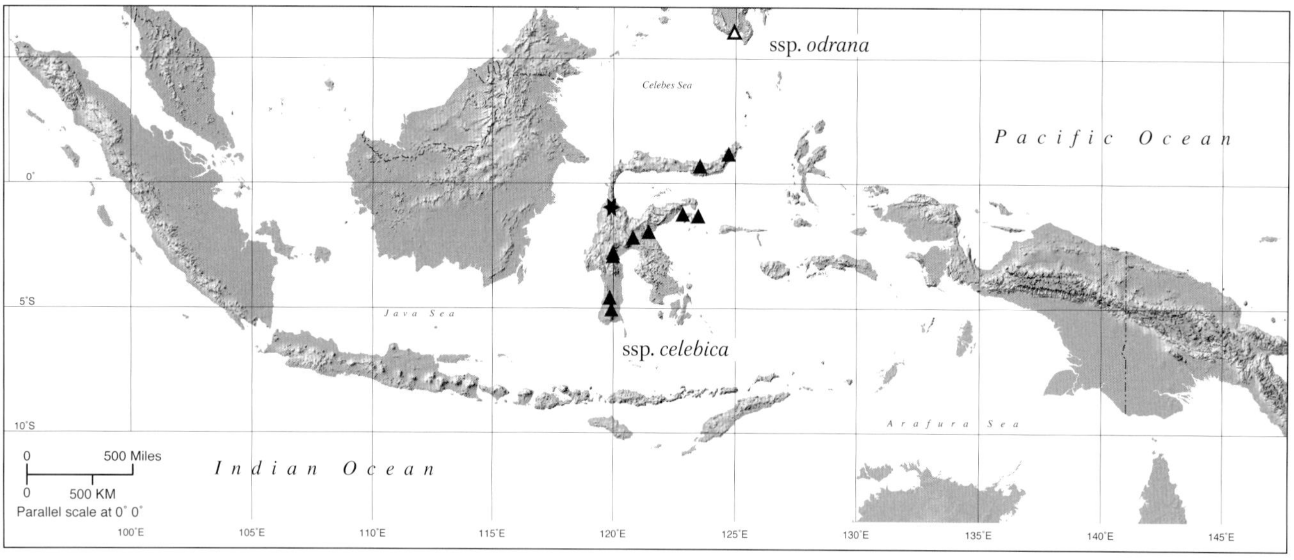

FIGURE 102 Distribution of *Saliocleta odrana.*

"window" of the 8th sternite with two pairs of small projections. Shape of uncus is variable. There are broad, narrow and sometimes also bilobed shapes of uncus tip. Also socii vary in length and thickness. Juxta strong sclerotized and juxta process vary in shape somewhat, but shape is diagnostic. 8th tergite notched at posterior margin. Lobes of display characteristically triangular sclerotization. Female genitalia have long apophyses posteriores with broad basal part. Ductus bursae short. The corpus bursae display a diagnostic signum, consisting from two separated reniform shaped parts.

Bionomics. Adults occuring commonly and were observed between 0–1790 m.

Distribution. Distributed in south Thailand, Sumatra and Borneo.

112. ***Saliocleta sordida*** (Roepke, 1943): 103 (*Turnaca sordida*).
Holotype: ♂, [Indonesia], Java, Tjibodas – NBC, Leiden.
Schintlmeister 2013: 391.

Diagnosis. This is the sister species of *nonagrioides* and similar by external appearance. Separable by shape of angled and not rounded tornus. Forewing shape variable in length; and most specimens are smaller than *nonagrioides.* Forewing pattern and individual variability are as in *nonagrioides*, but hitherto males with contrasting blackish medial spot not known yet. Hindwings are paler brown colored than in *nonagrioides.*

The male genitalia resembles *nonagrioides*, in shape of uncus and Socii. Juxta smaller and not waisted. Phallus due to range of variability not separable from *nonagrioides.* Best feature to separate *sordida* are 8th abdominal segments. 8th sternite have in the "window" only one pair of small projections and lobes of the notched 8th tergite lacks a triangular sclerotization. Female genitalia have not yet been dissected.

Bionomics. Most known specimens of *sordida* in collections are > 80 years old and only a very few males were collected recently. They were taken in the lowlands up to 600 m, but in Bali also at 1.500 m.

Distribution. Restricted to Java and Bali.

113. ***Saliocleta odrana*** (Schaus, 1928): 86 (*Pydna odrana*).
Lectotype: ♂, Philippine Islands, Surigao –USNM, Washington DC.
Schintlmeister 2013: 390.

Taxonomic note. *Saliocleta odrana* was described of Mindanao. In Sulawesi the ssp. *celebica* (Kiriakoff, 1970): 109; fig. 8, holotype: ♂, [Indonesia, Sulawesi], W Celebes, Mt. Tompoe, Paloe, 2.700 ft. – NHMUK, London. occur. Due to wide width variation, separation of ssp. *celebica* from ssp. *odrana* by external appearance is difficult. 30% of the examined specimens of ssp. *odrana* display a blackish medial blotch prominently developed, while in ssp. *celebica* only one of 45 males have this blotch present. Phallus of ssp. *celebica* differ by the swollen shape of the upper

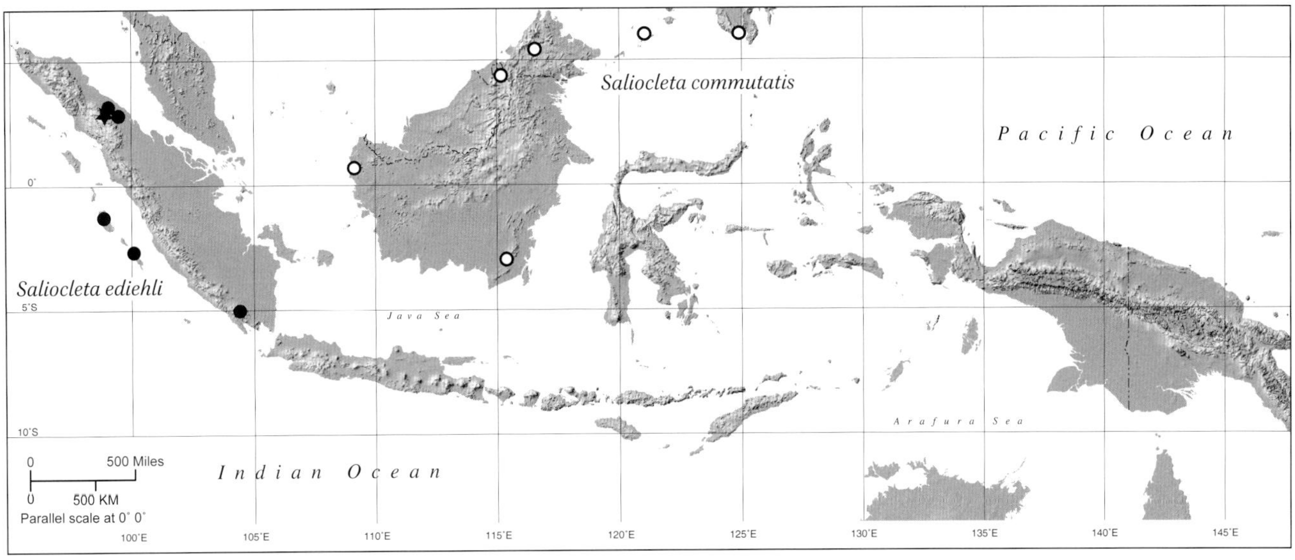

FIGURE 103 Distribution of *Saliocleta ediehli* and *commutatis*.

process from Mindanao populations. Different shapes of phalli from various Philippine islands suggests however, that in the Philippines several subspecies occur.

Diagnosis. *Saliocleta odrana* is extrem variable in size, forewing shape and forewing pattern, as the selection of figured adults show. The fuscous apical streak on forewing extends in most of examined specimens (n > 80) until the base in both sexes. Tornus of forewing is not angled but rounded. Hindwings vary from pale yellowish to yellowish brown. They are paler than in *nonagrioides*. Sexual dimorphism is limited. The few examined females have a more pointed apex of the forewings and paler hindwings, but width variation is seemingly like in males.

Male genitalia are very similar to *nonagrioides*. They differ from *nonagrioides* by a smaller juxta process. Phallus is shorter and thicker than in *sordida* and *nonagrioides*. The pair of processes at phallus tip and in particular the upper swollen arm separates *odrana celebica* from *sordida* and *nonagrioides*. 8th sternite very similar to *nonagrioides*, but the second pair of central projections reduced in size. 8th tergite notched at posterior margin. Female genitalia show a resemblance to *nonagrioides*, but base of apophyses posteriores less broad and lamella antevaginalis smaller. Ductus bursae short, corpus bursae circular and large. Signum appear undivided.

Bionomics. *S. odrana celebica* fly commonly from sea level up to 1400 m, also in urbanized areas.

Distribution. *Saliocleta odrana celebica* occur in Sulawesi and Peleng. The ssp. *odrana* (and probably further subspecies) inhabit almost all Philippine Islands including Palawan.

114. ***Saliocleta ediehli*** Schintlmeister, 1994: 223; pl. 3: 16; GU-pl. 4: 2.

Holotype: ♂, Indonesia, North Sumatra, Prapat HW 2, 1.300 m – coll. A. Schintlmeister, Dresden.

Schintlmeister 2013: 389.

Diagnosis. *Saliocleta ediehli* is the sister species of its eastern vicariant *S. commutatis*. Separable from *comutatis* by less reddish tinged wings, shorter forwings. Tornus is rounded, not strong angled. Holotype of *S. ediehli* represent an unusual fuscous brown colored male (fig. 114c). Further material currently collected appear yellowish and not tinged brown. A fuscous blotch in the medial area is weakly developed, but visible. Submarginal and marginal fasciae marked as rows of fine dots. Postmedial fascia often indistinct as row of blackish dots. Hindwing chcolate brown or yellowish brown. Female is unknown.

Male genitalia have long and tapered uncus; tip fine pointed. Socii are angeled and large. They differ from the similar genitalia of *commutatis* by the thicker uncus and shape of its tip. The 8th sternite display a smaller "window" as *commutatis*.

Bionomics. Adults are rare (n = 6 males are known) and prefer seemingly lower altitudes up to 300 m; holotype specimen was collected however at 1300 m altitude.

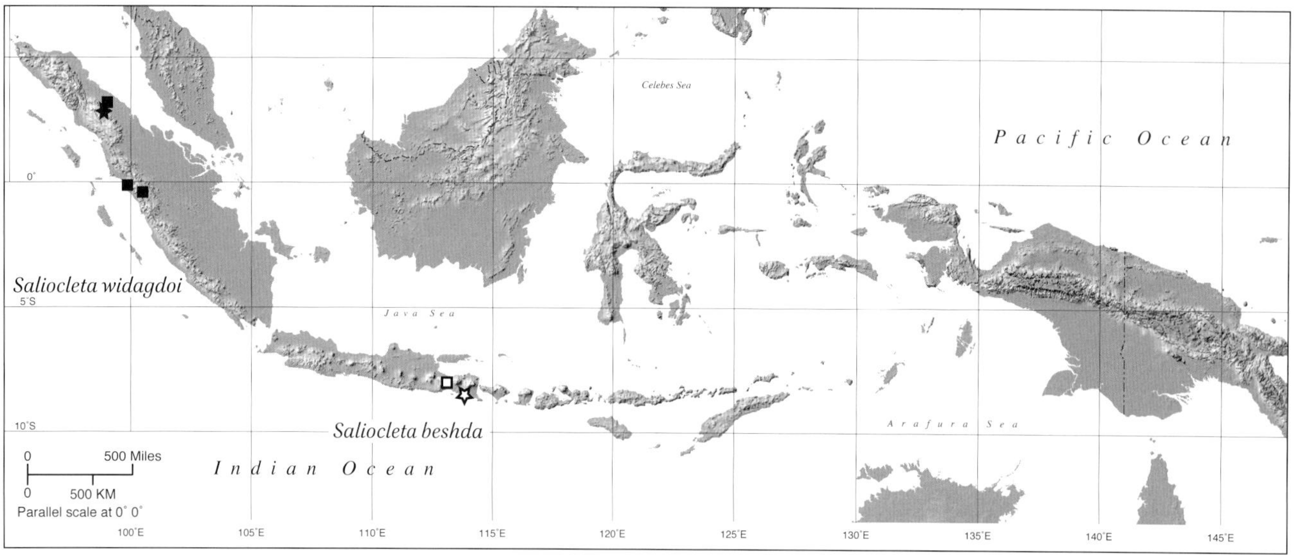

FIGURE 104 Distribution of *Saliocleta widagdoi* and *beshda*.

Distribution. Distributed in Sumatra and adjecent islands of Siberut and Mentawei.

115. ***Saliocleta commutatis*** Schintlmeister & Lourens, 2010: 116; figs. 194, 195, 196.
Holotype: ♂, Philippinen, Palawan, S. Vicente, 20 km NEE Roxas, 10°21'N, 119°10'E, 400 m – NHM, Wien.
Schintlmeister 2013: 389.

Diagnosis. Very similar to *S. ediehli*, but separable by more reddish tinged wings. Forewing length larger tornus strong angled and pronounced. Black forewing pattern sharp. A transverse band runs from base to apex. Postmedial, submarginal and marginal fasciae marked as rows of fine black dots. A fuscous blotch in the medial area is weakly developed, but visible. Hindwings pale reddish brown. A male from Tawi-Tawi Island (Philippines) is very small, but identity was confirmed by genitalia dissection. Sexual dimorphism is limited; female have broader forewing with pointed apex and more rounded than pointed tornus. Hindwings are pale yellowish slightly tinged reddish brown hindwings.

Male genitalia are separable from *ediehli* by the narrower and clearly pointed uncus and broader shape of valvae. Phallus is shorter than in *ediehli.* 8th sternite appear broader than *ediehli* and window is larger. Female genitalia are characterized by relatively reduced apophyses posteriores and massive but short apophyses anteriores. Ductus bursae long and well sclerotized. Corpus bursae circular; it display a striolate surface structure on lower half. Signum small and reniform shaped.

Bionomics. *S. commutatis* occur rare and appear in the Philippines between 50–950 m. From Borneo are only n = 5 specimens known, taken at 300 m and 800 m.

Distribution. Distributed in south Thailand, Borneo, Palawan, Tawi Tawi Isls. and Mindanao.

116. ***Saliocleta widagdoi*** Schintlmeister 1994: 222; pl. 3: 13; GU-pl. 4: 1.
Holotype: ♂, Indonesia, North Sumatra, Prapat HW 3 [1.500 m] – coll. A. Schintlmeister, Dresden.
Schintlmeister 2013: 391.

Taxonomic note. As already argued in Schintlmeister & Lourens, 2010: 114, *S. widagdoi* represent a complex of several similar, mostly undescribed species, which includes also the Taiwanese *S. virgata* (Wileman, 1914).

Diagnosis. *Saliocleta widagdoi* is after *Saliocleta dejoannisi* Schintlmeister, 1997, described from Vietnam, the largest species in the *widagdoi* group. Well characterized by broad forewing shape with a strong angled tornus. Postmedial fascia of forewings marked as a row of fine black dots which show in the middle between costa and dorsum a prominently enlarged and contrasting black dot. Submarginal fascia and marginal fascia appear as a double row of fine black dots. Hindwings are fuscous chocolate brown. Presumable female, which was found in the Malayan Peninsula, is much larger than male and display light yellowish brown hindwings. It resembles the female of *dejoannisi.* The latter species is display however whitish hindwings and not tinged brownish.

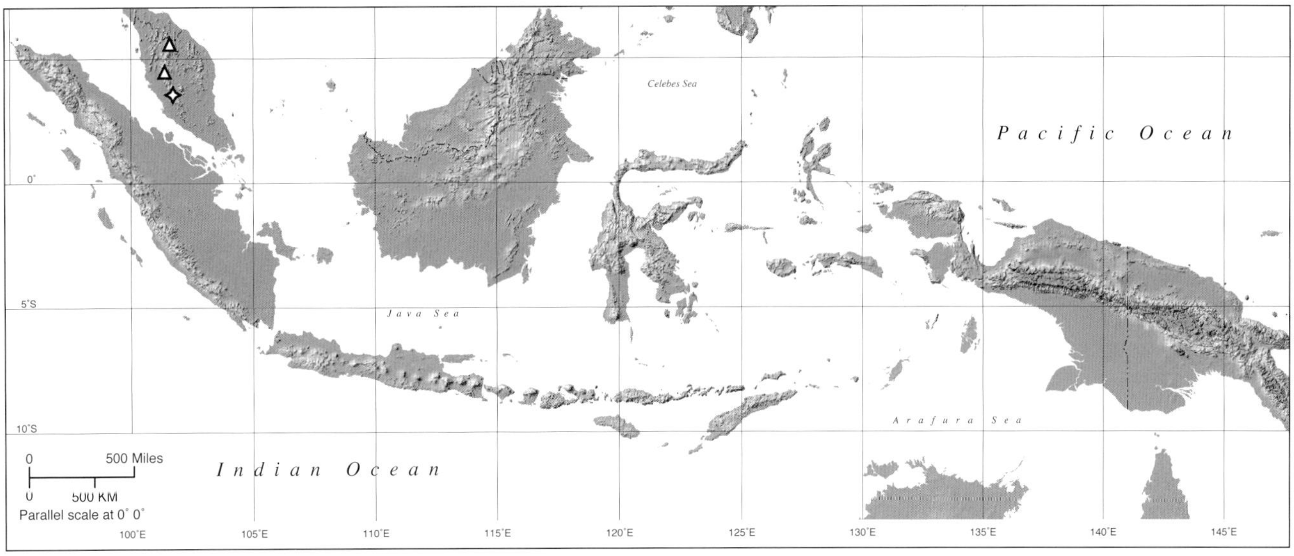

FIGURE 105 Distribution of *Saliocleta malayana*.

Male genitalia are characterized by the shape of the broad uncus and unique curved shape of socii with falcate tip. Socii have no straight margins at all three sides but convex curved. The phallus is relatively long and zigzag shaped. The 8th sternite differs in its shape and in particularly by the shape of the smaller window from the next species. Female genitalia have not yet been dissected.

Bionomics. A rare species (n = 10 specimens) that occur between 180–1300 m.

Distribution. *S. widagdoi* was found in Sumatra; a single female was collected in the Malayan peninsula (Genting Highlands).

117. ***Saliocleta beshda* spec. nov.** (figs. 115 a–b).
Holotype: ♂, Indonesia, Java, Meru Betiri National Park, 25 km S Kalibaru, 8°30' S, 114°03' E, 300 m–500 m, v.–vi. 1996, leg. St. Jakl – in coll. A. Schintlmeister, Dresden.

Paratypes (6 ♂♂):
Java: 3 ♂♂, Meru Betiri National Park, 25 km S Kalibaru, 8°30' S, 114°03' E, 300 m–500 m, v. –vi. 1996, leg. St. Jakl (GU 95-55, MWM 1.775); 1 ♂, Mt. Sempohan, 8°10' S, 113°40' E, 410 m, 28. i. 1998, leg. Jackl, Schintlmeister & Cervenka.

Diagnosis. Forewing length ♂♂ 21–23 mm. Similar by external apperance to *S. widagdoi,* but slightly (1–2 mm) smaller. Forewing fawn, postmedial and the marginal fasciae are marked as rows of fine black dots. *S. widagdoi* have one black dot of the postmedial fascia between costa and dorsum prominent developed, *beshda* not. Submarginal fascia consists as a row of fine black dots, but submarginal fascia, which well developed in *widagdoi* is absent. Along the costa of the forewings there are several fine black dots. The tornus of the forewings strong angled. The hindwings are chocolate brown but paler than *widagdoi.* Fringe of hindwing checkered fuscous yellow-dark brown. Female is unknown.

Male genitalia resembles to *widagdoi* but readily to separate by narrower uncus. Inner margin toward uncus of socii concave shaped. Valvae are narrower and longer than in *widagdoi.* Phallus characteristically broader at tip. It is shorter and not zigzag shaped as in *widagdoi.* Everted endophallus bear only a few cornuti. 8th sternite convex rounded at posterior margin and bilobed at anterior margin. Window smaller than in *widagdoi* and central processes are smaller. 8th tergit bilobed at posterior margin as in *widagdoi,* but central sclerotization reduced.

Bionomics. The few known specimens has been taken up to 500 m above sea level.

Distribution. The species is known only from West Java.

118. ***Saliocleta malayana*** (Schintlmeister 1994): 223; pl. 3: 19; GU-pl. 4: 3. (*Ceira malayana*).
Holotype: ♂, W Malaysia, Prov. W Pahang, Genting Tea Estate, 2.000 ft. – coll. A. Schintlmeister, Dresden.
Schintlmeister 2013: 389.

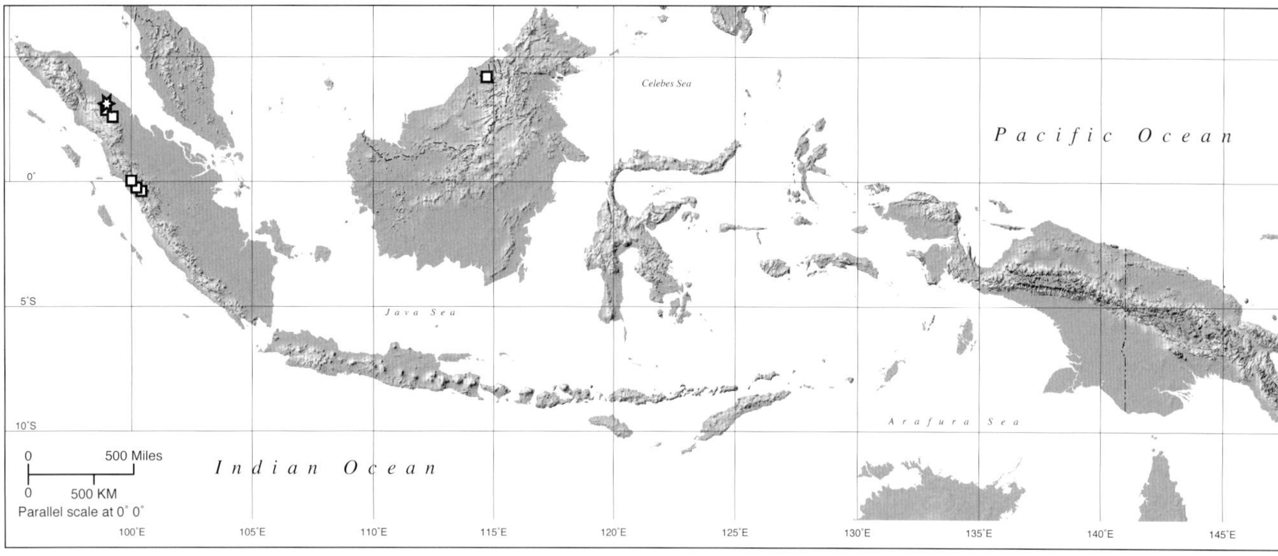

FIGURE 106 Distribution of *Saliocleta armata.*

Diagnosis. This larger species is readily to identify by its elongated forewing shape and rounded tornus. Forewing display a clearly marked submarginal fascia. A postmedial fascia, also marked by blackish dots occur occasionally (fig. 118d). Forewing bear many indistinct larger fuscous patches. Hindwing chocolate brown with a fine yellowish brown marginal fascia. Female sexualdimorphic show reduced fuscous pattern of the forewing, but marginalfascia of fuscous dots is present. Hindwings are light brownish yellow.

Uncus triangular shaped with large knob-shaped tip. Socii large, tapered and rectangular curved. Valvae rhomboid. A diagnostic feature is the serrated sacculus of valvae. Length of tooths is variable as illustrated. It is currently not known if a geographic correlation exists. Phallus broad and rather straight; less modified.

The Female genitalia have not yet been dissected.

Bionomics. Adults occur not rare in primary and secondary forests. Most specimens has been taken below 500 m above sea level, but one male from the Malayan Peninsula was collected at 1750 m.

Distribution. Distributed in southwestern China (Yunnan), Myanmar, Laos, Thailand, Vietnam and the Malayan Peninsula.

Subgenus *Biraia* Kiriakoff, 1962

Type species: *Ceira postica* Moore, 1879.

Small to the medium sized adults of warm yellow colored forewings. Hindwings yellowish, often tinged with reddish brown. Sexual dimorphism is limited. Male genitalia do not differ in its basic structures from subgenus *Saliocleta,* but 8th sternite have the "window" toward the margin not closed. The subgenus comprises nine species.

119. ***Saliocleta armata*** (Kiriakoff 1974): 383; pl. 2: 1 (*Ceira armata*).

Holotype: ♂, [Indonesia], NO-Sumatra, Dolok Merangir, 180 m – ZSM, Munich.

Schintlmeister 2013: 388.

Diagnosis. Forewing warm yellow to brownish yellow. Tornus of forwing angled. Forewing pattern brown. Transversal fascia from base to apex well developed. Submarginal fascia marked as row of small brown dots. Occasionally a brown medial fascia or a postmedial fascia consisting from brown dots occur (fig. 119e). A characteristic feature for identification is a large but indistinct brown dorsal blotch near tornus. Hindwing light yellow, slightly tinged brownish in the anal region. Female is unknown.

Umbrella shaped uncus tip of male genitalia variable as illustrated. Length of socii also variable. Valvae are simple. Phallus zig-zag shaped. Everted endophallus bear a field of spines in diverticulum. 8th abdominal segments long and bilobed at both margins. 8th sternite display stronger sclerotizations along the entire posterior margin. Due to enlargement of the central projections there are two larger windows (weak sclerotized area).

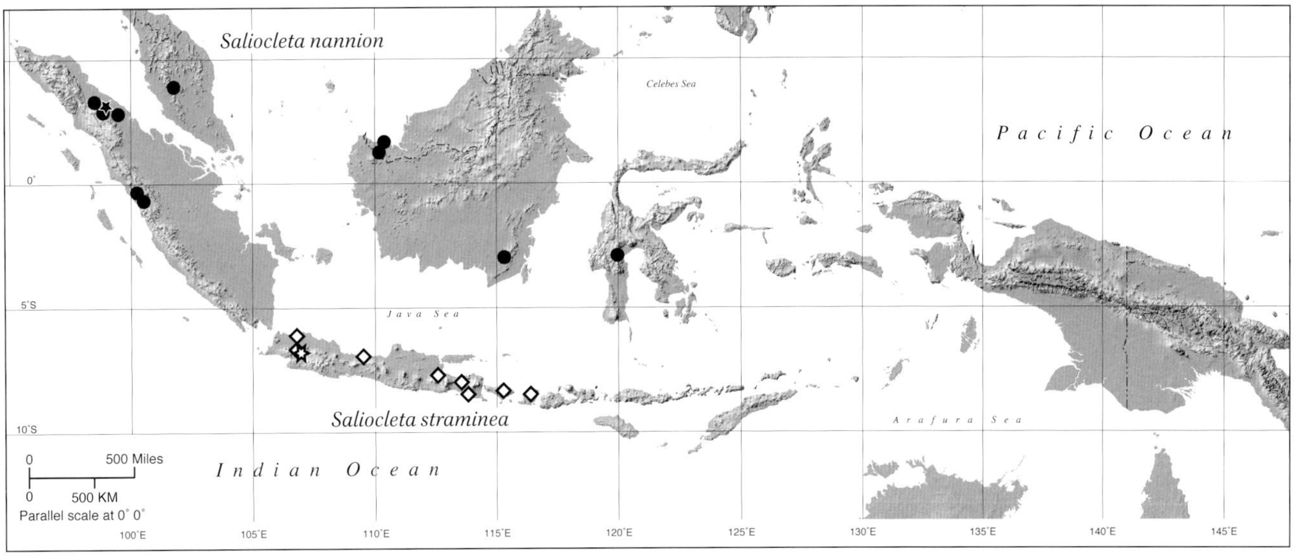

FIGURE 107 Distribution of *Saliocleta nannion* and *straminea*.

Bionomics. *S. armata* occur infrequent between 300 m-1300 m above sea level.

Distribution. Known from Borneo and Sumatra only.

120. ***Saliocleta nannion*** Kiriakoff 1974: 383; pl. 2: 2; fig. 8.
Holotype: ♂, [Indonesia], NO-Sumatra, Dolok Merangir, 180 m – ZSM, Munich.
Schintlmeister 2013: 388.

Diagnosis. Similar to *S. armata* and *S. straminea* but slightly (1–2 mm) smaller in forewing length. Not separable by forewing pattern from *straminea*. Brown forewings pattern as in *armata*, but much lighter and sometimes reduced. Brown dorsal shadow near tornus is often absent or very weakly developed. Female resembles the male, but hindwing lighter.

Male genitalia are similar to *armata*, but umbrella like uncus tip at average shorter. Individual variation of uncus tip and socii length are comparable as in *armata*. Best feature to separate *nannion* from other simlar congeners is a large hook of variable shape at phallus tip. Phallus slightly curved; everted endophallus bear a long field with strong sclerotized spines. 8th sternite differ from *armata* in shorter central projections; sclerotized band along posterior margin is open and appear as a third window. 8th tergit as in *armata* but central sclerotized part narrower.

Female genitalia have a large 8th abdominal segment with two pairs of robust apophyses. Ductus bursae very short. Corpus bursae large with striolate surface. Signum reniform and strong sclerotized.

Bionomics. *S. nannion* flies rather commonly in cultivated areas, but occur also in rainforests. Lower altitudes ubelow 500 m are seemingly preferred, but a few specimens were taken also at 1200m and 1500 m above sea level.

Distribution. Distributed in Sumatra, the Malayan Peninsula, Borneo and Sulawesi. The latter record, confirmed by genitalia dissection GU 44-76, extends the areal much toward east and is from biogeographic point of view somewhat exceptional.

121. ***Saliocleta straminea*** (Roepke, 1943): 103; figs. 16a, 16b (*Turnaca straminea*).
Holotype: ♂, [Indonesia, Java], Perwatee – NBC, Leiden.
Schintlmeister 2013: 391.

Taxonomic note. Two dissected males from Bali and Lombok differ by a longer process at phallus tip from Javan males (3 GU). Due to lack of further material it is currently not possible to draw taxonomic consequences.

Diagnosis. This species resembles externally much to *S. nannion* and represent its sister species. Separable from *nannion* by slightly longer (1–2 mm) forewings. Forewing pattern as in *nannion* and not separable, but occasionally a fuscous medial patch occur. Range of individual variability as in *nannion*. Yellowish hindwings are also not different from *nannion*. Sexual dimorphism is limited; hindwing lighter.

Male genitalia differ from *nannion* in having longer Socii. Uncus und valvae are as in *nannion*, but phallus differ

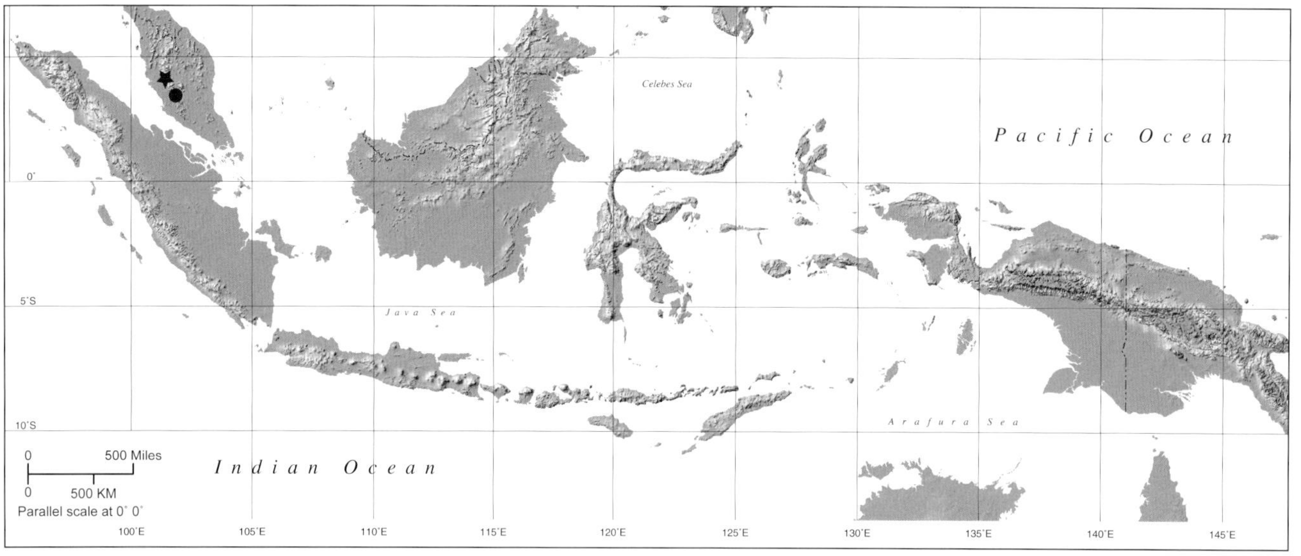

FIGURE 108 Distribution of *Chalepa pendleburyi.*

in a much smaller process at tip. Everted endophallus have no long field of many pines, but display only four or five strong sclerotized small thorns. 8th sternite have the posterior margin weaker sclerotized than in *nannion* and central projections are shorter. 8th tergite clearly waisted and deeply notched at posterior margin.

Bionomics. *S. straminea* is recorded mostly from cultivated areas, but occur also in forests. Adults has been taken between 100 m–1200 m; most specimens at medium altitudes around 1000 m.

Distribution. Distributed in Java, Bali and Lombok.

Chalepa Kiriakoff, 1959: 332.

Type species: *Chalepa binotata* Kiriakoff, 1959.

Schintlmeister 2013: 78.

Diagnosis. Small notodontids of fawn, reddish or pale brown color. Antennae of males are bipectinated with long rami. Female antennae looking like filiform, but they arev also bipectinated, but with very short rami. Forewing display a fine black discaldot and an indistinct indistinct bron apical streak. But not all species display this feature. Hindwings are in most species of *Chalepa* whitish or pale yellowish brown, but both hereunder described new species and some further hithertho undescribed species from Indochina having fuscous chocolate brown hindwings. Female sexualdimorphic, having a falcate apex of forewing. Forewings pattern display a fine black discal spot and in many species a pair of fuscous brown streaks in medial area.

Male genitalia are homologue to *Periergos* by divided and long valva. Lower arm narrow, strong sclerotized and curved. Uncus contrary to *Periergos* well developed and bear many setae. Socii absent. Phallus straight or slightly curved. At phallus tip there is a small and strong sclerotized knob-shaped projection. Everted endophallus bear many strong sclerotized triangular thorns. 8th sternite notched at posterior margin, which is stronger sclerotized. Anterior margin slightly bilobed. 8th tergite of triangular shape with rounded posterior margin. Female genitalia are characterized by long apophyses posteriores. Ductus bursae very long, less sclerotized and slender. Corpus bursae display two small signa.

Chalepa comprises currently 11 described species, but about further 8 species from mainland Asia awaiting their description. Distributed in Nepal, NE India, Indochina, S China, Malaya and Sumatra.

122. *Chalepa pendleburyi* spec. nov. (figs. 122a–d).

Holotype: ♂, Malaysia, Perak, F. M. S., Batang Padang, Jor Camp 1800 ft., ca. 4°20' N, 101°20'E, 26. v. 1923, H. M. Pendlebury (BM 10314005) – NHMUK, London.

Paratypes (2 ♀♀):

Malaya: 1 ♀, W. Pahang, Genting Tea Estate, 2000', 7. ii. 1978 leg. H. Barlow; 1 ♀, ibid 14. xi. 1981 (BM 10314006).

Etymology: Named for Henry Maurice Pendlebury (b. 1893–d. 1945 in Bangalore), who worked as an

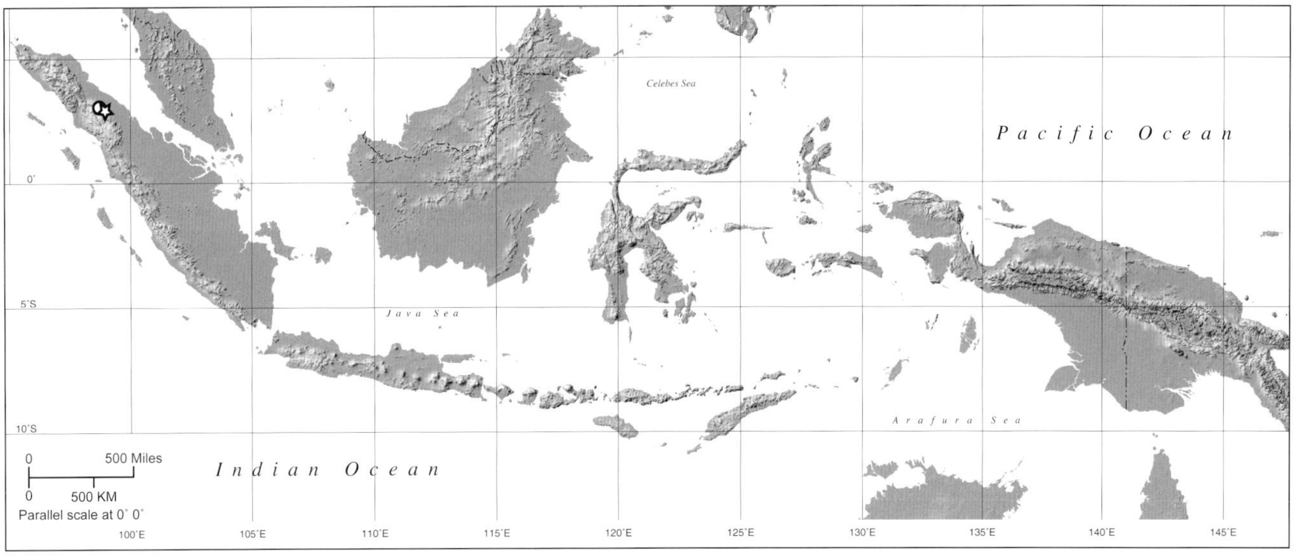

FIGURE 109 Distribution of *Chalepa takaloa.*

entomologist at the Federated Malay States Museum at Kuala Lumpur and did extensive collecting in Malaya of insects, including many moths.

Diagnosis. Forewing length ♂, 15 mm, ♀♀ 20 mm and 21 mm. Holotype male lacks the antennae; females antennae are brown and bipectinated with short rami. Thorax and frons fawn, abdomen fuscous brown colored. Forewing fawn. Postbasal, medial, submarginal and marginal fasciae marked as rows of fine black dots. Discal spot small black, accompanied by two indistinct blackish streaks. Hindwings of male uniform chocolate brown. Male is similar to *C. takaloa*, but the latter have very indistinct forewing pattern Female sexualdimorhic and entire body including abdomen is fawn. Black discal dot present, but no further black pattern is visible on forewing. Hindwing of same fawn color as forewing, but anal area tinged fuscous brown.

Male genitalia are as in most *Chalepa*. Uncus large, triangular shaped with many setae. Socii absent. Valva divided upper arm triangular with pointed tip; in *C. takaloa* shape of upper valva arm is only slightly conically shaped with knob-shaped tip. The lower valva arm of *pendleburyi* is longer than *takaloa* and very slender with knob-shaped tip. Phallus slightly arched with small projection at the tip. Everted endophallus short, of circular shape and armed with many small triangular shaped thorns. It differ by circular shape from *takaloa.* 8th sternite of holotype is demaged at posterior margin. It looks probably as in *takaloa* (fig. 122a). Female genitalia have large papillae analis. Apophyses robust and long. Postvaginal plate large, notched at anterior margin. Ductus bursae long, slightly curved. Corpus bursae circular shaped. It bear a butterfly-shaped signum and a second circular, smaller strong sclerotized signum at bottom.

Bionomics. The few known adults (n = 3) were taken at medium altitudes between 500 m and 600 m.

Distribution. Known from the Malayan Peninsula.

123. ***Chalepa takaloa*** spec. nov. (figs. 123a–c).

Holotype: ♂, Indonesia, N Sumatra, Simalungun, Parapat, Holzweg 2, 2°44' N 98°57 E, ca. 1050 m, 21.v.1993, leg. Dr. E.W. Diehl (GU 64-81) – coll. A. Schintlmeister, Dresden.

Paratype (1 ♂):

Sumatra: 1 ♂, 15 km SW Sindar Raya, 500 m, 3°10' N, 98°56' E, 20. v. 1991 leg M. Graul & A. Schintlmeister (GU 19-51).

Diagnosis. Forewing length ♂♂, 15 mm and 16 mm. Antennae of males brown and strong bipectinated with long rami. Forewing fawn, discal dot small and black. Postbasal, medial, submarginal and marginal fasciae marked as rows of indistinct brown dots. Male resemble by external appearance much to *C. pendleburyi,* but forewing of *takaloa* display many indistinct brown shadows, in particular near the apex. Hindwing fuscous brown as in *pendleburyi.* Female is unknown.

Male genitalia resembles *pendlebury.* Uncus triangular shaped with many setae. Upper arm of divided valvae long, curved and pointed at tip. Phallus very similar to *pendleburyi* but narrower. Everted endophallus long,

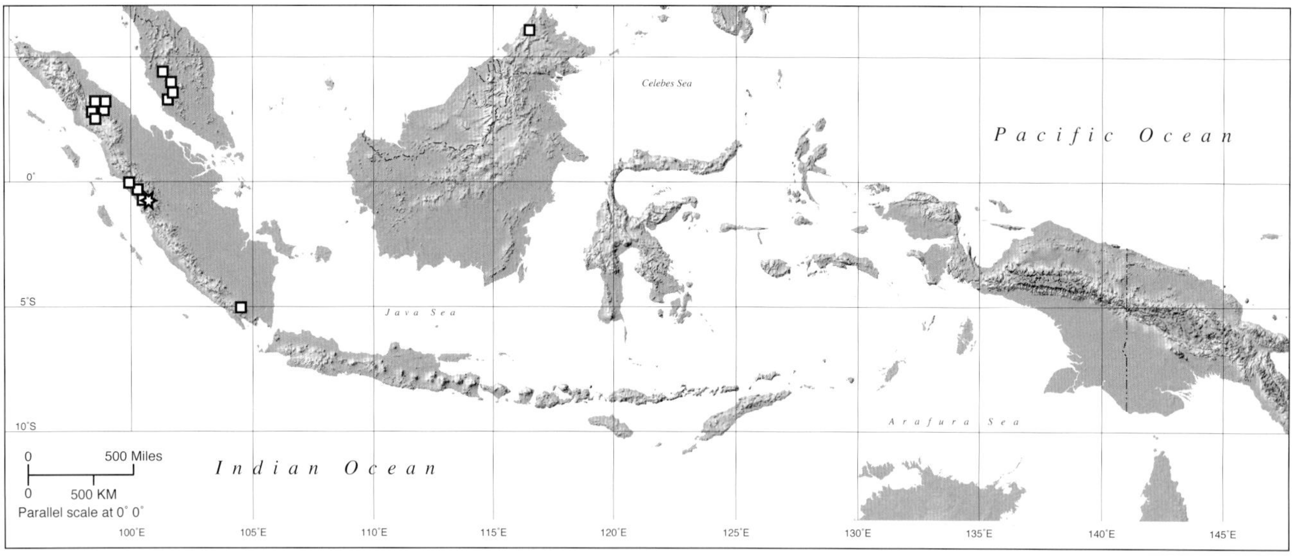

FIGURE 110 Distribution of *Periergos postrubra*.

rather than circular as in *pendleburyi*. 8th sternite at posterior margin stronger sclerotized and notched. 8th tergite similar as *pendleburyi,* but shape stronger conical.

Bionomics. The known two adults were taken at 500 m and 1050 m.

Distribution. Restricted to Sumatra.

Periergos Kiriakoff, 1959: 321.
Type species: *Periergos obsoleta* Kiriakoff, 1959.
Schintlmeister 2013: 322.

Diagnosis. Medium sized to larger notodontids of fawn, reddish-brow or bownish white color of wings and body. Antennae of males are brown and strong bipectinated with very long rami; those of the females have short or medium sized rami. Forewing bear a small but prominent contrasting black discal spot as in *Chalepa*. A submarginal fascia is marked as black patches (in some individuals absent). A brown band runs from apex of forewings to base, which is in many species indistinct developed. Hindwing ranges from chocolate brown, yellowish brown to white. Sexual dimorphism is strong developed. Females are much larger and have a pointed, mostly falcate apex of forewings (in males rounded). Hindwings are generally lighter than in males. The indistinct forewing pattern is often different from those of males, e.g. by presence of an larger fuscous streak or patch near the discal dot.

The male genitalia are very distinct by the deeply divided valves, where the upper arm is usually developed as a thinner and pointed spine. The usually divided uncus display in most species long processes on both sides. The socii are mostly reduced in size or absent, as in *P. hunyada*. Many species in Indochina and China show prominent developed juxta processes. The 8th abdominal segments are less modified; the 8th tergite usually slightly bilobed at the posterior margin.

The female genitalia have large papillae anales and a very short ductus bursae. The corpus bursae relatively small. No signum.

20 species are in the genus, which are concentrated in Indochina and China.

The genus occur in north India, Nepal, south China, Indochina, the Philippines, Borneo, Sumatra and Java.

124. ***Periergos postrubra*** (Swinhoe, 1903): 199 (*Pydna postrubra*).
Holotype: ♂, [Indonesia], Sumatra, Solok – NHMUK, London.
Schintlmeister 2013: 324.

Diagnosis. Forewing color vary from rusty brown to brownish white; gound color not yellowish tinged as in *P. hunyada* or *spinosa*. Variable also in size (forewing length). Males have a blackish submarginal fascia on forewings usually well developed. This separates *postrubra* from *P. hunyada* or *spinosa* and other congeners from Indochina. Females are very large, much larger than other congeners the hindwings are white, not yellowish tinged.

Male genitalia have a squarish and slightly notched uncus. Socii are subject of individual variation as illustrated; also asymmetrical socii pairs occur. Valve divided. *P.*

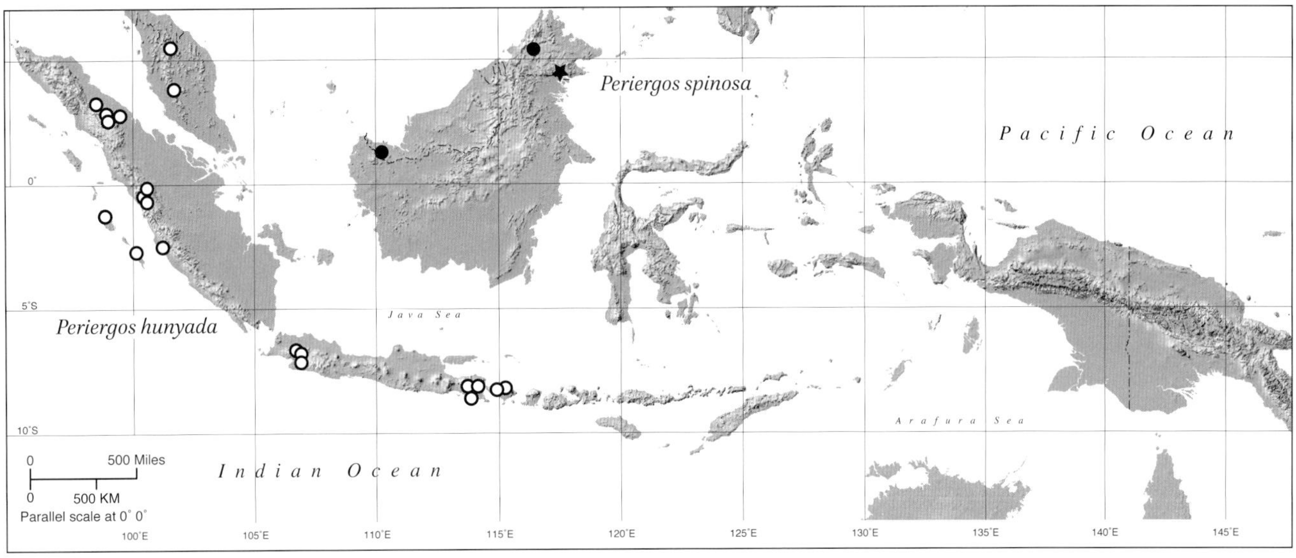

FIGURE 111 Distribution of *Periergos hunyada* and *spinosa*.

postrubra is unmistakable by the very large and strong sclerotized lower part. Both arms of valva vary individually in shapes. Phallus small and bear at tip a spine. Female genitalia have not yet been dissected.

Bionomics. Adults occur in Sundaland rather infrequent between 50 m–1700 m. Caterpillar was reared in Germany on various bamboo (Poaceae) and also on *Poa*. (Schintlmeister & Pinratana 2006: 103).

Distribution. Distributed in Indochina, the Malayan Peninsula, Borneo and Java.

125. ***Periergos hunyada*** (Swinhoe, 1903): 504 (*Pydna hunyada*).
Holotype: ♂, [Indonesia], S. Java, 1500 ft. – NHMUK, London.
Schintlmeister 2013: 323.

Diagnosis. A very variable species, which is not distinguishable by external characters from *P. spinosa* and also the range of individual variability is in both species almost identical. From other congeners *P. hunyada* may be distinguished by two brown blotches in medial area on the transverse fasciae of forewing. Postmedial fascia is marked by brown dots. Range of individual variability includes strong contrasting black blotches, a clearly pronounced fuscous brown medial area or forms with fuscous colored veins (fig. 125e). Sexual dimorphism is contrary to other congeners limited. Female antennae of both species have longer pectinations; all other congeners have very short rami (like filiform).

Male genitalia are distinct by a delicate built tegumen of rectangular shape and clearly divided valva. Uncus reduced and "shoulders" at base significant smaller than in *spinosa*. Valva divided. Upper valva arm pointed at tip. Lower valva arm slenderer than *spinosa*. The last part toward tip rectangular curved. Phallus long, slender and curved. 8th abdominal segments are less modified. 8th sternite almost not bilobed at anterior margin (in *spinose* bilobed). 8th tergit rather straight at posterior margin. Female genitalia are relatively small. Papillae anales large and long; apophyses reduced in length. Antevaginal plate notched at anterior margin (in *spinosa* not). Ductus bursae very short; corpus bursae ellipsoid. No signum.

Bionomics. The species occur in Sundaland commonly up to 500 m; but single individuals were taken also at 1200 m above sea level.

Distribution. Distributed in Indochina, Malaya, Java, Bali, Luzon, Negros and Panay.

126. ***Periergos spinosa*** (Holloway, 1983): 46; pl. 4: 5; fig. 32 (*Pydna spinosa*).
Holotype: ♂, [Malaysia], Borneo, E. Sabah, Kalabakan, 100 m – NHMUK, London.
Schintlmeister 2013: 324.

Diagnosis. A very variable species, which is not distinguishable by external characters from *P. hunyada* and also the range of individual variability is in both species almost the same (see under *hunyada*). Antennae of females have short pectinations; all other congeners (except *hunyada*) have antennae with very short rami.

Male genitalia have a reduced uncus. Shoulders at uncus base are much larger than in *hunyada*. A diagnostic feature are prominent setae on dorsal arms of valva, which are absent in *hunyada*. The phallus long, slender and curved. The 8th abdominal segments are less modified. 8th sternite deeper bilobed at anterior margin than *hunyada*. 8th tergit slightly notched at posterior margin. Female genitalia are relatively small and similar to *hunyada*. Papillae anales large, broader than in *hunyada*; apophyses reduced in length. Antevaginal plate not notched at anterior margin as in *hunyada*. Ductus bursae very short; corpus bursae circular. No signum.

Bionomics. The species occur in Borneo rarely, but in Palawan commonly up to 400 m.

Distribution. Distributed in Borneo and Palawan. A single male was taken in N Thailand (Schintlmeister & Pinratana, 206: 100).

Cerurinae

Note. *Cerura ejecta* Prout, 1920: 289 was described of "Key Is". The holotype ♂ in NHMUK, London lacks the abdomen but belongs doubtless to the neotropical genus *Tecmessa*, to a species complex of *Tecmessa lancea* (Schaus, 1905). *Tecmessa* does not occur in Oriental region; a case of wrong type locality of *Cerura ejecta*.

Neocerura Matsumura, 1929: 88 (key), 89.
Type species: *Cerura liturata* Walker, 1855.
Schintlmeister 2013: 273.

Diagnosis. Medium sized moths. Anennae bipectinated with long rami in both sexes. Forewing color shiny white with contrasting black pattern, which is as in *Kamalia*. Postbasal fascia well marked as a broad black band, often filled olive. Discal streak and postmedial fascia present. On costa near to apex a larger black blotch occur. Fringe of all wings contrasting black and white checkered. Hindwings white or black with transistional forms. A discal spot on hindwings, which is present in *Kamalia*, is absent. Sexual dimorphism is limited.

Male genitalia have a broad and round uncus, which is slightly notched at tip. Socii very robust of half size of uncus. Tegumen squarish and massive; saccus reduced. Valvae weakly sclerotized. *Neocerura* is distinguished from *Kamalia* by short phallus, with bifurcate tip. Both processes are long and pointed. 8th abdominal segement are contrary to *Kamalia* less modified. In particular the W-shaped strong sclerotized structure of 8th sternite is absent. Female genitalia of *N. liturata* have broad 8th abdominal segment, short and broad pairs of apophyses and a long ductus bursae. Corpus bursae is large and circular shaped. No signum.

Neocerura is represented by five species distributed in Nepal, entire India, Sri Lanka, China, Taiwan, Indochina, Sundaland, Java, the lesser Sundas, the entire Philippines, Sulawesi and – as a remote point – in Tanimbar.

127. ***Neocerura liturata*** (Walker, 1855): 988 (*Cerura liturata*).
Lectotype: ♀, [Bangladesh], Silhet – NHMUK, London by **present designation**.
Schintlmeister 2013: 274.

Taxonomic notes. *Cerura liturata* was described of 4 ♂♂, 1 ♀, which are according to the original description from "Silhet" and "Madras". The author traced in the holdings of NHMUK a ♀ marked as type. To stabilize the nomenclature this female is hereby designated as lectotype with the following labels: circular white label "Silhet", green ringed, circular label "Type", "10. CERURA LITURATA.", yellow label " Photo done by A. Schintlmeister #4315". A lectotype label was added. Lectotype is depicted here in fig. 127f.

Development of black scales is very variable. Specimens of dryer areas tends to produce smaller individuals with reduced black pattern, see also Schintlmeister & Lourens, 2010: 136. So specimens from the Lesser Sunda Islands, Bali and E Java are mostly whitish and relatively small; however – depending from altitude and microclimatic influences from special habitats also larger and more blackish specimens occur in these areas, as illustrated in fig. 127c. In these cases there is no evidence for subspecification. Populations from Sulawesi, Peleng, Salayer Isl. and the Philippines, however, display in majority forms with reduced black forewing pattern. But in the Phillippines also a few individuals with well developed black forewing pattern were found. From Sulawesi only whitish individuals are known, but *N. liturata* is known there only from a few specimens (n < 10).

Diagnosis. Adults are similar to several *Kamalia* species, but smaller. Forewing pattern contrasting black on shiny white forewing color. Hindwings display no blackish discal spot, which separates *Neocerura* from *Kamalia*. *N. liturata* show considerable variation in size and the development of black pattern. Sexual dimorphism is limited, but females are much larger than males.

Male genitalia are best to distinguish by the shape of bifurcate phallus from other congeners. Shape and length of the phallus processes vary individually. Also shape of socii is subject of individual variation. 8th abdominal segments are not modified. The female genitalia are as described in genus diagnosis.

Bionomics. Moth appear in rainforests rather rare but commonly in secondary vegetation including urbanized areas. Adults whas been observed in Indonesia at lower and medium altitudes between 100–1150 m, but in Sikkim up to 1900 m. Hostplants of the caterpillars are various Flacourtiaceae and *Terminalia* (Conbretaceae). Caterpillar was reared in Europe on Salicaceae (*Populus, Salix*).

Distribution. Distributed in entire India and Nepal, Sri Lanka, China, Taiwan, Indochina, Malaya, Borneo, the

 | DOI:10.1163/9789004354258_006

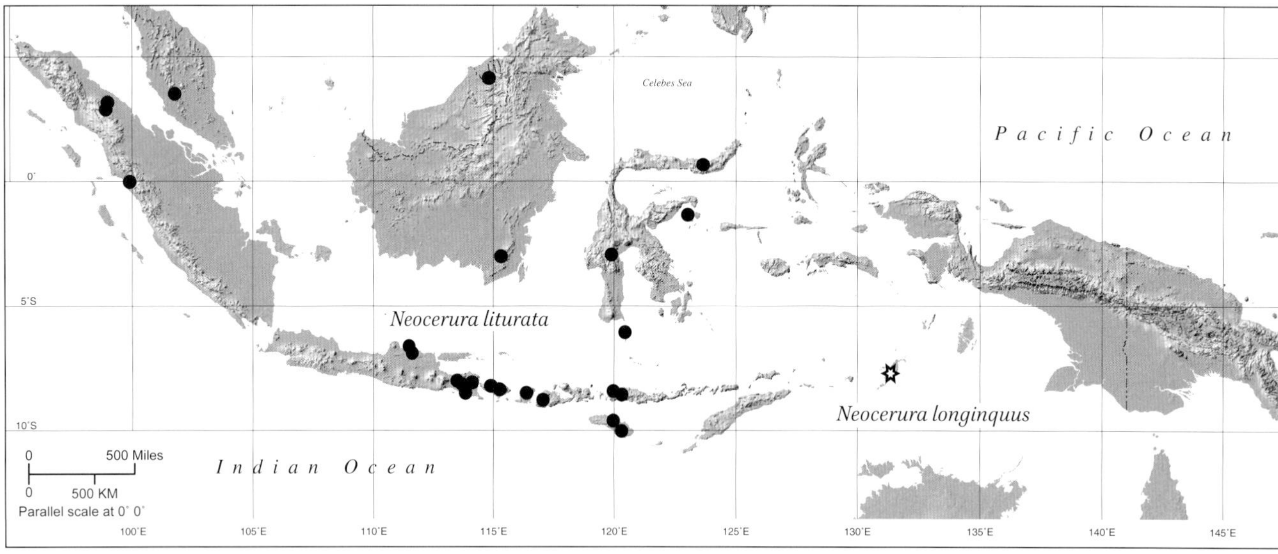

FIGURE 112 Distribution of *Neocerura liturata* and *longinquus*.

entire Philippines, Sumatra, Java, Bali, Lombok, Flores, Sumba, Sumbawa, Sulawesi, Peleng Isl. and Salayer Isl.

128. ***Neocerura longinquus*** **spec. nov.** (figs. 128a–c).
Holotype: ♂, Indonesia, Tanimbar, Yamdema Isl., 21 km N Saumlakki 7°47'S, 131°18'E, 150 m, 5.–28. xii. 2005, leg. St. Jakl (GU 92-67) – in coll. A. Schintlmeister, Dresden.

Paratypes (2 ♂♂, 3 ♀♀):
Tanimbar: 1 ♂, 1 ♀, Yamdema Isl., 21 km N Saumlakki 7°47' S, 131°18' E, 150 m, 27. xi.–11. xii. 2005, leg. St. Jakl 1 ♂, 1 ♀, Yamdema Isl., 20 km N Saumlakki, Loulun village, 7°47' S, 131°18' E, 150 m, 5.–28. xii. 2006, leg. St. Jakl; 1 ♀, Yamdema Isl., Loroklun vill., 20 km N Saumlakki, 150 m, 15.–30. xii. 2007, leg. St. Jakl.

Diagnosis. Forewing length ♂♂ 19 mm, ♀♀ 22–26 mm. Antennae in both sexes long bipectinate and black. Forewing color shiny white. Black pattern reduced in males. Postbasal fascia well developed, interrupted of yellowish-olive color. Postmedial fascia contrasting black near the costa; the remaining part of fascia toward the dorsum extinguished. Discaldot small but well visible. Fringe of forewing black and white checkered. Hindwings are pure white without black markings, except a few blackish dots on the fringe. Female is sexualdimorphic, displaying black forewing pattern more strongly developed. A black contrasting postmedial fascia runs until dorsum of forewing. Postbasal fascia surrounded in black. Basal fascia marked by a few black patches. Hindwings white with broad blackish postmedial area, fading white toward medial area. Contrasting black anal markings well visible. *Neocerura longinquus* resembles by external appearance, at the first glance somewhat to *Kamalia timorensis*, which is also a small species. The latter species display a rich black forewing pattern and a blackish discal spot on the hindwings, which is absent in *longinquus*.

Male genitalia are similar to *N. liturata*. Uncus broad and of rounded shape. Socii robust but not of triangular shape as in *liturata*. They show a less sclerotized projection. Valva weakly sclerotized; valva costa not curved as in *liturata*. Phallus short and straight, broader than in *liturata*. Processes of bifurcate phallus tip are broader and not pointed at tip. 8th abdominal segments are not modified. Female genitalia were not dissected.

Bionomics. Only five specimens are currently known, which were taken in three different years at 150 m above sea level, almost at the same locality.

Distribution. Endemic to Tanimbar.

Kamalia Koçak & Kemal, 2006: 3.
Type species: *Cerura tattakana* Matsumura, 1927.
Schintlmeister 2013: 228.

Diagnosis. Medium sized moths. Antennae are black and bipectinated with long rami in both sexes. Forewing color is shiny white with contrasting black pattern, that resembles *Neocerura*. But most *Kamalia* species are larger than *Neocerura*. Hindwings are white to black colored and display an indistinct blackish discal spot, which is not present in *Neocerura*. Sexual dimorphism is limited.

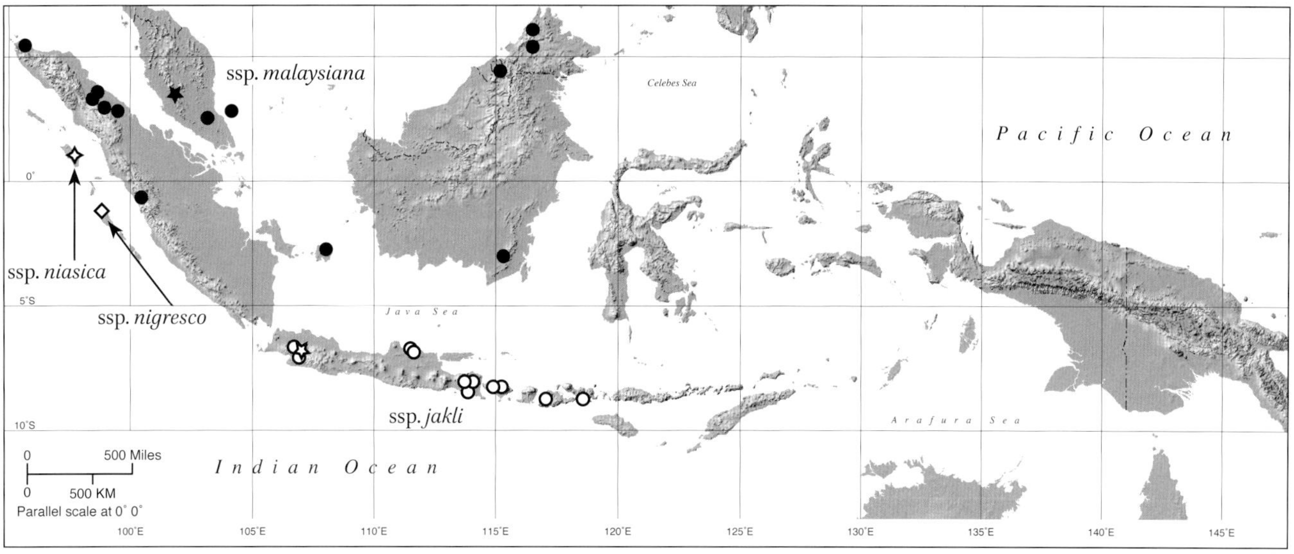

FIGURE 113 Distribution of *Kamalia malaysiana*.

Male genitalia of *Kamalia* resemble those of *Neocerura*. Uncus short with broad tip. Socii robust, with pointed tip. Valva weakly sclerotized and relatively small. Phallus shape divide *Kamalia* into three groups: rather straight, short and tapered phallus without further processes. Everted endophallus bears a serrated band with many spines. A second group have a bifurcate phallus, which bears further projections or processes. The third group have a long, slender and curved phallus; everted endophallus without sclerotizations and projections.

8th sternite displays a prominent and strongly sclerotized "W"-shaped structure, which is a characteristical generic feature. 8th tergite not modified. Female genitalia have long apophyses posteriores and a strongly sclerotized, large postvaginal plate. Ductus bursae is short. Corpus bursae is disc-shaped and large. No signum.

Caterpillar's hostplant are Flacourtiacea, but in Europe some species has been succesfully reared with Salicaceae (*Populus, Salix*). Caterpillars and behavior resembles much to those of *Cerura vinula* (Linnaeus, 1758) and other Cerurinae.

The genus *Kamalia* was reviewed by Schintlmeister, 2002. It comprises currently 14 species, which are distributed from S Japan, China, the Himalaya and entire oriental tropics including Sri Lanka, Australia, New Caledonia and Solomon isl.

129. ***Kamalia malaysiana*** (Holloway, 1982): 203; pl. 22: 1 (*Cerura malaysiana*).

Holotype: ♂, Malaysia, W. Pahang, Genting tea Est., 2.000 ft. – NHMUK, London.

Schintlmeister 2013: 229.

Taxonomic notes. The species tend to establish geographically subspecies.

The following subspecies occur in the area investigated here:

Ssp. *niasica* (Schintlmeister, 1994): 224; pl. 4: 10 (*Cerura malaysiana niasica*); holotype: ♂, Indonesia, Nias Isl., 10 km E Idanö Gawö – coll. A. Schintlmeister, Dresden. Characterized by deep black hindwings.

Ssp. *jakli* (Schintlmeister, 2002): 116; figs. 11, 14, 50 (*Paracerura malaysiana jakli*); holotype: ♂, Indonesia, W Java, Mt. Gedeh, Pangrango Nat. Park, 6°47'S, 107°' E, 1.250 m – coll. A. Schintlmeister, Dresden. Small adults with reduced black forewing pattern. Blackish hindwings mixed with white scales. Imagines of ssp. *jakli* becomes generally more pale in Sumba and Sumbawa compared with Java or Bali populations. From Java only pure white females of ssp. *jakli* are known.

A further subspecies is introduced here as:

Kamalia malaysiana nigresco **ssp. nov.** (figs. 129f, 129i).

Holotype: ♂, Indonesia, Sumatra, Siberut Isl., Mt. Malacan, 1°06' S 98°56'E, 300 m, i. 2004, leg. St. Jakl – in coll. A. Schintlmeister, Dresden.

Paratypes (3 ♂♂, 1 ♀):

Siberut: 1 ♂, Mt. Malacan, 1°06' S 98°56'E, 300 m, i. 2004, leg. St. Jakl; 2 ♂♂, 2 ♂♂, Labuhan Bajan, 100 m, 0°58' S 98°56'E, i.2004, leg. St. Jakl (GU 64-96); 1 ♀, Bojahan, 1°06' S 98°47'E, xi. 2004, leg. St. Jakl.

Diagnosis of ssp. *nigresco*. Forewing length ♂♂ 28 mm, ♀ 34 mm. This new subspecies resembles adults of ssp.

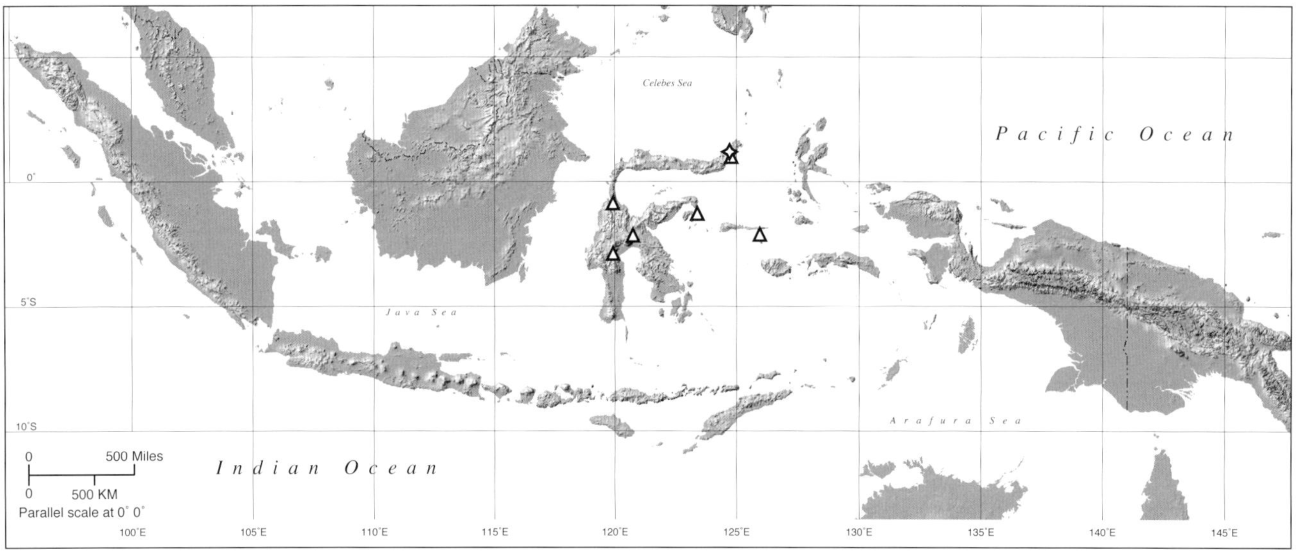

FIGURE 114 Distribution of *Kamalia minahassae*.

niasica rather than Sumatran populations. Hindwings are deeper black than in ssp. *niasica*. Forewing color is dull white, not shiny white as in the other known subspecies of *malaysiana*. Black postbasal fascia of forewing not brownish filled. A distinct character are the coalesced blackish area of upper thorax part. All other known *Kamalia* species display two or four black blotches, but not a single black field.

Male genitalia do not differ significantly from ssp. *malaysiana* and ssp. *niasica*.

Diagnosis. *Kamalia malaysiana malaysiana* is indistinguishable by external appearance from *K. minahassae* and *K. priapus* (Schintlmeister, 1997), described from Indochina.

Male genitalia apparatus as in other *Kamalia*. The shape of the broad and straight phallus separates *malaysiana* from other congeners. Everted endophallus bears a long serrated band consisting from many long spines. W–shaped sclerotisations of 8th sternites distinguish *malaysiana*. But there is some individual variation, as illustrated in fig. 126b. Female genitalia have large 8th abdominal segment and almost disc-shaped papillae anales. Postvaginal plate large, rectangular shaped and strongly sclerotized. A projection arises from postvaginal plate. Ductus bursae very reduced in length. Corpus bursae circular, where the upper half is more strongly sclerotized. A dissected female from Palawan (fig. 126a) displays this projection very different and also the postvaginal plate is slightly notched at posterior margin.

Bionomics. *K. malaysiana malaysiana* occur in Sundaland locally, preferring medium altitudes in rain forests between 600–1400 m and might be at some places common. Ssp. *niasica* and *nigresco* inhabiting lowlands (0–300 m). Ssp. *jakli* is common and occur also in cultivated land, but mostly at lower altitudes below 500 m. It was collected in Bali up to 1300 m above sea level. *K. malaysiana malaysiana* was reared several times in Germany on *Salix* and *Populus* (Salicaceae).

Distribution. *K. malaysiana malaysiana* occur in south Myanmar and south Thailand sympatrically with *K. priapus*. It is also distributed in the Malayan Peninsula, Tiomah Isl., Sumatra, Belitung Isl. and Borneo. *K. malaysiana palawana* (Schintlmeister, 2002) is restricted to Palawan; ssp. *niasica* and ssp. *nigresco* are endemics of Nias isl. and Mentawai isls. respectively. *K. malaysiana jakli* flies in Java, Bali and Sumbawa.

130. ***Kamalia minahassae*** (Holloway, 1982): 203 (*Cerura minahassae*).

Holotype: ♂, [Indonesia, Sulawesi], Celebes, Minahassa, Tomohon – NHMUK, London.

Schintlmeister 2013: 229.

Diagnosis. Indistinguishable by external appearance from *K. malaysiana*. A series of eight males from Peleng Isl. and also a single male from Sulu are in average 2–3 mm smaller in forewing length than males from Sulawesi.

Genitalia apparatus of the male does not differ significantly from *malaysiana*. But phallus is distinguished by

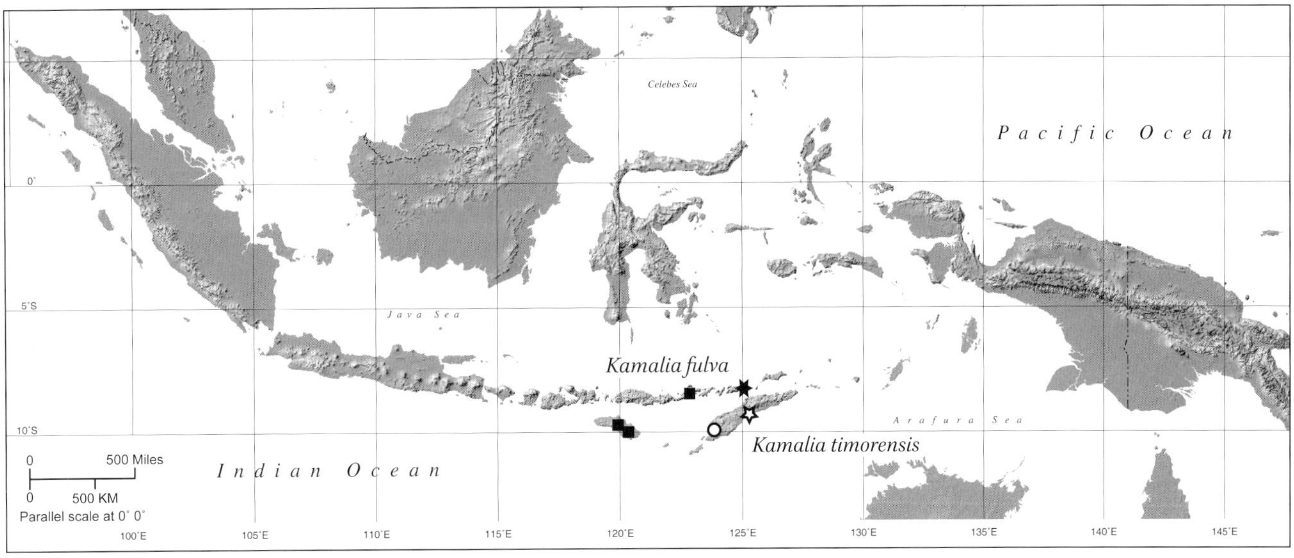

FIGURE 115 Distribution of *Kamalia fulva* and *timorensis*.

long and curved shape with sharp pointed tip. Everted endophallus displays the serrated band of fine spines shorter than *malaysiana*. The center of the W-shaped structure of 8th sternite bear a long central spine, the longest in all known *Kamalia* species. Both lobes of the "W" are weakly sclerotized. Female genitalia have not yet been dissected.

Bionomics. The species appear infrequent in Sulawesi at medium altitudes between 400–1400 m. In Peleng Isl. imagines were observed at 150 m.

Distribution. *K. minahassae* is distributed in Sulawesi, Peleng Isl. and Sula Isl.

131. ***Kamalia fulva*** **spec. nov.** (figs. 131a–f).

Holotype: ♂, Indonesia, Lesser Sunda Islands, Alor Isl., Moru, 8°17' S, 124°29' E, 500 m, 22. iii.–3. iv. xii. 2006, leg. St. Jakl – in coll. A. Schintlmeister, Dresden.

Paratypes (11 ♂♂, 1 ♀):

Alor: 8 ♂♂, Moru, 8°17' S, 124°29' E, 500 m, 22. iii.–3. iv. xii. 2006, leg. St. Jakl (GU 92-63, GU 92-64);

Sumba: 1 ♂, Nusa Tenggara Timur, E Sumba Isl., Luku Melolo, 10°00' S, 120°10' E, 500 m, 29. vi.–15. vii. 2005, leg. St. Jakl (GU 64-97), 1 ♂, Sumba, xi. 1998; 2 ♂♂, Mt. Langgarilu, ii. 2015, leg. Benny de Groof;

Flores: 1 ♂, 1 ♀, O. Flores, Nobo, 100 m, 28. i. 1954 (J.M.A. Groenendael).

Diagnosis. Forewing length ♂♂ 27–31 mm, ♀ 36 mm. Antennae in both sexes bipectinate and black with long rami. Forewing color yellowish white. Black forewing pattern less contrasting but similar to other congeners. Postbasal fascia of forewing filled orange-brown. All black markings of forewings are thinner and more sparsely in relation to *K. malaysiana* or *minahassae*. Hindwings are white, tinged yellow. Anal region mixed with blackish scales. Discal blotch of the hindwing indistinct blackish marked. All wings have a black-white checkered fringe. A pair from Flores have black pattern more reduced than specimens from Alor or Sumba and also hindwing are not blackish tinged.

Male genitalia are similar to those of *minahassae*. But uncus is slightly larger and socii broader. Valvae somewhat variable in shape: apex might be knob shaped or slightly pointed. Valvae of *fulva* generally are narrower and less rounded at apex than *minahassae*. Saccular process of valva well developed. Phallus strongly sclerotized, pointed and curved as in *minahassae*. Everted endophallus circular shaped. It arises seemingly from the central part of phallus, whereas in *minhasassae* endophallus is connected with phallus tip. 8th sternite have "W" shaped structure more similar to *malaysiana* rather than *minahassae*. Female genitalia have not yet been dissected.

Bionomics. *Kamalia fulva* occur not rare at lower altitudes up to 500 m.

Distribution. Distributed in Alor, Sumba and Flores.

132. ***Kamalia timorensis*** (Kiriakoff, 1970): 111; fig. 10 (*Cerura timorensis*).

Holotype: ♂, [East Timor], Portuguese Timor, Suai – NHMUK, London.

Schintlmeister 2013: 230.

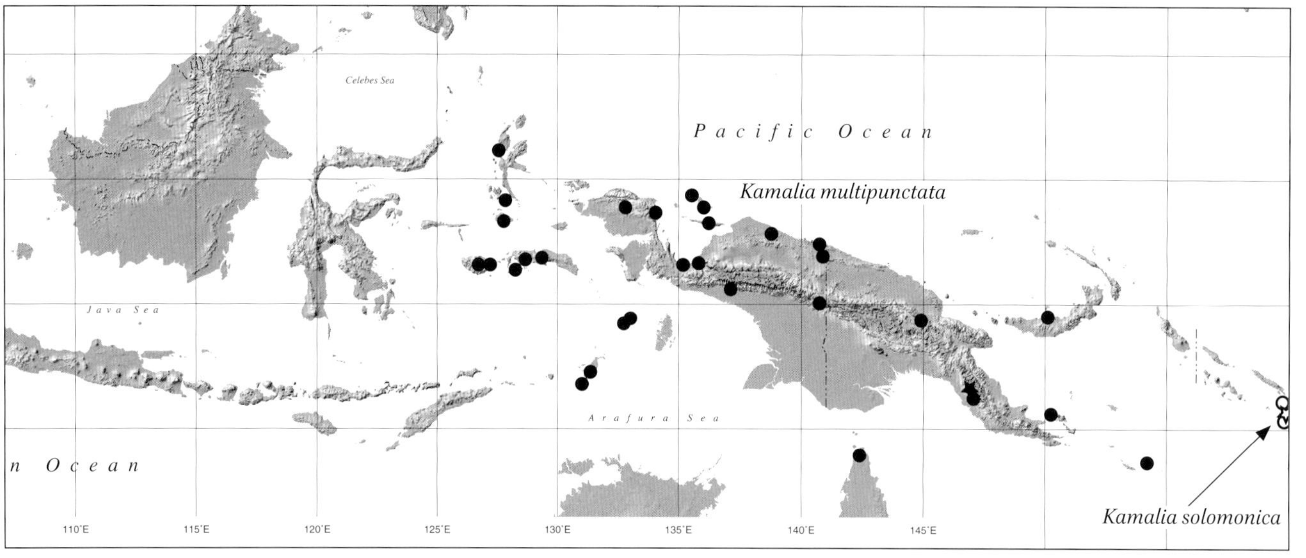

FIGURE 116 Distribution of *Kamalia multipunctata* and *solomonica*.

Diagnosis. By far the smallest species in *Kamalia*. Forewing length ♂♂ 21 mm, ♀♀, 30 mm. Forewings are whitish and slightly yellow tinged, but less than *fulva*. Forewing pattern, in particular the postbasal, yellowish-brown filled. Postbasal fascia, resembles to those of *fulva*. Black pattern in some individuals, including the holotype, is weakly developed. Hindwings are light, mixed with many black scales. Discal streak on hindwings is black and well visible. Sexual dimorphism is limited. Female have hindwings slightly more fuscous than male.

Male genitalia resembles *fulva*, but differ by smaller Socii. Valvae well distinguished from other congeners being narrow and kinked. Sacculus process of valva reduced. Phallus short and less curved, compared to *fulva*. 8th sternites are less tall then in *fulva* and there is a central spine on the "W", which is shorter than in *minahassae*. Female genitalia have long and slender apophyses posteriores, and triangular shaped, short apophyses anteriores. Postvaginal plate is ellipsoid with a pair of triangular lobes on both sides. Ductus bursae reduced in length; corpus bursae disc-shaped and without signum.

Bionomics. Adults has been collected commonly (n > 40 during 10 days) at 350 m above sea level.

Distribution. Endemic to Timor island.

133. ***Kamalia multipunctata*** (Betune-Baker, 1904): 381; pl. 6: 9 (*Cerura multipunctata*).

Holotype: ♂, [Papua New Guinea], Dinawa – NHMUK, London.

Schintlmeister 2013: 229.

Taxonomic notes. The species appear very variable in black forewing pattern. Even specimens from same geographical region might be very different in various years, see figs. 133j 133k, which are both from Arfak Mts. Also male genitalia are variable in shapes of socii and phallus. Apart from this, probably also geographical variability occur, as longer series from various islands from several years indicate. Due to overlapping individual and geographical variation it is currently not possible to recognize and separate taxonomically subspecies of *K. multipunctata*.

Diagnosis. Distinguished by the prominent black pattern of the shiny white forewings. A postbasal fascia is absent. Imagines are variable, including in wingspan. The black patches of the forewings are sometimes reduced in size and in other individuals prominently enlarged. Hindwings might be black or pure white with transistional forms. Female is much larger than the male and hindwings display contrasting black postmedial and marginal areas with a black medial fascia.

Male genitalia are distinguished by club-shaped valvae, where the saccular process is absent. Uncus and robust socii are as in other congeners. Phallus slender and curved; everted endophallus displays a small field of strongly sclerotized spines. W-shaped sclerotisations of the 8th sternite are slightly variable in shape, probably geographically correlated to various islands. 8th tergite broad, bilobed at anterior margin. Female genitalia have large and rounded papillae annales and long apophyses posteriores. Postvaginal plate is large and strongly sclerotized, with a pair of large lobes. Ductus bursae reduced in length. There is a prominent circular appendix bursae. Corpus bursae disc-shaped and without signum. Three dissected females from various geographical regions appear very different in their papillae anales, apophyses posteriore, postvaginal plate and also in shape of antrum, as illustrated in figs. 133c–e.

Bionomics. Males of *K. multipunctata* flies commonly in various habitats from lowlands, 50 m up to 1200 m above sea level. Single specimens were collected in Arfak Mts. up to 1700 m above sea level.

Distribution. Widely distributed in entire New Guinea Isl. including adjecent Islands such as: Goodenough Isl. Rossel Isl., Biak, Yapen, New Britain. Also found in Halmahera, Obi, Bacan, Ambon, Seram, Key, Tanimbar and in Australia (Queensland).

134. ***Kamalia solomonica*** **spec. nov.** (figs. 134a–c).
Holotype: ♂, Solomons Islands, Gualdalcanal isl., 6 km S Honiara, 2 km S Mbarana village, prim forest, 9°30' S, 160°00' E, 360 m, 1.–3. ii. 1998 leg. Dr. Ron Brechlin (barcoded NOTO-AS43.37) – in coll. A. Schintlmeister, Dresden.

Paratypes (5 ♂♂):
Solomons: 1 ♂, Guadacanal, Popomanaseu ascent Nuhu village gardens; 26. x. 1965 (Roy. Soc. Exped.); 2 ♂♂, Tulagi I. (Woodford) (BM 804); 3 ♂♂, Tugela Isl. (Woodford).

Diagnosis. Forewing length ♂♂ 26–29.5 mm. Closely related and similar by external appearance to forms with reduced black pattern of *K. multipunctata*. Sometimes small black anal markings occure on the white hindwings. Female is unknown.

Male genitalia resembles *multipunctata* but differ by the enlarged Socii, which are significantly longer than in *multipunctata*. Uncus and club-shaped valvae are as in *multipunctata*. Phallus is shorter, less curved and broader than in *multipunctata*. 8th sternite with "W" shaped sclerotizations resembles *multipunctata*, but a central projection is absent.

Bionomics. The holotype was collected at 350m above sea level in primary forest. Paratypes were presumable taken also at lower altitudes.

Distribution. Known from Solomon Isls.: Gualdalcanal, Tulagi and Tugela.

135. ***Kamalia amoa*** (Holloway, 1979): 359; pl. 68: 7; fig. 71: 2 (*Cerura amoa*).
Holotype: ♂, New Caledonia, Port Boisé, Site 74, 20 m – NHMUK, London.
Schintlmeister 2013: 229.

Diagnosis. Distinguished by dull white wings. Black pattern mixed with white scales. Submarginal fascia of all wings marked as rows of fine black dots. Hindwings white with blackish anal region and a medial shadow. Fuscous discal spot on hindwings is contrasting. Sexual dimorphism is limited.

Uncus of male genitalia and socii are narrower than in *multipunctata*. Valvae are narrow and display a small projection near apex. Phallus is slightly longer than in *multipunctata*. 8th sternite bears a central process on the sclerotized "W".

Bionomics. Males occur commonly at lower altitudes up to 300 m. Caterpillar was reared in New Caledonia on *Homalium deplanchei* in Flacourtiaceae (pers. comm. Thierry Salesne, Farino-New Caleonia 27. ii. 2013).

Distribution. Restricted to New Caledonia; also occuring in adjecent Maré and Lifou islands (no distribution map).

Dicranurinae

Liparopsis Hampson, 1893: 125 (key), 154.
Type species: *Liparopsis postalbida* Hampson, 1893
Schintlmeister, 2008: 137, Schintlmeister 2013: 237.

Diagnosis. Small notodontid moths. Males have bipectinate brown antennae and long rami. Rami of females are short. Last part until the tip of antenna, about 15% is simple. Forewing color is whitish and margenta-brown tinged. Discal spot is circular and whitish in all species, but often less contrasting. Forewing pattern consists of a small whitish basal area, accompanied by a large fuscous filled postbasal area. Near the apex there is a large magenta-brown costal patch and a further large fuscous area at the tornus. Postmedial fascia is weakly marked as a fuscous shadow. Hindwings are broad and whitish; an apical patch is contrasting fuscous. Postmedial fascia of forewing continues on hindwing. Sexual dimorphism is limited.

Uncus is long and slender; socii are absent. Valvae bear several projections of taxonomic value. Phallus is straight. 8th abdominal segments are less slerotized and less modified. 8th sternite bears a pair of central projections at posterior margin. Female genitalia have large papillae analis and reduced apophyses anteriores. Ductus bursae is large and thick; corpus bursae is circular shaped and relatively small. No signum.

Caterpillar's hostplants of *Liparopsis postalbida* is *Quercus* (Fagaceae).

All four *Liparopsis* species occuring in Sundaland. One species extends to the Himalaya, Myanmar, Indochina, China and Taiwan.

136. ***Liparopsis postalbida*** Hampson, 1893: 154.
Holotype: ♂, [India, Nagaland], Naga Hills, 6.000 feet – NHMUK, London.
Schintlmeister, 2008: 137, Schintlmeister 2013: 237.

Taxonomic note. Populations from Sumatra and S Thailand differ in its male genitalia from Himalayan, Thailand, Vietnam, southeastern China and Taiwan populations. They are described below as:

Liparopsis postalbida brevispinus **ssp. nov.** (figs. 136c–e).
Holotype: ♂, Indonesia, N Sumatra, Samosir Isl.-Lake Toba, Partunghoan, 2°37'N, 98°44'E, 1600 m, 20. ix. 1991, leg. Graul & Schintlmeister – in coll. A. Schintlmeister, Dresden.

Paratypes (24 ♂♂):
Sumatra: 1 ♂, Aceh, Gayo-Berge, 38 km N Blangkejeren, 4°15'N, 97°05'E, 1800 m, 6. ii. 1994, leg. Malicky; 1 ♂, N Sumatra, Prapat, Holzweg 2, 30 km S Pematang Siantar, 2°48'N, 98°55'E, 30. viii.–28. ix. 1991, 1050 m, leg. Graul & Schintlmeister; 1 ♂, ibid, 20. v. 1991, leg. Dr. E. W. Diehl; 3 ♂♂, Samosir Isl.-Lake Toba, Partunghoan, 2°37'N, 98°44'E, 1600 m, 20. ix. 1991, leg. Graul & Schintlmeister; 1 ♂, Dairi Mts., 2°52'N, 98°29'E, 28. viii. 1979, leg. Diehl & Schintlmeister; 1 ♂, Dairi-E, 1850 m, 25. iv. 1981, leg. Dr. Diehl; 1 ♂, 9 km SE Tele, Dairi Mts., Primärwald, 1600 m, 2°31'N. 98°38'E, 21. ix. 1991, leg. Graul & Schintlmeister; 3 ♂♂, Tele, 2°31'N, 98°38'E, 4. vi. 1990, leg. Dr. E. W. Diehl; 1 ♂, Mt. Talamau, Westseite, 12 km S Talu, 0°08'N, 99°59'E, 11. ix. 1991, leg. Graul & Schintlmeister; 5 ♂♂, 22 km E Padang, Bukit Subang, 0°40'S, 100°12'E, 1200 m, 19.–20. x. 1981 (GU 15-44); 3 ♂♂, E Padang, 2. ix. 1984; 1 ♂, 30 km SW Solok, Gunung Talang, 1°00' N, 100°38'E, 1600 m, 7.–9. xi. 1981, leg. A. Schintlmeister.
Mentawei: 1 ♂, N Pagai, Siliakaop, 26. ii. -7. iii. 1992, leg. Diehl (GU 65-32). **Thailand:** 1 ♂, Chumphon Prov., 5 km S Phato, 9°46'N, 98°47'E, 155 m, leg. Langer, Löffler & Naumann; 1 ♂, Doi Mussoe, road Tak to Mae Sot, 16°50'N, 98°51'E, 800 m, 5.–10. iii. 2002, leg. Ihle (GU 79-77).

Diagnosis of ssp. *brevispinus*. Forewing lengths ♂♂ 14–16 mm. Not separable by external appearance from *Liparopsis postalbida postalbida* Hampson, 1893, which was described of NE India. Female is unknown.

Male genitalia differ from the nominotypical subspecies by shape of the valvae. The lower arm of the divided valva is significantly shorter. Upper valva arm is of same length as ssp. *postalbida*, but apical spine and also a spine originating below the apex of the upper valva arm are shorter. Phallus of ssp. *brevispinus* is longer and slenderer than in dissected specimens from India, north Thailand, Vietnam, China or Taiwan.

Diagnosis. *Liparopsis postalbida* is the largest species within the genus. The border of the fuscous postbasal area is curved toward the medial area. Postmedial area toward the dorsum is violet-brown filled. There is individually variability in contrast of forewing pattern. Hindwing

 | DOI:10.1163/9789004354258_007

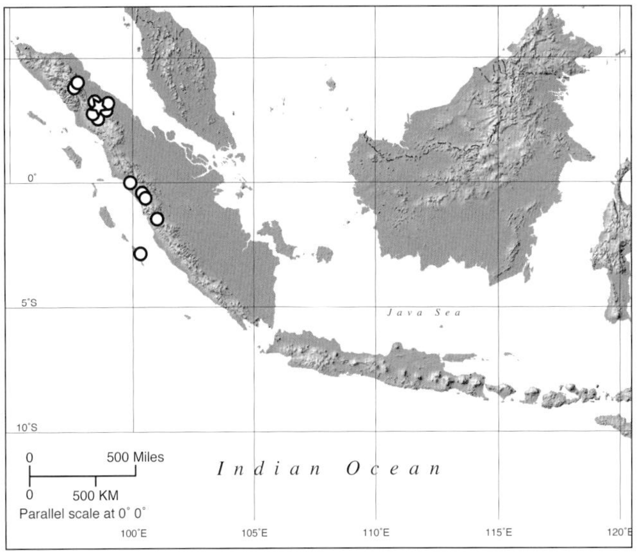

FIGURE 117 Distribution of *Liparopsis postalbida brevispinus.*

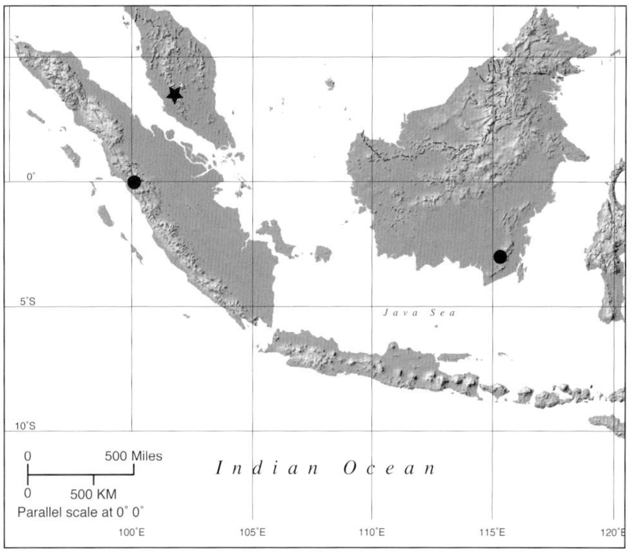

FIGURE 118 Distribution of *Liparopsis barlowi.*

displays a contrasting fuscous brown costal patch, which separates it from *L. barlowi.* Hindwings are not brownish tinged in the anal area as in *dierli* or *sundana.* Female of ssp. *postalbida* resembles the male, but they are much larger.

Uncus is long and slender and the tip is slightly pointed. Male genitalia differs from other congeners by shape of the divided valvae, which have in the upper arm three projections. One of them is a strongly sclerotized and bears a sharp spine. Female genitalia have large papillae analis and reduced apophyses. Ductus bursae is of medium slength. Corpus bursae is circular shaped. No signum.

Bionomics. Males occur in Sumatra commonly, mostly at altitudes between 800–1850 m. A male from Mentawei was collected below 200 m.

Caterpillars has been reared in Taiwan on *Quercus* (Fagaceae) (Sugi, 19 87: 286).

Distribution. *Liparopsis postalbida postalbida* is widely distributed in northern India, Nepal, Myanmar, Thailand Vietnam, Laos, China (Sichuan, Shaanxi, Yunnan, Hainan) and Taiwan. Ssp. *brevispinus* was found in S Thailand and Sumatra, but not yet in Malaysia.

137. ***Liparopsis barlowi*** Holloway, 1985: 110; fig. 22.
Holotype: ♂, Malaysia, W. Pahang, Genting Tea Estate, 2.000ft. – NHMUK, London.
Schintlmeister 2013: 237.

Diagnosis. External appearance is similar to *postalbida,* but moths are smaller in wingspan and paler in coloration. Forewing pattern is as in *postalbida,* but postmedial area toward the dorsum is less intensive brownish tinged. Most specimens display the white discal spot of forewing more contrasting than in *postalbida.* Hindwings are whitish, without margenta-brown scales. A fuscous costal patch, which is characteristic for *postalbida,* is absent. The female is unknown.

Male genitalia have the uncus with a knob-shaped tip. Valva bears a large triangular shaped projecion. Apical valva process is diagnostically rectangular angled. Phallus is slender and slightly curved with a pair of apical spurs at the tip. 8th sternite bears a pair of small central lobes at the posterior margin, which are smaller than in *sundana* or *dierli.*

Bionomics. Adults are rare in Sumatra (n = 1) and in Malaya (n = 4), but from South Kalimantan a good series (n > 20) was examined. The moth has been taken between 600 m and 1100 m above sea level.

Distribution. Distributed in Sumatra, Malaya and Borneo.

138. ***Liparopsis sundana*** Holloway, 1983: 48; pl. 4: 13; fig. 46.
Holotype: ♂, [Malaysia], Sarawak, Gunong Mulu Nat. Park, Site 15, Camp 2.5, Mulu, 1.000 m – NHMUK, London.
Schintlmeister 2013: 237.

Diagnosis. This is the smallest *Liparopsis.* It resembles by external appearance rather to *dierli* than *barlowi.* White

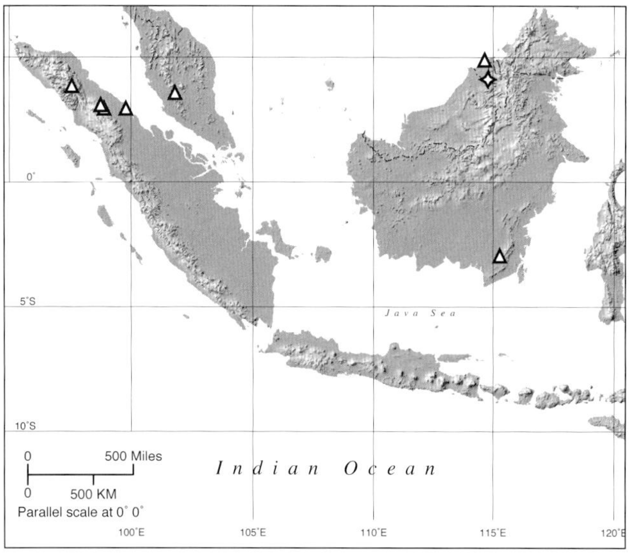

FIGURE 119 Distribution of *Liparopsis sundana*.

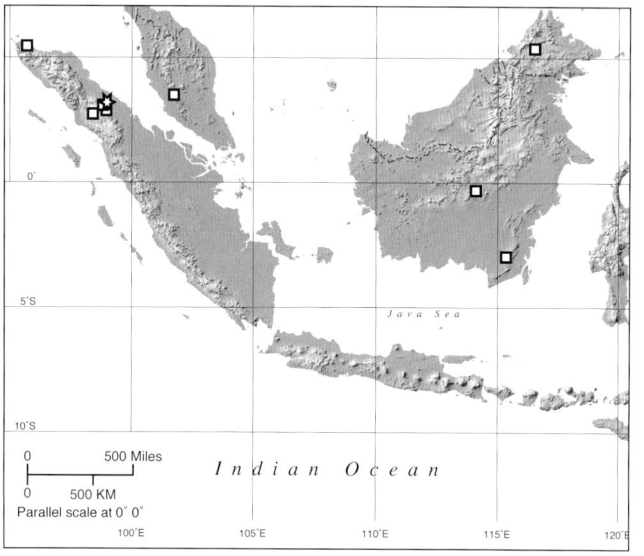

FIGURE 120 Distribution of *Liparopsis dierli*.

forewing color is less contrasting to the fuscous margenta brown pattern than *dierli*. Sometimes rather uniformly fuscous margenta-brown individuals (fig. 138d) and also transistional forms (fig. 138c) occure. Hindwings are whitish with a broad margenta-brown tinged anal area. In more fuscous forms entire hindwings may margenta-brown colored. Medial and postmedial fasciae are weakly marked as fuscous shadows on the hindwing. The presumable female of *sundana* is much smaller than *dierli* females. Hindwings displays an indistinct fuscous medial fascia, which is absent in *dierli*.

Male genitalia resembles *barlowi*, but the tip of the long uncus is less swollen. Valva has a third and pointed triangular projection, which is absent in *barlowi*. Phallus is curved, short and displays no spines at the tip. 8th sternite bears a larger pair of outwards curved central lobes at the posterior margin. Female genitalia have not yet been dissected.

Bionomics. Adults occur infrequently and were taken between 400–1685 m, most commonly at 1000 m above sea level.

Distribution. Distributed in Sumatra, Malaya and Borneo.

139. ***Liparopsis dierli*** Bender & Bender, 1984: 27; figs.
Holotype: ♂, [Indonesia], Sumatra, Dolok Merangir – Staatliches Museum für Naturkunde Karlsruhe (not located there in 2016).
Schintlmeister 2013: 237.

Diagnosis. Forewings display a contrasting white area in the upper part of the medial and marginal area. There are fuscous margenta-brown costal spots in postbasal and postmedial areas, which separates *dierli* from other congeners, in particular from the smaller *L. sundana*. Half of the hindwings are margenta-brown filled; the remaining part is white with a contrasting fuscous costal patch as in *postalbida*. Female is sexual dimorphic displaying reduced fuscous pattern in the white area of the forewings and weaker developed costal spots. Hindwings are whitish, but postmedial area light margenta-brown tinged.

Male genitalia have the uncus as in *barlowi*. Valvae are distinctive sinous shaped with only one smaller and rounded process. Phallus is curved with a pair of large curved processes at tip. 8th sternite bears a pair of large central lobes, which are not curved. Female genitalia have a large 8th abdominal segment. Papillae anales are very large and rounded. Postvaginal plate is squarish. Ductus bursae is slender and of medium length.. Corpus bursae is ellipsoid shaped. There is no signum but a fine band-like sclerotized structure resembles to females of *Pantanopsis diehli*.

Bionomics. The adults occur commonly up to 1200 m above sea level.

Distribution. Distributed in Sumatra, the Malayan Peninsula and Borneo.

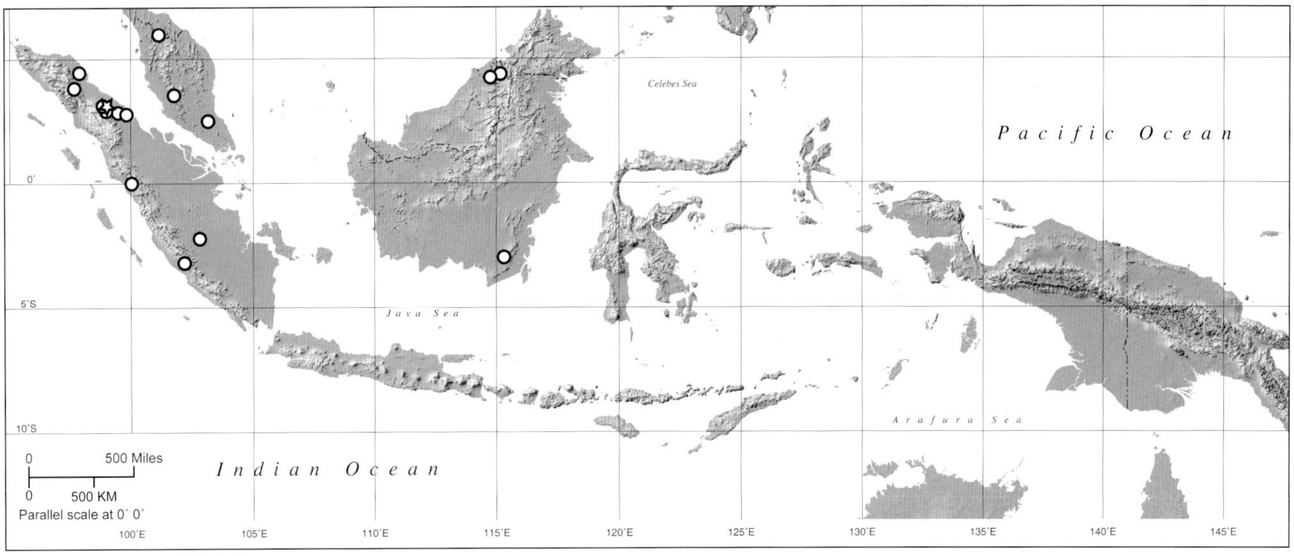

FIGURE 121 Distribution of *Pantanopsis diehli*.

Pantanopsis Kiriakoff, 1974: 387.
Type species: *Pantanopsis diehli* Kiriakoff, 1974
Schintlmeister 2013: 309.

Diagnosis. Small notodontids which are closely related to *Liparopsis*. Males have bipectinate black antennae with long rami until the tip. Antennae of females have shorter rami. Forewings are almost light greyish-white. Postbasal area fuscous margenta-brown or grey filled. Near the apex there is a large costal blotch of same color as the postasal area. Discal spot on forewing is not visible. Postmedial and marginal fasciae are blackish marked. Hindwing is whitish; on the costa there are two contrasting fuscous patches, divided by a contrasting squarish white field. Sexual dimorphism is limited, but female have darker hindwings.

Male genitalia have contrary to *Liparopsis* a short, broad and bilobed uncus. Socii are small and slender. Valvae are small and pointed at the apex. Saccus of tegumen displays a pair of projections. Phallus is straight with several smaller spurs. 8th abdominal segment are not specifically sclerotized. 8th sternite is heart shaped, while 8th tergite is of conical shape. Female genitalia have semicircular papillae annales, which are smaller than in *Liparopsis*. Both pairs of apophyses are reduced in length. Corpus bursae is large and of circular shape. It displays a long serrated band and central area of many strongly sclerotized micro thorns. No signum.

Pantanopsis comprises two species which are distributed in Indochina, Sundaland, the Philippines and Sulawesi.

140. ***Pantanopsis diehli*** Kiriakoff, 1974: 388; pl. 2: 3; fig. 10.
Holotype: ♂, [Indonesia], Sumatra sept., Deli [= Medan], Dolok Merangir, 180 m – ZSM, Munich.
Schintlmeister 2013: 309.

Taxonomic note. Two dissected females from Kalimantan and Sumatra are very different in shape of ductus bursae and development of the sclerotized band on the corpus bursae (figs. 140c, 140d). Several genitalia dissections of males (n = 8) from Sumatra, Malaya, Indochina and Borneo are, apart from individual variability in valvae, however not significantly different.

Diagnosis. The delicate species is readily to recognize by the contrasting margenta brown patches near the apex of all wings. Basal area of forewing is fuscous margentabrown filled. Postmedial fascia marked as a fine line. Marginal fascia on all wings is present as fine streaks. Hindwings of male have a slightly brownish tinged submarginal area, which vary individually in its darkness. Female is slightly larger than male and tend to have more fuscous hindwings, but a few females display, however lighter hindwings as male have.

Male genitalia are characterized by rather small valvae and shape of the bifurcate saccus with long projections. Uncus is less broad than *celebensis*. Valvae shape is variable as illustrated. Phallus is long and straight and bears triangular spines. 8th abdominal segments are not modified. 8th sternite is straight at posterior margin. Female genitalia are as described under genus diagnosis.

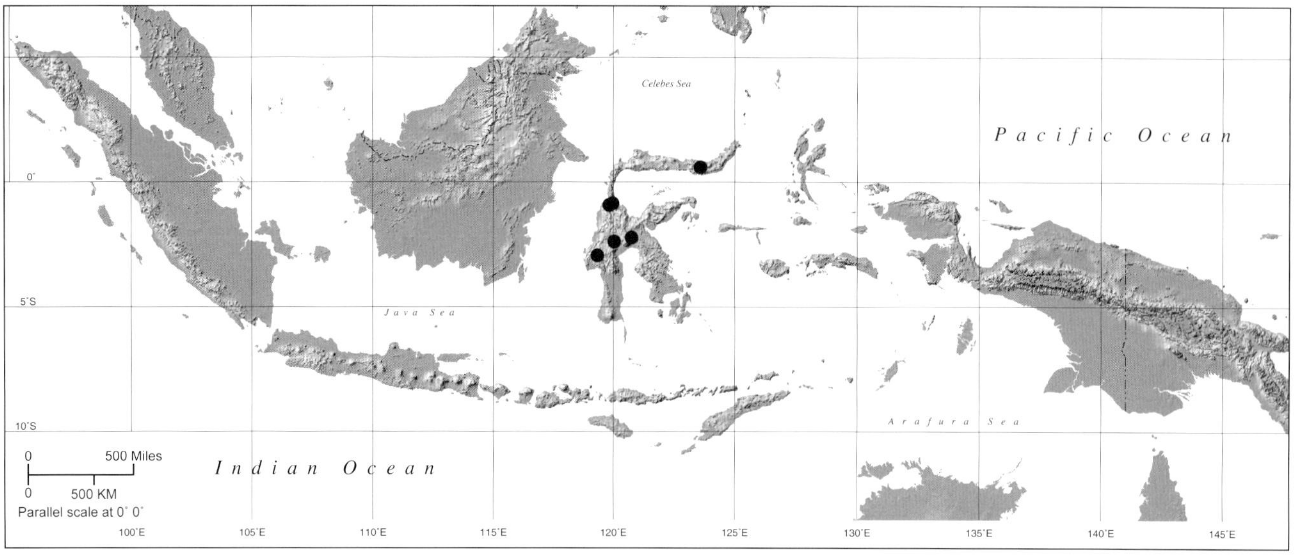

FIGURE 122 Distribution of *Pantanopsis celebensis.*

Bionomics. Imagines occur infrequent at lower altitudes up to 500 m, but a few males has been observed in Sumatra at 1200 m.

Distribution. Distributed in Thailand, Myanmar, Laos, Cambodia, Malaya, Sumatra, Borneo and the Philippines (Samar, Leyte and Mindanao).

141. ***Pantanopsis celebensis*** (Gaede, 1930): 632; pl. 84: a (*Liparopsis celebensis*).
Holotype: ♂, [Indonesia, Sulawesi], [North] Celebes – NHMUK, London.
Schintlmeister 2013: 309.

Diagnosis. *Pantanopsis celebensis* is not really similar to *P. diehli*. Forewing pattern is less contrasting and not margenta tingged as in *diehli*. Basal, postbasal and postmedial fasciae are doubled fuscous grey. Postbasal area is grey filled. Medial area of forewing is lighter than the remaining parts. Costal blotch near apex is as in *diehli*, but not margenta tingged. A fuscous grey discal dot is well visible. Hindwings are white and bear two contrasting grey costal patches. Sexual dimorphism is limited, but female have hindwings in postmedial area grey tinged.

Male genitalia resembles those of *diehli,* but the uncus is much broader and bilobed. Valva is not pointed at the apex. Saccus projections are short. Phallus is straight, long and slender as in *diehli*, but displays only one long spines at the tip and a further pocess in the everted endophallus. 8th sternite is contrary to *diehli* bilobed at posterior margin. Female genitalia have not yet been dissected.

Bionomics. Imagines occur infrequently between 1400–1800 m. A few males has been collected at 500 m and 2200 m above sea level.

Distribution. Endemic to Sulawesi.

Harpyia Ochsenheimer, 1810: 19.
Type species: *Bombyx milhauseri* Fabricius, 1775.
Schintlmeister 2008: 140; Schintlmeister 2013: 193.

Diagnosis. Medium sized to large insects. Antennae of males black and bipectinated with long rami in both sexes, except for the last 25% towards the tip which are filiform. Most species have a shiny silver or white forewing color with black pattern. Postbasal fascia is broad, often reduced for two blotches at costa and dorsum. Medial and postmedial fasciae of forewing are marked as a rows of black dots, with a larger black blotch at costa. Hindwings are whitish with prominent black anal spot. Fringe of hindwings is black-white checkered. Sexual dimorphism is limited.

Male genitalia have a large, squarish and bilobed uncus. At the uncus base a pair of small gnathoi arises. Socii are small. Valvae are club-shaped with a tapered costal process. Phallus is often slender with a spine of variable size at the tip. Posterior margin of 8th sternite is specifically modified by its shape (concave, convex or sinous) and areas of dense setae. Anterior margin is bilobed with a pair of lobes. 8th tergite is less modified. Female genitalia show a broad 8th abdominal segment. Papillae anales are large

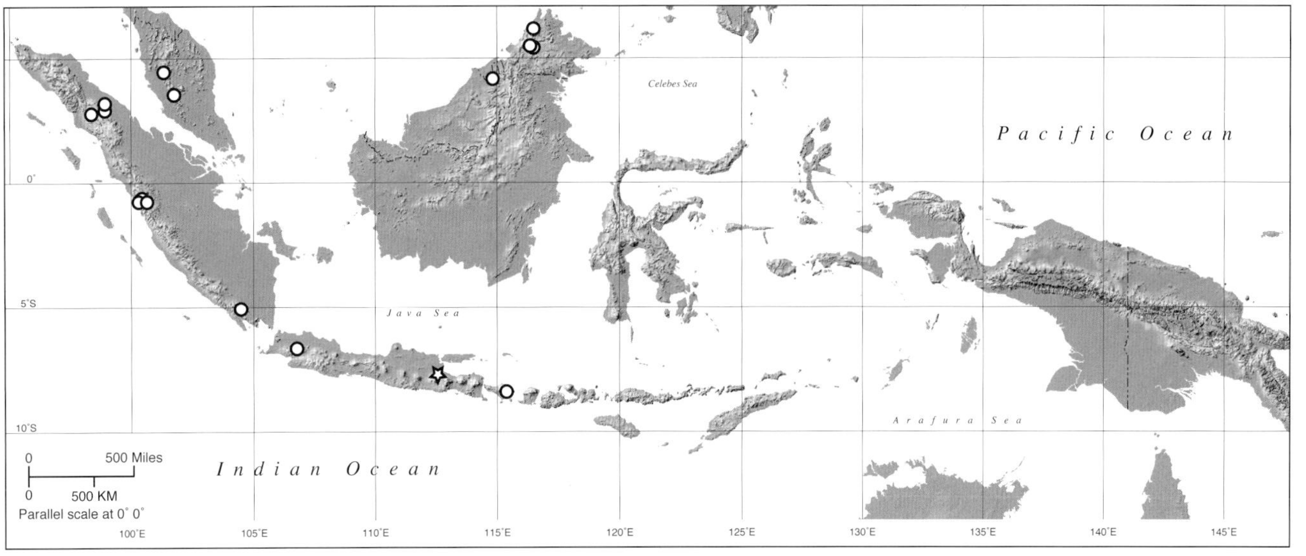

FIGURE 123 Distribution of *Harpyia kalisi.*

and rounded. Both pairs of apophyses are short. Antrum is long and strongly sclerotized. Ductus bursae is short. Corpus bursae is large with a knoted structure at the bottom. No signum.

Caterpillars are – as far as they are known – bizarre and similar in external appearance in all *Harpyia*. They feed on various Fagaceae, in particular *Quercus*.

Harpyia comprises 14 species. The genus is distributed throughout the palaearctic region, from northern Africa to Japan and Afghanistan, whole of India, Pakistan, Nepal, Indochina, Sundaland and Java.

142. ***Harpyia kalisi*** (Roepke, 1943): 79; fig. 8 (*Macrohoplitis kalisi*) **stat. rev.**
Holotype: ♂, [Indonesia], East Java, Djunggo-Ardjuno, 1500 m – NBC, Leiden.
Schintlmeister 2013: 194.

Taxonomic note. *Macrohoplitis kalisi* was placed by Nakamura, 1960: 36 as a junior synonym of *Harpyia microsticta* (Hampson, 1893). The latter was described of Naga Hills, 6000 ft. [NE India, Nagaland]. In Sundaland three *Harpyia* species occur and as genitalia dissection of the holotype of *microsticta* showed (fig. 142Aa), this species is quite distinct by the shape of the phallus and does not occur in Indochina nor Sundaland. China, Indochina and Taiwan are populated by a further different species, for which the oldest available name is *Damatoides baibarana* Matsumura, 1927, described from Taiwan. The taxon was formerly (Nakamura, 1960: 36) treated as a subspecies of *H. microsticta.* Due to the sympatrically occurence of three species of the complex in Sumatra, the taxonomic value of different phallus shapes and shapes of the 8th sternite must be evaluated newly. The conclusion is, that *H. kalisi* and *H. baibarana* must be reinstalled in the rank as genuine species (**stat. rev.**) due to distinct differences in these characters.

Sumatran and Bornean populations which are associated here with *kalisi,* match this taxon by external appearance (size, forewing shape, dorsal markings of forewing) well. The holotype female from Java of *kalisi* was, however not dissected.

Diagnosis. *H. kalisi* is the largest species within the group in Sundaland. It differ from *H. baibarana* by whitish rather than greyish forewing color. Forewing pattern is more similar to *baibarana* than to *kobesi* or *koeppeli.* Forewing shape have an angled outer margin, which is a good character to recognize the species. Outer marigin below the angle is slightly concave, but in *koeppeli* with a deeper dent and in *koebesi* rather straight. Dorsal streak in the medial area of the forewing is black and shorter than in *koeppeli.* Hindwings are whitish with black anal markings. Female is much larger than the male and postmedial area of hindwing is blackish tinged.

Male genitalia have a slightly notched and relatively long uncus, which is longer than in other species of the *microsticta* group. A pair of socii arising at the uncus base are small. Apart from socii there are a variable number of further small spines or projections at the uncus base. Valva is club shaped and bears a costal process. Phallus is variable in its shape, but never sinous curved as in *baibarana* or zig-zag shaped as in *microsticta.* Most of dissected males (n = 16) show the phallus process straight and with

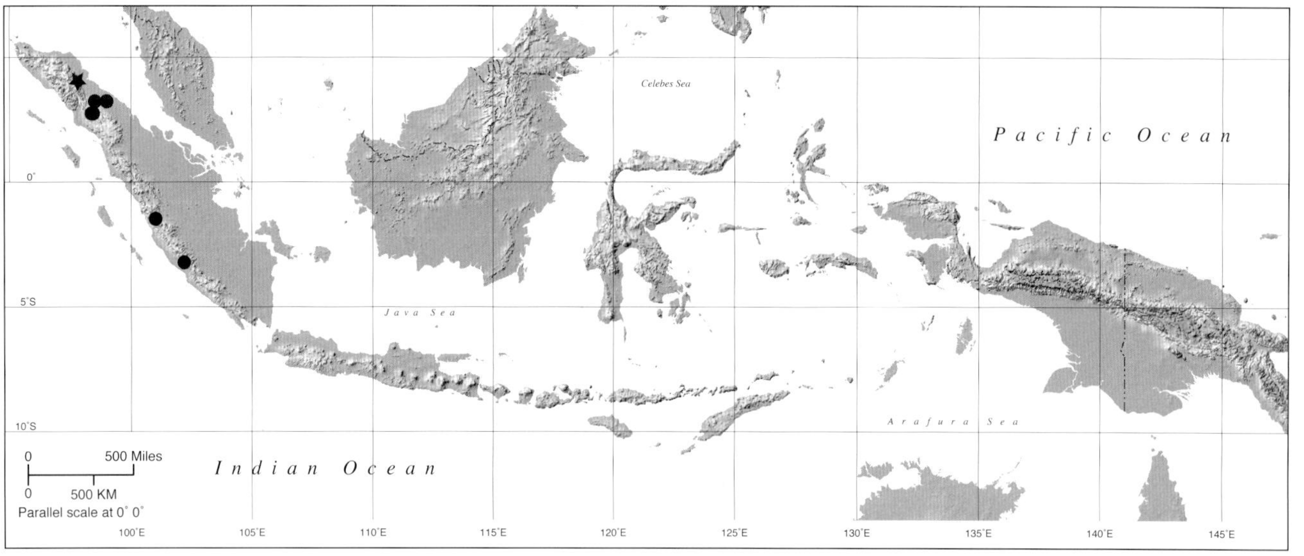

FIGURE 124 Distribution of *Harpyia kobesi*.

a knob-shaped tip, but one dissected male (fig. 142c) have the process strongly curved and pointed. 8th sternite have the posterior margin concave with a large area of dense setae. In *baibarana* the posterior margin is only slightly notched, see also Schintlmeister 2008, figs. 617, 619. Female genitalia have papillae anales rounded and falcate. Postvaginal plate is large and broad. Antrum and ductus bursae are short. Corpus bursae is the largest in the genus with a knoted structure at the bottom. No signum.

Bionomics. By far the most common *Harpyia* in Sundaland but not really frequent. Adults occuring from 100 m up to 1600 m above sea level.

Distribution. Known from Myanmar, Malaya, Borneo, Sumatra, Java and Bali.

143. *Harpyia kobesi* spec. nov. (figs. 143a–d).

Holotype: ♂, Indonesia, Sumatra, Aceh, Gayo-Berge, 44 km N Blankjeren, 4°17' N, 97°05' E, 1800 m, 2. ii. 1994, leg. Malicky – in coll. A. Schintlmeister, Dresden.

Paratypes (11 ♂♂, 2 ♀♀):

Sumatra: 2 ♂♂, Dolok Merangir, 9. viii. 1983 (GU57-77); 1 ♂, dto. 19. ii.–8. iii. 1981, leg. Dr. Diehl; 1 ♂, Dairi Mts., 2°52'N, 98°29'E, 1600 m, 28. viii. 1979, leg. Diehl & Schintlmeister (GU65-37); 2 ♂♂, Dairi Mont., 1200 m, ii. 1981, leg. Dr. Diehl; 1 ♀, Dairi Mts., 10. x. 1980, leg. Dr. E.W. Diehl (GU 65-40); 1 ♀, dto., 8. xi. 1970, 1 ♂, Berastagi, 1500 m, ii.1985 (GU 57-76), 1 ♂, Brastagi, 1800 m, 6. x. 1972, leg. E. W. Diehl; 1 ♂, Tele, 18. ix. 1976, leg. Dr. E. Diehl; 1 ♂, Sitahoan, 80 m, 24. xii. 1981, leg. Dr. E. Diehl; 1 ♂, Slopes of Mt. Korintji, 5000 ft., ix.–x. 1921, leg. C., F., & J. Pratt; 1 ♂, dto., viii.–ix. 1921.

Etymology. Named for the late Prof. Dr. Lutz Kobes (b. 26. viii. 1933 Crimmitschau d. 21. vii. 2016 Göttingen), former president of the Heterocera Sumatrana Society e.V. who devoted much of his life time for the exploration of Sumatran moths, in particular the Noctuidae.

Diagnosis. Forewing lengths ♂♂ 21–23 mm, ♀♀ 26 -28 mm. The smallest species of the *microsticta* group. About 3 mm shorter in forewing length than *kalisi*. Antennae are in both sexes black and bipectinated with long rami. The last 20% toward the tip is filiform. Forewing color is shiny white, lighter than in *kalisi* or *koeppeli*. Shape of the forewing marigin is gently concave or straight, not angled. Dorsal spot of the forewings is like a streak, but there is a blackish tinged squarish area below this streak. Postmedial area contains several wavy fasciae, mostly marked as rows of black dots. Hindwings are light greyish with a blackish-white checkered fringe and an inconspicous blackish anal spot. Female resembles the male, but hindwings are darker at their basal part.

Male genitalia are similar to *kalisi*. Uncus is broader and shorter than in *kalisi* and uncus tip is slightly bilobed. Socii are small. Apart from socii there are two ore more small projections at the base of uncus. Valvae are narrow and bear a costal process, which is slightly longer than in *kalisi*. There is a single male (fig. 143c), which differs in valva shape and its costal spines from other dissected genitalia of *kobesi*. Phallus is slightly curved and thick; the process at the phallus tip is reduced in length. 8th

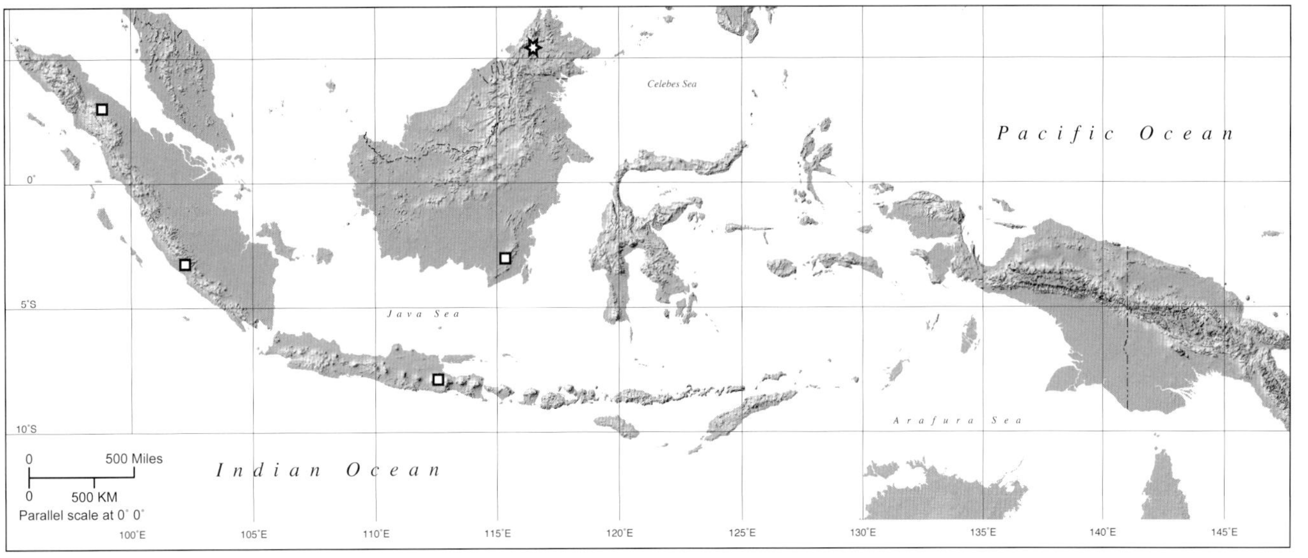

FIGURE 125 Distribution of *Harpyia koeppeli.*

sternites are very variable in shape of posterior margins as illustrated. But these different forms have, however, their anterior margin of the 8th sternite including their lobes are distinctly and homogenously shaped concave. 8th tergite is less modified. Female genitalia are small in size in comparison to other *Harpyia.* Apophyses are short and finer than in *kalisi.* Antrum is long and the ductus bursae is longer than in other congeners. Corpus bursae is ellipsoid rather than circular with a knoted structure at the bottom.

Bionomics. This beautiful species appears seemingly infrequent. Most specimens were taken between 1500–1800 m in primary forests, but three males has been observed at 80 m and 180 m above sea level in gardens.

Distribution. Endemic to Sumatra.

144. ***Harpyia koeppeli*** **spec. nov.** (figs. 144a–d).

Holotype: ♂, Malaysia, Borneo, Sabah, Tambunan reg., Trus Madi Mts., 1200 m, 5°35' N, 116°27' E, 12.-24.i.2007, leg. A. Sochivko (GU17-65a) – in coll. A. Schintlmeister, Dresden.

Paratypes (8 ♂♂):

Borneo: 2 ♂♂, Sabah, Trus Madi, 1150 m, 9. -14. iv. 1996 and iii. 2003, leg. Martini;

Kalimantan: 1 ♂, Kalimantan Selatan, 30 km E Kandangan, 15 km NE of Loksado, 1100 m, 2°52'S, 115°38'E, xii. 1997, leg. St. Jakl (GU 46-34);

Sumatra: 3 ♂♂, Sindar Raya, 330 m, 18. xi. 1976, leg. Dr. E. Diehl (13-01); 1 ♂, 30 km SW Siantar, Holzweg 2, 1050 m, 14. iv. 1969, leg. Dr. Diehl; Lebong Tandai, 18.–23. x. 1923, leg. C. J. Brooks (BM 1749); 1 ♂, Lebong Soelit [Bengkulu district], Java: 1 ♂, Tengger, Singolangoe, iv. 1934, leg. F. P. A. Kalis (BM 2017-04).

Etymology. Dedicated to Dr. Christian Köppel, Gaggenau, who is, as a lepidopterist, engaged for the protection of our nature. He initiated with his friends succesfully the founding of the Black Forest National Park in Germany (established 1. i. 2014), a project which is urgently needed also for many endangered areas in Indonesia.

Diagnosis. Forewing lengths ♂♂ 24–28 mm. Slightly (1–2 mm) smaller than *kalisi* and clearly (3–4 mm) larger in forewing length than *kobesi.* Antennae are bipectinate with long rami and black. The last 20% toward the tip are filiform. Forewing color is shiny silvery white as in *kalisi,* but less light than in *kobesi.* Shape of forewing marigin have a prominent dent in the lower part. Forewing margin is not angled as in *kalisi.* A contrasting black and long dorsal streak is longer than in any other known congeners. Postmedial area of forewing contains several indistinct wavy fasciae, which includes only a few black dots. Hindwings are whitish with an almost blackish fringe. There are less contrasting fuscous anal markings. Female is unknown.

Male genitalia are characterized by shape of the deepley notched uncus. Both lobes are pointed at their tips. Socii are slenderer and longer than in *kalisi* or *kobesi* and there are no further projections at the uncus base. Valvae are club shaped, and slenderer than those of *kalisi.* Phallus shape is broad and resembles those of *kobesi* but the tip is tapered. 8th sternite have a wavy or convex posterior margin with many dense setae. Shape of margins vary individually, but the center is always notched. 8th tergite is less modified.

Bionomics. Only a few males are currently known, which has been taken between 350–1200 m in rain forests.

Distribution. Known from Sumatra, Borneo, and Java.

Teleclita Turner, 1903: 45 (key), 53.

Type species: *Teleclita cydista* Turner, 1903 [= *Notodonta dryinopa* Dodd, 1902].

Schintlmeister 2008: 145; Schintlmeister 2013: 454.

Diagnosis. Large notodontids. Antennae of males are fuscous brown and bipectinate in both sexes except for the last 25% towards the tip. Forewings are elongate and narrow. Discal spot is indistinct developed and of circular pale brown. Characteristic pattern elements are a prominent black apical streak and three short black lines running rectangularly to the costa. Sometimes blackish postbasal and postmedial fasciae occur. Hindwings are whitish, having contrasting black anal markings. Hindwing fringe is black. Sexual dimorphism is on low level. The females have hindwing in the submarginal area blackish tinged, fading whitish toward the base.

Male genitalia are specially modified. Uncus is large and widely bilobed, sometimes it displays a small central projection. Further processes and projections arising from the uncus arms. Socii are large and of rounded shape. Tegumen is in the anterior part bulbous shaped. It bears a strongly sclerotized plate, from which a serrated, long process arises. Juxta is specifically and strongly sclerotized. Valva is large with rounded apex. Sacculus is large and bears a spine; digitus is well developed and have a serrated, triangular spine. Phallus is straight and relatively short. At the tip there is a pair of serrated and strongly sclerotized flaps. 8th sternite is large and unusual long with specifically sclerotizations. The anterior margins of both abdominal segments are bilobed. Female genitalia have two pairs of robust, medium sized apophyses. Antevaginal plate is large, and trapeze-shaped. Ductus bursae is slender and medium sized. Corpus bursae is ellipsoid. No signum.

Caterpillars of *Teleclita strigata dryinopa* resembles by external appearance and resting position remarkable to South African *Afroplitis bergeri sulaki* Schintlmeister & Witt, 2015 and both species are feeding on *Terminalia* (Combretaceae).

Teleclita comprises currently five species. Distributed in entire oriental region from India, China up to New Guinea and Australia. *Afroplitis* Kiriakoff, 1964 is a closely related African genus.

145. *Teleclita strigata* (Moore, 1879): 68; pl. 3: 13 (*Hoplitis strigata*).

Lectotype: ♂, [S India], Kanara – NHMUK, London by **present designation.**

Schintlmeister 2013: 455.

Taxonomic notes. *Hoplitis strigata* was described of 2 ♂♂, which are according to the original description both from "Kanara". In fact the syntype in ZMHU, Berlin from the Atkison collection comes from Darjeeling as its label show. To stabilize the nomenclature, hereby the ♂ from NHMUK with the following labels is designated as lectotype: "S India | Ward | 241", "Moore Coll. 94-106", red ringed, circular label "Type", yellow label " Photo done by A. Schintlmeister #4216". A lectotype label was added. The remaining male from Darjeeling becomes a paralectotype.

The taxa in the *T. strigata* complex show well developed geographically variation and trated here, except the Sulawesian populations, as one species because of the similarities of male and female genitalia. Thus *cathana, insignifica, cinnamomea* and *dryinopa* are all downgraded here as subspecies of *strigata* **(stat. nov.).**

Teleclita strigata strigata occurs in Sri Lanka and South India. Males have contrasting whitish and golden pattern on grey forewings. Populations from the Himalaya, southern China and Indochina belong probably to a further, undescribed subspecies with more uniformly greyish-brown colored forewings.

Teleclita strigata cathana (Schaus, 1928): 81 (*Neopheogia cathana*), **stat. nov.**

Lectotype: ♂, Philippine Islands, Luzon, Mt. Makiling – USNM, Washington.

This subspecies is characterized by pale greyish, golden tinged forewings in both sexes with contrasting black basal fascia and grey filled basal area. In Bali transitional forms occurs, where the males resembling already somewhat to the ssp. *mediocana*, but the females are as *cathana* from Sundaland.

The lectotype of *cathana* is an aberrative, fuscous male, but the paralectotype female – lecto- and paralectotype are illustrated in Schintlmeister, 2016: 566 – as well as the other examined specimens from the Philippines are matching well to Sundanian populations in its markings.

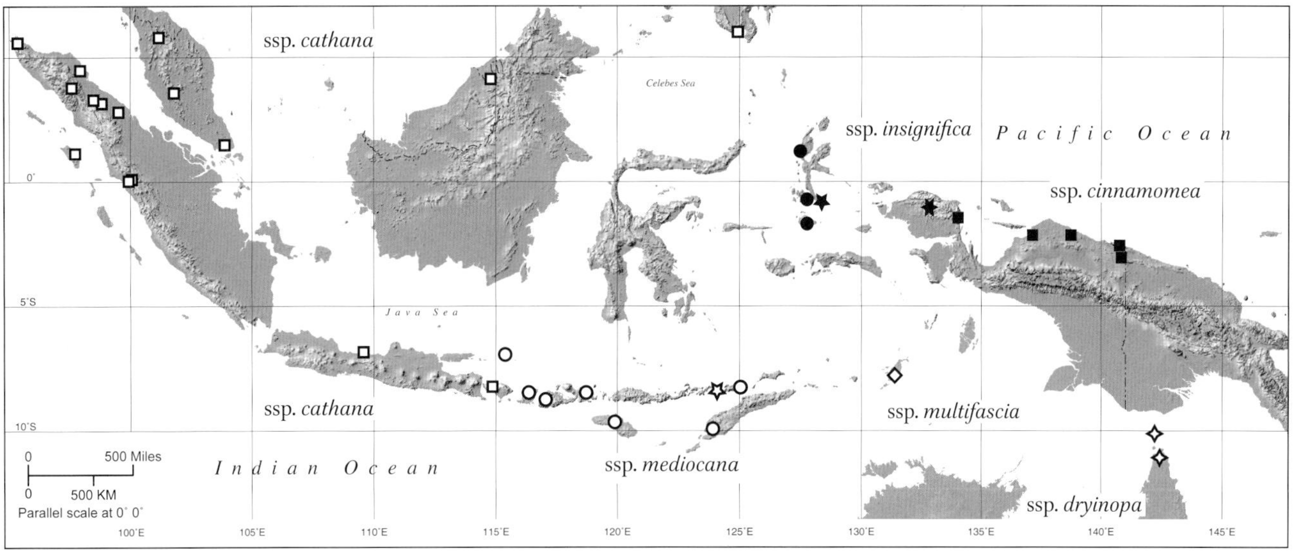

FIGURE 126 Distribution of *Teleclita strigata*.

Forewing color are in adults from Luzon, Panay and Mindanao however slightly darker.

Holloway, 1983 described from Singapore *Teleclita sundana*, which was synonymized by Schintlmeister & Lourens, 2010: 140 with *cathana*.

T. strigata cathana is distributed in the Philippines, Thailand, Myanmar, the Malayan Peninsula, Sumatra, Nias, Borneo, Java and Bali.

Teleclita strigata mediocana **ssp. nov.** (figs. 145d, 145e, 145l)
Holotype: ♂, Indonesia, Nusa Tenggara Timur, Pantar Isl., Tanah Labang, 350 m, 8°19'S, 124°16'E, 9.–21. iii. 2006, leg. Jakl – in coll. A. Schintlmeister, Dresden.

Paratypes (34 ♂♂, 8 ♀♀):
Pantar: 7 ♂♂, Tanah Labang, 350 m, 8°19'S, 124°16'E, 9.–21. iii. 2006, leg. Jakl (GU 94-30);
Alor: 8 ♂♂, 5 ♀♀, Moru, 500m, 8°17' S, 124°29'E, 22. iii. -3. iv. 2006, leg. Jakl (GU 93-18, GU 93-19, GU 93-20, GU 94-38); 2 ♂♂, dto., 1.–8. iii. 2006 (GU 94-37);
Timor: 2 ♂♂, 3 ♀♀, Buraen, 35 km E Kupang, 350 m, 10°15'S, 123°49'E, 26. i. -9. ii. 2006, leg. St. Jakl (GU 94-33, GU 94-34);
Sumbawa: 2 ♂♂, vii. 2004, leg. St. Jakl (GU 94-29); 1 ♂, Parado, 60 km to Bima, 60 m, 21.–30. xii. 1996, leg. Andang (GU 94-35);
Lombok: 1 ♂, Pusuk Hills, 700 m, 10°21'S, 123°39'E, 16.–22. iii. 2009, leg. St. Jakl (GU 94-03);
Kanguan: 3 ♂♂, Kanguan Isl., 100 m, 6°50'S, 115°15'E, xi. 2004, leg. Jakl (GU 46-58).

Diagnosis of ssp. *mediocana*. Forewing length ♂♂ 23–27 mm, ♀♀ 29–34 mm. A small subspecies of *strigata* – only ssp. *multifascia* is in average 2 mm smaller in forewing length. Forewings of male are fuscous ash-grey, much darker than ssp. *cathana*. Near the basal area there is an indistinct, but contrasting pale yellowish-brown broad streak. A second yellowish-brown circular to ellipsoid area in the medial area, near the costa is accompanied by a black discal streak. The other pattern elements are as in other subspecies. Hindwings are whitish with a pale brown marginal fascia. The sexual dimorphic female is characterized by a pale greyish contrasting medial area on forewing. In ssp. *multifascia* the medial area of forewings is less contrasting. The only known female of ssp. *insignifica*, the lectotype, resembles by the pale medial area much to ssp. *mediocana*, but *insignificata* displays a broader fuscous area on hindwing. Two of five females from Alor (fig. 145w) having the grey medial field of forewing darker and displays a contrasting broad golden transversal strip from the basal area toward the marigin.

Male genitalia differs due to individual variability not significantly from those of ssp. *cathana* or ssp. *multifascia*. Female genitalia are distinguished by a broadand wider bilobed vaginal plate from the other subspecies, but only three female genitalia of *mediocana* have been dissected.

The subspecies *mediocana* is distributed in Kanguan, Lombok, Sumbawa, Alor, Pantar and Timor.

Teleclita strigata multifascia **ssp. nov.** (figs. 145f, 145m)
Holotype: ♂, Indonesia, Tanimbar Isls., Yamdema isl., 21 km Saumlakki, 150 m, 7°47'S, 131°18'E, 27. i x.-11. xii. 2005, leg. Jakl – in coll. A. Schintlmeister, Dresden.

Paratypes (32 ♂♂, 18 ♀♀):

Tanimbar: 23 ♂♂, 12 ♀♀, Yamdema isl., 21 km Saumlakki, 150 m, 7°47'S, 131°18'E, 27. ix.–11. xii. 2005, leg. Jakl (GU 94-39); 5 ♂♂, 3 ♀♀, dto., 5.–28. xii. 2006 (GU 94-31, GU 94-32); 3 ♂♂, dto. 6. ii.-30. iii. 2007; 1 ♀, dto., 15.–30. xii. 2007; 1 ♂, 2 ♀♀, East Yamdena, Arui Bab, 120 m, ii. 1997 (GU 46-93).

Diagnosis of ssp. *mutifascia*. Forewing length ♂♂ 22–25 mm, ♀♀ 22–32 mm. This subspecies is by far the smallest in the group, 2 mm smaller in forewing length than ssp *mediocana* and 6 mm smaller than ssp. *cinnamomea* from New Guinea. Males display on the fuscous ash-grey forewings contrasting double post basal and double postmedial fasciae, consisting from thick blackish lines, which are unique in the *strigata* complex. The pale yellowish discal spot is relatively small and circular, mostly filled with a indistinct fuscous spot. Hindwings are whitish with a pale brown marginal fascia. The sexual dimorphic female resembles in facies ssp. *mediocana,* but the pale medial area is less contrasting.

Male genitalia differ not significantly from those of ssp. *multifascia.* Female genitalia differ seemingly by a broader and wider bilobed vaginal plate from the other subspecies, but only one female genitalia of *multifascia* have been dissected.

Teleclita strigata multifascia is endemic to Tanimbar.

Teleclita strigata insignifica (Rothschild, 1917): 250; pl. 4: 6 (*Hoplitis insignifica*), **stat. nov.**

Lectotype: ♀, [Indonesia, Halmahera], Dammer Island – NHMUK, London by **present designation**.

Hoplitis insignifica was described according to the original description from 2 ♀♀, but in fact the author traced in NHMUK 1 ♀ and 1 ♂ syntypes. To stabilize the nomenclature the ♀ with the following labels is designated as lectotype: "*Hoplitis insignifica* Rothsch.", "Dammer, 11. XI. 98 (H. Kühn).", blue label "Notodontidae genitalia prep. No. 43", red ringed, circular label "Type", yellow label " Photo done by A. Schintlmeister #9277". A lectotype label was added. The lectotype is illustrated on fig. 145ac. The remaining male with the data "Dammer, 28. XI. 98 (H. Kühn)." becomes the paralectotype.

The currently only known female (the lectotype) resembles by its pale medial forewing area remarkably in facies to females of ssp. *mediocana.* But males of *insignifica* are less similar to males of *mediocana* or *multifascia* by an indistinct costal markings and a less pronounced pale discal patch. Males of *T. strigata insignifica* resembles ssp. *cinnamomea* by forewing pattern, but they are smaller in forewing length.

Male genitalia of the dissected paralectotype from Dammer have shorter saccular spines of the valva than specimens from Halmahera, but since no further males from Dammer are available, it is currently not possibly to verify this. Female genitalia of the lectotype is not significantly different from ssp. *mediocana*, but the ductus bursae is shorter and the corpus bursae smaller than in other dissected females of *strigata,* perhaps due to the process of genitalia preparation.

T. strigata insignifica is distributed in Dammer, Halmahera, Obi and Bacan (genitalia of males from all islands have been dissected).

Teleclita strigata cinnamomea (Rothschild, 1917): 250; pl. 7: 6 (*Hoplitis cinnamomea*), **stat. nov.**

Holotype: ♂, [Indonesia, W Papua], Dutch New Guinea, Central Arfak Mountains, Ninay Valley 3.500 ft. – NHMUK, London.

This is the largest known subspecies of *strigata.* Males resembles by their fuscous color and the forewing pattern much to some males from Bali, but the grey filled basal area is less pronounced in *cinnamomea.* Female is sexual dimorphic and similar to ssp. *cathana,* but the medial area displays some fuscous shadows. A second female is much darker in color than the illustrated one.

Male genitalia have the tegumen process less serrated, but only two male genitalia were examined. The female genitalia are not dissected.

T. strigata cinnamomea is currently known only from W Papua.

Teleclita strigata dryinopa (Dodd, 1902): 43; 194 (*Notodonta dryinopa*), **stat. nov.**

ST: unspecified number of ♂♂, ♀♀, Australia, Queensland, Townsville – NHMUK, London.

This subspecies is characterized by the well developed black rectangular costal blotch in the postbasal area in both sexes. External appearance resembles somewhat to ssp. *strigata.*

Male genitalia are already more modified from the other subspecies of *strigata* by the shape of the bulbous plate and a thicker process arising from this plate. Apex of valvae is more acute than in other taxa of *strigata.* Sclerotizations of 8th sternite is centrally curved and not straight as in other subspecies. Female genitalia have a wider, posteriorly rather straight vaginal plate.

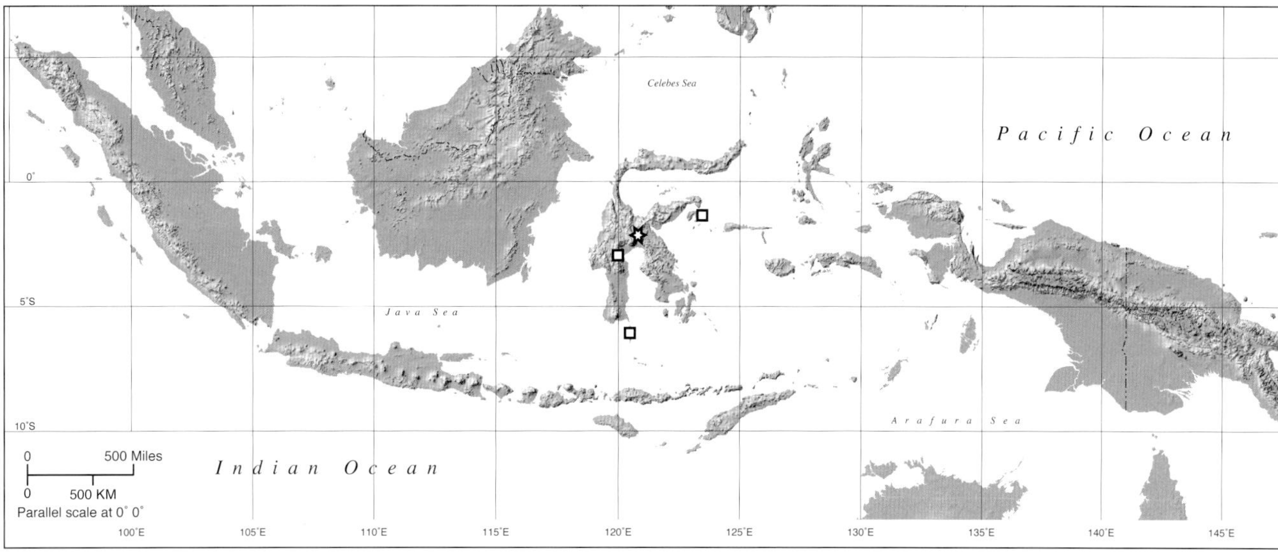

FIGURE 127 Distribution of *Teleclita rasilis.*

T. strigata dryinopa is distributed in Australia and Tasmania.

Diagnosis. *Teleclita strigata* is characterized by elongated forewings which have a black apical streak and a greyish costa. A whitish-grey basal area occur in all subspecies of *strigata* but not in the similar *T. rasilis*. Forewings vary in paleness and in most subspecies also fuscous reddish brown forms occur. Hindwing of male is white with contrasting black anal markings. Female is sexual dimorphic having the medial area of forewing paler and almost without pattern. Postmedial and submarginal areas of hindwing are fuscous grey.

Genitalia are as described in the genus diagnosis. Diagnostic features for identification of the species are a tapered valvae processes. Male genitalia showing individual variation in shape of the uncus structures – sometimes with triangular central projections – and the shape of the socii. Shape and length of tegumen- and valva processes as well as length of phallus are subjects of individual variation. Female genitalia have a trapeze-shaped vaginal plate.

Geographical variation and the many subspecies are dealt with above.

Bionomics. Adults prefer lowlands up to 500 m above sea level and occure mostly commonly. In W Papua *strigata*, where the species is seemingly rare, imagines has been taken at altitudes up to 1700 m.

Distribution. Widely distributed in several subspecies (see above): entire India, Nepal, southwestern China (Yunnan), Indochina, Sundaland, the Philippines, the lesser Sundas, Tanimbar, Moluccas (not yet found in Ambon and Seram), W Papua and Australia.

146. *Teleclita rasilis* spec. nov. (figs. 146a–d).

Holotype: ♂, Indonesia, Sulawesi, Mt. Sampuraga, 1400 m, 2°10' S, 120°45' E, 11.–12. i. 1995, leg. V. Siniaev & Tarasov (GU 51-48) – in coll. A. Schintlmeister, Dresden.

Paratypes (4 ♂♂, 2 ♀♀):

Sulawesi: 1 ♀, Puncak, Palopo, 900–1300 m, 3°00'S, 120°10'E, xii. 1996; 2 ♂♂, 1 ♀, Dumogo-Bone National Park, Site 14, 1040 m, 23.–25. ii. 1985, leg. J. D. Holloway (BM 1480); 1 ♂, Paloe [= Palu], Gunung Tompoe, 2700', |i. 1937, J. P. A. Kalis;

Peleng: 1 ♂, 2 km W Sambiut, 150 m, vii. 1998 (MWM 15.726).

Further material (no paratype):

1 ♂, **Saleyer**, Somarisi, 500 m, xii. 1938, leg. J. P. A. Kalis.

Diagnosis. Forewing lengths ♂♂ 27 and 25 mm, ♀ 36 mm. Forewing color is violet brown. A black apical streak is present. The whitish grey basal area lacks a black fascia and is not filled whitish grey, as it is characteristic for *T. strigata. T. rasilis* display from the base toward the tornus and along the dorsum a prominent black, indistinct transversal fascia. Postmedial area is paler than the remaining part of the forewing. Forewing pattern is as in *strigata*. Hindwings are whitish with black anal markings. Female is sexual dimorphic. Forewing pattern is indistinct developed; the black transversal fascia weakly visible.

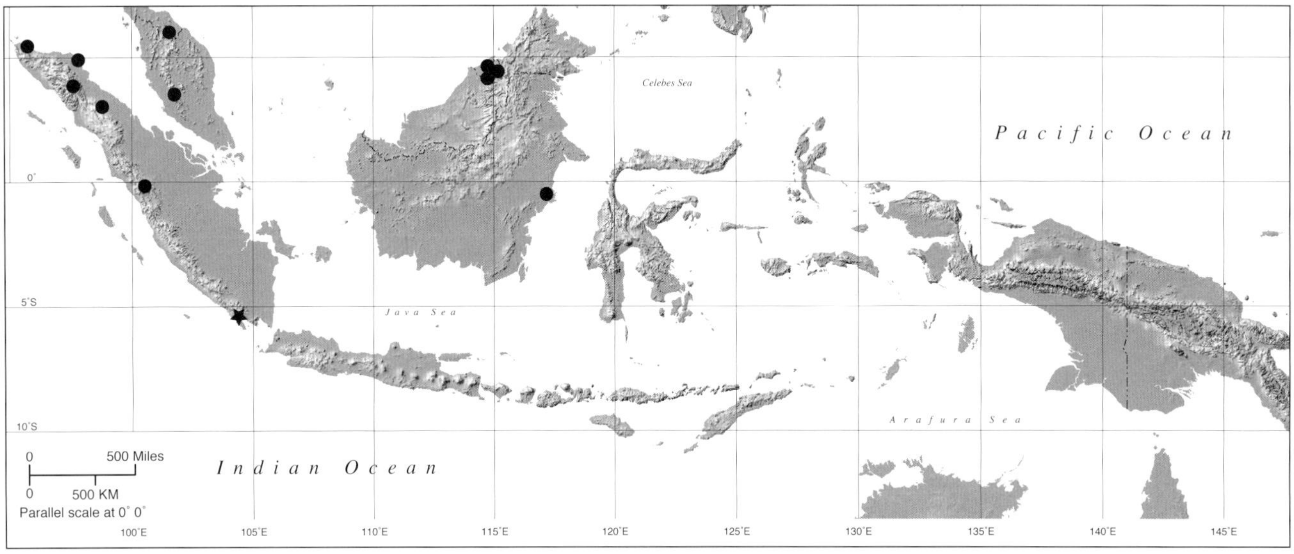

FIGURE 128 Distribution of *Teleclita flavisticta.*

Hindwings have a thin blackish marginal fascia, fading toward the medial area, while in *strigata* there is a broader fuscous band. *Teleclita rasilis* is best to separate from *strigata* by the pale area in the center of the postmedial area and the absent greyish basal area.

Male genitalia resembles *strigata*, but the arms of the widely bilobed uncus are much larger and rounded, not pointed as in *strigata*. Uncus bears a triangular central projection. Tegumen processes are as in *T. strigata dryinopa* and not serrated. Valvae are greatly different in shape and lacking a harpe. The saccular spine is in *rasilis* reduced in length. Phallus is bifurcate, straight and rather short. It bears on each of the processes a small strongly sclerotized area of spines. Female genitalia have not yet been dissected.

Bionomics. This rare species was collected at altitudes between 150 and 1400 m.

Distribution. Restricted to Sulawesi and Peleng.

147. ***Teleclita flavisticta*** (Gaede, 1930): 639.

Holotype: ♀, [Indonesia], Sumatra, Barisan-Gebirge, [„Western Slopes"] 800 m – NHMUK, London.

Schintlmeister 2013: 455.

Diagnosis. Rather umistakable by bluish grey colored forewings, with brownish tinged basal and submarginal area. Basal, postbasal and postmedial fasciae are fat and contrasting black. Discal spot is pale yellowish-brown, with a brown ring. Hindwings are whitish and display black anal markings. Individual variability as well as sexual dimorphism are limited. Females display a blackish marginal fascia on hindwing.

Male genitalia have not yet been dissected. Female genitalia resembles *strigata*, but the vaginal plate very reduced in size. Ductus bursae is much longer than in *strigata*.

Bionomics. Adults are rare in collections (n < 20) and the majority has been taken in lowlands below 600 m and in primary forests. One female was collected in Sumatra at 1350 m.

Distribution. Distributed in Thailand, Malaya, Sumatra and Borneo.

Pheressaces Turner, 1903: 45 (key), 56, **stat. rev.**

Type species: *Notodonta cycnoptera* Lower, 1894.

Schintlmeister 2013: 454.

Taxonomic note. Due to a misidentification of *P. cycnoptera* Schintlmeister, 2013 placed *Pheressaces* wrongly as a subjective synonym of *Teleclita.* In fact *cycnoptera* differs greatly from *Teleclita* species and is doubtless not congeneric with the latter.

Diagnosis. Medium sized. Antennae of males are grey and bipectinate except for the last 15% to the tip. Female antenna have very short pectinations.

Forewing color is light greyish. Forewing have blackish medial and postmedial fasciae accompanied by a whitish band. Discal streak is whitish and indistinct developed.

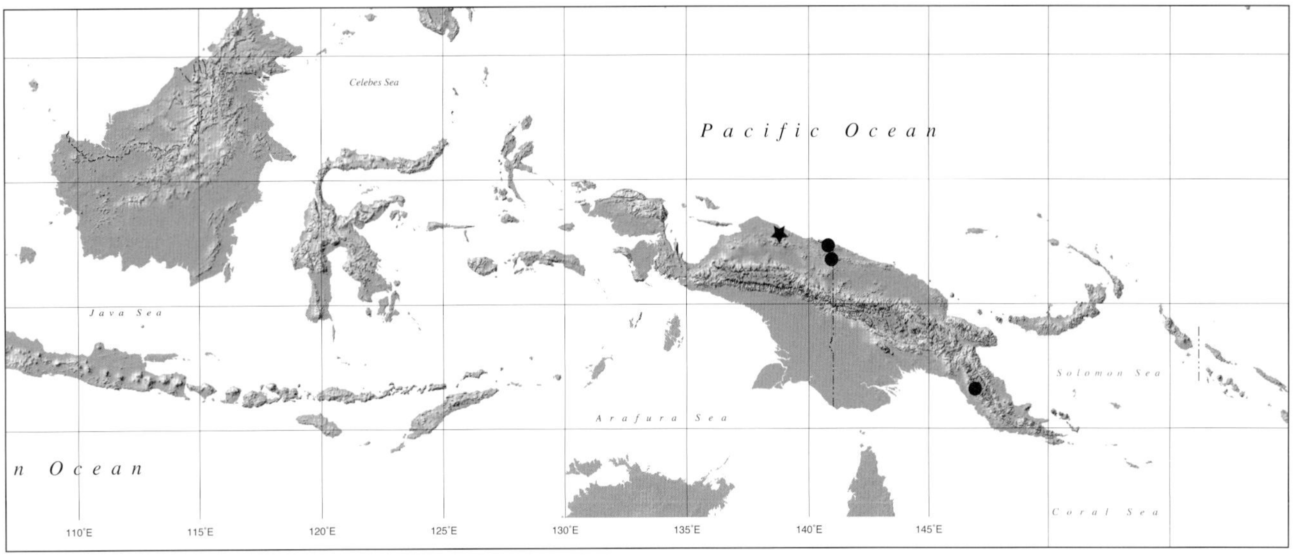

FIGURE 129 Distribution of *Pheressaces siegfriedihlei*.

Males have in the medial area along the dorsum a fuscous streak, which is absent in females. Veins are tinged black. Hindwing are white without anal markings. Female of *cycnoptera is* sexual dimorphic having the postmedial and submarginal area fuscous grey filled.

Male genitalia have a broad, deeply triangular notched uncus. Socii are much longer than the uncus and and massive. Valvae are rounded with a large and strong sclerotized sacculus. Harpe is well developed. Valva costa bears a long process. Phallus is slightly curved with a pointed tip. The everted endophallus bears a carina with many small and strong sclerotized thornes. 8th abdominal segments is less modified. Female genitalia of *cycnoptera* have the 8th abdominal segment of similar size as males, but the remaining parts are very small. Both paires of apophyses are long and slender. Postvaginal plate is of rather circular shape. Antrum is squarish and strongly sclerotized. Ductus bursae is long and slender. Corpus bursae bears a triangular signum.

The genus comprises two vicariant species and occurs in Australia and New Guinea.

148. ***Pheressaces siegfriedihlei*** spec. nov. (figs. 148a–c).
Holotype: ♂, Indonesia, W Papua, Dist. Sarmi, Gautier/ Foja Mountains, Bora Bora village, 2°20.448'S, 138°47.476'E, 11.–15. v. 2016, 60 Hm, leg. Bretschneider (genitalia slide Ihle 01-2017) – in coll. S. Ihle, Filderstadt.

Paratypes (9 ♂♂):
W Papua: 2 ♂♂, Bewani Hills, Om, 22. xi. 1934, W. Stüber; 1 ♂, Achterland, Hollandia [= Jayapura], 30. viii. 1938, leg. W. Stüber; 1 ♂, Waris District, Ampas, 30.viii. 1938, leg. W. Stüber;
PNG: 5 ♂♂, B. New Guinea, Mount Kebea, 6000 ft., vii. 1903, leg. A. E. Pratt (G.T.B. Baker Coll.).

Etymology. Named for Siegried Ihle, Filderstadt, who gave the author acess to his collection, which includes in particular interesting material from W Papua, leading to three descriptions of new species.

Diagnosis. Forewing lengths ♂♂ 18–22 mm, at average 20 mm. Forewings are light grey with many black spots. Basal, medial and postmedial fasciae near the forewing costa are well developed, fading toward the dorsum. Veins in the medial and postmedial area are black tinged. In the medial area along the dorsum there is a prominent black tinged vein (A1+2). The new species resembles in males to its vicariant *Pheresaces cycnoptera* (Lower, 1894): 78 (Syntypes: 2 ♂♂, [Australia], Queensland, Duaringa – Queensland Museum, Brisbane). The latter is readily separable by the blackish apical blotch of hindwing and the fuscous grey filled medial area of forewing. The female of *siegfriedihlei* is unknown.

Male genitalia are as described under genus diagnosis. The differ from *cycnoptera* by longer and more rounded uncus arms of the triangular notched uncus. Socii are broader and shorter than in *cycnoptera.* Valva bears a long and pointed costal process; harpe is much longer than in *cycnoptera.* Juxta is deeply notched.Phallus is less curved than *cycnoptera* and the tip less pointed. Carina of the everted endophallus bears a large area of small thorns, which is larger than in *cycnoptera.*

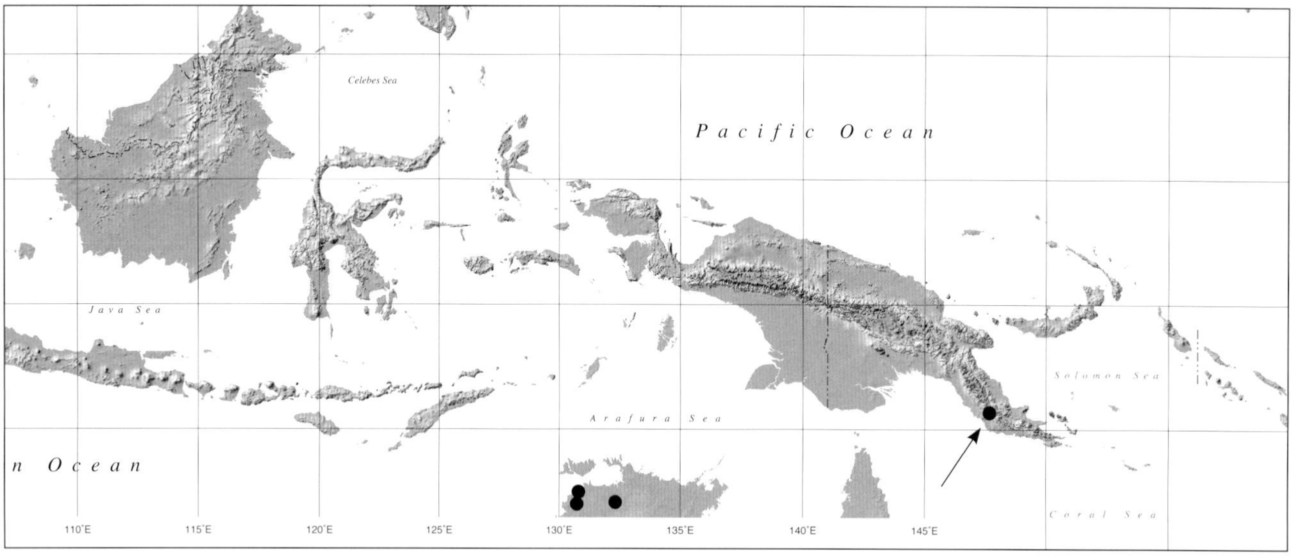

FIGURE 130 Distribution of *Destolmia lineata.*

Bionomics. The few known males (n = 6) has been taken between 60 m and 1830 m above sea level.

Distribution. Endemic to New Guinea in its entirety.

Destolmia Walker, 1855: 978 (key), 991.
Type species: *Destolmia lineata* Walker, 1855.
Schintlmeister 2013: 106.

Diagnosis. Medium sized moth of grey forewing color. Antennae are bipectinated on 50% of its length; the remaining part is filiform. Female are filiform. Forewing costa display a pair of indistinct blackish patches. Submarginal fascia is weakly developed, but a pair of fine black streaks on this fascia is a characteristic feature. Hindwings are chocolate brown or whitish without anal markings. Sexual dimorphism is limited.

Male genitalia have a large bifurcate uncus with long uncus arms. Socii are also bifurcate and large. Valva have a apical projection and a tapered basal process. Phallus is straight, with a serrated surface. 8th abdominal segments are less modified. Female genitalia are inconspicous. Apophyses are reduced; antrum is medium sized; ductus bursae is broad and of medium length; corpus bursae is circular without a signum.

The genus comprises two or more species and probably further seemingly closely related genera from Australia such as *Paradestolmia* Kiriakoff, 1967 or *Commnia* Kiriakoff, 1968 must be placed as synonyms of *Destolmia.*

The larva feeds in Australia on Eucalyptus (Myrtaceae). *Destolmia* is restricted to Australia and New Guinea.

149. *Destolmia lineata* Walker, 1855: 992.
Syntypes: 2 ♀♀, Australia, Sidney and [Western Australia] Swan River – NHMUK, London.
Schintlmeister 2013: 106.

Taxonomic note. The identity of *Destolmia lineata* is not clear. A syntype female from Sidney matches very well to the illustrated female from Queensland (fig. 149g). A dissected male from Sidney (fig. 149b) matches in its genitalia (with minor differences such as uncus shape, projection at the valva apex and phallus length) to the only known specimen of *Destolmia* from New Guinea (figs. 149a). Both specimens are however very different by external appearance. Individuals, which resembles externally to the male from PNG (fig. 149d) are also occuring in Australia, but they belong to a different species with very different male genitalia. For them the name *Destolmia lanceolata* (Walker, 1855) is applyable. The latter was described from a female from Australia, Queensland. Apart from this the author found apparently further species of *Destolmia* in Australia, but it was beyond the focus of this work to revise the species complex here.

Diagnosis. Most specimens from Australia appear grey with indistinct blackish markings and chocolate brown hindwings as described under genus diagnosis. The male from PNG have light brownish-grey forewings. A fuscous brown band extends from the middle of the dorsum to the middle of the outer margin. Hindwing are whitish without markings. Fringe of all wings is blackish brown/light brown checkered.

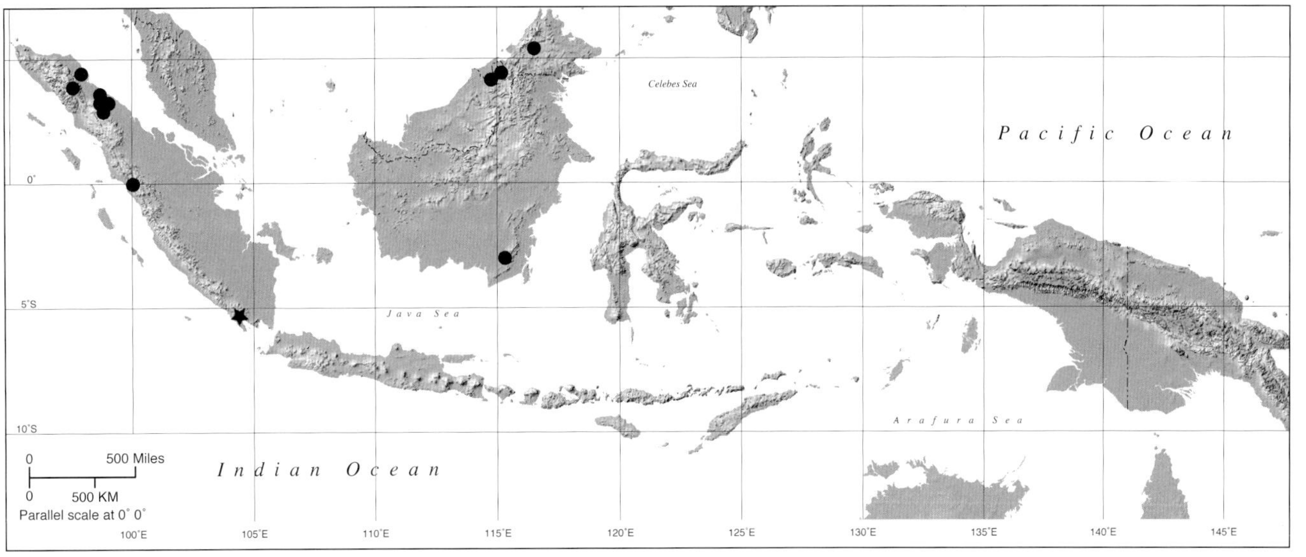

FIGURE 131 Distribution of *Stauroplitis annulata*

Male genitalia of *lineata* have a widely notched uncus with a pair of long uncus arms. Socii are large and bifurcate. Valva is elongate and bears a tapered basal processes at costa and the apical projection. Phallus is long and slender with a tapered tip. The everted endophallus displays a strongly sclerotized carina. 8th sternite is slightly notched at the posterior margin and displays circular sclerotizations. 8th tergite is slightly notched the posterior margin. Dissected specimens from Australia display the posterior margin slightly more rounded than the specimen from New Guinea. Female genitalia are as described under genus diagnosis.

Bionomics. From PNG only known from a single male, taken at 500 m above sea level. In Australia seemingly common and restricted to the lowlands, below 300 m.

The caterpillar was reared in Australia on *Eucalyptus caesia* (Myrtaceae).

Distribution. Known from Papua New Guinea (Kokoda Trails). Widely distributed in Australia (NT, Qsd, New South Wales, West Australia).

Stauroplitis Gaede, 1930: 639.

Type species: *Stauroplitis annulata* Gaede, 1930.

Schintlmeister 2008: 146; Schintlmeister 2013: 423.

Diagnosis. Larger notodontids. Antennae of males are fuscous brown and bipectinate except for the last 25% to the tip, which are filiform. Rami of males are long. Female have filiform antennae. Forewing color is light grey or fuscous brown. Discal spot of forewing is whitish and circular; apical blotch is black. Postbasal and postmedial fasciae are indistinct blackish. The submarginal fascia is whitish. Characteristic pattern elements are a two contrasting costal black streaks in the medial area and the apex and a blackish dorsal patch. Hindwings are chocolate brown. Sexual dimorphism is limited.

Male genitalia have a large uncus and a pair of robust socii. Valvae display a well developed sacculus with several processes. Phallus is short and bifurcate at the tip, where both lobes bear many strong sclerotized micro thorns. 8th sternite have a pair of prominent and large lobes at posterior marigin. 8th tergite is less modified. Female genitalia are relatively small, but their 8th abdominal segment is large. Apophyses posteriores are long, while apophyses anteriores are almost absent (reduced). Ductus bursae is reduced in legth. Corpus bursae is squarish and strongly sclerotized. No signum.

The genus comprise three species. Distributed in the Himalaya, S China, Indochina, Sundaland and the Philippines.

150. ***Stauroplitis annulata*** (Gaede, 1930): 639, pl. 84: g.

Holotype: ♂, [Indonesia], Sumatra, Barisan-Gebirge, 800 m – NHMUK, London.

Schintlmeister 2013: 423.

Taxonomic note. In middle and northern Thailand the ssp. *plumbalis* Schintlmeister, 2006 occur, which displays forewings greyish-white.

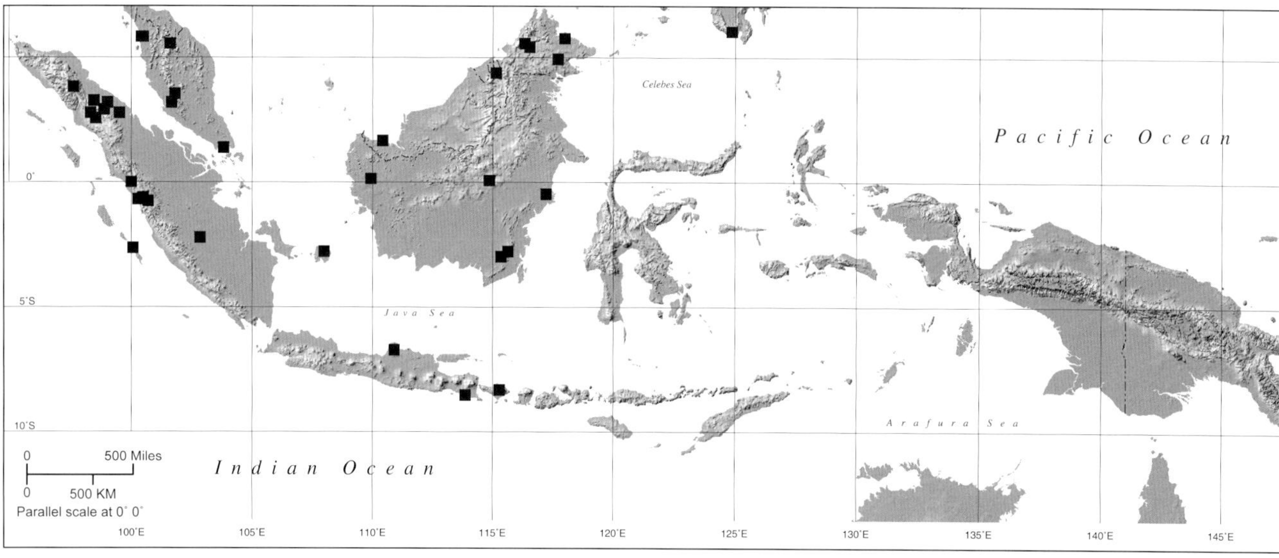

FIGURE 132 Distribution of *Oxoia smaragdiplena.*

Diagnosis. In Sundaland rather umistakable by size and the triangular forewing shape. Females vary individually in their forewing coloration, and also uniform pale grey individuals occure.

Male genitalia are as described above. Saccular process near the valva apex is a characteristic feature. Saccus appear variable in its serration and number of projections and both valvae are sometimes not symmetrically. Phallus is of an unique shape as illustrated. Both phallus lobes bear many small and strong sclerotized thorns. 8th sternite have apart from the pair of lobes at the posterior margin, a further central process. 8th tergite is slightly bilobed. Female genitalia are as described in the genus diagnosis above; the very long appophyses posteriores and the strong sclerotized corpus bursae are diagnostically features.

Bionomics. Moths occur infrequently in rain forests between 700 m and 1500 m.

Distribution. *S. annulata annulata* occur in Sumatra, Borneo and southernThailand. Central and north Thailand is populated by ssp. *plumbalis.*

Oxoia Kiriakoff, 1967: 53.

Type species: *Exaereta smaragdiplena* Walker, 1862.

Schintlmeister 2008: 146; Schintlmeister 2013: 308.

Diagnosis. Large adults. Antennae of males are fuscous brown and bipectinate; the last 50% to the tip are filiform. Rami of males are long; those of females are slightly shorter. Forewing color deep violet brown with intensively green pattern. Discal spot on forewing is large, circular and contrasting green. Characteristic pattern elements of forewings are a black basal streak, and a large black streak in the central part of the postmedial area. Hindwings are chocolate brown without markings. Females display less green pattern elements than males.

Male genitalia are in all species very large. They have a long uncus and a pair of small socii. Valvae are elongate and bear a large apical spine. The harpe is well developed. Phallus is long and straight. The everted endophallus bears near the phallus tip a field of many needle-like sharp spines. Carina bears a field of small thorns. 8th sternite is notched and 8th tergite is less modified. Female genitalia have a small 8th abdominal segment, which shows several notches. Both pairs of apophyses are long. Ductus bursae is extremely long and thick. Corpus bursae is rather small and of circular shape. No signum.

The genus includes three species, which are distributed from the Himalaya, China, Indochina, the Philippines, the Malayan Peninsula, Sumatra, Borneo, Java, Bali, Sulawesi up to New Guinea.

151. ***Oxoia smaragdiplena*** (Walker, 1862): 134 (*Exaereta smaragdiplena*).

Holotype: ♂, [Malaysia], Borneo, Sarawak – Oxford University Museum of Natural History.

Schintlmeister 2008: 147; Schintlmeister 2013: 308.

Diagnosis. The smallest species in *Oxoia*. It shows by far most green pattern of all congeners. Basal and postmedial fasciae of the forewings contains two prominent circular

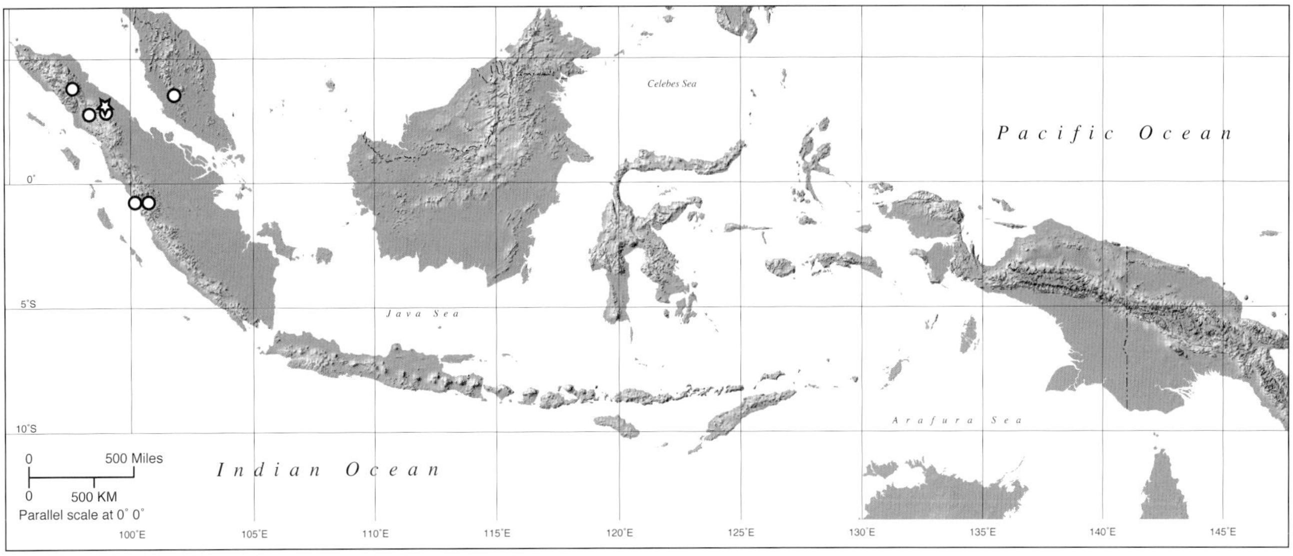

FIGURE 133 Distribution of *Oxoia viridipicta.*

black blotches. In other *Oxoia* species, there are streakes. Near the forewing apex there is a further, inconspicous black blotch, which is not seen in *viridipicta.* Hindwings vary from light brown to dark chocolate brown. Females of *smaragdiplena* show less green pattern.

Male genitalia are distinguished by the shape of the uncus tip, which is not rounded but bears a central projection. Socii are small. Valva have a long and straight costal process at apex. Harpe and sacculus are well developed and slightly stronger sclerotized than the valvae. Phallus is straight and long. The everted endophallus bears a field of cornuti and a stronger sclerotized carina. Shape of 8th sternite is diagnostically rhomboid shaped. 8th tergit displays centrally a characteristical heart shaped sclerotization. Female genitalia are as described under genus diagnosis. The are distinguished from *viridipicta* or *irrotativiridis* by the diagnostically triangular notched antevaginal plate and a longer ductus bursae.

Bionomics. *O. smaragdiplena* occurs commonly from the lowlands (100 m) up to 1600 m. In Thailand adults has been observed up to 2200 m above sea level.

Distribution. Distributed in Nepal, northeastern India, southwestern China (Yunnan), Indochina, Malaya, the Philippines in its entirety, Borneo, Sumatra, Belitung, Mentawei Isl. (Pagai), Java and Bali.

152. ***Oxoia viridipicta*** (Kiriakoff, 1974): 419; pl. 5: 4; fig. 30 (*Panteleclita viridipicta*).
Holotype: ♂, [Indonesia], N.-Sumatra, Dolok Merangir, 180 m – ZSM, Munich.
Schintlmeister 2008: 147; Schintlmeister 2013: 309.

Diagnosis. Similar by external appearance to *Oxoia irrotativiridis,* but distinguished by larger and more prominent developed black basal streak on forewing. *Oxoia smaragdiplena* is clearly smaller than *viridipicta* and both black markings are not streak like but circular. Along the forewing costa there are several blackish patches. Green pattern on forewings is reduced, and only postbasal, submarginal fasciae and discal area are green marked. A few males display more expanded green markings but they does not reach the greenish appearance as in *smaragdiplena.* Sexual dimorphism is limited.

Male genitalia have long uncus with rounded tip, which separate *viridipicta* from *smaragdiplena.* Socii are larger than in *irrotativiridis* and rounded at their tips. Valva is elongate and has a curved and pointed apical process. Together with the digitus process it displays a wrenche like shape. Shape and length of both processes vary individually. Phallus is straight and shorter than in other congeners. The everted endophallus bears two separate fields with many needle-like sharp spines (cornuti) and a strongely sclerotized carina. 8th sternite is of ellipsoid shape with a notched posterior margin. Sclerotizations of 8th tergite are similar to *smaragdiplena.* Female genitalia are distinguished by shape of the apostvaginal plate, which display contrary to other *Oxoia* a central projecetion. Ductus bursae is longer than in *smaragdiplena* and corpus bursae larger, as in *irrotativiridis.* Corpus bursae displays a strongly sclerotized structure (not a signum), but only one female was dissected.

Bionomics. Adults occur rarely (n < 10) in Sumatra up to altitudes of 1000 m. In Thailand, *viridipicta* occurs more

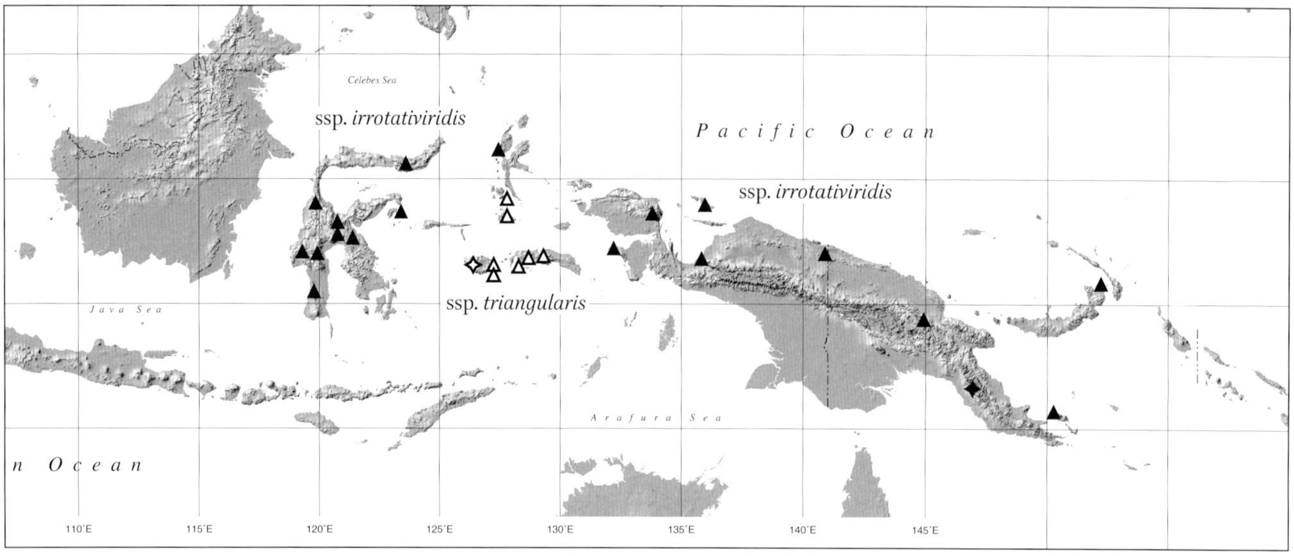

FIGURE 134 Distribution of *Oxoia irrotativiridis*.

commonly and has beeen observed until 2000 m above sea level.

Distribution. Distributed in northeastern India (Meghalaya), Vietnam, Thailand, Malaya, and Sumatra.

153. ***Oxoia*** (Bethune-Baker, 1904): 380; pl. 5: 8 (*Notodonta irrorativiridis*).
Holotype: ♂, [Papua New Guinea], Dinawa [4.000 ft.] – NHMUK, London.
Schintlmeister 2013: 308.

Taxonomic note. Popuations from Buru were described as *Somera triangularis* Gaede, 1930: 632; pl. 83: h. Holotype: ♂, Buru, 1.500 m [Gamoe, Mrapat] – NHMUK, London. Holloway, 1987: 116, placed *triangularis* as a synonym of *irrotativiridis*. Due to a wider range of individual width variation, *triangularis* does not differ significantly by facies from other populations, but male genitalia are different from Papuan material. Socii are shorter, the juxta flatter and the phallus is slenderer, significantly longer and curved (fig. 150ad- f). The 8th tergite have a rounded, not notched posterior margin. Female genitalia are less different, but differ slightly in shape of the postvaginal plate (but only two females from Buru and from W Papua were dissected). Three dissected males from Obi and Bacan matching well to three genitalized males from Buru, but two dissected males from Halmahera and two males from Sulawesi are surprisingly virtual identical with their male genitalia of New Guinean specimens. The morphological differences between these two taxa indicates the occurence of two different species, but from this anormal distribution together with the minor differences in female genitalia *triangularis* is classified here in the rank of a subspecies of *irrotativiridis* (**stat. nov.**).

Diagnosis. Similar to allopatric *O. viridipicta*, but at average 2 mm smaller in forewing length. Pectinations in male antennae extending more toward the tip than in *viridipicta*. Best feature to separate both species is a small black basal patch in *irrotativiridis*, which is replaced in *viridipicta* by a large streak in the postbasal area. *Oxoia irrorativiridis* vary much in size and the extension of the green forewing pattern as illustrated. Green discal blotch is sometimes white tinged. Hindwings vary from light to dark chocolate brown. Females display reduced green markings on forewing.

Male genitalia have a slightly broader uncus than *viridipicta*, which more rounded at the tip. Apical and costal valva processes are longer than in *viridipicta* and the costal process arises from the valva center. Both processes are variable in shape. Phalli and 8th sternites are in both subspecies of *irrotativiridis* different. 8th tergite resembles *viridipicta*. Postvaginal plate of female genitalia is centrally convex notched, in *smaragdiplena* the notch is triangular.

Bionomics. Adults are common. They occur up to 1700 m above sea lecel, but lower altitudes up to 500 m are apparently preferred.

Distribution. *Oxoia irrorativiridis irrotativiridis* occur in New Guinea, New Britain, Halmahera, Sulawesi and Peleng. Ssp. *triangularis* is distributed in Buru, Ambelau Isl., Obi, Bacan, and presumably Ambon and Seram.

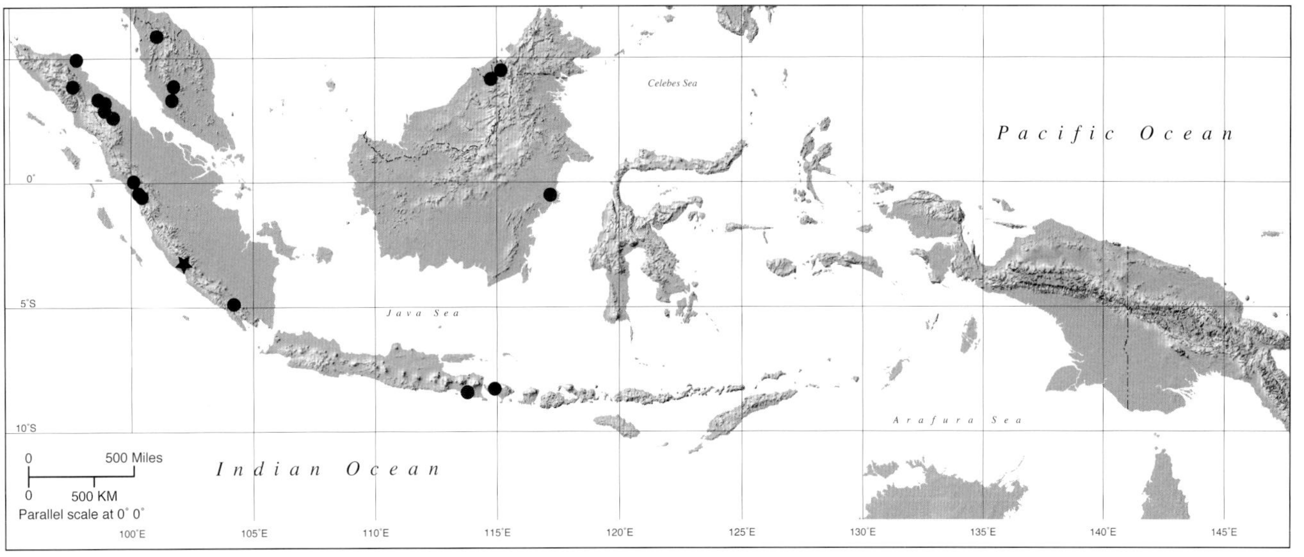

FIGURE 135 Distribution of *Rodneya caudata*.

Rodneya Kiriakoff, 1967: 63.
Type species: *Rodneya caudata* Kiriakoff, 1967.
Schintlmeister 2013: 380.

Taxonomic note. There are currently three heterogenous species in *Rodneya,* which are in fact representing in three different genera. The genus diagnosis refer therefore only to the type species, *Rodneya caudata.*

Diagnosis. Medium sized males, large females. Antennae of males pale brown and bipectinated with very long and tufty rami; last 30% to the tip are filiform. Antennae of females are as in males, but tufty bipectination extends only 50%. Forewing golden-brown, greenish yellow tinged in submarginal area. Postmedial fascia on forewing pale brown, submarginal and marginal fasciae fuscous brown. Discal spot small and black. In the apical area severeal white patches occur. There is a prominent circular postbasal blotch on forewing. Hindwings are pale to fuscous brown with yellow tinged anal area. Sexual dimorphism is limited.

Uncus is large and bifurcate. Socii robust and curved. A gnathos is visible, but very reduced. The large juxta is prominently strongly sclerotized, fan shaped. Valvae are elongated with rounded apex. Saculus well developed with a more strongly developed basal projection. Phallus straight with many micro thorns on surface. Both 8th abdominal segments are notched at the posterior marigin. Female genitalia have a large 8th abdominal segment with large and robust apophyses posteriores and reduced apophyses anteriores. Antevaginal plate is large and bilobed. Ductus bursae broad and medium sized. Corpus bursae very small. No signum.

For distribution see under *Rodneya caudata* below.

154. ***Rodneya caudata*** (Kiriakoff, 1967): 63; fig. 12.
Holotype: ♂, [Indonesia], W. Sumatra, Lebong Tandai – NHMUK, London.
Schintlmeister 2013: 381.

Diagnosis. Unmistakable by forewing shape, coloration and the violet postbasal markings. The species shows almost no individual variability.

Genitalia are as described under genus diagnosis. Juxta vary individually slightly in its shape.

Bionomics. Moth occur uncommonly between 300 and 1700 m, most specimes were collected in primary forests below 500 m above sea level.

Distribution. Distributed south Vietnam, Thailand, Myanmar, Malaya, Borneo, Sumatra, Java and Bali.

Stauropus Germar, 1812: 45.
Type species: *Phalaena fagi* Linnaeus, 1758.
Schintlmeister 2008: 151; Schintlmeister 2013: 424.

Taxonomic note. As shown by Schintlmeister, 2007 and 2008 this large genus can be divided into several subgenera. The various subgenera differ chiefly by differences of structures of 8th abdominal segments.

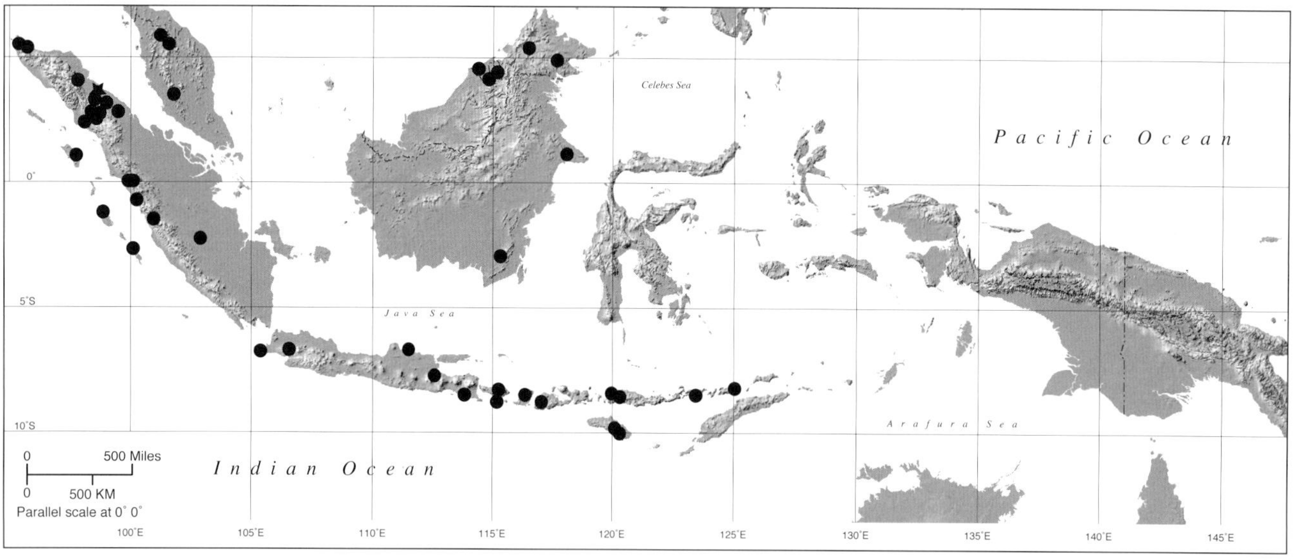

FIGURE 136 Distribution of *Stauropus major.*

Diagnosis. Medium sized adults. Males have bipectinate brown antennae with long rami. The last 25–35% to the tip are simple. Antennae of the female are filiform. Forewings are brown, greyish, or green. Species of subgenus *Benbowia* Kiriakoff, 1967 are all green colored. Discal spot of forewing marked as a fuscous brown or black vertical streak. Near the apex and forewing costa there is a small pale patch, which is present in almost all *Stauropus*, except those of the subgenus *Benbowia.* Basal area of forewing is separated by greenish tinged or paler coloration from medial area. There are also a well developed postmedial, submarginal and marginal fasciae. The latter consists of prominent brown dots. Hindwings display in their costal half the ground color and pattern as in the forewings. In resting position of the adult this part of the hindwings is visible. The remaining half of hindwings are pale brown to yellowish-brown or fuscous chocolate-brown. Sexual dimorphism is limited, but females of some species are much larger than males.

Uncus is absent in male genitalia; socii are large and developed like lobes. Valvae are often modified with processes or serrated structures and bear an elongated apical process. Phallus straight, very slim and long. In subgenus *Benbowia* slightly thicker and curved. 8th sternite is of taxonomic value in the genus. Shape and presence of a central "vein" resembles to *Syntypistis.* Female genitalia have a very large 8th abdominal segment and also papillae anales are large. Apohyses are robust but short. Ductus bursae is narrow and of moderate length; in *Benbowia* reduced in length. Corpus bursae is conterary to the size of the 8th abdominal segment very small. No signum.

Caterpillars are bizarre, ant-like and homologue by their extremely long developed thoracic legs and long caudal appendages. Hostplant of caterpillar are in Fagaceae (e.g. *Quercus*) although a few species, such as *Stauropus alternus,* are polyphagous. 36 *Stauropus* species are currently known.

Species of the subgenus *Stauropus* (12 species) are distributed in entire palaearctic region, India to Sri Lanka and in the oriental tropics including the Moluccas. The subgenus *Chlorostauropus* (16 species) is of Papuan origin, with a few species occuring in Australia and the lesser Sundas. The subgenus *Benbowia* (8 species) is distributed from Himalaya, China, Indochina, the Philippines and Sundaland until Bali.

Subgenus ***Stauropus*** Germar, 1812

Distinguished by the 8th tergites, which bear central projections and a pair of large lobes. Also females display a central projection on 8th tergite.

155. ***Stauropus major*** van Eecke, 1924: 175; pl. 14: 6.

Holotype: ♂, [Indonesia], Sumatra, Deli [= Medan] – NBC, Leiden.

Schintlmeister 2008: 155; Schintlmeister 2013: 428.

Diagnosis. This large species is best recognized by its broad forewing shape and dark chocolate brown wing color, sometimes mixed more intensively with dark green scales. The species is variable in size as illustrated. Postbasal, medial and submarginal fasciae whitish. A postmedial

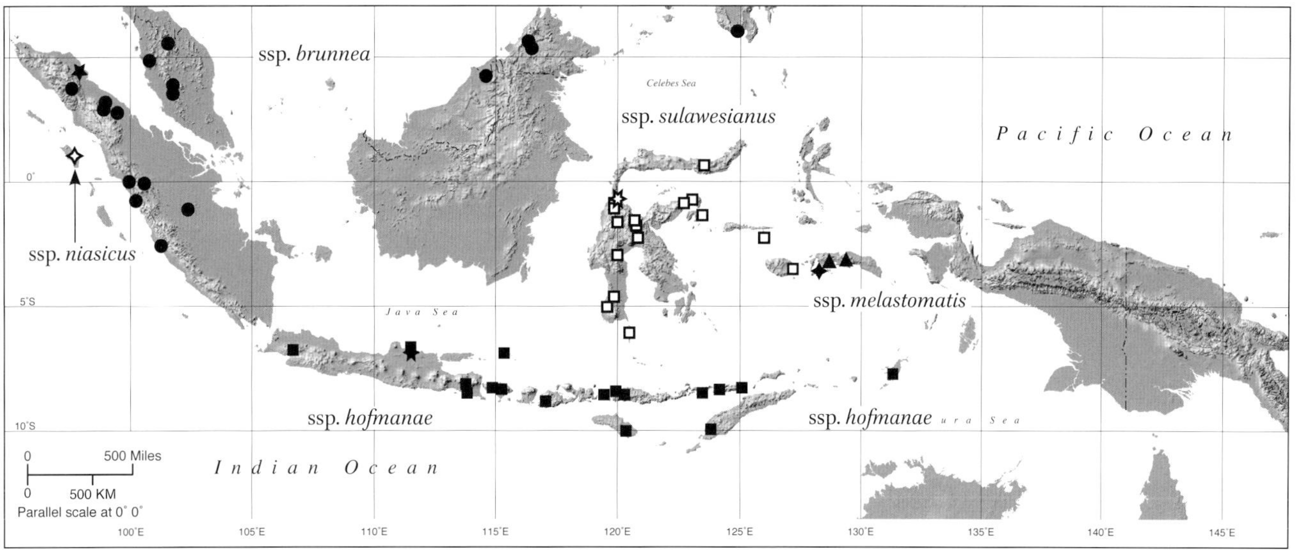

FIGURE 137 Distribution of *Stauropus alternus*.

fascia is marked as a row of blackish dots. Fringe have white patches. Discal spot very indistinct. Hindwings are of same color as forewings with paler anal region. Sexual dimorphism limited.

Genitalia apparatus is variable in length and thickness of the various processes and lobes of valva. However, no geographically correlation was found between the different forms. 8th tergite is unique by the pair of large projections. Shape of 8th tergite is somewhat variable at posterior marigin as illustrated. Female genitalia have very large papilles anales and a large 8th abdominal segment. Apophyses are very robust, but not long. Ductus bursae long and broad; corpus bursae small and of circular shape.

Bionomics. Adults are commonly in forests between 100 to 1700 m above sea level, but most specimens were observed below 700 m.

Distribution. Distributed in northeastern India, Andaman Isls., Myanmar, China, Indochina, Malaya, Borneo, Palawan, Sumatra, with adjecent islands of Nias, Siberut, and Pagai (Mentawei); Java, Bali, Lombok, Flores, Sumba, Sumbawa, Alor.

156. ***Stauropus alternus*** Walker, 1855: 16021.

Syntypes: 2 ♀♀, [Bangladesh], Silhet –NHMUK, London.

Schintlmeister 2008: 157; Schintlmeister 2013: 425.

Taxonomic notes. 8th sternite bears a separate sheet with an process and belong therefore not to the subgenus *Chlorostauropus*, where it was formerly placed by Schintlmeister, 2013.

Stauropus alternus occur in Indonesia in several subspecies, which was reviewed by Schintlmeister 2003.

Stauropus alternus brunnea (Schintlmeister, 1981): 285; figs. 2a, 2b; holotype: ♂, Indonesien, Sumatra, Aceh, 20 km NW Langsa, 4°32‘ N, 97°45‘ E, 80 m – coll. A. Schintlmeister, Dresden. External appearance as in ssp. *alternus*. Male genitalia of ssp. *alternus* differ from ssp. *brunnea* by slenderer apical part of the valves and a central notch of the 8th sternite, which have rounded central lobes. 8th sternite in ssp. *brunnea* is almost straight at posterior margin and not notched. The appendix of the 8th sternite of ssp. *brunnea* is also rather straight at the posterior margin. Female genitalia differ from those of ssp. *alternus* by thikker and shorter apophyses posteriores. Ducutus bursae is longer, straight and strongly sclerotized.

Stauropus alternus niasicus (Schintlmeister, 1981): 286; figs. 3a–c; holotype: ♂, [Indonesia], Nias, 10 km E Idanö Gawö, 0°46‘ N, 97°52‘ E, 3 m – coll. A. Schintlmeister, Dresden. This is the largest subspecies of *S. alternus* and the hindwings are slightly more fuscous than in ssp. *brunnea*. Male genitalia are not significantly different from ssp. *brunnea*. A male from Mentawei does not differ by external appearance and male genitalia from ssp. *brunnea*.

Stauropus alternus hofmanae Schintlmeister, 2003: 106; figs. 16, 20, 77, 82; holotype: ♂, Indonesien, W Java, Jatiraga, 28 km NE Blora, 6°46‘ S, 111°39‘ E, 150 m – coll. A. Schintlmeister, Dresden. Significantly, 2–4 mm, shorter forewing length than all other known populations. External appereance is generally paler than in other Sundanian *alternus*; in particular, the hindwings are whitish in their lower part. Male genitalia have a slenderer apical part of valvae than ssp. *brunnea*. Posterior margin of 8th sternite is not straight, but conical. 8th sternite appendix is more deeply notched. Populations from Lesser Sundas and Timor differ slightly from Java or Bali populations by slenderer valvae arms, slenderer socii and by shape of posterior margin of the 8th sternite. These differences are probably an expression of clinal variation. Female genitalia of *hofmanae* have shorter and slenderer apophyses than ssp. *alternus* or ssp. *brunnea*. The strongly sclerotized part of ductus bursae is curved, not straight as in ssp. *brunnea*.

Stauropus alternus sulawesianus Schintlmeister, 2003: 106; figs. 17, 18, 78, 83; holotype: ♂, Indonesien, Sulawesi Tengah, Mt. Salubaya, 0°45‘ S, 120°00‘ E, 1.000 m – coll. A. Schintlmeister, Dresden. Forewing length is on average 2–3 mm larger than ssp. *brunnea*. Forewing pattern displays a contrasting whitish basal area; the medial area of forewings is fuscous and contrasting. Hindwings are darker than in other subspecies, except ssp. *melastomatis*. Male genitalia apparatus does not differ significantly from ssp. *brunnea*, but the notch of 8th sternite and also the appendix as well are characeristically shaped as confirmed by n = 4 genitalia dissections of males from Sulawesi. Female genitalia resemble ssp. *brunnea*, but apophyses posteriores are slightly thicker.

Two dissected males from Buru, fig. 156j, differ in shape of 8th sternite as illustrated. Due to the limited material at the disposal of the author this character cannot presently be confirmed by further dissections, hence the Buru populations are provisionally placed in ssp. *sulawesianus*.

Stauropus alternus melastomatis Felder, 1861: 39; syntypes: unspecified number of ♂♂, ♀♀, [Ambon], Amboina – NHMUK, London. This subspecies is well characterized by its dark blackish-brown general appearance. Old material in collections, including the-150-year-old syntypes, tend to fade to pale reddish brown. Male genitalia are not yet dissected. Female genitalia, fig. 156k, have slenderer and longer apophyses posteriores than in other subspecies, and the apophyses anteriores are reduced in length.

Diagnosis. *Stauropus alternus* is best to recognized among the Indonesian *Stauropus* by the elongated forewing shape. Basal area of forewings is paler and does not display a black streak or spot as seen in many *Chlorostauropus*. Medial area of forewings darkened. Hindwings display a submarginal fascia, marked by whitish dots. Femalea are slightly sexual dimorphic and have mostly a darker wing coloration than males. White submarginal fascia of hindwings appear more contrasting.

Male genitalia are distinguished by a broad apical part of the valvae. 8th sternite bears a central appendix, which is a diagnostical character. Female genitalia are characterized by large apophyses posteriores and large papillae anales. Apophyses anteriores are short. Ductus bursae is in its anterior portion strongly sclerotized. Corpus bursae is small and bear no signum.

Bionomics. Adults are commonly in forests and urban landscapes, a synantrophic species. It occurs between 100 to 1700 m above sea level, but in the Himalayas up to 2500 m. Most specimens were observed at altitudes below 700 m.

Caterpillars are polyphagous and several rearing experiments in Germany was successfull with *Quercus* spp. (Fagaceae), *Tilia* (Tiliaceae), *Malus, Rubus, Rosa* (Rosaceae), *Rhododendron* (Ericaceae), *Syringus, Ligustrum* (Oleaceae).

Distribution. *S. alternus alternus* is distributed in Pakistan, Nepal, India, Andaman Isls., Myanmar, southern China, Taiwan, south Japanese islands and entire Indochina. The ssp. *albescens* Moore, 1879 occur in Sri Lanka and southern India. Ssp. *brunnea* flies in Malaya, Borneo, the Philippines, Sumatra, with the islands of Siberut, Pagai (Mentawei). Ssp. *niasicus* is endemic on Nias; ssp. *hofmanae* inhabits Java, Bali, Lombok, Flores, Sumba, Sumbawa, Alor, Timor, Tenimbar. Ssp. *sulawesianus*: Sulawesi, Peleng, Selayar, Sulabes, Buru and ssp. *melastomatis* occur in Ambon, Seram.

157. ***Stauropus pusilla* spec. nov.** (figs. 157a–e).

Holotype: ♂, Indonesia, W Timor, Amarassi, Buraen, 300 m, 10°21' S, 123°39'E, 7.–13. iv. 2007, leg. St. Jakl – coll. A. Schintlmeister, Dresden.

Paratypes (30 ♂♂, 2 ♀♀):

W Timor: 9 ♂♂, Amarassi, Buraen, 300 m, 10°21' S, 123°39'E, 7.–13. iv. 2007, leg. St. Jakl (GU 94-11, 94-12); 17 ♂♂, 2 ♀♀, Buraen, 35 km SE Kupang, 350 m, 10°15' S, 123°49'E, 26. i.–9. ii. 2006, leg. St. Jakl; 6 ♂♂, dto. 20.–27. ii. 2006.

Diagnosis. Forewing length ♂♂ 14.0 -15.5 mm, ♀ 21 mm. The smallest species within the genus. Antennae of males brown, bipectinated with long rami; the last 5% to the tip filiform. Female antennae are filiform. Wing color of male greenish-brown; female is less green tinged. Basal area of forewing whitish filled with a small black basal streak.

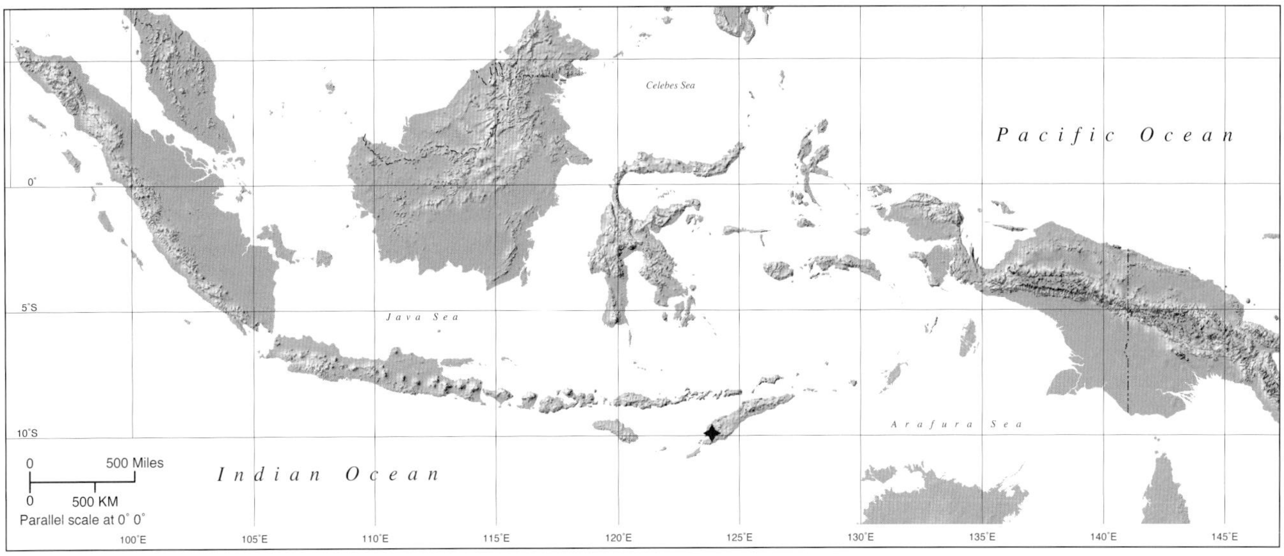

FIGURE 138 Distribution of *Stauropus pusilla*.

Medial area is more fuscous, but postmedial area is lighter. Near the apex there is an indistinct paler patch. Marginal fascia of forewing marked as a row of large brown patches. Costal area of hindwing display the forewing pattern and color, while the anal area ia paler brown filled. Female resembles the male, but postmedial area of forewing is darker.

Male genitalia are small. Ventral process of valva is broad and long. Valvae are broader than in other congeners. Phallus long and slender. Phallus base knob-shaped as in *S. alternus*. 8th sternite straight at posterior margin and notched centrally. It bear a central appendix as in *alternus*. 8th tergite slightly bilobed at posterior appendix with characteristic sclerotizations. Female genitalia have large papillae anales and large apophyses posteriores. Ductus bursae is slender and long. Corpus bursae is small and ovoid. No signum.

Bionomics. A common species that was collected at lower altitudes at 300 –350 m.

Distribution. Endemic on the Island of Timor.

Subgenus ***Chlorostauropus*** Kiriakoff, 1968: 135.

Type species: *Stauropus viridissimus* Bethune-Baker, 1904.

8th tergites are rather simply with convex rounded posterior margin. Forewings of most species are mixed with green scales.

158. ***Stauropus viridissimus*** Bethune-Baker, 1904: 378; pl. 4: 1.

Lectotype: ♂, [Papua New Guinea], Ekeikei [1.500 ft.] – NHMUK, London **by present designation.**

Schintlmeister 2013: 426.

Taxonomic notes. *Stauropus viridissimus* was described from an unspecified number of ♂♂ and ♀♀ from Dinawa and Eikeikei. 1 ♂ and 1 ♀ syntypes were traced by the author in the holdings of NHMUK. To stabilize the nomenclature the male with the following labels is designated hereby as lectotype: “Ekeikei | B. C. New Guinea. | 1500 ft. | Jan.–Feb. 1903 | A.E Pratt Coll.”; green circular label “Type”; “G. T. B. Baker Coll. | Brit. Mus. 1927 – 360”; yellow label “Photo done by A. Schintlmeister #4055”. A lectotype label was added. The remaining female, depicted in fig. 158j, also from Ekeikei, but March – April 1903, becomes a paralectotype.

The species tends to establish geographically subspecies, which are separable best by facies. But also male genitalia displaying differences between the various subspecies, such as ventral valvae processes or shape of the 8th sternites.

Subspecies *viridissimus* displays a splendid bluish green wing coloration. The ventral process of valva is broad and not knob-shaped at the tip. A single male from Key Isl. displays a very indistinct forewing pattern. Distributed in New Guinea, Biak, Yapen, Key and Waigeu.

Populations from Australia were described as *Stauropus habrochlora* Turner, 1922: 378; syntypes: 1 ♂, 1 ♀, Australia, N. Q[ueens]land, Kuranda near Cairns – ANIC Canberra. This taxon was placed in synonymy to

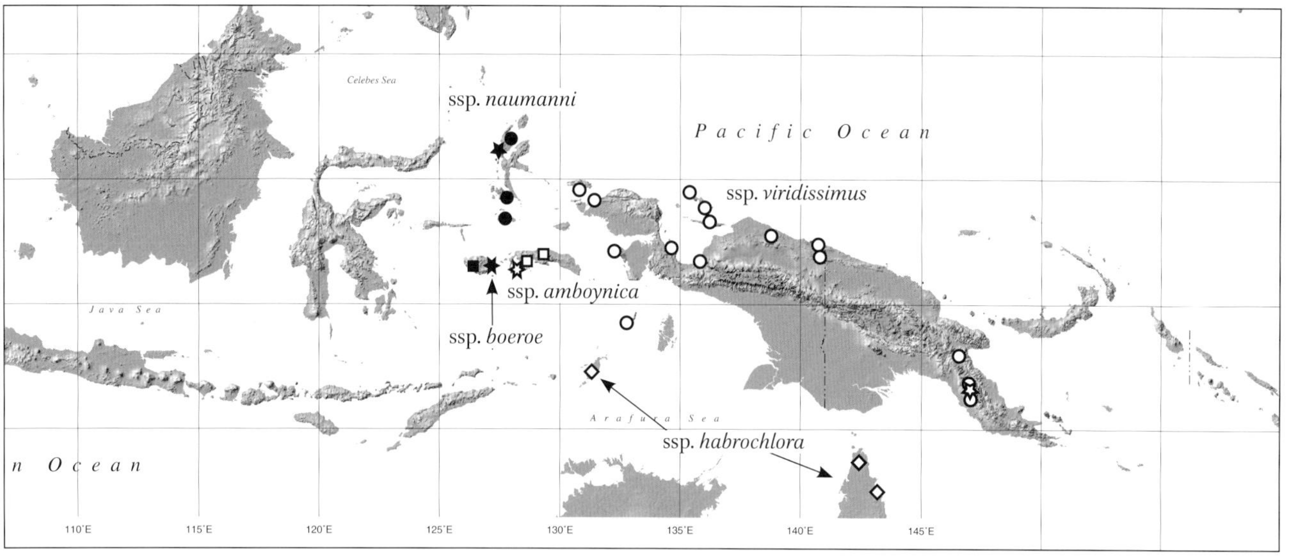

FIGURE 139 Distribution of *Stauropus viridissimus.*

Stauropus viridissimus by Edwards, 1996: 274. Examination of material from Queensland showed that *habrochlora* is conspecific with *viridissimus* and do not differ by external appearance from New Guinean populations. But male genitalia differ by the shape of a more bulbous shaped valva. 8th sternite has the strongly sclerotized central part at the posterior margin larger. Thus, the taxon is classified here as *Stauropus viridissimus habrochlora* **stat. nov.** A series of 4 ♂♂ from Tanimbar, fig. 158l, resemble males from Australia, Northern Territories, but the series is smaller in forewing length. Male genitalia are virtually identical, but 8th sternite of Tanimbar specimens (n = 3 GU) does not show the extended, strongly sclerotized area. The Tanimbar population is provisionally placed with *habrochlora* and could represent a further subspecies.

Stauropus viridissimus amboynica Oberthür, 1911: 323; pl. 66: 633; holotype: ♀, [Indonesia, Moluccas], île d'Amboine – NHMUK, London, differ by brownish tinged forewings and also by broader forewing shape from other known populations. They somewhat resemble by facies *S. talesea* from New Britain.

Male genitalia are distinguished from ssp. *viridissimus* by a smaller ventral valva process Phallus is the shortest among all *viridissimus* subspecies.

Stauropus viridissimus naumanni Schintlmeister, 2003: 116; figs. 52, 53, 56, 100; holotype: ♂, Indonesien, N Moluccas, Halmahera, Straße Baru-Basale, Gn. Talagarana, 600 m – coll. A. Schintlmeister, Dresden, is the largest subspecies in *viridissima,* on average about 1.5 mm larger than ssp. *viridissima*. Forewing pattern is indistinctly developed, in particular the postmedial fascia. Ssp. *naumanni* is the brightes of all subspecies, and the green wing color is slightly yellowish tinged. Anal area of hindwing is pale yellowish-brown. Males from Bacan (n = 6) and Obi (n = 2) display slightly darker hindwings than the type series from Halmahera (n = 20), but the other features match ssp. *naumanni* well.

Male genitalia have a slender ventral valva process, and the apical valva process is slenderer in comparison to ssp. *viridissimus* as well.

Stauropus viridissimus boeroe **ssp. nov.** (figs. 158d, 158r, 158s).
Holotype: ♂, Indonesia, SE Buru, Ilat, Remajah, 250 m, 3.–20. i. 2013, leg. St. Jakl – in coll. A. Schintlmeister, Dresden.

Paratypes (20 ♂♂):
Buru: 17 ♂♂, Ilat, Remajah, 250 m, 3.–20 i. 2013, leg. St. Jakl (GU 93-31, GU 93-32, GU 93-64); 2 ♂♂, dto. 5.–25. iii. 2013, leg. St. Jakl; 1 ♂, Central Buru, Kako Takalago, 700 ft., v. 1921, C. F., & J. Pratt.

Etymology: Boeroe is the former name for Buru Island.

Diagnosis of ssp. *boeroe*. Forewing lengths ♂♂ 18–19 mm; one male have forewing length of 21 mm. Smaller than ssp. *naumanni* or ssp. *amboynica.* Forewing pattern is sharp and disctinctly developed. Forewing color is greyish-green, not brownish-tinged as in ssp. *amboynica* or yellowish green as in ssp. *naumanni.* Most specimens of the type series display a blackish medial area; 4 ♂♂ (of n = 21) lack this fuscous medial area. The female is unknown.

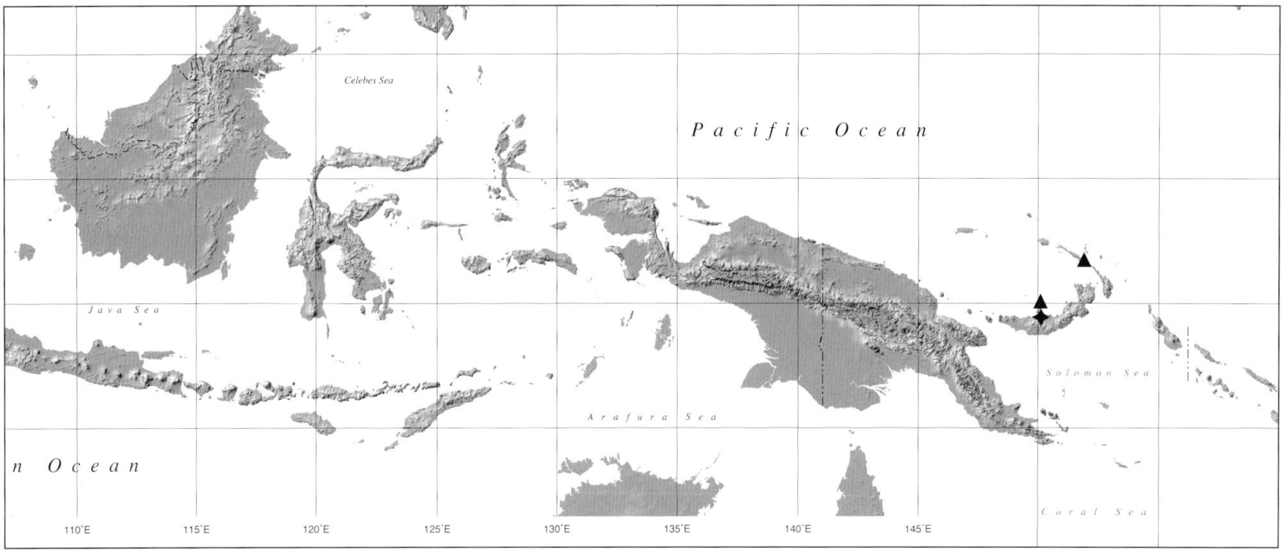

FIGURE 140 Distribution of *Stauropus talasea.*

Apical valva process is the broadest among all *viridissimus* subspecies and the 8th sternite has the posterior margin not rounded as in ssp. *amboynica* or ssp. *viridissimus,* but straight.

Diagnosis. *Stauropus viridissimus* is rather unmistakable by its greenish wing color in various shades. A characteristic pattern feature is a black dot near the forewing apex, which is sometimes weakly developed. A small black discal spot is present on forewing. In males the forewings display often a blackish medial area. This form occurs in all subspecies. Sometimes the basal area of forewings has a large black blotch or a long streak, sometimes reaching the medial area at the dorsum. The sexually dimorphic female is rather brownish by facies, with a reduced green pattern. The few known females (n = 6) vary individually somewhat in the intensity of the green pattern, but no correlation to geographical subspecies was found in external appearance.

Male genitalia are distinguished by valvae shape. The long and broad apical process bears a triangular projection near the tip. The various shapes of the large and strongly sclerotized ventral processes of valvae are taxonomically important in separating the different subspecies. Phallus is long and slender, as usual in *Stauropus*. 8th abdominal segments are less modified. Female genitalia have not yet been dissected.

Bionomics. Males occur commonly in lowland forests up to 500 m, but adults were observed occasionally also at altitudes up to 1670 m.

Distribution. *Stauropus viridissimus boeroe* is restricted to Buru; ssp. *naumanni* occurs in Obi, Bacan, Halmahera; ssp. *amboynica*: Ambon, Seram; ssp. *viridissimus* occurs in New Guinea in its entirety, Key, Biak, Waigeu, and ssp. *habrochlora* is distributed in Australia (Queensland, Northern Territories) and Tanimbar.

159. *Stauropus talasea* spec. nov. (figs. 159a–d).
Holotype: ♂, Papua New Guinea, W New Britain prov., Talasea district, Kulkelpun near Btalla, rainforest, 500 m, 5°46.257'S, 150°33.572'E, 21.-23.iii.2003, leg. Rudloff & Schaarschmidt (MWM 15.542) – Museum Witt, Munich.

Paratypes (25 ♂♂):
New Britain: 8 ♂♂, Talasea peninsula, Bulumuvi, Lake Dakataua, 5°01.330'S, 150°08379'E, 1.–4. iv. 2003, leg. Rudloff & Schaarschmidt (MWM 15.541); 2 ♂♂, Talasea district, Kulkelpun near Bialla, 500 m, 5°46.257'S, 150°33.572'E, 21.–23. iii. 2003, leg. Rudloff & Schaarschmidt; 1 ♂, Plotplun, S Lopun, 300 m, 5°50.756'S, 150°39.824'E, 23.–25. iii. 2003, leg. Rudloff & Schaarschmidt; 3 ♂♂, Talesea distr., Bereme, SE Kimbe, 350 m, 5°43.952'S, 150°36.720'E, 19.–26. iii. 2003, leg. Rudloff & Schaarschmidt; 3 ♂♂, Pomio district, Lopun, near Wakis, 100 m, 5°57.277' S, 150°42.763' E, 24.–29. iii. 2003, leg. Rudloff & Schaarschmidt.
New Ireland: 4 ♂♂, New Ireland, xi. 1923 (A. F. Eichhorn) (BM 2213);
New Hanover: 4 ♂♂, New Hanover, iii. 1923 (A. S. Meek); 4 ♂♂, Nunmiava vilage, 02°36.526'S, 150°25.560'E, 3.–5, iii. 2004, leg. J. P. Rudloff.

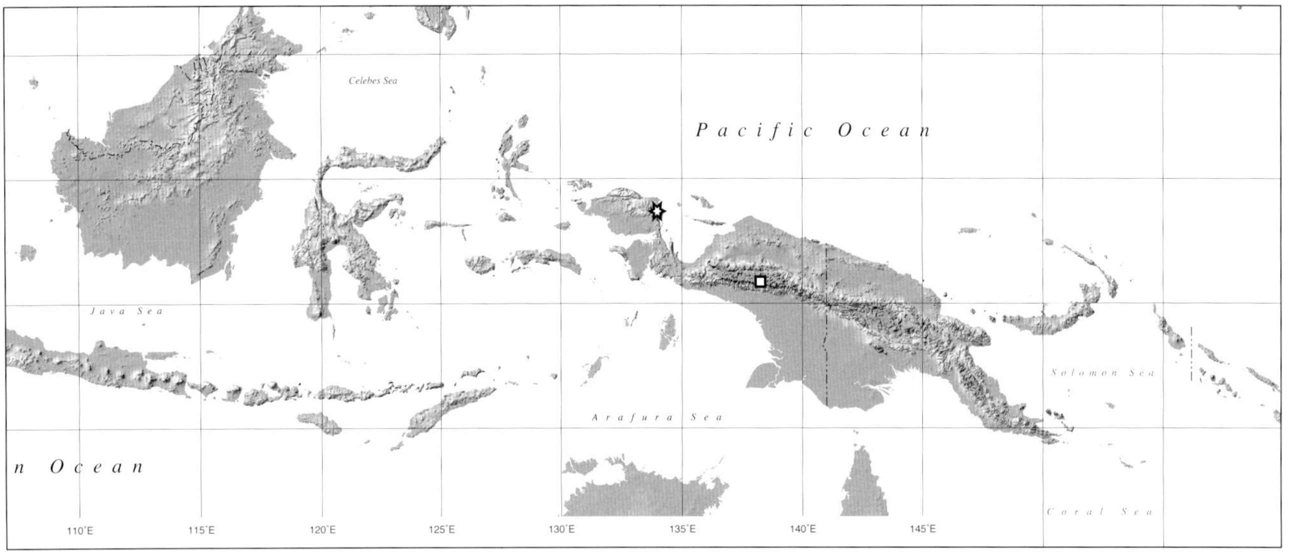

FIGURE 141 Distribution of *Stauropus vanmastrigti.*

Etymology: Named after the type locality.

Diagnosis. Forewing length ♂♂ 20 mm. The species is closely related and similar to *S. viridissimus.* Forewing shape is broader than in *viridissimus.* Forewing color is fawn, greenish tinged, and the pattern appears somewhat indistinct. Discal spot is in contrasting black and well visible. Marginal fascia is marked on all wings with a row of large brown dots. As in *viridissimus* a fuscous-brown medial area occurs as individual forms. Hindwing color and pattern continues from the forewing in the costal area, but medial and anal area are a light brownish-white, which distinguish *talesea* from *viridissimus.* Hindwing fringe is checkered whitish-brown. The female is unknown.

Male genitalia differ from *viridissimus* by the pointed and slender shape of the ventral valva process. The apical valva part is pointed at the tip. Sacculus displays an asymmetrically pair of processes. Phallus is long and slightly curved. There are two smaller projections near the tip; a second dissected male from New Britain (W 15.541) has only one projection at that position. A further dissected male from New Ireland, see fig. 155b, displays a longer, more strongly curved phallus than New Britain males and lacks the projections.

Bionomics. Adults have been collected in New Britain in primary forests at lower altitudes, up to 200 m.

Distribution. New Britain, New Ireland and New Hanover.

160. *Stauropus vanmastrigti* spec. nov. (figs. 160a–d)
Holotype: ♂, Indonesia, W Papua, Arfak Mts., Maibri vilage, 1670 m, 1°07'S, 133°54'E, 5.-20. xii. 2012, leg. St. Jakl – coll. A. Schintlmeister, Dresden.

Paratypes (13 ♂♂):
W Papua: 6 ♂♂, W Papua, Arfak Mts., Maibri vilage, 1670 m, 1°07'S, 133°54'E, 5.–20. xii. 2012, leg. St. Jakl (GU 94-08, GU 94-24); 5 ♂♂, dto. 1.–15. ii. 2013; 1 ♂, dto., 10.–30. iii. 2013; 1 ♂, road Wamena-Setani, km 50, 2000–2200 m, 26. - 29. i. 2000, leg. K. Cerny. **PNG:** 1 ♂, Morobe Prov., near Wau, Mt. Kaindi, 2360 m, 29.–30. viii. 1983, leg. S. E. & P. M. Miller.

Etymology: Named for the late Bruder Henk van Mastrigt (b. 15. i. 1946 Heerlen – d. 5. viii. 2015 in Jayapura), who lived in Jayapura from 1974 and established a great collection of Papuan butterflies, and later also moths. Before his death he sent the author a complete set of images of his notodontids, which could be identified, among them many hitherto undescribed species.

Diagnosis. Forewing lengths ♂♂ 19.5–20.0 mm. Forewing color is greyish green, fading somewhat over time to brownish. The basal area of forewing is contrastingly greenish-white filled. One male displays a contrasting black streak in the basal area. Postbasal area is fuscous greenish grey marked. The discal spot is fine black, surrounded in white. Postmedial fascia is whitish, and marginal fascia is marked as a row of black dots and fine streaks. Near the forewing costa and the apex a sharply

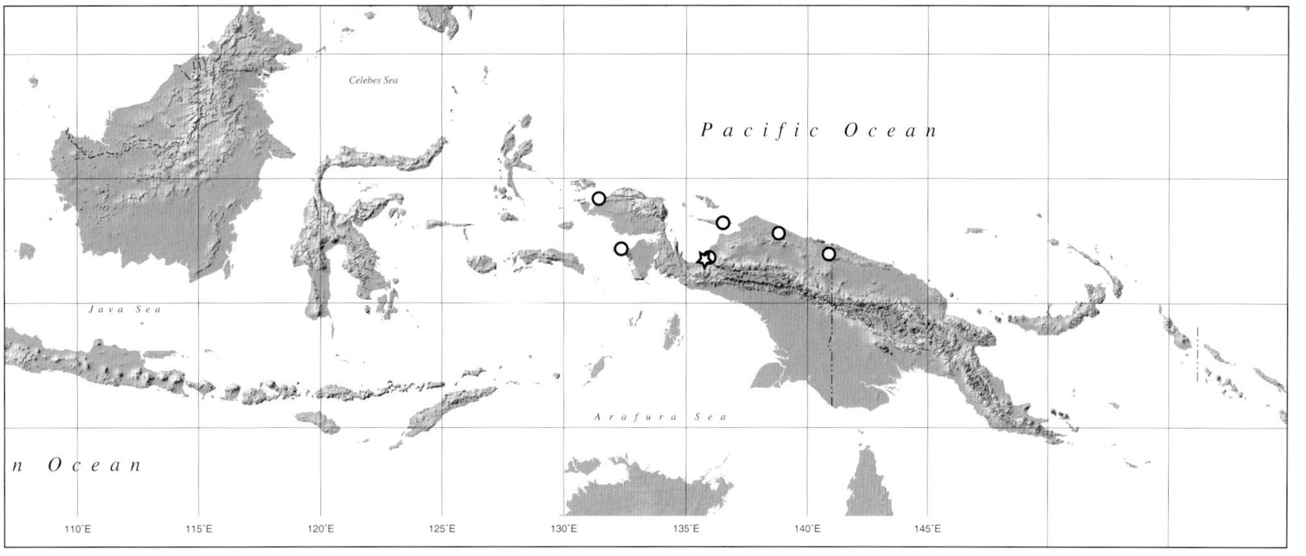

FIGURE 142 Distribution of *Stauropus caerulea*.

marked black dot is present, which is usual for most *Stauropus*. The hindwing in its costal part corresponds to the forewing coloration and pattern. Anal area of hindwing is fuscous reddish-brown, sometimes, in individually variations, pale brown. Between the thorax and the abdomen there is a brush of deep black hairs. Female is unknown. There is a resemblance to *S. viridissimus*, but the new species is clearly smaller and more fuscous greyish colored.

Male genitalia display a reduced uncus, as usual in *Stauropus*. The ventral valva process is rectangular shaped, with a knob shaped tip. The apical valva process is slender and squarish at the tip. There is no projection at the tip as in *viridissimus*. The saccular part of valva bears a pair of lobes, of which one is very reduced in length. Phallus is straight, slender, and relatively short and distinguished by a broad base. 8th sternite is narrow, the posterior margin concavely rounded. 8th tergite is of squarish shape with a slightly rounded posterior margin.

Bionomics. The species is mountainous, and adults were collected in primary forests at altitudes between 1700–2200 m.

Distribution. Known from W Papua and PNG.

161. *Stauropus caerulea* spec. nov. (figs. 161a–158d).

Holotype: ♂, Indonesia, W Papua, Straße Nabire-Ilaga, km 57, 710 m, 3°50437S, 135°72106E, 2. ii. 1998, leg. M. Schaarschmidt & F. Roick – coll. A. Schintlmeister, Dresden.

Paratypes (12 ♂♂, 1 ♀):

W Papua: 1 ♂, Nabire, 700–1000 m, vi. 1996, leg. Diufri; 3 ♂♂, 6 km NE Fak-Fak, 400 m, 2°55'S, 132°20'E, 10.–25. xii. 2008, leg. St. Jakl (GU 65-47; GU 94-67); 1 ♀, Bewani Hills, Ampas, 23. xii. 1938, leg. W. Stüber; 2 ♂♂, Nabire, 12 km E Sanabusa, Lagari, 50 m, 6.–8. xii. 1993, leg. K. Cerny (MWM 29.426); 1 ♂, Dist. Sarmi, Bora Bora village, Gauttier/Foja Mts., 2°20.488' S 138°47.476'E, 11.–15. v. 2016, leg. G. Bretschneider; 1 ♂, Sorong, Mt. Enau, 10 km of Baenkete, 7–800 m, vii. 2015 leg. Benny de Groof & Team

Yapen: 4 ♂♂, Dist. Windesi, 5 km from Asei village, 150 m, 14.–19. ix. 2013, leg. G. Bretschneider.

Diagnosis. Forewing lengths ♂♂ 18 -19 mm, ♀ 18 mm. A small species. Forewings are fuscous chocolate brown tinged with dark green scales. The fuscous green colored basal area bear two black spots. Discal spot is marked as a black streak surrounded by pale brown. Postmedial and the submarginal fasciae are a contrasting pale brownish color. A single male has a blackish-brown darkened medial area. In the costal part of hindwing the forewing color and pattern continues. A pair of broad and indistinct blackish lines at the costa of the hindwings distinguish *caerulea* from other congeners. The remaining part of hindwing is reddish-brown. Female is like the male, but the forewing color is tinged a more reddish-brown, perhaps due to the age of the only known specimen.

Male genitalia are characterized by a pointed and slightly bifurcate tip of the apical valva process. Uncus is absent. The ventral valva process of is slightly serrated and pointed at the tip. There is one long and robust saccular

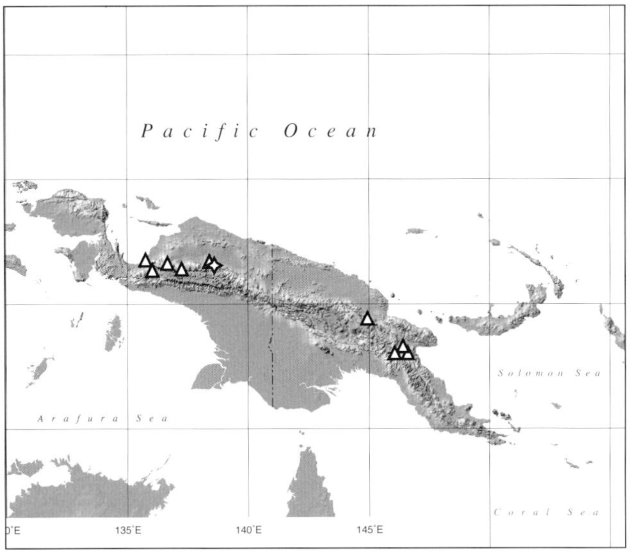

FIGURE 143 Distribution of *Stauropus affinis*.

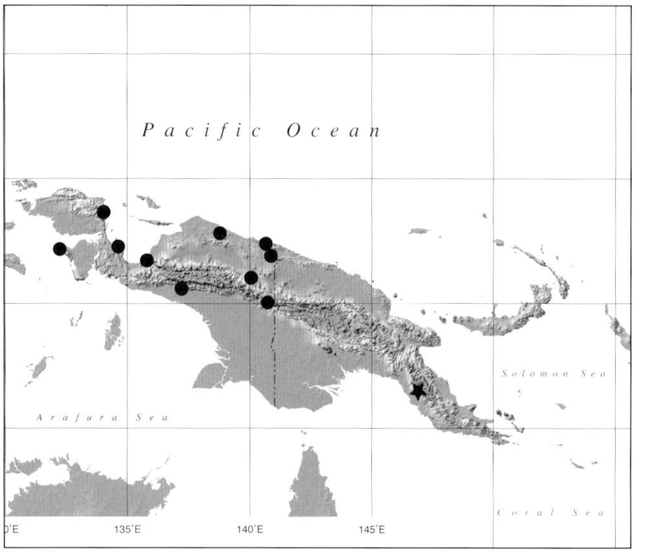

FIGURE 144 Distribution of *Stauropus pratti*.

valva process, which is also longer and broader than in *vanmastrigti*. The second process is rather inconspicuosly, weakly sclerotized, and short. Phallus is short and rounded, not pointed, at the tip. Base of phallus is characteristically club-shaped. 8th sternite is wavy at the posterior margin. 8th tergite is narrower than *vanmastrigti*.

Bionomics. The few known specimens (n = 13) were taken between 50–1000 m above sea level.

Distribution. Known from W Papua and Yapen Isl.

162. ***Stauropus affinis*** Rothschild, 1917: 245; pl. 8: 14.
Syntypes: 2 ♀♀, [Indonesia, W Papua], Central Dutch New Guinea, Mount Goliath 6.000-7.000 ft. – NHMUK, London.
Schintlmeister 2013: 425.

Diagnosis. An inconspicuous species. Forewings are brown, mixed with green scales. Basal area of forewing whitish green, not bordered sharply toward the fuscous postbasal area. Forewing pattern is indistinct and postmedial fascia are visibly only in the costal area. The blackish spot near the apex is visible in most specimens but not unprominently developed. A submarginal fascia is whitish marked. Female is slightly sexual dimorphic and displays before the black apical spot a whitish blotch, which is a good feature for identification.

Male genitalia resemble those of *viridissima*, but they are smaller and have a broad apical valva process with a triangular projection at the tip. Length of the saccular lobes varies individually, as illustrated. Phallus is very long, straight and pointed at the tip. 8th sternite is characteristically convex-shaped at the posterior marigin. 8th tergite is concave rounded at the posterior margin. Female genitalia have not yet been dissected.

Bionomics. *S. affinis* has been observed infrequently and mostly at medium altitudes between 700–1400 m), but a few males have been taken in mountain regions, e.g. Mt. Hagen, up to 2000 m.

Distribution. Distributed in New Guinea in its entirety.

163. ***Stauropus pratti*** Bethune-Baker, 1904: 380; pl. 4: 18.
Holotype: ♂, [Papua New Guinea], Ekeikei [1.500 ft.] – NHMUK, London.
Schintlmeister 2013: 426.

Diagnosis. The species is readily recognized by a contrasting white lower part of the postmedial fascia on forewing. Shape and length of forewings are as in *caerulea*. Forewing is brown, tinged with green scales. A blackish spot near the apex is weakly developed. Hindwings are in the anal part of a warm yellow, which distinguishes *pratti* from other congeners. The female is unknown.

Male genitalia display a pointed apical valva process. The ventral valva process is rather short. Sacculus bears a pair of projections. Phallus is of a conical shape at its base. 8th sternite is notched at the posterior marigin; 8th tergite is shaped as in *affinis*.

Bionomics. Adults are not rare and having been observed between 600 and 1700 m.

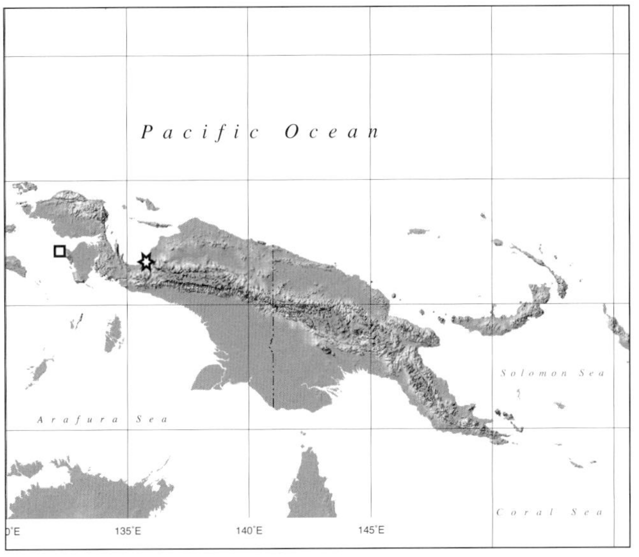

FIGURE 145 Distribution of *Stauropus schaarschmidti*.

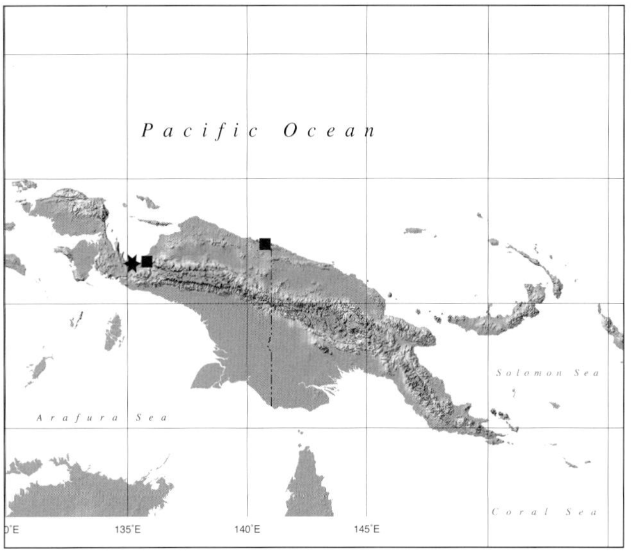

FIGURE 146 Distribution of *Stauropus evanescens*.

Distribution. Endemic to New Guinea.

164. ***Stauropus schaarschmidti*** Schintlmeister, 2003: 110; figs. 48, 98.

Holotype: ♂, Indonesia, Irian Jaya, Straße Nabire – Ilage, km 54, 700 m, 3°49'N, 135°73'E – coll. A. Schintlmeister, Dresden.

Schintlmeister 2013: 426.

Diagnosis. Forewings are elongated and narrow. Basal and medial area are prominently blackish-colored; postmedial and marginal area are paler. Pattern is indistinctly developed and less visible. By its wingshape and black colored wings the species is rather unmistakable. Hindwings are in the costal part blackish as in the forewings, but the anal area is light brown. The female is unknown.

Male genitalia are distinguished by a broad apical valva process. Phallus is relatively short and pointed at the tip. Phallus is distinguished by its broad and kinked base. 8th sternite is slightly bilobed at the central part and both lobes are strongly sclerotized.

Bionomics. The species is rare; n < 10 specimens are taken from the lowlands up to 745 m.

Distribution. Known only from W Papua.

165. ***Stauropus evanescens*** Gaede, 1930: 627; pl. 83: c.

Holotype: ♂, [Indonesia, W Papua], Neu Guinea [„Nomnagihé, 25 miles south of Wangaar"], 600 m – NHMUK, London.

Schintlmeister 2013: 426.

Diagnosis. Rather unmistakable by the pale greenish-grey forewings, with a contrasting fuscous greyish-green postbasal area. Hindwings have the costal part also fuscous greyish-green colored. The female is unknown.

Male genitalia are similar to most other *Chlorostauropus*, but the tip of the valva is pointed and bears a small triangular projection. Phallus shape is different to other *Stauropus,* being thicker and having a triangular base. 8th sternite is slightly notched andresembles *schaarschmidti*.

Bionomics. *Stauropus evanescens* occur uncommonly and was mostly observed in the lowlands, up to 100 m. The type specimen, however, was collected at 600 m above sea level.

Distribution. Distributed in New Guinea.

166. ***Stauropus pelz* spec. nov.** (figs. 166a, 166b).

Holotype: ♂, Indonesia, Nusa Tenggara Timur, Sumba, Gunung Wanggameti, 1050 m, 10°04' S, 120°13'E, 1.–10. vii. 2005, leg. St. Jakl – coll. A. Schintlmeister, Dresden.

Paratypes (6 ♂♂):
Sumba: 6 ♂♂, Gunung Wanggameti, 1050 m, 10°04' S, 120°13'E, 1.–10. vii. 2005, leg. St. Jakl (GU 65-45).

Diagnosis. Forewing lengths ♂♂ 18 -19 mm, 4 mm smaller at average than the similar *S. loki*. Basal area of forewing bears a small black patch. Medial area is fuscous chocolate-brown. Near the apex there is an indistinct paler patch. Marginal fascia is marked as a row of fine fuscous brown streaks. Hindwing displays as in *loki* one broad

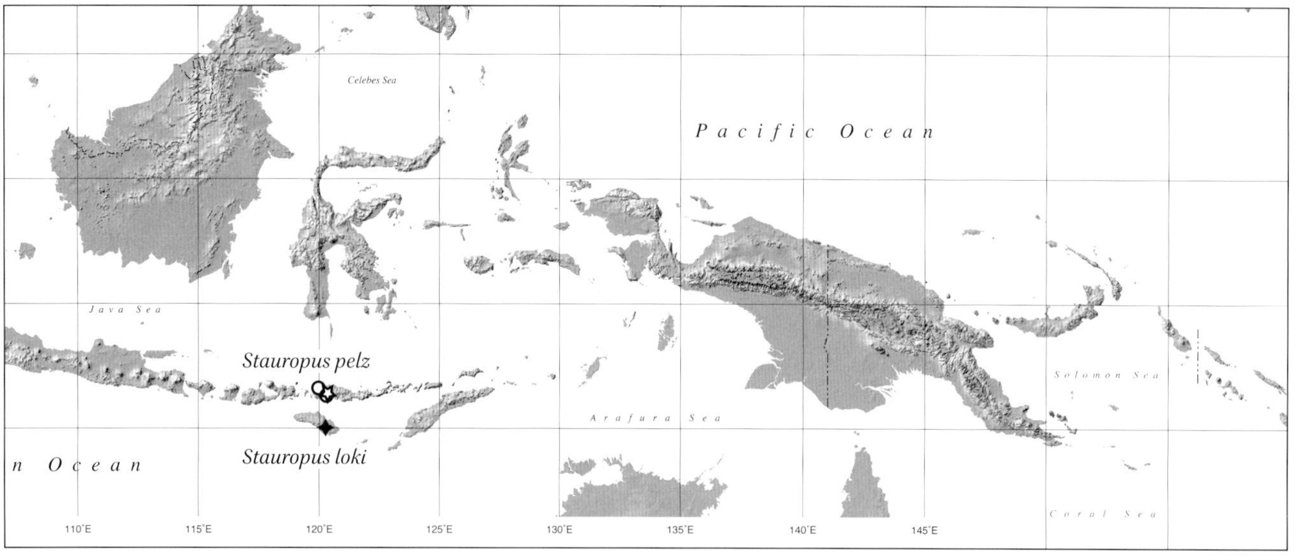

FIGURE 147 Distribution of *Stauropus pelz* and *loki*.

fuscous brown streak in the costal area. Hindwings are paler than in *loki*. The female is unknown.

Male genitalia are similar to *loki*. Uncus is very small with a small triangular central projection, which distinguishes the new species from other congeners. Socius is inconspicuous, arising from a bulbous base. Valva is broad and curved; the ventral process is enlarged, but is not strongly sclerotized as the remaining valva part. Phallus is long and straight. 8th sternite is triangularly notched at the posterior margin. The projection at the anterior margin is broad and arrow-like; shorter than in *loki*. 8th tergite has a concave rounded posterior margin.

Bionomics. The species is known only from the type locality and was taken at 1050 m above sea level.

Distribution. Endemic to Sumba.

167. ***Stauropus loki*** Schintlmeister, 2003: 108; figs. 42, 43, 44, 95.
Holotype: ♂, Indonesien, Prov. Nusa Tenggara Timur, Flores, Gunung Ranaka, 9 km E Ruteng, 1.140 m – coll. A. Schintlmeister, Dresden.
Schintlmeister 2013: 426.

Diagnosis. Similar in facies to *S. pelz*, but forewing lengths ♂♂ 3–5 mm larger. Forewings color is darker and more greenish-tinged than in *pelz*. Basal area displays a small black patch. A pale patch near the forewing apex is in most specimens weakly developed or invisible. Hindwings are darker brown than *pelz*.

Uncus reduced and of a squarish shape without a central projection. Socii small, weakly sclerotized and rectangularly angled. Valva is broad and large; ventral process is enlarged, but it is shorter than in *pelz*. Phallus is straight, long, and slender. 8th sternite notched at the posterior margin and more strongly sclerotized in this area. The arrow-like anterior projection is slenderer and longer than in *pelz*. Female genitalia have not yet been dissected.

Bionomics. The species was observed commonly in the mountains from 900 m up to 1820 m above sea level.

Distribution. Endemic to Flores.

Subgenus ***Benbowia*** Kiriakoff, 1967: 52.
Type species: *Stauropus virescens* Moore, 1879.

Forewings and costal part of hindwings are green. Phallus is much thicker than in other subgenera and often curved. Uncus is absent, but socii are enlarged with projections at the tip. 8th sternite of *Benbowia* has a pair of small central lobes. 8th tergite has a serrated posterior margin.

168. ***Stauropus virescens*** Moore, 1879: 404.
Holotype : ♂, [India], Darjiling [=Darjeeling] – NHMUK, London.
Schintlmeister 2008: 159; Schintlmeister 2013: 424.

Taxonomic notes. A variable species. Apart from external differences, many parts of male genitalia and 8th abdominal segments show great width variation, which led to the

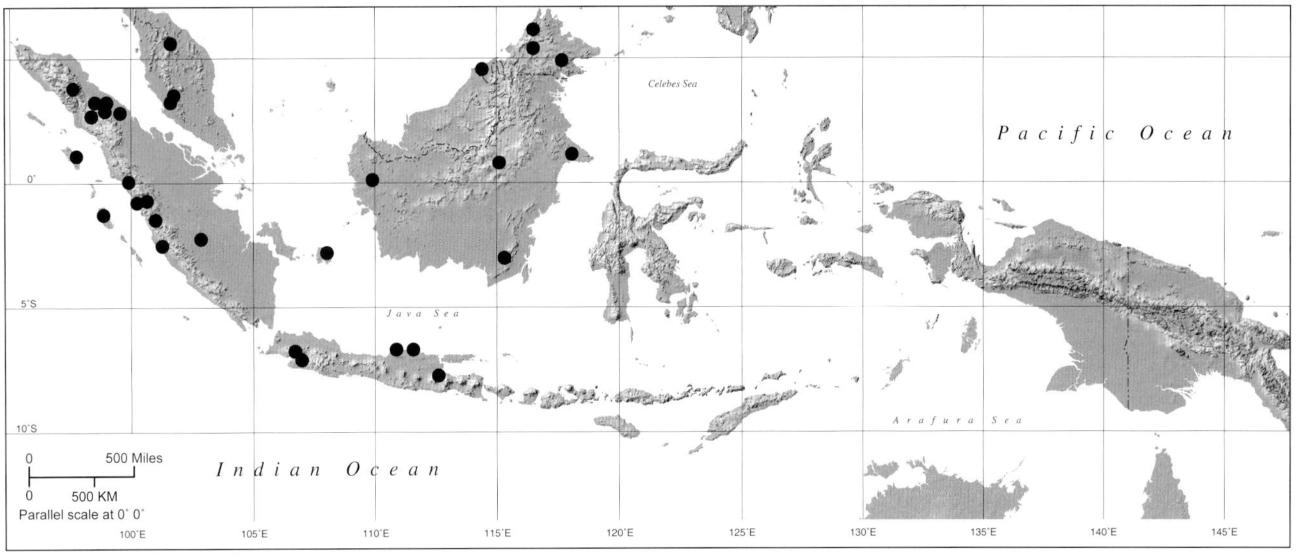

FIGURE 148 Distribution of *Stauropus virescens.*

descriptions of several taxa. However, after dissection and examination of many (n = 42) genitalia, no correalation to geographic localities was found. Populations from India, Nepal, Bhutan and Myanmar differ by their distinctly longer processes at phallus tip, see fig. 168f, from Sundaland specimens and also by shape of the serrated valva apex, without a bulbous, spine bearing projection.

The oldest available name for Sundanian populations is *Stauropus obscuratus* van Eecke, 1929: 165; pl. 14: 4; holotype: ♂, Sumatra, Fort de Kock [=Bukittinggi] – NBC, Leiden, which is placed here as a subspecies of *virescens* (**stat. nov.**). *Stauropus obscuratus* was described according to a melanic specimen, an individual form which occurs frequently in Sumatra and Borneo, but not known as yet in Java.

Benbowia dudgeoni Kiriakoff, 1967: 53; fig. 13; holotype: ♂, Sumatra, Lebong Tandai – NHMUK, London, is a junior synonym of *obscuratus* (**syn. nov.**) and both taxa were placed earlier as synonyms to *virescens* by Holloway, 1983: 59.

Diagnosis. Forewings and thorax are green. Basal, postmedial and marginal fasciae are marked as rows of black and white patches. Hindwing in its costal area is green with blackish markings. The remaining part is chocolate- to reddish-brown. Occasionally forms with blackish colored basal and medial area occur, including transistional individuals, compare fig. 168k. The green color varies from light yellowish-green, bluish-green to deep mossy-green; a male from Java displays yellowish forewing. Forewing fasciae varies also individually in their shapes. Variation width is widely overlapping with the only further species occurring in Sundaland, *S. elisabethae.* Both species are hard to separate according to external features. *S. viridescens* displays the black apical forwing patch smaller than in *elisbethae.* A good characteristic is the black band near the hindwing apex, running parallel to margin and hits almost at right angles (70–90°) the costa, while in *elisabethae* this streak is acute-angled (40–50°). Sexual dimorphism is limited, and individual variations do not differ between the sexes.

Male genitalia are best recognized by the narrowness of the ventral valva process, and its pointed tip, which displays strong serrations. Socii have a bulbous tip with serrations. Valva is curved and large with a bulbous projection at the apex. Phallus is curved and bear a pair of processes near the tip, which are short in ssp. *obscuratus.* Shapes of socii, valvae apex, ventral valva process are subjects of individual variation. 8th sternite convex at posterior margin with a central, notched projection. Margin of 8th tergite is concavely rounded at the serrated posterior margin, which distinguishes *virescens* from *elisabethae.* The latter displays a deeply convex notched margin. Female genitalia have rather less rounded papillae anales than *elisabethae.* Antevaginal plate displays a large bifurcate projection. The notch is much deeper than in *elisbethae.* Ductus bursae is short and narrow. Corpus bursae is disc-shaped and small. No signum. The most telling feature in differentiating *virescens* from *S. elisabethae* is the smooth, non-serrated shape of the 8th abdominal segment.

Bionomics. Moths occur commonly and fly in Sundaland from the lowlands (50 m) up to 1500 m in a wider range of

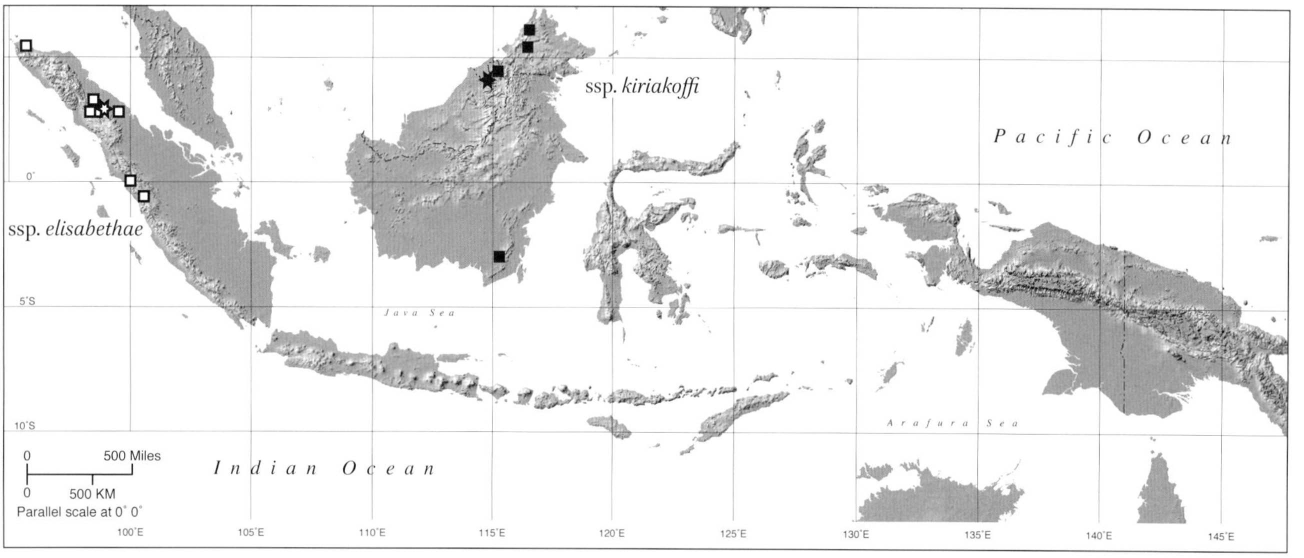

FIGURE 149 Distribution of *Stauropus elisabethae.*

habitats, including urban areas. In Sikkim *virescens virescens* was observed also at 2000 m above sea level.

Distribution. *B. virescens virescens* is distributed in Nepal, northeastern India, Myanmar and Indochina. Further eventually different subspecies flying in Andaman Isls. and southwestern China (Yunnan). In Malaya, Borneo, Palawan, Sumatra, Nias, Siberut, Belitung Isles, and Java the ssp. *obscuratus* occurs.

169. *Stauropus elisabethae* Dierl, 1981: 171; fig. 1, 2.
Holotype : ♂, [Indonesia], Sumatra, [Prapat, 1.000–1.200 m] – ZSM, Munich.
Schintlmeister 2013: 424.

Taxonomic notes. Dierl's original description provides no exact data of the holotype and states only: "Typen: Holotypus ♂, Sumatra, in Coll. Dr. Bender.", but the author traced the holotype in ZSM, Munich and added from the label, the collecting locality: "Prapat, 1.000–1.200 m, Sumatra, 27. 7. 1976, legit Dr. E. Diehl"

Male genitalia of the holotype are not yet dissected. The genitalia drawing provided in the original description by Dierl (fig. 2) show in fact *S. virescens.* It was made from a paratype of *elisabethae*. All other paratypes of *elisabethae* examined by the author (n = 5) are conspecific with *S. virescens* and only the holotype of *elisabethae* represents, according its external appearance a different species. The holotype is illustrated here for first time, see fig. 179f.

Benbowia kiriakoffi Holloway, 1983: 60; pl. 6: 5, 8; fig. 61; holotype ♂, Borneo, Sarawak, Gunong Mulu Nat. Park, Site 14, Camp 2.5, Mulu, 1.000 m – NHMUK, London, is conspecific with *elisabethae,* but differs from Sumatran populations in the shape of its triangular-shaped ventral valva process and a larger serrated projection at the valva apex. These characteristics were confirmed of n = 9 male genitalia dissections. The taxon *kiriakoffi* is placed here as a subspecies of *S. elisabethae* **(stat. nov.).**

Diagnosis. Similar to *S. virescens* by external appearance; for description of pattern see above. The variabilities of both species are overlapping, but in *S. eliabethae* there are apparently no melanic forms occuring. *B. elisabethae* distinguished from *virescens* by a more strongly developed black apical patch of the forewing and by the position of the last black fascia near the apex of the hindwing, which hits the costa acute-angled (40–50°). Sexual dimorphism is limited.

Male genitalia resembles *virescens* but its small uncus is triangularly shaped and the ventral valva process is broader. The serrated projection at valva apex is larger in *elisabethae* than in *virescens.* 8th sternite has a rather straight, not curved posterior margin. Central projection of 8th sternite, with a notch of variable shape, is as in *virescens.* 8th tergite is serrated and deeply bilobed at the posterior margin. Female genitalia are similar to *virescens*, but the 8th abdominal segment is serrated and the projection of the antevaginal plate is only slightly bifurcate.

Bionomics. Adults occur sympatrically with *virescens*, but they are much rarer and prefer medium altitudes between 600 m and 1500 m. From the lowlands (around 50–350 m) only a very few specimens are known.

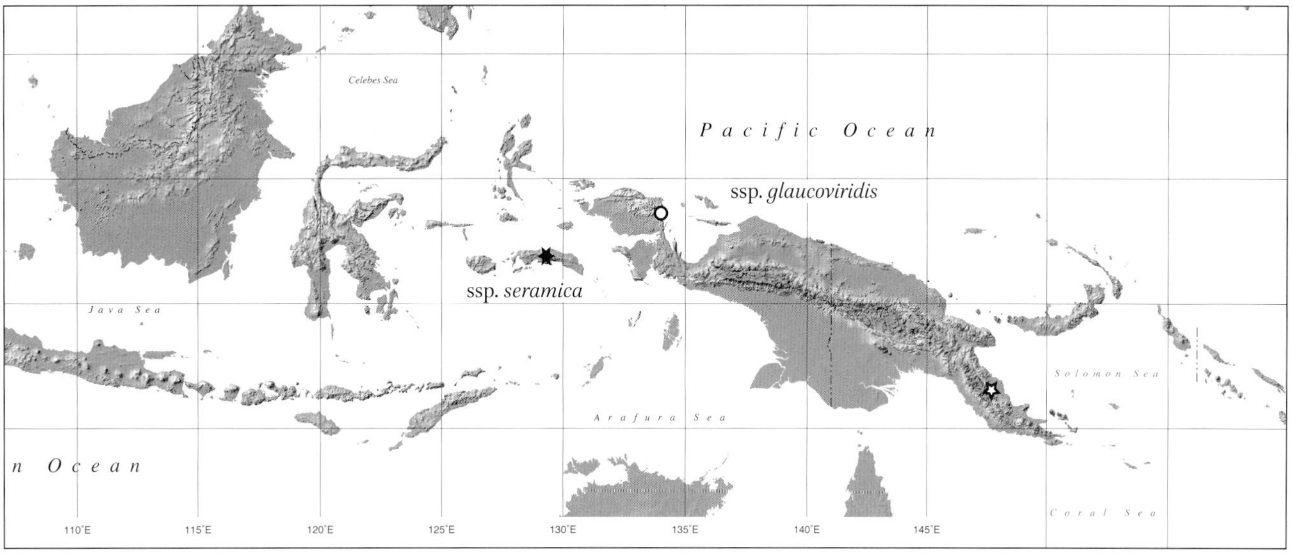

FIGURE 150 Distribution of *Kakostauropus glaucoviridis*.

Distribution. The ssp. *elisabethae* is endemic to Sumatra and ssp. *kiriakoffi* is restricted to Borneo.

Kakostauropus Gaede, 1930: 632.
Type species: *Stauropus glaucoviridis* Rothschild, 1917.
Schintlmeister 2013: 230.

Diagnosis. Medium-sized adults with bluish-green colored forewings. Antennae are black and bipectinate in both sexes almost until the tip. The last 3% of the antennae length are filiform. Rami are long. Forewing pattern is black and reduced to some costal dots and a prominent black postbasal fascia. Postmedial fascia is slightly darker green and marginal fascia marked as a row of black dots. Forewing fringe is of fuscous grey and bluish-green checkered. Hindwings are brownish with a fuscous-brown marginal line. Sexual dimorphism is limited.

Socii are slender, long and pointed at the tips. Valva is of elongated shape with a conical and rounded apex. The digitus process is large and on each side of valvae asymmetrically shaped. Phallus is straight. 8th sternite is notched at the posterior margin; 8th tergite is concavely rounded at the margin with specific sclerotizations in the central part. Female genitalia have long apophyses anteriores, but reduced apophyses posteriores. 8th abdominal segment is wide. A large central projection of the antevaginal plate is a diagnostic feature for the genus. Antrum is strongly sclerotized and of squarish shape. Ductus bursae is very reduced in length. Appendix bursae is large and disc-shaped. Corpus bursae is large and very broad. No signum.

The genus is monotypic and restricted to New Guinea and Seram.

170. ***Kakostauropus glaucoviridis*** (Rothschild, 1917): 242; pl. 3: 6 (*Stauropus glaucoviridis*).
Holotype: ♀, [Papua New Guinea], North-east British New Guinea, Mambare [River], Biagi, 5.000–7.000 ft. – NHMUK, London.
Schintlmeister 2013: 230.

Taxonomic note. Population from Seram are different from Papuan material and introduced as.

Kakostauropus glaucoviridis seramica **ssp. nov.** (figs. 170c, 170f).
Holotype: ♂, Indonesia, Central Ceram, Manusela, 6000ft., Oct. – Dec. 1919, C. F. & J. Pratt (BM 2158) – NHMUK, London.

No paratypes.

Diagnosis of ssp. *seramica*. Forewing length 19 mm. Indistinguishable in external appearance from nominotypical *K. glaucoviridis* from W Papua, but male genitalia are different: uncus is broader and shorter than ssp. *glaucoviridis;* digitus processes of valvae are narrower and not bilobed but rounded. Phallus is slenderer and displays near the tip two small projections. 8th sternite in ssp. *seramica* is triangularly notched, while in ssp. *glaucoviridis* the notch is convex. 8th tergite displays the central sclerotization more narrowly and the marginal sclerotizations are toward the center circularly bordered and not straight as in ssp. *glaucoviridis*.

Diagnosis. Unmistakable by the bluish green forewing color with black pattern for further details see under genus diagnosis. Individual variability is seemingly limited.

Male genitalia are distinguished by the shape of the large uncus and the digitus process. Female genitalia as described under genus diagnosis. They are readily recognized by the unusually broad shape of the corpus bursae and the large appendix bursae.

Bionomics. Adults occur rare, n < 10 specimens are known, in montan rain forests between 1500–2000 m.

Distribution. *Kakostauropus glaucoviridis glaucoviridis* is known hitherto from two localities only: upper Mambare river, (Papua New Guinea) and Arfak Mts. (W Papua). In Seram the ssp. *seramica* occurs.

Syntypistis Turner, 1907: 679.

Type species: *Syntypistis chloropasta* Turner, 1907. [= *Syntypistis kebeae* Bethune-Baker, 1904].

Schintlmeister 2008: 164; Schintlmeister 2013: 435.

Diagnosis. Medium-sized adults. Antennae are brown and bipectinate; the last 20–30% to the tip are filiform. Rami of males are long, those of females are slightly shorter. General appearance resembles *Stauropus*, but forewing shape is more elongated. Forewings of almost all species display green scales, in various shades or includes green fields. Only a few species have light greyish or brown ground color of the forewing. General appearance resembles *Stauropus*, and forewings have a postbasal fascia and a postmedial fascia and often a fine fuscous submarginal fascia occuring. Basal area of many species is of a contrasting whitish or green filled. Hindwings are whitish, pale brown or fuscous brown, except the costal region, which displays the corresponding forewing pattern and coloration. Many *Syntypistis* species are strongly sexually dimorphic.

Male genitalia have a large uncus and a pair of large socii. Valvae are slender and often strongly sclerotized with an apical process. Phallus is slender, and most species display a strongly sclerotized tip. Shapes and sclerotizations of 8th abdominal segments are in *Syntypistis* usual of taxonomic value. The anterior 8th sternite has a central projection at the anterior margin, and the posterior margin is often notched. Female genitalia have long and slender apophyses, where the apophyses anteriores are usually slightly shorter than the posterior apophyses. Antrum is strongly sclerotized and large, ductus bursae is in most species short. Corpus bursae displays a prominent, sclerotized long oscillating band, but no signum.

Caterpillars – if known – are green and of cylindrical shape. Their hostpants are in the families Fagaceae, Styracaceae and Junglandaceae. Many *Syntypistis* species prefers mountainous habitats above 1,500 m, in the Himalaya some species have been caught up to 4,200 m. Higher mountains often host endemic *Syntypistis*.

Syntypistis is rich in species; more than 80 species are known. The genus is distributed in Pakistan, India, China, Korea, Japan, Taiwan, Indochina, the Philippines, Indonesia in its entirety up to Australia, New Guinea, and the Solomon Isl. The highest diversity of species is found in New Guinea and in China.

171. ***Syntypistis kebeae*** (Bethune-Baker, 1904): 378; pl. 5: 52 (*Stauropus kebeae*).

Lectotype: ♂, [Papua New Guinea], Mount Kebea [6.000 ft.] – NHMUK, London.

Schintlmeister 2013: 439.

Taxonomic notes. The *kebeae* complex is extremely variable in both the coloration and the forewing patterns of its moths as well as in their genitalia. Comparison of larger series of adults from various localities (Arfak Mts., Maibri village n = 77, 7 GU; Halmahera, 15 km SW of Baru n = 51, 4 GU; Bacan, Mt. Sibela, n=64, 2 GU; Seram, Kamarian, n = 35, Buru, Ilat, n = 36, 4 GU) indicate, that there might be differences among various islands, but the very wide range of individual variability overlaps this by far. The author failed to find significant taxonomic features to discriminate among such populations. This leads to the following new synonymies and *Syntypistis kebeae* is treated here as a monotypic species:

Syntypistis chloropasta Turner, 1907: 679; holotype: ♂, Australia, N. Q[ueensland], Kuranda – ANIC, Canberra, **syn. nov.**

Desmeocraera frugilegus form *aeruginosus* Gaede, 1930): 629; holotype: ♂, Zentral-Buru [„Gamoe 'Mrapat, 5.000 ft."] – NHMUK, London, **syn. nov.**

Quadricalcarifera ceramensis Kiriakoff, 1967: 44; fig. 9; holotype: ♂, Central Ceram, Manusela, 6.000 ft. – NHMUK, London, **syn. nov.**

Stauropus dubiosus Bethune-Baker, 1904: 379; pl. 6: 39 was described after 1 ♂, 1 ♀, [Papua New Guinea], Mount Kebea [6.000 ft.] and Ekeikei and both syntypes are in the holdings of NHMUK, London. The ♂ with the following labels is designated here as a lectotype to stabilize the nomenclature: "Mt. Kebea | B.C. New Guinea |6.000 ft. |

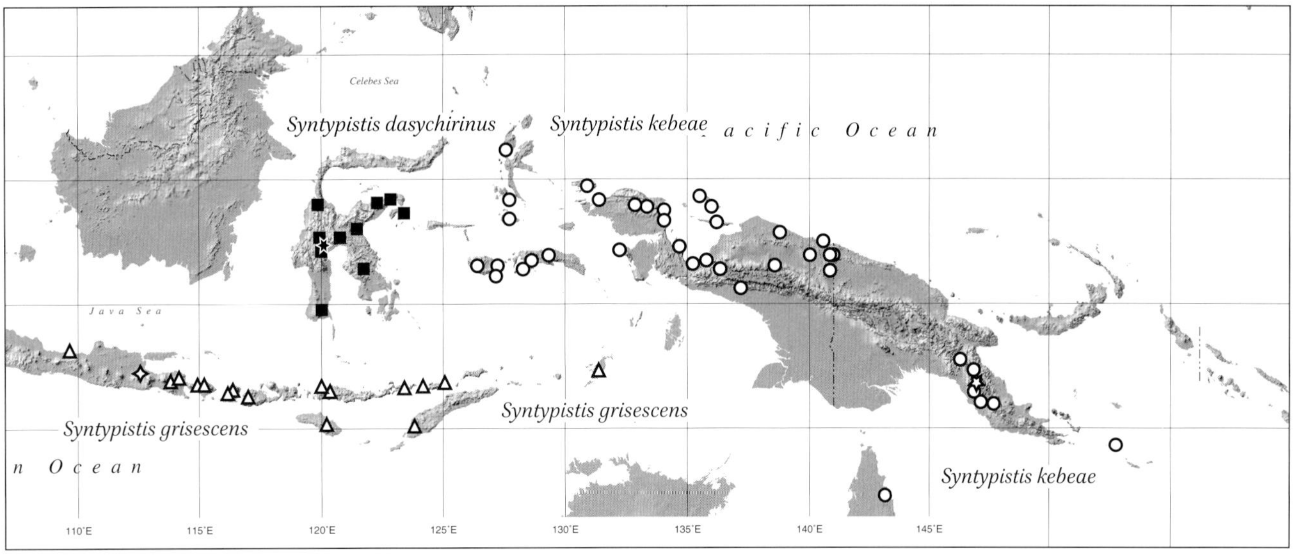

FIGURE 151 Distribution of *Syntypistis kebeae, dasychirinus* and *grisescens*.

March–April, 1903 | A.E. Pratt Coll."; green circular label "Type"; blue label "Notodontidae genitalia slide No. 1462, ♂"; yellow label "Photo done by A. Schintlmeister #4378". A lectotype label will be added. The remaining female becomes a paralectotype. The lectotype is conspecific with *S. kebeae* (**syn. nov.**) and resembles to fig. 171m; genitalia are examined. Both taxa were described in the same paper and as first revising author *dubiosus* is placed here as a synonym of *kebeae*.

The *kebeae*-complex includes two sister-species, *Syntypistis dasychirinus* and *grisescens,* with similar individual scopes of variation. But the fields of small spines at the apical half of the phallus and their placement differ in this taxa significantly as illustrated (figs. 171i, 173a, 174a), and are treated here as genuine species. However, it is possible, that *dasychirinus* and *grisescens* represent allopatrically subspecies of *kebeae*.

Diagnosis. The various individual forms of *kebeae* are never tinged with metallic deep green scales, which separates them from many other Papuan species. The illustrated forms of the adults reflecting in discrete steps the range of individual variability of *kebeae*. Most individual forms appear in both sexes. Sexual dimorphism is limited.

Male genitalia show in all parts, including shapes of uncus, socii and phallus, great variation scope. Valvae appear in some cases asymmetrical by length of apical process. Phallus bears from the middle part toward the tip many triangular smaller spines, which are arranged in two, parallel running rows, which is a characteristic feature. However, a few individuals show reduced numbers and reduced size of these spines. The female genitalia resemble most other Papuan congeners: large 8th abdominal segment, long and slender apophyses, corpus bursae with oscillating band. Specific characteristics are the short ductus bursae, the relatively large, triangular shaped anterior lobes of the antevaginal plate, and a strongly sclerotized large circular "shadow" in the central part of the vaginal plate.

Bionomics. The ecological valence is apparently very wide, and males occur in virtually any kinds of habitats, e.g. in cities, villages, rain forests, or dryer landscapes with shrubs. Most specimens are from lowlands; range of altitudes: 50–300 m. In Papua New Guinea specimens have been collected up to 2200 m above sea level. In the Arfak Mts. at 1670 m, the species was observed at artificial light quite commonly. *Syntypistis kebeae* is by far the most abundant *Syntypistis* in all areas where *kebeae* occur.

Distribution. Distributed in New Guinea in its entirety, Biak, Waigeu, Halmahera, Bacan, Obi, Ambon, Seram, Buru, Ambelau Isl. and Australia (Queensland).

172. ***Syntypistis dasychirinus*** (Roepke, 1944): 22; pl. 5: 52, **stat. rev.** (*Stauropodopsis dasychirinus*).

Holotype: ♂, [Indonesia, Sulawesi], Central Celebes, Todjambu – NBC, Leiden.

Schintlmeister 2013: 436.

Taxonomic note. *Stauropodopsis dasychirinus* was synonymized by Schintlmeister, 2013: 436 with *Stauropodopsis celebensis*. But based on phallus shape, *celebensis* is conspecific with *S. charistera*, and *dasychirinus* appear as a genuine species, which is closely related to *kebeae*.

Diagnosis. The species appears in three externally different forms: the most common form (fig. 169f) resembles in facies *kebeae* f. *aeruginosa*. Forewings are grey with a whitish pattern. Sometimes pale greenish scales appear in the apical area of forewing. The nominotypical morphe (figs. 172h, 172i) have whitish forewing color with contrasting brown pattern, including a prominent tornal patch. Similar forms occur also in *S. kebeae* (fig. 171l). In *S. grisescens* and *S. charistera* there are also individuals, which have a resemblance to *dasychirinus* but both species have a darker greyish forewing color than *dasychirinus*. A third form (fig. 172g) appears with fuscous brown forewings. All these forms occur also in females with corresponding pattern and coloration. Sexual dimorphism is limited.

Male genitalia resembles *kebeae,* but the unified socii is longer than the phallus. Uncus, and socii are variable in their shapes. Valvae are also variable but display a pronounced apical valva process. Phallus is slightly thicker than that of *kebeae* and displays in the middle part a double ring of small triangular spines. There are no spines from the middle portion toward the tip as seen in *kebeae.* The number of triangular spines in the ring part is sometimes reduced. Everted endophallus bears many fine but strongly sclerotized microthorns, which are not appear in *kebeae.* 8th sternite is notched at the posterior margin, and displays a comparable variability as in *kebeae.*Female genitalia are similar to those of *kebeae* and *grisescens,* differ by a longer ductus bursae and by shape of the relatively small antrum.

Bionomics. Imagines occur commonly between 700 m–2000 m. There are no records from the lowlands.

Distribution. Restricted to Sulawesi and Peleng Island.

173. ***Syntypistis grisescens*** (Roepke, 1944): 21; figs. 22a–d (*Stauropodopsis grisecens*).
Holotype: ♂, [Indonesia], E Java, Djunggo-Ardjuno, 1.300 m – NBC, Leiden.
Schintlmeister 2013: 436.

Diagnosis. This is the smallest species in the *kebeae* complex. Forewings are fuscous-brown, mixed with many fine deep green scales. Basal, medial and postmedial fasciae on forewing are present as fine black bands, but they are often less visible on fuscous forewing color. Medial area displays in many specimens indistinct, weakly whitish-marked circular discal, medial and postbasal blotches. In a few cases these patches are larger and prominent contrasting white. A few specimens appear alike nominatform of *S. dasychirinus*, but with darker greyish ground color of forewings. A few males (fig. 173n) appear with contrasting black medial area. These forms are known only from Bali, Lombok and Pantar. Females appear less variable than males. Variation scope includes forms with black basal and postmedial fasciae and various shades of darkness. Sexual dimorphism is limited.

At first glance male genitalia resemble *kebeae*, but apical processes are reduced in length and not pronounced as in *dasychirinus.* Uncus and socii are less variable in its shape; their tips are gently rounded and not bilobed. Valvae vary somewhat in their broadness. Phallus is variable in shape. The positions of spines are as in *dasychirinus,* but the spines are larger. Chitinized structures of the everted endophallus resembles *dasychirinus,* but they are denser and larger. 8th sternite is notched at the posterior margin and displays along the notch small lobes, which are not present in other species of the *kebeae* group. In the more strongly sclerotized part of 8th sternite several unsclerotized blotches occur in various numbers and sizes. Posterior margin of the 8th tergite is slightly bilobed at the rounded margin. Female genitalia resembles *dasychirinus*, but apophyses anteriores are of the same length as apophyses posteriores and ductus bursae is shorter. The oscillating band encloses almost the complete corpus bursae.

Bionomics. *Syntypistis grisescens* is a common species, except in Java, where it appears infrequently and fly at altitudes between 50 and 1820 m.

Distribution. Distributed in Java, Bali, Lombok, Flores, Sumba, Sumbawa, Loblein, Pantar, Alor, Timor and Tanimbar.

174. ***Syntypistis frugilegus*** (Rothschild, 1917): 243; pl. 3: 26 (*Stauropus frugilegus*).
Lectotype: ♂, [Papua New Guinea], New Guinea, Rook Island [= Umboi Isl.] – NHMUK, London, **by present designation.**
Schintlmeister 2013: 438.

Taxonomic notes. *Stauropus frugilegus* was described from 2 ♂♂, without designation of a holotype. In the holdings of NHMUK, London, the author was able to trace one male designated hereby as the lectotype to stabilize the nomenclature. The lectotype ♂ bears the following labels: "Rook Isl. | August 1913 |(A.S. Meek)", "*Stauropus* | *frugilegus* | Rothsch." in Rothschild's hand, red ringed circular label "Type", blue label "Notodontide genitalia slide No.

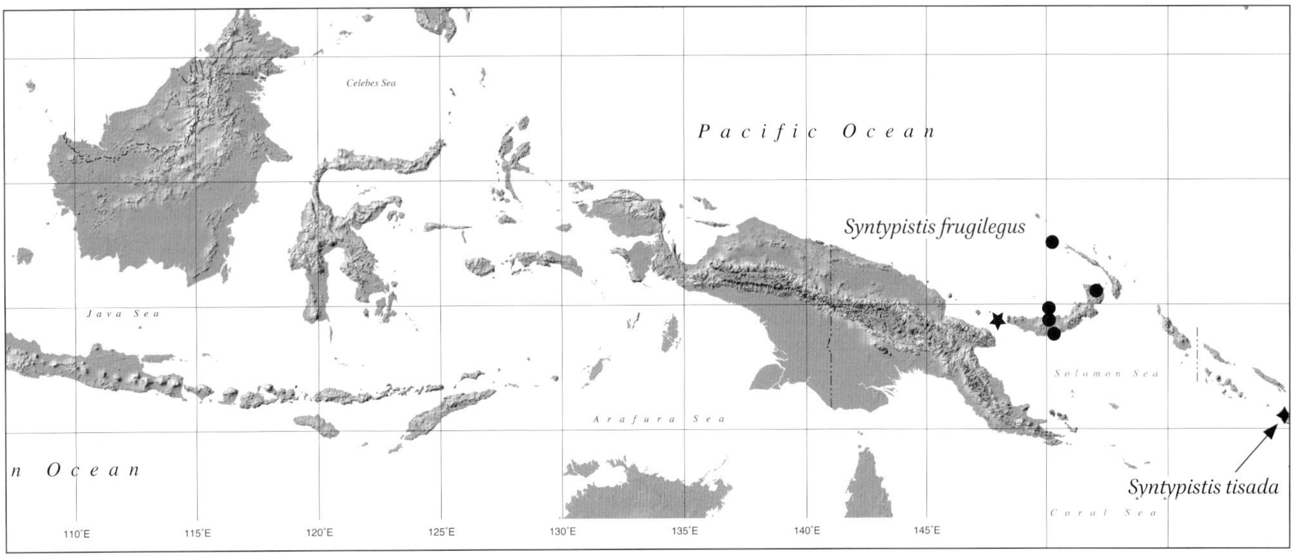

FIGURE 152 Distribution of *Syntypistis frugilegus and tisada.*

1445, ♂", yellow label "Photo done by A. Schintlmeister #4390". A lectotype label was added. The lectotype is illustrated here in fig. 174d.

Syntypistis frugilegus is treated here as a distinct species and not as a possible subspecies of *kebeae*, because of the smooth phallus and the shape of 8th sternite, which differs significantly from all of the many mounted genitalia slides of *kebeae*.

Populations from New Britain and New Ireland sharing these characters and *Quadricalcarifera eichhorni* (Kiriakoff, 1970): 111; fig. 15 (Holotype: ♂, [Papua New Guinea], New Ireland – NHMUK, London), is placed here as a junior synonym of *frugilegus* (**syn. nov.**). The holotype of *eichhorni* resembles by external appeareance much to an illustrated male (fig. 174e) from New Britain.

Diagnosis. 1–3 mm larger in forewing length than *S. kebeae*. Postbasal fascia of forewing are doubled and black. Submarginal fascia are marked as a fine black band. Occasionally a doubled postmedial fascia occur. Forewing color varies from dark brown, mixed with many green scales, shiny green, and grey. The hindwings are light brown or yellowish, whereas most specimens of *kebeae* display the anal part of the hindwings fuscous reddish-brown. Individual variabilitity of males is as illustrated; most specimens are of a series from New Britain (12 of n = 14 males) resembles externally however to the lectotype (fig. 174e). There are only two females at the author's disposal. One have reddish-brown hindwings but the second female is grey-colored and more similar to *kebeae*.

Male genitalia are similar to those of *kebeae*. They are distinguished by the smooth phallus, where no spines or thorns are present. At the phallus tip a knob-shaped, serrated projection is present. Everted endophallus displays a larger field of dense, chitinized microthorns as in *grisescens*. Uncus and socii are seemingly less variable and not distinguishable in its shape from *kebeae*. Valvae are like those of *kebeae* and variable in their shape as illustrated. 8th sternite has a wide central notch, of rather circular shape at the bottom; in *kebeae* the notch is pointed at the bottom. Female genitalia have not yet been dissected.

Bionomics. The species has been commonly observed in New Britain at 250 m above sea level.

Distribution. Known from Umboi Island, New Britain, New Ireland and New Hannover.

175. ***Stauropus tisada*** **spec. nov.** (figs. 175a–c)

Holotype: ♂, Solomon Islands, Guadalcanal, 6km S Honiara, 2 km S Mbarana village, 360 m, 1.–3. ii. 1998, leg. Dr. Ron Brechlin – coll. A. Schintlmeister, Dresden.

Paratypes (6 ♂♂):

Solomon Islands: 2 ♂♂, Solomon Islands, Guadalcanal, 6km S Honiara, 2 km S Mbarana village, 360 m, 1.–3. ii. 1998, leg. Dr. Ron Brechlin (GU 65-65); 3 ♂♂, Popomanaseu ascent, Nuhu village gardens, 28. x. 1965 (Roy. Soc. Exped.) (BM 02-2017, BM 03-2017).

Diagnosis. Forewing lengths ♂♂ 22 -23 mm. Similar by external appearance to *S. kebeae,* but 2 mm larger in

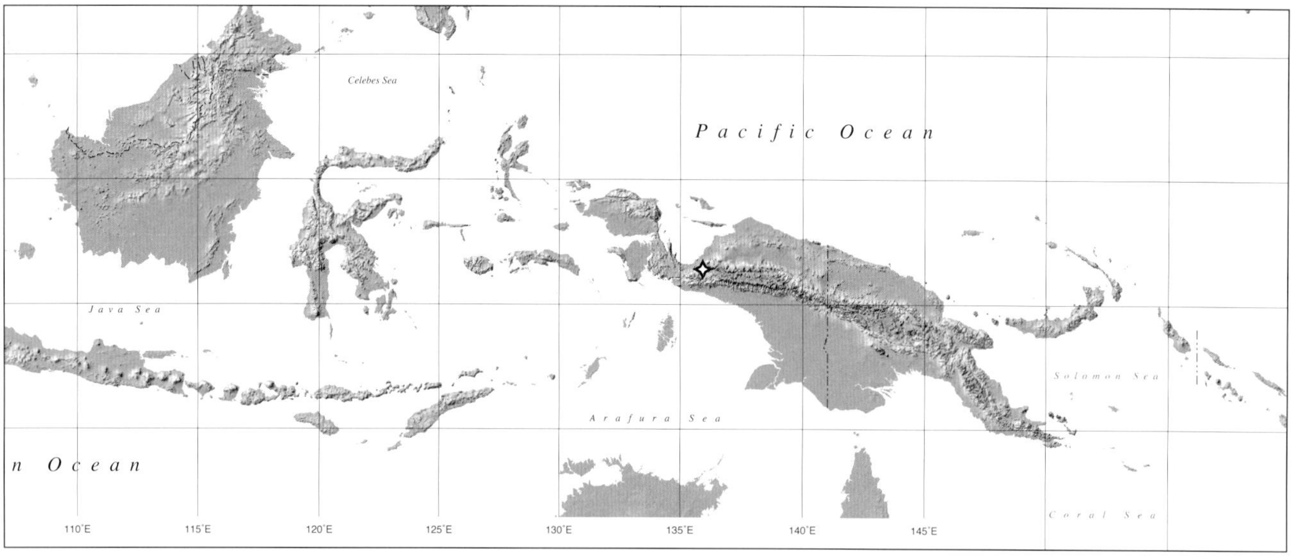

FIGURE 153 Distribution of *Syntypistis mapia*.

forewing length. Forewing color varies from dark brown, mixed with green scales, to shiny green, mixed with white scales to grey. Range of forewing variation width is as in *kebeae*, but forewing pattern is more indistinctly developed. Hindwings are pale brown with slightly more fuscous brown postmedial area. Female is unknown.

Male genitalia resembles *kebeae*. Uncus is rectangularly shaped and slightly bilobed. Unified socii are of round shape and also bilobed. Bases of uncus are squarish and less rounded than in other species of the *kebeae* complex. Valva is as in *kebeae*; apical processes pronounced but asymmetric at both valvae. Phallus is distinguished by a serrated knob at the tip. Its surface bears no spines or thorns. Everted endophallus displays no sclerotized microstructures. 8th sternite margin have numerous spines and posterior margin is notched. 8th tergite is less modified; posterior margin is concavely rounded.

Bionomics. The species occur frequently at lower elevations, up to 360 m.

Distribution. Endemic to the Solomon Islands, Guadalcanal.

176. ***Syntypistis mapia* spec. nov.** (figs. 176a–d)
Holotype: ♂, Indonesia, W Papua, Nabire distr., 10 km N Mapia, 1500–1700 m, 3°31'S, 135°44'E, viii. 2006, leg. local collectors – in coll. A. Schintlmeister, Dresden.

Paratypes (3 ♂♂, 1 ♀):
W Papua: 3 ♂♂, 1 ♀, Nabire distr., 10 km N Mapia, 1500–1700 m, 3°31'S, 135°44'E, viii. 2006, leg. local collectors (GU 65-74, GU 65-75, GU 66-06).

Other material: The late H. v. Mastrigt sent the author in 2016 for identification a photo of a male of this new species, which was collected presumably by him in W Papua, but the exact locality was not provided.

Diagnosis. Forewing lengths ♂♂ 22 mm–23 mm; ♀ 27 mm. A larger species with narrow forewing shape. It is clearly narrower than *uskwara,* which is the only comparable large Papuan *Syntypistis.* Antennae are in both sexes plumous. The last 20 % toward the tip is not bipectinated but filiform. Forewings are brown; basal and postmedial area up until the margin are tinged dark green. Basal, postbasal, postmedial and submarginal fasciae are marked indistinctly blackish. Discal spot in the brown medial area is developed as an indistinct paler brown ellipsoid patch. Hindwings are reddish brown. Costa of the hindwings are of a slightly darker brown, tinged with a few dark green scales. Sexual dimorphism is limited; the female looks exactly like the male, but its forewings are larger.

Male genitalia have a slender, rounded, and centrally notched uncus. Unified socii are of the same length as the uncus. Valva costa is swollen; apex bears a long and slender process. Phallus is straight, its phallus tip knob-shaped and serrated. 8th sternite is deeply notched at the posterior margin and more strongly sclerotized around the notch. 8th tergite is less modified but slightly bilobed at the posterior margin. Female genitalia were not dissected.

Bionomics. Known from four specimens which were taken between 1500 m and 1700 m altitude.

Distribution. Known from the Jayawijaya Mts. in W Papua.

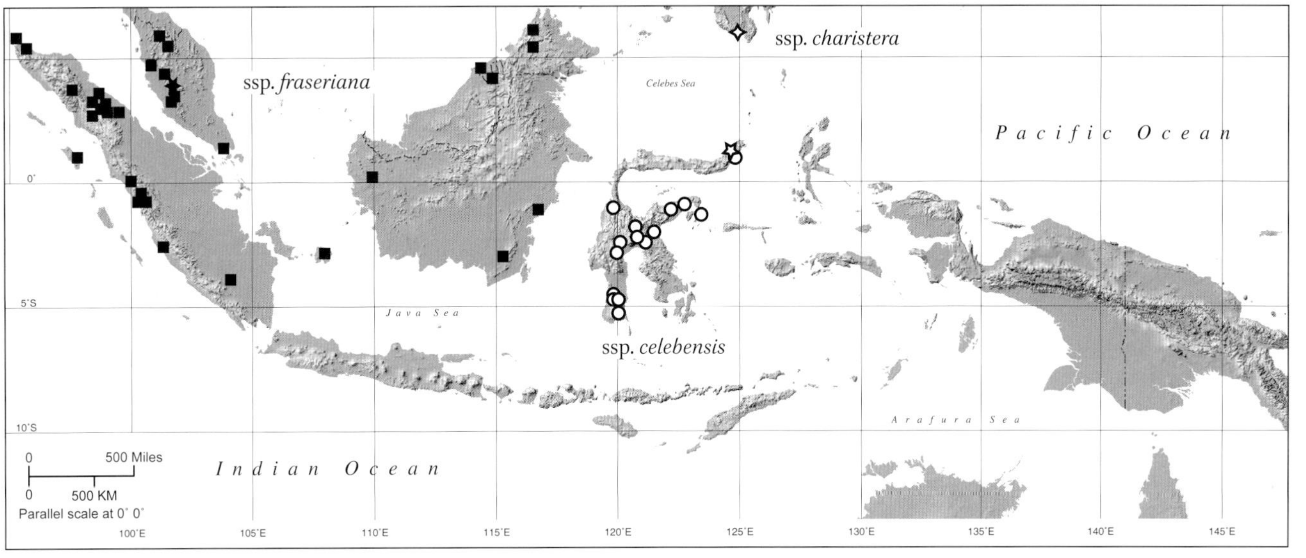

FIGURE 154 Distribution of *Syntypistis charistera*.

177. ***Syntypistis charistera*** (West, 1932): 211 (*Stauropus charistera*).
Holotype: ♂, Philippine Is., Mindanao I., subprov. Lanao, Kolambugan plains – NHMUK, London.
Schintlmeister 2008: ; Schintlmeister 2013: 436.

Taxonomic notes. *Syntypists charistera* is a polytypic species, with a wide range of individual variability. In the Indonesian Archipelago two subspecies occur.

Syntypistis charistera fraseriana (Kiriakoff, 1967): 43; fig. 7 (*Quadricalcarifera fraseriana*); holotype: ♂, Singapore, Fraser's Hill – NHMUK, London. The range of variation in adults is similar to ssp. *charistera*. But hindwings of ssp. *charistera* are fuscous reddish brown. In ssp. *fraseriana* or *celebensis* hindwings are rather yellowish white. Ssp. *fraseriana* is well distinguished by the shape of the 8th sternite. Its posterior margin is gently bilobed, whereas ssp. *charistera* display a large and deep notch (fig. 177f).

Syntypistis charistera celebensis (Roepke, 1944): 22 (*Stauropodopsis grisescens celebensis*). Holotype: ♂, [Indonesia, Sulawesi], N Celebes, Menado, Tonsea Lama – NBC, Leiden. This subspecies is readily discernible by external appearance. Adults have in most cases a large, blackish brown basal blotch, which is rarely seen in ssp. *fraseriana*, and three white stigma on the forewings. Forewing shape is significantly broader than in other subspecies. 8th sternite resembles to ssp. *fraseriana*. The taxon *celebensis* might represent due to the great externally differences a genuine species and not merely a subspecies.

Diagnosis. A highly variable species. The relatively small moth is distinguished in the medial area of forewings by three whitish stigmas, which are often weakly developed. Fuscous forms as depicted in fig. 177k, occur also, which resembles to *kebeae* or *grisescens*. Hindwings of the males are light brownish-white, sometimes pale reddish-brown tinged in the anal area. Costal area corresponds to the forewing pattern, as is usual in *Syntypistis*. Sexual dimorphism is limited, but females have darker brown hindwings. Individual forms seen in males also occur in females.

Male genitalia have socii longer than the squarish uncus, in *S. grisescens*. Valvae are longer and narrower than *grisescens* and bear also an apical process. Shapes of uncus, socii and valvae vary individually, but the range of variation is limited. Phallus is straight and larger than in *grisescens* or *dasychirinus*. Phallus tip bears one to four spines, which separates *charistera* from other members of the *kebeae* group. 8th sternite is slightly bilobed in Indonesian populations with a variable number of spines at the posterior margin. The strongly sclerotized margin displays a variable number of finer spines. 8th tergite is rounded, without specifically sclerotizations as in most congeners. Female genitalia show long and slender apophyses posteriores and shorter but broader apophyses anteriores. Antrum is squarish and in ssp. *celebensis* triangularly notched (fig. 177d). Ductus bursae is strongly sclerotized and short; corpus bursae large with an oscillating band, which is shorter than in *grisescens* or *dasychirinus*.

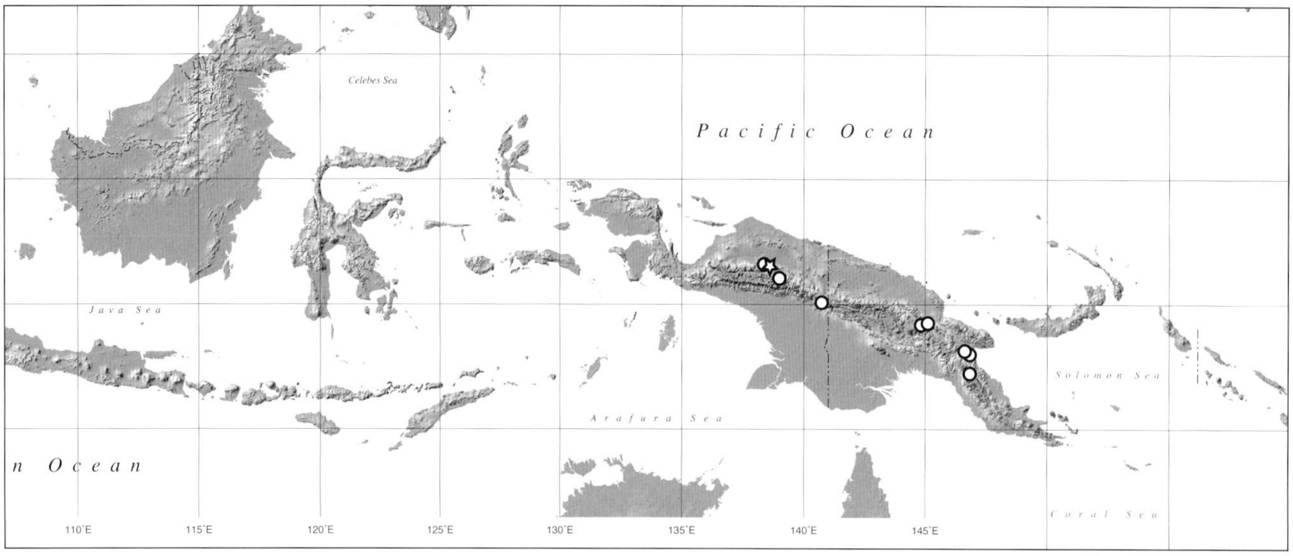

FIGURE 155 Distribution of *Syntypistis triplagosus*.

Bionomics. Males are common at light and occur in various habitates, also in urban areas. The species has been observed from the lowlands (10 m) up to 1200 m, but most adults has been found below 1000 m. In the Philippines adults has been collected also at 2200 m above sea level (Schintlmeister & Lourens, 2010: 182).

Distribution. *Syntypistis charistera charistera* inhabit entire Philippines, except Palawan. Ssp. *celebensis* is restricted to Sulawesi and Peleng. Ssp. *fraseriana* is distributed in southern Thailand, Myanmar, the Malayan Peninsula, entire Borneo and Sumatra with adjecent islands We, Nias and Belitung. In south Vietnam and Cambodia the ssp. *minima* Schintlmeister, 2006 occurs.

178. ***Syntypistis triplagosus*** (Rothschild, 1917): 243; pl. 8: 13 (*Stauropus triplagosus*) stat. rev.
Holotype: ♀, [Indonesia, W Papua], Central Dutch New Guinea, Mount Goliath 5.000 ft. – NHMUK, London.
Schintlmeister 2013: 439.

Taxonomic notes. Schintlmeister, 2013: 439 placed *S. triplagosus* as a synonym of *kebeae*. Since more material become available, it becomes evident, that male genitalia, e.g. uncus, valvae, phallus, of *triplagosus* are beyond the range of variation of *kebeae*. *Triplagosus* is reinstalled here as a genuine species.

Kiriakoff, 1967 described almost from same locality in the Jayawijaya Mts. in W Papua the following four species:

Quadricalcarifera nitidula Kiriakoff, 1967: 199; fig. 7, holotype: ♂, New Guinea, Araucaria Camp, [Jayawijaya Mts., 3°30'S, 139°11' E, 800 m] – NBC, Leiden.
Quadricalcarifera testacea Kiriakoff, 1967: 95; fig. 4; holotype: ♂, New Guinea, Top Camp, [Jayawijaya Mts., 3°30'S, 139°02' E, 2150 m] – NBC, Leiden.
Quadricalcarifera didyma Kiriakoff, 1967: 203; fig. 10; holotype: ♂, New Guinea, Top Camp, [Jayawijaya Mts., 3°30'S, 139°02' E, 2100 m] – NBC, Leiden.
Quadricalcarifera trivia Kiriakoff, 1967: 204; fig. 12; holotype: ♂, New Guinea, Top Camp, [Jayawijaya Mts., 3°30'S, 139°02' E, 2150 m] – NBC, Leiden.

All holotypes of these taxa are very similar by external appearance and are illustrated here, figs. 178l, 178m, 178n. Unfortunately the author was unable to locate any of Kiriakoff's original genitalia slides, including of the holotypes. They are probably lost. The genitalia drawings provided by Kiriakoff in the original descriptions are rather schematic and the description gives no differential diagnosis between these taxa. The author dissected and examined about n = 12 male genitalia in this complex and found a wider scope of seemingly individual variability, which does not correlate with external appearance in various forms of the dissected males. Furthermore, no separation based on male genitalia from the external different *triplagosus* was possible. In particular, the shape of uncus and socii, as well as the shape of the apical valvae processes, are variable in their forms. Figs. 178b and 178e illustrates a pair of asymmetric valvae processes. Juxta and phallus in all of the dissected males are each remarkably similar in

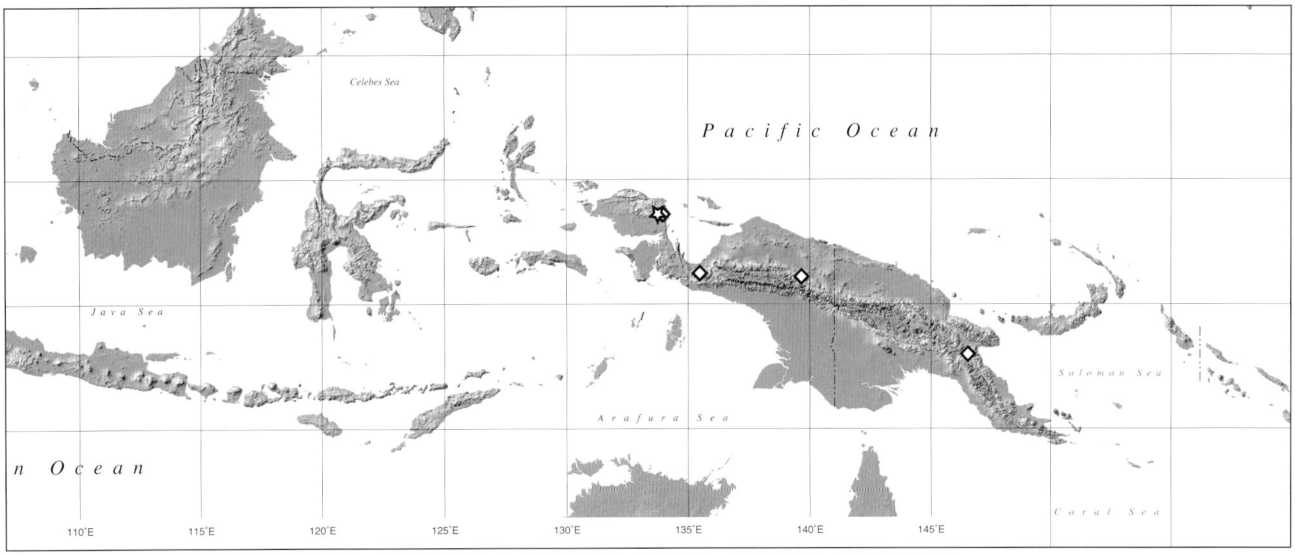

FIGURE 156 Distribution of *Syntypistis trisospylus.*

its shapes. As a taxonomically consequence all of these taxa are placed hereby as junior synonyms of *Syntypistis triplagosus* (**syn. novs.**).

Diagnosis. The nominat form of *triplagosus* displays light greyish forewing color with a large, blackish basal blotch, tornal patch, and a larger costal blotch near the apex. In the medial area three whitish, pale brown filled stigmas occur. The individual form *trivia* displays very reduced or absent black pattern elements on forewing, but the three white stigmas in the medial area are visible. Blackish basal and costal patches appear in a reduced form. Forewings are greyish, sometimes mixed with a few greenish scales. Hindwing are in both forms light brownish-white with an indistinct brownish postmedial fascia. Costal area displays near the apex a blackish patch. Sexual dimorphism is limited and both forms, including a transitional specimen, fig. 178r, occuring also in females.

Male genitalia have broad, rather rectangularly shaped valvae with an apical process. Gnathos is significantly narrower than in *kebeae.* Phallus is slightly curved. The curved basal part of the phallus seperates *triplagosus* from *didyma.* 8th sternite resembles *kebeae.* Female genitalia have not yet been dissected.

Bionomics. *Syntypistis triplagosus* occur commonly in mountainous regions from 800 m up to 3800 m. Most specimens were collected above 2600 m above sea level.

Distribution. Known from Arfak Mts. and the central mountain system of W Papua and Papua New Guinea.

179. ***Syntypistis seramica*** Kobayashi, 2011: 311; figs. 19; 42a–c

Holotype: ♂, Indonesia [Moluccas], Seram – National Science Museum, Tokyo.

Schintlmeister 2013: 444.

Taxonomic note. *Syntypistis seramica* was described from a single worn male without exact type locality. The identity is currently uncertain. The illustrated male genitalia, from the original description, suggests a relationship to *S. triplagosus* from New Guinea.

Diagnosis. Externally apparently similar to *S. kebeae.*

Male genitalia are distinct by valvae shape, with a smaller apical process. At the posterior margin the 8th sternite has a central projection which is different from other known *Syntypistis.* Phallus is smooth and does not bear the small thorns, which are diagnostic for the *kebae*-group. The female is unknown.

Bionomics. No information is available.

Distribution. Described from Seram (no map).

180. ***Syntypistis trisospylus*** (Joicey & Talbot, 1917): 60; pl. 2: 1. (*Stauropus trisospylus*).

Holotype: ♂, [Indonesia, W Papua], Arfak Mtns, Angi Lakes [„6.000 ft."] – NHMUK, London.

Schintlmeister 2013: 445.

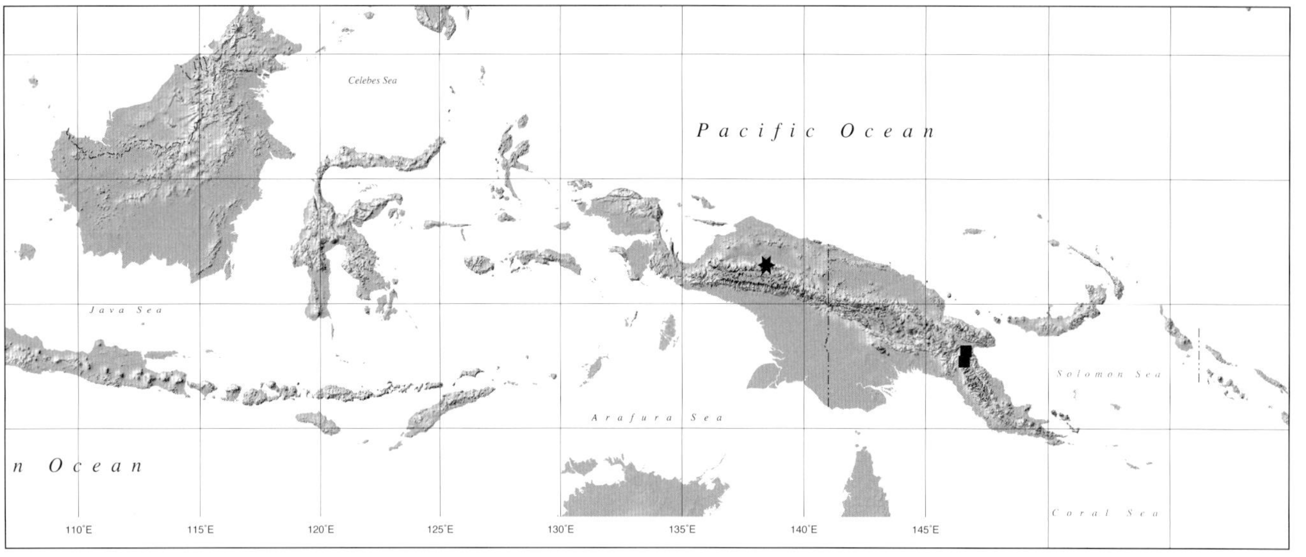

FIGURE 157 Distribution of *Syntypistis famelica.*

Taxonomic note. The holotype male of *trisospylus* lacks the abdomen and their genitalia remains unknown. The identity of *trisospylus* is therefore uncertain.

The holotype of the individual form *roseus* resembles much to the specimen illustrated here in fig. 180c. The genitalia of this male has been dissected and do not differ from specimens that matching externally the holotype of *trisopylus*.

Two females are placed tentatively to *trisopylus*, but it is also not clear, if they are conspecific with the holotype of *trisospylus.*

Diagnosis. Similar by external appearance to *S. triplagosus,* but larger in forewing length and forewings are more greenish tingged. There are three contrasting white and pale brown filled stigmas. The brown nucleus of these stigmas is much more contrasting than in *triplagosus.* There is no black apical patch on forewing as in *triplagosus.* Forewing pattern includes black basal fascia, a doubled postbasal fascia, a postmedial fascia and a fine submarginal line. Hindwings are brownish, darker than in *triplagosus*, with an indistinct slightly more fuscous submarginal area. Two females here attributed to *trisospylus* have a resemblance to *S. mixta*, but forewings of the latter are more elongated. Sexual dimorphism is slightly developed: the submarginal area of forewing is whitish and displays in its center a black blotch. A single female, see fig. 180f, has darker and more green-tinged forewings. Hindwings are darker brown than in males.

Male genitalia differ by a straight shape of valva apex from the other Papuan *Syntypistis* species. Apical valva process is large and blunt. Phallus resembles by its shape that of *triplagosus*, but it is slightly longer and rather straight. Posterior margin of 8th sternite is sinous with a central notch. Female genitalia have not yet been dissected.

Bionomics. Known are only a few specimens (n = 5) which were taken between 1670 m- 1830 m above sea level.

Distribution. Known from W Papua.

181. ***Syntypistis famelica*** (Kiriakoff, 1967): 204; fig. 11. (*Quadricalcarifera famelica*).
Holotype: ♂, [Indonesia], New Guinea, Mist Camp, [Jayawijaya Mts., 3°30'S, 139°02' E, 1800 m] – NBC, Leiden.
Schintlmeister 2013: 438.

Taxonomic note. Only two males have been dissected to examine their genitalia. Surprisingly both genitalia are very different. Fig. 181a is different from Kiriakoff's drawing in the original description, but the adult (fig. 181d) match very well the holotype by external appearance. Fig 181b match in its genitalia perfect Kiriakoff's drawing, but the worn imago has an enlarged white stigma on the forewing.

Diagnosis. A smaller species in the mountainous *triplagosus* – complex. Forewings of fresh specimens mixed with shiny green scales. Older material is fading into brown (181c, 181g). Basal area of the forewings is separated from

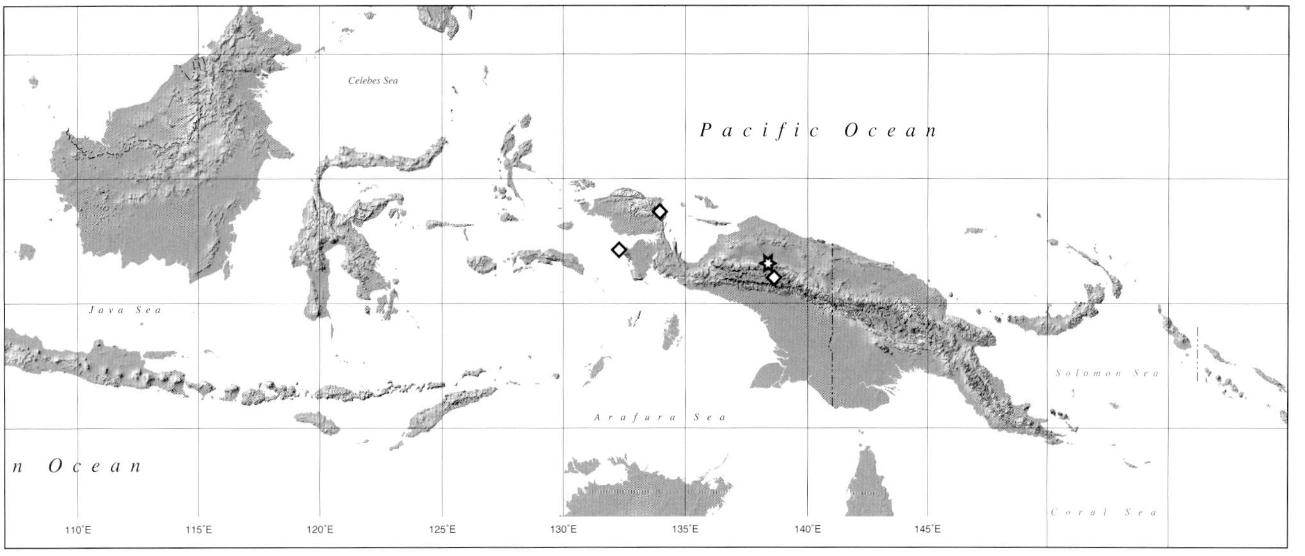

FIGURE 158 Distribution of *Syntypistis quadrivittata.*

the medial area by a broad blackish band. Three whitish stigmas on the forewings are indistinctly developed but visible. Submarginal fascia marked as a blackish line. In the postmedial area of forewings, a large fuscous dorsal patch occurs. Hindwing is light brownish-white with a broad, slightly more fuscous broad submarginal band. Near the apex a squarish fuscous costal patch occur. Sexual dimorphism is limited.

Male genitalia are occuring in two forms, as illustrated. Both forms share a rectangularly uncus with slightly smaller socii. Phallus is characterized by a longer process, which is hook-like at the tip. The everted endophallus bears in both forms many strongly sclerotized micro thorns. Posterior margin of 8th sternites are in both forms different. Female genitalia have not yet been dissected.

Bionomics. Only a few specimens (n < 10) are known, which has been taken between 1800 m- 2250 m above sea level.

Distribution. Known from two localities only: the Jayawijaya Mts. of W Papua and Mt. Kaindi, Morobe prov. in Papua New Guinea. The late Henk van Mastrigt sent several images of the species, depicted in figs. 181e, 181f, but unfortunately without collecting data.

182. ***Syntypistis quadrivittata*** (Kiriakoff, 1967): 197; fig. 5. (*Quadricalcarifera quadrivittata*).
Holotype: ♂, [Indonesia], New Guinea, Top Camp, [Jayawijaya Mts., 3°30'S, 139°02' E, 2150 m] – NBC, Leiden.
Schintlmeister 2013: 444.

Taxonomic notes. In the holdings of NBC, Leiden, there are two type specimens with identical data labels. They bear both a label in Kiriakoff's hand "Dr. S. G. Kiriakoff dét. | *Quadricalcarifera quadrivittata* sp.n.". But they lacks red type labels, as were usually used by Kiriakoff, and it is not clear, which of the males is the designated holotype and which is the paratype. In the original decription a holotype and a paratype specimens are mentioned. Therefore the better preserved and uncutted specimen is subsequentely selected and designated as the holotype specimen and illustrated in fig. 182b.

Quadricalcarifera nana Kiriakoff, 1967: 201; fig. 8; holotype: ♂, New Guinea, Top Camp, [3°30'S, 139°.02' E], 2.150 m – NBC, Leiden) has the same type locality as *quadrivittata* and does not differ from the latter, except by its smaller wingspan. Male genitalia of *nana*– as illustrated in the original description of Kiriakoff, 1967, fig. 8 – differ slightly in the shape of the uncus, the socii, and the apical valva process from mounted genitalia slides of the author, fig. 182a. Genitalia of *quadrivittata* as illustrated by Kiriakoff in its original description show valva apex less bifurcate. This difference falls in the individual width of variation and both taxa are brought therefore in synonymy. As first reviser the author places *nana* as synonym of *quadrivittata* (**syn. nov.**).

Diagnosis. *Syntypistis quadrivittata* is readily separated from the other Papuan congeners by the prominent blackish straight and broad postbasal fascia of the forewing. Postmedial fascia is developed similarly as a broad blackish band. Both fasciae vary individually in their broadness

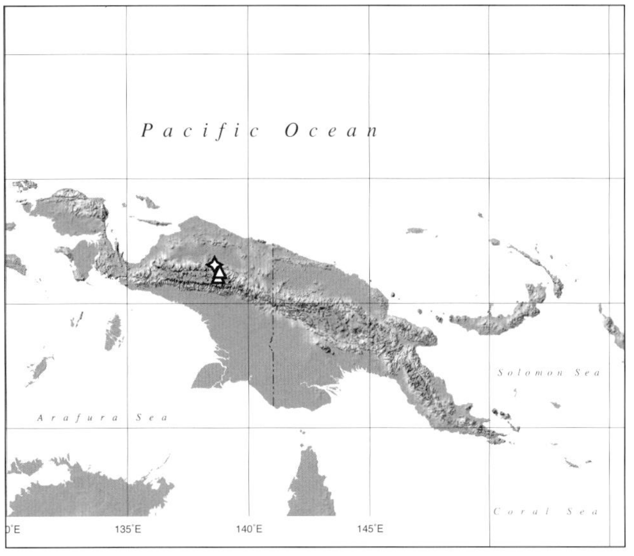

FIGURE 159 Distribution of *Syntypistis histrionica.*

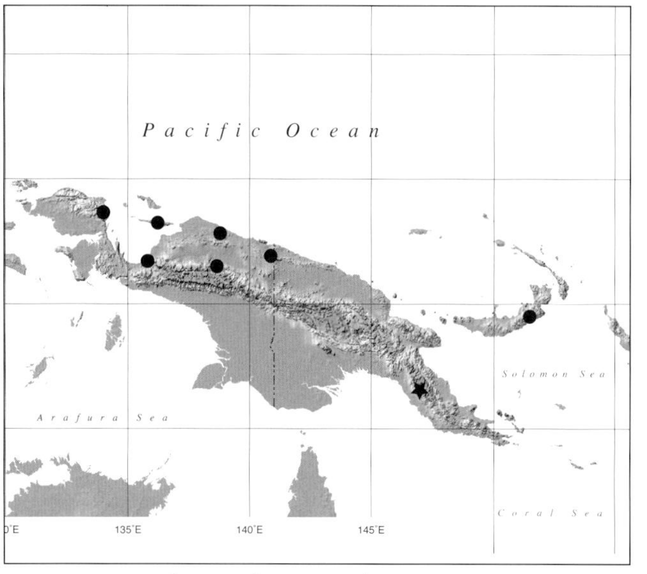

FIGURE 160 Distribution of *Syntypistis dinawa.*

and in contrast. The only available fresh (collected in 2012) male displays forewing an intense and shiny green forewing, while the remaining four old males (> 80 years), are more brownish faded. The three whitish stigmas in the medial area are indistinctly developed. Hindwings are of a light brownish-white to brown. Female is unknown.

Male genitalia resemble those of *famelica* and other Papuan congeners. They are best to recognize by the shape of the slightly conical uncus. Valva is convex-shaped on costa and sacculus. The apical process arises from the costa just before the apex and the entire apex appears therefore, bilobed. Phallus is similar to that of *famelica*, but the tip is knob-shaped, not like a hook. The everted endophallus bears many small microthorns. 8th sternite has a rather straight posterior margin; 8th tergite is not modified.

Bionomics. Six males (n = 6) are currently known, which were taken at 600 m, 1670 m, and 2100 m above sea level.

Distribution. Known from W Papua.

183. ***Syntypistis histrionica*** (Kiriakoff, 1967): 202; fig. 9 (*Quadricalcarifera histrionica*).

Holotype: ♂, [Indonesia], New Guinea, Mist Camp, [Jayawijaya Mts., 3°30'S, 139°02' E, 1800 m] – NBC, Leiden.

Schintlmeister 2013: 438.

Diagnosis. Rather unmistakable by the diagnostically contrasting greenish apical blotch on deep blackish-violet forewings. Three whitish stigmas are marked very indistinctly by a few white scales. A squarish dorsal patch near the tornus is of a contrasting blackish-violet in a green field. Fringes of all wings are checkered in white and fuscous-brown. Hindwing is light brownish-white with a brown costal area. Female is unknown.

Male genitalia – according to the drawing of Kiriakoff, 1967: fig. 9 – resembles *quadrivittata,* but the uncus is notched.

Bionomics. Only three males (n = 3) are known, which were taken between 1800–2200 m above sea level.

Distribution. Known from W Papua: Jayawijaya Mts. and Wamena.

184. ***Syntypistis dinawa*** (Bethune-Baker, 1904): 379; pl. 4: 17 (*Stauropus dinawa*).

Holotype: ♂, [Papua New Guinea], Dinawa [4.000 ft.] – NHMUK, London.

Schintlmeister 2013: 437.

Diagnosis. Forewings are black with shiny green scales. Older specimens (> 100 years) fading into brown and most of the green scales becoming less visible as well, see fig. 184c. In such old specimens a black, postbasal, postmedial, fasciae are visible. Basal area of forewing is more greenly tinged than the medial area. A characteristic feature is the submarginal fascia, marked as contrasting white patches, in particular prominently near the tornus. Hindwings are pale brownish, with darker brown submarginal area. Pattern of the costal part corresponds with the blackish and green forewing pattern. Sexual dimorphism is limited, but female hindwings are slightly darker.

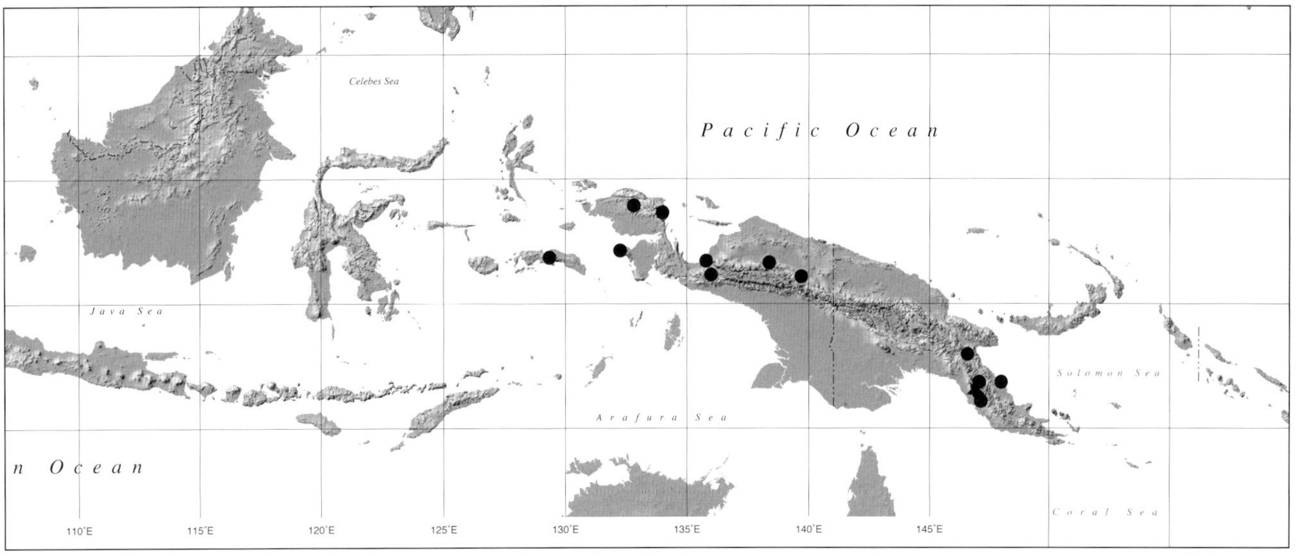

FIGURE 161 Distribution of *Syntypistis bella*.

Male genitalia are similar to those of *mediogriseus*, but phallus is broader and shorter. Uncus is broad, club-shaped, rounded and slightly notched. Valvae are rounded with a small triangular evagination at the apex. Phallus is straight and robust in comparison with the other Papuan congeners. It bears in its central part many microthorns. Phallus tip is not curved. 8th sternite is notched at the posterior margin and displays a pair of prominent lobes as in *mediogriseus*. Female genitalia have not as yet been dissected.

Bionomics. Adults occur infrequently from lower altitudes, e. g. 50 m, up to 1670 m above sea level.

Distribution. Distributed in New Guinea in its entirety, Yapen Island and New Britain.

185. ***Syntypistis bella*** (Bethune-Baker, 1904): 379; pl. 4: 14 (*Stauropus bella*).
Holotype: ♀, [Papua New Guinea], Ekeikei [1.500 ft.] – NHMUK, London.
Schintlmeister 2013: 436.

Diagnosis. Distinguished by the uniformly greyish-green colored forewings and a broad contrasting reddish-brown submarginal area of the hindwings. Some males display on the forewings indistinct fuscous postbasal and postmedial fasciae as well as a whitish discal spot, see fig. 185c. Females lack the whitish discal spot but display in variable numbers and sizes, larger fuscous violet-brown patches on their postmedial fasciae. Hindwings are in both sexes light brownish with a broad brown-filled submarginal area. Coloration in costal area of hindwings sets the forewing coloration forth.

Male genitalia have a rectangular or slightly conical uncus shape and club shaped unified socii. Valvae shapes are variable, as illustrated, and the apical process also varies individually in its size. Phallus is broad and curved, with a hook at the tip. 8th sternite is sinuous-shaped and has a more strongly sclerotized posterior margin. Female genitalia have very long apophyses posteriores and a pair of apophyses anteriores of half length. Ductus bursae is broad and strongly sclerotized. Corpus bursae is rather circular shaped. The oscillating band runs on the left half of the corpus bursae.

Bionomics. *S. bella* occurs commonly, in particular at lower altitudes of up to 500 m, but adults have been collected infrequently in mountain areas up to 2000 m as well.

Distribution. Distributed in New Guinea in its entirety and Seram.

186. ***Syntypistis mediogriseus*** (Gaede 1930): 62 9; pl. 4: 14 (*Desmeocraera trisopylus* f. *mediogriseus*).
Lectotype: ♂, [Indonesia, W Papua], Neuguinea ["Ninay valley, Central Arfak Mts., 3.500 ft"] – NHMUK, London.
Schintlmeister 2013: 441.

Taxonomic note. A smaller series of males from New Britain differ from adults from W Papua and are described below as:

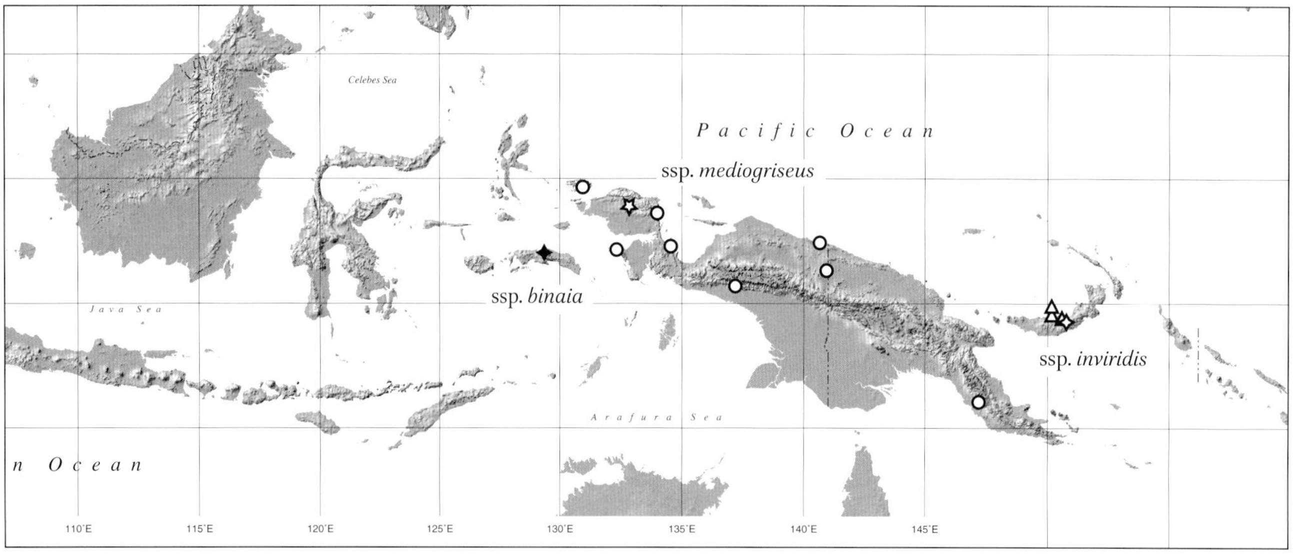

FIGURE 162 Distribution of *Syntypistis mediogriseus*.

Syntypistis mediogriseus inviridis **ssp. nov.** (figs. 186c, 186f)
Holotype: ♂, Papua New Guinea, New Britain, Talasea district, Kulkelpun near Bialla, 500 m, 5°46.257'S, 150°33.572'E, 21.–23. iii. 2003, leg. Rudloff & Schaarschmidt – in coll. A. Schintlmeister, Dresden.

Paratypes (5 ♂♂):
New Britain: 2 ♂♂, Plotplun, S Lopun, 300 m, 5°50.756'S, 150°39.824'E, 23.–25. iii. 2003 (GU 65-78), leg. Rudloff & Schaarschmidt; 1 ♂, Talesea peninsula, Lake Dakataua, Bulumivi, near bulu Murli, 250 m, 5°01.330'S, 150°08.379'E, 1.–4. iv. 2003 (GU 65-78), leg. Rudloff & Schaarschmidt; 2 ♂♂, Talesea distr., Bereme, SE Kimbe, 350 m, 5°43.952'S, 150°36.720'E, 19.–26. iii. 2003, leg. Rudloff & Schaarschmidt.

Diagnosis of ssp. *inviridis*. Forewing length ♂♂ 16–18 mm. The new subspecies is distinguished by a reduced number of green scales on the forewings. The external appearance is therefore greyish rather than greenish and also darker than ssp. *mediogriseus*. Basal area of forewing is blackish filled. Three stigmas in the medial area of forewing are less contrasting. Postmedial fascia appear stronger and more contrasting than in ssp. *mediogriseus*. Hindwing lacks the darkened submarginal area.

Male genitalia differ from ssp. *mediogriseus* by a less swollen valva costa, which bears a shorter apical process. 8th sternite lacks in *inviridis* the central lobes.

Specimens from Seram are different and introduced hereunder as:

Syntypistis mediogriseus binaia **ssp. nov.** (Figs. 186c, 186f)
Holotype: ♂, Indonesia Seram, Gunung Binaia, north slopes 1000 m, Secondary forest, Site 9: 2, viii.–ix. 1987, Operation Raleigh, J. D. Holloway et al. – NHMUK.

Paratype (1 ♂):
Seram: 1 ♂, Gunung Binaia, north slopes 1000 m, Secondary forest, Site 9: 1, viii.–ix. 1987, (genitalia slide BM 2017-05).

Diagnosis of ssp. *binaia*. Forewing length ♂♂ 18 mm. The new subspecies differ from ssp. *mediogriseus* by reduced fuscous pattern in the medial and postmedial area. These areas appear pale grey, except the postmedial fascia, which is developed as an indistinct fuscous greenish-grey shadow.

Male genitalia are distinct by rounded, club-shaped valvae with rounded apex. In other subspecies the valva apex is pointed. The apical valva projection is in ssp. *binaia* reduced in size. 8th sternite have the posterior margin concave with a central, bifurcate projection.

Diagnosis. Forewings are greyish, mixed with many green scales. There are three whitish, pale-brown filled stigmas in the light medial area. A characteristic feature is the fuscous dorsal spot near the tornus, accompanied by three whitish spots toward the medial area. The latter are part of an indistinctly developed submarginal fascia. Hindwings are light brownish-white in males and brown in the female. A postmedial fascia of the hindwings appears weakly and indistinct. Female displays the dorsal patch as a streak only. Hindwings are darker than in males.

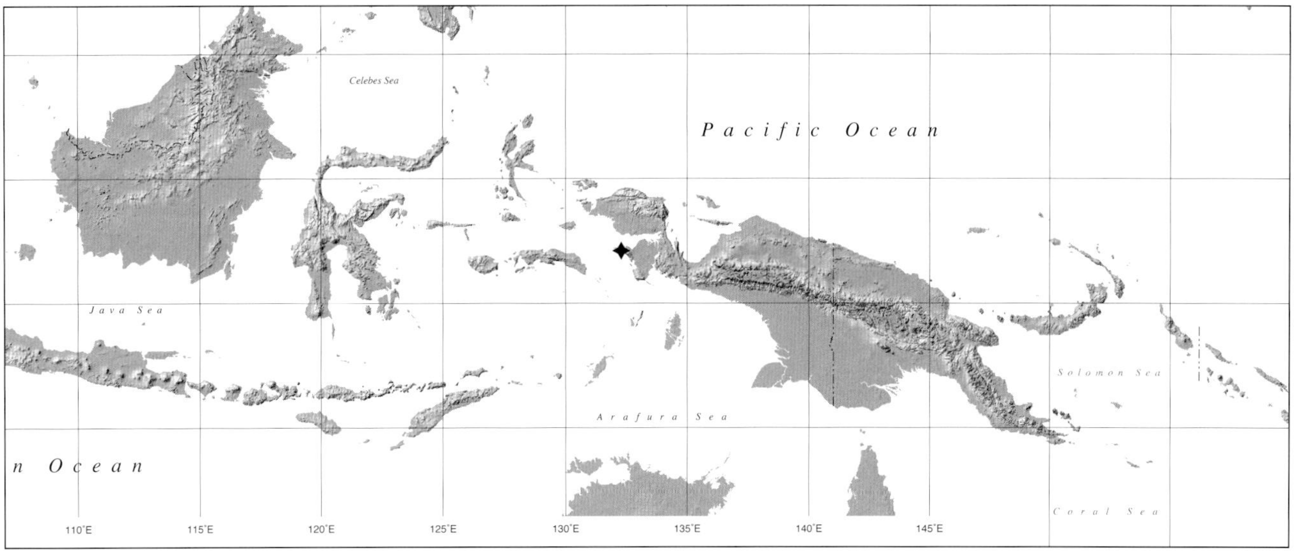

FIGURE 163 Distribution of *Syntypistis leucaspis*.

Male genitalia are similar to those of *S. dinawa*, but the uncus is narrower and less rounded at the tip. Socii are shorter than the uncus and broader than in *dinawa*. Valvae of ssp. *mediogriseus* are convex-rounded and display near the broad apex a long process. Phallus is slightly curved and of variable length. Phallus tip is tapered and less modified. 8th sternite concavely on the posterior margin with a small central notch. In ssp. *mediogriseus* a pair of strongly sclerotized central lobes occur. Female genitalia have not yet been dissected.

Bionomics. Males occuring frequently between 50 m and 1760 m above sea level.

Distribution. Ssp. *mediogriseus* is distributed in New Guinea in its entirety, while ssp. *inviridis* is restricted to New Britain and ssp. *binaia* to Seram.

187. ***Syntypistis leucaspis*** **spec. nov.** (figs. 187a–c)
Holotype: ♂, Indonesia, W Papua, Fak Fak, 600 m, 2°55' S, 132° 20' E, iv. 2009 – coll. A. Schintlmeister, Dresden.

Paratypes (3 ♂♂, 1 ♀):
W Papua: 3 ♂♂, 1 ♀, Fak Fak, 600 m, 2°55' S, 132° 20' E, iv. 2009 (GU 94-46).

Diagnosis. Forewing lengths ♂♂ 17 -18 mm; ♀ 20 mm. Antennae are black and in both sexes bipectinated with long rami. The last 25% toward the tip is filiform. The new species resembles *S. mediogriseus* at first glance, but the shape of the forewings is more rounded and less elongated. Forewing color is fuscous grey mixed with many metallic shining green scales in the basal and the postmedial area. There are three contrasting and prominent white stigmas, which are light-brown filled. Hindwings are a pale yellowish-brown with a broad fuscous-brown submarginal band, which extends also into the costal area. A squarish costal spot is mixed with many shiny green scales, accompanied by a contrasting white rectangular blotch. The female has a reduced light-brown nucleus of the stigma and displays a fuscous medial patch near the dorsum. Submarginal area of the forewing is broad and white-colored, where veins are contrasting blackish-tinged.

Male genitalia have a squarish uncus, which is slightly notched at the tip. Unified socii are club-shaped and notched. Valvae are rectangularly shaped. It bears an apical process as *mediogriseus*. Juxta is broader than most other Papuan congeners. Phallus is slightly curved and displays a knob-shaped tip as in *S. bella*. 8th sternite is strongly sclerotized and widely bilobed at the posterior margin. The trapezoid shape of the 8th tergite differs from most other Papuan congeners. Female genitalia have not yet been dissected.

Bionomics. *Syntypistis leucaspis* was collected in a smaller series (n = 5) together with *mediogriseus* at 600 m above sea level.

Distribution. Known only from the type locality, Fak Fak in W Papua.

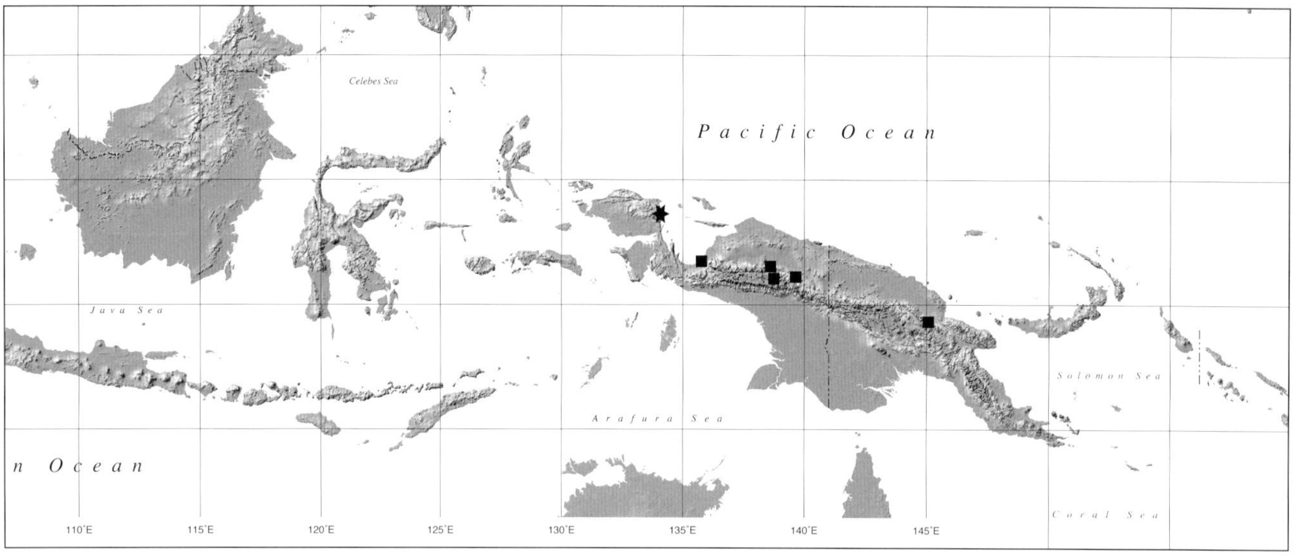

FIGURE 164 Distribution of *Syntypistis papuana*.

188. *Syntypistis papuana* spec. nov. (figs. 188a–g)
Holotype: ♂, Indonesia, W Papua, Arfak Mts., Maibri village, 1670 m, 1°07' S, 133° 54' E, 1.-15.iv.2013, leg. St. Jakl – coll. A. Schintlmeister, Dresden.

Paratypes (13 ♂♂, 1 ♀):
W Papua: 4 ♂♂, Arfak Mts., Maibri village, 1670 m, 1°07' S, 133° 54' E, 1.–15. iv. 2013, leg. St. Jakl (GU 93-79, GU 93-80); 1 ♂, 1 ♀, ibid but 5.–20. xii. 2012; 1 ♂, Straße Nabire-Ilaga km 53, 580 m, 3°49.460'S, 135°.7524'E, 22. ii. 1998, leg. M. Schaarschmidt & F. Roick; 2 ♂♂, Kecamatan Nipsan, Walmak, 1710 m, 4°07'S, 138°38'E, 31. i.–16. ii. 2005, UNZEN-ZMA Expedition 2005; 1 ♂, Baliem valley, Wika, 1600 m, 21. x. 1993, leg. A. J. de Beer, A. L. M. Rutten & R. de Vos; 1 ♂, Baliem valley, road Wamena – Sentani, km 48, 2150 m, 26. + 28. i. 2000, leg. Cerny & Hluchy, 1 ♂, Jayawijaya Mts., Rattan Camp, 3°30'S, 139°09' E, 1200 m, ii.–iii. 1939, leg. L. J. Toxopeus;
PNG: 2 ♂♂, West Highland, Kol Missionsstation, 61 km von Mt. Hagen, 1570 m, 5°72453' S, 144°84128'E, 22. iv. 1996, leg. M. Schaarschmidt (GU 94-49, GU 94-54); 1 ♂, Morobe pr. Wau, Wau Ecol. Inst., 1200 m, 1.–10. viii. 1983, S. E. & P. M. Miller.

Diagnosis. Forewing lengths ♂♂ 17–18 mm; ♀ 21 mm. Antennae are black and in both sexes bipectinated with long rami. The last 25% toward the tip is simple. Basal area of the forewings is contrasting filled in light green. Medial area usually not mixed with green scales. One male (fig. 188f) however, displays sparsely green scales on the entire forewing. Medial area varies from pale grey to fuscous violet- brown, with transitional forms. The postmedial area until the margin is grey colored, mixed with shiny green scales. Hindwing is light brown with a slightly darker brown, broad submarginal area. The apical costal region is colored as the dorsal part of the forewing and often mixed with green scales. Females display a whitish postmedial area of forewing and a prominent black blotch on the dorsum near to the tornus. Hindwings are darkener than in males.

Male genitalia have a conical uncus. One of the four dissected males (fig. 188b) has a notched uncus. Unified socii are narrower than in the previous three species. Valva tapered toward the apex with a small apical process. Phallus is slightly curved and relatively long, longer than in *S. mediogriseus* or *leucaspis*. Phallus tip is strongly curved and hook-like. 8th sternite have a sinous shaped and strongly sclerotized posterior margin, which resembles *S. bella*. Female genitalia have not yet been dissected.

Bionomics. *Syntypistis papuana* is a seemingly a common species. The late Henk van Mastrigt sent the author many images, unfortunately without exact collecting data. Specimens from the type series have been collected at altitudes between 580 m and 1700 m.

Distribution. Distributed in W Papua and PNG.

189. *Syntypistis mixta* (Bethune-Baker, 1916): 385 (*Stauropus mixta*).
Holotype: ♀, [Papua New Guinea], British New Guinea, Mount Kebea,3.000 ft. – NHMUK, London.
Schintlmeister 2013: 441.

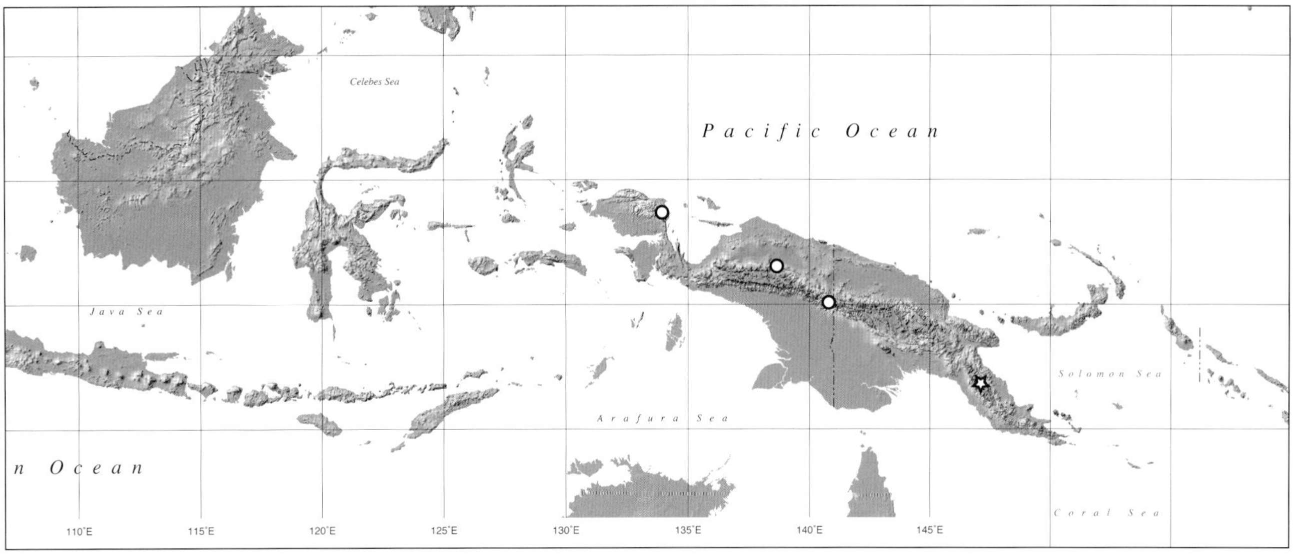

FIGURE 165 Distribution of *Syntypistis mixta*.

Taxonomic note. A direct comparison of the holotypes of *Stauropus mixta* and *Stauropus leucophaeus* Rothschild, 1917: 242; pl. pl. 8: 20; holotype: ♀, Central Dutch New Guinea, Mount Goliath 5.000-7.000 ft. – NHMUK, London, depicted here fig. 189e, does not show any differences by external appearance, respectively in their pattern. Therefore *leucophaea* is placed as a junior synonym of *mixta* (**syn. nov.**).

A series from the Arfak Mts. are tentatively assigned to *mixta*, but the only male (fig. 186e) and the females (fig. 189f) does not match the type of *mixta* fully. Genitalia figures (figs. 189a, 189b) and diagnosis are made from these specimens.

Diagnosis. Forewings are metallic shining green. Holotypes of *mixta* and *leucophaeus* and further old specimens from the Archbold expedition are less green colored, than recent material from the Afak Mts; faded brownish probably due to their age. The only male at the author's disposal has rather uniform greenish forewings, where three pale stigmas are weakly and indistinctly developed. Hindwings are reddish brown. Female is sexually dimorphic and displays a distinct forewing pattern. Postbasal and submarginal fasciae are marked as black lines. The postmedial fascia consists of several black blotches. In the medial area there are two whitish patches, with light brown nucleus. A third stigma is rather indistinctly developed. In the submarginal area there are several whitish fields. There are also females, with rather less contrasting forewing pattern.

Male genitalia resembles *papuana*, but the squarish uncus is shorter and the socii broader and club-shaped. Valvae are convex, rounded from both sides with tapered apex. Apical process is large. Phallus resembles *papuana*, but its process is narrower and has no knob-shaped tip. 8th sternit is distinguished by the shape of the strongly sclerotized posterior margin. At both sides small lobes arises. Female genitalia resembles *S. chloriolus* but the postvaginal plate toward the ostium is narrower. Ductus bursa is broad and medium sized; corpus bursae displays as in most congeners a long oscillating band.

Bionomics. A rather rare species in collections (n < 20) and females are in the majority. Known specimens has been taken between 1100–1670 m, but the holotype female of *leucophaea* was taken at 2100 m above sea level.

Distribution. Restricted to the Arfak mountains and the central mountain system in New Guinea.

190. ***Syntypistis chloriolus*** (Joicey & Talbot, 1917): 59; pl. 1: 15. (*Stauropus chloriolus*).

Lectotype ♀, [Indonesia, W Papua], Wandammen Mts., 3.000-4.000 ft. – NHMUK, London **by present designation.**

Schintlmeister 2013: 437.

Taxonomic notes. *Stauropus chloriolus* was described of two ♀♀ (syntypes). One of them the author traced in the holdings of NHMUK. To stabilize the nomenclature the female with the following labels is herewith designated as lectotype: "Wandammen Mts. | 3-4000 ft. D. N. Guinea. | Nov. 1914. A.C & F. Pratt."; red-ringed circular label "Type";

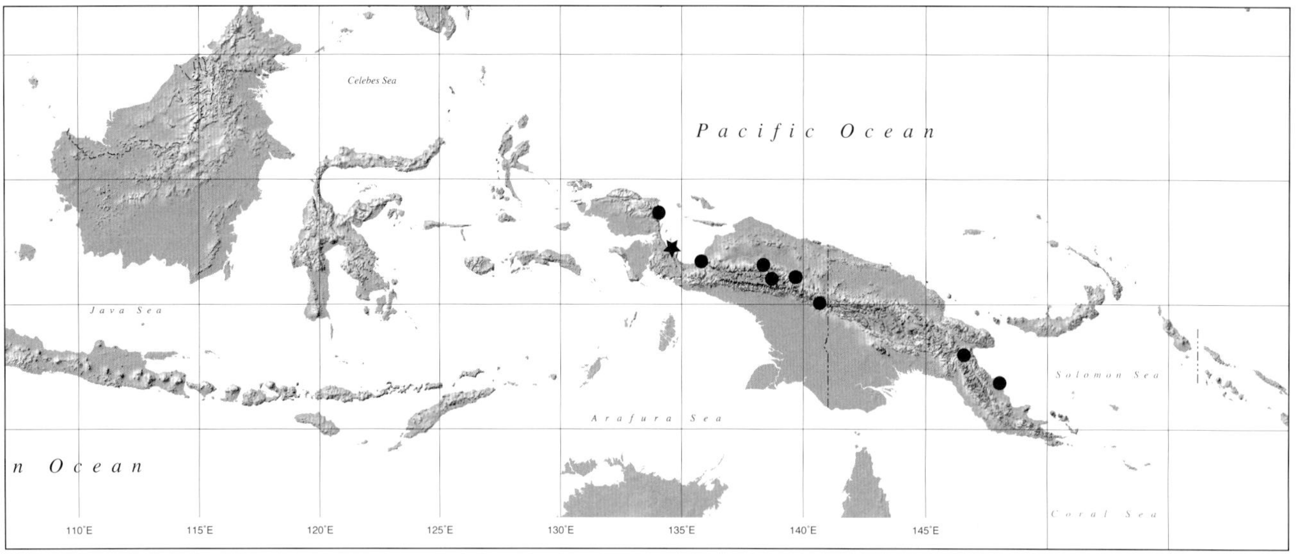

FIGURE 166 Distribution of *Syntypistis chloriolus.*

"*Stauropus* | *chloriolus* | J. v T. | TYPE ♀ (H.T.).''; yellow label "Photo done by A. Schintlmeister #4411''. A lectotype label was added. The lectotype is depicted here in fig. 190e.

A direct comparison of the lectotype of *chloriolus* with the ♀ holotype of *Stauropus viridinitens,* Rothschild, 1917: 242; pl. pl. 8: 21, holotype: ♀, North-east British New Guinea, Mambare River, Biagi 5.000-7.000 ft. – NHMUK, London, depicted in fig. 187i, showed, that both taxa are conspecific. The description of *Stauropus viridinitens* was published on 31. August 1917, the description of Joicey & Talbot appeared "July 1917" (according to the data imprinted on the wrappers of the issues) and *viridinitens* sinks therefore as a junior synonym (**syn. nov.**), while *chloriolus* takes the priority.

Diagnosis. Forewings shiny fuscous green and blackish brown with pattern. Older specimens in collections faded the green color into brown. The medial area displays three whitish stigma, which are in most males weakly developed and inconspicous. Postmedial fascia is blackish and bears in the middle part a larger black blotch, which is off-center toward the margin. Hindwings are reddish- brown with a slightly paler medial fascia. Female resembles the male, but a black postbasal fasciae is more strongly developed. Forewings of females are often mixed with more shiny green scales.

Male genitalia have a squarish uncus. Socii are variable in size, sometimes of same size as the uncus. Valvae are convexly rounded from both sides with a bifurcate apex. The upper valva process is large. Phallus bears a curved, knob-shaped process. Everted endophallus bears many strongly sclerotized microthorns. 8th sternite has an almost straight posterior margin. Female genitalia have long apophyses posteriores, a larger postvaginal plate, and a robust and strongly sclerotized ductus bursae. The "oscillating band" on the corpus bursae is almost rectangularly curved in its posterior part.

Bionomics. Moth occurs infrequently and females are in the majority. Only a few adults were observed at 1000 m, but most specimens has been collected from altitudes between 1670–2100 m.. One male was taken at 3300 m above sea level.

Distribution. Known from the the Arfaks and central mountains in W Papua and Papua New Guinea.

191. ***Syntypistis viridigriseus*** (Rothschild, 1917): 243; pl 3: 18 (*Stauropus viridigriseus*).
Holoptype: ♂, [Papua New Guinea], Mambare [River], Biagi, 5.000 ft. – NHMUK, London.
Schintlmeister 2013: 446.

Diagnosis. Forewings are in both sexes brownish grey; the basal area displays several green metallic shining scales. The holotype male displays less contrasting brownish stigma in the medial area. Both examined females have stigmas surrounded in contrasting white. Postbasal, medial, postmedial and submarginal fasciae are fuscous and grey-marked. Hindwings are pale brown with an indistinct whitish medial fascia.

Genitalia in both sexes are not yet dissected.

Bionomics. Only three specimens are currently known, which has been taken from 1000 m up to 2100 m.

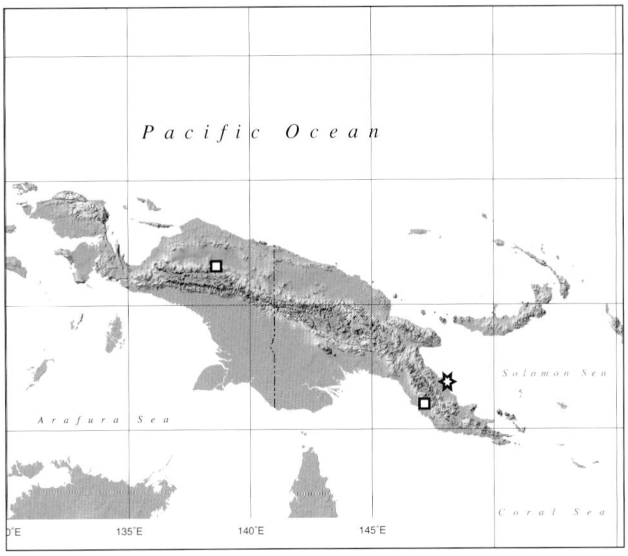

FIGURE 167 Distribution of *Syntypistis viridigriseus*.

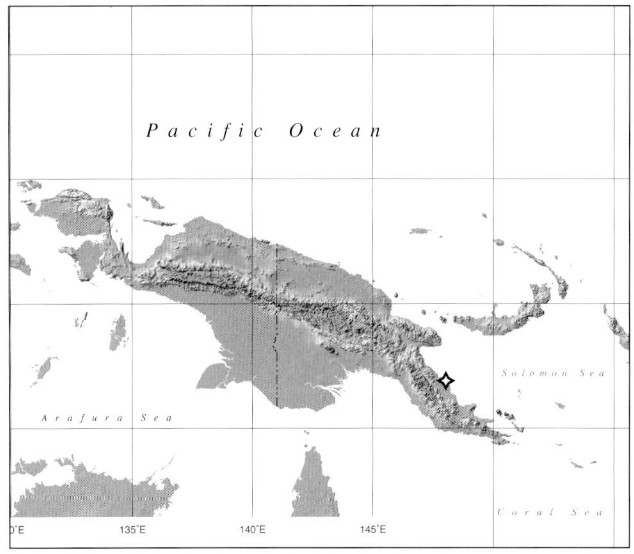

FIGURE 168 Distribution of *Syntypistis alboviridis*.

Distribution. Known from W Papua and Papua New Guinea including New Britain.

192. ***Syntypistis alboviridis*** (Rothschild, 1917): 244; pl. 8: 22 (*Stauropus alboviridis*).
Lectotype: ♀, [Papua New Guinea], Mambare [River], Biagi, 5.000 ft. – NHMUK, London **by present designation.**
Schintlmeister 2013: 435.

Taxonomic note. *Stauropus alboviridis* was described of two ♀♀ (syntypes). One of them the author traced in the holdings of NHMUK. To stabilize the nomenclature the female with the following labels is hereby designated as lectotype: "Biagi, Mambare | R. 5000 ft., B. N. | G., Jan. 06. | (A.S. Meeck)."; "*Stauropus* | *alboviridis* | R.."; blue label "Rothschild | Bequest | B.M. 1939–1"; yellow label "Photo done by A. Schintlmeister #34514". A lectotype label was added. The lectotype adult is figured in fig. 192b. The remaining female, whose genitalia were examined (BM 111), becomes a paralectotype.

Diagnosis. Forewing white, the pattern is metallic shining green, mixed with a few black scales. The fringe fuscous brown and white checkered. Among the Papuan *Syntipistis* species unmistakable by its distinct pattern. The only available female resembles by external appereance the female of *Syntypistis comatus* Leech, 1898 and its closely related species. It is likely, that the corresponding male of *alboviridis* is more strongly sexually dimorphic and would have also a resemblance to the male of *comatus.*

Female genitalia have long apophyses posteriores, a short ductus bursae. The oscillating band of the corpus bursae is distinguished by the larger amplitudes.

Bionomics. *S. alboviridis* was described from 2 ♀♀ which were taken at the same locality at 2100 m.

Distribution. Known only from the type locality in Papua New Guinea.

193. ***Syntypistis flavicollis*** (Rothschild, 1917): 241; pl. 3: 16 (*Stauropus flavicollis*).
Holotype: ♀, [Indonesia, W Papua], Central Dutch New Guinea, Mount Goliath 5.000-7.000 ft. – NHMUK, London.
Schintlmeister 2013: 438.

Taxonomic note. *Quadricalcarifera poecilochroa* Kiriakoff, 1967: 198; fig. 6, holotype: ♂, [Indonesia, W Papua. Jayawijaya Mts.], New Guinea, Mist Camp, [3°30'S, 139°.05' E], 1.800 m – NBC, Leiden is conspecific with *flavicollis* and sinks as a junior synonym (**syn. nov.**). Both type specimens share same forewing pattern and in particular the prominent pale frons and patagula, but both old (> 100 years) types of *flavicollis* show green color faded into yellowish brown. Male genitalia of *flavicollis* paratype (fig. 193a) and a male from type locality of *poecilochroa* (fig. 193b) sharing shapes of uncus, phallus and 8th abdominal segments, but socii and valvae are slightly different as illustrated. The genitalia drawing of Kiriakoff, 1967, fig. 6 show socii slightly notched, and apical valva process larger, but display the characteristic conical shape of uncus.

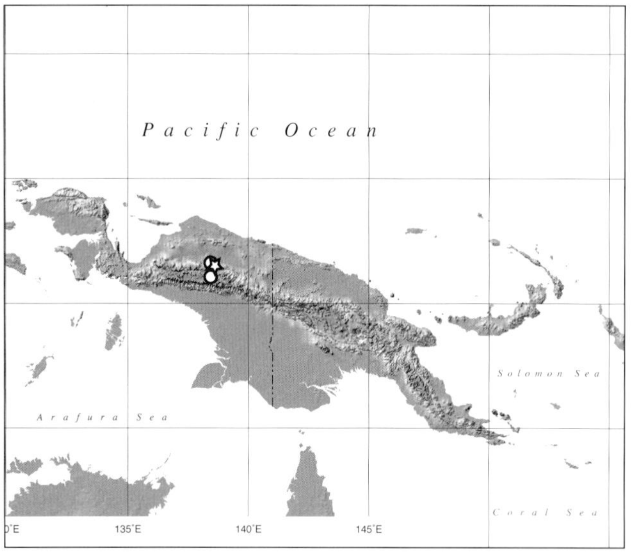

FIGURE 169 Distribution of *Syntypistis flavicollis*.

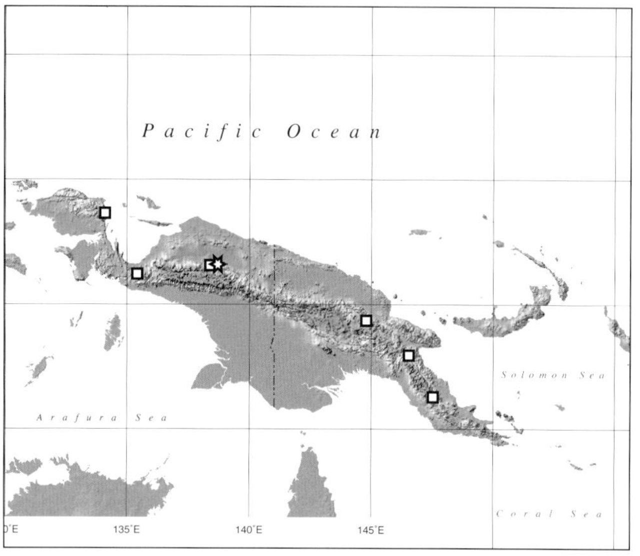

FIGURE 170 Distribution of *Syntypistis nitidus*.

Diagnosis. The pale yellow-greenish frons and patagula are diagnostic features. Forewings are fuscous brown mixed with metallic green scales, fading into yellowish-brown in older specimens. Basal area greenish filled; the medial area display a squarish paler green blotch. Most known specimens display three pale brownish stigmas in the medial area indistinctly. Submarginal fascia marked as a row of blackish and greenish patches. A larger circular black dorsal patch, accompanied by a greenish streak is a characteristic feature of *flavicollis*. Hindwings are of a light reddish-brown with an indistinct slightly darker broad submarginal band. Sexual dimorphism is is limited.

Male genitalia are distinguished by the broad – broadness varies individually – and conical uncus shape. Uncus and socii are variable in shape as illustrated (fig. 193a, 193b) and transistional forms occur. Valvae are ellipsoid with an apical process. Phallus robust and rarther straight. Phallus tip lacks a knob-shaped or hook-like projections. Endophallus bears many less strongly sclerotized microthorns. 8th sternite almost straight at the posterior margin and slightly bilobed centrally. Posterior margin displays centrally a small arrow-like projection. Female genitalia have not yet been dissected.

Bionomics. *S. flavicollis* occurs infrequently between 1800 m and 3300 m above sea level.

Distribution. Restricted to W Papuan central mountains, where it is known from a few localities.

194. *Syntypistis nitidus* (Rothschild, 1917): 241; pl. 3: 14, 15 (*Stauropus nitidus*).

Lectotype: ♂, [Indonesia, W Papua], Central Dutch New Guinea, Mount Goliath 5.000-7.000 ft. – NHMUK, London **by present designation.**

Schintlmeister 2013: 438.

Taxonomic note. *Stauropus nitidus* was described of 7 ♂♂, 2 ♀♀, (syntypes). In fact, as examination showed, the entire type series consists of males only. One male is marked in the holdings of NHMUK as "Type". To stabilize the nomenclature this male with the following labels is herewith designated as lectotype: "Mt. Goliath, | 5–7000 ft., Centr. | Dutch N Guinea | about 139° long., | Januar 1911 | (A. S. Meeck)."; red-ringed circular label "Type"; "*Stauropus* | *nitidus* | Rothsch."; yellow label "Photo done by A. Schintlmeister #4006". A lectotype label was added. The lectotype is depicted in fig. 194c.

Diagnosis. A variable species, which is best to recognized by postmedial and submarginal fasciae of the forewing, which are intersect at the center and form an "X". Postbasal fascia is accompanied in the anterior part often by a white line. From the apex of the forewings a greenish transverse hits the postmedial fascia. Forewing colors fuscous blackish-brown, mixed with many shiny green scales. Older specimens tend to a faded green color. In the medial area a pale brown stigma occurs, surrounded in cream. One male display three stigmas in the medial area in contrasting white (fig. 194e). Hindwings are light brown. Sexual dimorphism is limited, but females have hindwings of a darker brown than does the male.

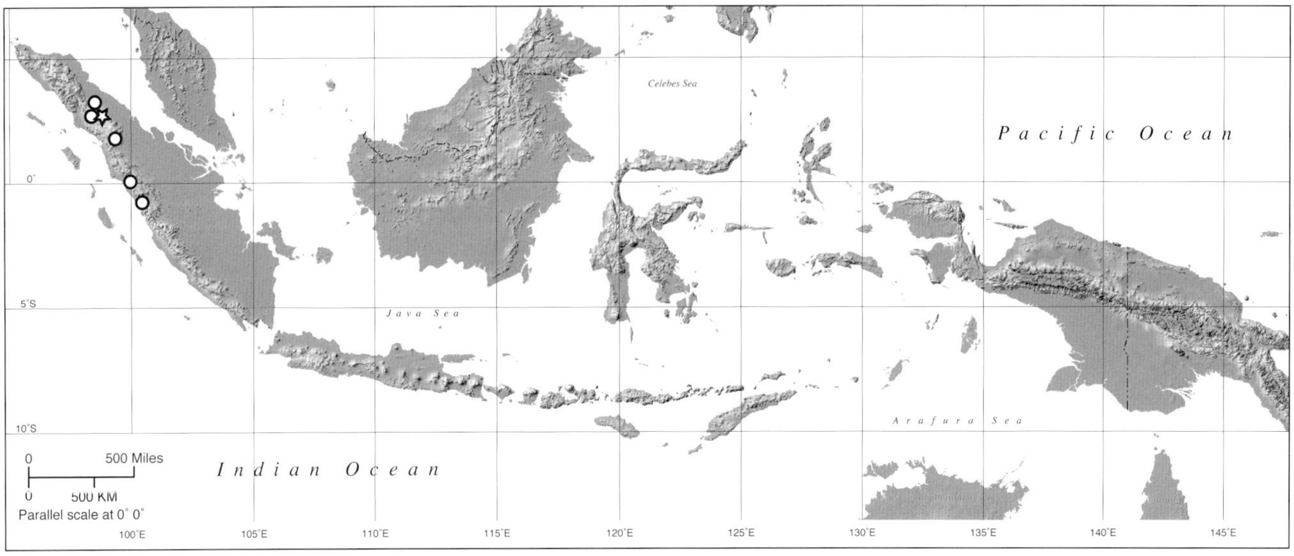

FIGURE 171 Distribution of *Syntypistis gemmea.*

Male genitalia have a squarish, non-conical uncus with a round tip. Unified socii are broad and club-shaped with a round tip. Valvae are convex rounded with apical process. Valvae have a rounded costa with an apical process. They are broader than in *flavicollis.* Phallus is rather straight and not curved at the tip. 8th sternite has a wave-shaped posterior margin. The arrow-like projection at the anterior margin is longer than in *flacicollis.* Female genitalia have a robust 8th abdominal segment of triangular shape. Apophyses anteriores are short and robust. Ductus bursae broad and weakly sclerozized. Corpus bursae is ovoid, its oscillating band weakly sclerotized.

Bionomics. *Syntypistis nitidus* occurs commonly at higher altitudes from 1670 m up to 3300 m; most specimens were observed above 2200 m.

Distribution. Distributed in central mountains in New Guinea in its entirety.

195. *Syntypistis gemmea* spec. nov. (figs. 195a–e)

Holotype: ♂, Indonesia, N. Sumatra, Samosir-Tobasee, Pattunghoan, 1600 m, 2°37' N, 98° 44' E, 20.ix.1991, leg. Graul & Schintlmeister – coll. A. Schintlmeister, Dresden.

Paratypes (21 ♂♂, 5 ♀♀):

Sumatra: 4 ♂♂, Indonesia, N. Sumatra, Samosir-Tobasee, Pattunghoan, 1600 m, 2°37' N, 98° 44' E, 20. ix. 1991, leg. Graul & Schintlmeister (GU 47-46); 1 ♂, N Sumatra, Bunga Bandar, near Sipirok, 1300 m, 19. xi. 1992, leg. Dr. E. W. Diehl; 1 ♂, 20 km NE Sipirok, Lake Marsabut, 1350 m, 1°40'E, 99°20'E, 8. i. 1994, leg. Z. Weidenhoffer; 4 ♂♂, 1 ♀, Dairi Mts., 1600 m, 2°52'N, 98°29' E, 28. viii. 1979, leg. Diehl & Schintlmeister; 1 ♀, Dairi, Silalahi, 1500 m, 4. ix. 1972, leg. Erber; 2 ♂♂, dto. 27. ix. 1970, leg. Dr. E. Diehl; 1 ♂, Nagaradja, 250 m, 20. -23. iv. 1974; 1 ♂, Berastagi, 1600 m, 3°01'N, 98°31' E, 21. viii. 1979, leg. Diehl & Schintlmeister; 3 ♂♂, dto., 5. vi. 1970; 2 ♀♀, Sumatra Barat, Mt. Singgalang near Padang Panjang, 2100 m, 10.–11. ii. 1996, leg. A. Kallies (GU 65-89); 1 ♂, Mt. Talamau, Westseite, 12 km S Talu, 500 m, 0°08'N, 99°59'E, 10. ix. 1991 leg. Graul & Schintlmeister; 1 ♂, 22 km E Padang, Bukit Subang, 0°40'S, 100°12'E, 1200 m, 4. i. 1985; 1 ♂, Siantar, Hozweg 2, 1050 m; 29. xi. 1968; 1 ♂, 1 ♀ North Korintji valley, 5000 ft., ix.–x. 1921, leg. C., F. & J. Pratt.

Diagnosis. Forewing lengths ♂♂ 19–21 mm; ♀♀ 21–24 mm. Forewings are black mixed with many metallic shining green scales. Basal fascia black, whitish marked on the dorsum. Submarginal fascia pale green, accompanied by a white streak near the dorsum. There are three contrasting white stigmas in the medial area, of which two are often merged y-shaped. Hindwings are reddish-brown. At the hindwing costa a green tinged area occurs. The individual variability concerns primarily the number and density of the green scales. Sometimes (fig. 195e) rather uniform greenish forewings occur. The species have an external resemblance to *S. nitidus,* but, as male genitalia showes, they are not closely related. Female resembles male; sexual dimorphism is limited.

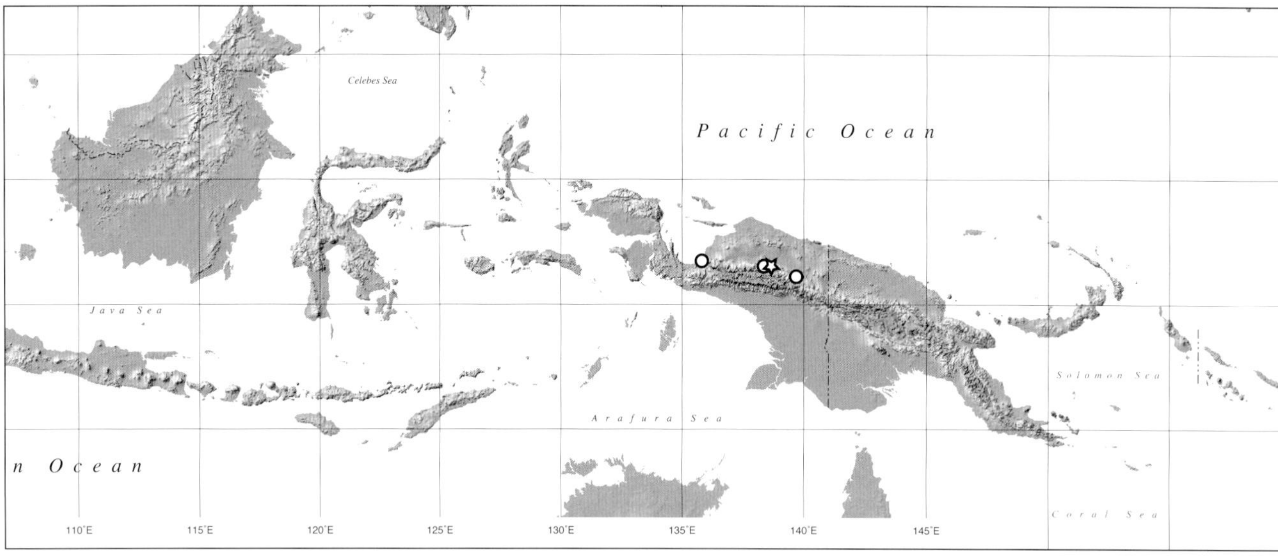

FIGURE 172 Distribution of *Syntypistis purpurascens*.

Male genitalia have a long, rectanguar uncus, rounded at the tip. Unified socii are of approximately same length as the uncus. Valvae shape is characterized by the broad apex, where both angles are pointed. Phallus is long, curved toward the tip. It displays a diagnostically wavy posterior margin. 8th tergite convex at posterior margin and less modified. Female genitalia have large apophyses posteriores with a robust base. They are twice as long as the apophyses anteriores. Antrum medium sized and bilobed. Ductus bursae broad and short. Corpus bursae long; the oscillating band weakly developed and rather short. It does not reach the base of ductus bursae.

Bionomics. *Syntypistis gemmea* prefers altitudes above 1200–2100 m, where it flies sometimes commonly. Only a few specimens were observed below 500 m.

Distribution. Endemic to Sumatra.

196. *Syntypistis purpurascens* (Rothschild, 1917): 242; pl. 3: 9, 10 (*Stauropus purpurascens*).

Lectotype: ♂, [Indonesia, W Papua], Central Dutch New Guinea, Mount Goliath 5.000–7.000 ft. – NHMUK, London.

Schintlmeister 2013: 438.

Taxonomic note. *Stauropus purpurascens* was described of 4 ♂♂, 2 ♀♀, (syntypes). One male is marked in the holdings of NHMUK as "Type". To stabilize the nomenclature, this male is herewith designated with the following labels as lectotype: "Mt. Goliath, | 5-7000 ft., Centr. | Dutch N Guinea | about 139° long., | Januar 1911 | (A. S. Meeck)."; red-ringed circular label "Type"; "*Stauropus* | *purpurascens* | Rothsch."; yellow label "Photo done by A. Schintlmeister #4016". A lectotype label was added. Lectotype is illustrated here in fig. 196c. The remaining five specimens, which are not marked yet as syntypes or types, were labelled as paralectotypes.

Diagnosis. Forewing color deep fuscous violet brown, mixed with a few green metallic shiny scales. Basal area and forewing costa displaying sometimes more green scales. *S. purpurascens* is readily to recognize by the contrasting green, submarginal forewing fascia, which is usually well developed. Older specimens tend fading green color into a paler brown. Hindwings vary from light to fuscous brown. Sexual dimorphism is limited.

Male genitalia have a rectangularly shaped uncus and broad unified socii as *nitidus*. Valvae are short and display a broad apex with a narrow apical process. Phallus shape resembles that of *nitidus*. 8th abdominal segments are as in *nitidus*, but anterior projection is longer. Female genitalia have not been dissected yet.

Bionomics. The species is common in mountainous regions and occurs at altitudes between 1700–2800 m.

Distribution. Known from the central mountains in W Papua.

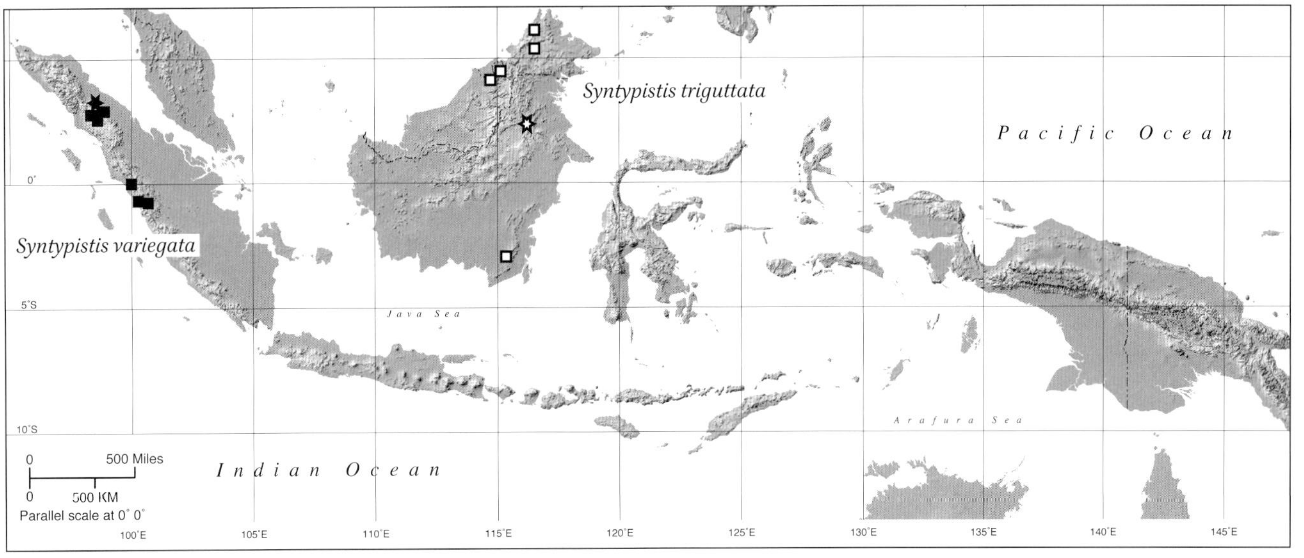

FIGURE 173 Distribution of *Syntypistis triguttata* and *variegata*.

197. ***Syntypistis triguttata*** (Kiriakoff, 1967): 46; fig. 20 (*Quadricalcarifera triguttata*).
Holotype: ♂, [Indonesia, Kalimantan], Midden Oost Borneo Expeditie, near L. Petak, 1.200 m – NBC, Leiden.
Schintlmeister 2013: 445.

Taxonomic note. Male genitalia of *triguttata* are as illustrated in the original description by Kiriakoff, 1967: fig. 20 but does not match his text-description. In fig. 15 in the same paper, wrongly named in the legend as *Quadricalcarifera ferrea*, match well to *triguttata*.

A good series (n = 58 specimens) from S Kalimantan, Loksado, appears shorter in forewing length. Forewings are darker and are less green tinged than specimens from N Borneo (Brunei, Sabah and type locality). Male genitalia, however, are not different.

Diagnosis. Forewing color fuscous violet brown, mixed with a few shiny green scales. *S. triguttata* is externally distinguished from *S. variegata* by the weakly developed or absent submarginal fascia. Males of *triguttata* occur in an individual form, where three contrasting white stigmas in the medial area are prominently developed. Hindwings are brown. Females appear sexually dimorphic and variable. They have more greenish and paler forewings. Forewing pattern, such as black postbasal and submarginal fasciae and tree stigma in the medial area are in most females well developed. The range of individual variation includes also forms of *variegata* females (198d, 198f), which is the sister species of *triguttata*.

Male genitalia have a rectangular shaped uncus and a tapered unified socii. They are similar to *variegata*, but differ in larger, narrower and club shaped valvae and bear a long apical spine. Phallus is very long, narrow and curved at the strongly sclerotized tip. 8th sternite concave shaped at the strongly sclerotized posterior margin. 8th tergite convex rounded at posterior margin. Contrary to *variegata* there is seemingly almost no individual variability in the male genitalia (n = 6 GU). Female genitalia have longer apophyses, than *variegata*. Apophyses anteriores are 60% of length of post apophyses. Antrum large, strongly sclerotized; ductus bursae broad and of medium length. Corpus bursae circularly shaped, bearing a long "oscillating band".

Bionomics. *Syntypistis triguttata* is a common species, which has been observed at altitudes between 800 m and 1800 m.

Distribution. Endemic to Borneo Island.

198. ***Syntypistis variegata*** (Kiriakoff, 1974): 391; pl. 3: 4; fig. 13 (*Quadricalcarifera variegata*), **stat. rev.**
Holotype: ♂, [Indonesia], Sumatra, Doulou bei Brastagi, 1.200 m – ZSM, Munich.
Schintlmeister 2013: 445.

Taxonomic note. The taxon *variegata* was placed by Schintlmeister, 2013: 445 as a subspecies of *triguttata* and both taxa are similar by external appearance. Structural differences of valvae, phallus and 8th sternit show,

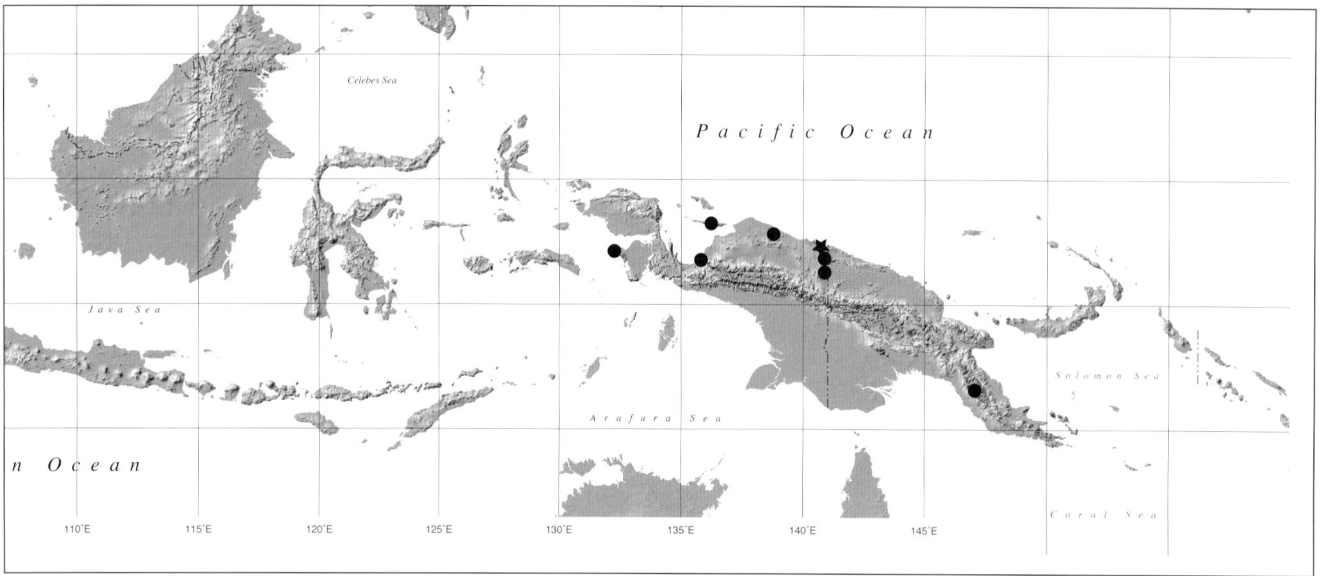

FIGURE 174 Distribution of *Syntypistis uskwara*.

however, that *variegata* must be classified as a distinct species (**stat. rev.**)

Diagnosis. Similar to *S. triguttata*, but the green submarginal fascia of the forewings is prominently green marked. Individual variability is overlapping with *triguttata* and forms, as illustrated for *triguttata* (fig. 197d), also occur. Females are sexually dimorphic and very variable as described under *triguttata*. Both species are not separable by external appearance.

Male genitalia are significantly smaller than *triguttata*. Uncus and socii are shaped as in *triguttata*. Socii are variable in length. Valva is shorter and rather straight. Valva apex bears a spine of variable length. Phallus is much shorter and thicker than in *triguttata*. Phallus tip is only slightly curved and shorter with several small spines. Posterior margin of 8th sternite wave-shaped with a small central notch. 8th tergite convex rounded at posterior margin. Female genitalia much resembles *triguttata*, but both pairs of apophyses are shorter, antrum is shorter and corpus bursae rather ovoid.

Bionomics. Adults occuring commonly between 150 m and 1800 m above sea level.

Distribution. Endemic to Sumatra.

199. ***Syntypistis uskwara*** (Kiriakoff, 1970): 112; fig. 13 (*Quadricalcarifera uskwara*).
Holotype: ♂, [Indonesia, W Papua], Dutch New Guinea, Humboldt Bay district [= Yos Sudarso Bay] – NHMUK, London.
Schintlmeister 2013: 446.

Diagnosis. This is the largest Papuan *Syntipistis*. The species is very variable. Forewing color is a fuscous chocolate-brown. Green patterns occur in variable parts, as illustrated, in extremely cases the entire forewings are green. Hindwings are pale reddish-brown. The costa bear two green areas with fuscous brown center. Sexual dimorphism is limited.

Male genitalia have uncus and gnathos of approximately equal size. Valvae display costa convex. Valva apex has no process. Phallus is straight. 8th sternite slightly bilobed at posterior margin. Female genitalia are not dissected yet.

Bionomics. *S. uskwara* occurs between 600 m and 1800 m above sea level.

Distribution. Distributed Papua New Guinea and W Papua.

200. ***Syntypistis miyanoi*** Kobayashi, 2011: 309; figs. 17, 18, 40a–h, 41a–b.
Holotype: ♂, Indonesia, Sulawesi, Palopo, Puncak, 1.100 m – National Science Museum, Tokyo.
Schintlmeister 2013: 445.

Diagnosis. Well distinguished by plumose male antennae and metallic shining deep green forewings. The intensity of the green color vary individually. Medial area of forewing bears a small fuscous brown discal spot, which is surrounded in a paler cream color. Doubled black postbasal and postmedial fasciae are present. Submarginal fascia is violet-brown and lighter green, accompanied by several black patches. Sometimes a prominent black transverse

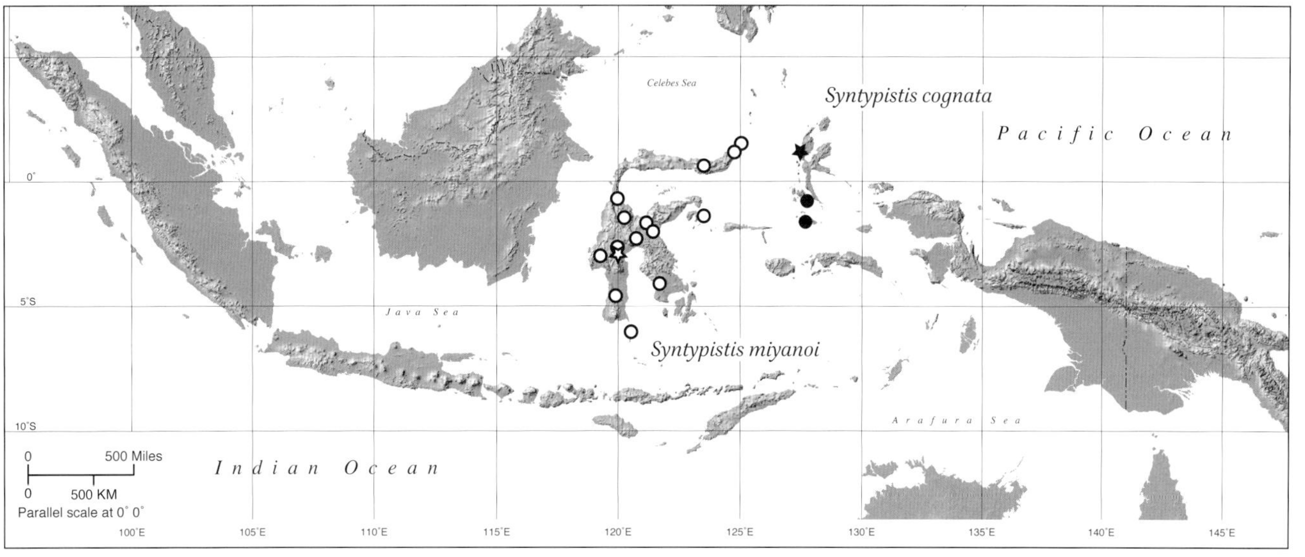

FIGURE 175 Distribution of *Syntypistis miyanoi* and *cognata*.

streak occurs parallel to the dorsum. Hindwings are reddish brown. Females resembles the males, but they have usually more black pattern elements and a paler colored medial area. *S. miyanoi* is similar and related to the following new species *cognata*; for differential diagnosis see there.

Uncus narrow, rounded at the tip and of equal length or longer than the curved socii. There is a shield at the base of uncus, which is much smaller than in *S. cognata*. Valvae are club-shaped. Phallus slightly curved with broad tip. Everted endophallus is distinguished by several diverticula. 8th sternite displays a central, gently rounded projection at posterior margin. 8th tergite has a straight posterior margin with prominent central notch. Female genitalia have medium sized and slender apophyses posteriores and reduced apophyses anteriores. 8th abdominal segment is bilobed. Antrum large, convex bilobed posteriorally. Ductus busae short and strongly sclerotized. Corpus bursae ovoid with a long, open oscillating band.

Bionomics. Adults occuring commonly between 700–2000 m. By far most specimens were collected at altitudes above 1000 m.

Distribution. Distributed in Sulawesi, Peleng Isl. and Selayar Isl.

201. ***Syntypistis cognata* spec. nov.** (figs. 201a–c)

Holotype: ♂, Indonesia, Moluccas, Obi Isl., 25 km N Tapaya, 1600 m, 1°38'S, 1°38' S, 127°° 48' E, 20. vi.–20. vii. 2008 leg. Jakl – coll. A. Schintlmeister, Dresden.

Paratypes (37 ♂♂, 3 ♀♀):

Obi: 2 ♂♂, 25 km N Tapaya, 1600 m, 1°38' S, 127°48' E, 20. vi.–20. vii. 2008 leg. Jakl (GU 93-85); 4 ♂♂, S.-coast, Seribu Mts., 22 km N Tapaya, 150 m, 1°38' S, 127° 48' E, 20. xi.–10. xii. 2008 leg. Jakl (GU 92-76, GU 92-77); 3 ♂♂, 1 ♀, dto. iii. 2009;

Bacan: 4 ♂♂, SE slopes Mt. Sibela, SE Makian, 200 m, 0°44'N, 115°01'E, 10. -30. viii. 2008 leg. Jakl (GU 65-90); 2 ♂♂, ibid 2.–12. v. 2008;

Halmahera: 23 ♂♂, 1 ♀, 15 km SW Baru, Mt. Talagarama, 600 m, 1°22' N, 127°32'E, 22. -31. i. 1996, leg. Sinjaev & Tarasov; 1 ♂, 1 ♀, Straße Baru-Basale, Gn. Talagarama, 600 m, 2. -7. iii. 1997, leg. Stefan Naumann (GU 42-66, GU 65-91).

Diagnosis. Forewing lengths ♂♂ 25 -26 mm; ♀♀ 27 and 29 mm, much larger, on average 2–3 mm, then the Sulawesian vicariant *S. miyanoi*. Antennae black and plumose in both sexes; female have slightly shorter rami than males. Forewings are deep green and shiny metallic, somewhat variable in the intensity of the green scales. Forewing pattern resembles *miyanoi*. Postbasal fascia doubled, black and more weakly developed than in *miyanoi*. A black streak in this fasciae is conspicuous and positioned at an angle of 45° positioned, while in *miyanoi* the angle is 75°. Postmedial and submarginal fasciae are indistinct, broad, blackish bands. Discal spot very indistinct, of brown color, surrounded in cream, but in most specimens of the type series a discal spot is invisible. Hindwings are pale reddish brown as in *miyanoi*. Female have more strongly developed black postmedial fascia and the medial area is paler colored. Sexual dimorphism is limited.

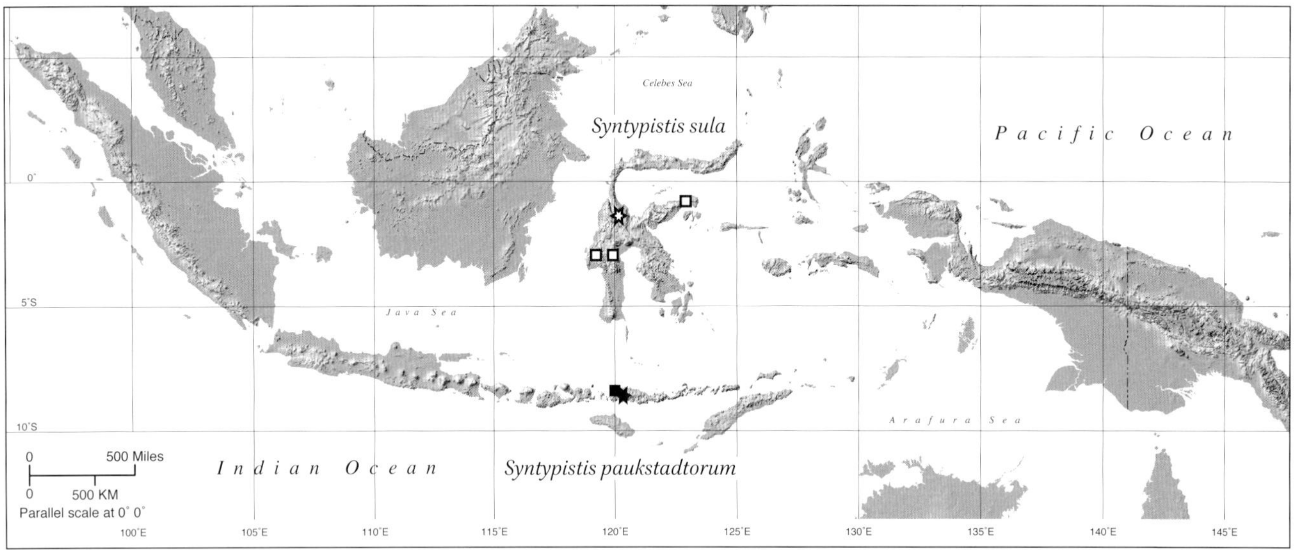

FIGURE 176 Distribution of *Syntypistis sula* and *paukstadtorum*.

Male genitalia are large and much larger than in *miyanoi*. Uncus and unified socii are of approximately equal length. At the base of the uncus a large fan shaped plate arises, which separates the new species from *miyanoi*. Valvae are curved and club shaped, with apex beeing broader than base. At valva base there is a projection. Juxta broad and pointed anteriorally. Phallus is significantly longer than *miyanoi* and of slightly sinuous shape. Uncus tip is narrower than in *miyanoi*. Posterior margin of the 8th sternite has the posterior margin straight, not wavy as in *miyanoi*. Anterior margin displays also a central process. Posterior margin of 8th tergite straight with a central notch. Female genitalia are similar to *miyanoi,* but larger. Apophyses posteriores long and slender, apophyses anteriores short and robust. Both pairs of apophyses are longer than in *miyanoi*. 8th abdominal segment is less deeply bilobed. Antrum is large, strongly sclerotized, and deeply triangularly notched. Ductus bursae short and strongly sclerotized. Corpus bursae ovoid with long open oscillating band.

Bionomics. *S. cognata* occurs commonly between 150 m and 600 m above sea level.

Distribution. Restricted to Obi, Bacan and Halmahera.

202. ***Syntypistis sula*** Kobayashi, 2011: 308; figs. 16, 39a–c.
Holotype: ♂, Indonesia, Sulawesi, Puncak Dingin, 1700 m – National Science Museum, Tokyo.
Schintlmeister 2013: 445.

Diagnosis. Forewings are dull green. Basal, postbasal and medial fasciae are doubled and black. Submarginal fascia marked as a blackish line, which fades toward the dorsum. Hindwings are brown. *S. sula* is the largest *Syntypistis* in Sulawesi and well distinguished by its broad forewing shape. Female is unknown.

Male genitalia resembles the Philippine *S. aswang* Schintlmeister & Lourens, 2010. They are characterized by an extremely swollen costa of the strongly sclerotized valva. Uncus and socii are narrow and of equal length. Phallus slender, rather straight and long. Phallus base is broad and tapered. Phallus tip narrow, curved, and tapered. 8th sternite is deeply concave notched. 8th tergite is distinguished by shape of the posterior margin with a pair of lobes.

Bionomics. *S. sula* was collected rarely (n = 3) at 440 m, 1700 m and 2000 m above sea level.

Distribution. Endemic to Sulawesi.

203. ***Syntypistis paukstadtorum* spec. nov.** (figs. 203a -f)
Holotype: ♂, Indonesia, Nusa Tenggara Timur, Flores, 9 km S Ruteng, Golo Lusang, 1820 m, 27. ii.–9. iii. 1992 leg. U. & L. Paukstadt – coll. A. Schintlmeister, Dresden.

Paratypes (45 ♂♂, 16 ♀♀):
Flores: 1 ♀, Mandgarai distr., Ruteng, Golo Lusang, 1500 m, 25. vii. 1991, leg. U. & L. Paukstadt; 18 ♂♂, 4 ♀♀, 9 km S Ruteng, Golo Lusang, 1820 m, 27. ii.–9. iii. 1992, leg. U. & L. Paukstadt (GU 47-85); 4 ♂♂, 7 km S Ruteng, Golo

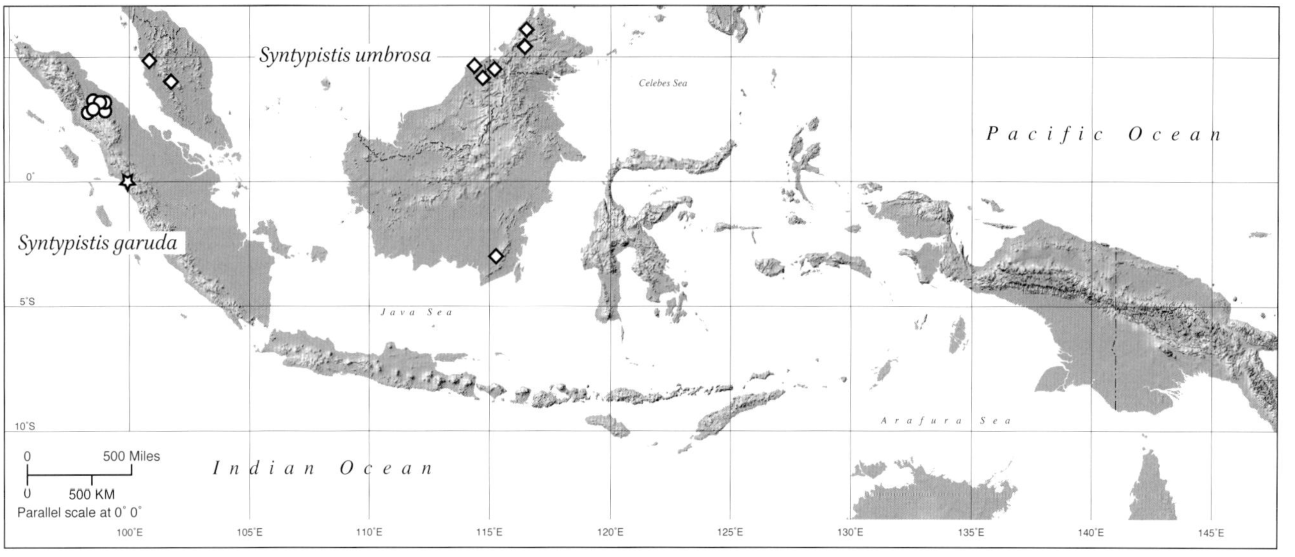

FIGURE 177 Distribution of *Syntypistis umbrosa* and *garuda*.

Lusang, 1750 m, 20. ix.–1. x. 1992, leg. U. & L. Paukstadt (GU 24-37); 12 ♂♂, 6 ♀♀, Gunung Ranaka, 9 km E Ruteng, 1140 m, 14.–15. iv. 1996 leg. Dr. Ron Brechlin (GU 42-65, GU 47-87, 54-86); 3 ♂♂, Rangawatu, Telekomstation, 33 km E Labuhanbajo, 900 m, 13. iv. 1996, leg. Dr. Ron Brechlin; 8 ♂♂, 5 ♀♀, Gunung Ranaka, 3 km S Mano, 18 km SE Ruteng, 1270 m, 17.–21. iv. 1996, leg. Dr. Ron Brechlin (GU 42-61).

Etymology: Dedicated to Cptn. Ulrich and Laela Paukstadt, Wilhelmshavn, who collected many years on various Indonesian islands Lepidoptera and left the Notodontidae for the author's collection.

Diagnosis. Forewing lengths ♂♂ 21–23 mm; ♀♀ 24–27 mm. Antennae of male black and plumose; the last 10% toward the tip is filiform. Female has shorter rami of the antennae. Forewings dull green; a few males have the postmedial area tinged a deep moss-green. Basal area slightly darker and separated toward the medial area mostly by a fuscous fascia. Postmedial fascia black, but in many specimens weakly developed. Hindwings are light brown with more fuscous brown colored anal region. Costal strip of forewing green with a large and indistinct blackish patch. Females are sexually dimorphic displaying a whitish medial and submarginal area. In the medial area three large and contrasting white stigmas occur, each with a small, pale-brown nucleus.

Male genitalia have unified socii, which are broad and slightly notched. They are slightly shorter than the uncus. Valvae are slightly curved and show a swollen sacculus. Apical valva process is long and slender. There is a resemblance to *S. umbrosa*, but the latter have broader valvae, in particular, a much broader valva apex. Phallus is slender and long. The rounded tip is more strongly sclerotized. 8th sternite is deeply tiangular notched at the posterior margin. 8th tergite convex rounded at posterior margin with a small central notch. Among the three dissected male genitalia one's genitalia display a short tegumen and a curved phallus (fig. 203c). Female genitalia have long and slender apophyses posteriores and short and robust apophyses anteriores. Antrum is relatively broad and not notched as *umbrosa* or *garuda*. Ductus bursae short. Corpus bursae lacks a oscillating band.

Bionomics. *S. paukstadtorum* was collected commonly in mountanous forests between 1100–1800 m above sea level.

Distribution. Endemic to Flores Island.

204. ***Syntypistis umbrosa*** (Matsumura, 1927): 6; pl. 1: 13 (*Quadricalcarifera umbrosa*).

Lectotype: ♀, [Taiwan], Formosa, Horisha – Entomological Institute of the Hokkaido University, Sapporo.

Schintlmeister 2013: 446.

Taxonomic notes. A small series (n = 7) from S Kalimantan appear significantly smaller than specimens from Sabah or mainland Asia, but male genitalia are not different.

Male genitalia from different geographically regions are diverse and the *umbrosa* complex, which includes also *S. tala* Schintlmeister & Lourens, 2010, described from

Palawan, and *garuda* (see below), must splitted into several further taxa, once the group is revised. For populations from Borneo and Malaya the name *hasegawai* Nakamura, 1976 would be applicable.

Diagnosis. Similar to *S. umbrosa*. Forewings dull, brownish-green and basal area darker brownish-green colored. Postbasal fascia fuscous brown and in contrast to *garuda,* weakly developed. In the costal area a pair of parallel fuscous brown streaks occur. The latter streak is a good character to separate *garuda* from *umbrosa*. Postmedial fascia is indistinct and whitish marked. Discal patch inconspicous, circularly shaped and of a light cream color. Submarginal fascia developed as a greenish line. Older specimens (>30 years old) fading green color into brown. Forewing fringe is checkered white and brown. Hindwings are light reddish-brown. Females are sexually dimorphic, have light grey colored forwings and a less pronounced basal area. The medial area bears three whitish stigmas, with pale grey nuclei. Distinguishable from females of *S. pallidadina* by the absence of metallic shining green scales. The species appear similar to *S. garuda* from Sumatra, but forewing pattern. including the stigma in the medial area, is more contrasting and hindwings are darker.

Male genitalia of Bornean populations have the unified socii longer than the uncus. Valva slightly curved and rectangularly shaped with broad apex. Apical valva process is long and pointed. Juxta is larger than *garuda* and a central anterior projection is drop-like and longer. Phallus long with curved distal part. Phallus tip is knob-shaped and serrated. 8th sternite notched at the posterior margin. 8th tergite has a convex posterior margin which is centrally notched. Female genitalia are distinguished by the large, deeply notched antevaginal plate. The notch is longer than in *S. kiriakoffi*. Apophyses posteriores are very long and slender, longer than in *garuda*. Apophyses anteriores are reduced in size. Corpus bursae features a long"oscillating band", which extends over 70% of the corpus bursae.

Bionomics. The species occurs in Malaya and Borneo according uncommonly and mostly at altitudes between 800 m and 1000 m above sea level. Holloway, 1983: 53, however reported *umbrosa* under the name *malayana* as a frequent species at lower altitudes in Sabah and Brunei. In Indochina and the Himalaya the species appear commonly and much more females than males has been observed. In noreastern India adults occur up to 3000 m.

Distribution. Widely distributed in Taiwan, southern China, Indochina in its entirety, Myanmar and the Himalaya. In Sundaland occuring on the Malayan Peninsula, Borneo and Palawan.

205. ***Syntypistis garuda*** (Schintlmeister, 1994): 225; pl. 4: 1, 2; GU-pl. 4: 4 (*Quadricalcarifera garuda*).
Holotype: ♂, Indonesia, W Sumatra, Mt. Talamau, Westseite, 12 km S Talu, 0°08' N, 99°59' E, 1.800 m – coll. A. Schintlmeister, Dresden.
Schintlmeister 2013: 438.

Diagnosis. Similar to *S. umbrosa*, but males have plumose antennae with longer rami in both sexes. General appearance rather brownish than greenish brown, but older specimens (>30 years old) are of fading green color into brown as *umbrosa*. Wing pattern is similar to that of *umbrosa*, but the darker basal area of the forewings is better distinguished. A strip along the dorsum is fuscous greenish-brown. Postbasal fascia strong fuscous brown. In the costal area near the basal area a parallel fuscous brown streak occurs. This prominently streak is a good characteristic to differentiate *garuda* from *umbrosa*. Postmedial fascia is indistinct whitish marked. Discal patch large, circular and light cream. Forewing fringe checkered white and brown. Hindwings are light reddish-brown in both sexes with darker anal area. Female is sexually dimorphic, has a light grey forewing color and less a pronounced basal area. The medial area bears three whitish stigmas, with pale grey nuclei. There are no metallic, shiny green scales as in *pallida*. Forewing pattern of females appear less contrasting and hindwings are paler than females of *umbrosa*.

Male genitalia resemble those of *umbrosa* but differ by valva shape. Valva apex is rather rounded than right-angled as in *umbrosa*. Juxta of *garuda* is narrower and the anterior projection is heart like and somewhat pointed. Phallus is rather straight, the distal part is not curved as in *umbrosa*. Phallus tip broader than *umbrosa* and serrated. 8th sternite widely bilobed; 8th tergite is deeper and more notched as a heart shape than *umbrosa*. Female genitalia are as in *umbrosa*, but apophyses posteriores are shorter. Antevaginal plate is deeply notched. The notch is deeper than in *kiriakoffi*.

Bionomics. The species occur infrequently. Less than n = 20 specimens has been sampled mostly in mountainous forests between 1000–1800 m above sea level. Two specimens came from altitudes of 550 m and 80 m above sea level.

Distribution. Endemic to Sumatra.

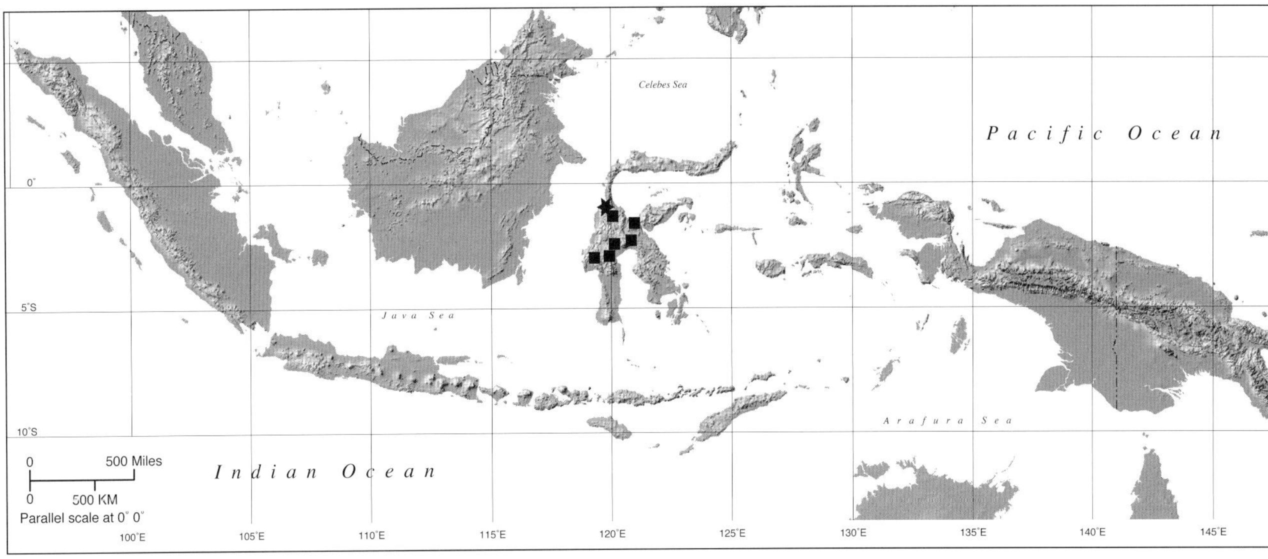

FIGURE 178 Distribution of *Syntypistis kiriakoffi*.

206. ***Syntypistis kiriakoffi*** (Schintlmeister & Lourens, 2010): 177.

Holotype: ♂, [Indonesia, Sulawesi], W. Celebes, Paloe district, Mt. Tompoe, 2.700 ft. – NHMUK, London.

Schintlmeister 2013: 440.

Diagnosis. Forewings of male display in most cases a shiny metallic-green basal area, costa, and dorsum, but forms without metallic scales are also occur, as depicted in fig. 206c. A prominent and contrasting white costal streak near the basal area is a characteristic feature. Discal spot is indistinctly cream-colored with a brown nucleus. Postmedial fascia present as a fuscous line. Fringe of forewing checkered brown and white. Hindwings are light brownish-white with a more fuscous brown anal region. Females are sexually dimorphic having white ore light grey forewings. Three white stigmas in the medial area are present in greyish specimens, but almost invisible in white females. Basal area, dorsum, submarginal area and a larger costal blotch near the apex are metallic green filled. *S. kiriakoffi* resembles females of *S. palladina*, but the green-filled submarginal area of the forewings is narrower.

Male genitalia of *kiriakoffi* are similar to those of *S. umbrosa*, but differ by the long and curved uncus and pointed unified socii. Uncus is also much larger than in *palladina*. Valvae are rectangularly shaped as in *umbrosa* with a long and tapered apical spine. Phallus narrow and curved at the distal part as in *umbrosa*. Phallus tip is strongly sclerotized and serrated. 8th sternite notched at posterior margin. 8th tergite is convex at posterior margin and also notched. Both notches are wider than in *umbrosa*. Female genitalia resembles *umbrosa*, but the antevaginal plate is less deeply notched and the antrum displays a pair of slenderer lobes. Apophyses, ductus bursae and corpus bursae with oscillating band are as in *umbrosa*.

Bionomics. Moths occur infrequently at altitudes between 650–2000 m.

Distribution. Endemic to Sulawesi.

207. ***Syntypistis palladina*** (Schaus, 1928): 76 (*Stauropus palladina*).

Lectotype: ♂, Philipine Islands, Mindanao, Surigao – USNM, Washington DC., designated by Schintlmeister, 2016: 578.

Schintlmeister 2013: 442.

Taxonomic notes. *Syntypistis palladina* tends to establish geographical subspecies. Populations from Borneo, the Malayan Peninsula and Sumatra belong to *Syntypistis palladina viridimargo* (Kiriakoff, 1967): 48; fig. 19. Holotype ♂, [Indonesia, Kalimantan], South East Borneo, Samarinda – NBC, Leiden. This subspecies is distinguished in both sexes by a whitish medial area of forewing. Uncus and unified socii in male genitalia are short and posterior margin of 8th sternite is straight.

In Java, Bali and Lombok a distinct subspecies occur, described below.

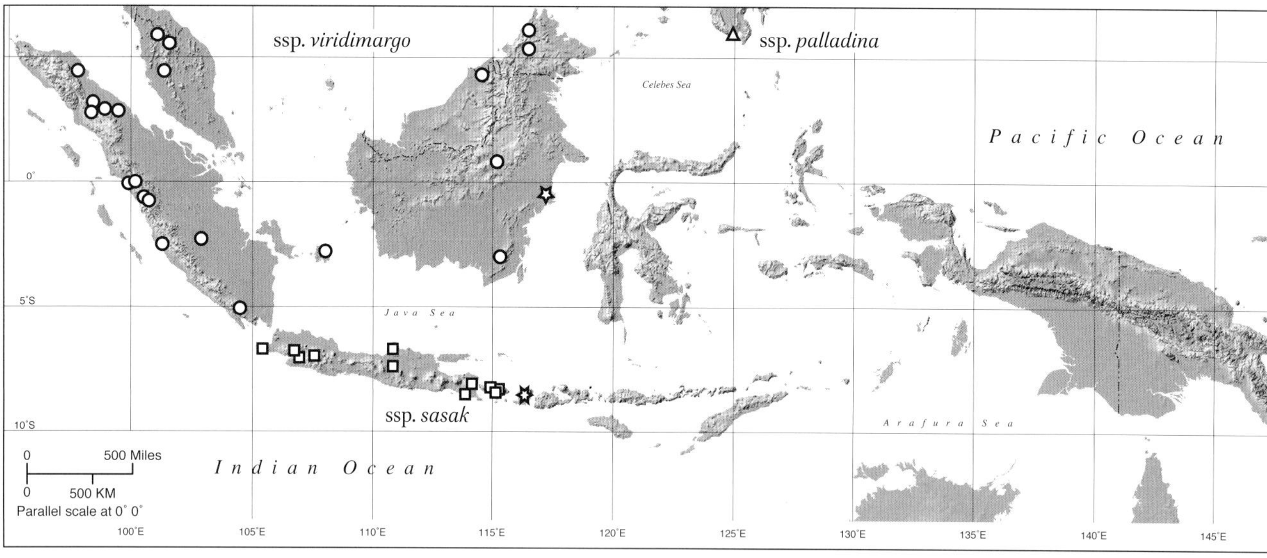

FIGURE 179 Distribution of *Syntypistis palladina*.

Syntypistis palladina sasak **ssp. nov.** (figs. 204b, 204e, 204i–m).

Holotype: ♂, Indonesia, Lesser Sundas, Lombok isl., Pusuk Hill, 700 m, 10°21'S, 123°39'E, 16.–22. iii. 2009, leg. S. Jakl – in coll. A. Schintlmeister, Dresden.

Paratypes (63 ♂♂, 38 ♀♀):

Lombok: 4 ♂♂, Pusuk Hill, 700 m, 10°21'S, 123°39'E, 16.–22. iii. 2009, leg. S. Jakl (GU 93-96, GU 93-97); 1 ♀, E Lombok, E Gunung Rinjani, 1550 m, 9. iv. 2000, leg. S. Naumann;

Bali: 9 ♂♂, 13 ♀♀, Tamblingan Lakes, 1300 m, 8°16'S, 115°06'E, vii. 2004, leg. St. Jakl; 9 ♂♂, 6 ♀♀ Bedugul distr., Tamblingan Nat. Park, 1200 m, 8°14'S, 115°08'E, i.–ii. 2004, leg. S. Jakl; 1 ♂, 4 ♀♀, Tamblingan Lakes, 1200 m, 8°15.47'S, 115°05.15'E, 25. xi.–10. xii. 2009, leg. St. Jakl (GU 94-66); 2 ♂♂, 2 ♀♀, dto., v. 2004; 5 ♂♂, 2 ♀♀, Bali centralis, Buyan Lake, 1300 m, 10. ii. 1997, leg. K. Cerny; 1 ♂, W. Bali Reserve, Titap, 650 m, 8°25'N, 114°58'E, 24. xii. 1998, leg. S. Naumann; 1 ♂, West Bali National Park, 680 m, Tabanan, Sepang, 18. iv. 2000, leg. S. Naumann; 1 ♂, Kuta, leg. U. & L. Paukstadt (GU 19-16);

Java: 2 ♂♂, Ujung Kulon, National Park, 6°50'S, 105°33'E, 100 m, viii. 1996, leg. S. Jakl; 13 ♂♂, 1 ♀, Mt. Pangrango, 30 km SE Bogor, 1625 m, 6°30'S, 107°10'E, 16. -20. ii. 1996, leg. V. Siniaev & E. Afonin (GU 42-62, GU 54-98, GU 79-68); 7 ♂♂, Mt. Gedeh, 1250 m, Pangrango Nat. Park, 6°47'S, 107° E, iv.–vii. 1996 (GU 47-50, GU 54-99); 1 ♂, 3 ♀♀, Sukanegara, Cianur reg., 1050 m, 7°06'S, 107°07'E, 18.–21. iii. 2007, leg. S. Jakl; 2 ♀♀, Mt. Solok, 600 m, ii.1981, leg. U. & L. Paukstadt (GU 16-34); 1 ♀, Mt. Baluran, 7°52'S, 114°23'E, 600 m–1000 m, iv. 1996 (GU 42-63); 1 ♂, Mt. Muria, 15 km N Kusdas, 1100 m, 6°35'S, 110°53'E, 21.–22. i. 1998, leg. Jakl, Schintlmeister & Cervinka; 1 ♂, Gunung Argopuro, 15 km SW Bondowoso, 8°04'S, 113°45'E, 7.–13. x. 2004, leg. S. Jakl; 1 ♂, 2 ♀♀, Meru Betiri National Park, 25 km S Kalibaru vill., 300 m–500 m, v.–vi. 1996, leg. S. Jakl (GU 51-02); 2 ♂♂, Buleleng district, Tamblingan, Danau Tambliungan, 8°16.1'S 115°05.5, 19.–21. ii. 2015, 1250 m, leg. J. Hájek & J. Sumpich.

Etymology. Named after the Sasak people from Lombok.

Diagnosis of ssp. *sasak*. Forewing lengths ♂♂ 21–22 mm, ♀♀ 23–27 mm. Male forewing rather uniform blackish brown with shiny metallic green scales. Often number of green scales is reduced as shown in the holotype (fig. 207g). Males of ssp. *sasak* are distinguished from ssp. *palladina* and *viridimargo* by lack of any whitish scales on the forewings. Hindwings of ssp. *sasak* are brown, darker than in ssp. *viridimargo*. Females are variable in pattern and size. The postmedial fasciae on forewing is more strongly developed than ssp. *palladina* or *viridimargo*. Medial area is brownish grey and darker than ssp. *viridimargo*. The three pale stigmas are therefore more contrasting and better visible. Hindwings are brown, sometimes with broad, fuscous, filled postmedial area. A single female from Java (fig. 207k) displays forewings with reduced brownish and green patterns and a rather uniform grey medial area.

Male genitalia of ssp. *sasak* having the uncus and the unified socii longer than in other subspecies. Base of the phallus is significantly narrower and tip less curved. The 8th sternite triangular notched and the 8th tergite more deeply notched than in *viridimargo* or *palladina*. Female genitalia does not differ from those of the other subspecies.

Diagnosis. *Syntypistis palladina* males have plumose antennae with long rami; females have shorter rami. Forewings of males are brown, mixed with shiny metallic green scales in basal area, along the costa, submarginal area and along the dorsum. Submarginal and marginal fasciae are marked as fine blackish lines. Hindwing light brownish white to brown with fuscous brown marginal line. Forewing pattern of female resembles the male's, but medial area is grey instead of brown. In the medial area three whitish stigmas occur. Females resembles females of *S. kiriakoffi*, but the green postmedial area of forewing is broader.

Male genitalia are distinguished from the other similar congeners by their short and robust uncus. Socii are unified and large. Valva bears a long apical process. Phallus is serrated at the tip. 8th tergite is notched. 8th sternite concave at the posterior margin. Female genitalia have long apophyses posteriores and short apophyses anteriores. Lamella postvaginalis is not notched as in *kiriakoffi* or *umbrosa*. Ductus bursae is longer than in *kiriakoffi* or *umbrosa*. The "oscillating band" is present at the large, ellipsoid corpus bursae.

Bionomics. The species occurs commonly in Sundaland, Java, and Bali from the lowlands up to 1600 m above sea level. Most specimens were taken below 1000 m.

Distribution. *Syntypistis palladina palladina* occur in the Philippines: Mindanao, Leyte, Samar and Mindoro. The ssp. *viridimargo* flies in Borneo, the Malayan Peninsula up to south Thailand, Sumatra and Belitung isl. *S. palladina sasak* is known from Java, Bali and Lombok.

208. ***Syntypistis comatus*** (Leech, 1898): 306 (*Stauropus comatus*).
Holotype: ♀, [China, Sichuan], Omei-shan – NHMUK, London.
Schintlmeister 2013: 437.

Taxonomic notes. The subspecific context is currently not clear yet. There are some indications, that *S. comatus* has established several subspecies in mainland Asia and for Sumatran populations the name *Syntypistis comatus bioculata* (Kiriakoff, 1967): 42; fig. 6; holotype: ♂, S. W. Sumatra, Barisan Range, 2.500 ft. – NHMUK, London, might be applicable, if Sundanian populations are taxonomically separable. A few dissected males from Kalimantan differ from Sumatran material by more compact tegumen and larger phallus, where the basal part is shorter and tapered.

Specimens from Java, however, differ significantly from all other known populations of *comatus* and are described below as:

Syntypistis comatus majapahit **ssp. nov.** (figs. 208d, 208h –j).
Holotype: ♂, Indonesia, Java, Sukanegara, Cianur reg., 1050 m, 7°06'S, 107°07'E, 18.-21 iii. 2007, leg. St. Jakl – in coll. A. Schintlmeister, Dresden.

Paratypes (22 ♂♂, 1 ♀):
Java: 3 ♂♂, Mt. Pangrango, 30 km SE Bogor, 1625 m, 6°30'S, 107°10'E, 16.–20. ii. 1996, leg. V. Siniaev & E. Afonin (GU 47-89); 10 ♂♂, Mt. Gedeh, 1250 m, Pangrango Nat. Park, 6°47'S, 107° E, iv.–vii. 1996 (GU 51-03, GU 54-87); 1 ♂, Gunung Kencana, Pangrango N. P., 350 m, 6°37'S, 106°06'E, 15.-18.iii.2007, leg. St. Jakl; 2 ♂♂, Mt. Salak, 6°42'S, 106°44'E, 1000–1500 m, vii. 1996; 2 ♂♂, Kelabang Bondowoso, St. Dongeng Sempol, 1150 m, 3. v. 1997, leg. S. & A. Saito; 2 ♂♂, 1 ♀, Meru Betiri National Park, 25 km S Kalibaru vill., 300–500 m, v.–vi. 1996, leg. St. Jakl (GU 43-06, GU 51-04); 1 ♂, Cikaniki Stn., Halimun-Salak N.P., 15.–18. x. 2015, leg. W. Mey.

Etymology. Named after an ancient kingdom in Java (from 1293–1527).

Diagnosis of ssp. *majapahit*. Forewing lengths ♂♂ 21–22 mm, ♀ 27 mm. On average 2 mm smaller than populations from Borneo and Sumatra. Subspecies *majapahit* is characterized by weakly developed or absent white stigma in the medial area of forewing, but one male (of n = 23) displays white stigmas as in ssp. *comatus*. The forewing shape is less elongated than in ssp. *comatus*. The sexually dimorphic female does not differ significantly from Sundanian females by external appearance.

Male genitalia are distinguished by shorter and thicker phallus. Other features, such as shapes of uncus, valva and apical valva processes are subject to individual variation. The only female have not yet been dissected.

Diagnosis. Males of *comatus* are readily identified by five contrasting white stigmas with grey nucleus on deep-green forewings. In ssp. *majapahit* forewings are often weakly tinged in white and such forms can be confused with *S. palladina sasak*. The latter, however, never displays white scales. The green forewing color fades in most specimens quickly into brown. A blackish postmedial fascia is mostly visible, but in a good series (n > 50) from Kalimantan this

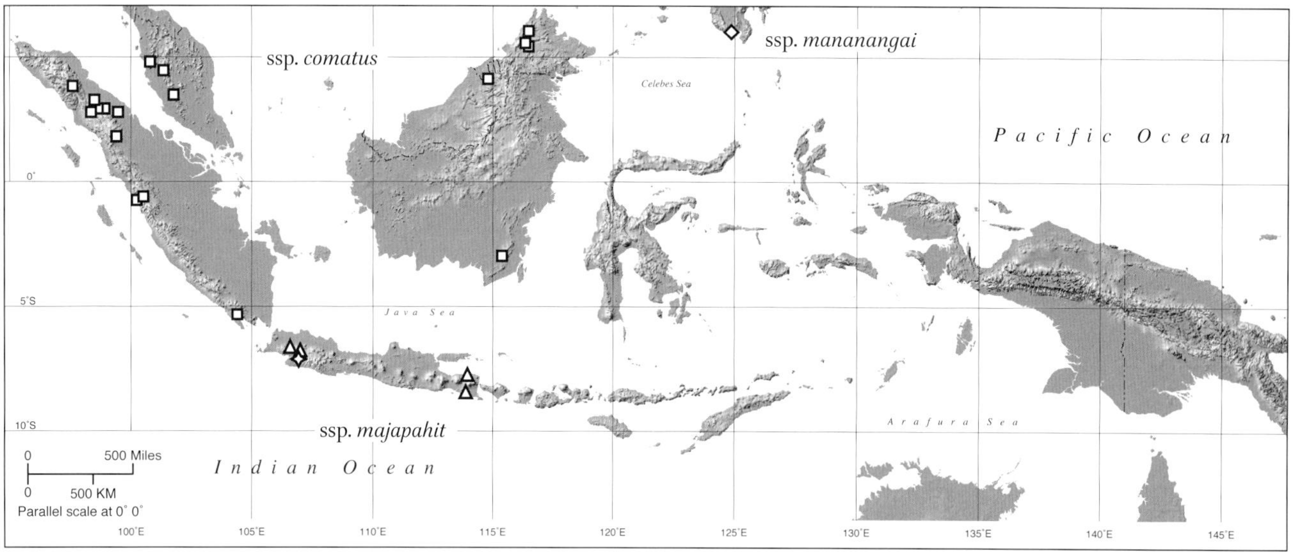

FIGURE 180 Distribution of *Syntypistis comatus*.

fascia is contrasting black developed, see fig. 208f. Individual variability is mainly seen in the extension of the white stigmas and sometimes they are confluent and lacking a grey nucleus. Hindwings are light reddish-brown with a more darkly colored anal area. The female is sexually dimorphic and displays white medial and submarginal area where four white stigmas are only to be recognized by the light greyish nucleus. Females of *comatus* are easy to separate from females of other congeners by a large green filled area along the dorsum.

Male genitalia have a narrow, long uncus and broad unified socii with rounded tip. Valva shape is narrow and curved. Shape of valva apex appears variable, but a large apical process is always curved like a hook and not as a spine. Phallus long and rather straight, distally kinked. Phallus tip is a more strongly sclerotisized knob. 8th abdominal segments are deep and broadly notched. 8th tergite is also notched at the posterior margin, but the notch is smaller. Female genitalia are well characterized by the broad antevaginal plate and the large antrum with a pair of large lobes. Apophyses posteriores long and slender; apophyses anteriores short, but longer than in *palladina*. Ductus bursae is longer than in members of the *umbrosa* group. Corpus bursae long ellipsoid with prominent developed "oscillating band".

Bionomics. Moth occurs commonly between 300–2600 m, in Nepal up to 3500 m above sea level. The species was most frequently observed between 800–1500 m. In Java the adults are apparently more rare.

Distribution. *Syntypistis comatus comatus* is distributed in China, Nepal, north India, Myanmar, Indochina, the Malayan Peninsula, the Philippines, Kalimantan, Borneo and Sumatra. The ssp. *sasak* is restricted to Java, and in Mindanao ssp. *mananangai* Schintlmeister & Lourens, 2010 occur.

209. *Syntypistis trioculata* (Holloway, 1976): 55; pl. 20: 238, 242; fig. 365 (*Quadricalcarifera trioculata*).
Holotype: ♂, [Malaysia], Borneo, Sabah, Mt. Kinabalu, Power Station, 1.930 m – NHMUK, London.
Schintlmeister 2013: 445.

Diagnosis. *S. trioculata* resembles by greenish rufous forewing color and the three white stigmas with grey nuclei to *S. comatus*. It might be distinguished from the latter by the whitish head and patagia. A whitish spot near forewing apex of males is also a good character to separate *trioculata* from *comatus*. Blackish postmedial fascia is indistinctly developed. Hindwings are reddish-brown and darker than *comatus*. Female is sexually dimorphic and has forewings colored almost white. Only the basal and marginal areas are green. Near the forewing apex there is a pair of black patches; a remnant of a large costal blotch of *comatus*.

Male genitalia have uncus and unified socii as in *comatus*. But valvae are short and broad. They bear at the apex a small triangular projection and not a hook- like process as in *comatus*. Phallus rather short, straight and kinked distally. Both abdominal segments are concavely at their posterior margin like *comatus*.

Female genitalia has not been examined.

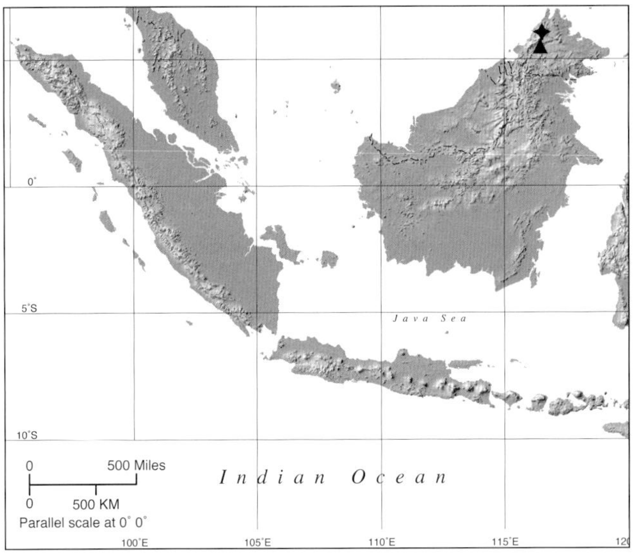

FIGURE 181 Distribution of *Syntypistis trioculata*.

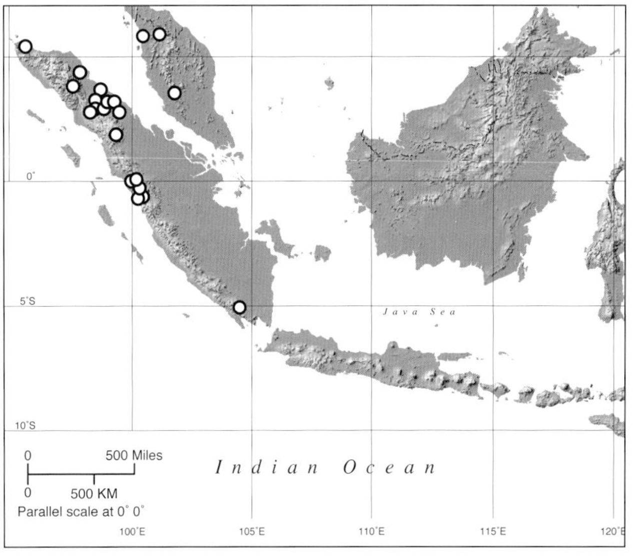

FIGURE 182 Distribution of *Syntypistis viridipicta*.

Bionomics. The species occurs commonly above 1100 m and has been observed up to 2000 m.

Distribution. Endemic to Sabah, Borneo. Not known from Kalimantan.

210. ***Syntypistis viridipicta*** (Wileman, 1910): 312 (*Stauropus viridipicta*).
Syntypes: 2 ♂♂, [Taiwan], Formosa, Kanshirei, 1.000 ft. – NHMUK, London.
Schintlmeister 2013: 447.

Diagnosis. Forewings fuscous brown, tinged with metallic green scales. They varying from deep moss-green to cyan-green; some specimens, but not all, are fading brown with time. Basal and submarginal fasciae are marked black, but most specimens have this fascia weakly and indistinctly developed. A few males displaying several additional fasciae in the medial area. Other males display contrasting white stigmas with brown nuclei in the medial area, see fig. 210f. About 30% of the males having these stigmas weakly and non-prominently developed (fig. 210d). There is also variability in broadness of forewings. Hindwings are reddish-brown. Sexual dimorphism is limited and females have the same wide range of individual variability as males. But most females have the blackish postmedial and submarginal fasciae better developed than do the males.

Uncus and unified socii are long and curved. Tegumen bear a diagnostically evagination, which is unique among the Indonesian congeners. Valvae are strongly sclerotized, with tapered apex and a smaller process of variable size. Phallus is distinctively zig-zag shaped, long and slender. Phallus base is strongly sclerotized and ends distally in an elongated process with serrations at the tip. 8th sternite is distinguished by the broad strongly sclerotized posterior margin which is notched. 8th tergite is deeply notched. Female genitalia have long apophyes posteriores and only slightly shorter apophyses anteriores. Antrum large, posterior slightly notched. Ductus bursae curved and medium sized. Corpus bursae long with prominent "oscillating band", which covers almost complete corpus bursae.

Bionomics. The moth is in Sumatra and Malaya among the most common notodontids and occur from the lowlands (10 m) up to 2000 m. In SW China, Yunnan up to 3000 m above sea level. Caterpillar's hostplant in Taiwan is *Styrax* (Stycaraceae).

Distribution. Widely distributed: Nepal, northeastern India, Myanmar, China, Taiwan, Indochina in its entirety, the Malayan Peninsula and Sumatra. Seemingly absent from the minor Islands around Sumatra and Malaysia.

211. ***Syntypistis stanleyi* spec. nov.** (figs. 211a–d).
Holotype: ♂, Indonesia, Sumatera Barat, Mt. Sanggul, 1200 m, 2°44'S, 100°23'E, 1700 m, iv. 2004, leg. S. Jakl (GU 94-23) – in coll. A. Schintlmeister, Dresden.

Paratype (1 ♂):
Borneo: 1 ♂, Mt. Kinabalu, Kalanggaan, 1400 m, 9.–13. viii. 2005, leg. W. Mey & K. Ebert (ZMHU 02-2017).

Etymology. Named for Stanislav ("Stanley") Jakl, Prague,

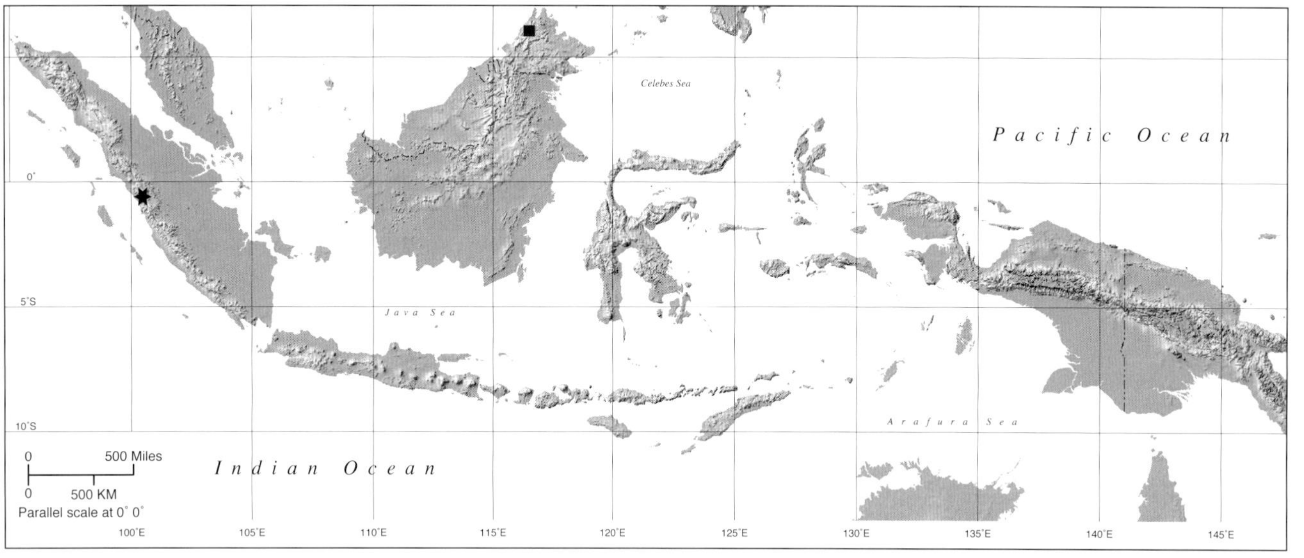

FIGURE 183 Distribution of *Syntypistis stanleyi.*

who collected – apart from many other valuable material – also the holotype of this new species.

Taxonomic note. Only two males are available, which are different in their genitalia (uncus, phallus, 8th sternite). It remains currently unknown if these differences are in the range of individual variability or indicate subspecific differences.

Diagnosis. Forewing length ♂♂, 20,5 mm. Antennae fuscous brown, bipectinated with long rami; last 20% toward the tip filiform. External appearance similar to *S. viridipicta*. Forewings are fuscous brown mixed with many shiny metallic green scales. Basal area of forewing displays green scales more densely. A black basal spot is present, which does not occur in *viridipicta*. Postbasal fascia is black and separates basal from from medial area. Submarginal fascia as in *viridipicta* well developed as a black line. Fringe of all wings slightly checkered brown and cream. Hindwings are as in *viridipicta* reddish-brown, with slightly darker anal area. The female is unknown.

Male genitalia of the holotype have short and narrow uncus and slightly longer socii. The paratype from Borneo displays uncus and unified socii clearly longer. Tegumen lacks an evagination, which is a characteristic feature for *viridipicta*. Juxta larger than in *viridipicta,* well sclerotized. It bears anteriorally a diagnostically drop-shaped projection. Valvae strongly sclerotized, narrow with bulbous costa as *viridipicta*. Valva apex tapered and without an apical process as in *viridipicta.* Phallus straight and slightly curved distally. Phallus tip is knob-shaped. 8th sternite notched centrally at posterior margin. The shape of margin differs between holotype and paratype from Borneo as illustrated. 8th tergite convexly notched an the posterior margin. The notch is less deep than in *viridipicta.*

Bionomics. The holotype was taken among several specimens of *viridipicta* (n = 6 GU) at 1700 m; the paratype has been collected at 1400 m above sea level. *S. stanley* must be a rare species. Among many dissected specimens of *viridipicta* only two males were traced.

Distribution. Known from West Sumatra and from Mt. Kinabalu in Borneo.

212. ***Syntypistis pallidifascia*** (Hampson, 1893): 151 (*Stauropus pallidifascia*).
Holotype: ♂, [India], Sikhim – NHMUK, London.
Schintlmeister 2013: 442.

Taxonomic note. Papuan populations, described as *Stauropus germana* Rothschild, 1917: 244; pl. 8: 18, 19, holotype: ♂, Dutch New Guinea, Central Arfak Mountains, Ninay Valley 3.500 ft. – NHMUK, London, does not differ in its wing pattern or genitalia from Sundanian specimens, but imagines are significantly, 2 -5 mm, on average 3 mm, larger in its forewing length. A series from New Ireland and a few specimens from New Britain are slightly smaller than Papuan specimens but also larger than Sundanian or Moluccan populations. Most females (n = 17) of *germana* have a blackish medial area; only one female from New Guinea appears with paler medial area. Among n = 17 females from other Indonesian islands, only one female from Sumatra displays a black medial area.

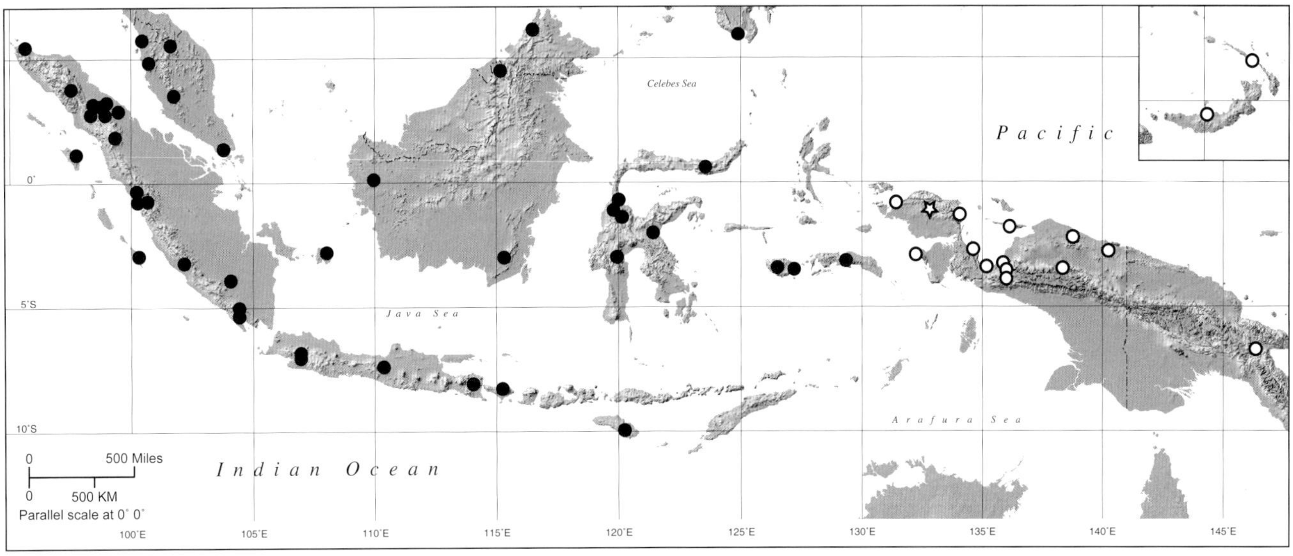

FIGURE 184 Distribution of *Syntypistis pallidifascia*.

Diagnosis. Readily recognized by the brown medial area of the forewings. Sometimes the brown medial area is less contrasting to the other parts of the wing and the forewings appear more uniform. Postbasal and postmedial fasciae consists of blackish and pale brown lines. Medial area displays a pair of circular, cream-colored stigmas with brown nuclei. A blackish submarginal fascia of forewing is often weakly developed. Females display postmedial fascia more strongly developed and hindwings are darker than in males.

Male genitalia have a curved uncus and a much longer curved unified socius. Valvae are asymmetric, bearing on their right arm a knob-shaped projection, which is absent on the left arm. This evagination varies individually in its shape. Valva costa convex shaped and valva apex bears a smaller process of variable size. In a few specimens this process is lacking. Phallus is diagnostically zig-zag shaped but variable in its length. Phallus tip is serrated. 8th sternite has a central triangular projection, which is rounded at the tip. The 8th tergite deeply notched with variable shape of the notch. Female genitalia are small in relation to male genitalia. 8th abdominal segment is broad, but not long. Apophyses posteriores long and slender; apophyses anteriores are short and robust. Antrum small and bilobed. Ductus bursae is strongly sclerotized and curved. Corpus bursae is large, and the "oscillating band" almost closed.

Bionomics. A common species in mainland Asia and in Indonesia, but apparently rare in Borneo island. Adults have been observed mostly between 800 and 1200 m above sea level, but a few specimens have been taken at 2000 m.

Distribution. Distributed in Nepal, N India, Myanmar, S China, Taiwan, Iriomote Isl in Japan (ssp. *iriomotensis* Nakatomi, 1980), Indochina, Philippines (in Mindoro: ssp. *juttamariae* Schintlmeister & Lourens, 2010), Malaya, Borneo, Sumatra, Belitung, Nias isl., Mentawei Isl., Java, Flores, Sulawesi, Buru, and Seram. The ssp. *germana* occur in New Guinea in its entirety, New Ireland and New Britain.

213. ***Syntypistis vaneeckei*** spec. nov. (figs. 213a–e)
Holotype: ♂, Indonesia, Sulawesi Selatan, Puncak, Palopo, 900–1300 m, 2°57' S, 120°03' E, iii. 1997 – coll. A. Schintlmeister, Dresden.

Paratypes (36 ♂♂, 1 ♀):
Sulawesi: 2 ♂♂, Puncak, 27 km W Palopo, 1000–1200 m, 2°57' S, 120°03' E, 10.– 19. v. 1984, leg. A. Schintlmeister (GU 19-14); 3 ♂♂, Puncak, Palopo, 900–1300 m, 2°57' S, 120°03' E, iii. 1997 (GU 47-59, GU 79-64); 1 ♂, ibid, vii. 1997; 8 ♂♂, ibid, x. 1997 (MWM 15.623, W 15624); 1 ♂, ibdi. xi. 1997; 2 ♂♂, ibid, xii. 1997; 2 ♂♂, ibid, xii. 1998 (MWM 15.605); 3 ♂♂, 1 ♀, Mamasa, 1500–2000 m, 2°56'S, 119°22'E, x. 1995; 5 ♂♂, Mt. Sampuraga, 1400 m, 2°14'S, 120°45' E, 11.–12. ii. 1995, leg. Siniajev & Tarasov (GU 43-30); 1 ♂, Mt. Salubaya, 1000m, 0°45'S, 120°00'E, 9.–10. ii. 1995, leg. Siniajev & Tarasov; 1 ♂, Quarles Mts., Polo Polo, N Rantepao, 2200 m, 26.–28. ix. 1995, leg. St. Naumann; 3 ♂♂, Dumogo-Bone National Park, Hill, Site 15, 1440 m, viii. 1985, leg. J. D. Holloway; 1 ♂ dto., Site 17, 1520 m, viii. 1985; 1 ♂ dto., Site 13, 1300 m, 5.–6. iii. 1985 (BM 1748).

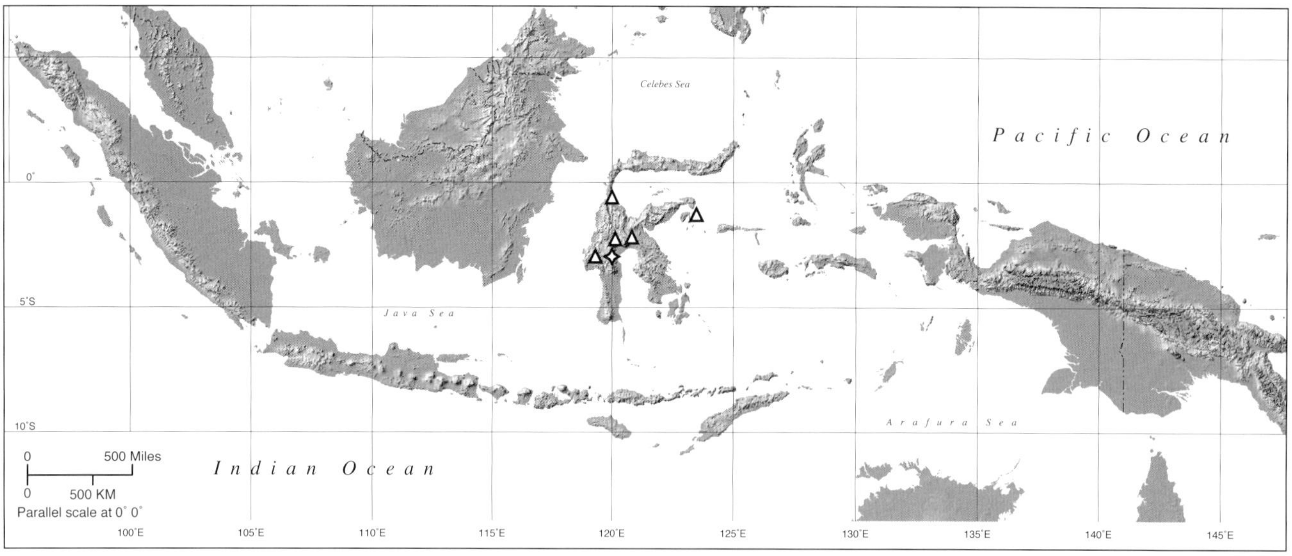

FIGURE 185 Distribution of *Syntypistis vaneeckei*.

Peleng: 1 ♂, Basiano, 700 m, x. 1999.

Etymology: Named for Rudolf van Eecke (b 23. x. 1886 in Buitenzorg/Java – 24. xii. 1975 in Leiden). He contributed much to the knowledge of Indo-Australian Lepidopoptera and published in 1930 at Brill an illustrated book "De Heterocera van Sumatra".

Diagnosis. Forewing lengths ♂♂, 21–22 mm; ♀ 24 mm. The new species is closely related to *S. pallidifascia*. Antennae in both sexes bipectinated with long rami; last 20% toward the tip are filiform. Forewing color deep bluish green. Sometimes darker specimens occur, where many of the green scales replaced by black (fig. 213d). Postbasal fascia less visible, but postmedial fascia mostly well developed as a black double line. Submarginal fascia black, often indistinct. Medial area is fuscous blackish brown, less green tinged. It contains a brown discal patch, which is surrounded in light brown. The forewing pattern resembles *pallidifascia*, but a black basal streak and a large blackish patch in the submarginal area separates it from the latter. Hindwings are light brown, but along the costa the forewing pattern continues as in most *Syntypistis*. Sexual dimorphism is limited.

Male genitalia resembles *S. pallidifascia* by their uncus shape, with "head" and setae, and curved unified socii. Valvae are strongly sclerotized, elongated and curved. Each valva arm has in its middle part a knob-shaped projection with dense setae. In *palidifascia* only one valva arm has such an evagination. Valva apex bear a long and slender process, which is shorter than in *pallidifascia*. Juxta is strongly sclerotized and narrower than in *pallidifascia*. Phallus straight, shorter and broader than in *pallidifascia*. Phallus tip rounded with small serrations. 8th sternite rather straigh at posterior margin. 8th tergite notched, but less deeply in comparison to *pallidifascia*. Female genitalia are not dissected yet.

Bionomics. *S. vaneeckei* is common in Sulawesi. It flies sympatrically with *pallidifascia*, but *S. vaneeckei* prefer seemingly higher altitudes. Adults has been observed between 900 and 2200 m above sea level.

Distribution. Known from Sulawesi and Peleng isl.

214. ***Syntypistis nigribasalis*** (Wileman, 1910): 289 (*Stauropus nigribasalis*).
Holotype: ♂, [Taiwan], Formosa, Rantaizan, 7500 ft. – NHMUK, London.
Schintlmeister 2013: 441.

Taxonomic note. The nominatform is restricted to Taiwan. All other populations outside Taiwan belong to *Syntypistis nigribasalis tropica* (Kiriakoff, 1974): 388; pl. 2: 4; holotype: ♂, Sumatra, Dolok Merangir – ZSM, Munich, which have the phallus tip strongly curved and the central evagination of the 8th sternite squarish. In the nominotypical subspecies 8th sternite is trapezoid.

Diagnosis. Grey to whitish forewings with contrasting brown and black basal area pattern. In the medial area three white stigmas with grey nuclei are visible. Postbasal and postmedial fasciae are black and white. A fine black submarginal fascia occurs. Forewing veins are tinged

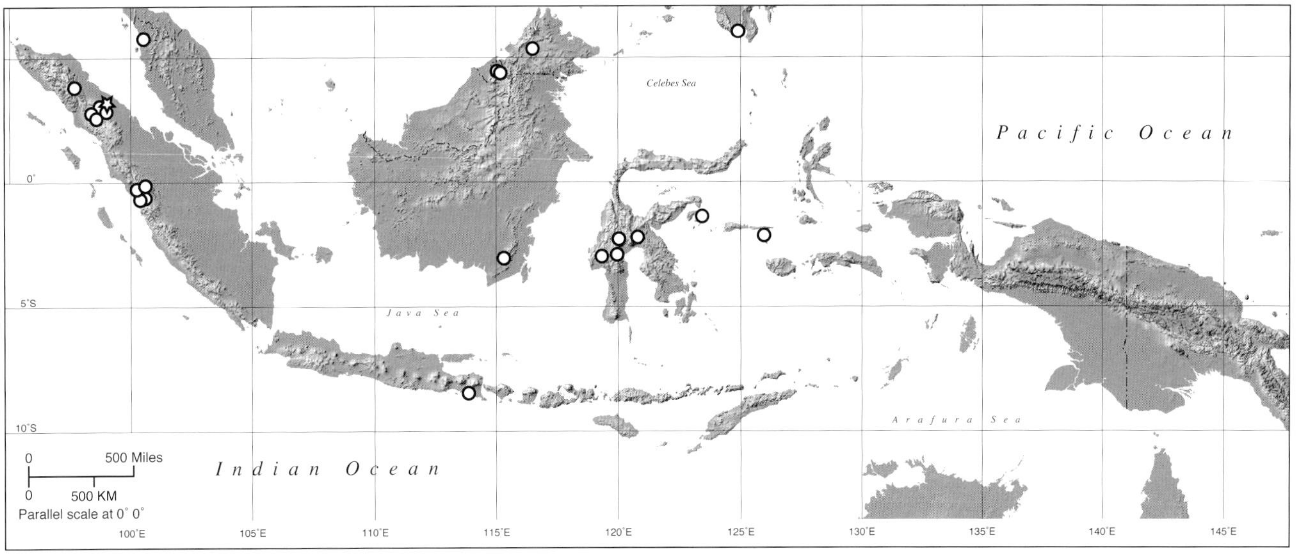

FIGURE 186 Distribution of *Syntypistis nigribasalis tropica.*

black. Sexual dimorphism is limited.

Male genitalia are characterized by long and compact uncus with squarish tip. Socii are reduced in size. They are present as a pair and not unified as in most other congeners. Valvae rather short and characteristically shaped with broad apex. Tegumen and saccus are large and compact. Phallus very long and rectangular curved distally. 8th sternite displays a squarish and broad central projection at posterior margin. 8th tergite concave. Female genitalia are elongated with narrow 8th abdominal segment. Antrum is strongly sclerotized, large and triangular shaped. Ductus bursae is strongly sclerotized and curved. Corpus bursae is ellipsoid and displays four oscillating bands. No signum.

Bionomics. *Syntypistis nigribasalis* prefer montainous habitats above 900 m and occur in Indonesia infrequently up to 1700 m. The holotype of *tropica*, however was collected at 150 m above sea level; the only specimen, which is known from Sumatra at such a low altitude. Two males from Java were taken at 300 m.

Distribution. *Syntypistis nigribasalis nigribasalis* is endemic to Taiwan. Ssp. *tropica* is distributed in S China, Indochina, the Malayan Peninsula, Sumatra, Borneo, Java, the Philippines, Sulawesi, Peleng isl. and Sanana isl. (Sula).

215. ***Syntypistis alleni*** (Holloway, 1983): 51; pl. 4: 17, 21; fig. 50 (*Quadricalcarifera alleni*).

Holotype: ♂, Brunei, Bukit Retak, 1.618 m – NHMUK, London.

Schintlmeister 2013: 435.

Taxonomic note. Comparison of *alleni* from Borneo, including the type series, with material from China, made clear, that *alleni* is more widely distributed than expected.

Specimens from south eastern China were described as *Syntypistis aspera* Kobayashi & Kishida, 2004 in Kobayashi, Kishida & Wang, 2004: 203; figs. 3, 8, 8a, 8b; holotype: ♂, China, Guangdong, Shaoguang, Nanling, 1100m – South China Agricultural University, Guangzhou. Specimens from China are not different in wing pattern and male genitalia from Bornean material, and both taxa are brought, therefore into synonymy (**syn. nov.**). The name *alleni,* as earlier described, takes priority.

Diagnosis. *Syntypistis alleni* resembles at first glance to *S. nigribasalis,* but forewing color is more blackish-brown tinged. The black basal area extends distally along the costa. There is a long and contrasting black patch at the tornus of the forewing. Veins are not tinged black, which is a good character to distinguish both species. Submarginal fascia is indistinctly developed. Hindwings are fuscous brown; the female have the hindwings slightly darker than the male. Sexual dimorphism is limited.

Male genitalia have broad, conically shaped uncus, which is shorter than in *nigribasalis* and notched at the tip. Socii are reduced in size. They are present as a pair and not unified. Valvae curved, with an acute apex. Saccus is shorter than in *nigribasalis.* Phallus short and broad. It bears distally a field, armed with many micro-thornes. Posterior margin of 8th sternite rather straight with a small central elevation. Female genitalia have not been examined yet.

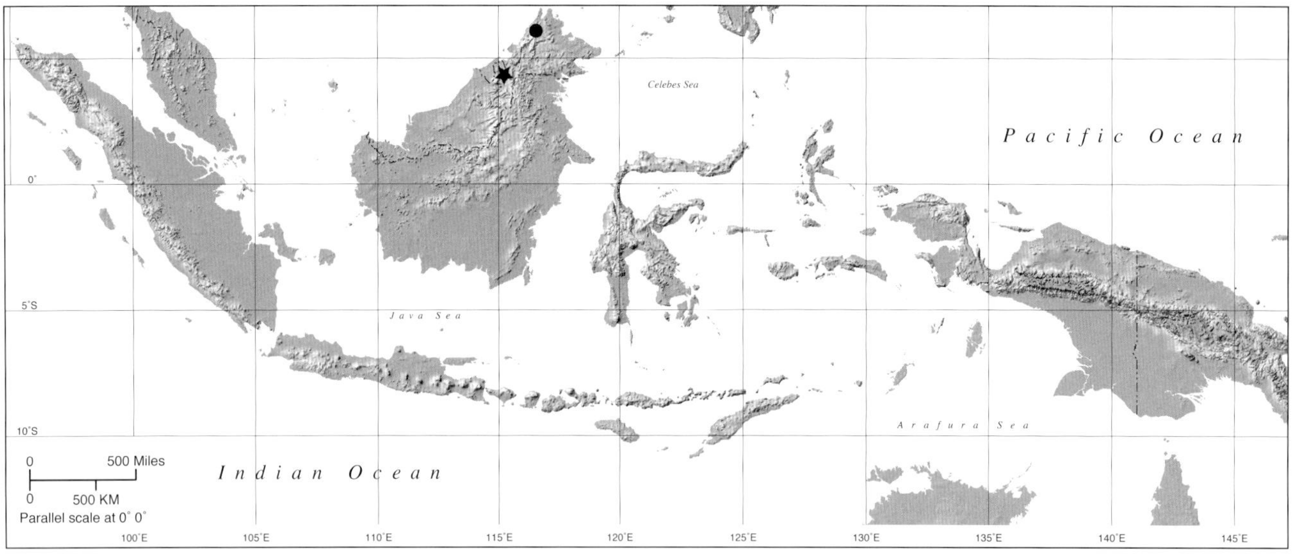

FIGURE 187 Distribution of *Syntypistis alleni.*

Bionomics. A rarer species, which is known in Borneo from six specimens only, which were taken in montainous habitats between 1600–1900 m. In China adults were observed between 1100–1600 m.

Distribution. Known from Borneo and – disjunct – southeast China (Fujian, Guangxi, Guangdong).

Parasinga Kiriakoff, 1967: 51.
Type species: *Somera lichenina* Butler, 1880.
Schintlmeister 2013: 313.

Diagnosis. Medium sized adults. Antennae bipectinate in both sexes except for the last 5% toward the tip. Imagines much resemble by external appearance to *Syntypistis*, but forewing shape is broader, in particular at the apex. Forewings are bluish-green or grey with variable fuscous brown pattern. Hindwings are chocolate to lights brown and forewing pattern continues in the costal area, as in *Syntypistis*. Sexual dimorphism is limited.

Male genitalia have a long, and slender uncus and gnathos-like unified socii (comparable as in *Syntypistis*) of same length. The juxta is caudally notched, and the tips of both lobes bear many and densely distributed setae. Further characteristically features for *Parasinga* are a very long saccus and a strongly sclerotized tip of the curved valva. Phallus long, straight and slender, as in *Stauropus*. 8th sternite characteristically trapezoid shaped and centrally notched. 8th tergite less modified with rounded posterior margin. Female genitalia have a large 8th abdominal segment in relation to the rather small and ellipsoid or ovoid corpus bursae. Both pairs of apophyses are long and slender. Antrum and ductus bursae are short and weakly sclerotized. No signum.

Parasinga contains five species which are distributed in India, Myanmar, Indochina, Sundaland, Java, Bali, Sulawesi and the Philippines.

216. *Parasinga lichenina* (Butler, 1880): 67 (*Somera lichenina*).
Holotype: ♂, [Malaysia], Borneo – NHMUK, London.
Schintlmeister 2013: 313.

Taxonomic note. The ssp. *penatus* Schintlmeister & Lourens, 2010 described from Palawan, is distinguished from ssp. *lichenina* by narrower shape of valva apex, more compact saccus and more deeply notched 8th sternite.

Diagnosis. A variable species. Forewing color vary from light brown to bluish green. Postbasal and postmedial fasciae occuring occasionally and are marked indistinctly blackish. Submarginal fascia is developed as a sharp and fine black line. Forewings may be uniform ore have a fuscous brown filled medial area. Hindwings are fuscous or light brown. Sexual dimorphism is limited, but females display a brown discal streak on forewing, surrounded in a whitish color. All individual forms occurring in males are also seen in females. Females of *lichenina* are in collections sometimes confused with females of *Syntypistis viridifusca*, but they are separable by their broader forewing shape from *Syntypistis*.

Male genitalia have a slim and pointed uncus. The unified socii are equal in length to the uncus. But they are

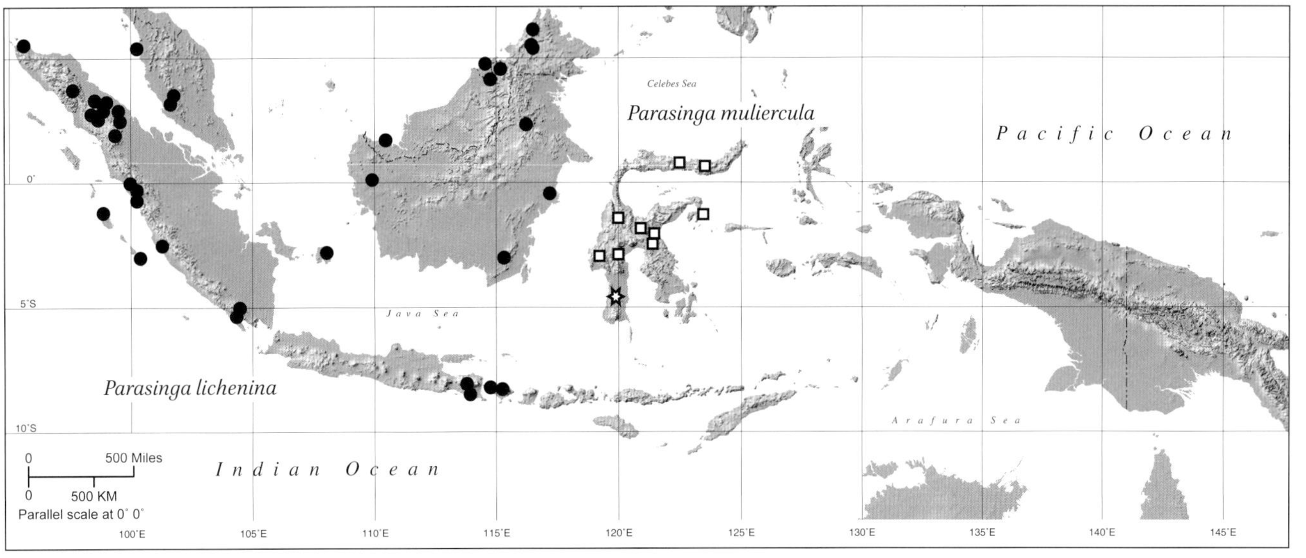

FIGURE 188 Distribution of *Parasinga lichenina lichenina* and *muliercula*.

broader and notched at the tip. Valvae display the most strongly sclerotized apex of *Parasinga* and costa have also more strongly sclerotized evagination. Phallus long and slightly curved in the middle part; all other congeners have the middle part of the phallus straight. 8th sternite is broad, triangularly notched. Female genitalia are long and narrow. Postvaginal plate bears a pair of strongly sclerotized long lobes, which is a diagnostic feature. Ductus bursae reduced in length and ellipsoid corpus bursae is long and narrow.

Bionomics. *Parasinga lichenina* is a common species, occuring from 100 m up to 2100 m above sea level.

Distribution. *S. lichenina lichenina* is distributed in southern Thailand, southern Myanmar, Cambodia, Malaya, Borneo, Sumatra, Belitung, Mentawei and Siberut islands, Java and Bali. In Palawan the ssp. *penatus* Schintlmeister & Lourens, 2010 occur.

217. ***Parasinga muliercula*** **spec. nov.** (figs. 217a–c)

Holotype: ♂, Indonesia, Sulawesi Selatan, 40 km NNE Ujung Padang, Bantimurung, 200 m, 4°56'S, 119°39'E, Talkessel, Primärwald auf Kalk, 2.–9. v. 1984, leg. A. Schintlmeister – in col. A. Schintlmeister, Dresden.

Paratypes (26 ♂♂, 16 ♀♀):

Sulawesi: 1 ♂, 1 ♀, 40 km NNE Ujung Padang, Bantimurung, 200 m, 4°56'S, 119°39'E, 2.–9. v. 1984, leg. A. Schintlmeister (GU 17-42); 1 ♂, dto., 7.–8. i. 1986; 1 ♀, Mamasa, 1500–2000 m, 2°56'S, 119°22'E, x. 1995; 1 ♂, 1 ♀, Puncak, Palopo, 2°57' S, 120°03', x. 1998; 1 ♂, 3 ♀♀, ibid, 1100 m, 25. ix. 1995, leg. St. Naumann (GU 43-22); 1 ♀, Namo, Straße Palu-Gimpu, 650 m, 21.–22. ix. 1995, leg. St. Naumann; 1 ♀, Straße Pendolo–Tentea, 2 km N Taripa, Pospangeo Mts., 1000 m, 4. viii. 1994, leg. St. Naumann; 1 ♂, Tambusisi, x. 1997; 1 ♂, Pangean [near Maros], 600 m, iii. 1938; 1 ♂, 2 ♀♀, Pangean (near Maros), 600 m, iii. 1938; 1 ♀, Soroako, 2°35'30"S, 121°22'30", 9. x. 1993, leg. M. J. & J. P. Duffels; 9 ♂♂, N. Sulawesi, Dumoga-Bone Natl. Park, 11. – 15. x. 1985, 220 m, leg. J. B. Heppner; 4 ♂♂, dto., 16. – 20. x. 1985; 1 ♂, dto. 6. – 10. x. 1985 (FLMNH-MGCL specimen 289680 – 28694); 6 ♂♂, Maros region, Camba area, 7-800 m, vi. 2017, leg. Benny de Groof;

Peleng: 1 ♂, 5 ♀♀, Basiano, 700 m, x. 1999 (GU 66-11).

Diagnosis. Forewing length ♂♂ 22 -23 mm; ♀ 27 mm. The new species belongs together with *P. lichenina* and *P. viridescens* Schintlmeister & Lourens, 2010 described from the Philippines, to a group of closely related species. Externally not separable from *lichenina*, but the width variation of the forewing markings is much more limited. Brownish fuscous forms such as illustrated in figs 216f, 216i for *lichenina* are not known in *muliercula*. Females display as in *lichenina* also a brown discal streak on forewing, but whitish scales are absent.

Male genitalia resembles those of *viridescens* rather more than *lichenina*. Uncus is much thicker than in *lichenina* and longer than in *viridescens*. The unified socii are of same length as the uncus. Valva of *muliercula* have a triangular digitus process, which is absent in *lichenina*. S. *viridescens* also display the triangular digitus process, but it is situated more toward the valva apex. Juxta notched in its posterior portion, with dense setae on both lobes. An

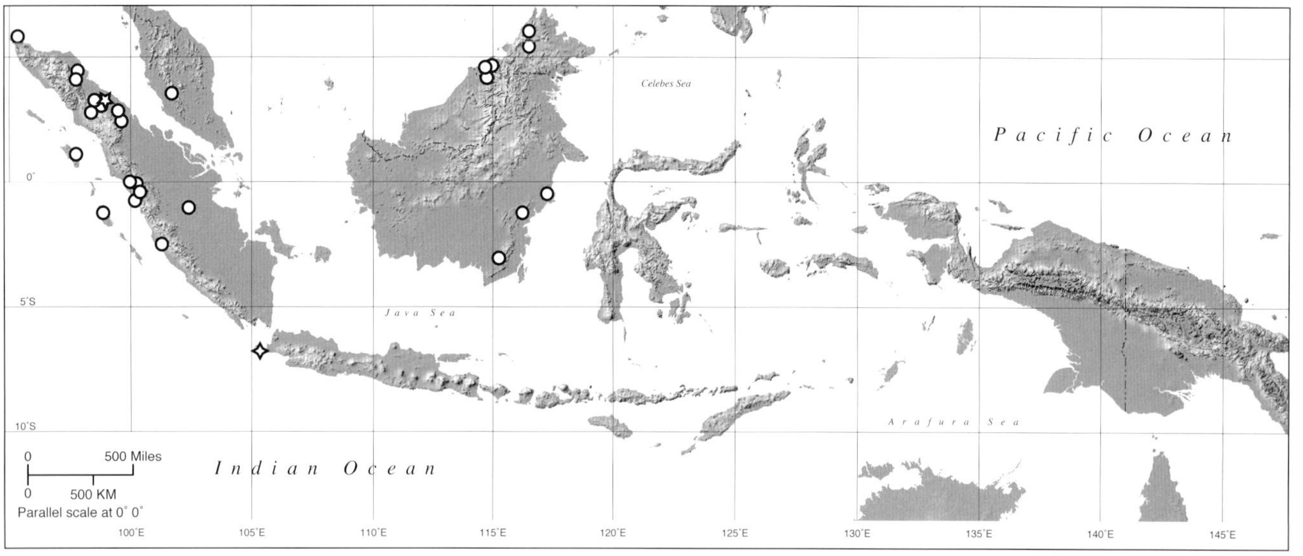

FIGURE 189 Distribution of *Parasinga cinerascens.*

anterior projection as in *lichenina* is absent. Phallus straight and significantly shorter than *viridescens.* 8th sternite resembles in shape *viridescens,* but posterior margin is straight and not notched. Female genitalia resemble *viridescens* by their broad 8th abdominal segment. 8th tergite posteriorally rounded and less modified as usual in *Parasinga.* Female genitalia have both pairs of apophyses very long. Apophyses anteriores are longer than in *lichenina.* Projections of vaginalplate are shorter than in *lichenina.* Ductus bursae short and corpus bursae relatively small. Corpus bursae is broader and shorter than in *lichenina.*

Bionomics. The species occurs infrequently and prefers lower altitudes below 1100 m; one female was taken between 1500–2000 m.

Distribution. Restricted to Sulawesi and Peleng Isl.

218. ***Parasinga cinerascens*** Kiriakoff, 1974: 395; pl. 3: 5; fig. 16.
Holotype: ♂, [Indonesia], Nord-Sumatra, Dolok Merangir, 180 m – Staatliches Museum für Naturkunde Karlsruhe.
Schintlmeister 2013: 313.

Diagnosis. A variable species. Forewings are silver-grey, sometimes dark bluish grey. The species is best recognized by a dark brown basal streak toward the forewing dorsum and fuscous brown hindwings, which are in both sexes darker than in any other congener. Other fuscous brown forewing pattern may vary, as illustrated, and also uniform individuals without visible forewing pattern occur. Sexual dimorphism is limited. The various individual forms of the males are also seen in the females.

Male genitalia resemble most other congeners, but uncus is club shaped and broader. Valvae shape is distinguished by smooth, curved and narrower shape at apex. A digitus projection is absent. Phallus long, curved in its distal portion, and pointed at the tip. 8th sternite concave at posterior margin, which separates it from other *Parasinga.* The female genitalia have no processes at vaginal plate. Ductus bursae is broader than in most congeners. Corpus bursae broad and long.

Bionomics. The species occurs frequently, and mostly below 1000 m above sea level. A few specimens have been taken between 1300 to 1700 m.

Distribution. Distributed in the Malayan Peninsula, Borneo, Nias, Siberut and Java.

219. ***Parasinga harmani*** Holloway, 1983: 57; pl. 5; fig. 63.
Holotype: ♂, Brunei, Temburong; Base Camp LP283, 350 m – NHMUK, London.
Schintlmeister 2013: 313.

Diagnosis. The species is unmistakable by the irregular, blue-suffused, blackish patches on its forewings. Forewing color varies from pale reddish, red brown to fuscous chocolate brown. A whitish discal patch occurs occasionally. Sexual dimorphism is limited.

Male genitalia resemble *P. cinerascens,* but apical part of valva is broader and valva shape is rather kinked than

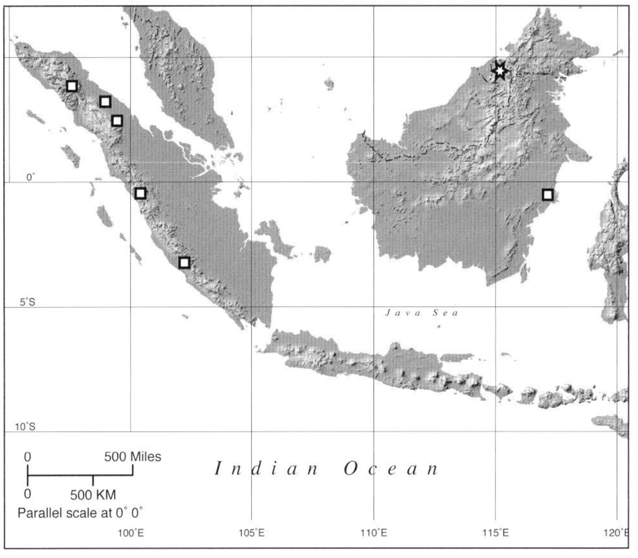

FIGURE 190 Distribution of *Parasinga harmani.*

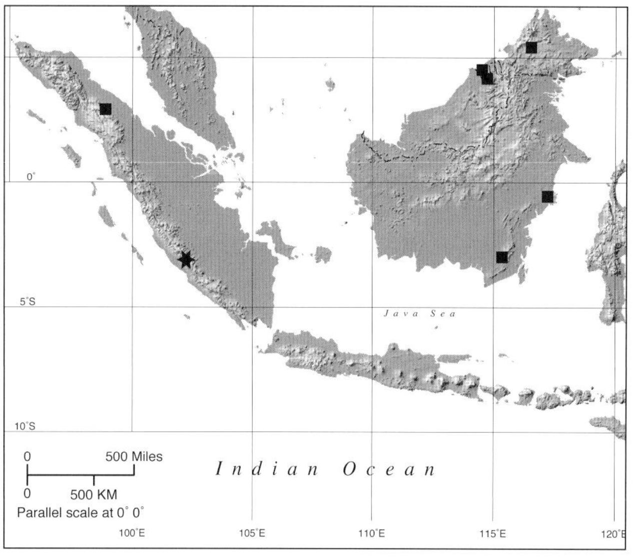

FIGURE 191 Distribution of *Parasinga insufficiens.*

curved. Juxta is characteristically shaped with a small anterior projection. Phallus gently curved distally. 8th sternite concavely shaped at posterior margin. Female genitalia have not yet been dissected.

Bionomics. *Parasinga harmani* is a rare species, eleven specimens are known currently. They were taken in lowland forests up to 700 m above sea level.

Distribution. Restricted to Borneo and Sumatra.

220. ***Parasinga insufficiens*** (Gaede 1930): 629; pl. 83: f. (*Desmeocraera insufficiens*).
Holotype: ♂, [Indonesia], Sumatra, Barisan-Gebirge, 800 m – NHMUK, London.
Schintlmeister 2013: 313.

Diagnosis. Forewing color dull olive green; one male appears bluish green colored. Forewing pattern fuscous brown very indistinct. On the costa there are 3 to 4 brown dots, which are sometimes less visible. Hindwings pale yellowish brown. Female are more brownish and display a fuscous brown medial area of the forewings, but currently only two females are known.

Male genitalia have a straight and pointed uncus. The unified socii are of same length as the uncus but broader. Valva are distinguished by a costal evagination. Juxta small without anterior projection. Phallus straight and slender. It is the shortest within the genus. 8th sternite distinguished by its conical shape and notched posterior margin. Female genitalia have reduced apophyses anteriores, which is unusual in the genus. Apophyses posteriores are long as in other *Parasinga.* Lobes of the vaginal plate are large and bear a small projection. Ductus bursae slenderer and longer than in other congeners. Corpus bursae is small, the smallest of all examined *Parasinga.*

Bionomics. A rare species, n < 20 specimens are known. The moth has been taken between 800–1600 m above sea level.

Distribution. Known from Borneo and Sumatra.

Cascera Walker, 1865: 460.
Type species: *Cascera muscosa* Walker, 1865.
Schintlmeister 2013: 69.

Diagnosis. Medium sized adults. Antennae of males are short bipectinated; the last 20%-segment toward the tip is filiform. Female antennae are filiform. Antennae are in both sexes very long, about 3/4 of the forewing length. Forewings are usually green – in *C. callima* fuscous violet brown, but suffused with a few green scales. Forewing pattern of all species is relatively homogenous: blackish basal patch, discal spot present as a broad fuscous area; medial area brownish. A pale postmedial fascia bears near the dorsum a contrasting white patch. Submarginal fascia whitish. Forewing fringe is checkered. Hindwings are reddish-brown. The type species of *Cascera, C. muscosa,* display in the anal angle of the hindwings a green tinged, whitish and blackish pattern. Other species in the genus

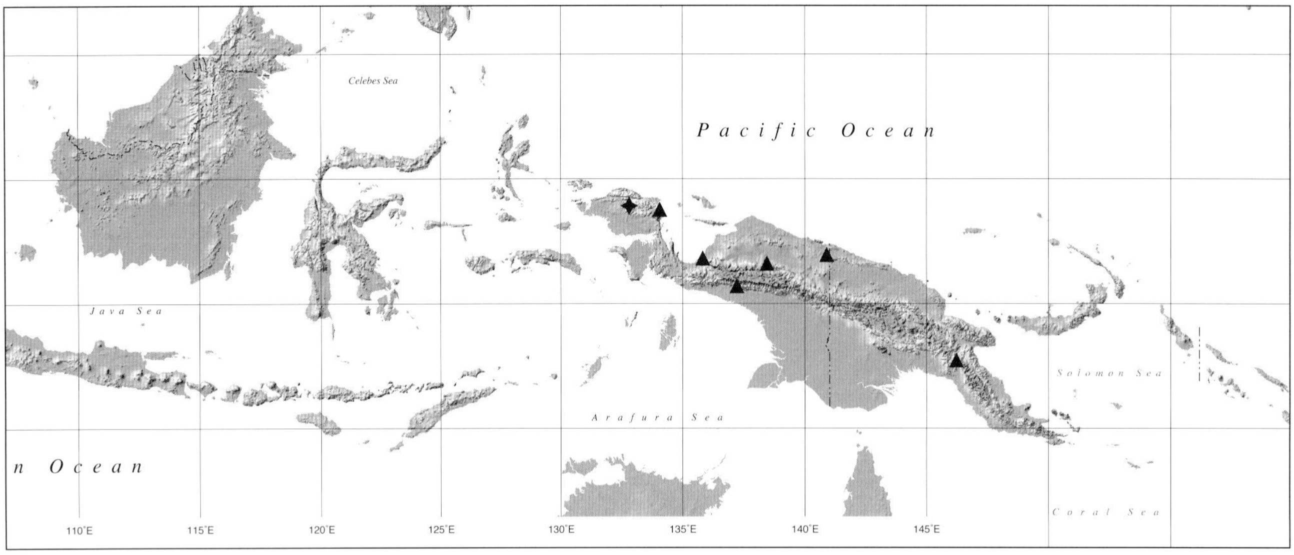

FIGURE 192 Distribution of *Cascera marginata.*

have no anal markings on their hindwings. Sexual dimorphism is limited.

Male genitalia are very similar within the genus and differences between the species are subtle. There is also a remarkable resemblance of the genitalia to those of *Omichlis.* Uncus is bifurcate; a pair of socii is large with several forks. Valvae are elongate with an apical process. A digitus process is present. Along the sacculus several evaginations, asymmetrically on both valvae arms, occuring. At valva base a brush with long and dense setae arises, which often drops out during to the genitalia dissection. Phallus straight and rather short. Everted endophallus is large, with a large and strongly sclerotized carina. 8th abdominal segments are specifically sclerotized. 8th sternite usually deeply notched at the posterior marigin; 8th tergite convexly rounded at posterior margin and occasionally slightly notched. Female genitalia have long and slender apophyses posteriores and short apophyses anteriores. Antrum rather small and inconspicous; ductus bursae strongly sclerotized, long and strongly curved. Corpus bursae disc-shaped with one or two reniform signa.

Cascera comprises currently ten different species, which are mostly montanous. The genus is restricted to New Guinea and Eastern Australia.

221. *Cascera marginata* Rothschild, 1917: 247; pl. 7: 3.

Holotype: ♀, [Indonesia, W Papua], Dutch New Guinea, Central Arfak Mountains, Ninay Valley 3.500 ft. – NHMUK, London.

Schintlmeister 2013: 70.

Diagnosis. Forewings are green in fresh specimens, fading later to yellow-brown. *C. marginata* is readily identifiable by a large black shadow with indistinct borders in the submarginal area of forewing. From basal area to tornus along the dorsum a broad blackish band runs, interrupted in the medial area. Medial fascia marked pale brown. Hindwings are brown, slightly pinkish-tinged at base in fresh specimens, see fig. 221c. Sexual dimorphism is limited.

Male genitalia are not yet dissected due to lack of material. Female genitalia resemble those of other congeners as described above. But the postvaginal plate is broader with longer, upwards-orientated lobes. Apophyses anteriores are slightly longer than in *inconcisa.* Anterior margin of the 8th abdominal segment is sinuous. Ductus bursae is broader and shorter than in other congeners. Corpus bursae bears one reniform signum and a more strongely sclerotized, streak-like structure, which was also found in a second dissected female from Australia.

Bionomics. Rare in collections, n < 20 specimens, and mostly represented by females. The moth flies in New Guinea in a band between 1000–1500 m above sea level, but two females were taken in Australia around 400 m.

Distribution. Distributed in New Guinea in its entirety. Also occurs in Australia, Queensland: 2 ♀♀ from Kuranda SF and Wooroonooran NP are the first records for Australia.

222. *Cascera inconcisa* Swinhoe, 1892: 299; pl. 7: 6.

Holotype: ♂, [Indonesia, W Papua], Mysol [= Misol Isld.] – Oxford University Museum of Natural History.

Schintlmeister 2013: 70.

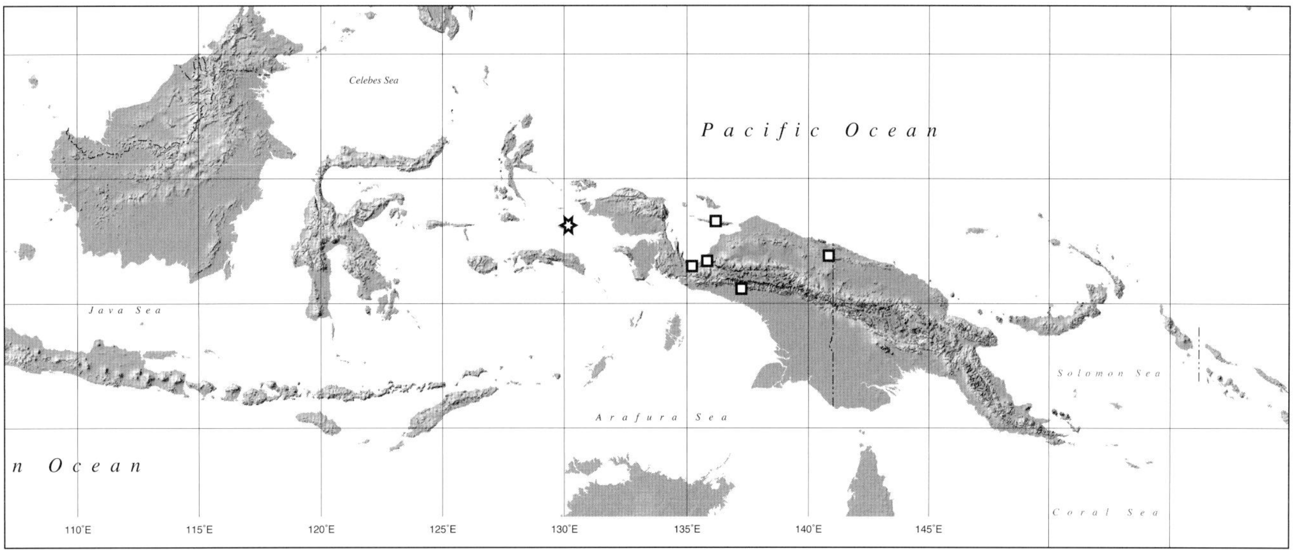

FIGURE 193 Distribution of *Cascera inconcisa*.

Taxonomic note. The holotype of *inconcisa* is a worn and somewhat brown faded male, which is well illustrated in the original description. Swinhoe's illustration matches fig. 222b, including the small white postmedial spot. Forewing pattern matchs also to *Cascera flavovirens* Rothschild, 1917: 247; pl. 7: 2; holotype: ♀, Dutch New Guinea, Snow Mountains, near Oetakwa River 3.500 ft. – NHMUK, London. The holotype, illustrated in fig. 222d, do not differ from *inconcisa* and placed therefore as a subjective, junior synonym of *inconcisa* (**syn. nov.**). Since Misol is an island and the holotype of *inconcisa* is not yet dissected in its genitalia, the synonymy remains somewhat uncertain.

Diagnosis. Forewings pale yellowish-brown, but since only old material (< 80 years) was at the author's disposal, it is assumed that fresh specimens are probably more greenly tinged. Rothschild, 1917: 247 mentioned in the original description "mossy yellowish sea-green forewings". Apart from the pale forewings *inconcisa* is distinguished by the shape of the black postmedial fascia with two sharp teeths pointing toward the apex. Submarginal fascia of the forewings of *inconcisa* marked as a row of blackish-white spots; sometimes weakly developed. The basal blotch is as in *marginata*, but due to the pale ground color of the forewings more contrasting. Hindwings are pale brown.

Male genitalia are not yet dissected. Female genitalia are similar to *marginata*, but the postvaginal plate is narrower. Ductus bursae longer and slenderer. Corpus bursae lacks the sclerotized streak-like structure of *marginata* but bear also a reniform signum.

Bionomics. Only a few specimens (n = 7) are currently know, mostly females. They have been taken from 350–1000 m above sea level.

Distribution. Known from W Papua and Misol Isl.

223. ***Cascera bella*** Bethune-Baker, 1904: 374; pl. 5: 44.
Lectotype: ♂,[Papua New Guinea], Mount Kebea [6.000 ft.] – NHMUK, London, **by present designation.**
Schintlmeister 2013: 69.

Taxonomic notes. *Cascera bella* was described from an unspecified number of ♂♂, and ♀♀ (syntypes) from "Dinawa"; "Mount Kebea"; "Eikeikei" and "Aroa River". In the holdings of NHMUK the author traced a pair of syntypes. To stabilize the nomenclature the ♂ with the following labels is designated hereby as lectotype: "Mt. Kebeae. | B.C. New Guinea | 6000 ft. | March–April 1903 | A.E Pratt Coll. "; green circular label "Type"; "G.T.B. -Baker Coll. | Brit. Mus. 1927–360; yellow label "Photo done by A. Schintlmeister #4096". A lectotype label was added. The lectotype is depicted in fig. 223f. The remaining ♀ from Eikeikei becomes a paralectotype.

Cascera irrorata irrorata Rothschild, 1917: 248; pl. 6: 6, holotype: ♂, Central Dutch New Guinea, Mount Goliath 5.000–7.000 ft. – NHMUK, London, matches in external appereance of its holotype the lectotype of *bella* exactly and is synonymized here (**syn. nov.**). The name *bella* takes priority.

Cascera irrorata pallida Rothschild, 1917: 248, holotype: ♂, Dutch New Guinea, Central Arfak Mountains, Ninay

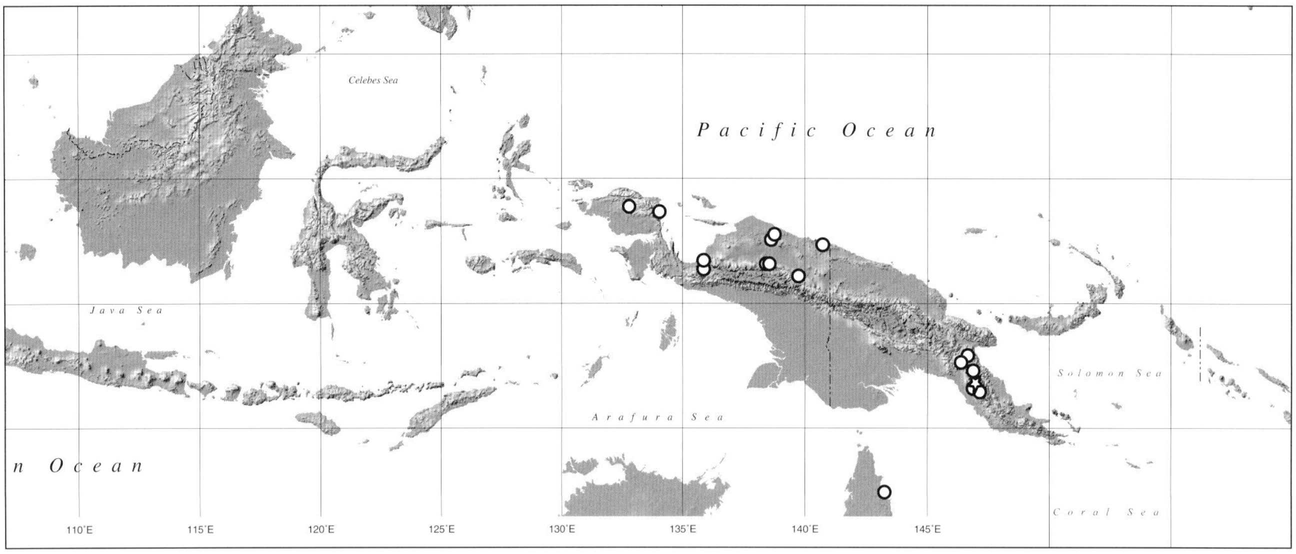

FIGURE 194 Distribution of *Cascera bella*.

Valley 3.500 ft. – NHMUK, London, is an individual form, that has more whitish scales on the forewings and not a geographical subspecies as further material from the Arfak mountains shows. The taxon *pallida* is therefore infrasubspecific and placed as a junior synonym of *bella* (**syn. nov.**).

Cascera variegata was described from 3 ♂♂, and 3 ♀♀ (syntypes). To stabilize the nomenclature the ♂ with the following labels is designate hereby as lectotype: "Ninay Valley | Central Arfak Mountains | Dutch New Guinea | 3500 ft. | Nov'08–Jan'09"; green circular label "Type"; "Type | *Cascera* | *variegata* | Rothsch."; yellow label "Photo done by A. Schintlmeister #4100". The lectotype is figured here in fig. 223g. A lectotype label was added. The remaining 2 ♂♂ and 3 ♀♀ becomes paralectotypes. The lectotype of *Cascera variegata* is conspecific with *bella* and represents another junior synonym of the latter (**syn. nov.**).

It is currently not clear, if the *bella* complex bears more than one species. Differences by external appearance and also in their genitalia such as shape of the valves, everted endophgallus, shape of ostium, ductus bursae, and the second signum, suggest this. But due to transistional forms it is impossible to separate different taxa taxonomically. Australian specimens are externally not different from the lectotype of *bella,* but a dissected male differ greatly in its genitalia by their shorter and slightly thicker uncus forkes, valva shape and a shorter phallus.

Diagnosis. Forewings green, older material fading into brown. Forewing pattern is contrasting. Discal patch large, accompanied by a smaller circular spot. Medial area separated by two fasciae, sometimes reddish-brown or whitish filled. Separable from *marginata* or *concisa* by a whitish contrasting submarginal fascia, and several fuscous blackish-green patches in the submarginal area. Hindwings are brown, in fresh specimens with a slightly pink-tinged base.

Male genitalia are as described under the genus diagnosis. Digitus process rather small. Socius bear three rounded projections. Sacculus of right valva arm displays one exvagination; the left valva arm bears two. In *C. latifasciata* the right valva arm bears two exvaginations of the sacculus. Everted endophallus is curved with a strongly sclerotized carina. 8th abdominal segments are characteristically sclerotized; the notch of the 8th sternite is wider and deeper than in *latifasciata.*

Male genitalia are variable in shape of the socii, digitus process, shape of endophallus, and in the sclerotizations of the 8th abdominal segments. Female genitalia are as in the other congeners, but corpus bursae bears two reniform signata. Ductus bursae longer and slenderer than in the previous species. Anterior margin of 8th abdominal segment rather more straight than sinuous. There is variability in shapes of the postvaginal plate and signum as illustrated.

Bionomics. This is the most common *Cascera.* Most adults were observed in New Guinea from 1000 m up to 2000 m above sea level. One male was collected in the lowlands at 70 m. In Australia adults occuring at lower altitudes below 500 m.

Distribution. Distributed in New Guinea in its entirety and Australia (Qusd).

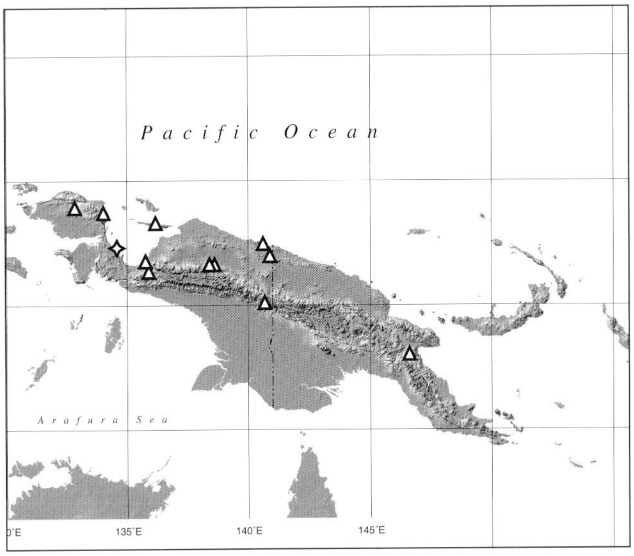

FIGURE 195 Distribution of *Cascera latifasciata.*

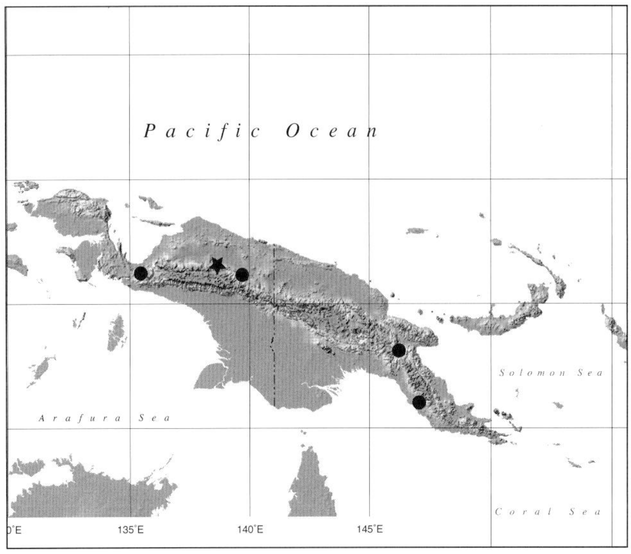

FIGURE 196 Distribution of *Cascera albiscripta.*

224. ***Cascera latifasciata*** Gaede, 1930: 633.
Holotype: ♂, [Indonesia, West Papua], Neu-Guinea, Wandammen-Gebirge – NHMUK, London.
Schintlmeister 2013: 70.

Diagnosis. Similar to *Cascera bella,* distinguished by narrower forewings and by olive green colored forewings. Older specimens of *latifasciata* faded into yellowish brown. Medial area brown, white postmedial spot is small and blackish submarginal shadow absent. These characters are however variable and sometimes overlapping with the individual variability of *bella*.

Male genitalia resemble *bella,* but uncus arms are broader. Digitus process of the valva broader and longer. Sacculus of the both valvae are symmetric and display on each valvae three exvaginations. A third triangular process is lacking in *bella*. Everted endophallus is strongly curved with a sclerotized carina. 8th abdominal segments characteristically sclerotized. A sclerotized notch of 8th sternite is wider than in *bella*. The "lobes" reach the posterior margin. Female genitalia resembles *bella*, but apophyses anteriores are longer. Anterior margin of 8th abdominal segment is sinuous. Corpus bursae disc-shaped. It bears two signata areas, of which the smaller signum differs by its rather circular shape from *bella.*

Bionomics. Moths occur infrequently between 800 m up to 1700 m above sea level.

Distribution. Distributed in New Guinea and Yapen isl.

225. ***Cascera albiscripta*** Rothschild, 1917: 246; pl. 7: 4.
Holotype: ♀, [Indonesia, W Papua], Central Dutch New Guinea, Mount Goliath 5.000-7.000 ft. – NHMUK, London.
Schintlmeister 2013: 69.

Taxonomic notes. *Cascera albiscripta virens* Rothschild, 1917: 247, holotype: ♀, British New Guinea, Angabunga River, affluent of St. Joseph's River – NHMUK, London, was described as a paler subspecies. Differences are within the range of individual variability and also comparison of further material from Papua New Guinea does not differ from W Papua specimens. *Cascera albiscripta virens* is a subjective synonym of *Cascera albiscripta* (**syn. nov.**).

Diagnosis. Forewings is colored olive-green. Older specimens fading into brownish green, but also fresh specimens sometimes appear "bronzy olive green" (from Rothschild's original description of *virens*) as illustrated in fig. 225c. *C. albiscripta* is readily to identifiable by a prominent broad white medialfascia of forewing, which displays a sharp tooth. Discal spot black and accompanied by a larger black patch. A white basal fascia on forewing is well developed. Hindwings are pale brown. Sexual dimorphism is limited, but females have slightly broader shaped forewings.

Male genitalia are distinguished by their uncus shape. Both uncus arms are short and the notch is not deep. Socii are compact, long and strongly curved. They bear no further processes. Valvae are rather short with long digitus process. Phallus is the shortest within *Cascera*. The

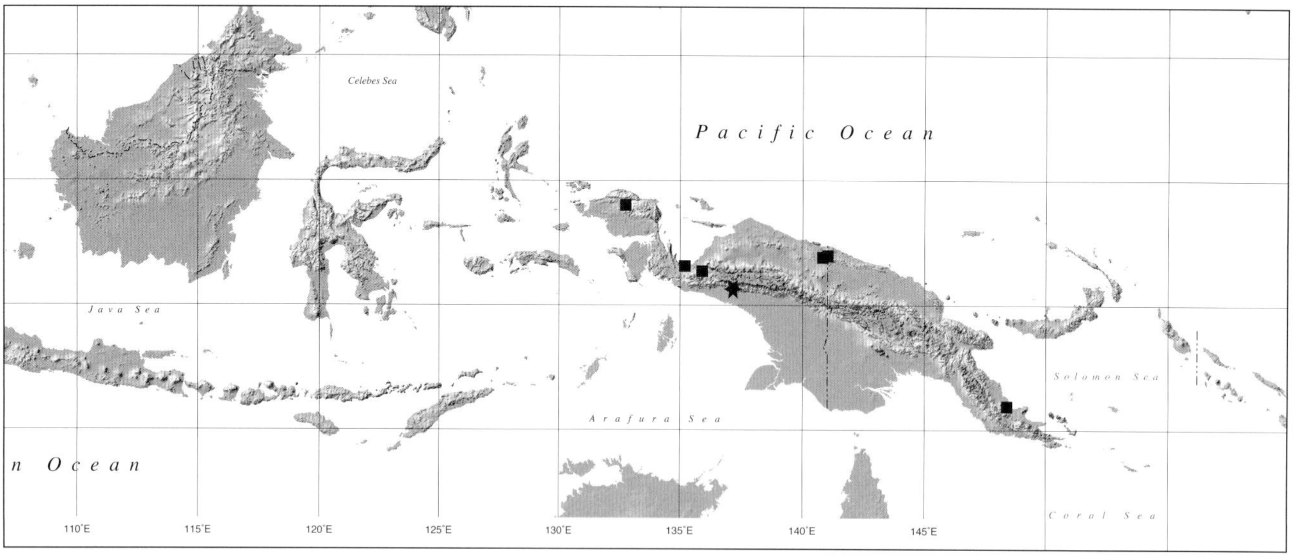

FIGURE 197 Distribution of *Cascera olivacea*.

everted endophallus is almost 300° rounded. 8th sternite concave, but posterior margin of 8th tergit is convex. Sclerotizations of 8th abdominal segment are specifically as illustrated. Female genitalia have not yet been dissected.

Bionomics. The species is rare in collections (n < 20) and has been collected at altitudes between 1700–2000 m.

Distribution. Restricted to New Guinea.

226. ***Cascera olivacea*** Rothschild, 1917: 247; pl. 6: 18.
Lectotype: ♂, [Indonesia, W Papua], Dutch New Guinea, Snow Mountains, near Oetakwa River, 3.500 ft. – NHMUK, London **by present designation.**
Schintlmeister 2013: 70.

Taxonomic notes. *Cascera olivacea* was described of a pair. To stabilize the nomenclature the ♂ with the following labels is designated hereby as lectotype: "nr. Oetakwa R. | Snow Mts., Dutch | N. G., up to 3500 ft., | x. xii. 1910 (Meek)"; red circular label "Type"; white labvel in Rothschild's hand "Type | *Cascera* | *olivacea* | Rothsch."; yellow label "Photo done by A. Schintlmeister #4112". The lectotype is figured in fig. 226g. A lectotype label was added. The remaining ♀ becomes a paralectotype.

Cascera olivacea flavolavata Rothschild, 1917: 247; holotype: ♂, Dutch New Guinea, Central Arfak Mountains, Ninay Valley 3.500 ft. – NHMUK, London, was described as a subspecies. It is yellowish and represent a paler color variant of *olivaceae*, which was found also in the eastern part of W Papua (fig. 226g). A transistional male was collected in Arfak Mts. and differs in its genitalia not from a male of the nominatform originated from same locality, see fig. 226h. The taxon *flavolavata* is treated therefore as an infrasubspecific individual form of *olivacea* and placed here as a subjective synonym (**syn. nov.**) of the latter.

Diagnosis. Forewing color olive-green. Best feature to distinguish *olivacea* from other similar congeners is the broad, indistinctly whitish submarginal forewing fascia. Other forewing pattern is as in *C. bella*. Hindwings are pale brown. *C. olivacea* appear in a color morphe (f. *flavolavata*), having orange-brown forewings and reddish brown hindwings. Sexual dimorphism is limited.

Male genitalia are large, much larger than all other congeners. Fig. 226d is slightly reduced in scale to fit the page width. Uncus bifurcate with slender and parallel.running uncus arms. Socii are large and display several projections of variable shapes. Valvae have a large digitus process of variable shape, as illustrated. Saccular process is bifurcate and a diagnostically feature of *olivacea*. Phallus as in *perscripta* but longer. Everted endophallus curved with a strongly sclerotized carina. 8th sternite notched and somewhat variable in shape. Two dissected female genitalia are different, also in size, but both females are by external appearance almost identical and were collected in the hydrographer Mts. in PNG. Female genitalia, illustrated in fig. 223a, display a pair of long apophyses anteriores. Ductus bursae long and strongly sclerotized. Corpus bursae disc-shaped and bears a large reniform signum, which is the largest within the genus. A second signum appears very small. A further female genitalia (fig. 223c) display

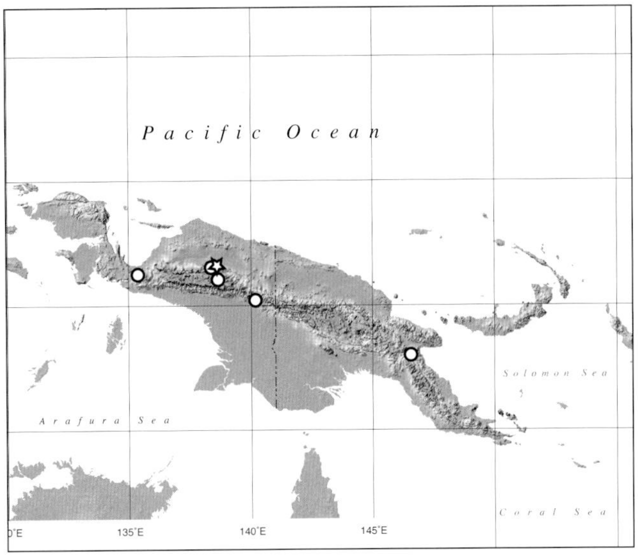

FIGURE 198 Distribution of *Cascera perscripta*.

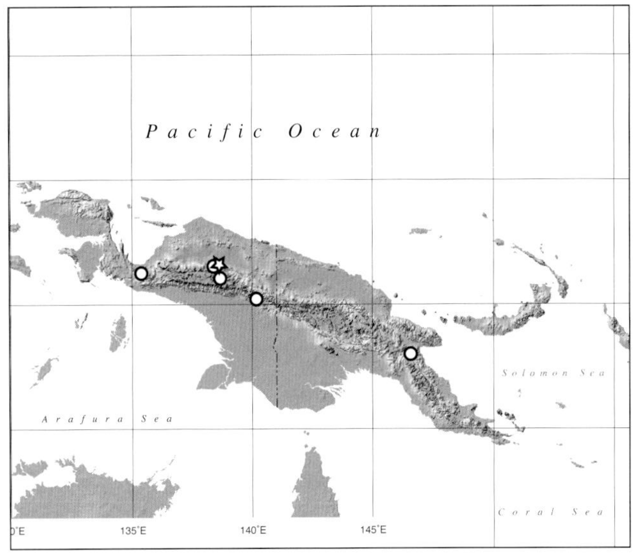

FIGURE 199 Distribution of *Cascera violetta*.

short apophyses anteriores and a smaller signum. The second, small signum of this specimen is partly hidden by the large signum.

Bionomics. The adults occur infrequently and appear between 700–1100 m.

Distribution. Restricted to New Guinea.

227. ***Cascera perscripta*** Rothschild, 1917: 248; pl. 6: 17.
Lectotype: ♂, [Indonesia, W Papua], Central Dutch New Guinea, Mount Goliath 5,000-7,000 ft. **by present designation.**
Schintlmeister 2013: 70.

Taxonomic note. *Cascera perscripta* was described of 2 ♂♂ and 6 ♀♀. To stabilize the nomenclature the ♂ with the following labels is hereby designated as lectotype: "Mt. Goliath, | 5–7000 ft., Centr. | Dutch N. Guinea, | about 139° long.,| Februar 1911 | (A. S. Meek)"; red circular label "Type"; "*Cascera* | *perscripta* | Type Rothsch."; yellow label "Photo done by A. Schintlmeister #4111". A lectotype label was added. The lectotype matches in coloration and pattern fig. 227c exactly. The remaining ♂ and 6 ♀♀ becomes paralectotypes.

Diagnosis. Forewing velvety olive-green. Basal and medial areas are pale green, mixed with many brown scales. There are no white scales or fields, which distinguish *perscripta* from other similar congeners. Discal patch is deep green, accompanied by a larger deep green blotch. Postmedial fascia of forewing is contrasting and characteristically in its shape. Veins in the postmedial and submarginal areas are pale green tinged. Hindwings are pale reddish-brown. Sexual dimorphism is limited.

Uncus is bifurcated. Socii are serrated and display several projections. Digitus process of valva is large. Phallus long, straight and slender; everted endophallus is small. 8th sternite has a gently concave posterior margin and is not notched. Female genitalia display a broad 8th abdominal segment with an almost straight anterior margin. Ductus bursae curved and strongly sclerotized in its entirety. Corpus bursae bears one signum only.

Bionomics. Adults are not rare and flying between 1700–2100 m above sea level.

Distribution. Restricted to New Guinea.

228. ***Cascera violetta*** Gaede, 1930: 633; pl. 84: c.
Holotype: ♂, [Indonesia, West Papua], Neu-Guinea, Mt. Kunupi, 2000 m [Weyland Mts., Menoo Valley] – NHMUK, London.
Schintlmeister 2013: 70.

Diagnosis. The holotype is rather unmistakable by the contrasting black discal spot on greenish-yellow forewing. Most other specimens, however, displaying indistinct markings and the forewing color varies from dull green to yellowish-orange. Hindwings are pale reddish-brown. Female is unknown.

Genitalia have not been examined by the author, but Kiriakoff, 1967: 209 described them as follows: "Differs

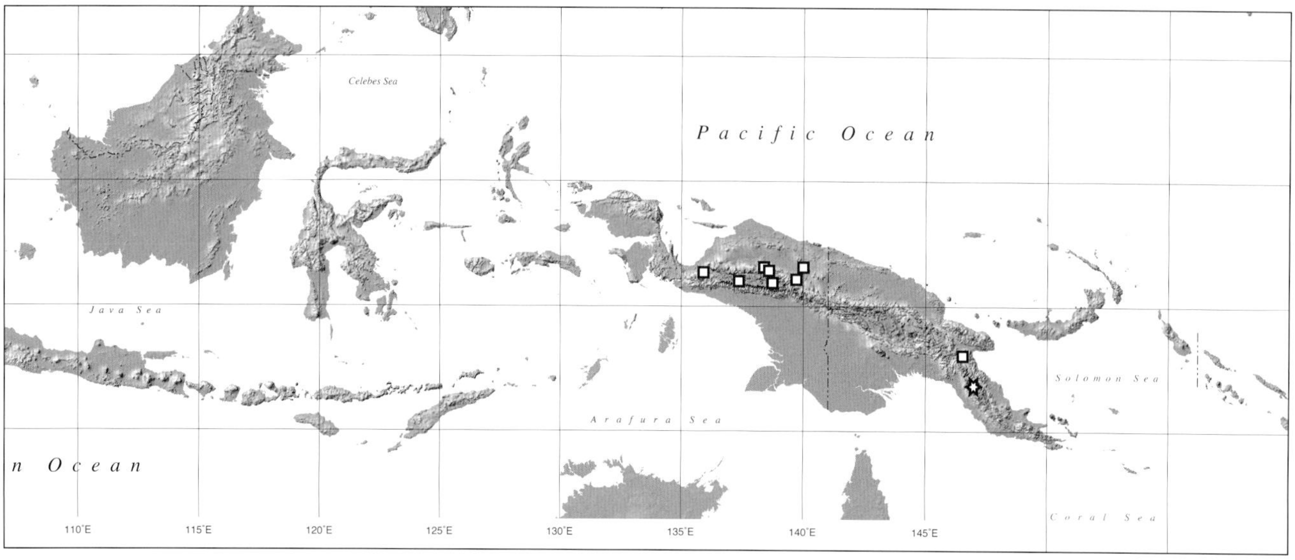

FIGURE 200 Distribution of *Cascera callima.*

from all the other species by the narrow uncus and the valva broadening at the tip."

Bionomics. Only a few males (n < 10) are currently known, all taken above 2000 m up to 2250 m above sea level. The late Henk van Mastrigt sent the author several images of specimens for identification, but unfortunately without geographical data.

Distribution. Known only from New Guinea.

229. *Cascera callima* Bethune-Baker, 1916: 384.
Holotype: ♀, [Papua New Guinea], British New Guinea, Mount Kebea, 6.000 ft. – *NHMUK, London.
Schintlmeister 2013: 69.

Taxonomic notes. *Cascera archboldiana* Kiriakoff, 1967: 206; fig. 13; holotype: ♂, New Guinea, 9 km E Lake Habbema [Baliem Valley], Moss Forest Camp, 2.800 m – NBC, Leide, was separated in the original description from *callima* by "the presence of a white discal spot". Curiously, the holotype of *callima* displays also such a spot, but most other specimens not. Apart from this feature the author was unable to find further characters to separate the white spotted form from uniform fuscous specimens and also their male genitalia are not different. *C. archboldiana* sinks therefore as a junior synonym of *callima.*

Diagnosis. Forewings dark violet brown, mixed with yellowish-green scales. Basal blotch, discal spot, shadows in the submarginal area of the forewings, and the submarginal fascia are black marked. Submarginal fascia accompanied by whitish or paler yellowish-green dots. The species displays often a white medial area of variable size, which is helpfull for quick identification. Hindwings pale reddish-brown. Sexual dimorphism is limited.

Male genitalia are best identified by unique shape of the large socius which bear several projections and a bulbous shape of valva costa. Digitus projection is large with pointed or rounded tip. Valva sacculus bears a small exvagination and is symmetrically on both avalvae arms. Phallus large and slightly curved. Everted endophallus is strongly curved and displays a large strongly sclerotized cornutus. 8th sternite has a squarish notch in the center of posterior margin. Both abdominal segments are specifically sclerotized as illustrated. Female genitalia have not yet been dissected.

Bionomics. *C. callima* is montainous and occurs commonly at higher altitudes. It has been observed at altitudes between 1800–2800 m. One male was taken at 1000 m abvove sea level.

Distribution. Endemic to the central mountain systems of W Papua and Papua New Guinea.

Omichlis Hampson, 1895: 279.
Type species: *Omichlis rufotincta* Hampson, 1895.
Schintlmeister 2008: 187; Schintlmeister 2013: 303.

Diagnosis. Medium sized adults. Male antennae bipectinated, except the filiform tip, 10–20% of the antennae length. Female antennae are filiform. Forewings are

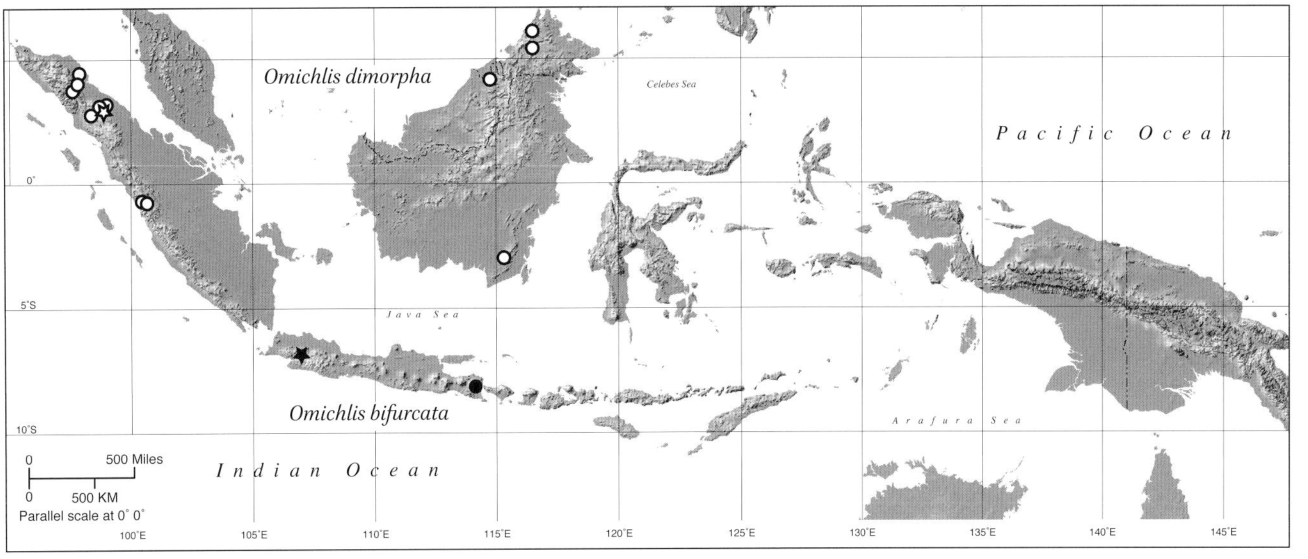

FIGURE 201 Distribution of *Omichlis dimorpha* and *bifurcata*.

reddish or yellowish brown. The angular distal forewing margin is a characteristic feature of *Omichlis*. A small black discal spot or a pale circular patch surrounded in fuscous brown and a small black basal spot are visible in most species. Many species are polymorphic. Often a black transversal band on the veine M2 from the base to the distal margin angle of the forewings occur. Hindwings are brown with small black anal markings. Sexual dimorphism is limited.

Male genitalia have a remarkable resemblance to *Cascera*. Uncus is long and bears a gnathos. Sometimes the gnathos is deeply bifurcated. Socii are long and slender, but less prominent. Valvae are elongated with an apical process. Valva costa is often swollen and also sacculus is enlarged in many species. A digitus process is prominently developed. Saccus of tegumen reduced in size. Phallus is small and bears taxonomically important spines, processes or serrated projections at the tip. The everted endophallus displays a more strongly sclerotized band. 8th abdominal segments are less modified. 8th sternites are mostly concave at the posterior margin. Female genitalia have short apophyses posteriores and – contrary to *Cascera* – enlarged apophyses anteriores. Ostium bursae and lamella postvaginalis are strongly sclerotized and specifically modified in shape. Ductus bursae is strongly sclerotized and relatively short. Corpus bursae is disc-shaped or ovoid. It bears one or two signa, which resembles in shape to *Cascera*.

The genus is currently comprised of 38 species. Most of them occuring in New Guinea. But *Omichlis* is in a few species widely distributed in the Himalaya, S China, Indochina, Sundaland, the Philippines in its entirety, Java, Sulawesi, Moluccas, New Guinea and Australia.

230. ***Omichlis dimorpha*** Kiriakoff, 1974: 399; pl. 4: 1; fig. 19.
Holotype: ♂, [Indonesia], Nord-Sumatra, 30 km SW [Pematang] Siantar, Holzweg 2, 1.050 m – ZSM, Munich.
Schintlmeister 2013: 303.

Diagnosis. A variable species, which occur in several morphes. The nominatform is as illustrated in fig. 230e, having several black veins on forewing. A pale yellowish brown discal spot on forewing is present, which is in males often indistinctly developed. A black basal spot occur in 80% of the specimens (n > 100). A blackish medial fascia appear occasionally. Specimens without black markings on forewing (fig. 230g) belong to f. *pallida* Kiriakoff, 1974. Female display the discal spot on the forewings more contrasting and sharply marked than the male, but sexual dimorphism is limited. *Omichlis dimorpha* is indistinguishable by external appearance from the Javan *bifurcata* (see below) and the Himalayan *O. rufotincta* Hampson, 1895. The latter displays analogous individual variability as *dimorpha*.

Male genitalia of both species are, however very different and *O. dimorpha* is distinguished by narrow, long, undivided, and tapered valva shape. Digitus process is prominent and worm-like, but vary in thickness as illustrated. Uncus long with variable shape of fan-like to circular tip. Below the tip a pair of large gnathoi arises. Socii are long and slender. and display at base a sickle-shaped lobe. Phallus become broadener toward the tip. The everted endophallus bear a stronggly sclerotized hook. Female

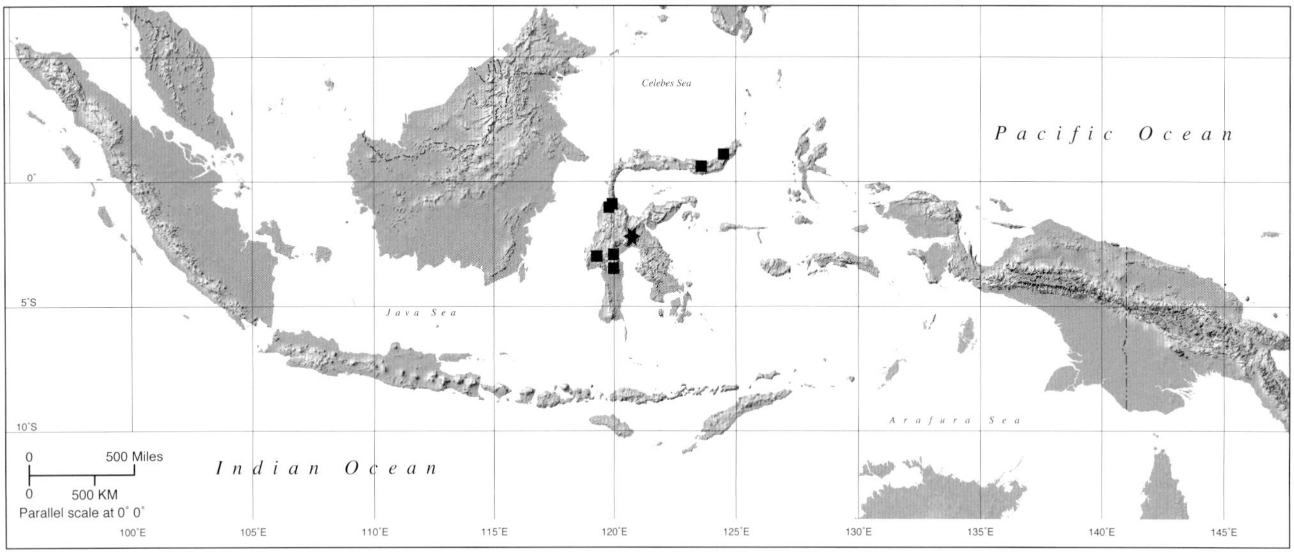

FIGURE 202 Distribution of *Omichlis polymorpha.*

genitalia are inconspicous. Apophyses posteriores shorter than in most other congeners. Shape of the postvaginal plate is diagnostic. Ductus bursae short and broad. Corpus bursae ovoid with a broad and bilobed signum.

Bionomics. *O. dimorpha* occur frequently, but locally. It was observed mainly below 1100m, but a few males has been taken at 1700 m above sea level.

Distribution. Distributed in Sumatra and Borneo.

231. ***Omichlis bifurcata*** **spec. nov.** (figs. 231a–c).
Holotype: ♂, Indonesia, West Java, Mt. Gede, Pangrango National Park, 1250 m, 6°47' S, 107° E, iv.–vii. 1996 (GU 51-44) – in coll. A. Schintlmeister, Dresden.

Paratypes (3 ♂♂):
Java: 1 ♂, Mt. Gede, Pangrango National Park, 1250 m, 6°47' S, 107° E, iv.–vii. 1996; 1 ♂, Mt. Gede, 4500 ft., ix. (US 127526); 1 ♂, Gunong Argopuro, 15 km SW Bondowoso8°04'S, 113°45'E7.-13. x. 2004, leg. St. Jakl.

Diagnosis. *Omichlis bifurcata* is the sister species of *dimorpha.* Forewing lengths ♂♂ 19.5–20.0 mm. Male antennae bipectinated; the last 30% toward the tip filiform. Externally not distinguishable from *O. dimorpha.* Four males at the author's disposal belonging to two different forms with black transverse and without a black transverse band as also seen in *dimorpha.* A black basal spot occurs in two specimens. Female is unknown.

Male genitalia resemble *O. dimorpha*, but differ greatly by the bifurcate valve apex. Uncus anvil-shaped and slightly bilobed. Uncus tip is not fan-shaped as in *dimorpha.* A notched gnathos is present. Socii are robust and curved. There are no sickle-shaped lobes as in *dimorpha.* Digitus process of valva is smaller than in *dimorpha.* Phallus broad at the tip, but slim at the base. Everted endophallus bears no hook, but a fork-like sclerotization is present. 8th sternite slightly bilobed.

Bionomics. The species is known from 4 males only, which were taken between 1250 and 1370 m.

Distribution. The species is endemic to Java.

232. ***Omichlis polymorpha*** **spec. nov.** (figs. 232a–h).
Holotype: ♂, Indonesia, S Sulawesi, Mt. Sampuraga, 1400 m, 2°10'S, 120°45'E, 1250 m, 1.–6. ii. 1995, leg. Sinjaev & Tarasov – in coll. A. Schintlmeister, Dresden.

Paratypes (33 ♂♂, 6 ♀♀):
Sulawesi : 1 ♂, 1 ♀, Mt. Sampuraga, 1400 m, 2°10'S, 120°45'E, 1250 m, 11.–12. ii. 1995; 1 ♂, Puncak, Palopo, 2°55' S, 120°05'E, 1000 m, 25.–31. i. 1995, leg. Sinjaev & Tarasov (GU 51-41); 1 ♂, dto., 13.–14. i. 1995; 3 ♂♂, Puncak, Palopo, 900 m- 1300 m, 3° S, 120°10'E, iii. 1997; **1** ♀, Puncak, Palopo, 900 m- 1300 m, x. 1997 (MWM 15.568); 1 ♂, dto., x. 197 (MWM 8.667); 1 ♂, dto. ii. 1998 (MWM 15.597); 1 ♂, Mamasa, 1500–2000 m, 2°56'S, 119°22'E, x. 1995; 1 ♂, Mamasa, 2000 m, 2°57'S, 119°24'E, 17.–18. x. 1995, leg. Gala; 1 ♀, Kulawi, 1000 m, 1°26'S, 120°00' E, 7.–8. ii. 1995, leg. Sinjaev & Tarasov; 1 ♀, Paloe [=Palu], Sidaonta, 1500 m, vii. 1937, leg. Kalis; 4 ♂♂, Loda, 4000', v. 1937 (J. Pratt); 1 ♂, Dumoga Bone Nat. Park, road Duloduo-Molibagu,

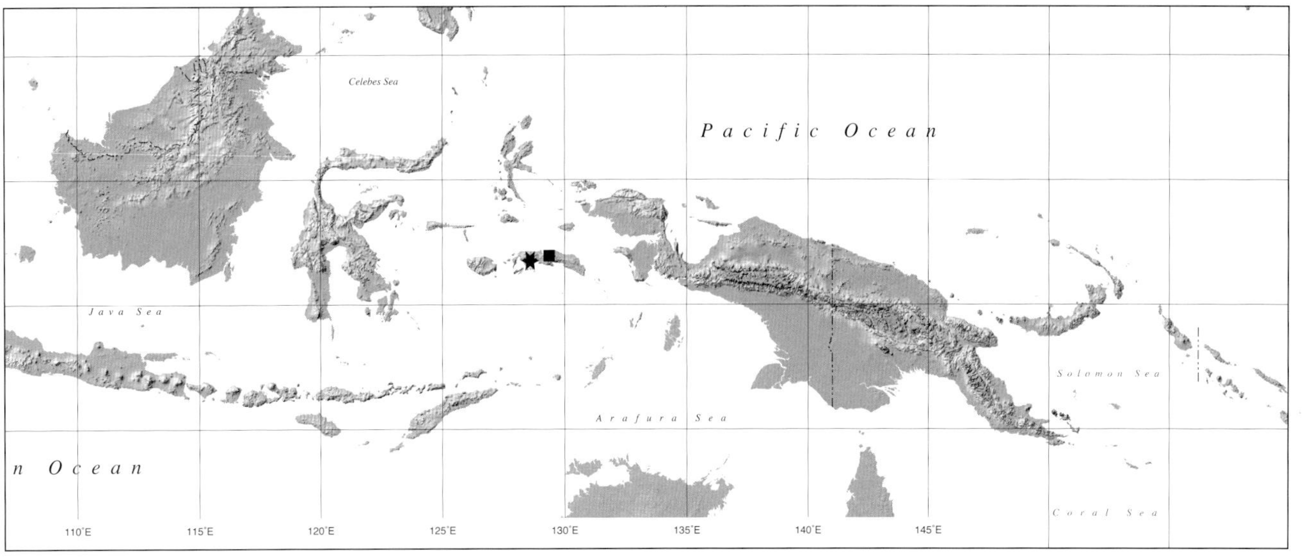

FIGURE 203 Distribution of *Omichlis dubia*.

250 m, 0°44430'N, 123°96781'E, 17. i. 2001, leg. M. Schaarschmidt (MWM 15.596); 1 ♂, Gunung Ambang Reserve, Road Kotamobagu-Motoling, 850 m, 0°86308' N, 124°44708'E, 15.-18. i. 001,leg. M. Schaarschmidt; 1 ♂, Dumoga Bone NP, Hog's back camp, N-side of ridge, 600 m, 10.–11. ii. 1985, J. P. Duffels & J. D. Holloway; 2 ♂♂, Dumoga Bone N.P., Tumpah Transect, 1040 m, 23.–25. ii. 1985, leg. J. D. Holloway; 5 ♂♂, 1 ♀, dto. ii. 1985; 1 ♂, dto. i. 1985; 2 ♂♂, Mt. Tambusisi, 4000', 1°39'S, 121°21'E, 3.–13. iv. 1980; 1 ♀, SW Sulawesi, 30 km NE of Enrekang, Gn. Rantemario, Gowa Camp, 1800 m, 3°24'45"S, 120°00'00"E, 7. xi. 1993, leg. J. P. & M. J. Duffels; 4 ♂♂, Dumoga-Bone Natl. Park, 11. – 15. x. 1985, 220 m, leg. J. B. Heppner; 2 ♂♂, dto., 16. – 20. x. 1985 (FLMNH-MGCL specimen 289682–289687).

Diagnosis. Forewing lengths ♂♂ 20–22 mm, ♀♀ 22–24 mm. Male antennae bipectinate; last 30% toward the tip are filiform. Female antennae are filiform. The species is polymorphic. There are specimens with black-tinged vein M2, and individuals without forewing markings. Both forms are indistinguishable by external appearance from corresponding forms of *O. dimorpha*. Holotype and a paratype female (figs. 232d, 232e) have a prominent contrasting fuscous brown discal spot with fuscous medial line and black basal spot. 3 ♂♂ resembles externally *O. pratti falcigera* from Halmahera (illustrated fig. 236). Other specimens (fig. 232f) having dark brown fore- and hindwings.

Several genitalia dissections (n = 5) ensure the conspecifity of all of these forms to one species. Male genitalia of *polymorpha* have a resemblance to the Philippine *Omichlis diversa* Schintlmeister, 1993 and *O. rufotincta*. Uncus long with a fan shaped tip. It may be slightly notched or rounded at the margin. Gnathoi lobes are longer than in *O. diversa* (fig. 232Aa). Socii as in *dimorpha*, e.g. slender and curved, but thinner than in *diversa*. Valvae have an enlarged saccus, which forms a rectangular angle – a diagnostic feature for *O. polymorpha*. Digitus process large, somewhat variable in shape. Tip of digitus process knob-shaped. Valva apex ends in a long apical process. Juxta is strongly sclerotized and displays a centrally projection, which is absent in *diversa*. Phallus is small and resembles much to *diversa*. In the middle part there is a curved process. Phallus tip bears a smaller process and a serrated plate in *diversa* and *polymorpha*. Sclerotisations of everted endophallus are different in both species as illustrated. 8th abdominal segments are as in *falcigera* with gently bilobed 8th sternite. Female genitalia have relatively small papillae anales and very long apophyses posteriores. Antevaginal plate bears two long projections, which are unique within the genus. Corpus bursae display a large signum.

Bionomics. The species flies infrequently and most specimens have been taken above 900 m up to 2000 m. One male was collected at 250 m.

Distribution. Endemic to Sulawesi.

233. ***Omichlis dubia* spec. nov.** (figs. 233a–e).

Holotype: ♂, Indonesia, Moluccas, W Seram, Kamarian, 0-400 m, 3°20' S, 128°38'E, 16. i.–ii. 1986, leg. A. Schintlmeister – in coll. A. Schintlmeister, Dresden.

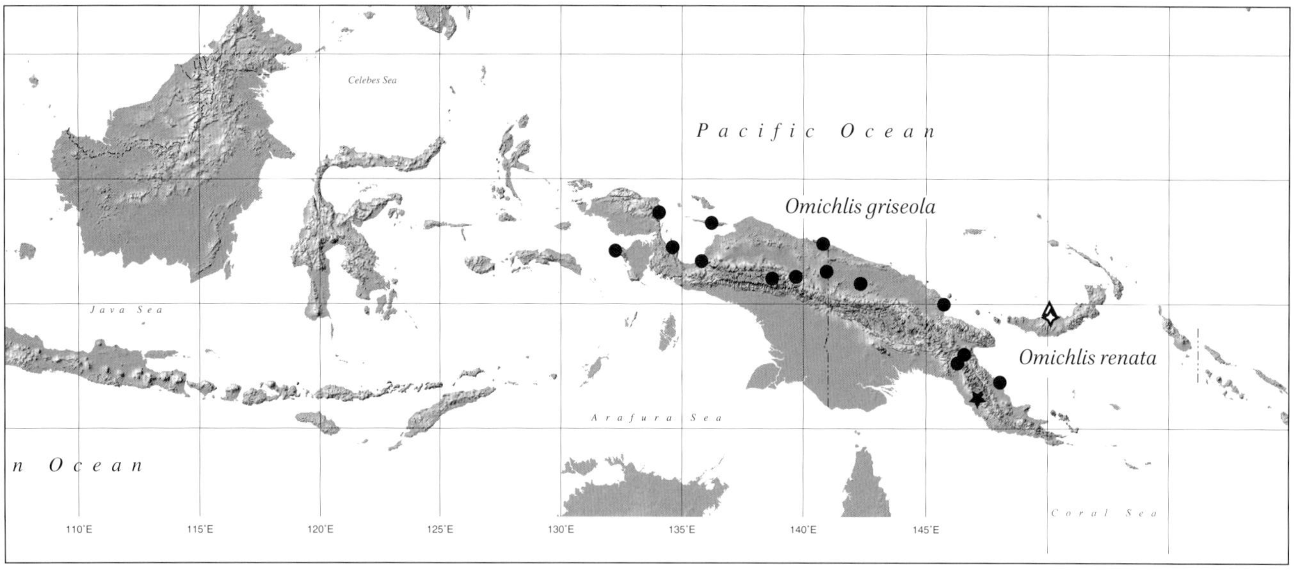

FIGURE 204 Distribution of *Omichlis griseola* and *renata*.

Paratypes (16 ♂♂):
Seram: 3 ♂♂, Kamarian, 0-400 m, 3°20' S, 128°38'E, 16.i.–ii. 1986, leg. A. Schintlmeister (GU 16-48); 9 ♂♂, Gunung Kobipoto, summit ridge, 1470 m, viii.–ix. 1987, leg. J. D. Holloway, D. T. Jones et al. (BM 1810); 1 ♂, dto., 900 m; 1 ♂, Solea, 75 m, leg. J. D. Holloway, D. T. Jones et al.; 3 ♂♂, Gunung Binaia, north slopes, 1000 m, leg. J. D. Holloway, D. T. Jones et al. (BM 1809).

Diagnosis. Forewing length ♂♂ 19–20 mm. Male antennae bipectinate, 30% toward the tip are filiform. Wings are reddish brown. Forewing shape and pattern are as in *griseola*. Discal spot on forewing is small, ellipsoid, and fuscous brown or yellowish colored. There are several slightly fuscous and indistinct fasciae on forewings. On the dorsum there is a fuscous reddish-brown shadow. At the gently angled distal margin (vein M2) of forewing there is a fuscous reddish brown shadow. Female is unknown.

Male genitalia have a long uncus, variable in length and a broad, slightly bilobed tip. A small gnathos process arises directly above the socii. Socii are shovel-shaped and curved. Valvae narrower than in *O. pratti falcigera*. Valva costa rather straight to concave, not convex and swollen as in *O. griseola*. Sacculus is large, larger than in *griseola*. Harpe bear two tapered projections. Valva apex bears a straight process. Phallus short and straight with broad circular ring at the tip. Endophallus displays a serrated band as in *griseola*. 8th sternite have a convex posterior margin.

Bionomics. Males have been taken between 400 and 2000 m above sea level.

Distribution. Endemic to Seram.

234. ***Omichlis griseola*** Bethune-Baker, 1904: 377; pl. 5: 7.
Holotype: ♂, [Papua New Guinea], Ekeikei [1.500 ft.] – NHMUK, London.
Schintlmeister 2013: 303.

Taxonomic notes. The *Omichlis griseola*-complex appears to be taxonomically difficult. There is a great variability in external appearance of adults as well as in the male genitalia. Unfortunately, not much material is available for studies, and thus it is currently not possible to decide whether there is simply a wide variation or several species are mixed in this complex. Provisorially such unique specimens, as illustrated in figs. 234d, 234h, are attributed to *griseola*.

Diagnosis. Inconspicuous species of reddish-brown color with a black dorsal patch in the basal area of forewing. In some specimens this black patch is absent. Discal spot is oval, pale brown, surrounded in fuscous brown. The holotype male displays three fasciae, marked as rows of small black dots. A short black streak at the distal angle of vein M2. Distal margin kinked at vein M2 and not rounded. This characters might be helpful in separating *griseola* from other similar taxa. Sexual dimorphism is limited. Females occure in similar individual forms as males.

Male genitalia have a relatively short uncus, which is shorter than in *dubia*. Uncus tip of the holotype is shaped like a reversed triangle. A pair of gnathal lobes arrises near the uncus tip. A Seram specimen has these lobes larger than Papuan material, see fig. 234b. Socii small and curved.

Valvae have a convex costa and a large digitus process of variable shape. Valva apex bears a longer process of variable shape. Sacculus at the base of the valva is bulbous and rounded. Harpe have near the digitus process a triangular projection. Phallus ends in a serrated tip and the everted endophallus displays a curved and serrated carina. Both structures are also visible in the holotype phallus (fig. 234a), which was not removed from the genitalia apparatus. 8th abdominal segments are less modified as illustrated.

The female genitalia are not yet dissected.

Bionomics. Adults occur infrequently (n < 25) and have been taken between 150 and 1700 m above sea level.

Distribution. Restricted to New Guinea Island.

235. ***Omichlis renata*** Gaede, 1930: 610; pl. 79: g. (*Omichlis rufotincta* form *renata*).
Holotype: ♀, [Papua New Guinea], Neu-Pommern [New Britain, Talesea] – NHMUK, London.
Schintlmeister 2013: 304.

Diagnosis. Similar to *griseola* and *pratti*, but the discal spot is circular and of a contrasting, fuscous-brown. *O. renata* is distinguished by a fuscous dorsal shadow near the base. Postmedial fascia marked as in *griseola* by a row of fine black dots. Submarginal fascia consists of a row of fine, black, U-shaped figures. A single male (fig. 235c) displays many fuscous brown fasciae on forewing. Hindwings are brown with fine, blackish anal markings. Sexual dimorphism is limited.

Male genitalia are separable from *griseola* by their longer uncus. Uncus tip shaped as a reversed sickle. Socii large and curved. Valva are distinguished from other similar species by a serrated process of the harpe, which is asymmetric on both valva arms. Apical valva process is very long. Digitus projection displays a knob-shaped tip. The everted endophallus bears a long and serrated carina, which is shaped like a chainsaw chain. Female genitalia are distinguished by shape of the antevaginal plate. Papillae anales are large as in *pratti* and apophyses anteriores reduced. Ductus bursae is robust and thicker than *pratti*. Corpus bursae is disc-shaped and the signum is heart-shaped.

Bionomics. Known from a few specimens (n = 7) which has been taken at 250 m above sea level.

Distribution. Endemic to New Britain.

236. ***Omichlis pratti*** Bethune-Baker, 1904: 377; pl. 5: 5.
Holotype: ♂, [Papua New Guinea], Ekeikei [1.500 ft.] – NHMUK, London.
Schintlmeister 2013: 304.

Taxonomic notes. *Omichlis parallelistriga* Rothschild, 1915: 80, holotype: ♂, Southern Dutch Guinea, Snow mountains, Utakawa River, 2.500–3.000 ft. – NHMUK, London, represents an individual form of *pratti*, as genitalia dissections of similar specimens, illustrated in fig. 236i, also in their genitalia fig. 236b, shows. They matching other forms of *pratti*. Holotype of *parallelistrigata* have more narrowly shaped forewings, but it displays a blackish shadow at the apical area of the forewings, which is seen only in *pratti*. *O. parallelostriga* is placed here as a junior synonym of *pratti* (**syn. nov.**).

O. pratti has been found to be common also in Australia, Queensland, confirmed by n = 3 genitalia dissections. *Omichlis hadromeres* Turner, 1922: 387, syntypes: 4 ♂♂, Australia, N. Q[ueens]land, Kuranda near Cairns – ANIC, Canberra, could be a junior synonym of *pratti*. But a further 3 dissected specimens from Queensland and Western Australia revealed another two Australian species, similar to *pratti* by external appearance. Thus, taxonomic changes should be undertaken, after the identity of *hadromeres* is fixed.

Differences in the size of the genital apparatus, of specimens from Fergusson Isl., Rossel Isl. and also Australia, are interpretad here as expression of individual variability and not of geographical subspecification.

Populations from Halmahera, Bacan and Obi differ however significantly from New Guinean specimens and introduced here as:

Omichlis pratti falcigera **ssp. nov.** (figs. 236a–d).
Holotype: ♂, Indonesia, Moluccas, Halmahera, Mt. Talagaranu, 15 km SW Baru, 600 m, 1°21'N, 127°32'E, 1250 m, 22.–31. i. 1996, leg. Sinjaev & Tarasov – in coll. A. Schintlmeister, Dresden.

Paratypes (12 ♂♂, 1 ♀):
Halmahera: 6 ♂♂, Mt. Talagaranu, 15 km SW Baru, 600 m, 1°10'N, 127°32'E, 1250 m, 22.–31. i. 1996, leg. Sinjaev & Tarasov (GU 51-42); 1 ♂, 3 km SW Baru, 1°10'N, 127°32'E, 20.i.1996, leg. Sinjaev & Tarasov (GU 51-39); 1 ♀, Road Baru-Basale, Gn. Talagarana, 600 m, 2.–7. iii. 1997, leg. S. Naumann;
Bacan: 3 ♂♂, SE slopes Mt. Sibela, SE Makian, 1200 m, 0°44' N, 115°01' E, 10.–30. viii. 2008, leg. Jakl (GU 94-82); 1 ♂, dto., 2.–12. v. 2008;

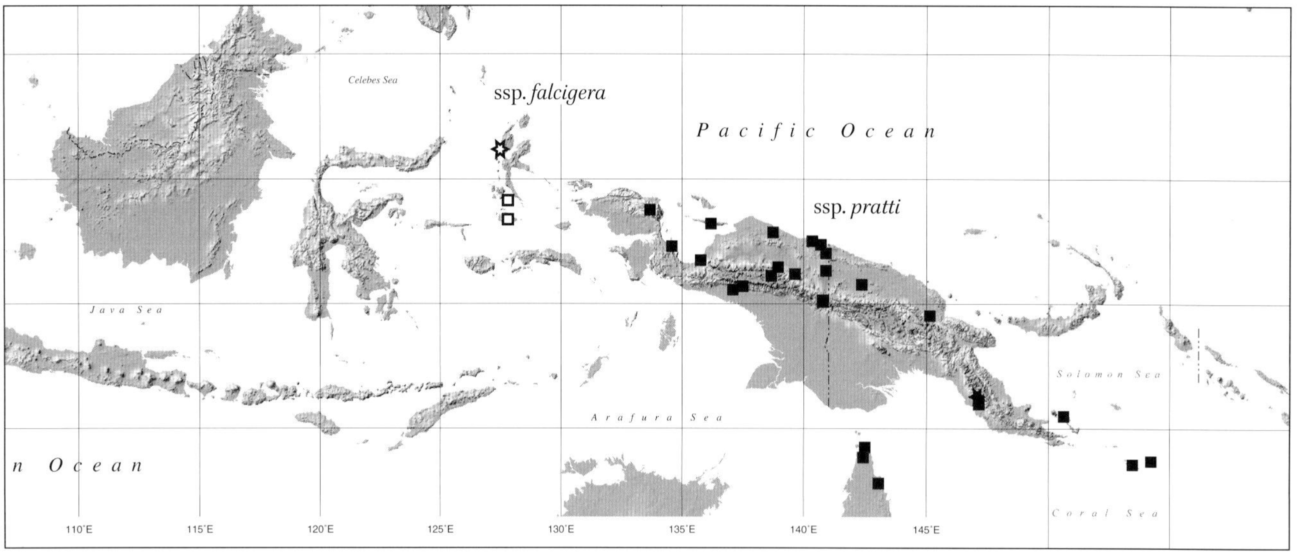

FIGURE 205 Distribution of *Omichlis pratti*.

Obi: 1 ♂, S Coast, Seribu Mts., 22 km N Tapaya village, 1200 m, 1°38'S, 127°48'E, iii. 2009, leg. St. Jakl (GU 92-90).

Diagnosis. Forewing length ♂♂ 20–22 mm, ♀ 24 mm. Forewing color is reddish-brown. All specimens of the type series have a small white discal dot on the forewings, not a streak as in ssp. *pratti*. Fuscous medial fascia weakly developed, but in most specimens visible. Postmedial and submarginal fasciae are marked by rows of fine black dots. Occasionally a black basal patch occurs. Apart from almost uniform forewings a morphe with black transversal line occur, but the line is broader and indistinct in comparison to ssp. *pratti* or to *polymorpha*. The variability of the pattern is more limited than in *pratti*. A blackish shadow in the apical area of the forewings does not occur. The only known female (fig. 236q) is fuscous brown colored.

Male genitalia differ from *pratti* by the shorter apical process and a more strongly curved costa of the valve. Phallus tip is bifurcated, where the upper process is not knob-shaped as in ssp. *pratti*.

Diagnosis. *O. pratti* is similar to *O. griseola*. Wing color varies from yellowish brown to reddish brown. Forewing shape rather more gently rounded than angled at M2 of the distal margin. The species appears very variable. A good character for identification is a blackish shadow in the apical area of the forewing, which appears in most specimens examined. The submarginal fascia is marked as fine black U-shaped figures. Discal patch of ssp. *pratti is* indistinct pale yellowish. Sometimes discal spot appear as a fine, contrasting white streak. A few individuals have a large reddish-brown discal patch, see fig. 236n. Forewings display occasionally a fuscous transverse band on the forewings, a blackish basal patch near the dorsum or a black tornal spot. Fringe at M2 is in almost all specimens blackish marked. Sexual dimorphism is limited; most individual forms of males also occur in females.

The sickle shaped uncus tip is a diagnostical character of male genitalia. Below the uncus tip there is a swollen portion instead of a gnathos. Socii are long, curved and slender, accompanied by a pair of broad lobes. Juxta have a central projection of variable shape. Valvae have a long apical process. Valva costa convexly curved. Digitus process is large, clearly larger than in *griseola*. Sacculus of the valva large and rectangular, distally slightly concave. There are are no further processes or projections as seen in *griseola*. Phallus differs greatly from *griseola* by a pair of strongly sclerotized portions. Everted endophallus bears a stronger carina and a small diverticulum. 8th abdominal segments are as in *griseola*. Female genitalia have long apophyses anteriores and short apophyses posteriores. Lamella postvaginalis squarish. Ductus bursae short. Corpus bursae display a large reniform signum, but in *renata* heart-shaped. A dissected Australian female (fig. 236f) is distinctively smaller, than other dissected females (n = 5).

Bionomics. Adults appear commonyly from the lowlands (50 m) up to 2200 m above sea level.

Distribution. *Omichlis pratti pratti* is distributed throughout New Guinea in its entirety and also on Fergusson Isl., Rossel Isl., Sudest Isl. and NE Australia (Queensland). In Halmahera, Bacan and Obi the ssp. *falcigera* occurs.

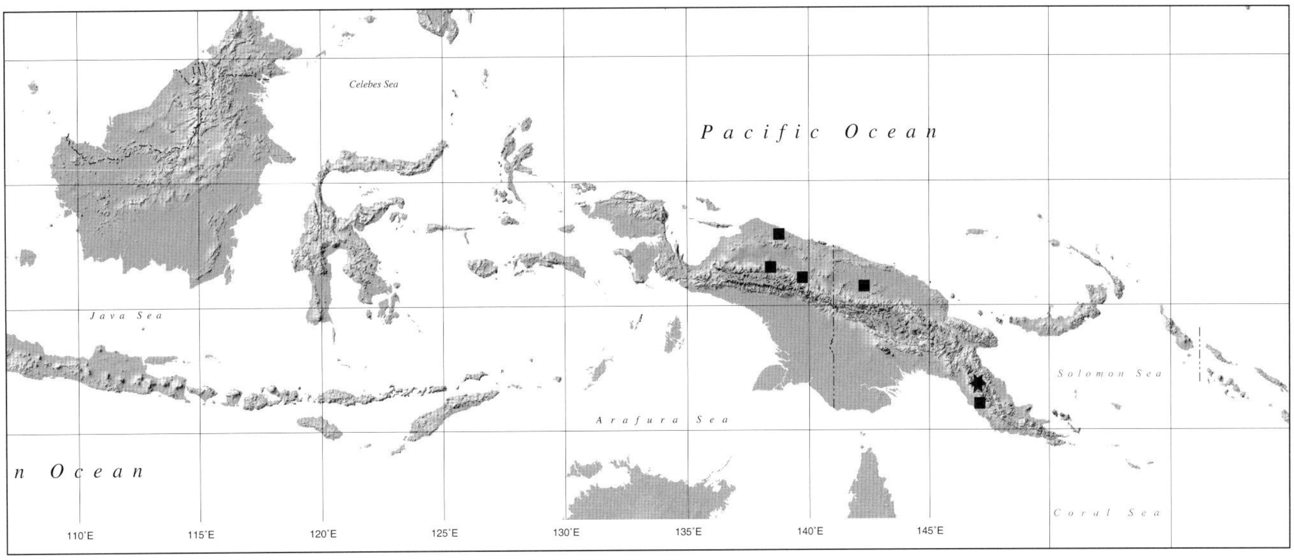

FIGURE 206 Distribution of *Omichlis erythra.*

237. ***Omichlis erythra*** Bethune-Baker, 1916: 383.
Holotype: ♂, [Papua New Guinea], Kebea Range, 6.000 ft. – NHMUK, London.
Schintlmeister 2013: 303.

Diagnosis. Adults are cinnamon brown. Forewing apex is pointed. Distal margin clearly angled at vein M2. Postbasal fascia indistinct whitish; medial fascia marked as a fuscous shadow. Postmedial fascia appear near the forewing costa as a row of white dots. Discal spot is fuscous brown, but in most specimens almost invisible. A black tornal spot occurs occasionally. Hindwings are dark reddish-brown with blackish anal markings. Sexual dimorphism is limited.

Male genitalia have a triangular shaped uncus tip. A pair of compact gnathoi arises just below the tip. Socii are well developed and hammer shaped. Juxta bears a pair of very long lobes, which are unique within the genus. Valva elongated with a long and curved apical process. Harpe lacks a digitus process. Sacculus of valva bears a sharp process and basally a squarish projection. Phallus have two processes at the tip. Everted endophallus has a strongly sclerotized carina. 8th abdominal segments less modified, most similar to *griseola*, but more deeply bilobed at their posterior margins.

Bionomics. Adults are rare in collections. n = 8 specimens are currently known, including an image of the late Henk van Mastrigt. They have been collected between 1200 m and 1800 m above sea level.

Distribution. Endemic to New Guinea Island.

238. ***Omichlis paragriseola*** **spec. nov.** (figs. 238a–c).
Holotype: ♂, Indonesia, W Papua, Nabire 6 km E Samabusa, Lagari, 50 m, 9.–13. xi. 1993, leg. R. Brechlin & K. Cerny (GU 94-84) – in coll. A. Schintlmeister, Dresden.

Paratypes (1 ♂, 1 ♀):
W. Papua: 1 ♀, Nabire 6 km E Samabusa, Lagari, 50 m, 9.–13. xi. 1993, leg. R. Brechlin & K. Cerny, 1 ♂, Nabire 12 km E Samabusa, Lagari, 50 m, 6.–8. xii. 1993, leg. R. Brechlin & K. Cerny (GU 66-21).

Diagnosis. Forewing lengths ♂♂ 21 mm, ♀ 23 mm. External appearance resembles to *O. griseola*, but the new species differs in both sexes by a serrated and rounded forewing margin. Discal spot of forewing is pale yellowish and less prominent. Medial fascia weakly developed as a blackish shadow. Postmedial and submarginal fasciae are marked as rows of black dots. The second paratype female (not illustrated here) show more forewing fasciae. Holotype male has a prominent black basal patch, but a dissected paratype lacks this black patch. Hindwings are brown with mottled fringe and a blackish anal stigma.

Male genitalia have a fan-shaped, slightly notched uncus tip. A pair of arrow-like gnathoi arises directly below the tip. Socii are long and broad. Juxta strongly sclerotized with a rectangular central projection. Valva costa concavely shaped. Apical valva process rather straight and shorter than in *griseola* or *pratti*. *O. paragriseola* is well distinguished from all other congeners by a serrated harpe and a small projection at the tip. Phallus straight, enlarged toward the circular tip. The everted endophallus bears a small, strongly sclerotized carina. 8th abdominal

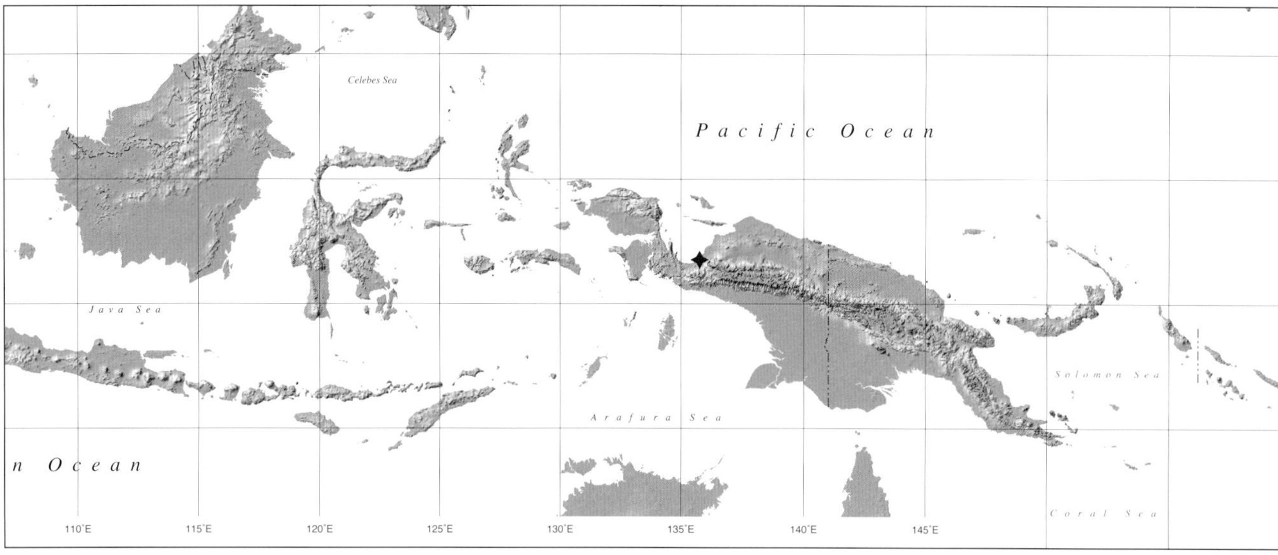

FIGURE 207 Distribution of *Omichlis paragriseola.*

segments less modified most similar to *griseola.* Female genitalia has not been dissected yet.

Bionomics. Only 3 specimens are known so far, all of which were taken in primary lowland forest at 50 m above sea level.

Distribution. Known only from W Papua.

239. ***Omichlis opacitas* spec. nov.** (figs. 239a–c).
Holotype: ♂, Papua New Guinea, [D. N. Guinea], Kaiserin-Augusta-Fluß-Expedition, Hauptlager bei Malu, Sepik river, 400 km upwards, ii. 1913, leg. S. G. Bürgers – in coll. A. Schintlmeister, Dresden.

Paratypes (10 ♂♂, 4 ♀♀):
Papua New Guinea: 1 ♂, Morobe, Bulolo Gorge, 800 m, 21. x. 1992, V. O. Becker Col.; 1 ♂, Mt. Kokoda, 1200 ft., ix. 1933, leg. L. E. Cheesman.
W. Papua: 1 ♀, Nabire 12 km E Samabusa, Lagari, 50 m, 6.–8. xii. 1993, leg. R. Brechlin & K. Cerny; 3 ♂ ♂, Nabire, Irian-Jaya-Highway, km 45, 745 m, 3°29'52"S 135°43'84" E, 22.–24. xi. 1997, leg. K. Cerny (MWM 8.639); 1 ♀, Snow Mts., nr. Oetakwa R., up to 3500 ft., x.–xii. 1910 (A. S. Meek); 2 ♂♂, Cyclops Mts., Sabron, Camp 2000 ft., vi. 1936, leg. L. E. Cheesman (BM 10314108); 1 ♂, Humboldt Bay dist.,Wembi, 8. vii. 1937; 1 ♂, Fak Fak, 1700 ft., i.–ii. 1908, J. Pratt; 1 ♂, Jafi District, Affi, 2350', 14. vi. 1939, leg. W. Stüber; 1 ♀, Mt. Goliath, about 139° E, 5000–7000 ft., ii. 1911 (A. S. Meek).
Waigeu: 1 ♂, Camp Nok, 5000 ft., xi. 1938, L. E. Cheesman.

Diagnosis. Forewing lengths ♂♂ 17–21 mm, ♀♀ 21 and 23 mm. Male antennae bipectinate, last 30% toward the tip are filiform. Female antennae filiform. Forewing yellowish-brown. Forewing pattern consists of an inconspicuous, pale yellow discal streak and a fuscous dorsal shadow near the base. Medial fascia fuscous-brown and well contrasting. Basal, postbasal and postmedial fasciae are indistinctly marked by brownish dots. Submarginal fascia consists of fine, sharp, fuscous-brown streaks. Hindwings are reddish-brown; fringe of all wings is whitish. Males display a prominently fuscous brown transverse fascia from base to margin along vein M2. Females lacks this transverse band. There is a resemblance to *O. lineata*, but the latter is readily distinguished by the absent dorsal shadow and a more angled forewing margin.

Male genitalia are characterized by a semicircular uncus tip with central spike. A pair of larger gnathoi processes arises near the uncus tip. Socii are slender and rectangularly kinked. Juxta displays a long central projection. Valva costa convexly shaped. Apical valva process much shorter than in *rudloffi.* Digitus process large and knob-shaped. Harpe displays two rounded projections. Sacculus bears a tapered process. Phallus straight, centrally waisted. Phallus tip has a serrated ring. The everted endophallus bears a strongly sclerotized carina and a serrated sclerotized diverticulum. 8th sternite wide triangular notched at the posterior margin. Female genitalia have not yet been dissected.

Bionomics. *O. opacitas* seems to occur frequently, as several images, unfortunately transmitted without data, by the late Henk van Mastrigt shows. Imagines of the

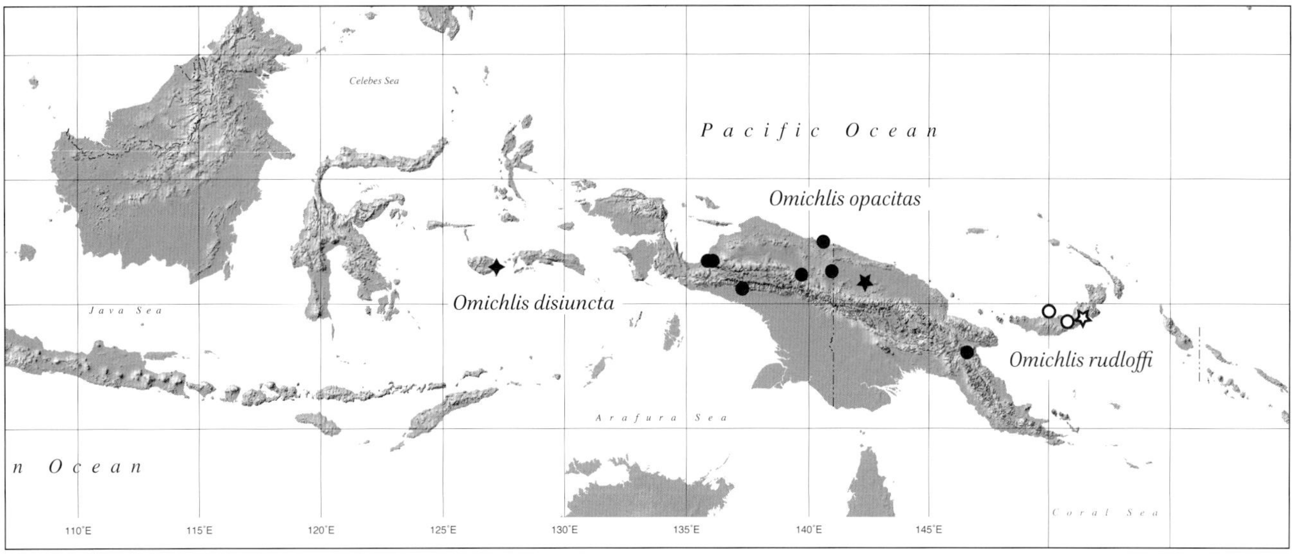

FIGURE 208 Distribution of *Omichlis opacitas, rudloffi* and *disiuncta*.

typeseries have been taken from 50 m up to 1070 m above sea level.

Distribution. Endemic to New Guinea Island.

240. ***Omichlis rudloffi*** **spec. nov.** (figs. 240a–c).
Holotype: ♂, Papua New Guinea, E. New Britain, Pomio district, Lopun, near Wakis, 100 m, 5°57.277' S, 150°42.763' E, 24.–29. iii. 2003, leg. Rudloff & Schaarschmidt (MWM 24.320) – in Museum Witt, Munich.

Paratypes (2 ♂♂):
New Britain: 1 ♂, Talasea peninsula, Bulumuvi, near bulu Murti, Lake Dakataua, 250 m, 5°01.330' S, 150°08.379' E, 1.–4. iv. 2003, leg. Rudloff & Schaarschmidt (MWM 24.319); 1 ♂, Talasea district, Bereme, SE Kimbe, 350 m, 5°43.952'S 150°36.720'E, 19.–26. iii. 2003, leg. Rudloff & Schaarschmidt;

Etymology: Named for Jan-Peter Rudloff, Roßlau, who collected the types series of this new species.

Diagnosis. Forewing lengths ♂♂ 20 and 21 mm. This species resembles *O. opacitas*, but forewings are narrower shaped and wing color is less reddish-tinged. Antennae are in both sexes as in *opacitas*. Discal spot is of forewing color, but surrounded fuscous-brown. Forewing fasciae and also the transverse band are as in *opacitas*. Holotype and one paratype have no dorsal shadow, but the remaining paratype (fig. 240c) displays a similar shadow as in *opacitas*. Hindwings are brownish and in the anal region is yellowish-brown. Female is unknown.

Male genitalia resemble those of *opacitas*, but uncus tip is anvil-shaped and gnathos is absent. A pair of socii is curved and much broader than in *opacitas*. Juxta bears a central projection, which is shorter than in *opacitas*. Valva have a long and slender apical process. Digitus projection knob-shaped and small. Sacculus of valva displays a pointed projection. Phallus straight with a serrated ring at the tip. Everted endophallus bears a diverticulum and a strongly sclerotized carina. 8th abdominal segments resembles *opacitas*, but posterior margin is wider bilobed.

Bionomics. A few males (n = 3) has been taken at lower altitudes between 100 m and 350 m.

Distribution. Endemic to New Britain.

241. ***Omichlis disiuncta*** **spec. nov.** (figs. 241a–d).
Holotype: ♂, Indonesia, SE Buru, Ilat, Remajah, 250 m, 3.–20. i. 2013, leg. St. Jakl – in coll. A. Schintlmeister, Dresden.

Paratypes (13 ♂♂):
Buru: 13 ♂♂, Ilat, Remajah, 250 m, 3. -20. i. 2013, leg. St. Jakl (GU 93-34, GU 93-35).

Diagnosis. Forewing lengths ♂♂ 19–20 mm. This species is doubtless closely related to *O. opacitas* and *rudloffi*. Antennae in both sexes are as in *opacitas*. Forewings are yellowish-brown and their pattern is similar to the previous two species. Discal spot is of forewing color but surrounded in fuscous-brown. Transverse band on forewings is absent. Fasciae are, except the submarginal fascia, weakly

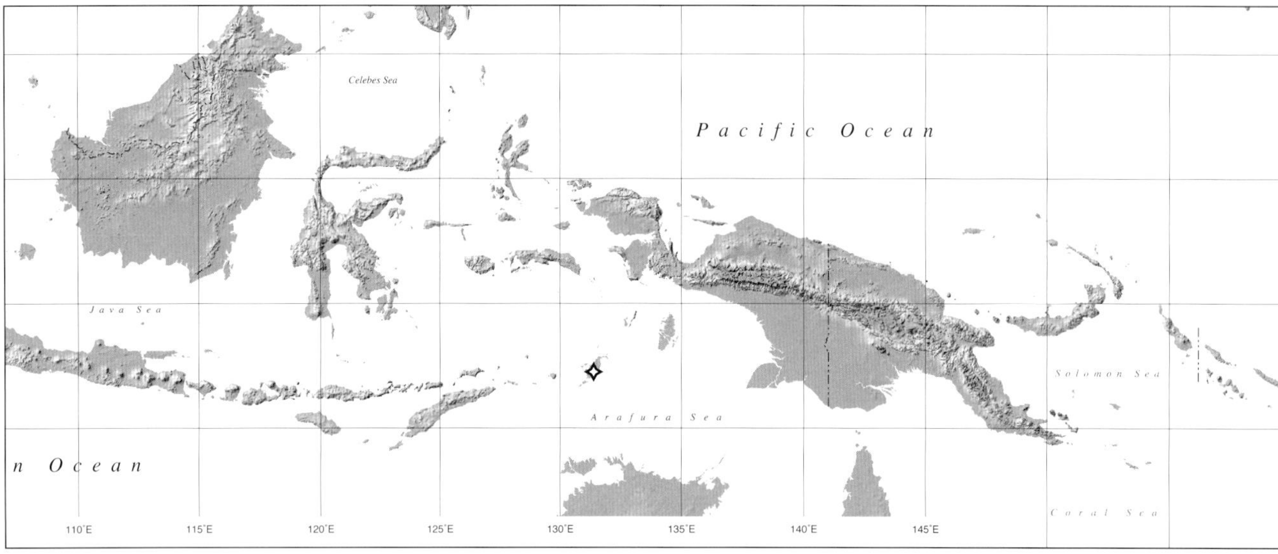

FIGURE 209 Distribution of *Omichlis priskae*.

and indistinctly developed. Hindwings are reddish-brown with blackish anal markings. Forewing fringe fuscous reddish-brown; hindwing fringe reddish-brown and yellowish-brown checkered. Female is unknown.

Male genitalia show a resemblance to *opacitas* and *rudloffi*, but they are significantly smaller. Uncus short, uncus tip rather rectangularly shaped and broad with a small central spine. Gnathos reduced to a swollen part of the uncus. Socii are broad and bulbous-shaped. Juxta of similar shape as *rudloffi* and with a small central projection. Valva shape resembles *rudloffi*, but apical processes is shorter. Valva costa displays a projection that distinguish *disiuncta* from other congeners. Harpe bears two smaller, but stronggly sclerotized pointed spines. Phallus straight with circular shaped tip. Everted endophallus has a strongly sclerotized band, but shape is not closed as in *rudloffi* or *opacitas*. Diverticulum is present, but much smaller than in the previous species. 8th sternite triangularly notched at posterior margin as in *opacitas*.

Bionomics. *O. disiuncta* was collected frequently in one lcality at 250 m above sea level.

Distribution. Endemic to Buru.

242. *Omichlis priskae* spec. nov. (figs. 242a–c).
Holotype: ♂, Indonesia, Tanimbar Isls., Yamdena, 21 km N Saumlaki, 150 m, 7°47' S, 131°18'E, 27. xi.–11. xii. 2005, leg. St. Jakl – in coll. A. Schintlmeister, Dresden.

Paratypes (2 ♂♂):
Tanimbar: 1 ♂, Yamdena, 21 km N Saumlaki, 150 m, 7°47' S, 131°18'E, 5.–28. xii. 2005, leg. St. Jakl (GU 93-04); 1 ♂, Yamdena, 20 km NE Saumlaki, village Loulun, 50 m, 7°47' S, 131°18'E, 5.–28. xii. 2006, leg. St. Jakl (GU 93-97).

Etymology. Named for Priska Erjani Jakl, wife of Stanley Jakl, who collected from 1996 on many islands of Indonesia Scarabaeidae, Coleoptera. During his many and long expeditions, also on remoted islands as well, he always took for the author's collection Notodontidae, which are now an important basis for this study.

Diagnosis. Forewing lengths ♂♂ 16–17 mm; the smallest known *Omichlis* species. Antennae bipectinated; the last 15% toward the tip is naked. Wing color pale yellowish-brown; forewing pattern is indistinct. Discal spot indistinct pale yellowish and hard to see. Submarginal fascia of forewing marked as fine blackish U-shaped streaks. The holotype displays a fuscous-brown medial fascia and a finer postmedial fascia. One of the paratypes has almost uniform forewings, where only a fuscous streak from the distal angle to the medial area is visible. The second paratype (fig. 242d) have uniform forewings with the fuscous distal streak and an additional black tornal spot. Female is unknown.

Uncus tip of male genitalia fan-shaped; a pair of gnathoi lobes arises directly below the uncus tip. Socii are large and curved with pointed tip. Valvae have a large costal projection, which bears a sharp spine at the tip. Digitus process reduced, the valva apex bears a short process. Phallus relatively long and curved. At the tip there is a characteristically hook. The everted endophallus shows a strongly sclerotized carina as in *pratti*. 8th sternite less

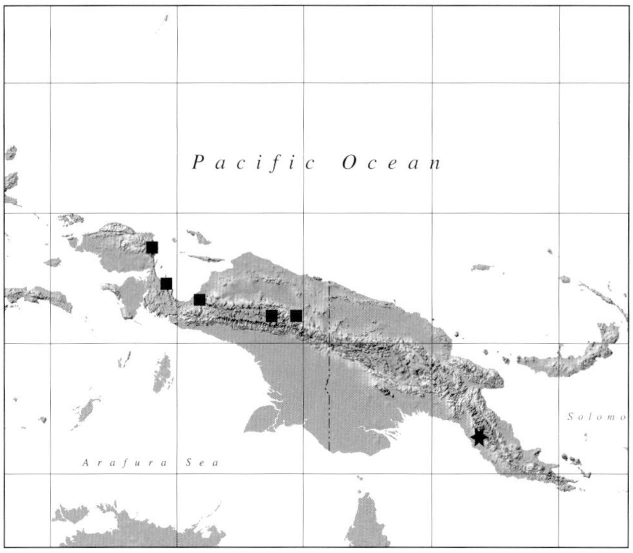

FIGURE 210 Distribution of *Omichlis lineata.*

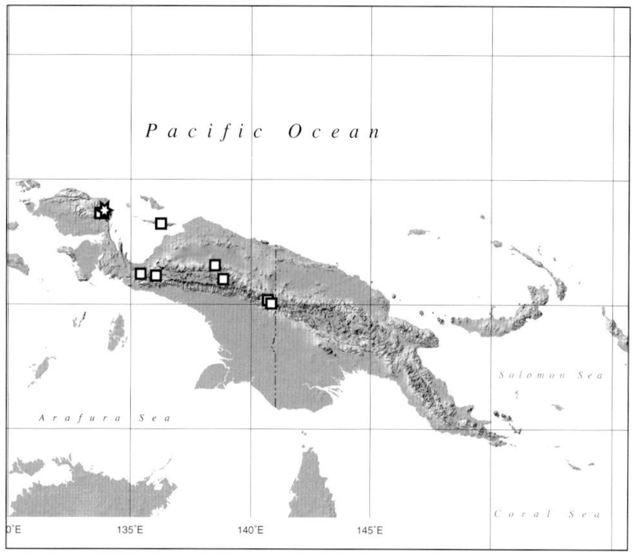

FIGURE 211 Distribution of *Omichlis plagiosa.*

deeply notched at the posterior margin than most other *Omichlis* species.

Bionomics. Only three males were collected during three visits to Yamdena. The collecting locality is situated at 150 m above sea level.

Distribution. Endemic to Tanimbar (Yamdena).

243. ***Omichlis lineata*** Bethune-Baker, 1904: 178.
Holotype: ♂, [Papua New Guinea], Kebea Range, 6.000 ft. – NHMUK, London.
Schintlmeister 2013: 304.

Diagnosis. This species is similar to *O. opacitas* and *rudloffi* but readily distinguished by the angled margin and the pointed forewing apex. Forewings are colored yellowish-brown. There is no dorsal shadow as in *opacitas.* The fuscous brown transversal band from the base to the distal margin on vein M2 is thin and slightly curved. The only known worn female has this band more weakly developed. On forewings several fasciae are visible: basal fascia, post basal fascia, and a curved medial fascia, which is in *opacitas* rather straight. Postmedial fascia marked as black dots and submarginal fascia by fine black streaks. Forewing fringe contrasting fuscous reddish-brown. Hindwings are reddish-brown. Sexual dimorphism is limited.

Male genitalia have a knob-shaped uncus tip and a pair of large gnathoi lobes below the tip. Socii are compact and of a distinctive shape. Juxta bears a central and tapered projection. Valvae have a large digitus. Harpe bears two pointed projections. Sacculus of valva is bulbous and large. Phallus straight with a pair of large serrated processes. Everted endophallus displays a strongly sclerotized carina. Diverticulum bears two small spines. 8th sternite deeply notched at posterior margin. Female genitalia have long apophyses anteriores, but apophyses posteriores are relatively short. Postvaginal plate squarish. Ductus bursae strongly curved. Corpus bursae is ovoid and displays a reniform signum.

Bionomics. Only a few (n < 10) specimens are currently known, which has been collected at altitudes between 750 m and 1800 m.

Distribution. Endemic to New Guinea Island.

244. ***Omichlis plagiosa*** Joicey & Talbot, 1917: 62; pl. 2: 3.
Holotype: ♀, [Indonesia, W Papua], Arfak Mtns., Angi Lakes, 6.000 feet. – NHMUK, London.
Schintlmeister 2013: 304.

Diagnosis. Forewings prominently violet-brown, a few specimens are yellowish-tinged. Discal spot of forewings small, of ellipsoid shape, and surrounded in fuscous-brown. Transverse band prominently broad and contrasting. Five fuscous-brown fasciae are clearly visible, but more weakly developed than in *O. trilineata.* Occasionally a blackish tornal blotch occurs. Hindwings are sometimes brown, or yellowish-tinged. Sexual dimorphism is limited.

Male genitalia are similar to those of *trilineata.* Uncus tip is fan-shaped; below the tip a pair of ganthoi arises. Socii are large and curved. Valvae have a long and curved apical process. Digitus process is of variable shape. Harpe

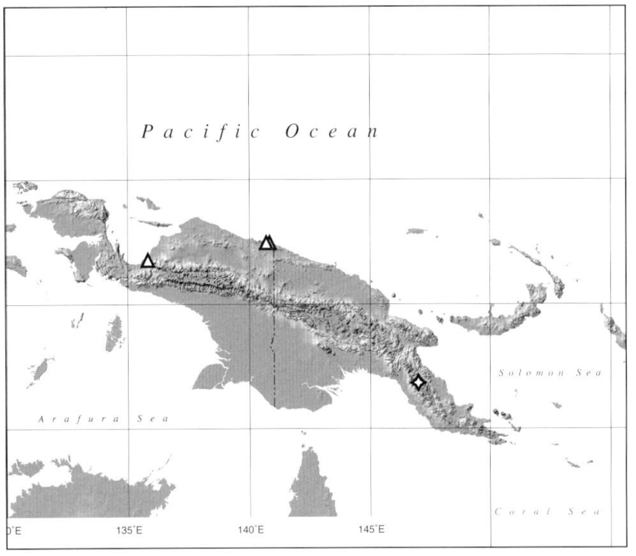

FIGURE 212 Distribution of *Omichlis trilineata.*

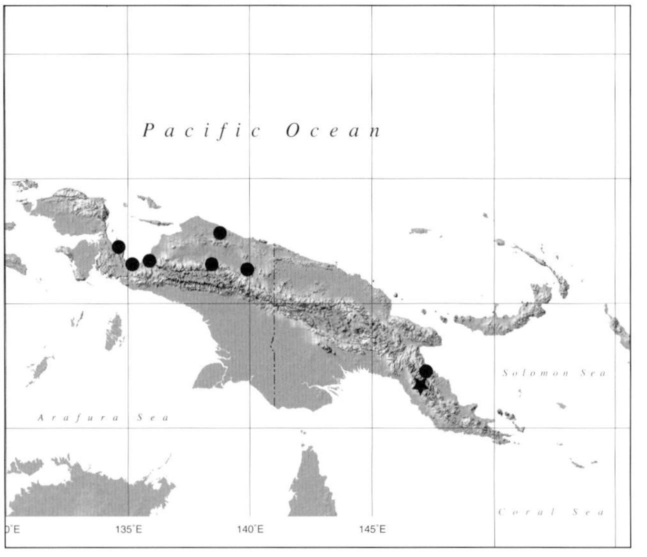

FIGURE 213 Distribution of *Omichlis dinawa.*

bears one tapered process. Phallus, somwehat variable in shape, has a pair of processes at the tip, as seen in *lineata* or *trilineata*. Everted endophallus has a small diverticulum and a relatively small, more strongly sclerotized carina. 8th sternite bilobed at the posterior margin, but 8th tergite rather straight. Female genitalia have not yet been dissected.

Bionomics. The moth occur infrequently, and has been observed between 1500 m and 2000 m above sea level..

Distribution. Endemic to New Guinea Island.

245. ***Omichlis trilineata*** Bethune-Baker, 1908: 178.
Holotype: ♂, [Papua New Guinea], Kebea Range, 6.000 ft. – NHMUK, London.
Schintl*meister 2013: 304.*

Diagnosis. *O. trilineata* differs from *plagiosa* by its more reddish-tinged forewing color. All fasciae are much stronger and more contrastingly developed. Postbasal fascia appear occasionally doubled. Discal spot is yellowish-brown filled, and rather disc-shaped. It is larger than in *plagiosa*. Hindwings are violet brown with more fuscous-brown fringe.

Male genitalia have a resemblance to those of *plagiosa*. Uncus tip fan-shaped and deeply notched. It bears a large central spine. Gnathoi lobes below the tip are large. Socii are of similar shape than in *plagiosa*, but larger. Central projection of the juxta is large and tapered. Valvae differ from *plagiosa* by downwardly curved apical process. Digitus process large; harpe bears three small spines. Phallus bears a pair of long, serrated processes at the tip. Everted endophallus as in *plagiosa*. 8th sternite distinctively shaped; where the notched posterior margin is rather narrow. Female genitalia have not yet been dissected.

Bionomics. *O. trilineata* is rare in collections; n < 10 specimens are known. The moth, in contrast to *plagiosa* occurs in the lowlands, up to 200 m. Holotype, however, was taken at 1800 m above sea level.

Distribution. Endemic to New Guinea Island.

246. ***Omichlis*** Bethune-Baker, 1904: 376; pl. 5: 3.
Holotype: ♂, [Papua New Guinea], Dinawa [4.000 ft.] – NHMUK, London.
Schintlmeister 2013: 303.

Diagnosis. This is the largest of three similiar species. Forewing margin gently rounded, not as angled as the following two species. Forewings orange brown with an indistinct, yellowish, apical blotch. A further yellowish patch occurs in the basal area near the dorsum. Discal spot is absent. Basal, postbasal, postmedial and submarginal fasciae marked as rows of black dots. In the medial area, near the costa a contrasting black dot usually occurs, accompanied by a whitish area. The holotype specimen lacks this black dot. A black marginal patch is present at vein M2. Hindwings are pale reddish-brown. Sexual dimorphism is limited.

Male genitalia are by far the largest in this group of three similar species. Uncus anvil-shaped without central projection. Gnathoi lobes below the tip are very large and

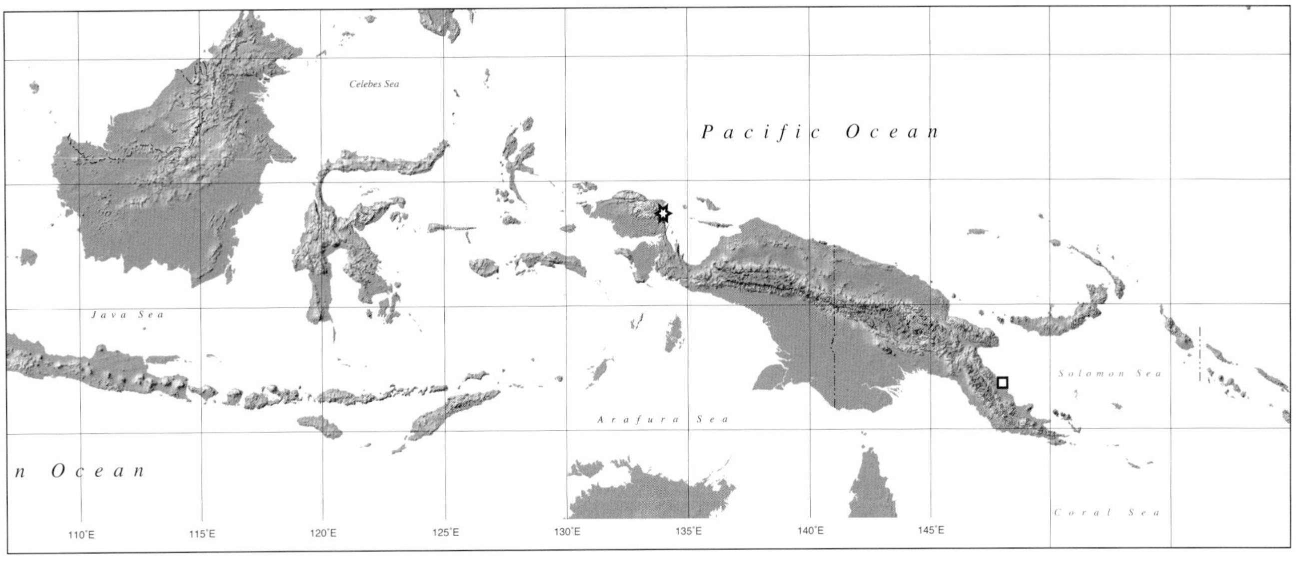

FIGURE 214 Distribution of *Omichlis ochromacula*.

bulbous. Socii are small, slender and curved. Valvae broad with a medium-sized apical process. Digitus projection large. Juxta bears a slender central process. Phallus straight; at the phallus tip there are three sharp pointed processes, like a trident. Everted endophallus bears a large serrated, circular diverticulum. Carina developed as a more strongly sclerotized plate. 8th sternite at posterior margin semicircularly notched. Female genitalia have not yet been dissected.

Bionomics. Adults occuring infrequently, from 600–1200 m above sea level.

Distribution. Endemic to New Guinea Island.

247. ***Omichlis ochromacula* spec. nov.** (figs. 247a–d).
Holotype: ♂, Indonesia, W Papua, 30 km S Manokwari, Arfak Mts., Ngat Biep, river Ngat valley, 850 m, 18.–19. xii. 1993, leg. R. Brechlin & K. Cerny (GU 94-83) – in coll. A. Schintlmeister, Dresden.

Paratypes (2 ♂♂):
Papua New Guinea: 1 ♂, Biagi, Mambare River, 5000 ft., iv. 1906, leg. A. S. Meek (BM 10314126); 1 ♂ dto. iii. 1906.

Further material:
1 ♂, **W. Papua**, image sent by H. v. Mastrigt (but without further data), illustration see fig. 239d.

Diagnosis. Forewing lengths ♂♂ 17 mm and 19 mm, smaller than *O. dinawa* Antennae of the male bipectinate, last 40% toward the tip filiform. *O. ochromacula* resembles *dinawa,* but the forewing shape is narrower and the margin at vein M2 angled. Basal area of forewing yellowish and bearing a prominent black patch, which is absent in *dinawa* and *baliema*. The yellowish area continues along the costa from the basal area to the postmedial area. In the apex a distinct, yellowish blotch occur. Discal spot of forewing distinct, large yellowish with a small brownish nucleus. Forewing fasciae and also a black spot at the margin (M2) are as in *dinawa*. Hindwings are fuscous-brown, darker than in *dinawa* and *baliema*. Female is unknown.

Male genitalia are smaller than those of *dinawa*. Uncus medium-sized, somewhat variable in its length. Distinguished by the unique shape of the uncus tip, which displays a broad central projection. Below the tip two smaller triangular gnathoi lobes arises. Socii are compact; the last part strongly curved and very slim. Juxta bears a small central projection. Valva shape narrower than *dinawa,* and the apical process longer. Digitus process large and square-shaped. Sacculus of valva reduced in size in relation to *dinawa* or *baliema*. Phallus short and straight. Phallus tip bears a large hook-like central process but two further projections are reduced in size. Diverticulum is small, serrated and not circular. Carina as in *dinawa*. 8th sternite rather triangularly notched at posterior margin, not rounded as in *dinawa* or *baliema*.

Bionomics. The species is recognized from 3 males to date. The specimens were collected at 850 and 1520 m above sea level.

Distribution. Endemic to New Guinea.

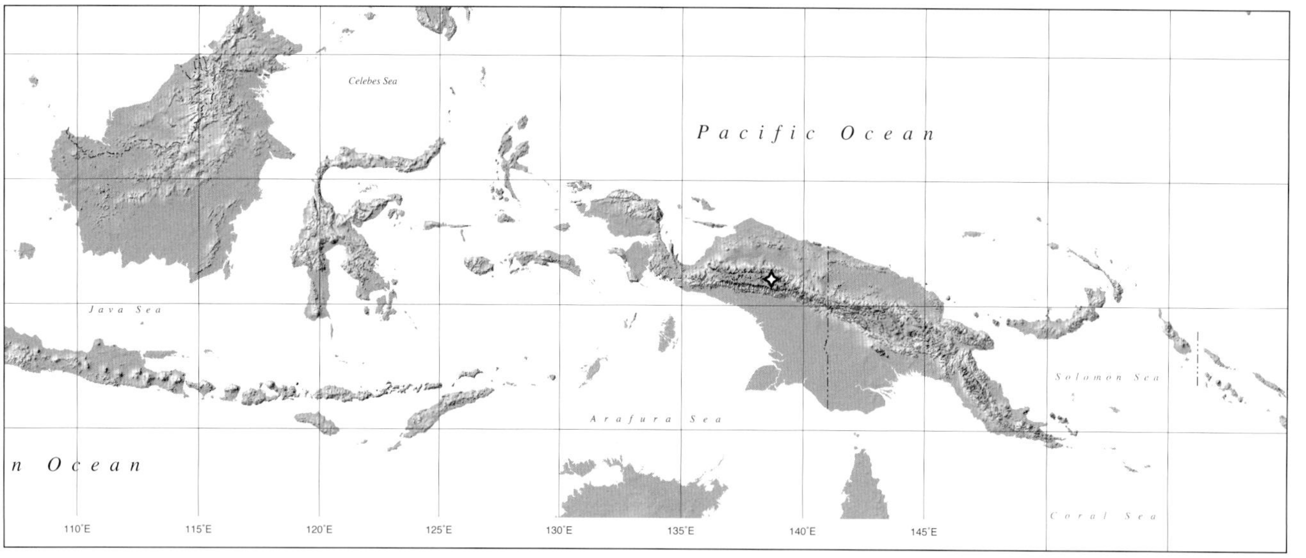

FIGURE 215 Distribution of *Omichlis baliema*.

248. *Omichlis baliema* spec. nov. (figs. 248a–b).

Holotype: ♂, Indonesia, W Papua, Baliem valley, road Wamena-Sentani, km 48, 2150 m, 26 + 29. i. 2000, leg. Cerny & Hluchy (MWM 24.316) – Museum Witt, Munich.

No paratypes.

Diagnosis. Forewing length ♂ 19 mm. Male antenna and wingshape as in *ochromacula*, but general appearance is paler and yellowish rather than ocherous. *O. baliema* differs from *ochromacula* by its lack of a black basal patch, the less contrasting yellowish discal spot, and the absent black marginal patch at vein M2. Hindwings are yellowish, not brown as in *ochromacula*. Female is unknown.

Male genitalia are slightly larger than those of *ochromacula.* Uncus tip resembles those of *ochromacula,* but there is no central projection. Gnathos below the uncus tip of circular shape. Socii less robust than *ochromacula* but of similar shape. Central projection of juxta is larger than in *ochromacula,* but shorter than in *dinawa.* Valvae are larger than *ochromacula.* Apical process is large and club-shaped. Digitus process long and tapered, not squarish as in *ochromacula.* Sacculus bears a small process. Harpe is larger than in *ochromacula* and bears two tapered projections. Phallus much larger than in *dinawa* or *ochromacula* and broader toward the tip. Phallus tip displays only a pair of small processes. Everted endophallus bears at the base a stronger, furcation-like sclerotized area, which resembles *calyptis*. Diverticulum is small; carina bears a circularly serrated structure. 8th sternite bilobed on both margins.

Bionomics. The holotype was taken at 2150 m above sea level.

Distribution. Known only from the type locality in the Baliem Valley of W Papua.

249. *Omichlis calyptis* (Swinhoe, 1892): 294 (*Pydna calyptis*).

Holotype: ♂, Indonesia, W Papua], Dorey [Bay] – Oxford University Museum of Natural History.

Schintlmeister 2013: 303.

Taxonomic notes. *Omichlis hampsoni* Bethune-Baker, 1904: 375; pl. 5: 2, was described of 2 ♂♂ from "Dinawa". To stabilize the nomenclature the ♂ with the following labels is hereby designated as lectotype: "Dinawa. | B. New Guinea. | 4000 ft. | March–July 1902 | Coll. A.E Pratt"; green circular label "Type"; "G.T.B. -Baker Coll. | Brit. Mus. 1927–360; yellow label "Photo done by A. Schintlmeister #3907". A lectotype label will be added. The lectotype is depicted here in fig. 249d. The remaining ♂ become a paralectotype.

The lectotype of *Omichlis hampsoni* is conspecific with the holotype of *calyptis* and *hampsoni* is placed as a junior synonym of the latter (**syn. nov.**).

Two dissected males from Ambon and Seram differ by their narrower and more pointed digitus processes from other examined specimens (n = 4 GU).

Australian males sometimes show a black patch in the tornus of forewing. Unfortunately it was not possible to dissect male genitalia from New Guinea or Australia to evaluate the taxonomy.

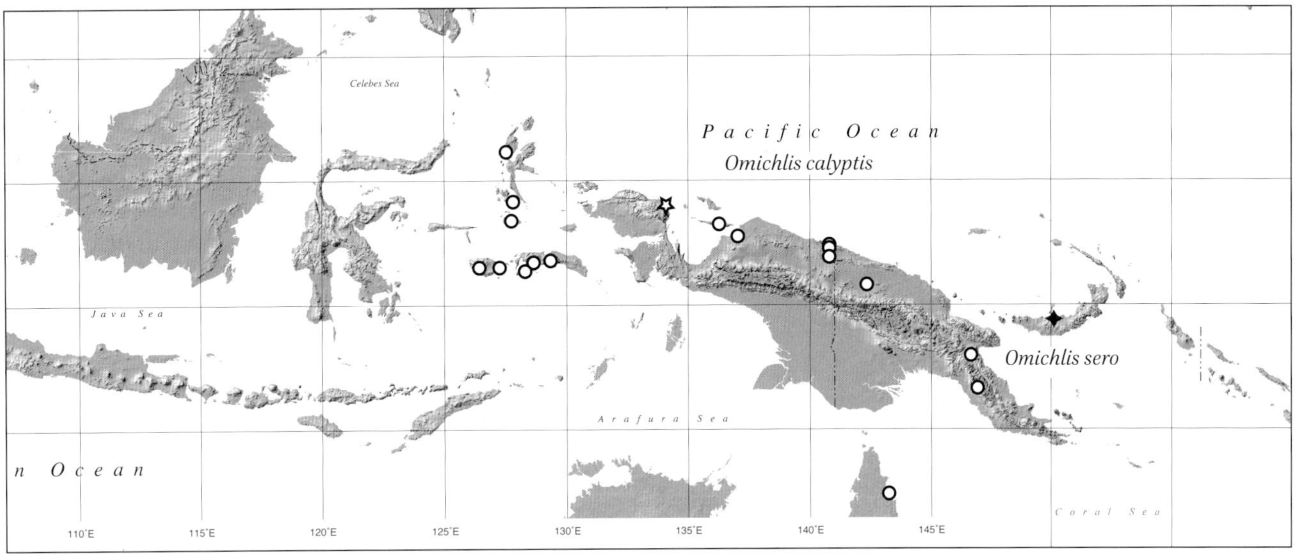

FIGURE 216 Distribution of *Omichlis calyptis* and *Omichlis sero*.

Diagnosis. A rather unmistakable species, which is best characterized by the golden-brown areas along the costa of the forewings. Remaining part of forewing is brown, but the darkness is some variable. Discal patch of forewing ellipsoid and surrounded in pale yellow. Fasciae are in the costal part well developed and fading toward the dorsum. Two whitish, contrasting fasciae in the postmedial area are significant. Hindwings are reddish-brown. Adults are variable in size. Females resembles males, but the distal margin is gently curved, not angled.

Male genitalia have a circular-to-ellpisoid uncus tip. Gnathos absent, but socii are large and characteristically curved downwards. Juxta bears a broad central projection. Valvae have a large apical process. Digitus process pointed, variable in shape. Harpe large, but without processes. Phallus straight, broader at the tip. From the tip a bifurcated sclerotized part arises, which is part of the everted endophallus and resembles *O. baliema*. Diverticulum is larger than in *baliema* and not circular. Carina bears a small spine and does not display serrated structures, as does *baliema*. 8th sternite slightly notched at posterior margin. Female genitalia have a large and long 8th abdominal segment. Apophyses posteriores are reduced in length; apophyses anteriores are very long. Ductus bursae curved and compact. Corpus bursae bear a large reniform signum.

Bionomics. Adults are common, but occur in in New Guinea infrequently. Most findings are from lower altitudes up to 600 m. A few specimens, including the lectotype of *hampsoni*, have been taken at 1200 m above sea level.

Distribution. Distributed in Buru, Ambelau Isl., Halmahera, Bacan, Obi, Ambon, Seram, New Guinea and Australia, Qsd.

250. ***Omichlis sero*** spec. nov. (figs. 250a, 250b).
Holotype: ♂, Papua New Guinea, W New Britain prov., Telasea district, Bèreme, SE Kimbe rainforest, 350 m, 05°43.952 150°36.720, 19.–26. iii. 2003, leg. Rudloff & Schaarschmidt (MWM 29.420) – Museum Witt, Munich.

No paratypes.

Diagnosis. Forewing length ♂ 21 mm. Male antennae are brown and last 25% to the tip are filiform. There is a resemblance to *O. baliema* and *calyptis*, but general appearance is more brownish and forewing markings are less contrasting. Forewing stigmas are present but indistinct. *O. sero* is distinguished by a broad and fuscous-brown median shadow and a submarginal fascia, which is marked as a row of black dots. Apical patch is brownish-yellow as in *baliema*. Hindwings are fuscous-brown. Female is unknown.

Male genitalia have a long uncus with disc-shaped tip, similar to those of *calyptis*, but there is a small and pointed central projection. A gnathos arises at the uncus tip. Socii are large and bear a projection with a spine at its tip. Central projection of juxta is as in *baliema*. Valva shape, where the apical processes are absent, and costa is curved, distinguish *sero* from other congeners. Digitus process is small, squarish, and serrated. Sacculus bears a small process toward the valva apex. Phallus is straight and bear a

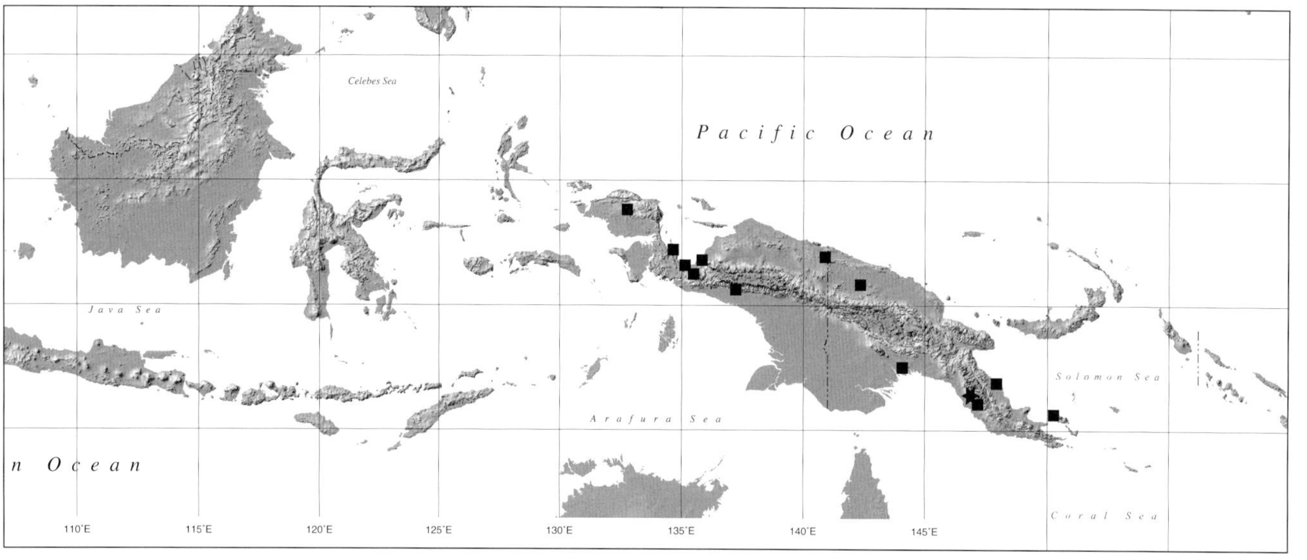

FIGURE 217 Distribution of *Omichlis ochracea*.

central process at the tip. Everted endophallus bears a strongly sclerotized carina but no cornuti. 8th sternite slightly bilobed on posterior margin, 8th tergite squarish.

Bionomics. The holotype was taken in a rainforest at 350 m above sea level.

Distribution. Known only from the type locality in Talesea, New Britain.

251. ***Omichlis ochracea*** Bethune-Baker, 1904: 376; pl. 5: 6. Lectotype: ♂, [Papua New Guinea], Ekeikei, 1.500 ft. – NHMUK, London **by present designation**
Schintlmeister 2013: 304.

Taxonomic notes. *Omichlis ochracea* was described of a pair from "Ekeikei, 1.500 ft." and "Mount Kebea [6.000 ft.]". To stabilize the nomenclature the ♂ with the following labels is hereby designated as lectotype: "Ekeikei. | B. C. New Guinea. | 1500 ft. | Jan.–Feb. 1903 | A.E Pratt. Coll."; green circular label "Type"; "G.T.B. -Baker Coll. | Brit. Mus. 1927–360; yellow label "Photo done by A. Schintlmeister #3915". A lectotype label was added. Lectotype is figured here in fig. 251d. The remaining ♀ becomes a paralectotype.

Omichlis macrospila Kiriakoff, 1970: 117; fig. 3, holotype: ♂, Dutch New Guinea, 25 miles S of Wangaar, Nomnagihé, 2.000 ft. – NHMUK, London, displays a prominent, ocherous forewing blotch on forewing. But it is merely an individual form of *ochracea* as several genitalia dissections (n = 5) showed. The illustration of holotype male genitalia of *macrospila* in Kiriakoff, 1970 does not show a digitus process. However, as illustrated (fig. 251c), forms with reduced digitus process occur as an expression of individual variability; the corresponding specimen of genitalia in fig. 251c resembles externally fig. 251f. *Omichlis macrospila* sinks as a junior synonym of *ochracea* (**syn. nov.**).

Diagnosis. The yellow forewing color distinguishes this species from most other congeners. Forewing pattern is very variable. There are forms with brown transversal band along vein M2, black tornal spot, and black spot at the angled margin (vein M2). Sometimes forms display a brown blotch in the postmedial area, near the apex. All these forms appear also in any combination and in both sexes. Furthermore an indistinct paler discal spot and a brown medial fascia occur occasionally in all forms. Hindwings are reddish-brown. Sexual dimorphism is limited.

Male genitalia have a rather rectangular, slightly conical uncus. A pair of gnathoi arising below the tip and are downwardly directed. Socii are rather small and curved. Juxta bears a central projection of variable shape. Apical valva process variable in its length and tip is in most cases knob-shaped. Digitus process varies in its length. Harpe bears a smaller process, which sometimes display fine serrations. Phallus is straight and waisted in the central part. Near the tip a pair of processes arises; one of them is large and hook like. Everted endophallus bears a sclerotized carina, as is usual in most *Omichlis*. 8th sternite gently bilobed at the posterior margin. Female genitalia resemble *calyptis*. Apophyses posteriores are reduced in length; apophyses anteriores are very long. Postvaginal plate is circularly shaped and have no lobes as *calyptis* do. Ductus bursae compact and curved. Corpus bursae is disc-shaped and displays a reniform, waisted signum.

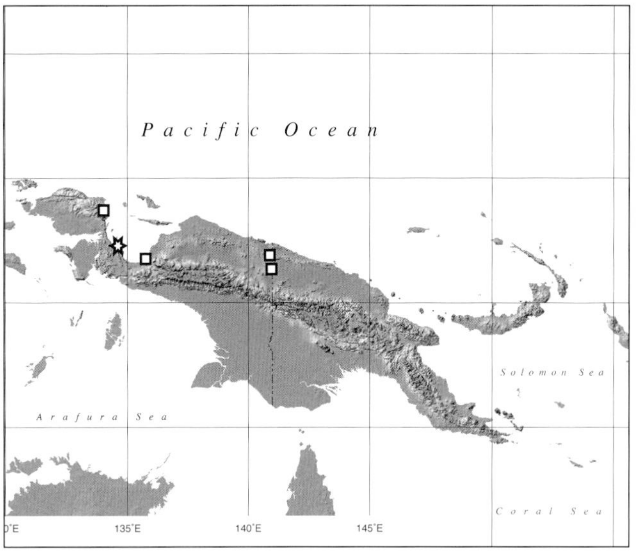

FIGURE 218 Distribution of *Omichlis leucosticta.*

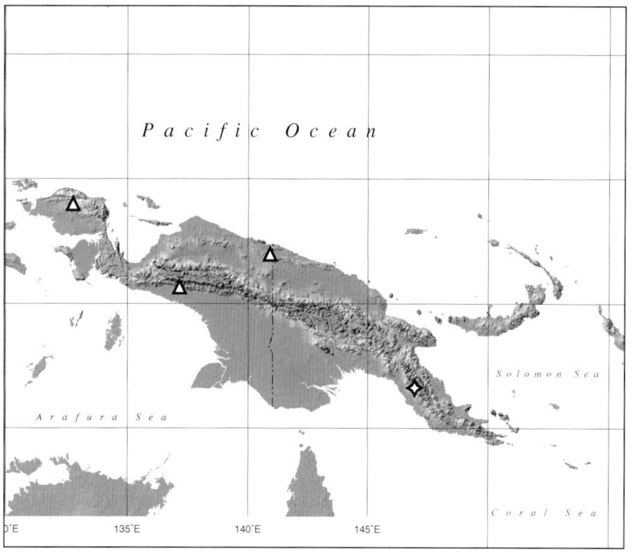

FIGURE 219 Distribution of *Omichlis transversa.*

Bionomics. Adults occur frequently and have been observed from 50 up to 1200 m above sea level.

Distribution. Distributed in New Guinea in its entirety and on Goodenough Isl.

252. ***Omichlis leucosticta*** Joicey & Talbot, 1917: 62; pl. 2: 4.
Lectotype: ♂, [Indonesia, W Papua], Wandammen Mtns. 3.000-4.000 feet. – NHMUK, London **by present designation**
Schintlmeister 2013: 304.

Taxonomic note. *Omichlis leucosticta* was described from a pair. To stabilize the nomenclature the ♂ with the following labels is hereby designated as lectotype: "Wandammen Mts. | 3–4000 ft. D. N. Guinea | Nov. 1914 | A. C & F. Pratt"; red circular label "Type | HT"; "*Omichlis* | *leucosticta* | J. + T. | Type ♂. (H. T.)"; yellow label "Photo done by A. Schintlmeister #3938". A lectotype label was added. The lectotype is depicted in fig. 252a. The remaining ♀ becomes a paralectotype.

Diagnosis. Forewing color dark brown-violet. Contrasting white pattern distinguishes this species from other congeners. Forewing pattern consists of many fasciae, marked as rows of white dots. Discal spot is ellipsoid and whitish with two deep black nuclei. Hindwings are fuscous-brown with black and contrasting white anal markings. Fringe of all wings is cream. The only known female is in poor condition (paralectotype) but resembles the male.

Genitalia are not yet dissected.

Bionomics. The species is known from n = 7 specimens. It has been collected from 750 m up to 1650 m above sea level.

Distribution. Restricted to New Guinea.

253. ***Omichlis transversa*** Bethune-Baker, 1908: 178.
Holotype: ♀, [Papua New Guinea], Kebea Range, 6.000 ft – NHMUK, London.
Schintlmeister 2013: 304.

Diagnosis. Forewings yellowish-brown. Forewing margin has a pronounced projection at vein M2, which is unique in *Omichlis*. Postmedial fascia straight and in a contrasting fuscous-brown. Postbasal fascia weakly developed. Discal spot small and pale yellowish-brown. A black spot near the dorsum toward the forewing tornus is helpful for identification, but it is weakly developed. Medial area surrounded by two straight, fuscous-brown fasciae. Hindwings are of same color as forewings. Female has narrower forewings than male.

Male genitalia are characterized by an umbrella shaped uncus tip. No gnathos, but a pair of slender and strongly curved socii are present. Valvae have an apical processes. Digitus processes long and curved; it arises from valva costa. Harpe bears two small, tapered processes. Phallus long and straight. Everted endophallus has a circular, strongly sclerotized carina. 8th sternite deeply notched at posterior margin. Female genitalia have short apophyses posteriores and long apophyses anteriores. Ductus bursae is curved, short and broad. Corpus bursae disc-shaped and

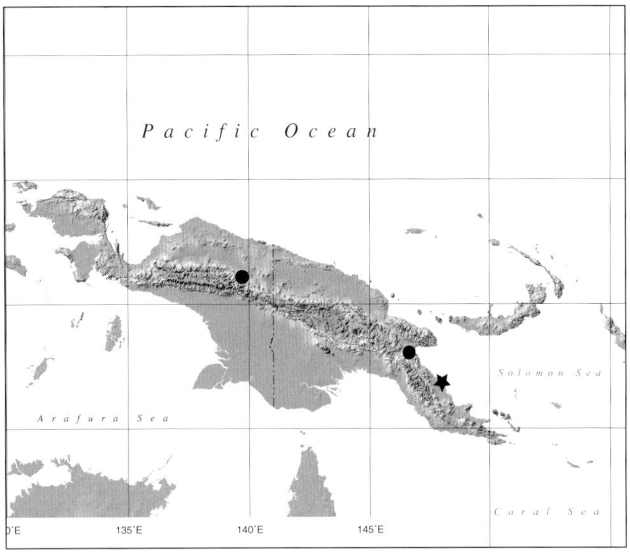

FIGURE 220 Distribution of *Omichlis strigata*.

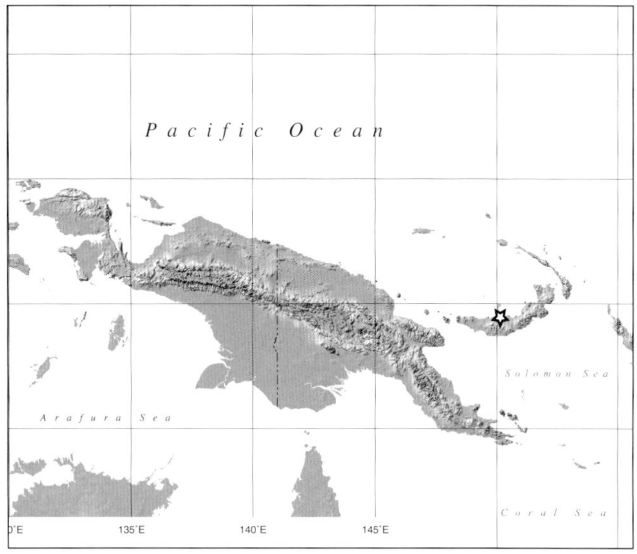

FIGURE 221 Distribution of *Omichlis geminata*.

bears a signum, which is not reniform as in most *Omichlis*, but fish-shaped.

Bionomics. A rare species (n < 10 known specimens). Samples have been collected at altitudes between 1000 m and 1200 m above sea level.

Distribution. Endemic to New Guinea.

254. ***Omichlis strigata*** Bethune-Baker, 1908: 179.
Holotype: ♂, [Papua New Guinea], Biagi, Mambaré River – NHMUK, London.
Schintlmeister 2013: 305.

Diagnosis. Forewings yellowish-brown, sepia toned. Forewing margin rounded, without angle at vein M2. Postbasal, postmedial and submarginal fasciae of forewing are fuscous-brown, accompanied by a yellowish contrasting band. Discal spot reniform, yellowish, and fuscous-brown filled. Veins are yellowish-tinged and this contrasting lines continues on the thorax. Forewing fringe fuscous-brown. Hindwings of same color as forewing, but fringe is yellowish brown, except the anal region, where fringe is of fuscous-brown. Sexual dimorphism is limited.

Genitalia are small in both sexes. Male genitalia are characterized by a fan-shaped uncus tip and a pair of long gnathoi lobes. Socii curved and broad. Valva bears a long curved apical process. Digitus process pointed. Sacculus large with two projections, of which one of them is pointed. Phallus slightly curved with a broader ring-shaped tip. Everted endophallus bears a carina and a strongly sclerotized, serrated band. 8th sternite is of circular shape, with a bilobed posterior margin. Female genitalia have weakly sclerotized apophyses posteriores of medium length, whereas apophyses anteriores are strongly sclerotized and long. Postvaginal plate bilobed. Ductus burse thick. Corpus bursae ovoid with reniform signum.

Bionomics. Only a few (n < 10) specimens are known, which has been taken at altitudes of 1500 and 1700 m.

Distribution. Endemic to New Guinea Island.

255. ***Omichlis geminata*** Gaede, 1930: 611; pl. 79: f.
Holotype: ♂, [Papua New Guinea, New Britain], Neu-Pommern, Talesea – NHMUK, London.
Schintlmeister 2013: 303.

Diagnosis. Wing color reddish-brown. Forewing margin slightly angled at vein M2. *O. geminata* is well characterized by a prominent, rather straight fuscous-brown postmedial fascia. Many further fasciae in basal, postmedial, and submarginal areas are indistinct and weakly developed. Discal spot small and yellowish-filled. At vein M2, near to the margin, a weakly developed blackish patch occurs. Hindwings show black and yellowish contrasting anal markings. Sexual dimorphism is limited; the female is more distinctively marked than the somewhat worn holotype male.

Genitalia have not yet been dissected.

Bionomics. A pair (n < 2) has been taken at lower altitudes up to 500 m.

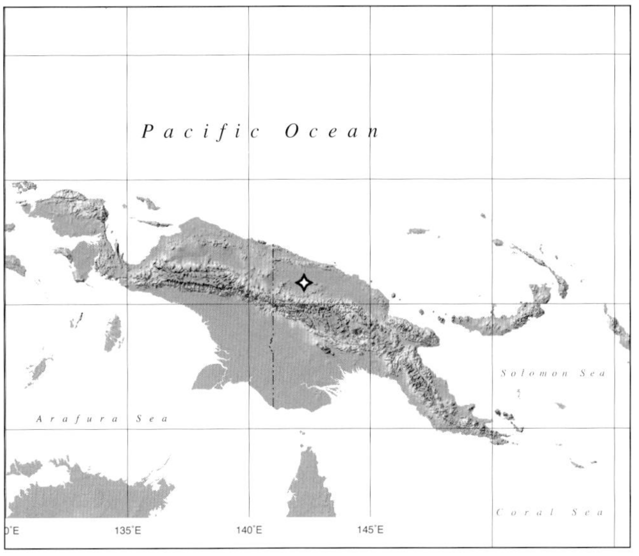

FIGURE 222 Distribution of *Omichlis buergersi*.

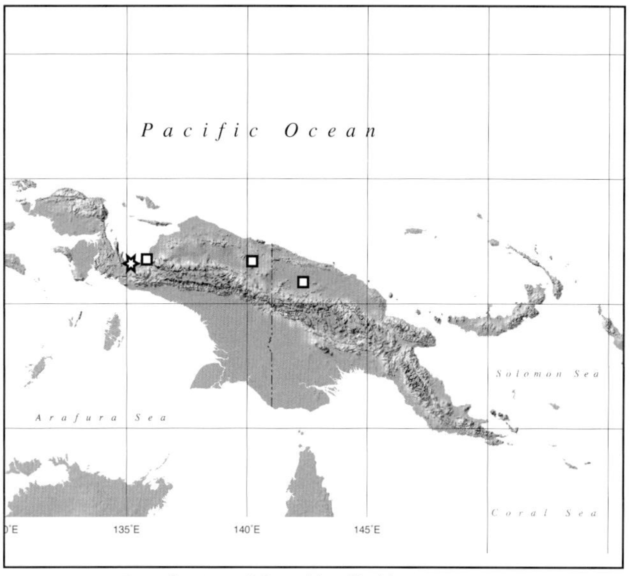

FIGURE 223 Distribution of *Omichlis albidilinea*.

Distribution. Known from New Britain,Talesea.

256. ***Omichlis buergersi*** **spec. nov.** (figs. 256a, 256b).
Holotype: ♂, Papua New Guinea, (D. N. Guinea), Kaiserin-Augusta-Fluß-Expedition, Etappenberg, 850 m (Sepik river, 400 km upwards), 2. x. 1912, leg. S. G. Bürgers (GU 66-25)– in coll. A. Schintlmeister, Dresden.

No paratypes.

Diagnosis. Forewing length ♂ 17 mm. Antennae bipectinate, the last 30% toward the tip filiform. Forewing color rusty brown, slightly angled at vein M2 at the margin. Unmistakable by pale greyish forewing wing pattern. Postbasal fascia white, accompanied basally by three black dots. Postmedial fascia curved and white. Apical region and neighbouring marginal area whitish filled. Discal spot rusty-brown, surrounded in pale grey. Forewing fringe rusty brown Hindwings are of same color as forewing. Hindwing fringe is pale grey, except the anal region. Anal markings blackish and grey. Female is unknown.

Male genitalia are distinguished by the anvil-shaped uncus tip, with a pair of long arms. Uncus is short. Gnathoi lobes are triangularly shaped and medium sized. Socii curved and slender. Juxta bears a small anterior projection and a pair of long lobes, as seen also in *O. erythra*. Valva elongated with pointed and relatively short apical process. Apical process is shorter than in most other congeners. Digitus projection large and broad with rounded tip. Harpe strongly sclerotized and developed as a toothed plate. Sacculus is large and of rounded shape. Phallus straight and short. Phallus tip of circular shape, serrated with a pair of short tapered projections. Carina moderately sclerotized. Everted endophallus bears a pair of serrated, more strongly sclerotized bands. 8th sternite bilobed with rounded lobes on the sides. 8th tergite have a convex posterior margin, which is in *Omichlis* an exception.

Bionomics. The holotype was collected at 850 m above sea level.

Distribution. Known only from the type locality at the Sepik river in Papua New Guinea.

257. ***Omichlis albidilinea*** Gaede, 1930: 611; pl. 79: g.
Holotype: ♂, [Indonesia, West Papua], Neu-Guinea, 700 m [Nomnagihé, 25 miles south of Wangaar, 2.000 ft.] – NHMUK, London.
Schintlmeister 2013: 303.

Diagnosis. Forewings violet-brown. A white and fuscous violet- brown, longitudinal band runs from base to margin along vein M2. Basal, antemedial, and postmedial fasciae are white. Medial fascia of forewing colored fuscous violet-brown. Discal streak small and pale yellowish. Submarginal fascia consists of fuscous, violet-brown fine streaks. Fringe of all wings half fuscous reddish-brown and contrasting white. Hindwings are reddish-brown becoming paler toward the base. Anal markings white and fuscous-brown. Female is unknown.

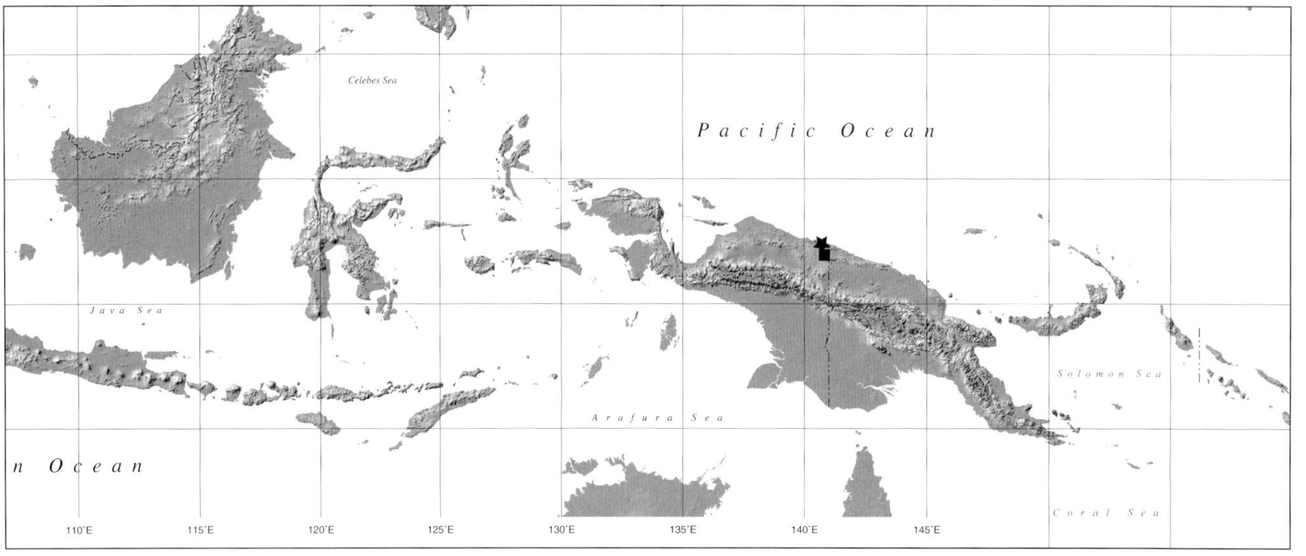

FIGURE 224 Distribution of *Omichlis stueberi.*

Male genitalia are well distinguished from those of other congeners by shape of the uncus tip, which has a pair of smaller lobes and a small central projection. Gnathoi lobes below the uncus tip are reduced in size. Socii broad, curved and rounded at the tip. Juxta displays a large central projection. Valvae have a convex shaped costa, a curved apical process and a relatively small knob-shaped digitus process. Harpe bears one or two small tapered projections. Phallus curved with a pair of small and rounded projections at the tip. Carina squarish and diverticulum reduced in size.

Bionomics. Only a few specimens (n = 8) are currently known, which have been collected 700–900 m above sea level.

Distribution. Endemic to New Guinea Island.

258. ***Omichlis stueberi* spec. nov.** (figs. 258a, 258b).
Holotype: ♂, Indonesia, W Papua (Ned. NW Guinea, Hollandia), Jayapura, Achterland, 20. vi. 1937, leg. W. Stüber – NBC, Leiden.

Paratypes (2 ♂♂):
W Papua: 1 ♂, Arso District Uskwar, 24. ix. 1938 leg. W. Stüber; 1 ♂, Hollandia, Uskwar, Bewani Mts., 7. iii. 1937.

Etymology. Named for Wilhelm Carl Julius Stüber (b. 13. viii. 1877 in Wollin d. 19. i. 1942 on SS Van Imhof near Nias). Stüber went 1917 to Hollandia (= Jayapura) and there collected chiefly dragonflies for Lieftinck, who described from this material more than 100 species as new to science. Later he worked until 1939 also for J. M. A. v. Groenendael and sent him and others numerous moths, which are now storaged in NBC, Leiden. His rich material was an important source in writing this book. Stüber died tragically as an internee.

Diagnosis. Forewing lengths ♂♂ 22–23 mm. A large species, larger than most other congeners. Antennae are covered with whitish scales. They are bipectinate, the last 30% toward the tip are filiform. Forewings distinguished by a small projection of the margin at vein M2 and a concave shape above this projection. Wing color pale brown. Postbasal and postmedial fasciae whitish, accompanied by a fuscous-brown shadow. Beyond the postmedial fascia a row of white spots occurs. Medial fascia fuscous-brown. Discal spot small and whitish, surrounded in black. Submarginal fascia marked as a row of blackish spots; sometimes weakly developed. Fringe of forewings fuscous brown. Hindwings are of same color as forewing, but fringe is paler brown. Female is unknown.

Unfortunately it was not possible to dissect and examine the male genitalia, but *stueberi* is by external appearance unmistakable.

Bionomics. Only three males (n = 3) are known so far. They have been collected at lower elevations, < 500 m.

Distribution. Known only in W Papua in the Jayapura area.

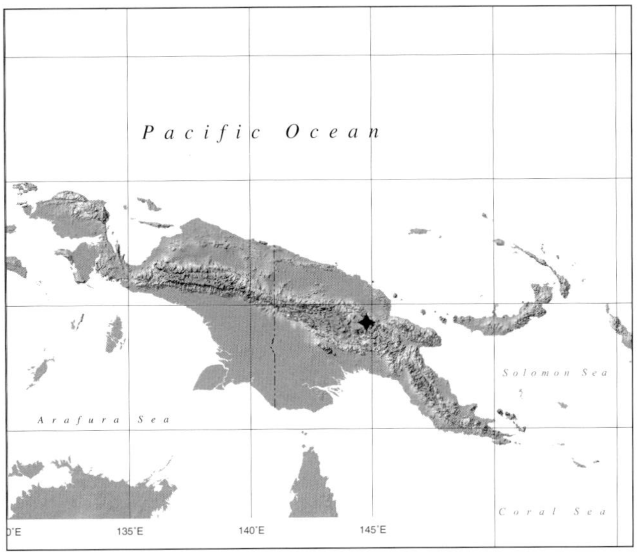

FIGURE 225 Distribution of *Omichlis superne*.

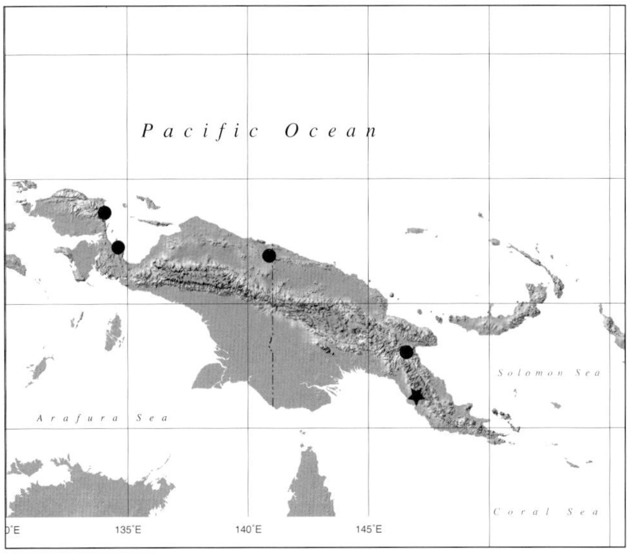

FIGURE 226 Distribution of *Omichlis plagata*.

259. *Omichlis superne* spec. nov. (figs. 259a, 259b).
Holotype: ♂, Papua New Guinea, Chimbu prov. Mt. Wilhelm, 3137 m, 6°00924' S, 144°58566' E, 16. ii. 2007, leg. M. Hoffmann (MWM 15.586) – Museum Witt, Munich.

No paratypes.

Diagnosis. Forewing length ♂ 22 mm. Antennae bipectinate, the last 28% toward the tip filiform. Outer margin of forewing rounded, not angled. Forewing color whitish-brown. Medial, postmedial and submarginal fasciae of a somewhat indistinct fuscous-brown. Discal spot yellowish. In the postmedial area veins are tinged black. A blackish brown apical streak is a good character for identification. Hindwings brown. Hindwing fringe is fuscous-brown and marginally whitish. The species is unmistakable in *Omichlis* by the pale external appearance. Female is unknown.

Male genitalia have a medium sized uncus with gently bilobed tip. A pair of small gnathoi lobes arises from the middle part of the uncus. Socii long and broad with rounded tip. Tegumen in the upper part compact and triangular shaped. Juxta posteriorally convex without a central projection, which is characteristic for many other *Omichlis* species. Valvae short and rounded. Apical valva process reduced in length. Digitus process tapered and strongly sclerotized at the pointed tip. Harpe have a strongly sclerotized process. Sacculus rather small with a pointed projection. Phallus curved with a characteristical shaped projection at the tip.

Bionomics. Holotype male was taken at 3100 m above sea level in subalpine forest.

Distribution. Known from Mt. Wilhelm in PNG.

260. *Omichlis plagata* Bethune-Baker, 1908: 179.
Holotype: ♀, [Papua New Guinea], Ekeikei, [1.500 ft.] – NHMUK, London.
Schintlmeister 2013: 304.

Taxonomic notes. A single specimen from PNG, Wau (adult fig. 260d, genitalia fig. 260b) differs by external appearance, e.g. discal spot, shape of the distal margin of the forewing, slightly from the illustrated holotype specimen. Male genitalia of this specimen display great differences in shape of uncus tip, socii, phallus tip bears a spine, smaller carina at endophallus, and shape of the 8th sternite. greatly from a dissected male from the Arfak Mts., see figs. 260a, 260c. Due to lack of material the differences in the male genitalia cannot confirmed by further dissections. This specimen is provisionally placed in *O. plagata*.

Diagnosis. Forewings fawn and margin is angled at vein M2. Postmedial and submarginal fasciae marked as rows of fine black dots. Best recognized by the dark chocolate-brown area along the dorsum, divided by a contrasting yellowish-brown line. Discal patch large and dark-brown filled. Hindwings are chocolate-brown.

Male genitalia occur in two forms: long and rectangularly shaped uncus with socii, longer than uncus length. The other form has a long uncus with a cruciform tip and medium-sized compact socii, which are shorter than the uncus. Valvae squarish having a long and curved apical process. Digitus projection knob-shaped. Sacculus bears a pointed process near base and a second pointed, larger

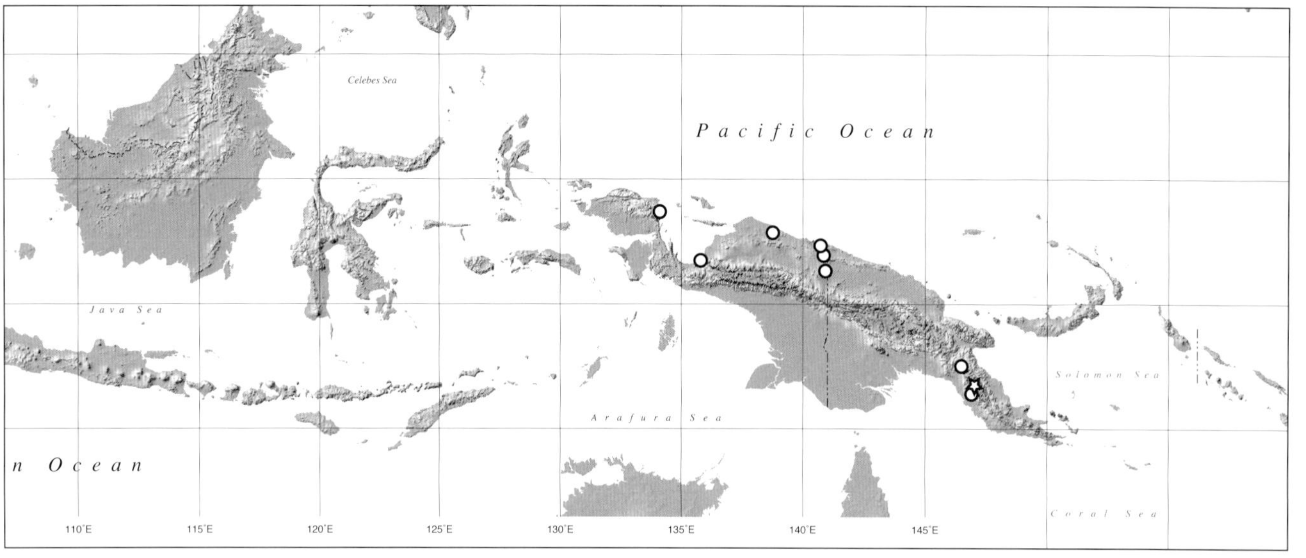

FIGURE 227 Distribution of *Omichlis rufofasciata*.

process distally. Harpe square-shaped. Phallus straight with a strongly sclerotized ring at the tip. Everted endophallus shows diverticulum with a serrated margin and two star shaped cornuti. Carina visible as a more strongly sclerotized rectangular plate. 8th sternite widely bilobed and 8th tergite slightly concave at posterior margin. Female genitalia have not yet been dissected..

Bionomics. The moth was observed rarely (n < 10 specimens currently known) at altitudes between 500–1700 m above sea level.

Distribution. Endemic to New Guinea.

261. ***Omichlis rufofasciata*** Bethune-Baker, 1904: 377; pl. 6: 17.

Lectotype: ♂, [Papua New Guinea], Mount Kebea [6.000 ft.] – NHMUK, London **by present designation.**

Schintlmeister 2013: 305.

Taxonomic note. *Omichlis rufofasciata* was described from a pair from "Mount Kebeae" and "Ekeikei". To stabilize the nomenclature the ♂ with the following labels is hereby designated as lectotype: "Mt. Kebeae. | B.C. New Guinea | 6000 ft. | March–April 1903 | A.E Pratt Coll. "; green circular label "Type"; "G.T.B. -Baker Coll. | Brit. Mus. 1927–360; yellow label "Photo done by A. Schintlmeister #3932". A lectotype label was added. The lectotype is depicted here in fig. 261c. The remaining ♀ from Ekeikei becomes a paralectotype.

Diagnosis. Forewings cream-colored, slightly greenish tinged. The prominent dark chocolate-brown medial area distinguishes the species from other congeners. The basal fascia consists of two fuscous-brown spots near the costa. Discal streak in the fuscous-brown medial area contrasting cream. Postmedial and submarginal fasciae marked as curved fuscous-brown streaks. At vein M2 a black margin patch occur. Hindwings reddish-brown with fuscous-brown and cream anal markings. Sexual dimorphism is limited.

Male genitalia have a resemblance to those of *O. plagata*, but uncus tip bears a pair of long arms. A central projection ends in a pair of lobes. A pair of short gnathos lobes arises directly above the curved and short socii. Valva elongated with medium-sized apical process. Digitus process large of rectangular shape. Harpe displays two small projections. Sacculus rather small. Phallus straight with a small untapered process at the tip. Everted endophallus more strongly sclerotized, bears a serrated diverticulum. Carina weakly sclerotized. 8th sternite at posterior margin triangularly notched. 8th tergite posteriorally concave shaped. Female genitalia have reduced apophyses posteriores, but apophyses anteriores are long and well sclerotized. Postvaginal plate deeply notched. Ductus bursae very compact, curved, and short. Corpus bursae ellipsoid with a medium sized reniform signum.

Bionomics. Adults occur infrequently and mostly in the lowlands between 50–700 m, but the lectotype was taken at 1800 m above sea level.

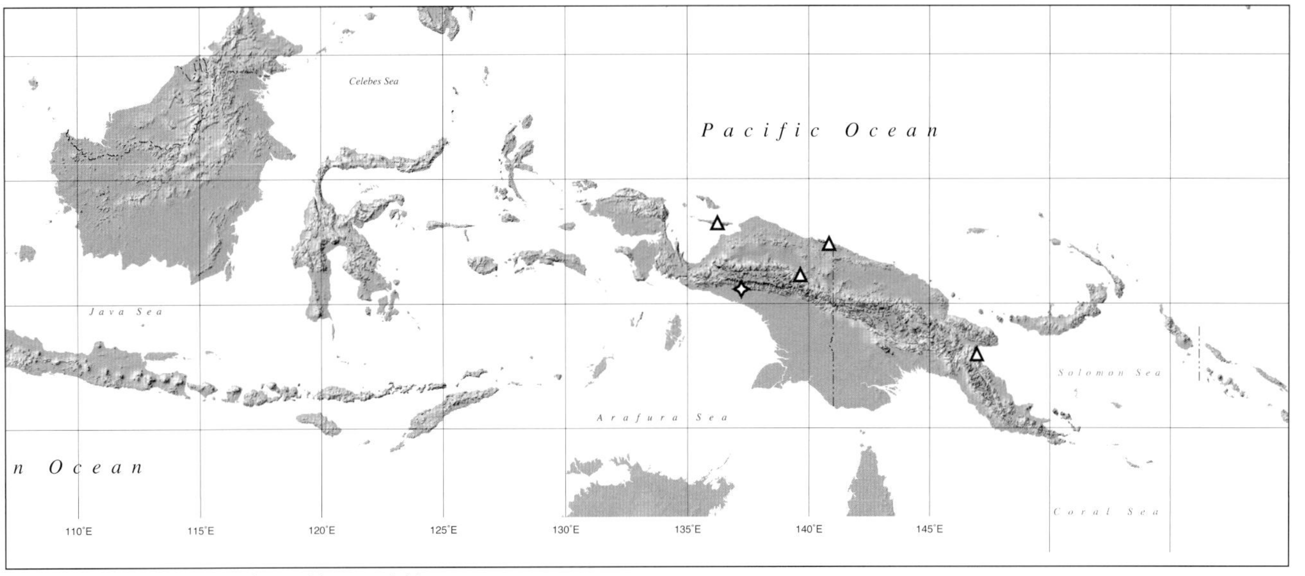

FIGURE 228 Distribution of *Omichlis pseudolibatrix*.

Distribution. Restricted to New Guinea Island.

262. ***Omichlis pseudolibatrix*** Rothschild, 1917: 259; pl. 6: 16.

Holotype: ♀, [Indonesia, W Papua], Dutch New Guinea, Snow Mountains, near Oetakwa River, up to 3.500 ft. – NHMUK, London.

Schintlmeister 2013: 304.

Diagnosis. Margin of forewings angled at vein M2 and forewing apex acute. Forewings of males are light brownish yellow, variable in paleness. Medial fascia inconspicous pale yellow, accompanied by a fine brownish line. In the postmedial area several fasciae occur of which on of them is prominently broad fuscous-brown. Contrasting white and submarginal fascia are unique characters and helpful for identification. Along vein M2 sometimes a fuscous-brown shadow of variable intensity occur. Discal spot ellipsoid, surrounded in dark brown. Hindwings are brown, with light yellowish-white medial and anal areas. The sexually dimorphic female is of a brown wing color, only in the medial area of forewings does a yellow dorsal patch appear. The margin of the forewing is convex above the vein M2. Forewing pattern resembles the male, but hindwings are brown. Anal markings consist of few small black dots.

Male genitalia have a large squarish shaped uncus, where a pair of gnathoi lobes arises from its central part. Socii are long, slender, kinked and pointed at their tips. Valva short and broad. Apex process very compact and short. Digitus process large; there is a second smaller projection. Harpe large and circular with a tapered projection. Sacculus small and inconspicuous. Phallus straight and short. At the phallus tip there is a large serrated and strongly slerotisized plate. Everted endophallus bears a large carina and a small serrated diverticulum. Female genitalia have a pair of pointed, strongly sclerotized lobes at the postvaginal plate. Ductus bursae long. Corpus bursae of circular shape with a large but weakly sclerotized signum.

Bionomics. Only a few specimens (n = 6) are known, which were taken at altitudes between 400 m and 1700 m above sea level.

Distribution. Distributed in New Guinea Island and Yapen Island.

263. ***Omichlis mediofasciata*** Rothschild, 1917: 260; pl. 6: 15.

Lectotype: ♂, [Indonesia, W Papua], North-east New Guinea, Mambare River, Biagi 5.000 ft. – NHMUK, London. **by present designation.**

Schintlmeister 2013: 304.

Taxonomic note. *Omichlis mediofasciata* was described of 2 males. To stabilize the nomenclature the ♂ with the following labels is designated hereby as lectotype: "Biagi, Mambare | R., 5000 ft., B. N. | G. April 06 | (A. S. Meek)"; red circular label "Type"; "*Omichlis* | *mediofasciata* | Rothsch."; yellow label "Photo done by A. Schintlmeister #3918". A lectotype label was added. Lectotype is figured here in fig. 263b. The remaining ♂ becomes a paralectotype.

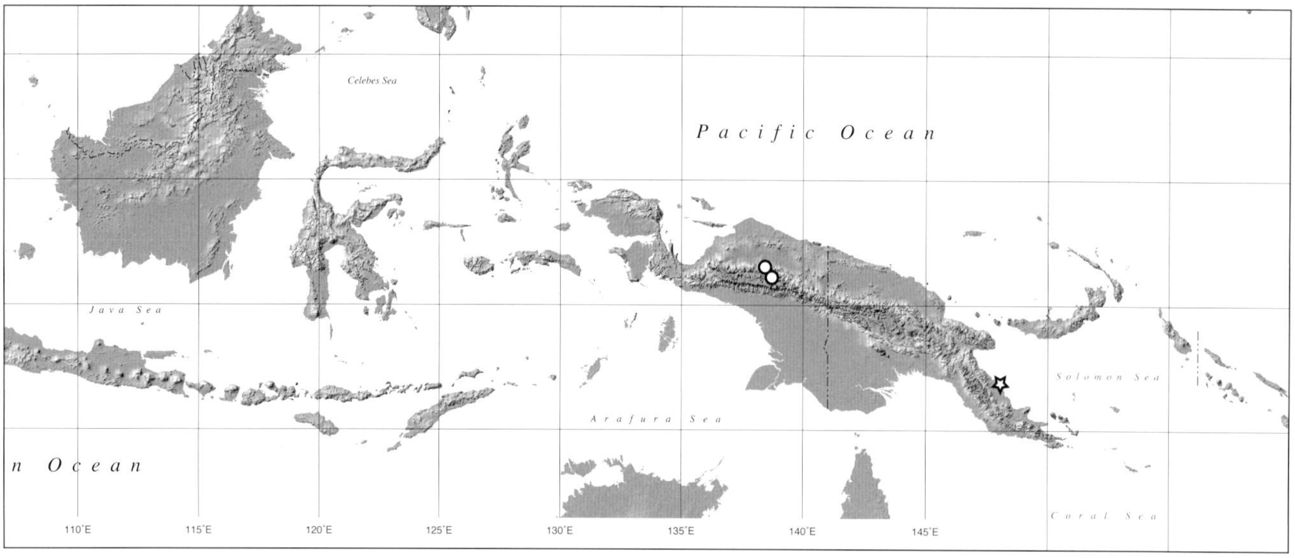

FIGURE 229 Distribution of *Omichlis mediofasciata*.

Diagnosis. Forewings brown to golden-brown. Forewing margin angled at vein M2. Postbasal, postmedial and submarginal fasciae are doubled, distinctly whitish and brown colored. Medial fascia is indistinct and brown colored. Discal spot ellipsoid of golden or brown color, surrounded by a pair of brown and whitish colored lines. Apex of forewing displays a diffuse yellowish-brown patch, which is a characteristic feature of *mediofasciata*. The lectotype and paralectotype display a longitudinal brown band along M2 to the margin. A further male (fig. 263b) and the female lacks this band, but having the vein yellowish-tinged. Hindwings are brown, fading toward the anal and basal region to yellowish. Anal markings are large and blackish-brown. Sexual dimorphism is limited.

Male genitalia are among the smallest within the genus. Uncus is very short with a broad ellipsoidal uncus tip. Gnathoi are not visible, but a pair of socii are well developed. Tegumen is in its posterior part unusually long. Valva has a convex costa and a short apical process. Digitus projection knob-shaped. Harpe displays a large projection. Sacculus small and inconspicuous. Phallus straight with strongly sclerotized ring at the tip. Everted endophallus bears a rather small carina but diverticulum is large and serrated. 8th sternite slightly bilobed at posterior margin. 8th tergite shape conical toward the straight posterior margin.

The author was unable to trace the genitalia slide of the only known female in NBC, Leiden.

Bionomics. Four specimens (n = 4) are currently known, which were taken at altitudes of 1500 m, 2140 m and 2600–2800 m above sea level.

Distribution. Restricted to New Guinea Island.

Omestia Bethune-Baker, 1908: 177.

Type species: *Omestia bella* Bethune-Baker, 1908.

Schintlmeister 2013: 302.

Diagnosis. Medium-sized moths. Male antennae are bipectinated with short rami, except for the last 30% toward the tip. Female antennae are filiform. Forewing color is fawn, those of the hindwings is reddish to chocolate-brown with reduced blackish anal markings in both sexes. Forewing have a rounded outer margin. Discal spot ellipsoid. Several fasciae are visible on the forewings: basal, medial, postmedial, submarginal fasciae. The occurrence of a contrasting white streak in the medial area is a characteristical genus feature. *Omestia bella* is slightly sexually dimorphic. Females have darker colored wings with acute apexes, while males have rounded apexes. Black pattern is reduced to some costal dots and the postbasal fascia. Postmedial fascia slightly darker, marked green. Fringes are fuscous grey and checkered bluish-green. Hindwings are grey.

Male genitalia are not known yet. Female genitalia have contrary to *Omichlis*, posterior and anterior apohyses long and of equal length. Papillae anales are large and strongly sclerotized. Antrum large, moderately sclerotized. Antevaginal plate rather small with a pair of lobes. Ductus bursae is long and curved, corpus bursae disc-shaped with a reniform signum on the bottom.

The genus is monotypic and restricted to New Guinea Island.

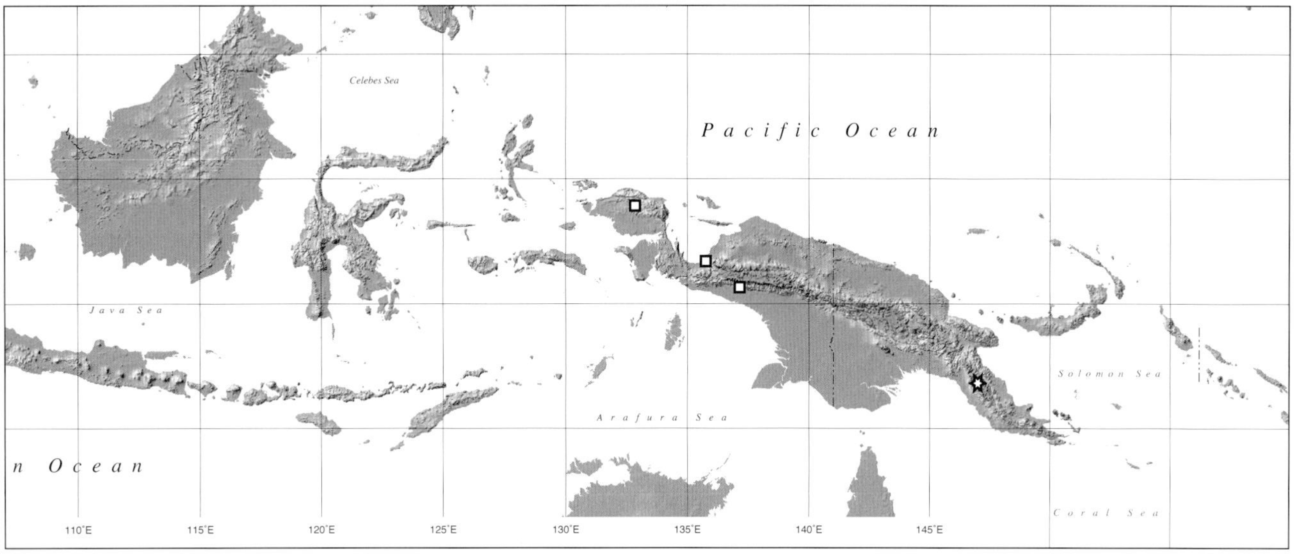

FIGURE 230 Distribution of *Omestia bella.*

264. *Omestia bella* Bethune-Baker, 1908: 177.
Holotype: ♂, [Papua New Guinea], Kebea Range, 6000 ft. – NHMUK, London.
Schintlmeister 2013: 303.

Taxonomic notes. The holotype of *Omichlis plusiotis* Rothschild, 1917: 259; pl. 4: 3., holotype: ♀, Dutch New Guinea, Snow Mountains, near Oetakwa River, up to 3.500 ft. – NHMUK, London, was originally described as a male, but in fact, like the other two paratypes, they are all females.

Comparison with the holotype ♂ of *bella* led to the conclusion, that *plusiotis* is the sexually dimorphic female of *bella,* and the latter takes priority (**syn. nov.**).

Kiriakoff, 1967 illustrated as "*Omestia bella*" on pl. 7: 51 the imago of the holotype female of *Syntypistis bella.* It is questionable, to which species the accompanying illustration of a male genitalia in Kiriakoff, 1967: fig. 112 really belongs. The holotype male is not dissected yet.

Diagnosis. Best recognized by the contrasting white streak in the medial area near the dorsum. Costa of the forewings prominently fuscous-brown colored. Sometimes an indistinct fuscous blotch near the tornus occurs. Female is sexually dimorphic as described above. Among the few examined females greater differences in forewing length occur.

Male genitalia are unknown; for description of female genitalia see under genus diagnosis.

Bionomics. Only a few specimens (n = 6) are known, which were taken at 700 m and 1100 m above sea level.

Distribution. Restricted to New Guinea Island.

Netria Walker, 1855: : 1388 (key), 1504.
Type species: *Netria viridescens* Walker, 1855.
Schintlmeister 2008: 181; Schintlmeister 2013: 281.

Diagnosis. Large adults. Antennae in both sexes brown and bipectinated. Last 8% toward the tip are filiform. Forewings are deep green to pale bluish green colored. They display brown basal, postbasal, postmedial and submarginal fasciae. Marginal fascia consists of fine, black, U-shaped parts. Hindwings are uniformly chocolate-brown except for a strip along the costa, where the green color and brown postmedial fascia of forewings continues. Hindwings does not show anal markings. Females are sexually dimorphic having fuscous medial area of a darker brown than the basal or postmedial area. Green forewing color appears often paler and more bluish-tinged than in the corresponding male.

Male genital apparatus relatively small. Uncus reduced; a pair of weakly sclerotized and slender socii are present. Valvae are small and bulbous. They bear a curved apical process, which is pointed at the tip. Juxta is large, strongly sclerotized, and centrally deeply notched. Tegumen bears often a pair of long processes; saccus developed long or very long. Phallus long, slender and slightly curved. Near the tip a long serrated plate arises. Everted endophallus approximaterly of same length as uncus but bears no sclerotized structures. A characteristic feature of the genus and taxonomically important are the modified 8th abdominal segments, which are visible by brushing the underside of the abdomen tip from its hairs. 8th sternite have an anterior central projection and the posterior margin bears spines, processes or other structures, which are often asymmetric. 8th tergite is concave at the anterior margin and posterior margin bear often a central projection. Female genitalia are much larger in relation to male genitalia size. They have a pair of long apophyses posteriores, while apophyses anteriores are shorter. Antevaginal plate strongly sclerotized, large and either posteriorally bilobed or deeply notched. Antrum small. Ductus bursae slender, curved and of variable length. Corpus bursae circular, sometimes with a smaller circular or ellipsoid appendix. No signum.

Hostplants of caterpillars are *Bassia* (Amaranthaceae), *Mimusops* (Sapotaceae), and *Achras* (Sapotaceae) (Holloway, 1983: 61).

Netria was revised by Schintlmeister, 2006. Currently it includes 12 species and many subspecies. Species are distributed from Himalaya, India in its entirety, Sri Lanka, Myanmar, S China, Taiwan, Indochina, the Philippines, Sundaland, Java, Bali, lesser Sundas, Timor, Wetar, Tanimbar, Sulawesi, Moluccas, New Guinea and Australia (Queensland).

265. ***Netria viridescens*** Walker, 1855: 1504.
Holotype: ♀, [Indonesia], Java – NHMUK, London.
Schintlmeister 2008: 182; Schintlmeister 2013: 282.

Taxonomic note. The species tends to establish geographical subspecies. In Indonesia the ssp. *viridescens* occur. Dissected males from Sumatra and Kalimantan (n = 7 GU) differ seemingly from Javan males (n = 5) by a thicker phallus. But the entire series from Sumatra and Kalimantan is inhomogeneous by external appearance and also male genitalia vary individually as well.

Diagnosis. Males are as described in the genus diagnosis and hard to separate by external appearance from most other congeners. A blackish dorsal shadow occurs near the tornus in both sexes. But there are also specimens, where this patch is very weakly developed and hard to see. Hindwings are chocolate-brown; on the costa there is a slender green strip, interrupted by two blackish patches. A blackish hairbrush on the first abdominal segment in both sexes is sometimes a helpful character in differentiation from the some sympatrically congeners, such as *N. carentis* or *N. multispinae*. Female is sexually dimorphic having the green forewings tinged with white scales and a fuscous-brown medial area. An indistinct whitish postmedial fascia on forewing distinguish the female from other congeners. The discal spot of ssp. *viridescens* is indistinctly developed pale yellowish-brown and variable in its visibility. Medial area is suffused in brown.

Genitalia capsula and phallus are as described in genus diagnosis. There is some individual variation in valvae shape, its apical processes as well as in the thickness of the saccus (fig. 265b). Phallus is shorter than in *longisacci* or *insularis*. *N. viridescens* is best identified by shape of its 8th abdominal segments. 8th sternite bear one large central spine accompanied by a 1–3 smaller triangular spurs. 8th tergite conically shaped and straight at the smooth posterior margin. Female genitalia are distinguished by a deeply notched postvaginal plate. Antrum is the largest among *Netria* species. Ductus bursae long and straight. Corpus bursae circular with an appendix.

Bionomics. A common species, which occurs in Sundaland between 0 m and 1700 m, but medium altitudes of around 1000 m are preferred. In Nepal the ssp. *continentalis* Schintlmeister, 2006 was observed up to 3000 m above sea level.

Distribution. *Netria viridescens viridescens* occur in Malaya, Borneo, Sumatra, Belitung, Java and Bali. In Mainland Asia (Himalaya, Myanmar, China, Taiwan, entire Indochina) the ssp. *continentalis* Schintlmeister, 2006 fly; in SW China (Yunnan) ssp. *suffusca* Schintlmeister, 2006 and in Palawan ssp. *pallidabasis* Schintlmeister, 2006 occur.

266. ***Netria torajae*** Schintlmeister, 2006: 70; pl. 2: 1, 2, 4; pl. 7: 1, 2.
Holotype: ♂, Indonesia, N-Sulawesi, Mt. Sampuraga, 2°10' N, 123°45' E, 1.400 m – Museum Witt, Munich.
Schintlmeister 2013: 282.

Diagnosis. Distinguished by the contrasting discal spot surrounded in white and a reniform stigma, also surrounded in white, on the forewing. Apart from this, a few white scales are seen only at the fuscous-brown basal fascia and occasionally at the postmedial fascia. Fuscous medial fascia developed as an indistinct shadows. The chocolate-brown hindwings display a fuscous medial shadow, which does not occur in other congeners. Individual variability in external appearance is limited in males. There are only two females at the author's disposal, neither of which shows sexual dimorphism. The medial area of the female does not differ in coloration from the other forewing parts.

Male genitalia resemble those of *N. viridescens*, but the strongly sclerotized valvae are broader and curved. Apical

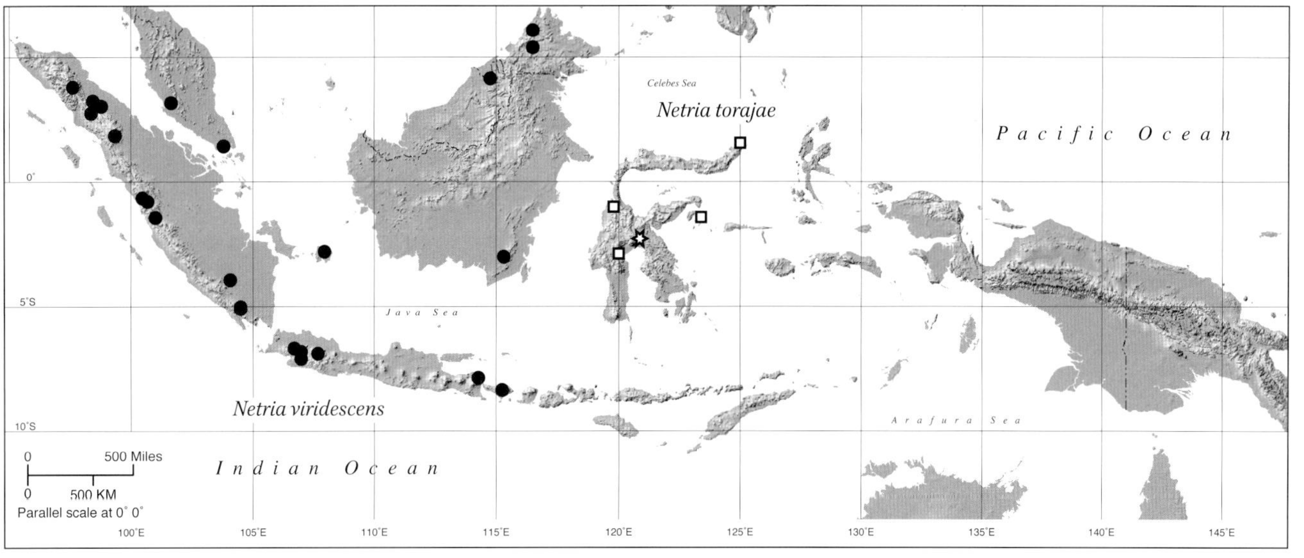

FIGURE 231 Distribution of *Netria viridescens* and *torajae*.

valva process very short. Saccus variable in shape. Phallus short and straight. The clubbed phallus tip distinguishes *torajae* from other congeners. 8th sternite bears one large central spine, but contrary to *viridescens* the entire posterior margin is broad and strongly sclerotized. The posterior margin of the 8th tergite straight, but wider than in *viridescens*. Female genitalia are similar to *viridescens*, but postvaginal plate is rather straight and shallower notched at posterior margin.

Bionomics. Males occur frequently, in Sulawesi between 900–1700 m, but on Peleng Isl. between 150–700 m above sea level.

Distribution. Restricted to Sulawesi and Peleng Isl.

267. ***Netria longisacci*** Schintlmeister, 2006: 71; pl. 2: 8, 11; pl. 7: 4.
Holotype: ♂, Indonesia, Nusa Tenggara, Sumbawa, Gunung Takan, 800 m – coll. A. Schintlmeister, Dresden.
Schintlmeister 2013: 281.

Diagnosis. Discal spot and reniform stigma of forewings rown tinged, contrasting, medial area. The forewing color of the female varies from light bluish-green, as illustrated, to deep green, as seen in the males. Females of *longisacci* display a large projection from postbasal fascia toward the medial area of forewing.

Male genitalia are distinguished by a pair of long and strongly sclerotized tegumen processes. In the closely related *N. insularis* these tegumen processes are absent, but in *arisemna* they are present. Valva costa is sinuous and bears a medium sized, curved apical process. Saccus appears very long, but saccus length is subject of individual variation, see fig. 267d. Juxta is strongly sclerotized and deeply notched. Both lobes of juxta are sharply pointed at their tip. Phallus is, together with *insularis*, among the longest within the genus and gently curved. 8th sternite is asymmetric and characteristically shaped. Posterior margin is strongly sclerotized and bears a prominent spine of variable shape; occasionally a second or third small spine occurs. 8th tergite is conical with a short process at the triangular tip. Male genitalia and 8th abdominal segements are variable in their shapes, but no geographical correlation was found between dissected material from various islands (n = 10 GU). Female genitalia have apophyses anteriores of equal length to the apophyses posteriores. Postvaginal plate gently bilobed, narrower than in *insularis*. Ductus bursae straight and corpus bursae disc-shaped.

Bionomics. Moths have been collected frequently between 350–1820 m above sea level.

Distribution. Distributed in the Lesser Sundas: Lombok, Sumbawa, Sumba, Flores, Timor, Pantar, Alor and Wetar.

268. ***Netria insularis*** Schintlmeister, 2006: 71; pl. 2: 3, 5; pl. 7: 3.
Holotype: ♂, Indonesia, Tanimbar isl., East Yamdena, Arui Bab, 120 m – coll. A. Schintlmeister, Dresden.
Schintlmeister 2013: 281.

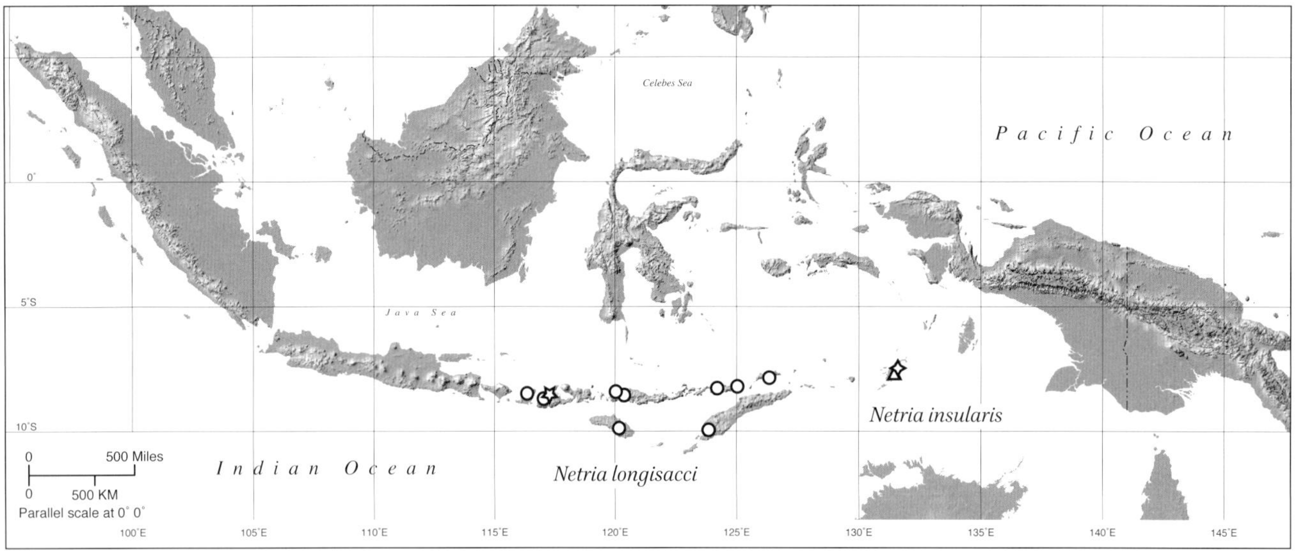

FIGURE 232 Distribution of *Netria longisacci* and *insularis*.

Diagnosis. Closely related and similar to the prevous species. Due to their overlapping individual variability, males are externally indistinguishable from *N. longisaccus*. Sexually dimorphic females (n= 4 ♀♀) show a greasy appearance in the medial area of forewings, a very special effect, which does not occur in any other *Netria*. The projection from postbasal fascia toward the medial area of forewing is shorter in *insularis* in comparison to *longisacci*.

Male genitalia are similar to those of *longisacci*, but well distinguished by the absent tegumen processes, longer socii, and less curved apical valva process. Juxta is more widely notched. Phallus is thicker than in *insularis*, and basal projection is reduced. 8th sternite less waisted, also with several spines of variable shape at the posterior margin. 8th tergite differs by wider bilobed anterior margin and lacks a process at the triangular shaped posterior tip. Anterior margin of *insularis* is wider and both lobes are more pointed. Female genitalia differ from *insularis* by wider 8th abdominal segment, shorter and thicker apophyses anteriores and a a curved corpus bursae.

Bionomics. Adults have been observed several times frequently at lower elevations of 150 m above sea level.

Distribution. Endemic to Tanimbar, Yamdena Isl.

269. *Netria arisemna* (Turner, 1931): 328 (*Syntypistis arisemna*).

Holotype: ♂, Australia, North Queensland, Cap York – Queensland Museum, Brisbane.

Schintlmeister 2013: 281.

Taxonomic notes. The species tends to establish geographical subspecies, which are best characterized by the shapes of the central process of the 8th tergite, in particular their thickness and also by shape of the juxta.

Schintlmeister (2006) described *Netria bipartita* as a genuine species with several subspecies. *Netria arisemna* was treated in that paper as a different species from Australia. Now, since more material became available, several transitional forms have been found, and it is evident, that *bipartita* and *arisemna* are conspecific.

So *N. arisemna* is represented by the following subspecies:

Netria arisemna bipartita Schintlmeister, 2006: 72; pl. 3: 1, 2, 5; pl. 7: 5, 6; pl. 10: 3, **stat. nov.**
Holotype: ♂, Indonesia, Sulawesi, Mt. Tambusisi – Museum Witt, Munich;

Netria arisemna andreasmuelleri Schintlmeister, 2006: 73; pl. 3: 3; pl. 7: 9, **stat. nov.**
Holotype: ♂, Indonesia, Maluku, W-Seram, Kamarian, 3°20'S, 120°30' E, 200 m – coll. A. Schintlmeister, Dresden;

Netria arisemna tarasovi Schintlmeister, 2006: 73; pl. 3: 4; pl. 7: 7, 8, **stat. nov.**
Holotype: ♂, Indonesia, Halmahera, 15 km SW Baru, 0°10' N, 127°32' E, 600 m – coll. A. Schintlmeister, Dresden.

Netria arisemna philippina Schintlmeister, 2006: 73; pl. 3: 6, 7, 9; pl. 8: 1, 2, **stat. nov.**
Holotype: ♂, Philippinen, Negros, Mt. Canlaon, W-Route via Mambucal, 10°22' N, 123°12' E, 600 m – Museum Witt, Munich.

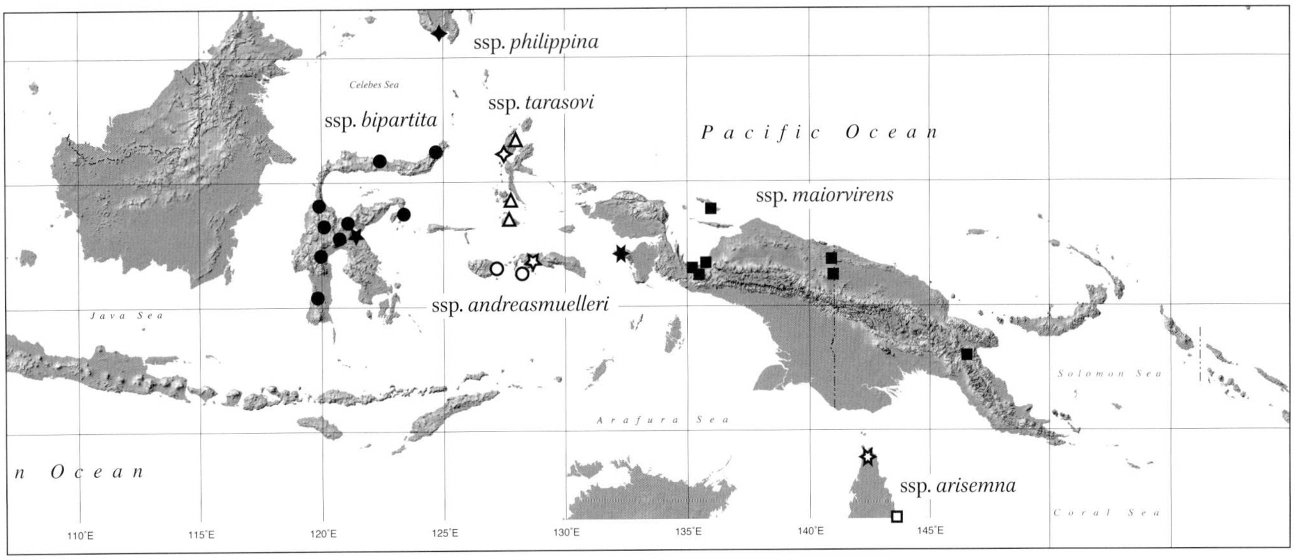

FIGURE 233 Distribution of *Netria arisemna*.

Populations from New Guinea representing a further, hitherto undescribed subspecies. They are described below as:

Netria arisemna maiorvirens **ssp. nov.** (figs. 269a, 269b, 269m–o).

Holotype: ♂, Indonesia, West Papua, NE Fak-Fak, 600 m, 2°55'S, 132°20'E, iv. 2009 – coll. A. Schintlmeister, Dresden.

Paratypes (9 ♂♂, 1 ♀):

W Papua: 2 ♂♂, NE Fak-Fak, 600 m, 2°55'S, 132°20'E, iv. 2009 (GU 66-33); 1 ♂, Arso district, Uskwar, 28. x. 1937, leg. W. Stüber; 1 ♂, Jafi district, Affi, 16. vi. 1939, leg. W. Stüber; 1 ♂, Nabire, Irian Jaya Highway, km 45, 745 m, 3°29'52" S, 135°83'85" E, 22.–24. xi. 1997, leg.K. Cerny (GU 81-32); 1 ♂, 1 ♀, Nomnagihé, 25 m South of Wangaar, 2000 ft., i.–ii. 1921, leg. C., F., & J. Pratt; 1 ♂, Mt. Kunupi, Menoo Valley, Weyland Mts., 6000 ft., xii. 1920–i. 1921, leg. C., F., & J. Pratt;

Biak: 1 ♂, Biak, 29. x. 1953;

Papua New Guinea: 1 ♂, Morobe prov., near Bulolo, Mt. Susu Nat. Res., 975 m, 27.–28. viii. 1983, leg. S. Miller.

Diagnosis of ssp. *maiorvirens*. Forewing lengths ♂♂ 30–38 mm; ♀ 42 mm. Much larger in forewing length, on average 4 mm in both sexes, than ssp. *arisemna* or other subspecies of *arisemna*. Forewing color of ssp. *arisemna* is greyish-green, and in the other subspecies deep green, but in *maiorvirens* pale green, sometimes yellowish-green, see fig. 269o. Postbasal area mixed with whitish scales. Hindwings are as in ssp. *arisemna* but darker than in other populations. Female is sexualdimorphic and resembles those of ssp. *bipartita*. Australian populations have paler and more greyish tinged females.

Male genitalia, resembles to those of ssp. *arisemna*. Valva and its apical process rather straight. Juxta is the largest among all known subspecies of *arisemna*. Saccus compactly developed and longer than those of any other Indonesian populations. 8th tergite bear the largest central process of all known subspecies. Female genitalia have not yet been dissected.

Diagnosis. Postbasal area of the forewings mixed with white scales. There is a black dorsal spot near to the tornus, as in *N. viridescens*. Three black costal dots in the postmedial area of the forewings distinguish *arisemna* from *insularis* or *longisacci*. Sexually dimorphic female have light blue-green forewings with a contrasting, fuscous-brown medial area. Postbasal fascia is deeply notched toward the outer margin.

Male genitalia resemble *viridescens*, but juxta and saccus are larger. Phallus long and slightly curved or straight. Endophallus of same length as phallus, but without sclerotized structures. *N. arisemna* is best to distinguish from other congeners by shape of 8th abdominal segment. 8th sternite bears a long double spine of different length. Posterior margin broad and strongly sclerotized with a pair of smaller central spines. 8th tergite has a very large and prominent central process. From east (New Guinea) to west (Sulawesi) these projections become slenderer. Female genitalia are distinguished by the shape of a large and strongly sclerotized postvaginal plate. Apophyses anteriores are short, shorter than in other congeners. Ductus

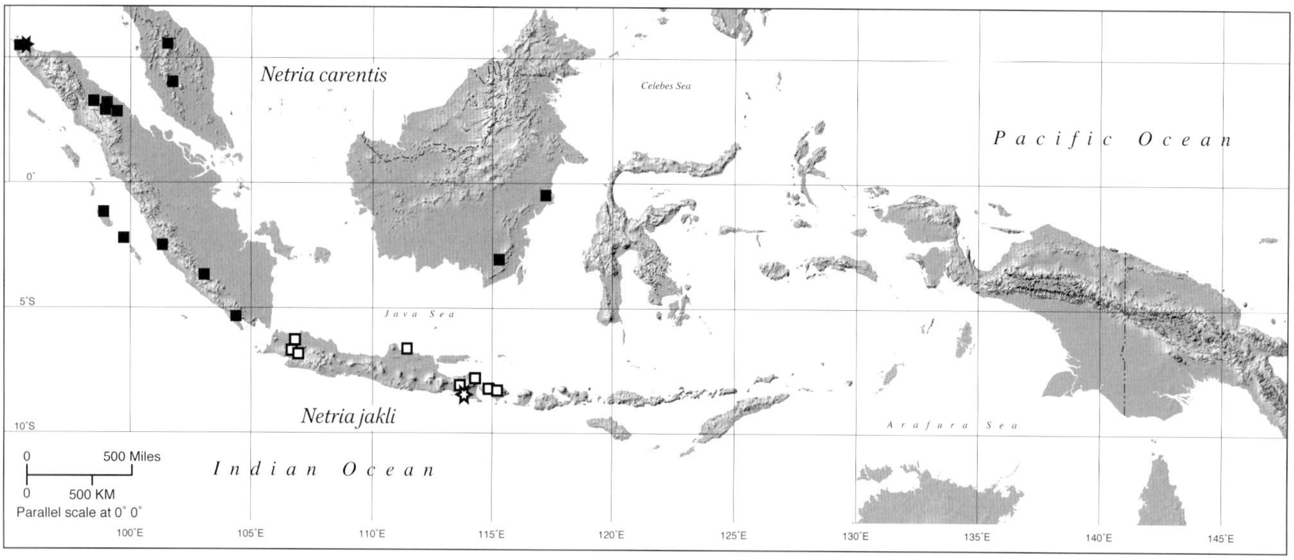

FIGURE 234 Distribution of *Netria jakli* and *carentis*.

bursae is the broadest within *Netria* and rather straight. Corpus bursae circular with a tendency to establish an appendix as in *viridescens*.

Bionomics. Adults occur frequently, except in New Guinea and Australia, where this species is apparently rare. Moths occur at lower and medium altitudes up to 1200 m; a single male was taken at 1800 m.

Distribution. *N. arisemna arisemna* occur in NE Australia (Queensland); ssp. *maiorvirens* is distributed in New Guinea in its entirety and Biak; ssp. *andreasmuelleri* inhabits Buru, Ambon and Seram. Ssp. *tarasovi* is found in Halmahera, Bacan and Obi, whereas ssp. *bipartita* occur in Sulawesi and Peleng. The entire Philippines (except Palawan) are populated by ssp. *philippina.*

270. ***Netria jakli*** Schintlmeister, 2006: 77; pl. 4: 1-3; pl. 8: 5-7.
Holotype: ♂, Indonesia, E Java, Meru-Betiri Nat. Park, 25 km S Kalibaru, 8°13'S, 113°45' E, 1.400 m – Museum Witt, Munich.
Schintlmeister 2013: 281.

Diagnosis. Similar in external appearance to *N. carentis* but smaller. Discal spot and reniform stigma on forewing marked by white scales. Most males of *jakli* display the green forewings, tinged fuscous-brown; the illustrated holotype, fig. 270d, do not. Sexually dimorphic female differs from sympatrically *viridescens* by its smaller size and more fuscous colored medial area of forewing.

Valva strongly sclerotized and strongly curved at the tip; apical process short. Juxta and saccus significantly smaller than in *carentis*. Phallus straight and slender. 8th sternite bear a pair of spines, which are absent in *carentis*. Posterior margin strongly sclerotized. Shape of the 8th tergite resembles *carentis* with a large central projection. Female genitalia differ from *carentis* by short apophyses anteriores. Postvaginal plate bears a pair of large projection with rounded tip, which is unique within *Netria*. Ductus bursae short.

Bionomics. Adults occur infrequently and has been observed mostly at medium altitudes between 500 m up to 1500 m.

Distribution. Restricted to Java and Bali.

271. ***Netria carentis*** Schintlmeister, 2006: 80; pl. 4: 4-6; pl. 9: 1, 2; pl. 10: 6.
Holotype: ♂, Indonesia, Sumatra; NW Aceh, 34 km E Banda Aceh, Mt. Silawa, 600 m – coll. A. Schintlmeister, Dresden.
Schintlmeister 2013: 281.

Diagnosis. Males are distinguished from other congeners by relatively narrow forewing with more prominently and broad developed fuscous-brown postbasal and postmedial fasciae. A characteristic feature of *carentis* is an additional brown medial fascia. A blackish tornal patch on forewings as occurring in *viridescens* is absent. Female is sexually dimorphic by pale blue-green forewing color contrasting fuscous-brown medial. Both features distinguish females from *jakli* or *viridescens.*

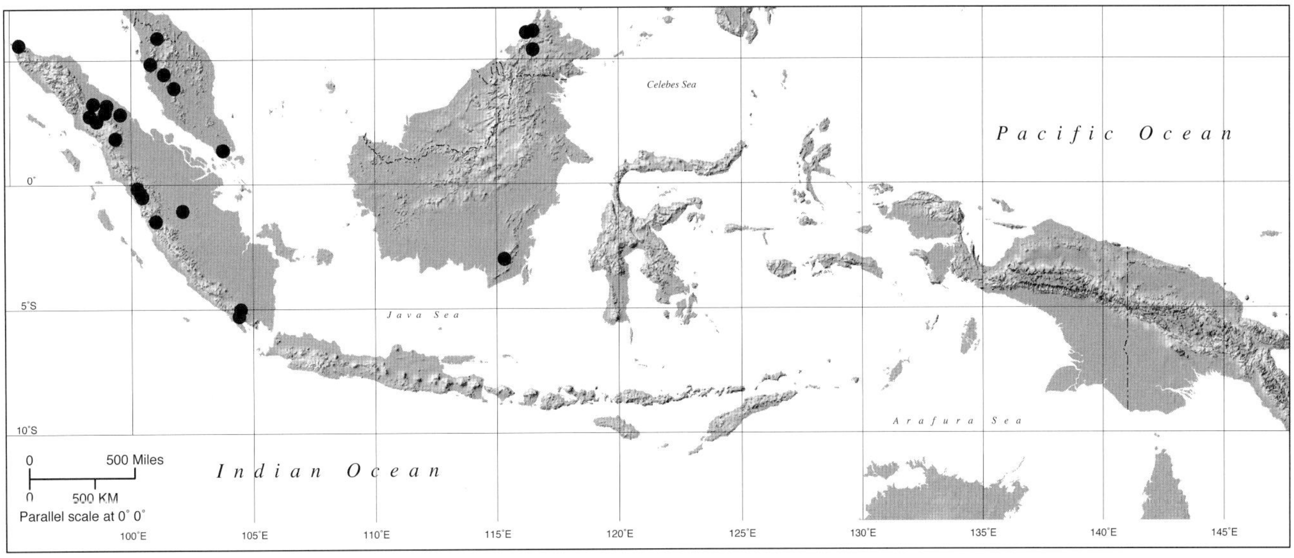

FIGURE 235 Distribution of *Netria multispinae.*

Male genitalia are large and characterized by relatively small valvae, where the distal part is rectangularly angled. Juxta is deeply notched and saccus very long. Phallus straight or slightly curved, compact and large. Phallus base is elongated like a slender process. 8th sternite diagnostically shaped and without processes or spines on the strongly sclerotized posterior margin. Shape of the 8th tergite is as in *jakli,* but the posterior projection triangular notched at the tip. Female genitalia resemble *N. insularis.* Both pairs of apophyses are robust and of equal length. Postvaginal plate deeply notched. Corpus bursae of circular shape.

Bionomics. Adults are frequent and has been observed mostly at altitudes below 1000 m above sea level. A few adults has been taken at higher altitudes up to 1500 m.

Distribution. Distributed in Sumatra, Mentawei Isls. (Siberut and N Pagai), Belitung, Borneo and the Malayan Peninsula.

272. ***Netria multispinae*** Schintlmeister, 2006: 81; pl. 5: 1-5, 10; pl. 9: 5.
Holotype: ♂, N Vietnam, Fan-si-pan, N-Seite, 22°17' N, 103°44' E, 1.600 m – Museum Witt, Munich.
Schintlmeister 2008: 182; Schintlmeister 2013: 282.

Taxonomic note. The melanic *Netria multispinae nigrescens* Schintlmeister, 2006 was described from SE China (Jiangxi). Outside Cina and in Sundaland the nominotypical subspecies *multispinae* occur.

Diagnosis. This is the largest *Netria* in Sundaland and well characterized by narrow forewing shape of males with somewhat pointed apex. Along the postbasal fascia whitish scales occur and the discal spot is surrounded by several fine white lines. A blackish tornal spot, as present in *viridescens,* is absent. Female is sexual dimorphic having a conrasting furcous green filled basal and medial area. Forewing shape is narrower than in other female congeners.

Male genitalia have socii longer than in other congeners. Valvae are variable in shape, but more bulbous than in other *Netria.* Saccus short. Phallus compact, slightly curved and also of variable shape. Best feature to recognize *multispinae* are many spines on the posterior margin of the 8th sternite. They are variable in number and shapes, but usually 2 or 3 symmetrically pairs of spines occur. 8th tergite less modified of conical shape with a straight posterior margin. Female genitalia have a resemblance to *virididescens,* in particular in shape of corpus bursae and its circular shaped appendix. Apophyses anteriores are short but more robust than in *viridescens.* Postvaginal plate almost not bilobed at posterior margin. Antrum relatively large and strongly sclerotized. Ductus bursae rather straight and compact.

Bionomics. Moth is very common in mainland Asia but rather infrequent in Sundaland. Adults has been recorded in Sumatra and Borneo from 150 m up to 2000 m; most specimens were observed at medium altitudes between 800 m and 1200 m. In the Himalaya the species occur up to 2600 m above sea level.

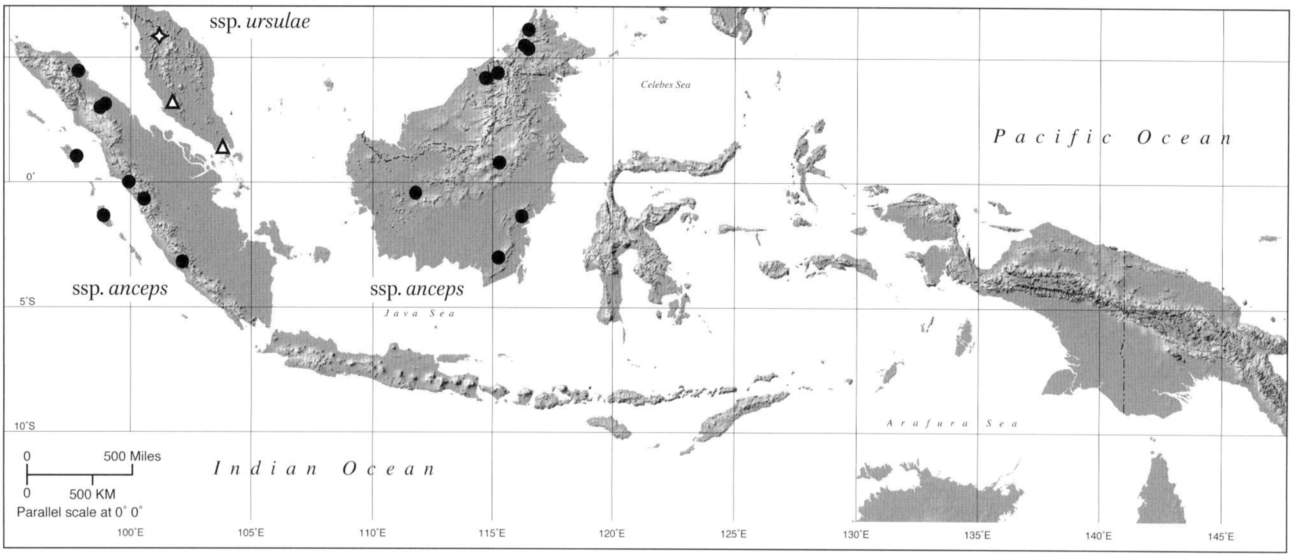

FIGURE 236 Distribution of *Cerasana anceps*.

Distribution. Widely distributed in Nepal, northern India, Myanmar, entire Indochina, Taiwan, Sumatra, Borneo and the Malayan Peninsula. In mainland China including Hainan the ssp. *nigrescens* occur.

Cerasana Walker, 1862: 123.

Type species: *Cerasana anceps* Walker, 1862.

Schintlmeister 2008: 183; Schintlmeister 2013: 71.

Diagnosis. Medium-sized adults. The genus was originally errected in the Arctiidae, later transferred to the Lymantriidae and 1930 placed by Gaede in the Notodontidae.

Antennae in males brown, strong bipectinated until the tip. Females have rami slightly shorter. External appearance resembles to Arctiidae or Lymantriidae by the "tigered" abdomen, where four rows of black dots on the intensive yellow abdomen occur. Forewings are yellow or orange-yellow. Forewing pattern is blackish-brown. Submarginal fascia marked as a row of spotty fuscous dots. Other fasciae on forewings are not visible. Discal spot fuscous-brown, well developed. A pair of contrasting brown spots are present in the basal area. Hindwings are yellowish-brown, often reddish-brown tinged in the anal area. Hindwings display a broad brownish postmedial band and a brown discal spot. Sexual dimorphism is limited.

Male genitalia have a long uncus and a large curved gnathos. Valvae are narrow, reduced in size and strongly asymmetrically in length. Phallus zig-zag shaped with a central projection and a hook shaped process the tip. 8th abdominal segments are highly modified. 8th sternite deeply notched and of asymmentric shape. 8th tergit widely bilobed at posterior margin. Anterior margin bear two long and slender projections. Female genitalia have a large and strongly sclerotized 8th abdominal segment. Both pairs of apophyses are rather short, and sometimes very reduced. Antrum large and strongly sclerotized. Ductus bursae short; corpus bursae large and pyriform or disc-shaped. Signum large, curved and strip-like.

The genus was revised by Schintlmeister 2004. There are currently six species known, which are distributed in south China, Indochina, Sundaland, Java and the Philippines. *Cerasana* is closely related to the African genus *Catarctia* Holland, 1892.

273. ***Cerasana anceps*** Waker, 1862: 123.

Holotype: ♂, [Malaysia], Borneo, Sarawak – Oxford University Museum of Natural History.

Schintlmeister 2013: 71.

Taxonomic note. *Cerasana anceps* is polytypic:

Cerasana anceps lutea (Pagenstecher, 1890): 14.

Holotype: ♂, [Philippines], Palawan [Puerto Princessa] – ZMHU, Berlin.

Cerasana anceps ursulae Schintlmeister, 2004: 198; figs. 4, 27.

Holotype: ♂, S Thailand, Yala, Pa Hala, 6°90' N, 101°16' E, 800 m – coll. A. Schintlmeister, Dresden.

Ssp. *ursulae* is 2–3 mm smaller in forewing length than ssp. *anceps* and display less contrasting pattern in the light yellowish forewings. Brown pattern is reduced.

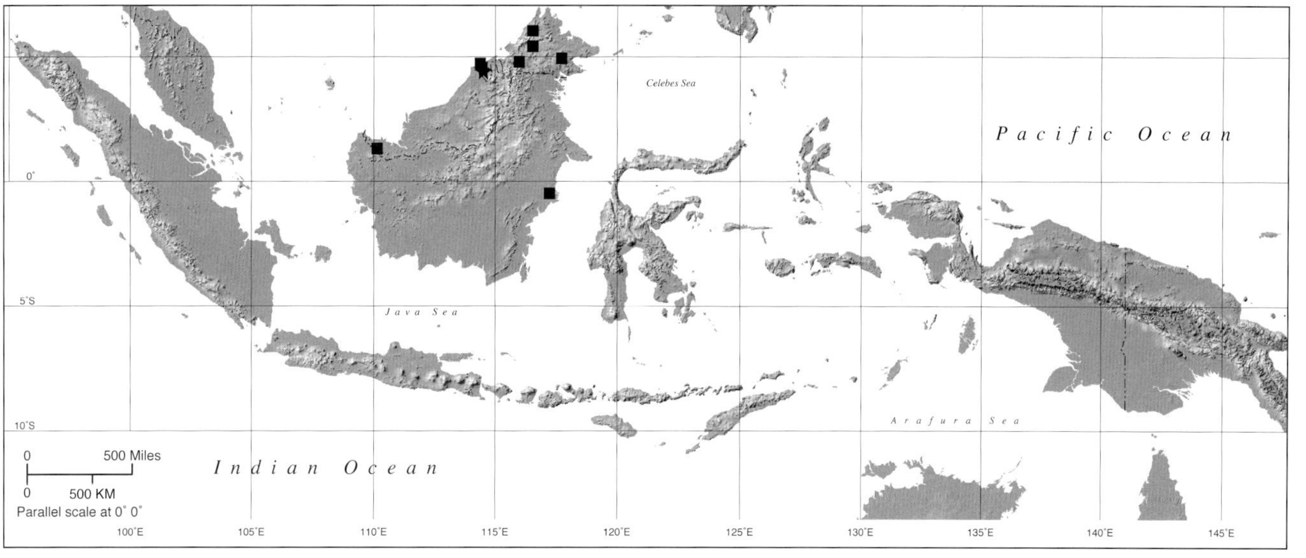

FIGURE 237 Distribution of *Cerasana alleni.*

Both subspecies, *ursulae* and *lutea* differ in their male genitalia (uncus length, phallus shape, shapes of 8th abdominal segments) from ssp. *anceps* as it is shown in detail by Schintlmeister, 2004.

Diagnosis. *Cerasana anceps* is the largest species in the genus. It is similar to *C. alleni*, but forewing shape is narrower in both sexes and the margin angled at vein M2, not rounded as in *alleni. C. anceps* is variable in its brown forewing pattern and development of the submarginal fascia. The latter is sometimes reduced to two spots only. Postbasal and postmedial fasciae are light yellowish. Veins are tinged light yellowish. Forewing costa displays a variable number of brown dots. Hindwings vary from pale yellowish to reddish-brown. Basal area often tinged reddish-brown. Medial band broad and light yellowish. Most specimens display a fuscous postmedial area. Sometimes also the submarginal area is fuscous blackish-brown filled.

Uncus is the longest in *Cerasana.* Gnathos curved and pointed, larger than the curved uncus tip. Valvae are asymmetric and of reduced length. Left valva arm is more than twice longer than the right arm. Phallus zig-zag shaped with a serrated plate in the middle part, which is for each of the subspecies specifically shaped. Phallus tip is pointed, but in *alleni* knob-shaped. 8th sternite deeply notched and posterior margin asymmetrically shaped. Anterior margin of 8th sternite also notched. 8th tergite displays at posterior margin a pair of long projections. Female genitalia are the largest within the genus. 8th abdominal segment very large. Antrum and ductus bursae are reduced. Corpus bursae disc-shaped. Signum-band long and gently curved, in *alleni* this band is rather V-shaped.

Bionomics. Adults are infrequent and prefer seemingly the lowlands up to 500 m altitude. A few specimens has been taken above 1000 m up to 1200 m.

Distribution. *Cerasana anceps anceps* is distributed in Sumatra, Nias, Mentawei Isls. (Siberut), Belitung and Borneo. Popultions from the Malayan Peninsula and Thailand belong to ssp. *ursulae.* In Palawan (including Balabac Isl.) the ssp. *lutea* occur.

274. ***Cerasana alleni*** Holloway, 1983: 98; fig. 22.
Holotype: ♂, Brunei, Rempayoh, 75 m – NHMUK, London.
Schintlmeister 2013: 71.

Diagnosis. Very similar by external appearance to *C. anceps*, but slightly smaller. Forewings less narrow and outer margin rounded, not angled as in *anceps*. Submarginal fascia on forewing consists of two brown spots only. The remaining brown pattern less contrasting and more indistinct than in *anceps.* Hindwings display a large, contrasting costal patch. The sexual dimorphism is limited.

Male genitalia have the uncus tip and gnathos as in *anceps*, but length is significantly shorter. Valvae are slender and assymmetric in length. Juxta bears, in contrast to *anceps,* a pair of small projections of different length. Phallus is curved but not zig-zag-shaped. There is no central serrated plate as in *anceps*, but toward the base a bulbous projection occurs. Everted endophallus is large and of circular shape. Posterior lobes of the 8th sternite is large. A pair of anterior processes are curved and long. Female genitalia resemble *anceps,* but they are smaller and 8th

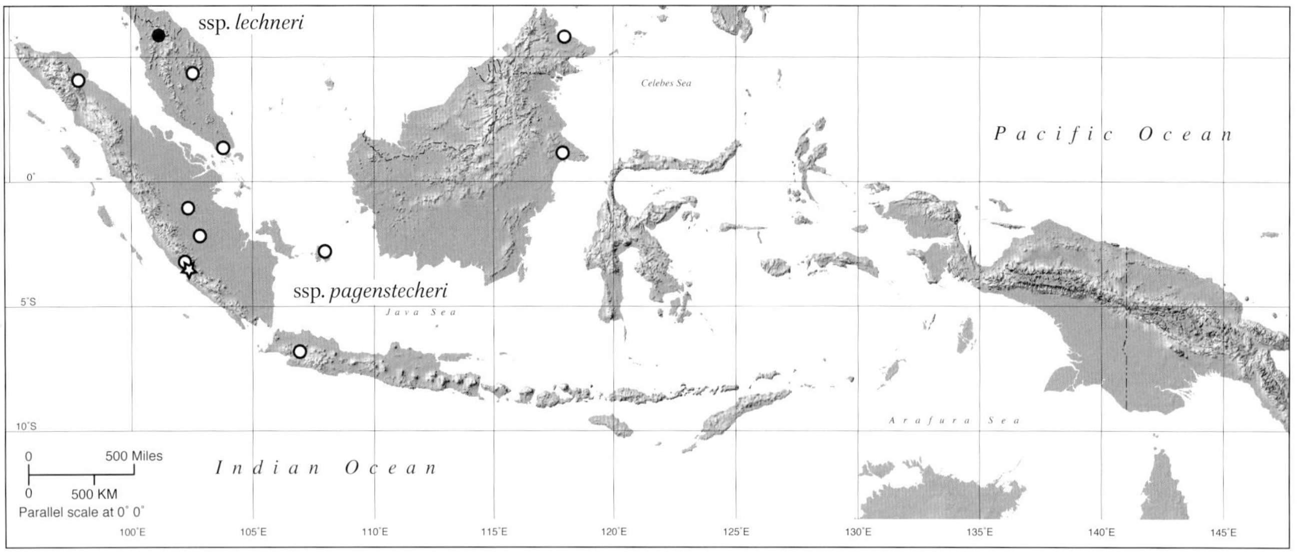

FIGURE 238 Distribution of *Cerasana pagenstecheri.*

abdominal segment less robust. Antrum and ductus bursae are longer than in *anceps*. Corpus bursae is pyriform and the signum band is rather V-shaped than U-shaped as in *anceps*.

Bionomics. The species is rare (n < 15 specimens are currently known). Adults has been collected from 3 m up to 1100 m above sea level.

Distribution. *Cerasana alleni* is restricted to Borneo. A female from Singapore is probably mislabelled (Holloway, 1983: 98).

275. ***Cerasana pagenstecheri*** Schintlmeister & Lourens, 2010: 206; figs. 355, 356.
Holotype: ♂, Indonesia, Sumatra, 35 km N Bengkulu, 3°38'S, 102°29' E, 500 m – coll. A. Schintlmeister, Dresden.
Schintlmeister 2013: 72.

Taxonomic note. The ssp. *lechneri* Schintlmeister & Lourens, 2010: 207; figs. 355, 356 (holotype: ♂, S Thailand, Chumpon Prov., 5 km S Phato, 9°46' N, 98°47' E, 155 m – coll. A. Schintlmeister, Dresden), differ from ssp. *pagenstecheri* by reddish-brown hindwings and in particular, by the different shape of the phallus with its long central spine and non hook-shaped tip.

Populations from Java differ by their weakly developed brown pattern from Sumatran material and could represent a further subspecies. Since it was as yet impossible to dissect the genitalia of Javan specimens, these populations are provisionally placed under ssp. *pagenstecheri.*

Diagnosis. Smaller in forewing length than *C. anceps* or *C. alleni*. Forewings are narrower than in *alleni.* Outer margin is rounded, not angled as in *anceps.* Forewing color more intensely yellow. Forewing pattern and its individual variation is similar to *alleni.* Hindwings vary from light yellow to reddish-brown. On the costa of hindwings a doubled blackish spot occur, which separates *pagenstecheri* from its other congeners. Discal spot on hindwings weakly and indistinct developed. Female resembles the male, but forewing length is larger.

Male genitalia are distinguished by a long, curved gnathos, which is much longer than the curved uncus tip. Both valvae are of equal length. They bear many setae in the entire valva. Phallus straight and compact. It bears a small spine in the central part and a large hook-shaped projection at the tip. Everted endophallus small. 8th sternite symmetric, bilobed at the posterior margin and also at the anterior margin as well. There is a pair of long and slender processes as in *anceps*. 8th tergite of conical shape with a small central spine. Female genitalia have the postvaginal plate triangularly notched. Ductus bursae and the corpus bursae are, due to a technical mistake, absent on the only dissected female.

Bionomics. Adults appear infrequently, having been observed at lower alttudes up to 500 m above sea level.

Distribution. Distributed in Sumatra, Belitung, the Malayan Peninsula, Borneo, and Java. In Thailand, Cambodia and Vietnam the ssp. *lechneri* occurs.

Somera Walker, 1855: 1778 (key), 882.

Type species: *Somera viridifusca* Walker, 1855.

Schintlmeister 2008: 184; Schintlmeister 2013: 419.

Diagnosis. Medium-sized adults. Antennae in both sexes black and bipectinated until the tip. Forewings are green and display a doubled black postbasal band, and a black postmedial fascia. Near the tornus there is a prominent black spot. Discal streak brown, small and inconspicuous. Fringe of forewings fuscous-brown and cream checkered. Hindwings vary from chocolate to reddish-brown. Sexual dimorphism is limited.

Male genitalia have a long and pointed uncus; socii and gnathos are absent. The genus is characterized by a pair of prominently developed tegumen processes and a further pair of juxta processes. Valvae are ellipsoid and toward the apex pointed. Saccus very small. Phallus straight with a serrated process at the tip. 8th abdominal segments are unmodified and weakly sclerotized. Female genitalia have large and strongly sclerotized papillae annales. Both pairs of apophyses are long. Ductus bursae long; corpus bursae bear a reniform signum.

Caterpillar's hostplant in Taiwan was reported by Lin & Shen, 1996: 87 as *Glochidion* (Euphorbiaceae).

Four species are currently placed in *Somera*. The genus is distributed in Nepal and NE India (Himalaya), S China, Taiwan, Myanmar, Indochina, the Malayan Peninsula, Borneo, Sumatra, Java, Bali, the Philippines and Sulawesi.

276. ***Somera viridifusca*** Waker, 1855: 882.

Holotype: ♂, [Bangladesh], Silhet – NHMUK, London.

Schintlmeister 2008: 185; Schintlmeister 2013: 419.

Taxonomic notes. The species show extensive geographical variation:

Somera viridifusca sumatrana Schintlmeister, 1993: 113; pl. 6: 5.

Holotype: ♂, Indonesien, Nordsumatra, Pakkat, 2°10' n. Breite, 98°33' ö. Länge, 1.400–600 m – coll. A. Schintlmeister, Dresden.

is paler in external appearance than other subspecies of *viridifusca* and green color of forewings is somewhat bluish tinged. The genitalia are not different from mainland Asian populations.

Somera viridifusca celebica Schintlmeister, 1993: 114; pl. 6: 8; pl. 7: 2.

Holotype: ♂, Indonesien, Sulawesi, W Celebes, G. Tompoe, Paloe, 2.700' – NHMUK, London.

is larger in forewing length and slightly darker colored than ssp. *viridifusca*. Male genitalia are distinguished by short uncus and long phallus. The serrated projection at phallus tip is shorter than ssp. *sumatrana* and display a knob-shaped tip. Female genitalia are larger than ssp. *viridifusca* or ssp. *sumatrana* and both pairs of apophyses are more compact.

Populations from Java and Bali represent a further subspecies, which is described below as:

Somera viridifusca javana **ssp. nov.** (figs. 276c, 276k–m).

Holotype: ♂, Indonesia, Java, Sukanegara, Cianur reg., 1050 m, 7°06'S, 107°07'E, 18.-21.iii.2007, leg. St. Jakl – in coll. A. Schintlmeister, Dresden.

Paratypes (32 ♂♂, 3 ♀♀):

Java: 5 ♂♂, 1 ♀, Sukanegara, Cianur reg., 1050 m, 7°06'S, 107°07'E, 18.–21. iii. 2007, leg. St. Jakl (GU 66-43) 2 ♂♂, Mt. Pangrango, 30 km SE Bogor, 1625 m, 6°30'S, 107°10'E, 16.–20. ii. 1996, leg. V. Siniaev & E. Afonin (GU 43-45); 1 ♀, Mt. Muria, 1055–1100 m, 6°35'S, 110°53'E, 21.–22. i. 1998, leg. Jakl, Schintlmeister & Cervenka; 1 ♂, Meru Betiri National Park, 25 km S Kalibaru vill., 300–500 m, v.–vi. 1996, leg. St. Jakl; 1 ♂, East Java, Nongkodjadjar, 4000', iii. 1934, (A. M. R. Wegner) (GU 07-61); 3 ♂♂, dto., i. 1934 (BM 854); 1 ♂, vi. 1934; 2 ♂♂, dto., ix. 1934; 1 ♀, Ardjoeno, 4500 ', v. 1934, leg. J. P. A. Kalis; 1 ♂, 1 ♀, Mt. Gedeh, viii. 1926; 1 ♂, Tengger, Singolangoe, 3000' 'vi. 1934, leg. J. P. A. Kalis: **Bali:** 2 ♂♂, West Bali Reserve, Titap near Papuan, 550 m, 8°25'S, 114°58'E, 29.- 30. i. 1998, leg. Jakl, Schintlmeister & Cervenka (GU 66-42); 2 ♂♂, Tamblingan Lakes, 8°15.47'S, 115°05.15'E, 25. xi.–10. xii. 2009, leg. Jakl; 1 ♂, Mondokpaenpang, x. 1934 (BM 856), leg. J. P. A. Kalis; 1 ♂, Git-Git, 5000', v. 1936, leg. J. P. A. Kalis.

Diagnosis of ssp. *javana*. Forewing lengths ♂♂ 19.0–20.5 mm; ♀♀ 22.0 and 26.5 mm. The males are on average about 1 mm smaller in forewing length than *S. viridifusca sumatrana*. Forewing color slightly darker green than ssp. *sumatrana*. Forewing pattern is the same.

Male genitalia differ from ssp. *sumatrana* by narrower and shorter phallus. Lateral process at the phallus tip is significantly shorter than in all other known subspecies. Female genitalia have not yet been dissected:

Diagnosis. *S. viridifusca* is distinguished from *virens* by the larger brown blotch in the medial area of the forewings. Due to individual variability in *viridifusca* the brown patch is sometimes smaller and in such cases both species are hardly to separate. A further brown patch near to the apex of the forewings includes the margin of the forewings,

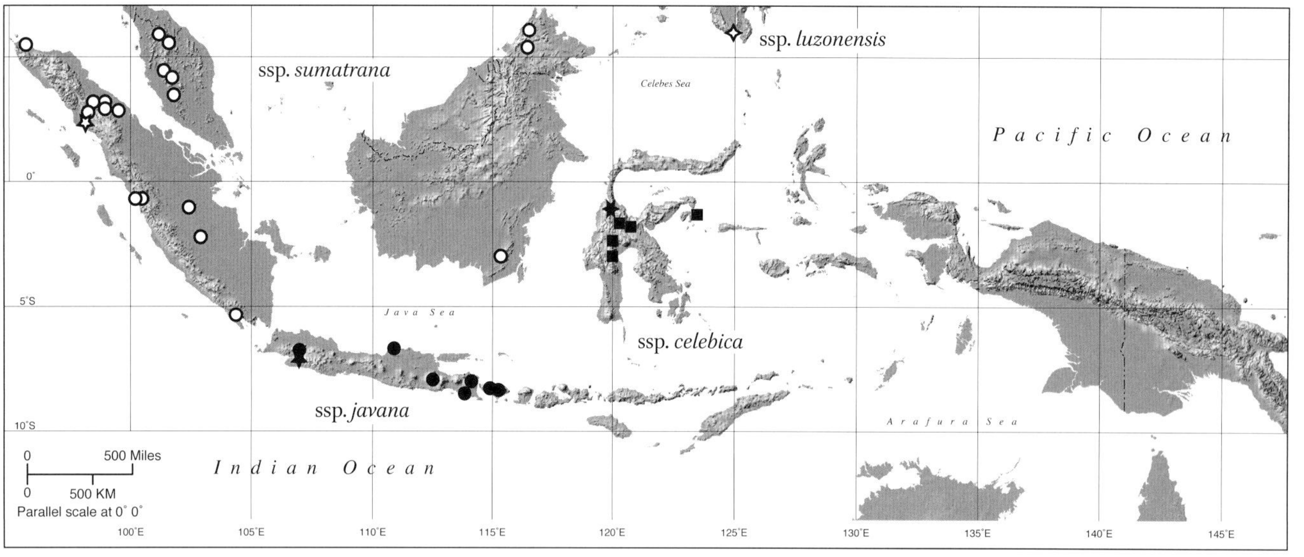

FIGURE 239 Distribution of *Somera viridifusca*.

whereas in *virens* it is markedly interrupted by a green band. Hindwings are reddish-brown. Sexual dimorphism is limited. Females of *S. virens* are readily differentiated from *viridifusca* by the longer rami of the antennae.

Male genitalia are characterized by a long and pointed uncus. Tegumen process is clearly longer than juxta process, which separates *viridifusca* from *virens*. Valvae are ellipsoid and slightly pointed at the apex. Phallus is straight with diagnostic, serrated projection at the tip. Female genitalia have boths pairs of apophyses long and apophyses anteriores are longer than in *virens*. Ductus bursae long and straight. Corpus bursae more circular shaped rather than ellipsoid as in *virens*. It bears a reniform signum, which is significantly larger than in *virens*. Corpus bursae display a pair of more strongly sclerotized bands, which are not seen in other congeners.

Bionomics. Adults are common in Sumatra, mainland Asia, the Phillippines and Sulawesi, but apparently infrequent in Borneo. Moths have been observed at lower altitudes up to 500 m as well as in montainous areas up to 2000 m; in Sikkim occurring also at 3800 m above sea level.

Distribution. *Somera viridifusca viridifusca* is widely distributed in Nepal, northern India, Myanmar, China, Hainan, Taiwan, and Indochina. The Philippines, except Palawan, are populated by ssp. *luzonensis* Schintlmeister, 1993. Ssp. *sumatrana* occur in the Malayan Peninsula, Sumatra, Borneo and Palawan. Ssp. *javana* is restricted to Java and Bali and ssp. *celebica* is endemic to Sulawesi including Peleng Isl.

277. ***Somera viriviri*** Kobayashi, 2012: 6; figs. 1, 2, 6a–h.
Holotype: ♂, Indonesia, Sulawesi – National Museum of Nature and Science, Tokyo.

Diagnosis. The smallest species within the genus, forewing length 18–19 mm, which much resembles *S. virens* externally. A brown discal streak on forewings is well developed and distinguishes *viriviri* from *virens* and *viridifusca*. A black tornal patch, which is characteristic of *virens* is absent in *viriviri*. Female is unknown.

Male genitalia resemble *viridifusca* by shapes of uncus, valvae, tegumen and juxta processes. They differ by phallus shape, which is bulging toward the tip, and the thick and compact projection at the phallus tip.

Bionomics. A rare species, which is currently known from 5 males (n = 5) only, which were collected between 900 m and 1300 m above sea level.

Distribution. Endemic to Sulawesi.

278. ***Somera virens*** Dierl, 1976: 84; fig. 2.
Holotype: ♂, [Indonesia], Sumatra, [Pematang] Siantar, Holzweg 2, 1.050 m – ZSM, Munich.
Schintlmeister 2008: 185; Schintlmeister 2013: 419.

Taxonomic note. In Sundaland the nominotypical ssp. *virens* occur.

From northeastern India (Assam) the fuscous ssp. *watsoni* Schintlmeister, 1997: 99; pl. 19: 9, 10; fig. 87 was described.

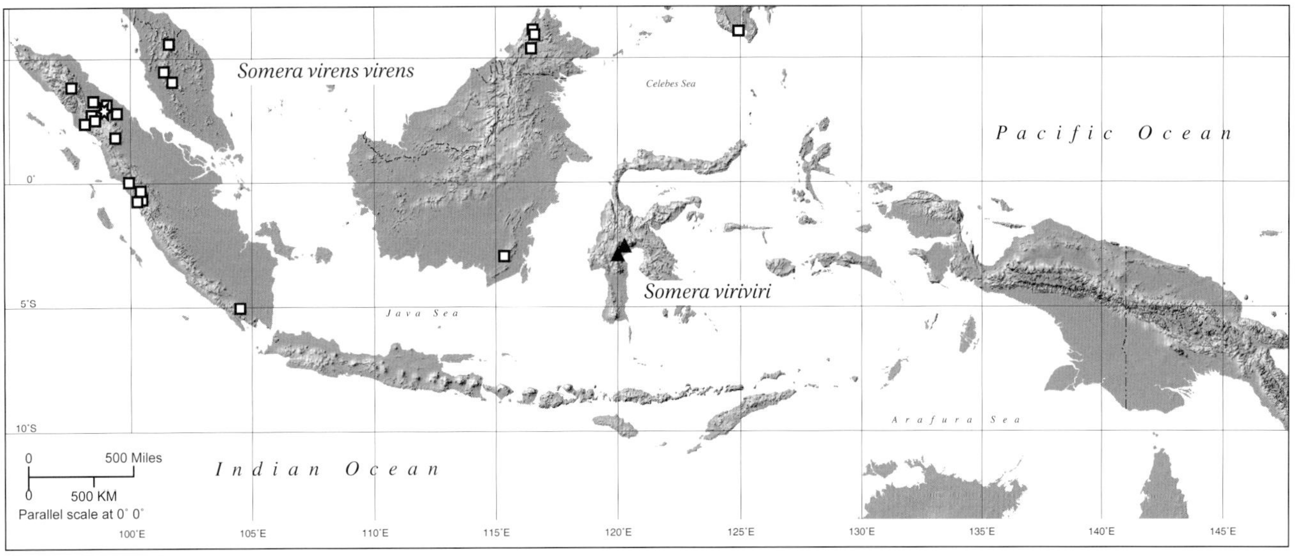

FIGURE 240 Distribution of *Somera viriviri* and *virens virens*.

Diagnosis. Similar by external appearance to *S. viridifusca*, but the brown blotch in the medial area of forewing is smaller. The green marginal band is not interrupted as in *viridifusca*. There is also a resemblance to *S. viriviri*, but the latter is clearly smaller. Females of *virens* are readily differentiated from *viridifusca* by much longer rami, comparable in their lengths to the males. Sexual dimorphism is limited.

Male genitalia have a long, slender and pointed uncus as in *viridifusca*, but both pairs of the tegumen and juxta processes are approximately of equal length. Valvae are more rounded and less narrow than in *viridifusca*. Phallus is straight and thick. Phallus tip is serrated, but without appendix. Female genitalia bear finer apophyses and the apophyses anteriores are shorter than in *viridifusca*. Ductus bursae curved, long and slender. Corpus bursae is ellipsoid with a small reniform signum, much smaller as seen in *viridifusca*.

Bionomics. Moths are common and occur mostly at lower altitudes up to 900 m. A few specimens have been taken up to 2000 m above sea level.

Distribution. *Somera virens virens* is distributed in south Myanmar, south Thailand, the Malayan Peninsula, Sumatra, Borneo, and Palawan. The ssp. *watsoni* occur in the Himalayas, southwestern China (Yunnan), Hainan, Myanmar, Vietnam Laos, and Thailand.

279. ***Somera brillians*** Gaede, 1930: 631; pl. 84: a.
Holotype: ♀, [Indonesia], Sumatra, Mt. Korintji, 2.500 m – NHMUK, London.
Schintlmeister 2013: 419.

Diagnosis. Forewing color shining metallic green, which distinguishes *brillians* from all other congeners. There is a resemblance to *Syntypistis gemmea*, but *Syntypistis* have a naked tip of the antennae and *Somera* are bipectinated until tip. The number of green scales on forewing varies individually, and sometimes specimens with very reduced number of metallic green scales occur as depicted in fig. 279d. There are also individuals with almost uniformly green forewings. Forewing pattern is similar to *S. virens*, but the discal streak is marked light brown. Sexual dimorphism is limited. Females have very short rami, the shortest within the genus.

Male genitalia have a shorter uncus, than other *Somera*. Juxta process arise from juxta base. Valvae are broader than *viridifusca*, and not rounded at the apex as in *virens*. Phallus straight and slenderer than other *Somera*. Phallus tip is serrated and curved. 8th sternite is slightly notched at posterior margin. Female genitalia display pairs of long apophyses of equal length. Ductus bursae is slender and curved. Corpus bursae ellipsoid. No signum.

Bionomics. *S. brillians* occurs infrequently and prefers higher altitudes between 1200 and 2500 m above sea level. Only one male was observed at 750 m.

Distribution. Endemic to Sumatra.

Psegmaphora Gaede, 1930: 636.
Type species: *Psegmaphora tripunctata* Gaede, 1930.
Schintlmeister 2013: 356.

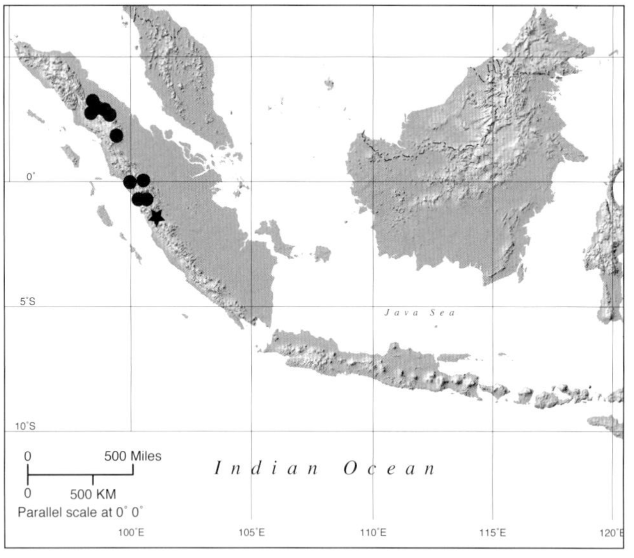

FIGURE 241 Distribution of *Somera brillians.*

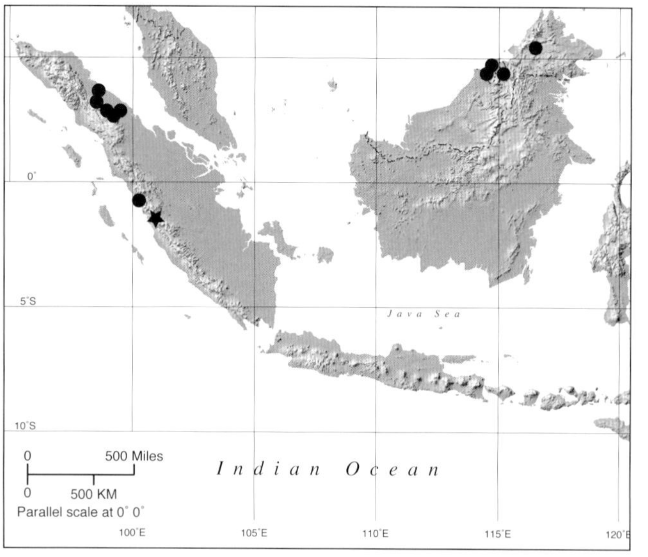

FIGURE 242 Distribution of *Psegmaphora tripunctata.*

Diagnosis. Medium-sized adults of greyish brown color. Antennae are in both sexes brown and bipectinated; the last 40 % to the tip are filiform. Characteristic pattern elements are a fine black basal streak, a smaller black costal-spot in the medial area and a broad blackish blotch area in the discal area. Postmedial fascia blackish and doubled; submarginal fascia marked as a row of black dots. Fringe of all wings fuscous-brown and cream-checkered. Sexual dimorphism is limited.

Male genitalia have a reduced and broad uncus. Socii club-shaped and long. Valvae simple, slightly curved and with costal projection. Phallus slightly curved and compact. Everted endophallus bear many needle-like spines. 8th abdominal segments are not modified.

The genus comprises only two species both of which are distributed in Indochina, Sumatra and Borneo.

280. ***Psegmaphora tripunctata*** Gaede, 1930: 637; pl. 84: f.
Holotype: ♂, [Indonesia], Sumatra, Mt. Korintji, 1.600 m – NHMUK, London.
Schintlmeister 2013: 356.

Diagnosis. See under the genus diagnosis. The underside of the forewings displays a well developed fuscous-brown medial fascia in *P. tripunctata*, which is very weakly visible or absent in the Indochinese distributed *P. yursvet* Schintlmeister, 2013.

Male genitalia of *tripunctata* have a pointed uncus tip, which is in *yursvet* slightly bilobed. Female genitalia of *Psegmaphora* are not yet dissected.

Bionomics. *P. tripunctata* occurs infrequently and is known only from a few localities, which are situated between 150–1200 m above sea level.

Distribution. Distributed in Sumatra and NE Borneo.

Shachia Matsumura, 1919: 75.
Type species: *Shachia subrosea* Matsumura, 1919
[= *Drymonia circumscripta* Butler, 1885].
Schintlmeister 2008: 186; Schintlmeister 2013: 417.

Diagnosis. Medium-sized and smaller adults. Antennae in both sexes bipectinated; the last 15% -40% toward the tip are filiform. Forewings triangle-shaped, violet colored, with greenish pattern elements. Black basal streak of the forewings prominently fat. Discal streak cream, embedded in a large whitish field. Sexual dimorphism is limited. Marginal fascia on all wings fine blackish. Fringes are checkered-cream and blackish. Hindwings whitish or brown. Discal spot indistinct and brownish.

Male genitalia are simple. Uncus straight, slender with rounded tip. Socii small, curved and slender. Valvae deeply divided. The upper arm is curved and bear many smaller teeth. Phallus curved; everted endophallus does not display csclerotized structures. 8th abdominal segments are not modified. Female genitalia have a broad 8th abdominal segment and rather short apophyses. Antrum very long and strongly sclerotized. Ductus bursae short and slender. Corpus bursae pyriform. It bears a large but weakly sclerotized signum.

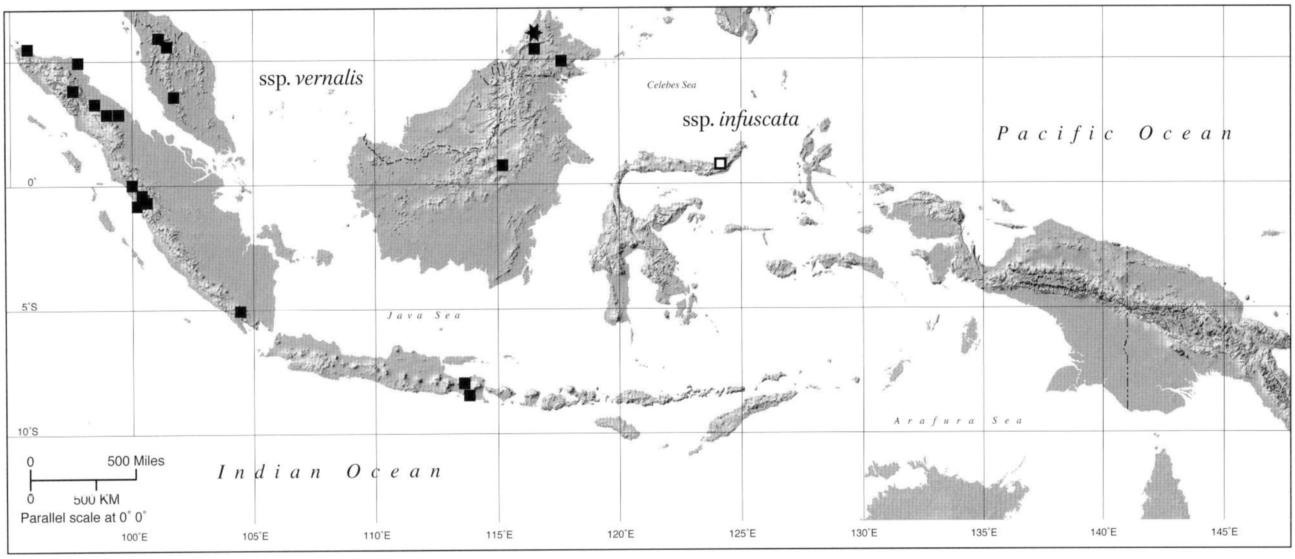

FIGURE 243 Distribution of *Shachia vernalis*.

Caterpillar of Japanese *S. circumscripta* (Butler, 1885) feeds on *Juglans* and *Pterocarya* (Juglandaceae).

Shachia comprised of 3 species, distributed in Japan, China, Thailand, Sundaland, Java and Sulawesi.

281. ***Shachia vernalis*** (Gaede, 1930): 640; pl. 84: h (*Pseudohoplitis vernalis*).

Holotype: ♂, [Indonesia, Sulawesi], Nord-Celebes – NHMUK, London.

Schintlmeister 2008: 186; Schintlmeister 2013: 417.

Taxonomic notes. Populations from Borneo have been described in the sense of a geographical subspecies as

Pseudohoplitis vernalis form *infuscata* Gaede, 1930: 640, holotype: ♂, Borneo, Kina-Balu – ZMHU, Berlin in the sense of a geographical subspecies, which differs from ssp. *vernalis* by brownish hindwings and enlarged blackish pattern, thicker black basal streak, large black apical blotch on the forewings and black tinged veins. It is likely, that *infuscata* must be classified as a genuine species and not as a subspecies, when the male genitalia of *vernalis* from Sulawesi become known.

Diagnosis. Unmistakable by wingshape and green pattern of the forewings. Anal tuft in both sexes green. Discal streak fine and brown, surrounded in white. Hindwings whitish in ssp. *vernalis,* but brownish with blackish brown tinged veins in ssp. *infuscata*. Sexual dimorphism is limited.

Male genitalia are as described under the genus diagnosis. Upper valva arm not pointed as in both the other congeners, but thick and rounded at the tip. It bears toward the tip many dense small spines. Lower arm of the valva unusually rectangle-shaped. The phallus rectangularly curved. Female genitalia are distinguished by the very long antrum. Corpus bursae pyriform with sclerotized microstructures in the upper part. Signum consist of two parallel bands.

Bionomics. Males of *P. vernalis infuscata* appear commonly, but females are rare in collections. Ssp. *vernalis* is known from a few and old males (n < 10) only.

Adults have been recorded mostly at lower altitudes up to 500 m. In Sumatra some males were observed between 1200 m to 1600 m and a single male from Borneo was collected at 2200 m above sea level.

Distribution. *P. vernalis vernalis* is restricted to N Sulawesi. Ssp. *infuscata* is distributed in Borneo, Java, Sumatra, the Malayan Peninsula, Thailand and southwestern China (Yunnan).

Pheraspis Turner, 1903: 46 (key), 61.

Type species: *Pheraspis polioxutha* Turner, 1903.

Schintlmeister 2013: 344.

Diagnosis. Medium-sized to smaller adults. Antennae are brown. They are in both sexes bipectinated; the last 5% until the tip 20 -25% are filiform. Female have short rami. Forewings are in most species grey. Two species display cream forewings. Postbasal and postmedial fasciae are

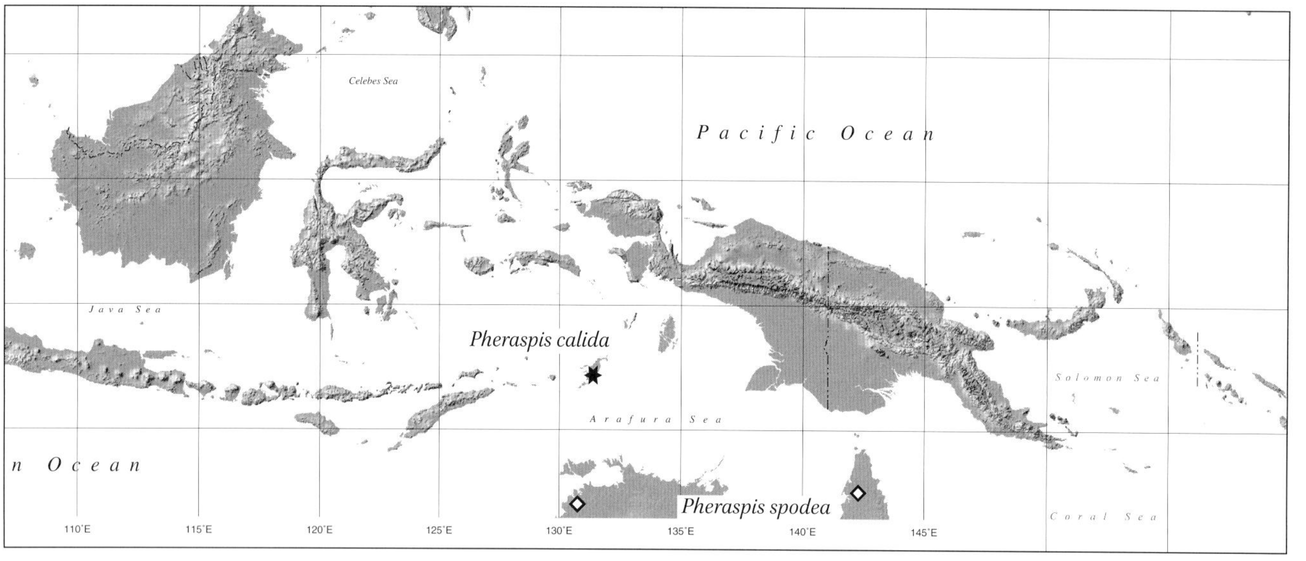

FIGURE 244 Distribution of *Pheraspis calida* and *spodea*.

fuscous-brown, but in some species less visible. Hindwings are cream or yellowish-brown with more yellowish or reddish-brown tinged anal area. Sexual dimorphism is limited.

Male genitalia have a robust uncus, which is bilobed at the tip. A gnathos occurs occasionally. Socii are long, curved and slender. Valvae of rhomboid shape with rounded apex. Costa bulbous in its central portion. Sacculus well developed but without projections. Phallus bears a small spine at the tip. Everted endophallus displays a more strongly sclerotized carina but no spines. 8th abdominal segments less modified. 8th sternite on both margins concave. Caterpillar of the Australian *P. symmetra* Turner, 1917 was reported feeding *Atalaya hemiglauca* (Sapindaceae).

The genus is Australian and only one species was found in Tanimbar. Currently six species are described.

282. ***Pheraspis calida*** spec. nov. (figs. 282a–d).
Holotype: ♂, Indonesia, Tanimbar Isls., Yamdena Isl., 20 km NE Saumlaki, 7°47'S, 131°19'E, 150 m, 6. ii.–30. iii. 2007, leg. St. Jakl – in coll. A. Schintlmeister, Dresden.

Paratypes (4 ♂♂, 1 ♀):
Tanimbar: 1 ♂, Yamdena Isl., 20 km NE Saumlaki, 7°47'S, 131°19'E, 150 m, 6. ii.– 30. iii. 2007, leg. St. Jakl (GU 92-89); 1 ♂, 1 ♀, Yamdena Isl., 21 km NE Saumlaki, 7°47'S, 131°19'E, 150 m, 27. ix.–11. xii. 2005, leg. St. Jakl (GU 66-29); 2 ♂♂, Yamdena Isl., 20 km NE Saumlaki, 7°47'S, 131°19'E, 150 m, 5–28. xii. 2006, leg. St. Jakl.

Diagnosis. Forewing lengths ♂♂ 17 mm; ♀, 20 mm. Body and wings cream-colored. Forewings rather uniform and brown pattern weakly developed. A small brown dot on the costa, near the basal area and a further dot on the medial vein in the postbasal area are visible. Postmedial fascia indistinct brownish, slightly darker than remaining forewing color. A postbasal fascia is sometimes developed as a broad brown shadow. Fringe of all wings cream and white checkered. Hindwings are of same color as forewing, but in the anal area is reddish-brown tinged. Female is slightly sexually dimorphic and displays darker brown forewing pattern. *P. calida* appears as the Australian *Pherastis spodea* Turner, 1903 and is inseparable by external appearance, see fig. 282Ab.

Male genitalia have a broad uncus, which is slightly bilobed at the tip and both lobes are rounded. A small gnathos is present near the uncus tip, but absent in *spodea*. Socii are curved and shorter than in *spodea*. Valvae display near the apex a small and slender process, which is not present in *spodea*. Harpe is strongly sclerotized and displays a pointed spine. In *spodea* harpe is much larger and bears two longer and pointed processes. Juxta strongly sclerotized and trapeze-shaped. Phallus rather straight; at the tip a small spine occur, which also present in *spodea*. Everted endophallus is curved in its more strongly sclerotized distal part (carina). 8th sternite gently rounded at the posterior margin; 8th tergite rectangularly shaped. Female genitalia have a strongly sclerotized 8th abdominal segment. Antrum short, ductus bursae long and curved. Corpus bursae disc-shaped. A small circular signum is visible.

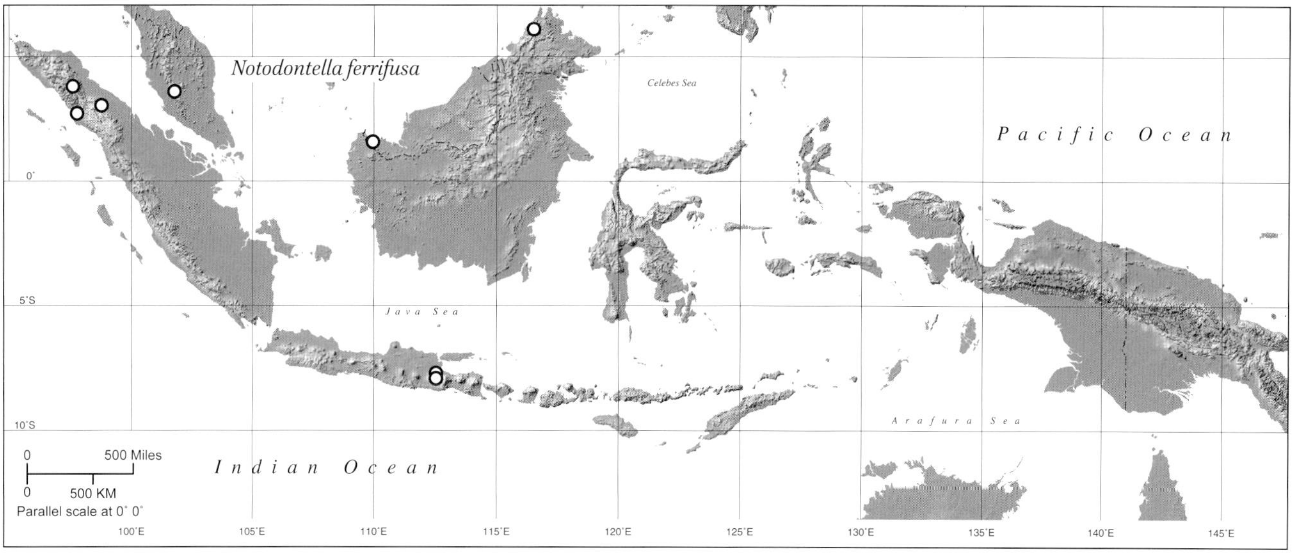

FIGURE 245 Distribution of *Notodontella ferrifusa*.

Bionomics. The few specimens (n = 5) of the type series from different years has been collected at 150 m above sea level.

Distribution. Endemic to Tanimbar.

Notodontella Roepke, 1943: 78.

Type species: *Notodontella nieuwenhuisi* Roepke, 1943 [= *Fentonia ferrifusa* Dudgeon, 1898].

Schintlmeister 2013: 289.

Diagnosis. Small notodontids. Antennae of the males strong bipectinated, females have short rami. Last 6–10 % toward the tip are filiform. Forewing color fuscous violet-brown with characteristically fuscous-brown dorsal blotch near to the base. Postmedial and submarginal fasciae marked as rows of black dots. Discal spot not visible. Hindwings are brown.

Male genitalia have a short uncus with knob-shaped tip. Socii well developed and curved. Valvae smooth, with pointed costal and apical processes. Phallus straight, slightly curved with a long sclerotized band at the tip. 8th sternite slightly bilobed at posterior and broad bilobed at anterior margin. 8th tergite not modified. Female genitalia have short apohyses anteriores; apophyses posteriores are three times longer. The antrum small; ductus bursae long and curved. The corpus bursae large and of circular shape. Signum large but weakly sclerotized.

The genus includes two species, which are distributed in India, the Andaman Isls., Myanmar, Indochina, Luzon, Sundaland, Java and Flores.

283. ***Notodontella ferrifusa*** (Dudgeon, 1898): 634 (*Fentonia ferrifusa*).

Lectotype: ♂, [India], Sikhim, 1.800 feet – NHMUK, London.

Schintlmeister 2013: 289.

Diagnosis. Small species. A fuscous-brown dorsal patch in the postbasal area surrounded by a whitish area is a diagnostic character. The fringe of all wings fuscous-brown and white checkered. The sexual dimorphism is limited.

For differentiation from its allopatric sister species *N. floresiensis* see below.

Male genitalia are variable in the shape of the short uncus and in the size and thickness of socii. Valvae bear a needle-like apical spine and a large tapered central costal process. Valvae vary individually in length and broadness. Phallus large, gently curved with a small spine at the tip. Endophallus displays a small strongly sclerotized spine. 8th tergite is widely bilobed at the anterior margin. Female genitalia are as described in genus diagnosis. Postvaginal plate have a characteristically, unique shape.

Bionomics. Adults are infrequent but occur locally sometimes in large numbers. The species prefers lower altitudes up to 500 m; only a few specimens are known from above 1000 m up to 1500 m. In Sikkim one male was taken at 2000 m.

Distribution. Known from north India, south India (Kerala), the Andaman Isls., Myanmar, Thailand, Laos, Vietnam, Luzon, Malaya, Borneo, Sumatra, and Java.

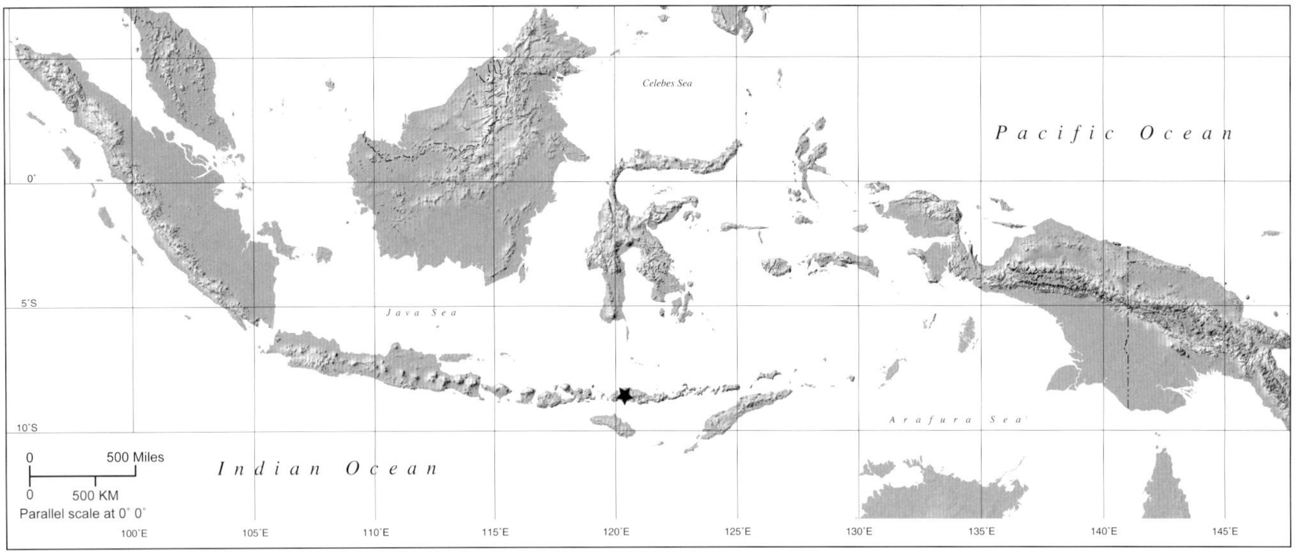

FIGURE 246 Distribution of *Notodontella floresiensis.*

284. ***Notodontella floresiensis*** **spec. nov.** (figs. 284a–c).
Holotype: ♂, Indonesia, Nusa Tenggara Timur, W Flores, Gunung Ranaka, 3 km S Mano, 18 km SE Ruteng, 1270 m, 17.–21. iv. 1996, leg. Dr. Ron Brechlin – in coll. A. Schintlmeister, Dresden.

Paratypes (2 ♂♂, 1 ♀):
Flores: 2 ♂♂, 1 ♀, Gunung Ranaka, 3 km S Mano, 18 km SE Ruteng, 1270 m, 17.–21. iv. 1996, leg. Dr. Ron Brechlin (GU 43-57).

Diagnosis. Small species. Forewing lengths ♂♂ 12–13 mm, ♀ 16 mm, 1–2 mm larger in forewing lengths than *N. ferrifusa*. *Notodontella floresiensis* resembles much to *ferrifusa*, but differ by an enlarged fuscous-brown basal patch, which extends along the dorsum to the forewing base. Forewings are generally paler in color than *ferrifusa*. Hindwings of *floresiensis* are light whitish brown, but brown in *ferrifusa*. Female resembles males, but the hindwings are light brownish with a fuscous-brown medial shadow.

Male genitalia are significantly smaller than those of *ferrifusa*. Uncus short with knob-shaped tip as in *ferrifusa*. Socii slender and curved. Valvae are narrower than in *ferrifusa,* beyond the range of individual variation. Valva apex bear a needle like process. Costal process is smaller than in *ferrifusa* and situated closer to valva base. Phallus is shorter and slenderer in comparison of *ferrifusa*. 8th abdominal segment as in the latter species. Female genitalia have not yet been dissected.

Bionomics. The type series was taken at 1270 m above sea level.

Distribution. Endemic to Flores.

Chloroceramis Kiriakoff, 1968: 237 stat. rev.
Type species: *Fentonia viridinota* Hampson, 1896
Schintlmeister 2013: 289.

Taxonomic note. *Chloroceramis* was placed by Schintlmeister & Pinratana, 2007: 147 as a junior synonym of *Notodontella*. Due to structural differences in antennae, male genitalia (uncus) and in particularly in the female corpus bursae *Chloroceramis* is reinstated herewith as a genuine genus.

Diagnosis. Small species. Antennae are in both sexes bipectinate, except for the last 40% toward the tip, which are filiform. Forewings grey or dark reddish-brown. Postmedial fascia marked as a row of black spots. Hindwings are brown. Sexual dimorphism is limited.

Male genitalia have a long uncus. The knob-shaped tip bears a pair of fine, small spines. Socii are well developed and long. Valvae smooth with rounded apex and without an apical process as in *Notodontella*. Phallus straight, rectangularly shaped. Female genitalia display strongly sclerotized papillae anales. Apophyses are short. 8th abdominal segment is narrow. Ductus bursae short. Corpus bursae ellipsoid and covered with numerous strongly sclerotized spines.

The genus is comprised of two species, which are distributed in the Himalaya, Myanmar, Indochina, Sundaland, Bali, Sulawesi and the Philippines.

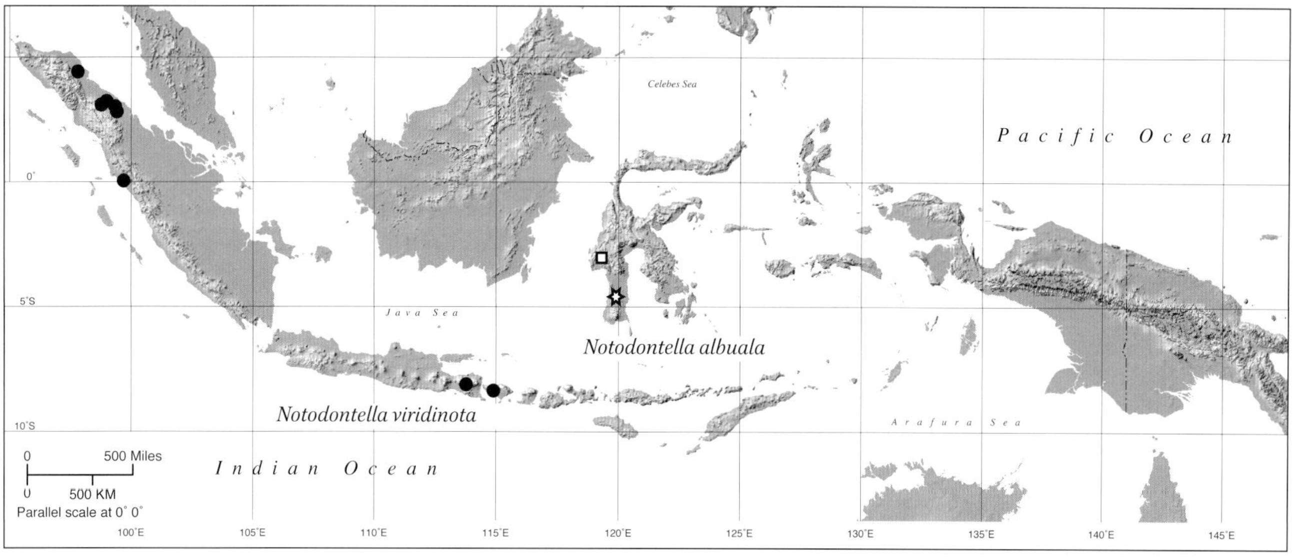

FIGURE 247 Distribution of *Notodontella viridinota* and *albuala.*

285. ***Chloroceramis viridinota*** (Hampson, 1896): 459 (*Fentonia viridinota*).
Holotype: ♂, Bhutan – NHMUK, London.
Schintlmeister 2013: 289.

Diagnosis. This small species is readily recognized by green areas along dorsum and costa of forewing. Thorax and anal tuft mixed with green setae. Basal area pale brown filled. Postbasal fascia doubled. Most specimens display at costa of medial area a white "<" shaped stigma. Sometimes also the entire, usually greenish colored costal area, is tinged in white. Postmedial fascia is marked as a row of black spots, sometimes accompanied by white patches. Marginal fascia fuscous blackish brown. Forewing fringe indistinctly fuscous and checkered in pale brown. Hindwings are fuscous-brown. Sexual dimorphism is limited. For separation from its allopatric sister species *C. albuala,* see below.

Uncus tip is knob-shaped and bears a pair of fine and small spines. Uncus varies individually in its shape and length. Socii are slender, curved and sharply pointed at the tip. Valvae are curved and smooth with an inconspicuous costal spine at base. Female genitalia have a broad ductus bursae. Corpus bursae is large; it bear in its central part numerous strongly sclerotized spines. No signum.

Bionomics. Adults may occur common in certain localities, but generally the species appears infrequently and has been recorded mostly below 500 m altitude. Only a few specimens have been taken in Sumatra at 1100 m above sea level.

Distribution. Distributed in Nepal, northern India, Bhutan, Myanmar, Thailand, Laos, Sumatra, Java, Bali and the Philippines.

286. ***Chloroceramis albuala* spec. nov.** (figs. 286a, 286b).
Holotype: ♂, Indonesia, Sulawesi Selatan, Bantimurung, 40 km NNE Ujung Padang, 4°56'S 119°39'E, Talkessel, 200 m, 2.–9. v. 1984, leg. A. Schintlmeister – in coll. A. Schintlmeister, Dresden.

Paratypes (9 ♂♂):
Sulawesi: 2 ♂♂, Bantimurung, 40 km NNE Ujung Padang, 4°56'S 119°39'E, Talkessel, 200 m, 2.–9. v. 1984, leg. A. Schintlmeister – (GU 43-58); 1 ♂, Mamasa, 1500–2000 m, 2°56'S, 119°22'E, x. 1995 GU 51-97); 1 ♂, Dumogo-Bone National Park, Tumpuh Transect, 1040 m, 23.–25. ii. 1985, leg. J. D. Holloway; 1 ♂, dto., 1140 m, 6.–8. viii. 1985; 1 ♂, dto., 540 m, 9.–11. ii. 1985; 1 ♂, Paloe [= Palu], G. Tompoe, 2700 ', i. 1937 leg. J. P. A. Kalis (BM 1471); 2 ♂♂, Landa Paloe. 3700', iv. 1937, leg. J. P. A. Kalis.

Diagnosis. Forewing lengths ♂♂ 12–13 mm. The small species is similar to *C. viridinota*. It differs in reduced green colored areas, in particular at the forewing dorsum. Apical area whitish and larger than the corresponding area in *viridinota.* Basal area paler brown, in comparison to *viridinota.* Postmedial fascia marked as an indistinct row of fuscous-brown patches. Hindwings are whitish, not brown as in *viridinota*, with pale brownish costal and anal region. The female is unknown.

Male genitalia are generally smaller than *viridinota* and uncus is shorter. Uncus tip is knob-shaped and bears a

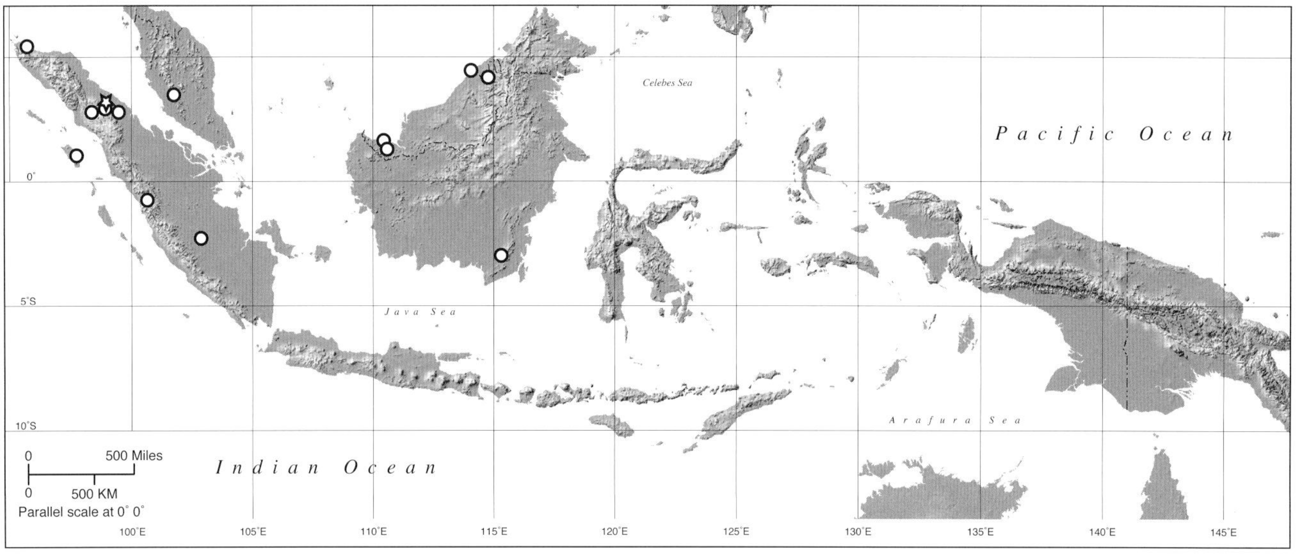

FIGURE 248 Distribution of *Medanella subterminalis.*

pair of fine spines. Socii are strongly curved, slender, and tapered. They are of same length as uncus. Valvae are narrower than *viridinota*. Near valva base a small costal spine arises. Phallus straight and base is broader than *viridinota*. 8th abdominal segments are less modified. They are bilobed at posterior and anterior margins.

Bionomics. *Notodontella albuala* occur infrequently between 200 and 2000 m.

Distribution. Endemic to Sulawesi.

Medanella Kiriakoff, 1974: 401.
Type species: *Medanella subterminalis* Kiriakoff, 1974.
Schintlmeister 2013: 256.

Diagnosis. Small notodontids. Antennae of males are brown and bipectinated, females have short rami. Last 45 % toward the tip are filiform. Body and wings are fuscous blackish brown. Forewing markings are deep black. Discal streak deep black. Postmedial and submarginal fasciae marked as rows of pale brown or cream patches. Sexual dimorphism is limited.

Male genitalia have a short and strongly curved uncus. Socii are curved and longer than uncus. Valvae narrow and curved. Sacculus is large with many dense setae. Saccus reduced. Phallus straight and long. Everted endophalus is very long, bearing several sharp, and large spines. Female genitalia display two pairs of apophyses of equal length. 8th abdominal segment is narrow and on its margins strongly sclerotized. Corpus bursae bears in its middle part many fine microspines. A signum is present.

Medanella is monotypic and distributed in the North East India, Indochina, Sundaland and Palawan.

287. ***Medanella subterminalis*** Kiriakoff, 1974: 401; pl. 4: 2; fig. 20.
Holotype: ♂, [Indonesia], Nordsumatra, Dolok Merangir, 180 m – ZSM, Munich.
Schintlmeister 2013: 256.

Diagnosis. An inconspicuous fuscous and small species that resembles at first glance to a noctuid moth. There are specimens with uniform blackish brown forewings, as well as with deep black markings, as described under genus diagnosis. Hindwings are paler brown than on forewing. Most females display in the center of postmedial fascia a larger black patch.

Male genitalia are as described under the genus diagnosis. Female genitalia have a large antrum and a large, curved ductus bursae. Signum of corpus bursae weakly sclerotized and reniform.

Bionomics. Adults appear frequently below 500 m above sea level. A few males have been taken in Sumatra at 1300 m altitude.

Distribution. Distributed in northeastern India, Myanmar, Thailand, Vietnam, Malaya, Sumatra, Nias, Borneo and Palawan.

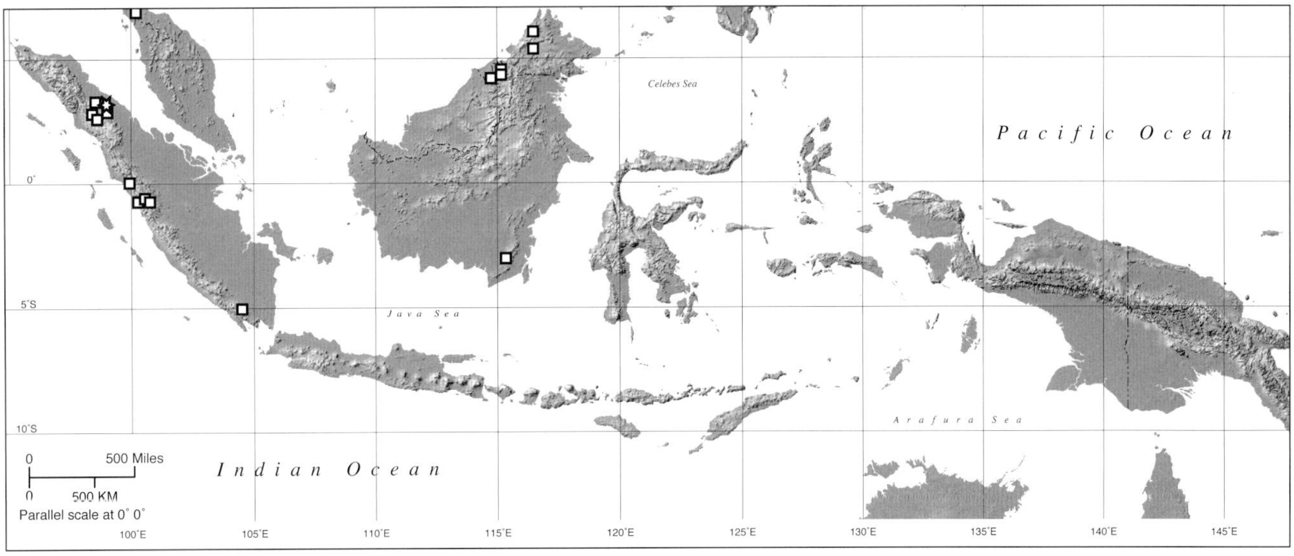

FIGURE 249 Distribution of *Antiphalera sumatrana*.

Antiphalera Gaede, 1930: 614.
Type species: *Phalera bilineata* Hampson, 1896.
Schintlmeister 2008: 188; Schintlmeister 2013: 40.

Diagnosis. Medium sized adults. Antennae plumose bipectinated; the last 30–40% toward the tip are filiform. Females have filiform antennae. Forewings are reddish-brown and of narrow shape. Wing pattern are in all known species homologous: brown discal streak and reniform stigma. Both are surrounded in cream or white. Below the stigma sometimes a prominent white blotch occures. Basal, postbasal, medial, postmedial and submarginal fasciae on forewing are black. Submarginal area is whitish, toward the apical area fuscous-brown colored. Hindwings are fuscous-brown with a paler medial shadow. Sexual dimorphism is, apart from the antennae morphology, limited.

Male genitalia have a compact uncus, which is broad at its base. From there a pair of large and curved socii arises. Tegumen is compact and broad. Valvae are less modified and weakly sclerotized. Phallus short and compact. It bears a large serrated wreath near the tip. 8th sternite show a central projection at posterior margin. Female genitalia have long apophyses posteriores. 8th abdominal segment is long. Postvaginal plate large and well sclerotized. Corpus bursae circular shaped. It bears a large, prominently sclerotized sickle-shaped signum.

The genus is comprised of five species, which are distributed in the Himalaya, south China, Indochina and Sundaland.

288. ***Antiphalera sumatrana*** (Kiriakoff, 1974): 378; pl. 1: 3; fig. 3 (*Grangulina sumatrana*).
Holotype: ♂, [Indonesia], Sumatra sept., Deli, Dolok Merangir, 180 m – ZSM, Munich.
Schintlmeister 2013: 41.

Taxonomic note. Females from Sumatra and Kalimantan show great differences in signum shape. In male genitalia no significant variability has been found, apart in shape of the uncus.

Diagnosis. Forewing pattern are as described before. The moth varies individually in size and development of whitish pattern, which is often replaced by brown; see fig. 288e. Males are best recognized in Indonesia by their plumose antennae. Sexual dimorphism is limited.

Male genitalia have long widely spaced socii, which distinguish *sumatrana* from other congeners. 8th sternite is unmistakeable by the long, triangular central projection on posterior margin. Female genitalia are strongly sclerotized and have pointed papillae annales. Corpus bursae disc-shaped and large. It bears a prominent signum of variable shape as illustrated.

Bionomics. Adults are common and fly between 800–2000 m above sea level.

Distribution. Distributed in southern Thailand, Sumatra, Borneo and Palawan.

Formofentonia Matsumura, 1925: 396.
Type species: *Formofentonia rotundata* Matsumura, 1925 [= *Stauropus orbifer* Hampson, 1893].
Schintlmeister 2008: 190; Schintlmeister 2013: 166.

Diagnosis. Medium sized adults. Antennae bipectinate in both sexes; the last 35% toward the tip are filiform. Female have much shorter rami. *Formofentonia* is unmistakable in its contrasting white areas along the dorsum and at the apex on fuscous violet-brown background. Except for the submarginal fascia, consisting of whitish spots, no fascia are visible. Discal spot is present. Hindwings are reddish-brown. The sexual dimorphism is limited.

Male genitalia have a bifurcated uncus, where uncus arms are slightly shorter than the slender pair of socii. Saccus small, but remarkeably strongly sclerotized. Valvae are smooth and bear an apical projection. Sacculus toward valva apex enlarged. Phallus long and slightly curved. Everted endophallus bears a more strongly sclerotized carina. 8th abdominal segment less modified. Posterior margins are gently bilobed. Female genitalia are characterized by two pairs of short and slender apophyses of equal length and a very long and slender ductus bursae. Corpus bursae is long ellipsoid. The prominent signum extends the entire width of the corpus bursae and differ greatly in shape between the different subspecies.

Formofentonia is monotypic, but there are several very distinct taxa, treated here as subspecies. *Formofentonia orbifer* is distributed in Himalaya, entire Indochina, China, Taiwan, Sundaland, Java, Sulawesi and Mindanao.

289. ***Formofentonia orbifer*** (Hampson, 1893): 152 (*Stauropus orbifer*).
Holotype: ♀, [India], Sikhim – NHMUK, London.
Schintlmeister 2008: 190; Schintlmeister 2013: 166.

Taxonomic notes. The species show extensive geographical variation. The following subspecies are recognized:

Formofentonia orbifer rotundata Matsumura 1925: 396; pl. 6: 10.
Lectotype: ♂, [Taiwan], Formosa, Horisha – Entomological Institute of the Hokkaido University, Sapporo.

Distinguished by its fuscous-brown general appearance, where in particular all cream or whitish fields of forewing are darkened. Male genitalia are as ssp. *orbifer*, but female genitalia display a long, interrupted rather straight band.

Populations from Java are classified as:

Formofentonia orbifer madena (Schaus, 1928): 81 (*Chadisra madena*).
Lectotype: ♂, [Indonesia], Java, Mount Gede, Tjibodas – USNM, Washington.

This subspecies is externally well distinguished by reduced white areas along dorsum and apex of forewing. Postbasal patch is fuscous-brown instead of cream or light brown as in other subspecies.

Male genitalia resemble by long apical valva process and rectangularly angled apical part of the sacculus ssp. *orbifer* rather than Sumatran populations. Phallus unique by long club-shaped projection on phallus tip.

Populations from Sundaland representing a further subspecies:

Formofentonia orbifer cremea **ssp. nov.** (figs. 289b, 289k, 289n, 289o).
Holotype: ♂, Indonesia, W Sumatra, Mt. Talang, 12 km E Padang, 1600 m, 0°57'S, 100°24'E, 13. ix. 1991, leg. M. Graul & A. Schintlmeister – in coll. A. Schintlmeister, Dresden.

Paratypes (28 ♂♂, 5 ♀♀):
Sumatra: 1 ♂, Prapat, Holzweg 2, 1050 m, 2°48'N, 98°55'E, 13. -28. vii. 1990, leg. E. W. Diehl; 1 ♂, dto., 13.–31. iii. 1993, leg. R. Brechlin; 2 ♂♂, 15 km NE Prapat, Holzweg 4, 1150 m, 2°46' N, 99° E, 3.–7. i. 1994, leg. Z. Weidenhoffer; 1 ♂, Sindar Raya, 350 m, 3°09'N, 98°57' E, 11.–14. viii. 1979, leg. Diehl & Schintlmeister; 1 ♂, Talu, 300 m, 0°14'N, 99°59'E, 7. ix. 1991, leg. A. Schintlmeister; 1 ♂, Bukit Subang, 22 km E Padang, Microwave Stat., 1200 m, 0°40'S, 100°32' E, 19.–20. i. 1991, leg. A. Schintlmeister (GU 16-80); 1 ♂, Gunung Rajabasa, 7 km E Kalianda, Microwave Stat., 200 m, 5°46'S, 105°36'E, 26. x. 1981, leg. A. Schintlmeister;
West Malaysia: 1 ♂, Selangor, vic. Ulu Gombak, FStC, 300–400 m NN, 3.–10. ii. 1986, leg. W. Nässig; 1 ♀, Genting Highland, 13. v. 1974 (GU 66-50);
Borneo: 2 ♂♂, Sabah, Golden Hill/KK, 1400 m, 14. iv. 1993, leg. Martini; 1 ♂, Mt. Trus Madi, Apin-Apin, 950 m, 17.iv. 1993, leg. Martini; 1 ♂, Sipulut, 800 m, 3. i. 1992; 1 ♂, Kinabalu Park, Headquarter, 1500 m, 13.–25. ii. 2000 leg. P. Spona; 1 ♀, Gunung Kinabalu, Sayap, 1000 m, 6° 10' N, 116°34' E, 11. iii. 2001;
Kalimantan: 2 ♂♂, 1 ♀, Kalimantan Selatan, 30 km E Kandangan, 15 km NE Loksado, 1100 m, 2°52' S, 115°38'E, 3.–22. ix. 1997, leg. S. Jakl; 1 ♂, dto., xii. 1997 (GU 72-39).
Palawan: 1 ♂, Mt. Mantalingajan, 600–800 m, 2.–12. viii. 2000, leg. D. Mohagan.

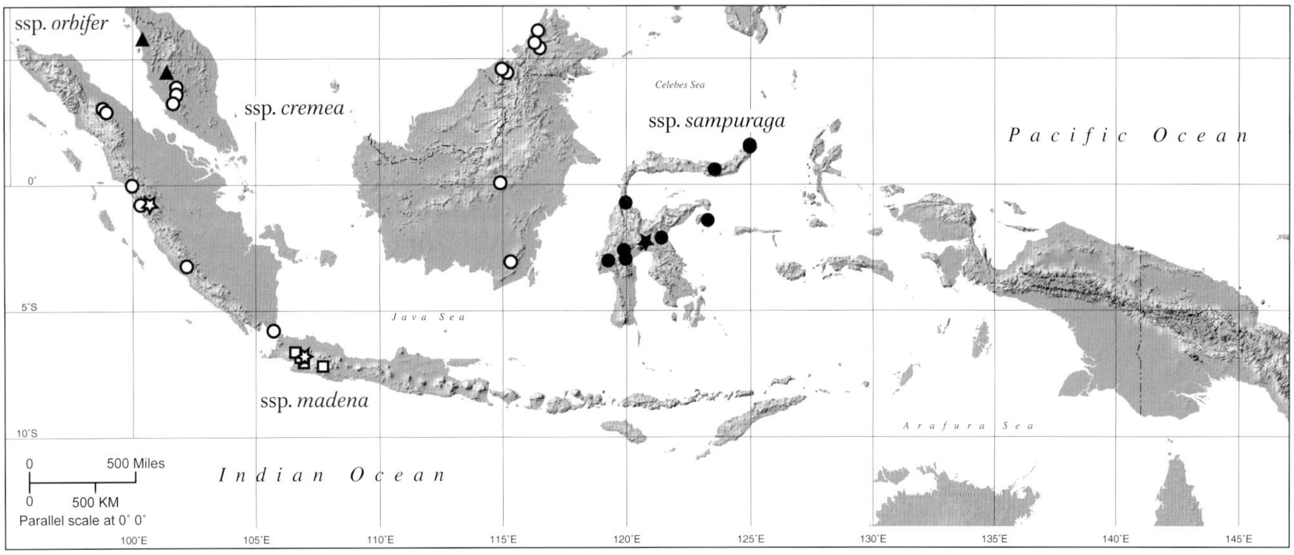

FIGURE 250 Distribution of *Formofentonia orbifer.*

Diagnosis of ssp. *cremea*. Forewing lengths ♂♂, 19–21 mm, ♀♀, 24–26 mm. This subspecies is distinguished differently from other subspecies by significantly paler brown hindwings in both sexes. The fuscous-brown costal strip in the basal area is narrow, the slimmest of all known subspecies.

Male genitalia differ by anvil-shaped apical part of valva from other subspecies. Margin of valva apex is straight, not buckled or rounded. Phallus as in ssp. *orbifer.* Female genitalia display a wing shaped, large signum.

Sulawesian populations are also distinct and described hereunder as:

Formofentonia orbifer sampuraga **ssp. nov.** (figs. 289e, 289l, 289p, 289q).

Holotype: ♂, Indonesia, Sulawesi, Mt. Sampuraga, 1400 m, 2°10' S, 120°45' E, 1.–6. ii.1995, leg. V. Siniaev & Tarasov – in coll. A. Schintlmeister, Dresden.

Paratypes (28 ♂♂, 5 ♀♀):

Sulawesi: 2 ♂♂, Mt. Sampuraga, 1400 m, 2°10' S, 120°45' E, 1.–6. ii. 1995, leg. V. Siniaev & Tarasov; 3 ♂♂, dto., 11.–12. ii. 1995; 6 ♂♂, Puncak, Palopo, 1000 m, 2°55'S, 120°05'E, 25.–31. i. 1995, leg. V. Siniaev & Tarasov; 1 ♂, dto. 10. 19. v. 1984, leg. A. Schintlmeister (GU 17-56); 5 ♂♂, 1 ♀, dto., 900–1300 m, iii. 1997, leg. R. Brechlin (GU 72-38); 1 ♀, dto. xi. 1997; 1 ♂, dto. xii. 1998; 1 ♂, Mamasa, 1800–2000 m, 2°56'S, 119°22'E, x. 1995; 2 ♀♀, Puncak, Palopo, 1100 m, 25.ix.1995, leg. S. Naumann (GU 66-49); 2 ♂♂, Mt. Salubaya, 0°45'S, 120°00'E, 1000 m, 9.–10. ii. 1995, leg. V. Siniaev & Tarasov; 2 ♂♂, Mt. Tambusisi, x. 1997; 2 ♂♂, Gunung Klabat,, ca. 20 km ESE Manado, 1700 m, 12.–13. ix. 2000, leg. C. Zorn; 1 ♂, Quarles Mts., Polo Poplo, N Rantepao, 2200 m, 26.–28. ix. 1995 leg. S. Nauman; 2 ♂♂, 1 ♀, Dumogo-Bone National Park, Site 14, 1440 m, 3.–8. iii. 1985, leg. J. D. Holloway; 1 ♂ dto., Site 8, 540 m, 12.–13. ii. 1985; 1 ♂ dto., Site 7, 540 m, 9.–11. ii. 1985; 1 ♂, Dumoga-Bone Natl. Park, 31. x. 1985, 220 m, leg. J. B. Heppner (FLMNH-MGCL specimen 289688); 1 ♂, Lake Moat, 20 km NE Cotamobagu, 1050 m, 26. – 30. x. 1985 leg. J. B. Heppner (FLMNH-MGCL specimen 289689).

Peleng: 3 ♂♂, 1 ♀, Basiano, 700 m, x. 1999 (GU 66-51).

Diagnosis of ssp. *sampuraga*. Forewing lengths ♂♂, 18–21 mm, ♀♀, 20–26 mm. External appearance resembles ssp. *cremea*, but blackish brown forewing pattern extended. The fuscous costalstrip at the basal area is broader. A submarginal row of white dots contrasts, as in ssp. *orbifer*. Ssp. *sampuraga* differ also by a straighter margin of the forewings from all other known subspecies.

Male genitalia are large. Space between the two uncus arms is wider than in the other subspecies. Uncus arms and socii are shorter. Sacculus at valva apex large and not rectangularly angled. Apical valva procjection is small; the shortest of all known subspecies. Phallus is thicker and the distal projection long, but without knob-shaped tip as in ssp. *cremeae*. Signum of female genitalia shaped as a quadrant and not as an interrupted band.

An seemingly isolated population from Mindanao is described hereunder as:

Formofentonia orbifer binansilang **ssp. nov.** (figs. 289h, 289r).
Holotype: ♂, Philippines, Mindanao, prov. Bukidnon, Mt. Binansilang, 45 km NW Maramag, 1200 m, 7°55' N, 124°40' E, 2. x. 1988, leg. K. Cerny & A. Schintlmeister – in coll. A. Schintlmeister, Dresden.

Paratypes (6 ♂♂, 1 ♀):
Mindanao: 2 ♂♂, Philippines, Mindanao, prov. Bukidnon, Mt. Binansilang, 45 km NW Maramag, 1200 m, 7°55' N, 124°40' E, 2. x. 1988, leg. K. Cerny & A. Schintlmeister (GU 17-57, GU 17-76); 1 ♂, 60 km NW Maramag, Talakag, 4 km N Dominorog, Kalatungan Mts., 1255 m, 7°36.029' N, 124°38.694' E, 2. x. 2011, leg. S. Naumann & J. Lourens; 1 ♂, 1 ♀, Prov. Bukidnon, Mt. Kitanglad, S Seite, Intavas, 8°07 N, 15. viii.–15. ix. 1998, leg. V. Sinjaev; 1 ♂, Provinz Davao del Norte, Mt. Caragan, vii. 1998; 1 ♂, Dalongdong, iv. 2000.

Diagnosis of ssp. *binansilang*. Forewing lengths ♂♂, 19–20 mm, ♀, 22 mm. External appearance similar ssp. *cremea*, but generally slightly more fuscous. Submarginal row of forewing weakly developed; whitish dots are contrary to ssp. *cremeae* hardly visible.

Male genitalia resemble mostly to those of ssp. *cremea*. The space between the two uncus arms is the closest in *orbifer*. Apical part of valva, e.g. sacculus and apical process is not straight as in *cremea* but curved. Phallus is shorter than in *cremea* and slightly curved.

Schintlmeister & Lourens, 2010: 232 already supposed that Mindanao populations may represent a distinct subspecies, but the available material at that time was insufficient.

Diagnosis. Adults of *F. orbifer* are unmistakable in their contrasting forewing pattern (as described in the genus diagnosis). The paleness of the white pattern varies slightly from shiny white to light brown.

Genitalia are as described in the genus diagnosis.

Bionomics. Adults are usually common and occur in Indonesia between 300–2000 m above sea level. In Mindanao, where the species is rare, it was recorded up to 2600 m.

Distribution. *Formofentonia orbifer rotundata* is endemic to Taiwan; the ssp. *orbifer* occur in mainland Asia: Himalaya (NW India, Nepal, NE India), Myanmar, S China, Indochina up to N Malaya. Ssp. *cremea* is distributed in Sumatra, Borneo, Palawan and the middle and southern parts of the Malayan Peninsula. Ssp. *madena* is endemic to Java.; ssp. *sampuraga* inhabits Sulawesi and Peleng isl., while ssp. *binansilang* is restricted to Mindanao.

Betashachia Matsumura, 1925: 398.
Type species: *Betashachia angustipennis* Matsumura, 1925.
Schintlmeister 2008: 191; Schintlmeister 2013: 51.

Diagnosis. Medium-sized imagines of grey forewings. Antennae of males bipectinated until the tip, but rami are short; those of the females are not filiform, but have very short rami. Last 45 % toward the tip are filiform. Forewing pattern consist of from several, indistinct black and white marked fasciae, such as basal, postbasal, medial, postmedial, submarginal fasciae. Discal spot on the forewings developed as a small black dot. Hindwings brown with indistinct, fuscous-brown anal markings. Sexual dimorphism is limited.

Male genitalia have a compact uncus and a pair of small socii. Valvae are broad with a costal process. Costal area of valva and large sacculus more strongly sclerotized. Phallus straight and long. Near the tip there are more strongly sclerotized structures, such as spines, hooks or plates. Everted endophallus bears several cornuti. 8th sternite large and squarish with a pair of projections near the posterior margin. 8th tergite notched at the posterior margin and bilobed at the anterior part. Female genitalia are characterized by a very long, compact, and strongly sclerotized antrum. Ductus bursae short. Corpus bursae circular with a large triangular or circular signum.

Caterpillar of *B. angustipennis* has been reared in Taiwan on *Styrax* (Styracaceae).

The genus is comprised of four species, which are all occur in eastern China. Only *B. angustipenis* extends to Taiwan, Indochina and Sumatra. One species occurs also in Korea.

290. *Betashachia angustipennis* Matsumura, 1925: 398, 407.
Holotype: ♂, [Taiwan], Formosa, Horisha – Entomological Institute of the Hokkaido University, Sapporo.
Schintlmeister 2008: 191; Schintlmeister 2013: 51.

Taxonomic note. From Sumatra ssp. *tropica* Kiriakoff, 1974: 407; pl 4: 3; holotype: ♂, Sumatra, Deli [=Medan], Dolok Merangir, 180 m – ZSM, Munich, has been described, which differ by an angled marigin of forewings at vein M2 and contrasting, whitish-developed forewing pattern from ssp. *angustipnnis*. Male genitalia are, however, not very different from mainland Asia or Taiwanese

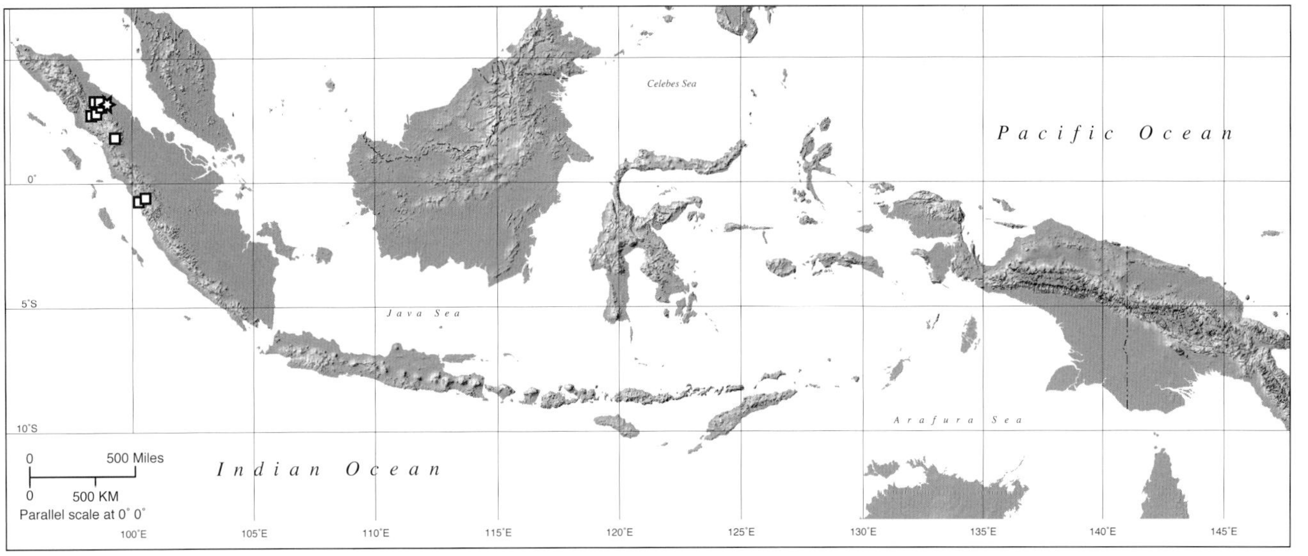

FIGURE 251 Distribution of *Betashachia angustipennis tropica*.

populations. Uncus is shorter, and costal valva process distinctly longer than in ssp. *angustipennis*. The length of the costal process varies geographically: Taiwan populations have the shortest process and specimens from Sumatra the longest. Phallus of Sumatran populations is thicker than in ssp. *angustipennis*.

Diagnosis. Ssp. *tropica* is readily recognized by its diagnostic white marginal forewing markings and contrasting black-white checkered fringe. Discal spot is small and deep black. Postbasal and postmedial fasciae are white, more contrasting toward the dorsum. Sexual dimorphism is limited.

Male genitalia have a characteristically shaped uncus and large, somewhat bulbous saccus. Valvae processes are long, tapered and curved. Phallus long and straight. At the tip there is a fan-like plate with serrated margin. 8th sternite large with a pair of projections below the posterior margin. Everted endophallus bears several cornuti. Female genitalia of ssp. *tropica* are not yet dissected. Ssp. *angustipennis* display a large antrum, which is posteriorly triangularly notched. Corpus bursae is disc-shaped and bear a large triangular signum.

Bionomics. Males are common. They occur in Sumatra mostly above 1000 m up to altitudes of 1800 m. Only a few males were collected at 300 m and 500 m above sea level.

Distribution. Nominotypical subspecies *angustipennis* is distributed in Taiwan, southern China including Hainan, and Yunnan, Vietnam, Laos, Thailand and southeastern Myanmar. In Central China (Shaanxi, Sichuan) ssp. *punctata* Schintlmeister, 2002 occurs. Ssp. *tropica* is endemic to Sumatra.

Fentonia Butler, 1881: 20.

Type species: *Fentonia laevis* Butler, 1881
[= *Harpyia ocypete* Bremer, 1861]

Schintlmeister 2008: 193; Schintlmeister 2013: 164.

Diagnosis. Medium-sized imagines of grey or violet-brown forewings. Antennae of males long bipectinated; the last 28–45% until the tip are filiform. Female antennae are filiform. Forewing pattern displays a well developed black submarginal fascia, often accompanied by a white band. Members of the violet-brown colored *ocypete* group have a black streak from base to the postmedian area, while the grey species, like *talboti,* have the basal area fuscous grey filled. Fringe of all wings are checkered. Hindwings varies from fuscous-brown to light- brown; anal markings are blackish and inconspicuous.

Male genitalia are similar in all *Fentonia* species. Uncus long and cylindric with dense setae. Socii slender slightly curved. Juxta inconspicuously sclerotized; sacculus reduced. Valvae ellipsoid, costa near the base is bulbous curved. Phallus straight with a serrated plate at the tip. 8th sternite are of taxonomic value for separation of *Fentonia* species. Posterior margin straight, curved or bilobed. In the central portion several projections or processes occur occasionally. 8th tergite broad and less modified; the posterior margin at th center often slightly notched. Female

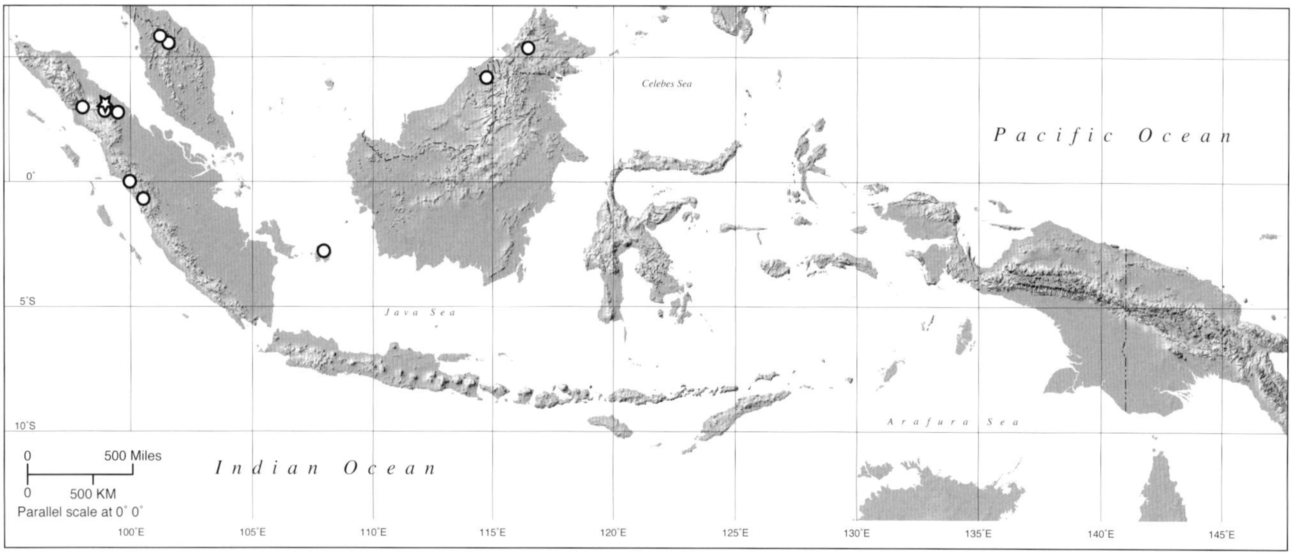

FIGURE 252 Distribution of *Fentonia sumatrana*.

genitalia have long apophyses posteriores, which are at their base strongly sclerotized and broad. Apophyses anteriores reduced in length. Antrum large and strongly sclerotized. Ductus bursae long and slender. Corpus bursae disc-shaped. It bears a strongly sclerotized triangular signum, mostly of irregular shape.

Caterpillars of *F. ocypete* or *baibarana* Matsumura, 1929 feeds in Japan and Taiwan on various Fagaceae, like *Quercus*.

The genus includes 12 species, which are distributed in Pakistan India, Nepal, China Russia Far East, Japan, Korean Peninsula, Taiwan, the entirety of Indochina, Sundaland, Palawan and Java. The greatest diversity of species is found in China.

291. *Fentonia sumatrana* Kiriakoff, 1974: 412; pl. 5: 1. (*Fentonia ocypete sumatrana*).

Holotype: ♂, [Indonesia], N.-Sumatra, Deli, Dolok Merangir, 180 m – ZSM, Munich.

Schintlmeister 2013: 166.

Taxonomic note. The females of *sumatrana, helena* and *hollowayi* are externally similar looking. It is currently doubtful whether the illustrated females of *sumatrana,* genitalia fig. 291b and corresponding imago fig. 291e, are really conspecific to the illustrated males.

Diagnosis. Similar to *helena* and *hollowayi* by external appearance and all three species occuring in Sumatra sympatrically. Forewings fuscous violet-brown; submarginal area slightly paler. The species have a contrasting white dorsal patch in the medial area of the forewings. Hindwings are of a paler brown than in *helena* or *hollowayi.* The males of *sumatrana* are distinguishable as follows: forewings are more fuscous and rather uniform. Black basal streak rather slim. Hairs on the thorax are contrasting whitish grey. *F. helena* has the forewing color more reddish-tinged and the black basal streak is thick and short, like a patch. Thorax hairs of *helena* are brown and not contrastingly developed. *F. hollowayi* have forewing pattern contrasting developed. The white dorsalspot is the largest among these three species. Basal streak rather weakly developed. Sexual dimorphism of *sumatrana* is limited.

Male genitalia are as described above. They are distinguished from *helena* and *sumatrana* by shorter socii and sacculus flaps of the valvae. Phallus tip bears a pair of strongly sclerotized and serrated projections. 8th sternite have posterior margin contrary to other congeners straight. In the center of 8th sternite a large semicircular projection and a pair of small lobes occur. Female genitalia have a long and strongly sclerotized antrum and a rather short ductus bursae. Corpus bursae is large and ovoid, bearing a large trapezoid signum. In *F. hollowayi* the signum is triangular.

Bionomics. *Fentonia sumatrana* occur in Sumatra infrequently and mostly between 900–1000 m above sea level. In Borneo it was reported as commonly (Holloway, 1972: 74) and flies also in the lowlands.

Distribution. Known from south Thailand, the Malayan Peninsula, Sumatra, Belitung Isl. and Borneo.

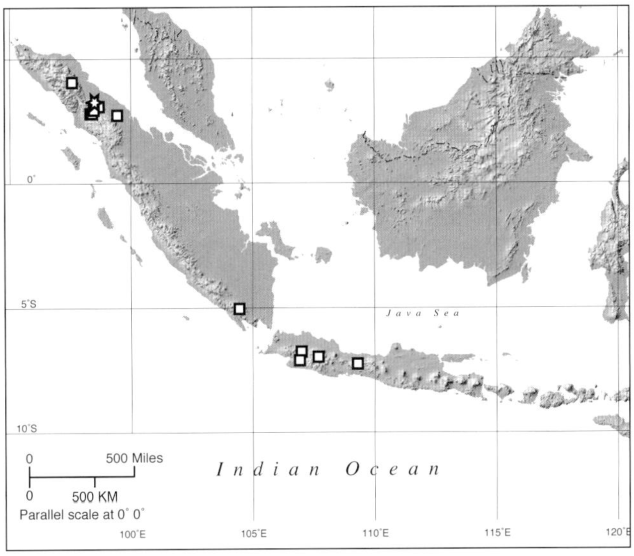

FIGURE 253 Distribution of *Fentonia helena*.

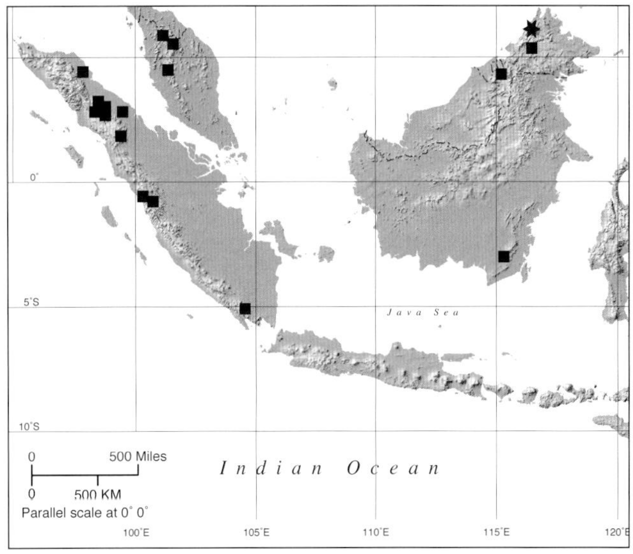

FIGURE 254 Distribution of *Fentonia hollowayi*.

292. ***Fentonia helena*** Kiriakoff, 1974: 413; pl. 5: 2; fig. 27.
Holotype: ♂, [Indonesia], N-Sumatra, Brastagi, 1.200 m – ZSM, Munich.
Schintlmeister 2013: 165.

Diagnosis. Similar to *F. sumatrana*, but hairs of the thorax rather brown than whitish grey. Forewings less dark than *sumatrana,* and more brownish-tinged. A single female from Java, where *sumatrana* reportedly does not occur, are very similar to the female attributed here to *sumatrana*, but the white dorsal spot on the forewings is absent.

Male genitalia have the uncus as in *sumatrana*, but socii are longer. Valvae have a straight sacculus. At valva base the costa is slightly bulbous as in *sumatrana,* but the bulb is smaller than in *hollowayi.* It seems, that valvae of Javanian males are slightly broader than in Sumatran populations. Phallus straight and much thinner than in *sumatrana* and *hollowayi.* 8th sternite is specifically shaped. Posterior margin slighly convex; in the central part a projection occurs, but there are no further processes or lobes as in *sumatrana* or *hollowayi.* Female genitalia have not yet been dissected.

Bionomics. Infrequent in Sumatra, but common in Java on Mt. Gede. Most specimens have been taken above 1000 m up to altitudes of 1600 m. Two males have been collected at 100 m and 500 m above sea level.

Distribution. Distributed in Sumatra and Java.

293. ***Fentonia hollowayi*** Schintlmeister, 2007: 157; pl. 24; fig. 125.
Holotype: ♂, Ostmalaysia, Sabah, Kinabalu Park Kg. Kiau, 1.000 m – coll. A. Schintlmeister, Dresden.
Schintlmeister 2013: 165.

Taxonomic note. Females of *sumatrana, helena* and *hollowayi* are very similar looking. The identity of the corresponding males is currently doubtful.

A good series (n = 162 ♂♂) from S Kalimantan differs from Kinabalu or Sumatran populations, being 2 mm smaller in forewing length and having slightly paler wings.

Diagnosis. Forewings fuscous violet-brown with contrasting pattern. Distinguishable from the similar *sumatrana* and *helena* by following characters: antennae of males have by far the longest rami; forewing has a white dorsal patch; large, orange-brown patch in the postmedial area broader and more contrasting; hindwings are fuscous-brown. Female have less contrasting, rather uniform colored forewings and is indistinguishable by external appearance from *sumatrana* or *helena.*

Male genitalia bear strongest resemblance to those of *helena*, but socii are thicker and shorter. Valvae broad with an apical flap. Costa bulbous, with large bump toward the base. Phallus large and thick, similar to *sumatrana*, but shorter. It bears near the conical tip a fan-shaped, strongly sclerotized appendix. 8th sternite convex at the posterior margin. In the center a pair of long and slender

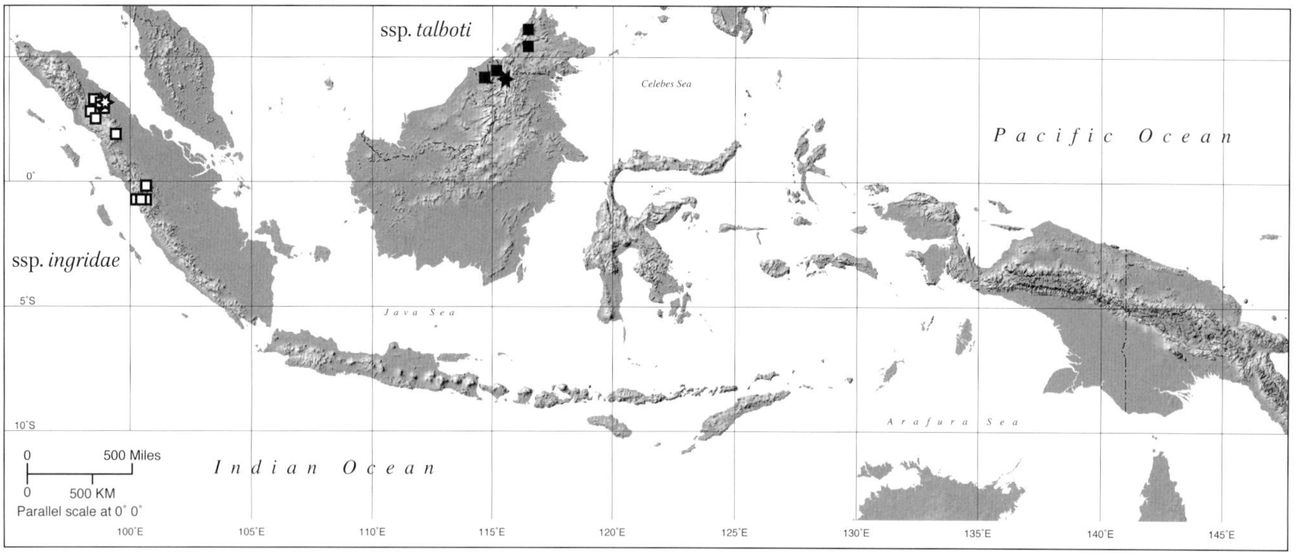

FIGURE 255 Distribution of *Fentonia talboti*.

lobes arises, which are a significant feature. Female genitalia resemble to those of *sumatrana*. They differ by a squarish, broad antrum and longer ductus bursae. Corpus bursae is smaller and circle-shaped. Signum large and triangular shaped.

Bionomics. This is in Sundaland the most common *Fentonia*, which occurs from the lowlands (80 m) up to 1700 m. Most specimens have been collected between 1200–1600 m above sea level.

Distribution. Distributed in Sumatra, the Malayan Peninsula and Borneo.

294. ***Fentonia talboti*** Gaede, 1930: 627; pl. 83: d.
Holotype: ♀, [Malaysia, Borneo], Sarawak – NHMUK, London.
Schintlmeister 2013: 166.

Taxonomic note. The species is represented in Sumatra in the subspecies

Fentonia talboti ingridae Holloway & Bender, 1985 in Bender, 1985: 80, 107; pl. 8: 7;
holotype: ♂, Sumatra sept., Deli [= Medan], Dolok Merangir 180 m – NHMUK, London.

Ssp. *ingridae* differ from ssp. *talboti* by less stronger defined medial and postmedial fasciae. Hindwings are darker in both sexes than in ssp. *ingridae*. The other features, given in the original description are subject of individual variability; the genitalia of both subspecies are virtually identical.

Diagnosis. Forewings of the male are grey, elongate and narrow. Basal area fuscous grey filled. Black postbasal fascia of forewing doubled. Medial fascia prominently black, in particularly in the costal area. Black postmedial fascia accompanied by a whitish fascia. Hindwing pale brown or whitish. The female is sexually dimorphic with broader and more rounded forewings. Hindwings of female are dark brown.

Male genitalia are relatively small. Socii are club-shaped and slightly curved. Valvae are sinuously shaped at costa and without any flaps or projections. Phallus short and gently curved. Phallus tip ends in a spine. Endophallus bears many needle-like cornuti. 8th sternite of symmetric shape, posterior and anterior margins convex. The central portion bears a curved projection. The female genitalia have compact apophyses posteriores with a broad base. 8th abdominal segment sinuous. Antrum small, ductus bursae long and strongly sclerotized. Corpus bursae circular and large. Signum strongly sclerotized and irregularly shaped.

Bionomics. An infrequent mountain species which was observed between 1200 and 1800 m. Bender (1985) reported it also from 180 m and 250 m above sea level.

Distribution. *Fentonia talboti talboti* is restricted to Borneo, whereas the ssp. *ingridae* occur in Sumatra.

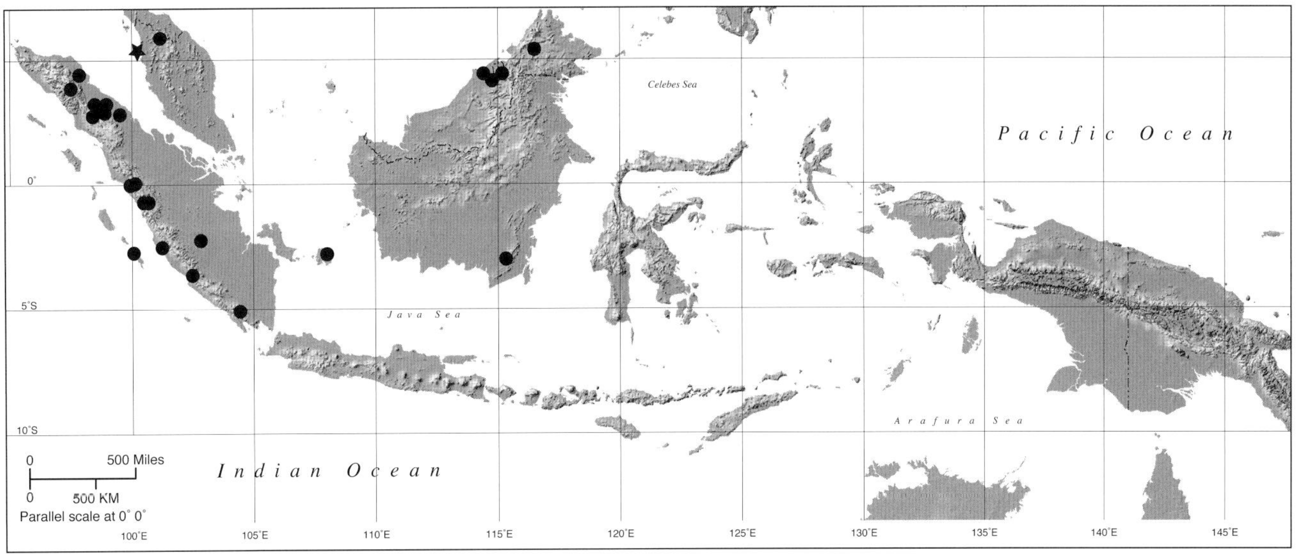

FIGURE 256 Distribution of *Fentonia bipunctus.*

295. ***Fentonia bipunctus*** (Rothschild, 1917): 246; pl. 5: 2 (*Stauropus bipunctus*).
Holotype: ♀, [West Malaysia], Penang – NHMUK, London.
Schintlmeister 2013: 164.

Diagnosis. Forewings ash grey and elongated as in *talboti.* Forewing pattern black and resembles *talboti* by strongly defined black postmedial fascia, the darker filled basal area and the greyish brown hindwings. Two prominent black spots, in the medial area are a good character to identify the species readily. Both spots vary individually in size and occasionally a third black dorsal spot occur. Sexual dimorphism is limited; the female have slightly darker hindwings.

Male genitalia are similar to those of *talboti,* but valva is broader, and there is a small flap at the harpe and a second flap at the sacculus. Phallus compact, slightly curved, and pointed at the tip. Endophallus bears many cornuti. 8th sternite convex, rounded at both margins. 8th tergite not bilobed as in *talboti,* but convex rounded. Female genitalia have strongly sclerotized and bilobed 8th abdominal segment. Papillae anales are much smaller than in other congeners. Apophyses posteriores relatively short and robust. Apophyses anteriores longer than in other *Fentonia.* Antrum small; ductus bursae medium-sized. Corpus bursae disc-shaped; signum small, but strongly sclerotized.

Bionomics. Adults are common and occur from 350 m up to 1700 m above sea level, more commonly above 1000 m.

Distribution. Distributed in Palawan, Borneo, Myanmar, Vietnam, Thailand, Malaya, Sumatra including Belitung and Mentawei (Pagai) islands.

Neopheosia Matsumura, 1920: 147.
Type species: *Pheosia fasciata* Moore, 1888
Schintlmeister 2008: 196; Schintlmeister 2013: 278.

Diagnosis. Medium-sized imagines. Closely related to *Fentonia.* Antennae are long bipectinated; the last 45% to the tip are filiform. Female antennae are filiform, except the *N. atrifusa* Hampson, 1897-group. Forewings are grey, black or orange brown. Forewing pattern have a whitish submarginal fascia. Discal spot inconspicuously black or brown. Fringe of all wings checkered. Hindwings brown; anal markings blackish. Sexual dimorphism is limited.

Male genitalia of *Neopheosia* are characterized by long uncus, a pair of large socii. Juxta inconspicous sclerotized. Valvae ellipsoid, sometimes with small basal processes and saccular projections. Phallus straight; endophallus bears star-shaped cornuti. 8th abdominal segment not modified. Female genitalia have medium-sized apophyses posteriores, but compact and short apophyses anteriores. 8th abdominal segment large and wave-shaped. Antrum large and strongly sclerotized. Ductus bursae medium sized in length, but broad. Corpus bursae disc-shaped with strongly sclerotized signum.

Caterpillar of *N. fasciata* feeds in Far East oligophag on Fagaceae (*Quercus*) and Rosaceae (*Prunus*).

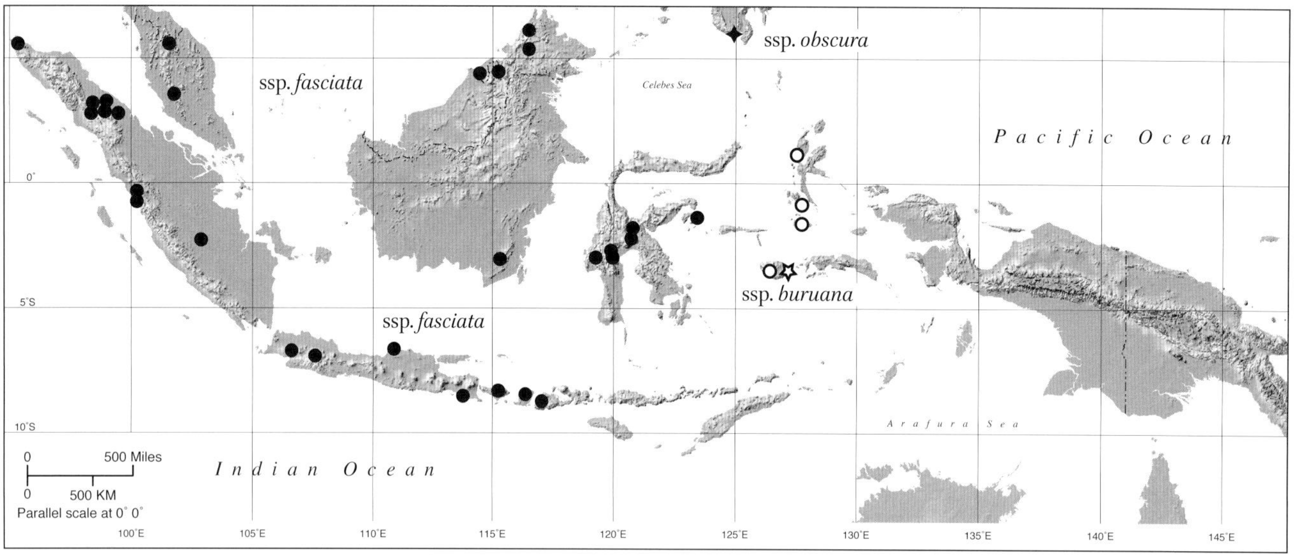

FIGURE 257 Distribution of *Neopheosia fasciata*.

Neopheosia comprised of five species. Distributed in Pakistan, N India, Nepal, China, Russia Far East, Japan, Korean Peninsula, Taiwan, all of Indochina, Sundaland, the Philippines, Java, the lesser Sunda Islands, Sulawesi, and Moluccas.

296. *Neopheosia fasciata* (Moore, 1888): 401 (*Pheosia fasciata*).
Lectotype: ♂, [India, Himachal Pradesh], Dharmsala, 6.200 feet – NHMUK, London **by present designation**.
Schintlmeister 2008: 196; Schintlmeister 2013: 278.

Taxonomic notes. *Pheosia fasciata* was described from an unspecified number of ♂♂, ♀♀ from Dharmsala, 6200 feet. The author traced in NHMUK a pair, both marked by red ringed circular label as „Type" and a further female with same data label as the previous mentioned pair. To stabilize the nomenclature the ♂ from NHMUK with the following labels is designatd as lectotype: "Kangra | 85-26 | 68.13", underside in Moore's hand: "*Pheosia* | *fasciata* ♂ | Type-Moore"; red ringed, circular label "Type", blue label „Notodontidae genitalia slide 1072 ♂", yellow label " Photo done by A. Schintlmeister #4117". A lectotype label was added. The lectotype is illustrated here as adult in fig. 296j and in its genitalia in fig. 296a. Both remaining females become paralectotypes.

Neopheosia fasciata is a variable insect. Populations from NW India and also Nepal are in general appearance darker than the series from Sundaland. But both forms cannot be separated seriously due to overlapping individual variability. Among NW Indian adults are also pale specimens, like the lectotype of *fasciata*. On other hand among a good series (n > 100) from Sumatra also a few dark specimens occur. Fuscous individuals appear occasionally among populations from Buru and Sulawesi, while in Mindanao (ssp. *obscura* Schintlmeister, 1993) apparently all specimens are fuscous-brown.

Populations from Buru are distinguished by a significantly short gnathos from Sundaland material as already pointed out by Holloway (1983: 69). Further dissections confirms this and a new subspecies is described below as:

Neopheosia fasciata buruana **ssp. nov.** (figs. 296e, 296f, 296o, 296p).
Holotype: ♂, Indonesia, SE Buru Isl., Ilat, Mt. Rernajah, 250 m, 3.–20. i. 2013, leg. S. Jakl (GU 93-29)– in coll. A. Schintlmeister, Dresden.

Paratypes (25 ♂♂):
Buru: 2 ♂♂, Ilat, Mt. Rernajah, 250 m, 3.–20. i. 2013, leg. S. Jakl (GU 93-30); 13 ♂♂, dto., Gamoe Mrapat, Central West Buru, 5000 ft.,iii.–iv. 1922; C., F. & J. Pratt (BM 1075); 1 ♂, ibid, 3. iv. 1922,
Bacan: 2 ♂♂, SE slopes Mt. Sibela, SE Makian, 1200 m, 0°44'N, 115°01'E, 2.–12. 2006, leg. S. Jakl; 1 ♂, dto. 10.–30. viii. 2008, leg. S. Jakl;
Obi: 1 ♂, S coast, Seribu Mts., 1500 m, 1°36'S, 127°48'E, 20. xi.–10. xii. 2008, leg. S. Jakl (GU 92-79);
Halmahera: 2 ♂♂, 15 km SE Baru, 300 m, v. 1998, leg. Gala; 3 ♂♂, „Halmahera", vi. 1998.

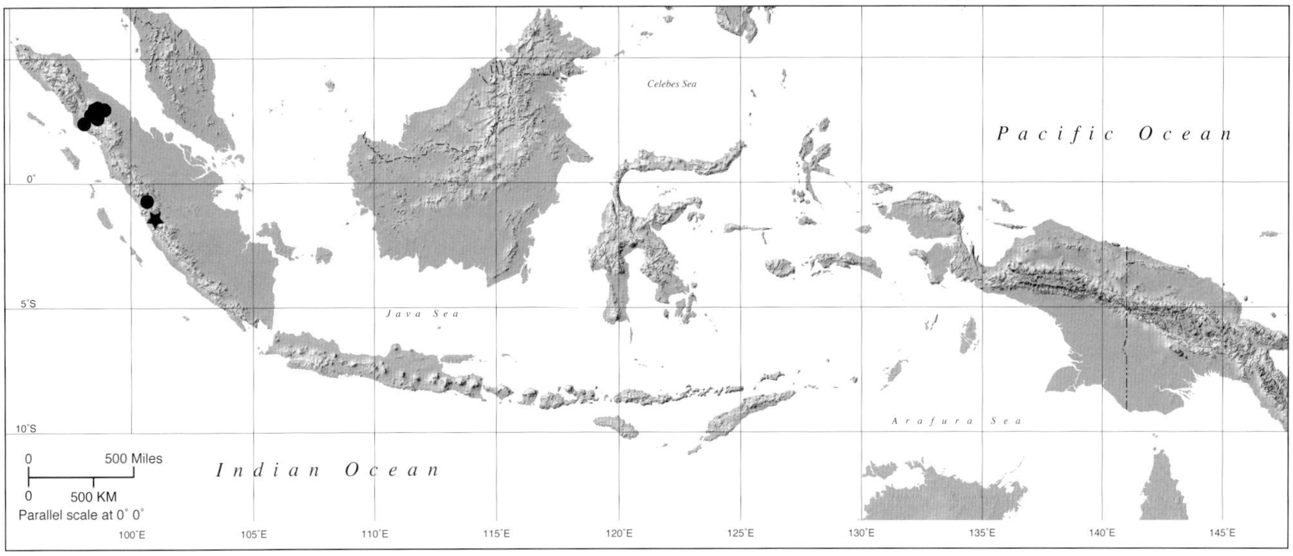

FIGURE 258 Distribution of *Neopheosia albiplaga*.

Diagnosis of ssp. *buruana*. Forewing lengths ♂♂, 22–23 mm, slightly, on average 1 mm, smaller than populations from Sumatra, Borneo or Java. There are specimens with pale yellowish-colored forewings as well as fuscous violet-brown individuals. But in fact, and due to individual variability ssp. *buruana* is not distinguishable by its forewing pattern. Females are still unknown.

Ssp. *buruana* is distinguished from other populations by a significantly shorter gnathos in male genitalia. For comparison n = 18 male genitalia from other regions has been dissected.

Diagnosis. Forewings are usual orange-brown with apical streak, brown discal streak, large brown dorsal line and checkered fringe. Adults are variable but unmistakable and sometimes fuscous forms occure. Range of individual variability includes also reducing of the brown pattern including the discal streak. Hindwings are whitish-brown with darker brown anal region. Sexual dimorphism is limited but females have broader forewing shape.

Male genitalia display a deeply bifurcated uncus of variable length and a central gnathos. Valvae are of ellipsoid shaped and free of processes or projections. Phallus straight; endophallus bears many star-like cornuti. 8th abdominal segments are less modified. 8th tergite slightly bilobed at the posterior margin. Female genitalia have a large 8th abdominal segment. Antrum and ductus bursae are compact and strongly sclerotized. Corpus bursae is disc-shaped and bear an ellipsoid, band-like signum of variable shape.

Bionomics. Adults are common and occur in Indonesia at altitudes between 300 m up to 1700 m. In the Himalaya the moth has been observed up to 2900 m above sea level. Hostplant of the caterpillar in Japan and Taiwan are various *Quercus* (Fagaceae), *Pyracantha* and *Prunus* (Rosaceae). The latter plant ("Cherry") was also provided in the original description by Moore, 1888 as hostplant.

Distribution. Distributed in the Himalaya from Pakistan to Sikkim, Assam, Myanmar, China, Hainan, Japan, Taiwan, entire Indochina, Malaya, Borneo, Sumatra, Java, Bali, Lombok, Sumbawa, Sulawesi, and most islands of the Philippines. In Mindanao and Panay the ssp. *obscura* Schintlmeister, 1993 occur. Ssp. *buruana* is distributed in Buru, Halmahera, Bacan and Obi.

297. ***Neopheosia albiplaga*** Gaede, 1930: 638; pl. 84: h.
Holotype: ♂, [Indonesia], Sumatra, Mt. Korintji, 2.400 m – NHMUK, London.
Schintlmeister 2013: 278.

Diagnosis. Forewings ash grey. Post basal and post medial fasciae are light grey. Most specimens are unmistakable by a contrasting blackish-grey filled medial area. But occasionally specimens occur, which have the medial area of same color as the basal and postmedial area. Discal spot indistinct and black. Fringe of all wings is checkered light and fuscous grey. Hindwings pale grey with blackish-grey anal markings. Sexual dimorphism is limited.

Male genitalia resemble those of *N. fasciata*, but socii are shorter. Gnathos is of half the length of socii. Valvae rather rectangularly shaped with well developed, pronounced sacculus. At valva apex there is a small flap. Phallus short and tapered at the tip. Endophallus bears many

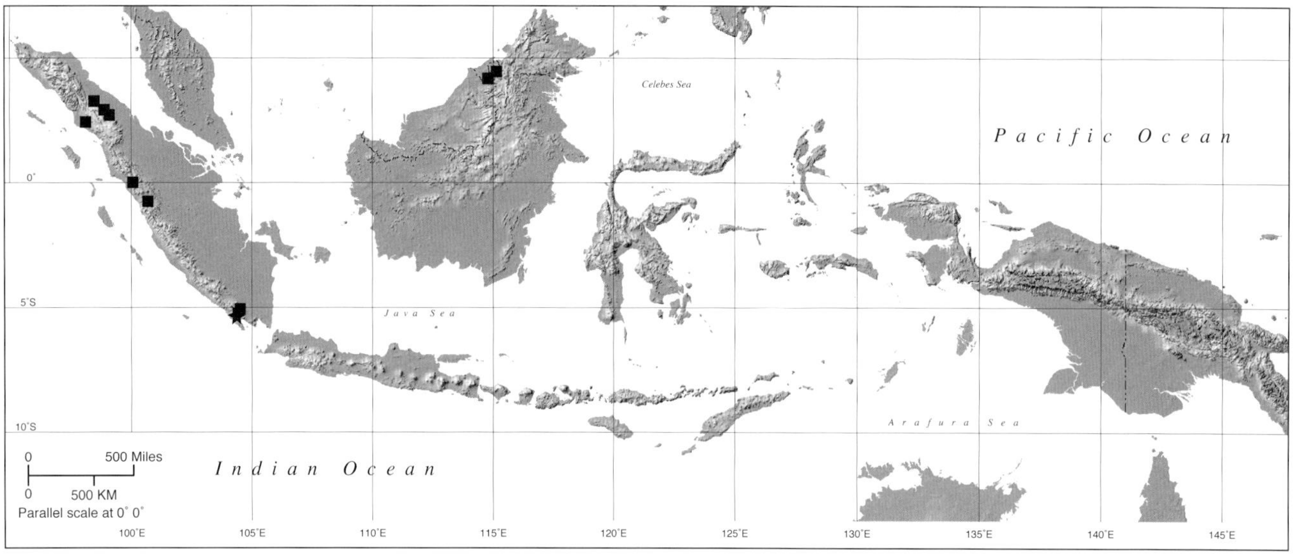

FIGURE 259 Distribution of *Fusadonta albipuncta*.

star shaped cornuti. Female genitalia have not yet been dissected.

Bionomics. The species is infrequent and has been observed between 350 m and 1700 m above sea level. Most specimens were collected at altitudes above 1200 m.

Distribution. Endemic to Sumatra.

Fusadonta Matsumura, 1920: 146.

Type species: *Notodonta basilinea* Wileman, 1911.

Schintlmeister 2008: 198; Schintlmeister 2013: 173.

Diagnosis. Medium-sized to large imagines. Antennae of the males bipectinated; the last 45–55 % until the tip are filiform. Female antennae are filiform. Wings and body are umber-brown and forewings display a dorsal scale tooth. Discal spot is pale brown with darker brown nucleus. Basal streak black and long, in most species strongly developed, but in *albipuncta* rather fine. Postmedial fascia indistinct, like a fuscous shadow. Hindwings display a fuscous or sometimes pale brown, medialfascia. Discal spot of hindwing indistinct, marked as a fuscous brown spot. Sexual dimorphism is limited.

Male genitalia are characterized by a bifurcated uncus and a pair of rather short, curved, and slender socii. Valvae are narow with a characteristic jaw spanner shaped apical part. Phallus compact with an central spine or projection. Endophallus bears many star shaped cornuti. A characteristic feature of *Fusadonta* are the many setae on the notched posterior margin of 8th sternite. 8th tergite less modified. Female genitalia have two pairs of long and slender apophyses. Postvaginal plate deeply notched. Antrum very large and strongly sclerotized. Ductus bursae short; corpus bursae circular shaped. Signum large like a broad band with characteristical central notch.

The genus is represented by four species. Distributed in China, Korea and Japan; one species occurs in Sundaland.

298. ***Fusadonta albipuncta*** (Gaede, 1930): 641; pl. 80: d (*Notodonta albipuncta*).

Holotype: ♂, [Indonesia], Sumatra, Barisan-Gebirge, 800 m – NHMUK, London.

Schintlmeister 2013: 173.

Diagnosis. Large species. Rather unmistakable by size and forewing shape with well developed dorsal scale tooth. Forewings are umbra, but often tinged yellowish-brown. Forewing base displays a fine black basal stripe. Hindwings with contrasting dark brown/cream doubled medial fascia.

Male genitalia display a bifurcated uncus and large socii. Apical part of valvae jaw spanner like shaped. 8th sternite very deep notched with many. Phallus has centrally a bump. A dissected single male from Palawan bears instead of the bump a pointed spine. Female genitalia are as described above.

Bionomics. *F. albipuncta* is seemingly a rare species, n < 20 specimens are known. Adults occur between 350 m and 1700 m above sea level.

Distribution. Distributed in Sumatra, Borneo and Palawan.

Genitalia Figures

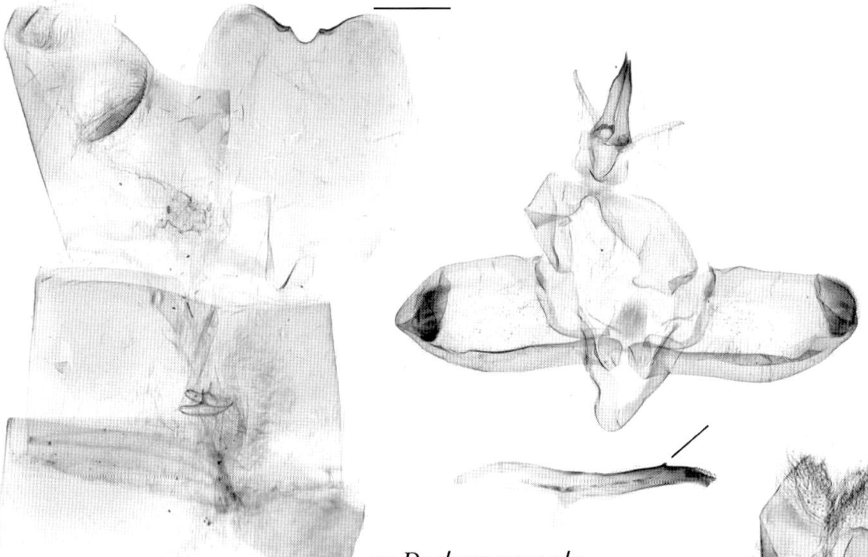

1a *Dudusa synopla*
Sumatra, GU 18-09

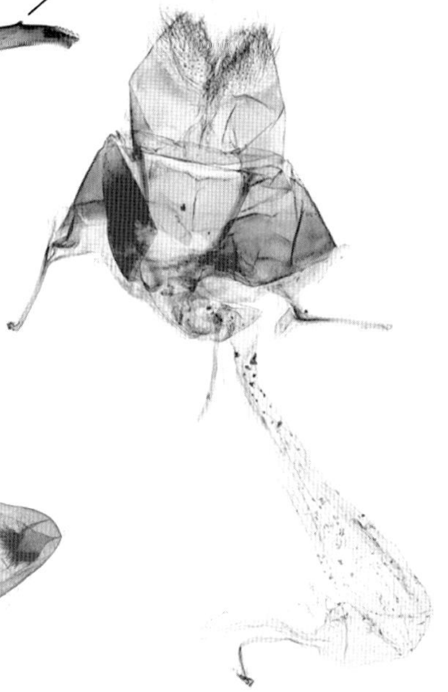

1b *Dudusa synopla*
Taiwan, MWM 4.872

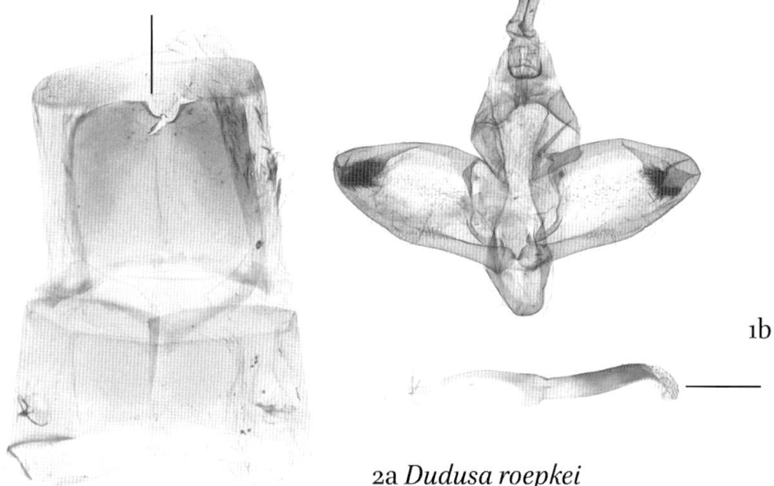

2a *Dudusa roepkei*
Java, GU 59-23, paratype

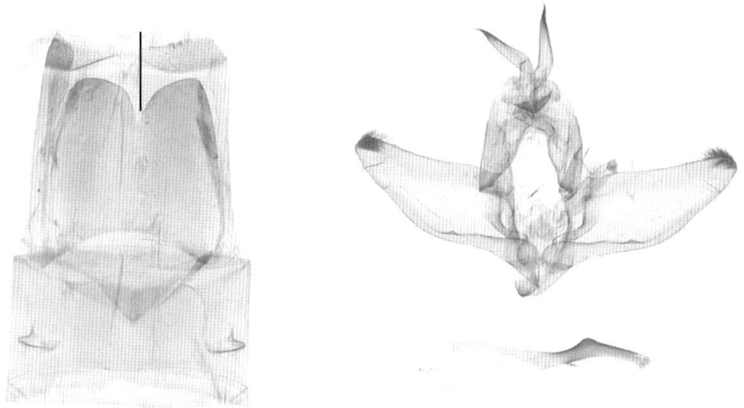

1 mm

3a *Dudusa celebensis*
Sulawesi, GU 59-24

 | DOI:10.1163/9789004354258_008

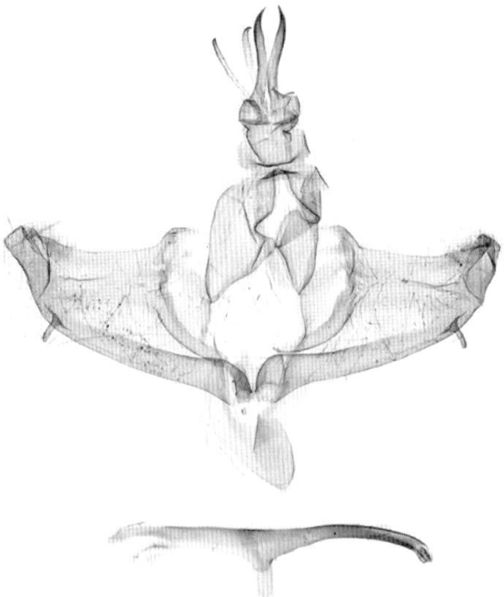

4a *Dudusa vethi vethi*
Nias, GU 18-10

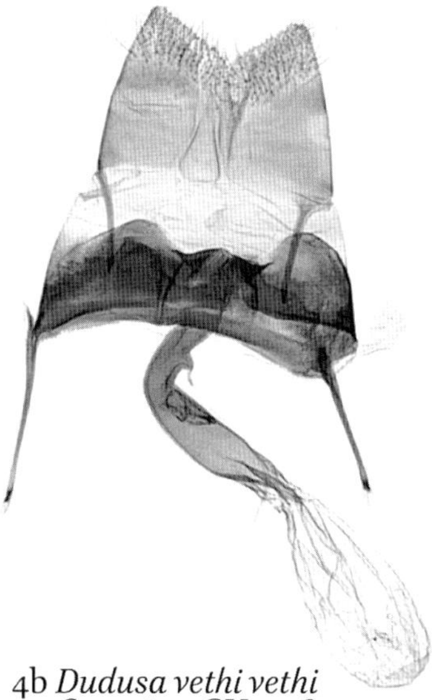

4b *Dudusa vethi vethi*
Sumatra, GU 95-85

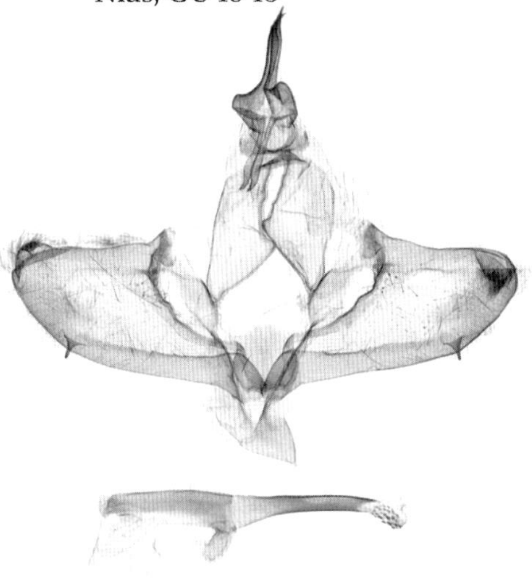

4c *Dudusa vethi javana*
Java, GU 59-22

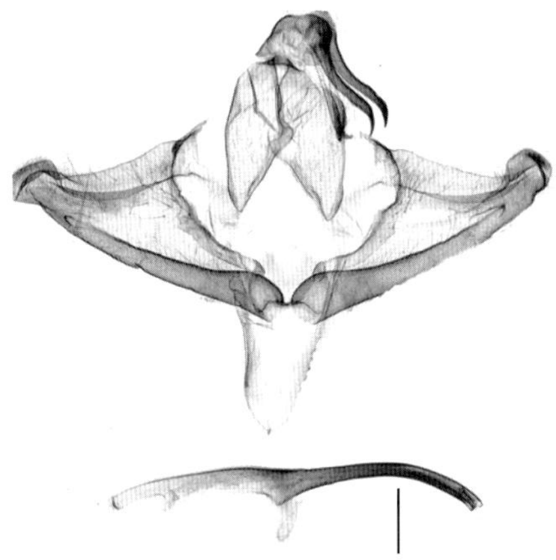

4d *Dudusa vethi borneensis*
Borneo, GU 42-07

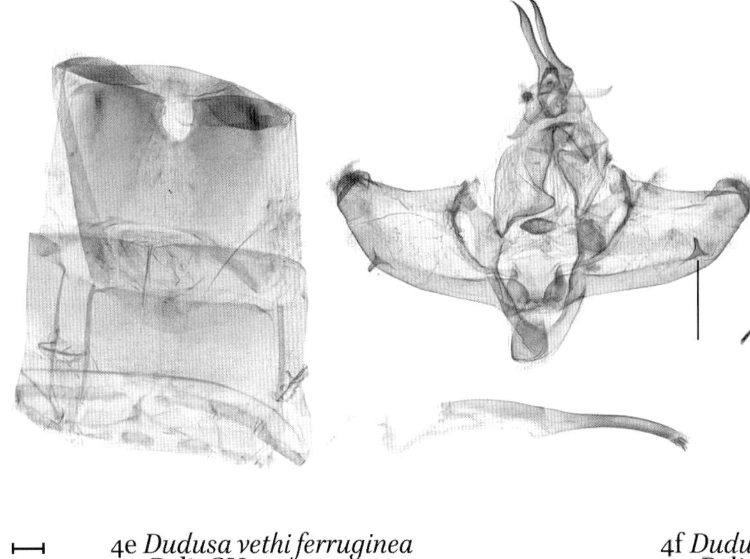

1 mm

4e *Dudusa vethi ferruginea*
Bali, GU 59-21, paratype

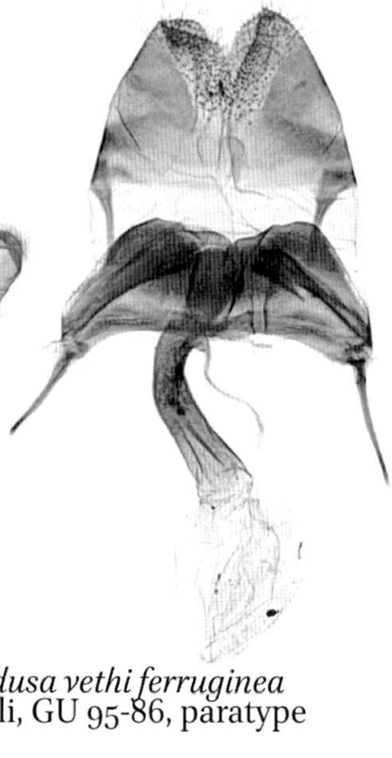

4f *Dudusa vethi ferruginea*
Bali, GU 95-86, paratype

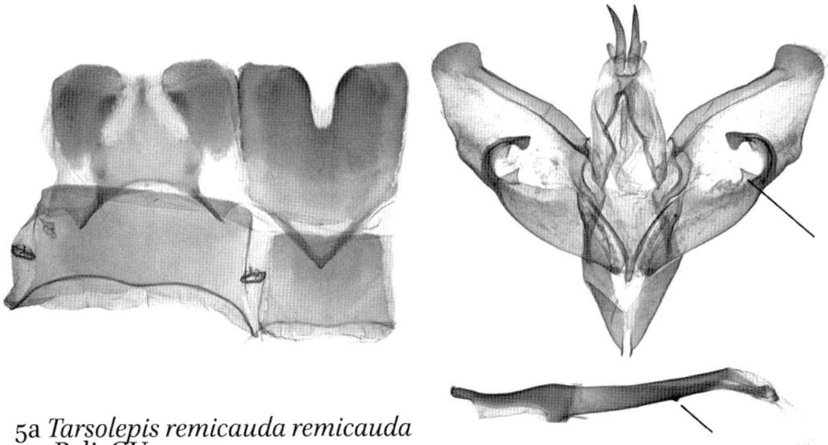

5a *Tarsolepis remicauda remicauda*
Bali, GU 95-02

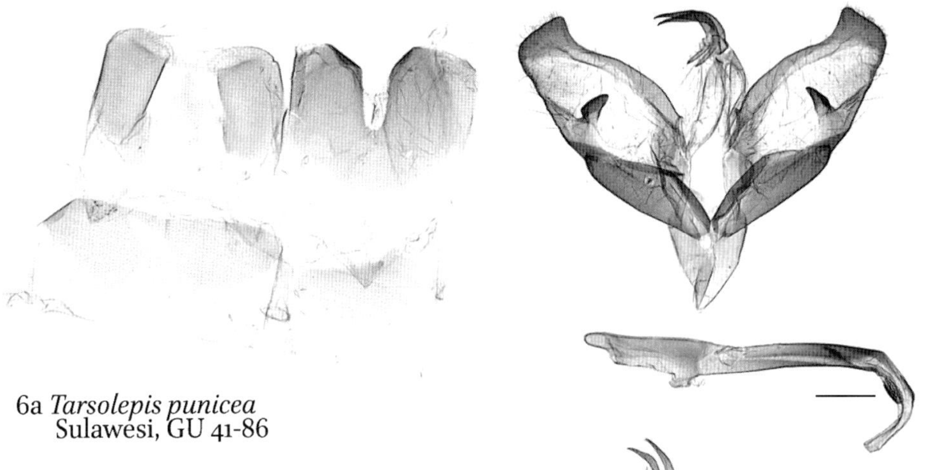

6a *Tarsolepis punicea*
Sulawesi, GU 41-86

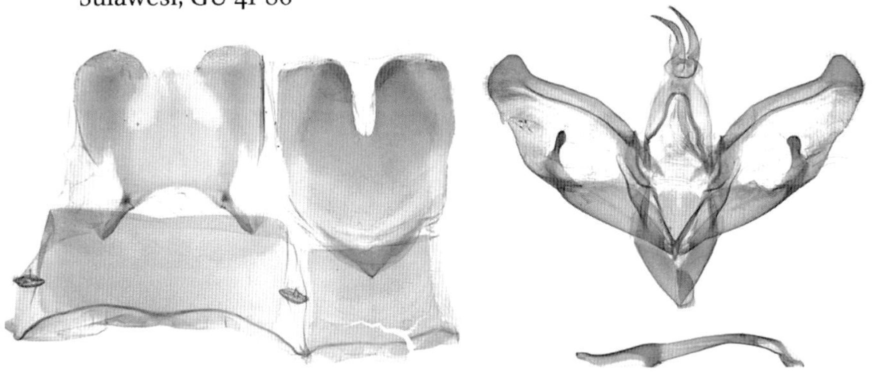

7a *Tarsolepis adultus*
Bacan, GU 95-01, paratype

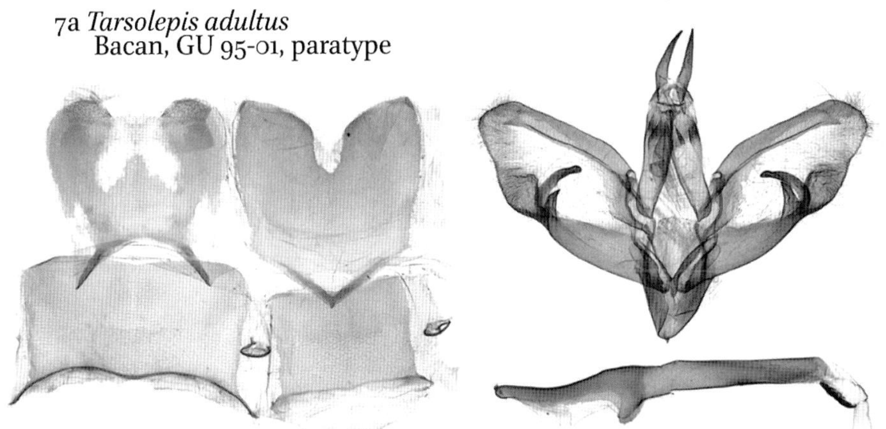

1 mm

8a *Tarsolepis sommeri*
Sumatra, GU 95-07

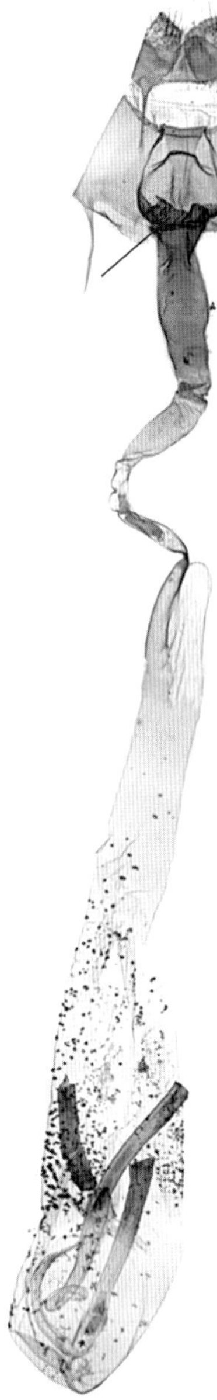

5b *Tarsolepis remicauda remicauda*
Sumatra, GU 95-96

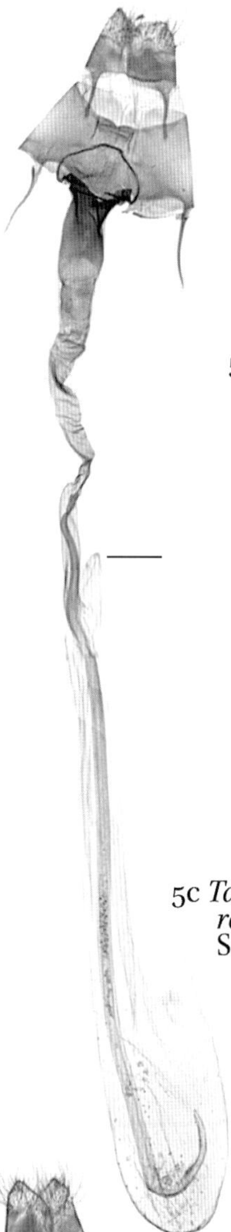

5c *Tarsolepis remicauda remicauda*
Siberut, GU 95-95

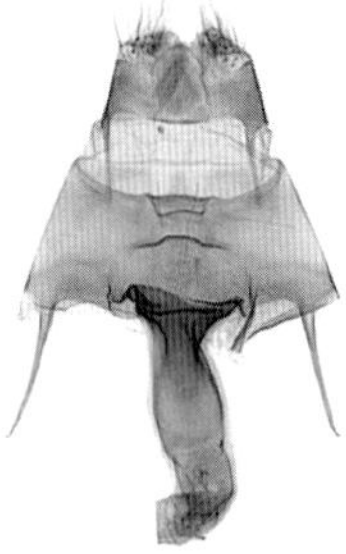

5d *Tarsolepis remicauda remicauda*
Siberut, GU 95-06

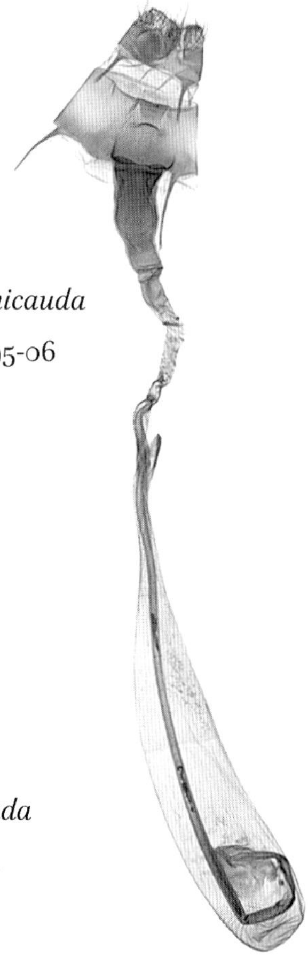

8b *Tarsolepis sommeri*
Bali, GU 95-92

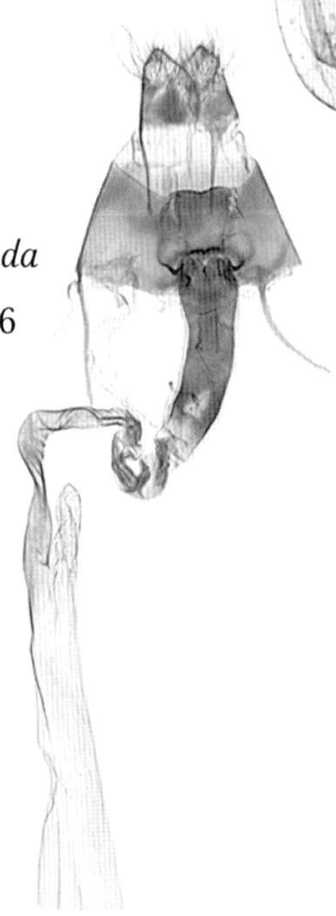

6b *Tarsolepis punicea*
Sulawesi, MWM 1.755

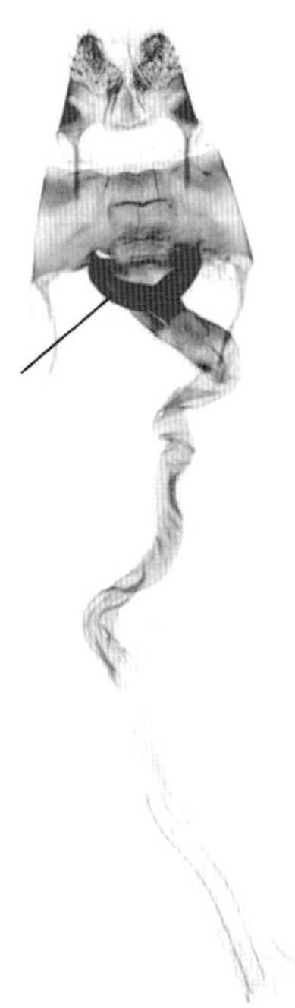

9a *Tarsolepis dinawensis extera*
Halmaheira, GU 73-42, paratype

1 mm

8c *Tarsolepis sommeri*
Bali, GU 95-08

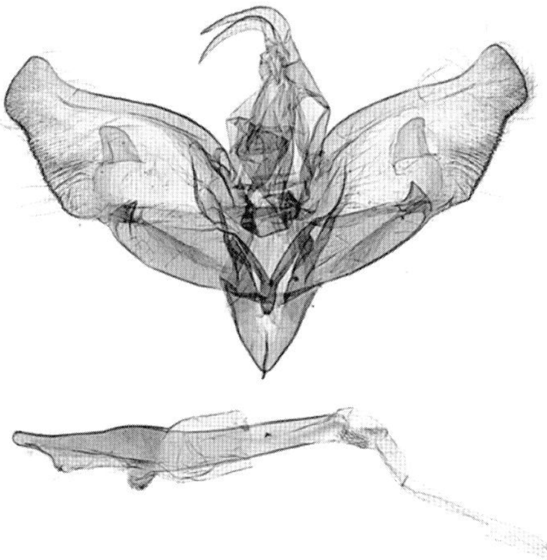

9b *Tarsolepis dinawensis dinawensis*
Papua New Guinea, BM 1195, holotype

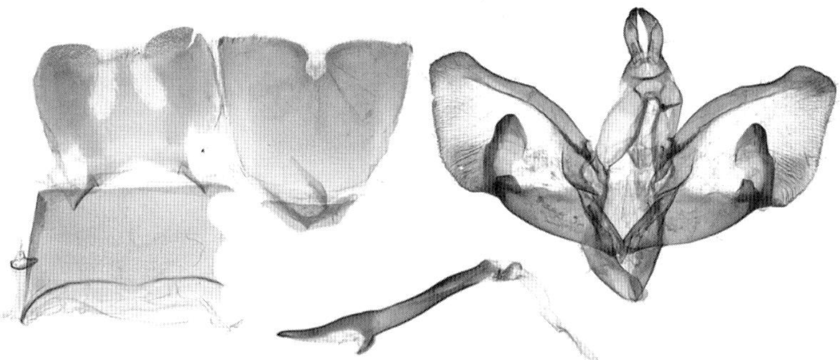

9c *Tarsolepis dinawensis dinawensis*
W. Papua, GU 95-05

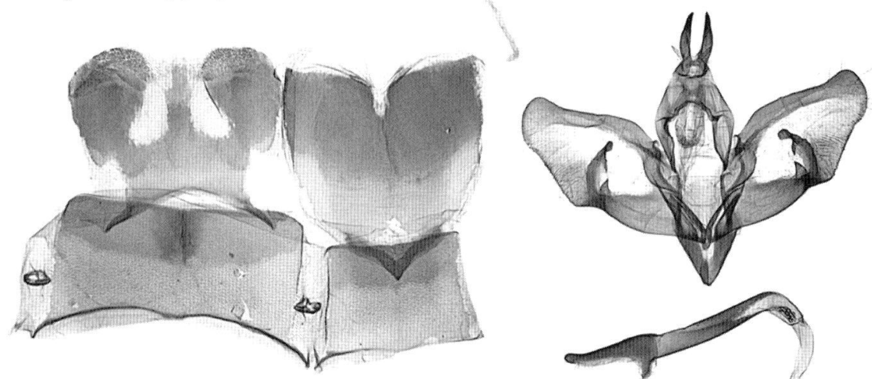

9d *Tarsolepis dinawensis dinawensis*
New Britain, GU 95-93

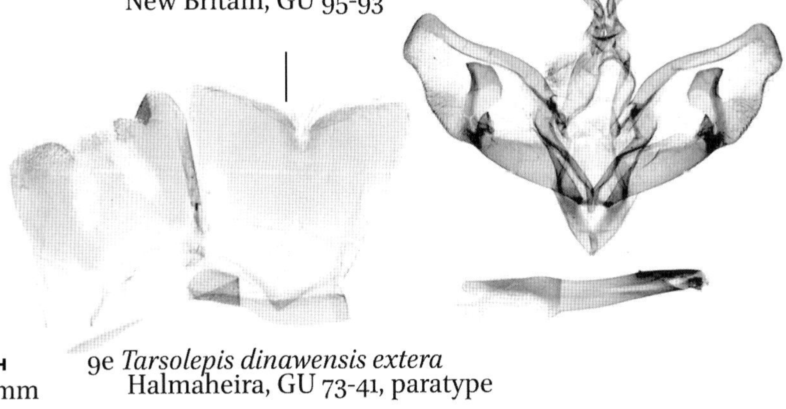

1 mm

9e *Tarsolepis dinawensis extera*
Halmaheira, GU 73-41, paratype

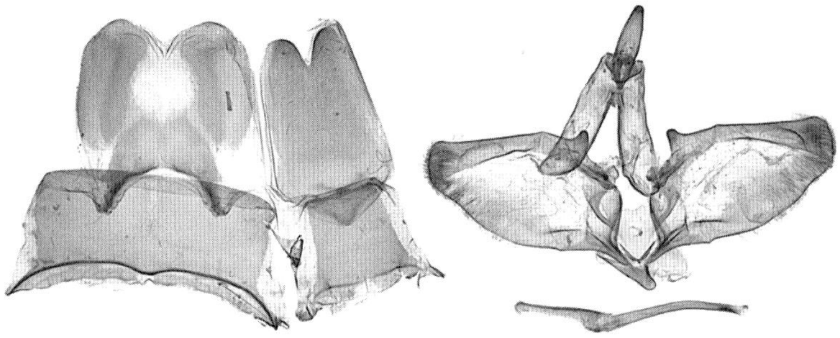

10a *Tarsolepis malayana*
Kalimantan, GU 95-88

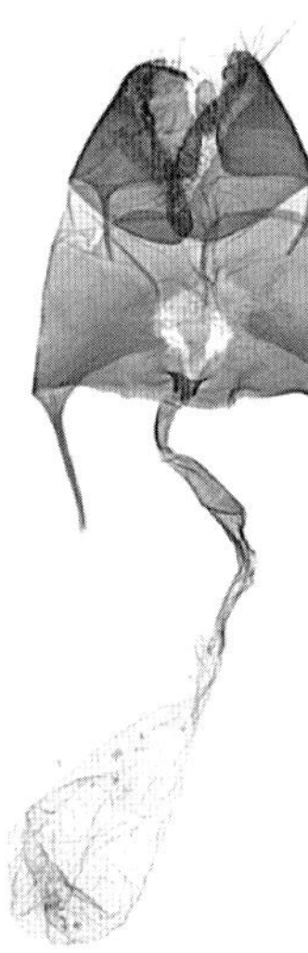

10b *Tarsolepis malayana*
Borneo, GU 95-89

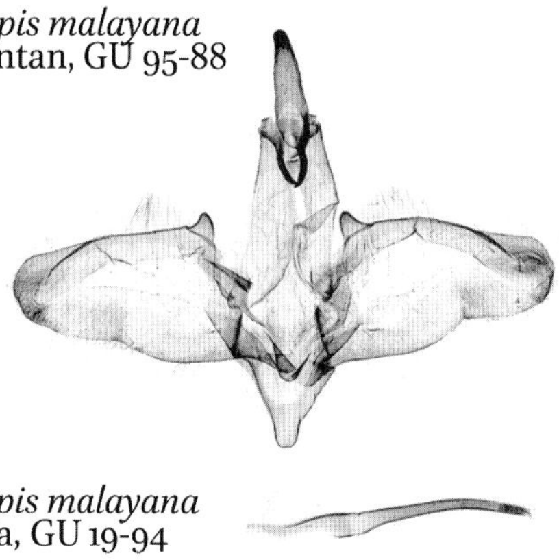

10c *Tarsolepis malayana*
Sumatra, GU 19-94

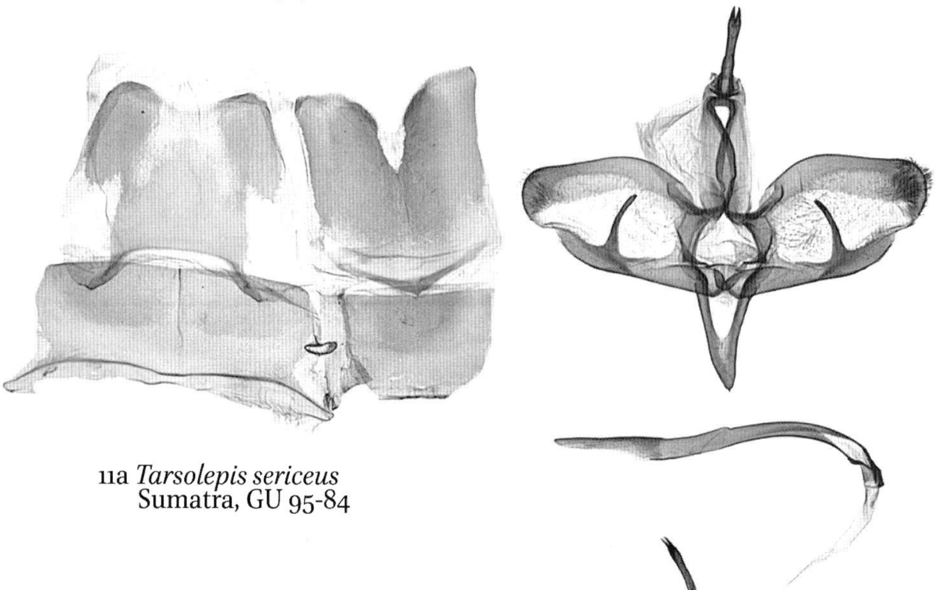

11a *Tarsolepis sericeus*
Sumatra, GU 95-84

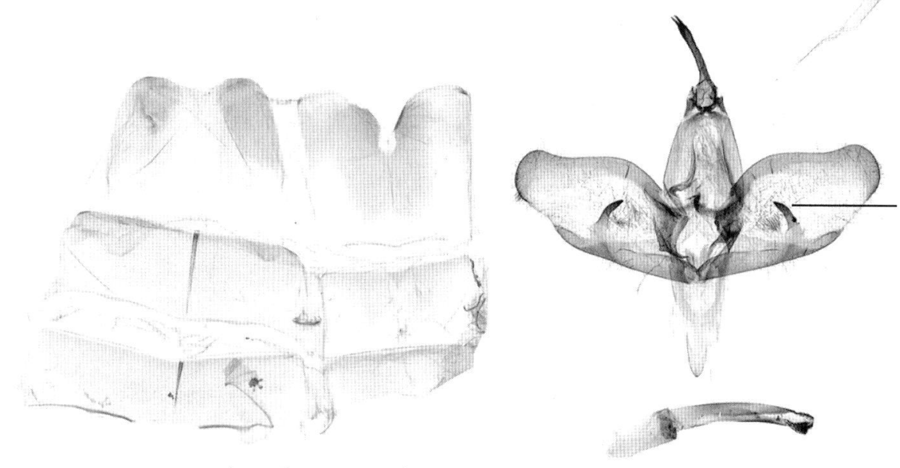

1 mm

11Aa *Tarsolepis hammamelis*
China, Hubei, GU 70-83

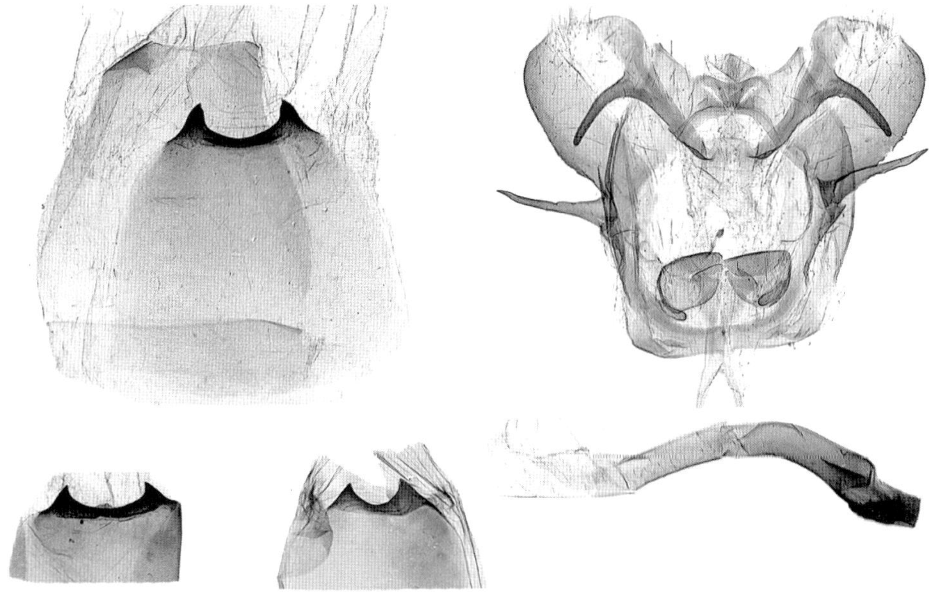

12a *Gangarides rosea*
Sumatra, GU 19-99
8th sternite

12b *Gangarides rosea*
Borneo, GU 44-88
8th sternite

12c *Gangarides rosea*
Sumatra, GU 16-56

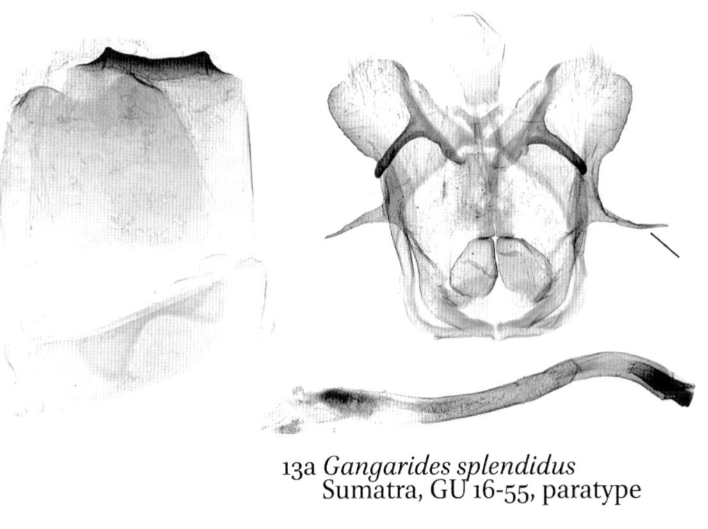

13a *Gangarides splendidus*
Sumatra, GU 16-55, paratype

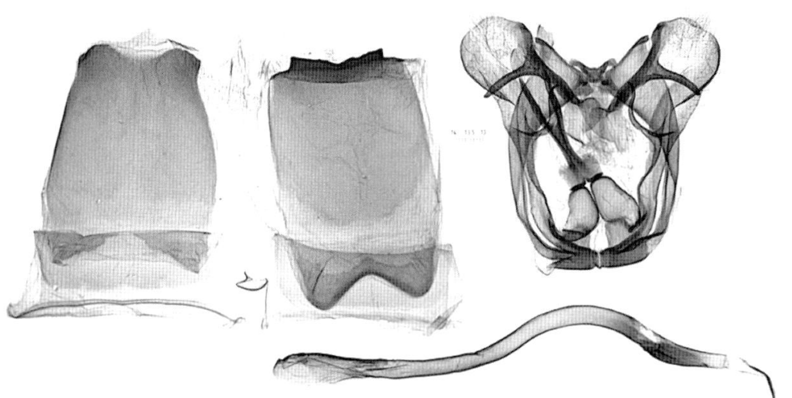

1 mm

13b *Gangarides splendidus*
Sumatra, Siberut Isl., GU 96-11

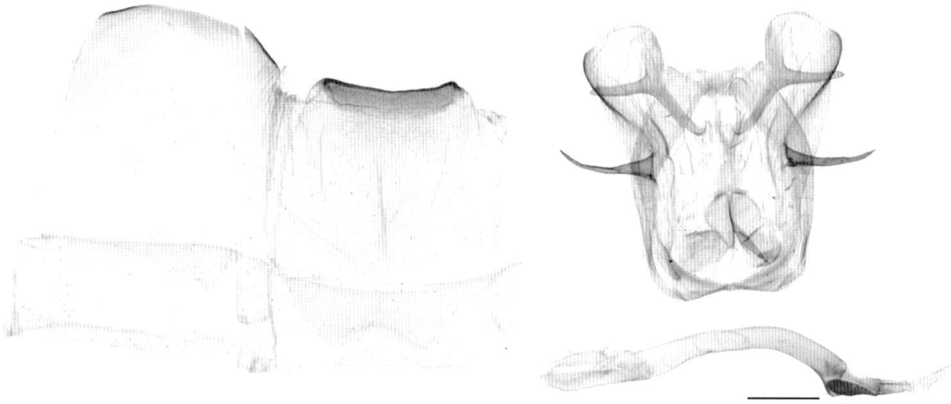

14a *Gangarides wallacei*
Sulawesi, GU 42-36

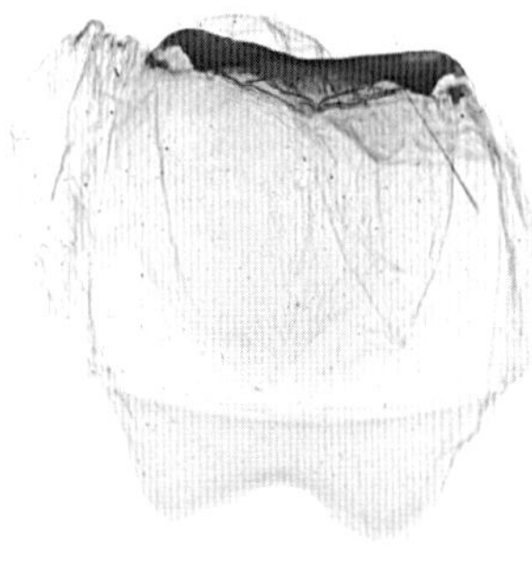

14b *Gangarides wallacei*
Halmaheira, GU 42-37
8th sternite

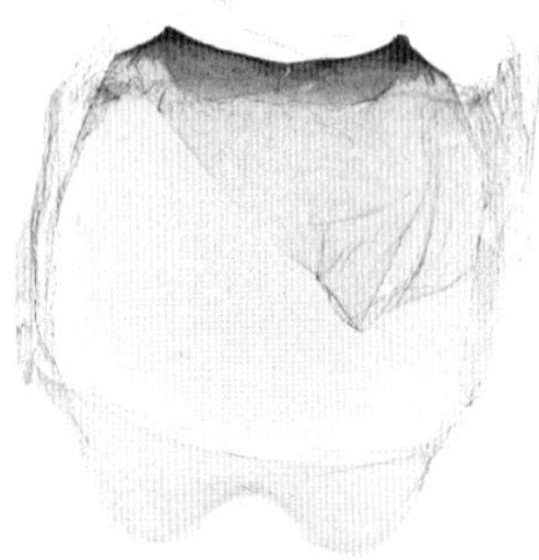

14c *Gangarides wallacei*
Sulawesi, MWM 4.896
8th sternite

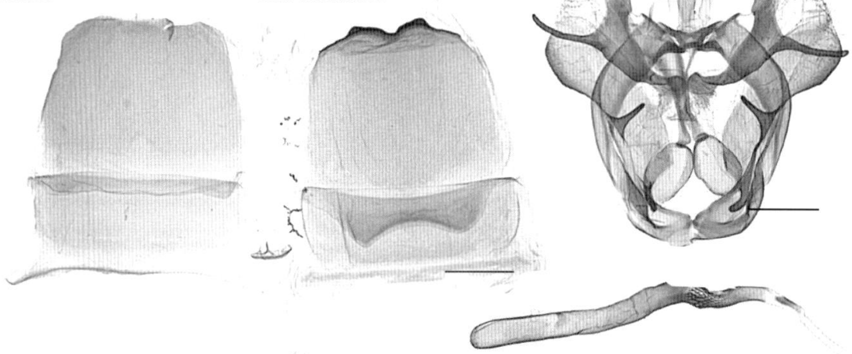

15a *Gangarides brunneus*
Sumatra, GU 96-15

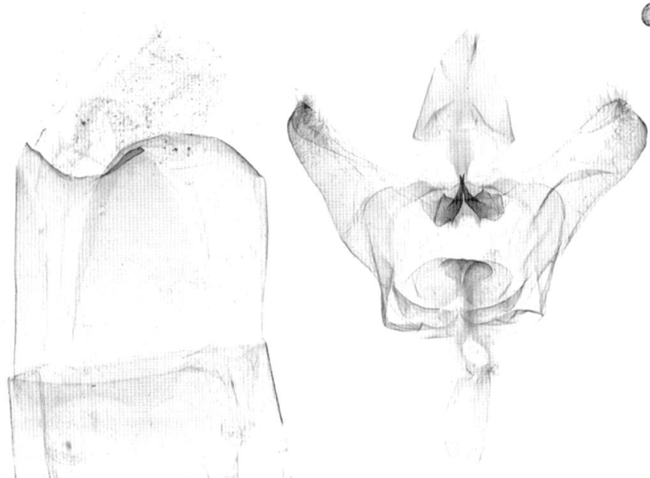

1 mm

16a *Gangarides vittipalpis*
Malaya, GU 16-60

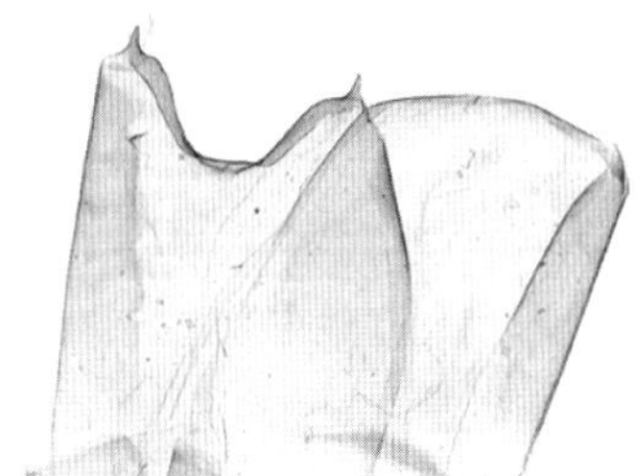

16b *Gangarides vittipalpis*
Sikkim, GU 19-96
8th abdominal segments

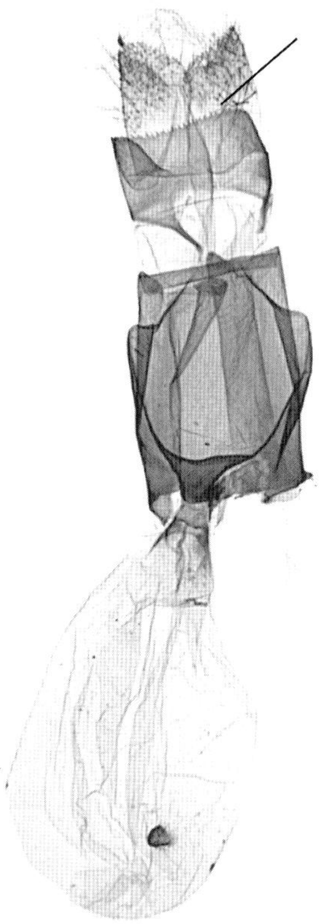

12d *Gangarides rosea*
Thailand, GU 96-10

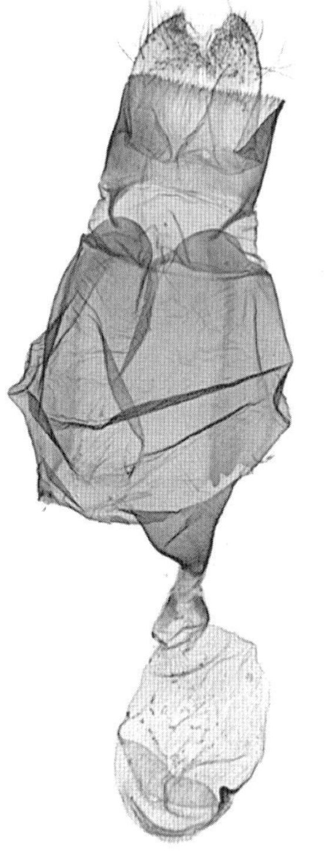

13c *Gangarides splendidus*
Sumatra, GU 96-07

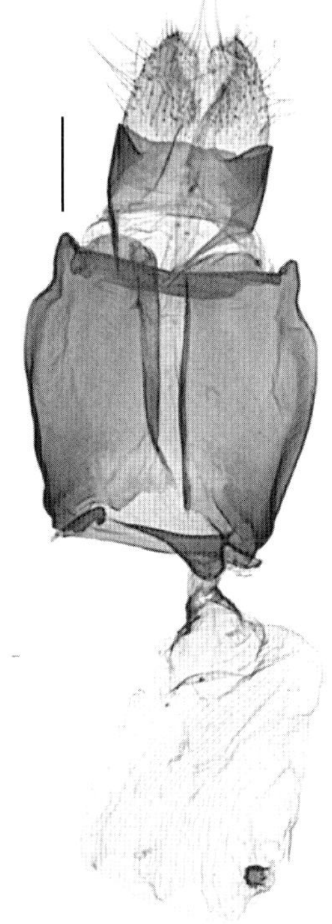

14d *Gangarides wallacei*
Seram, GU 92-06

1 mm

15b *Gangarides brunneus*
Sumatra, GU 42-35, PT

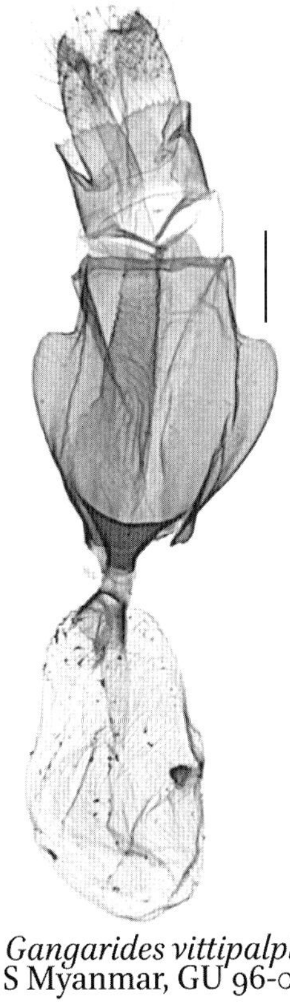

16c *Gangarides vittipalpis*
S Myanmar, GU 96-09

17a *Gangarides vardena*
Sumatra, GU 42-45

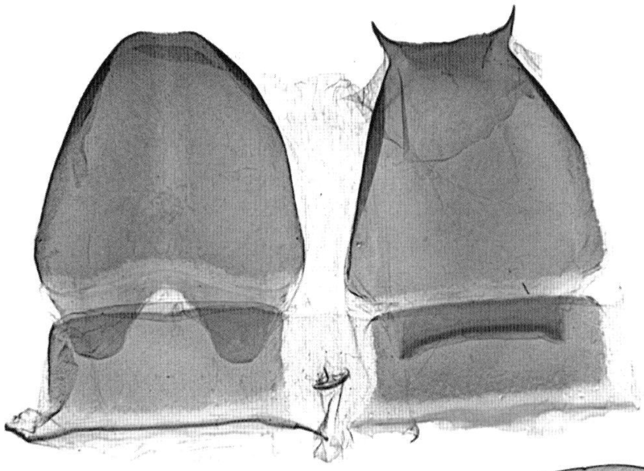

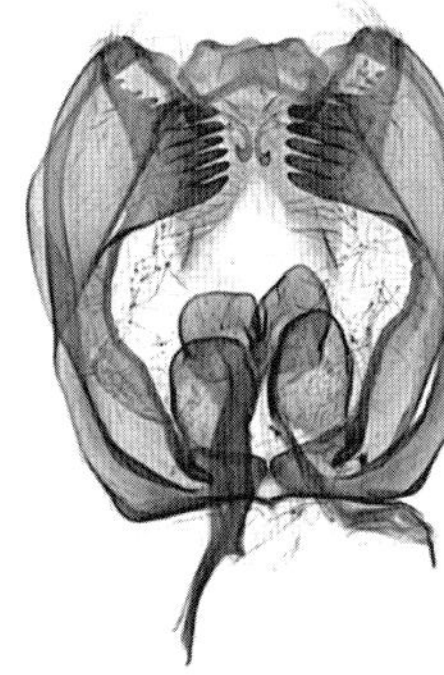

17b *Gangarides vardena*
Sumatra, GU 95-83

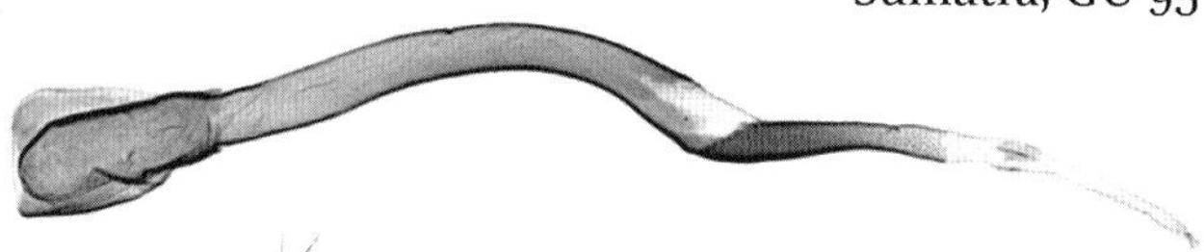

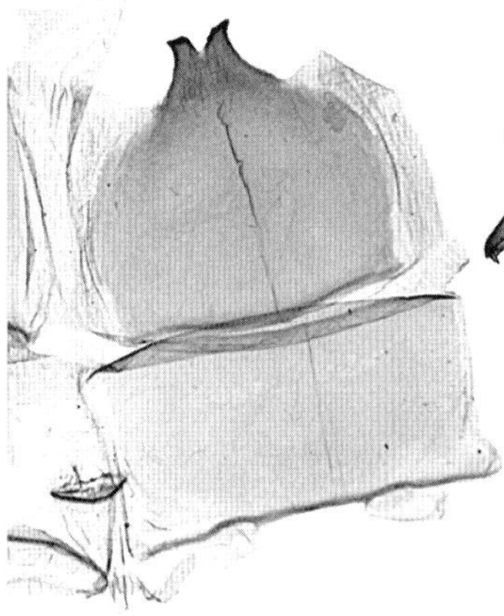

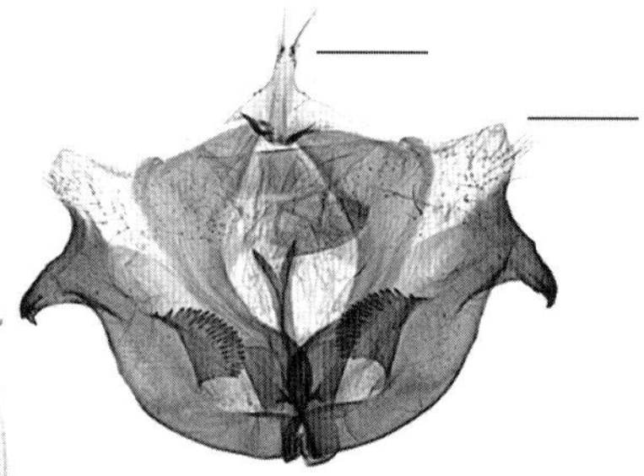

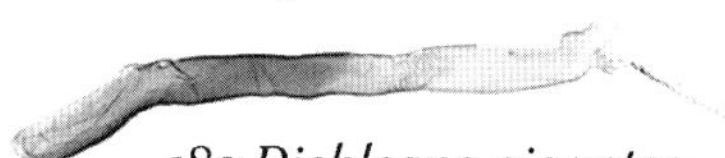

18a *Diehleana gigantea*
W Papua, GU 96-04

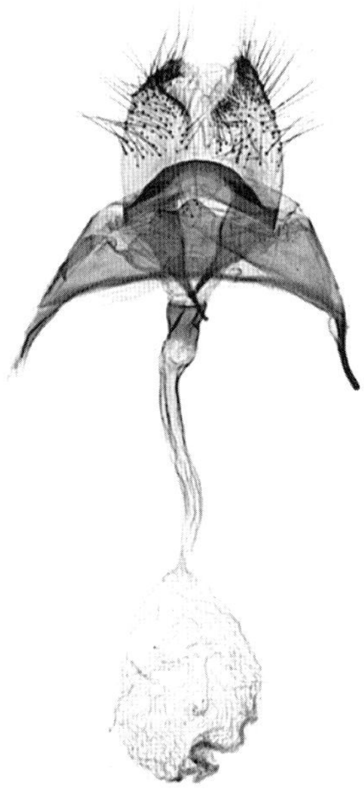

18b *Diehleana gigantea*
W Papua, GU 96-03

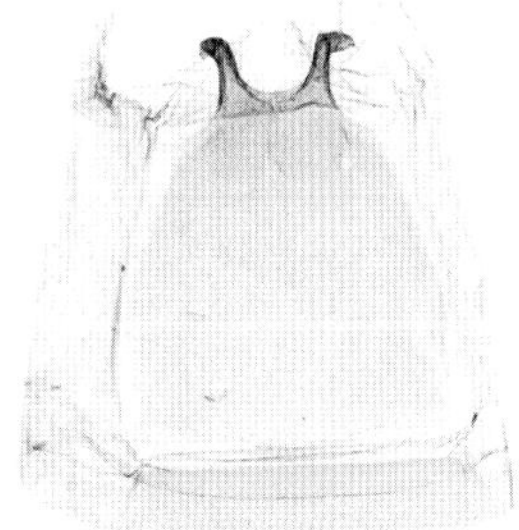

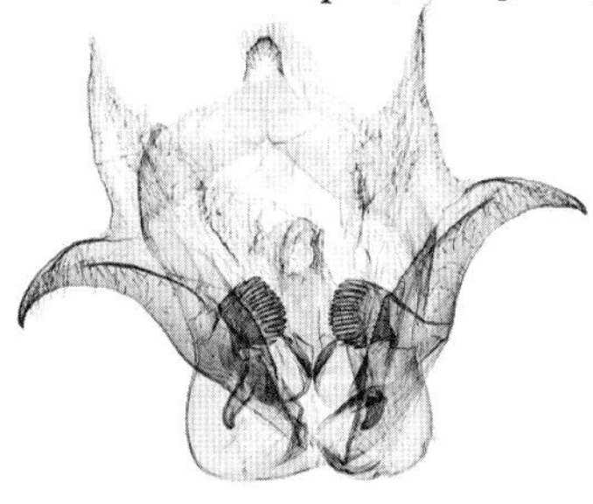

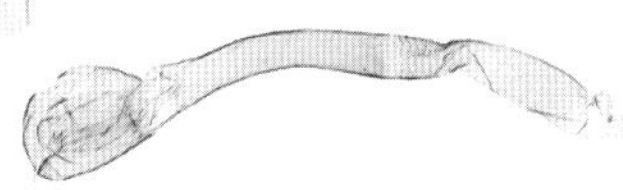

19a *Diehleana gilesi*
Sumatra, GU 42-25

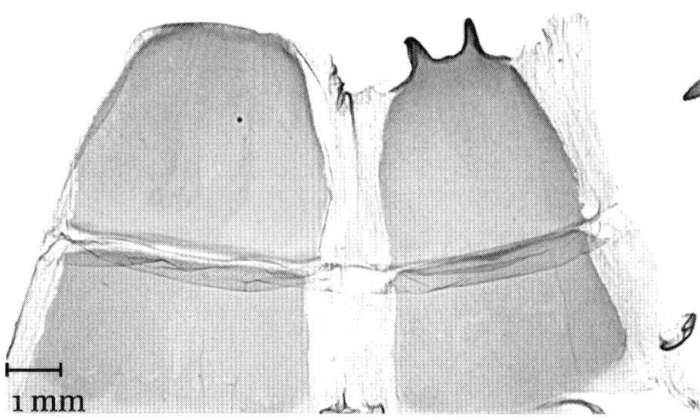

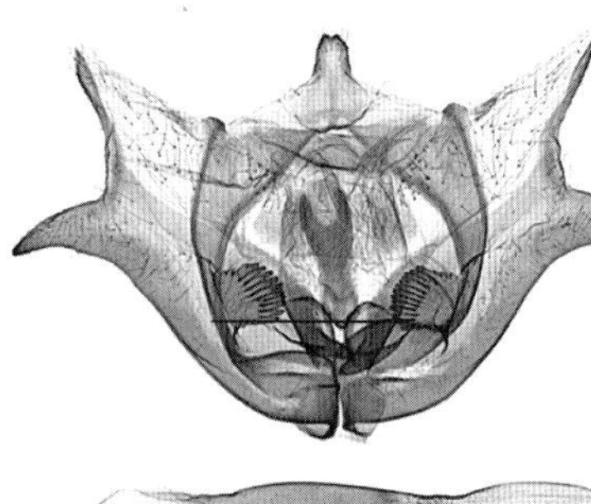

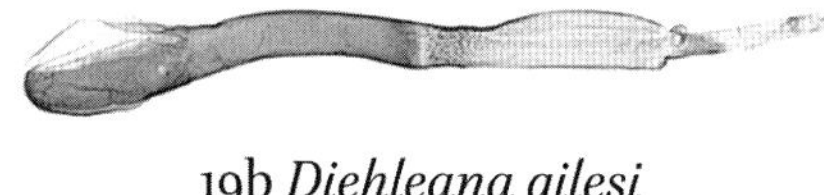

19b *Diehleana gilesi*
Thailand, MWM 29.398

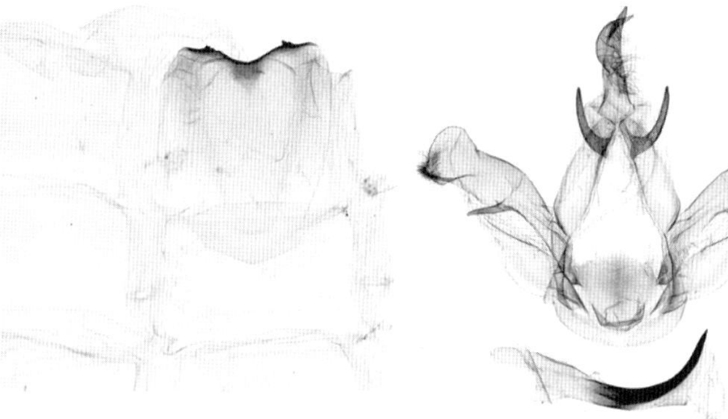

20a *Euhampsonia roepkei roepkei*
Brunei, GU 42-29

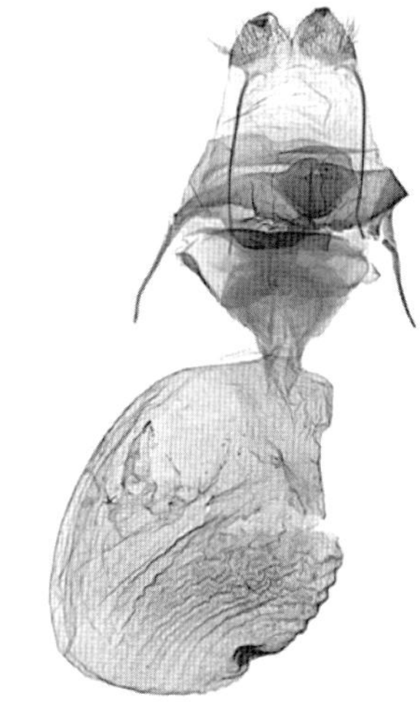

20b *Euhampsonia roepkei roepkei*
Brunei, GU 95-98

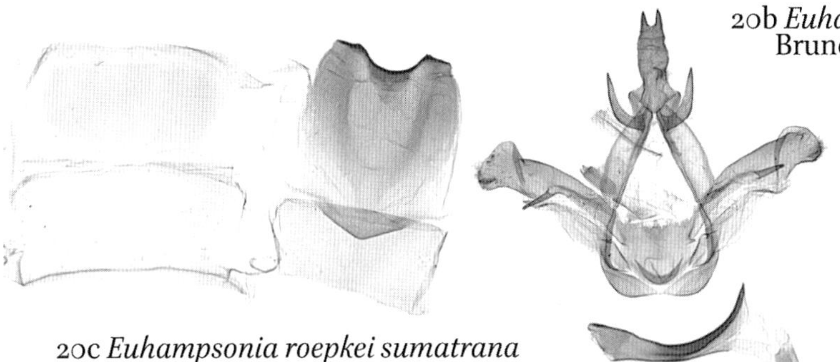

20c *Euhampsonia roepkei sumatrana*
Sumatra, GU 96-01, paratype

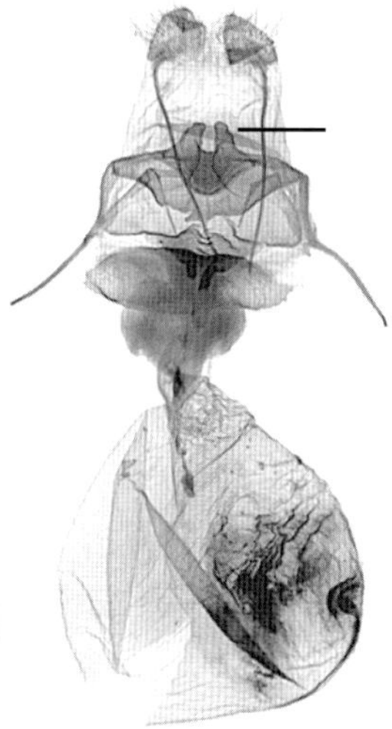

20d *Euhampsonia roepkei sumatrana*
Sumatra, GU 95-97, PT

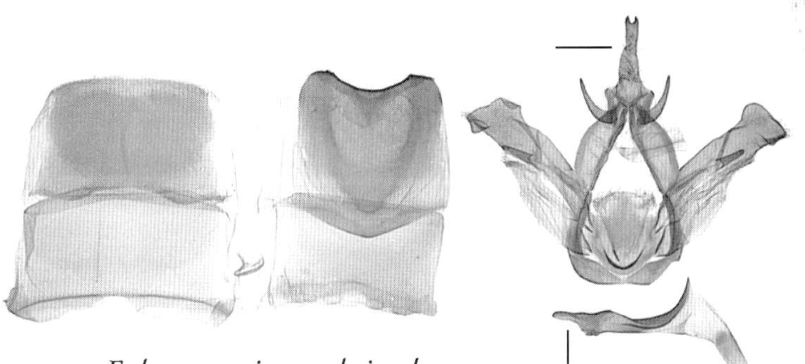

20e *Euhampsonia roepkei palawana*
Palawan, GU 95-100, paratype

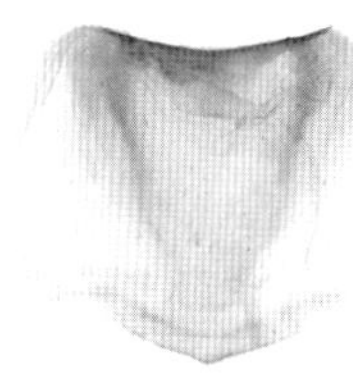

20f *Euhampsonia roepkei sumatrana*
Sumatra, GU 23-61
8th sternite, paratype

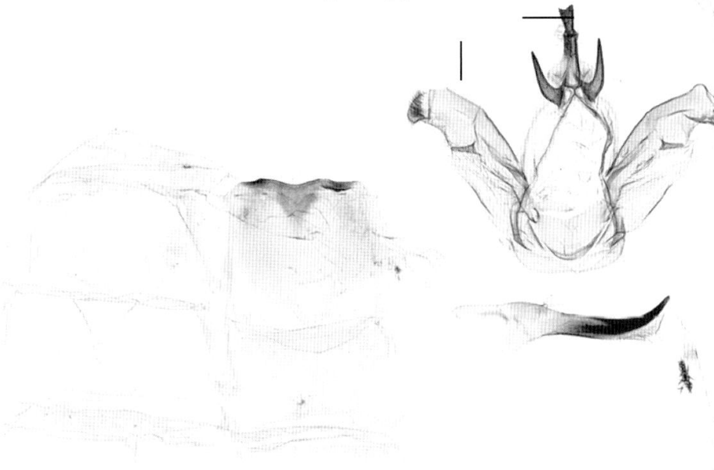

20h *Euhampsonia roepkei javana*
Java, GU 95-99,
phallus, paratype

1 mm

20g *Euhampsonia roepkei javana*
Java, GU 42-30, Paratype

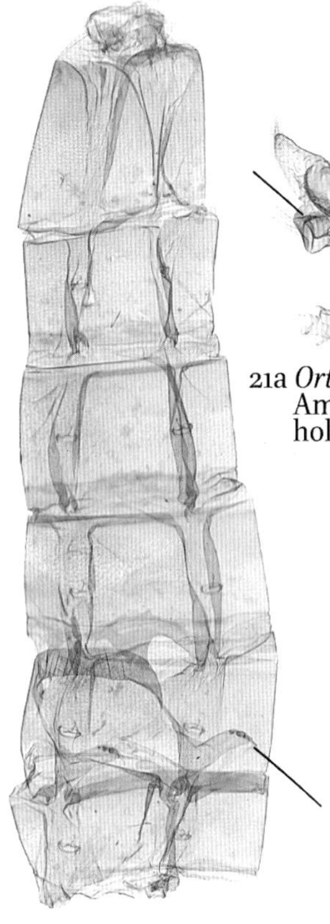

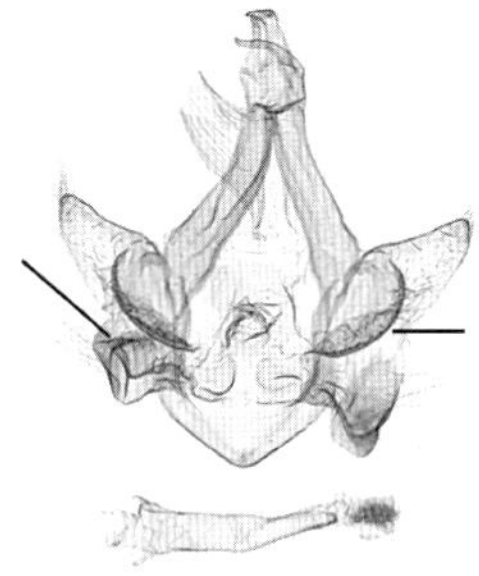

21a *Ortholomia moluccana*
Ambon, BM 329, Ambon, holotype

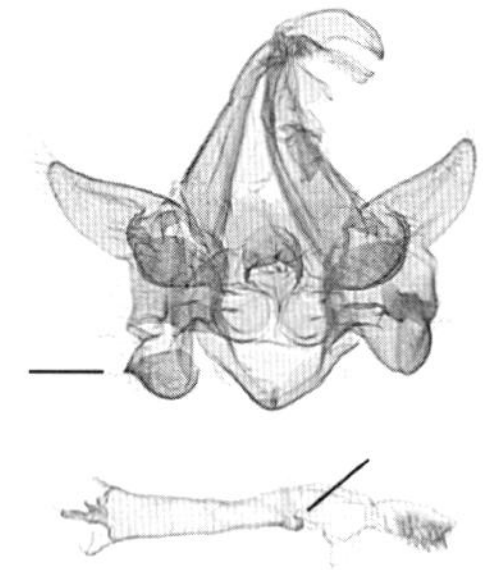

21A *Ortholomia glauca*
Australia-Queensland, BM 330

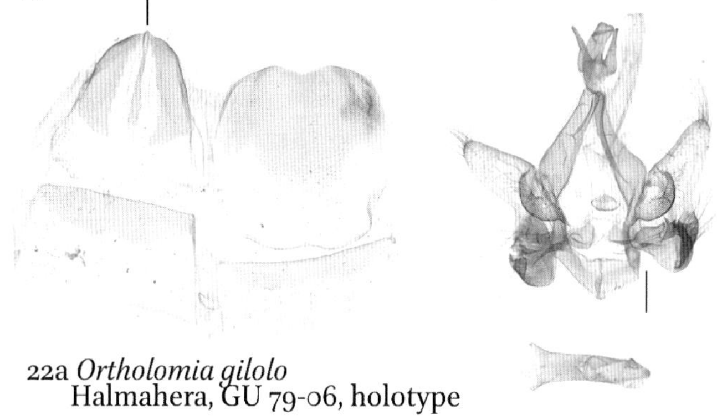

22a *Ortholomia gilolo*
Halmahera, GU 79-06, holotype

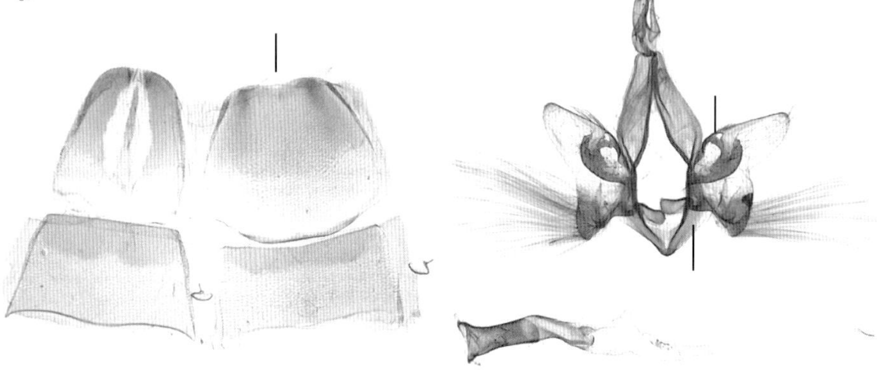

23a *Ortholomia subtilis*
Alor, GU 92-83, paratype

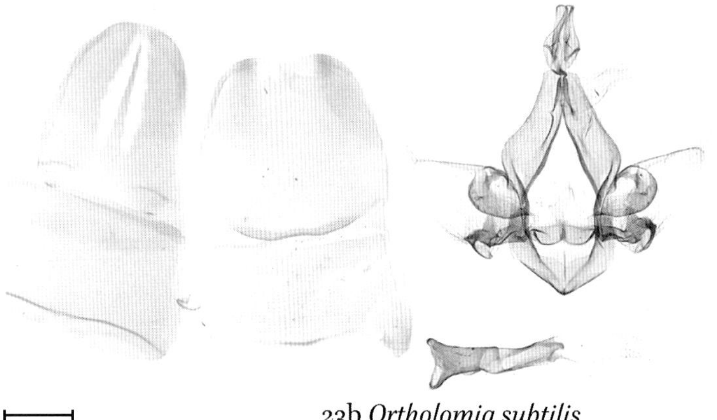

23b *Ortholomia subtilis*
Timor, GU 93-14, paratype

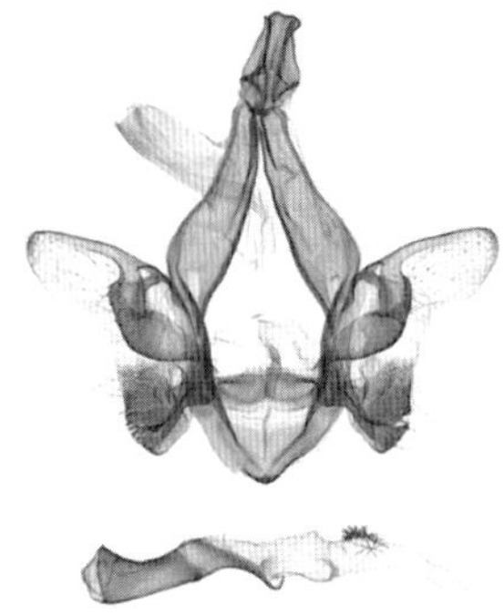

23c *Ortholomia subtilis*
Sumba, GU 95-20, PT

1 mm

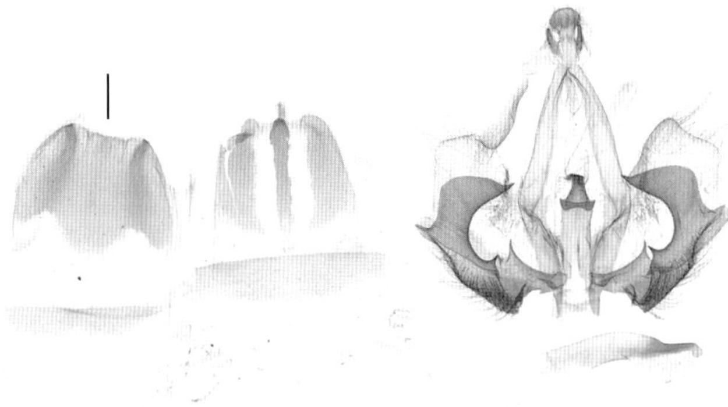

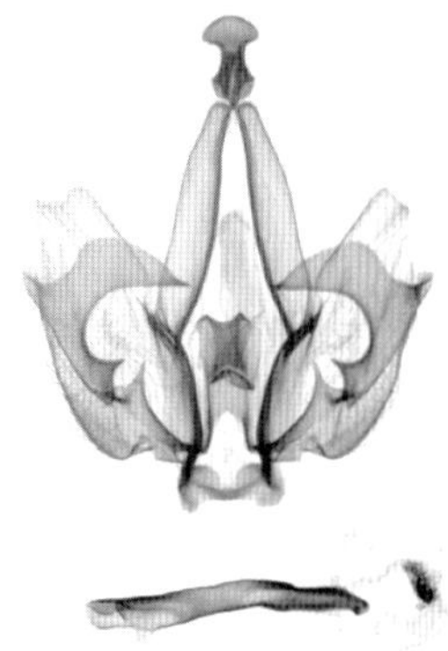

24a *Ortholomia xylinata*
Bali, GU 79-25

24b *Ortholomia xylinata*
Java, MWM 14.966

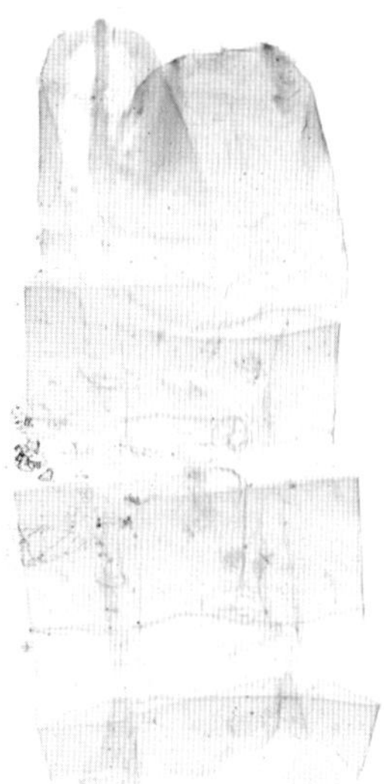

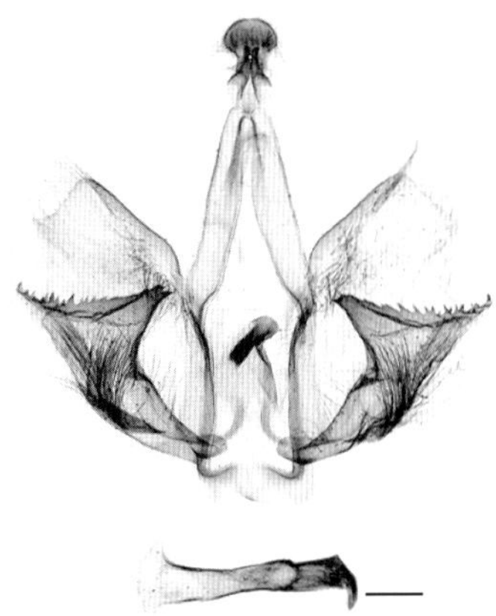

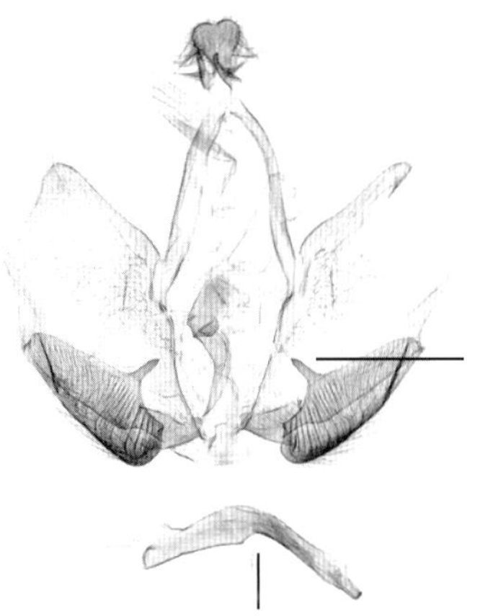

25a *Ortholomia besida*
Sumatra, GU 21-54,
holotype

26a *Ortholomia osica osica*
Mindanao, MV 17.565,
Paratype

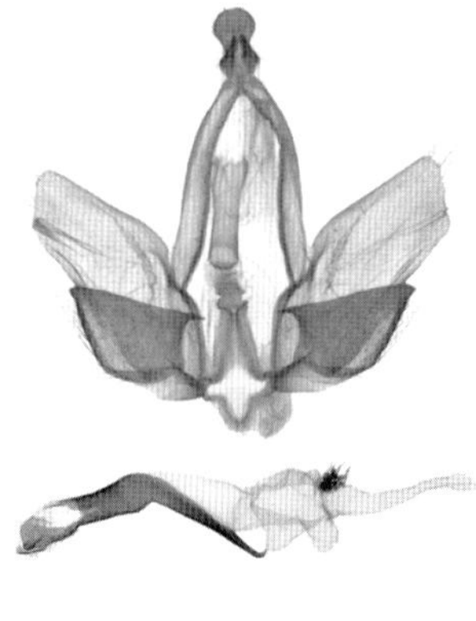

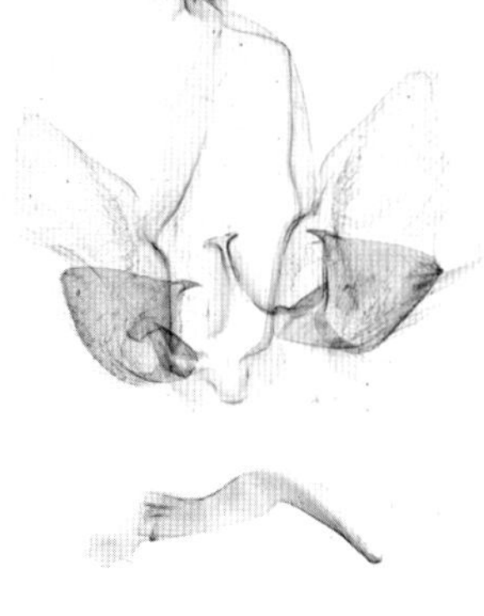

26b *Ortholomia osica sulawesiensis*
Sulawesi, MWM 14.964

1 mm

26c *Ortholomia osica sulawesiensis*
Sulawesi, GU 19-41

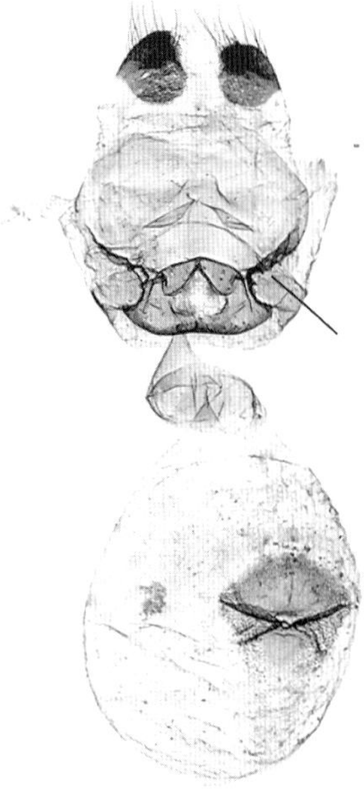

21b *Ortholomia moluccana*
Key Isl., GU 81-06

23d *Ortholomia subtilis*
Alor, GU 92-84,
paratype

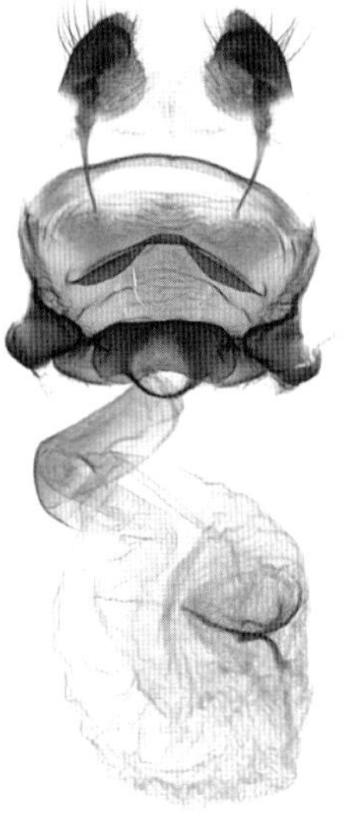

23e *Ortholomia subtilis*
Pantar, GU 92-86,
paratype

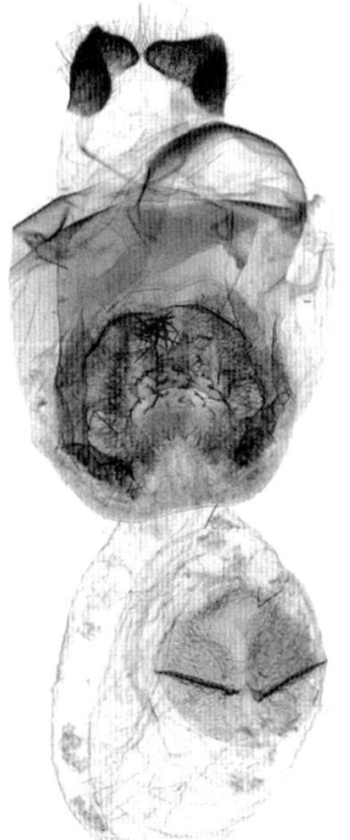

24c *Ortholomia xylinata*
Palawan, MV 17.536

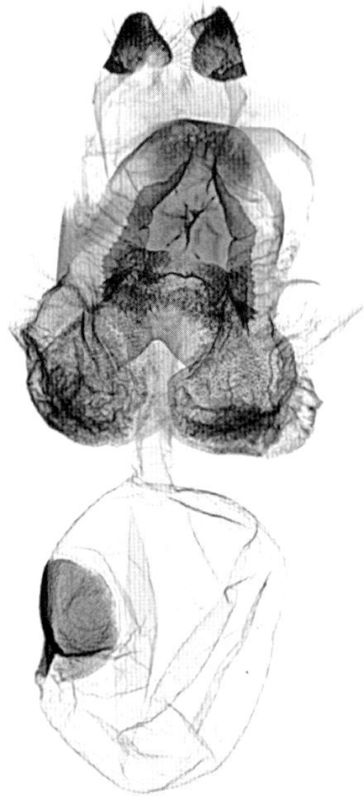

25b *Ortholomia besida*
Sumatra, GU 96-20

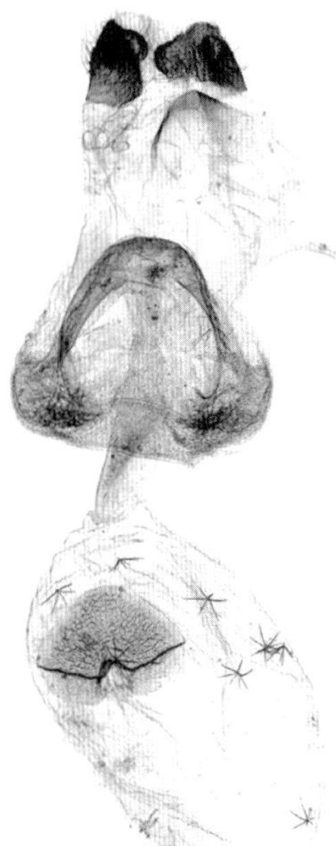

26d *Ortholomia osica*
Luzon, MV 17.859,
paratype

1 mm

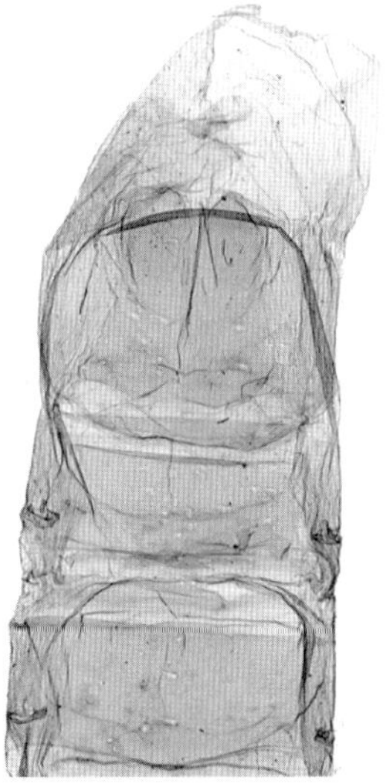

27a *Pseudogargetta diversa*
PNG, BM 34, lectotype

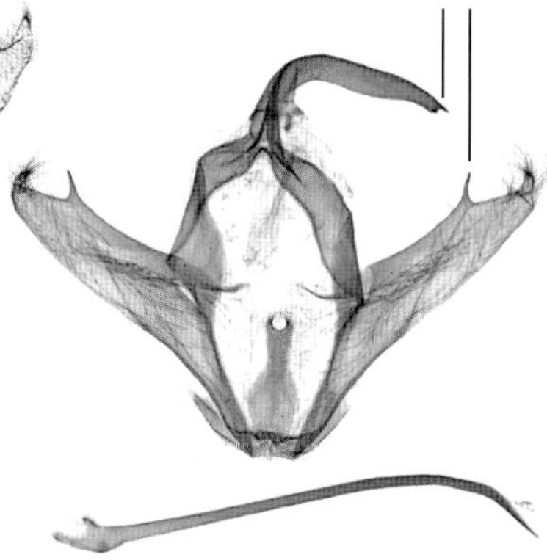

27b *Pseudogargetta diversa*
W Papua, GU 79-07

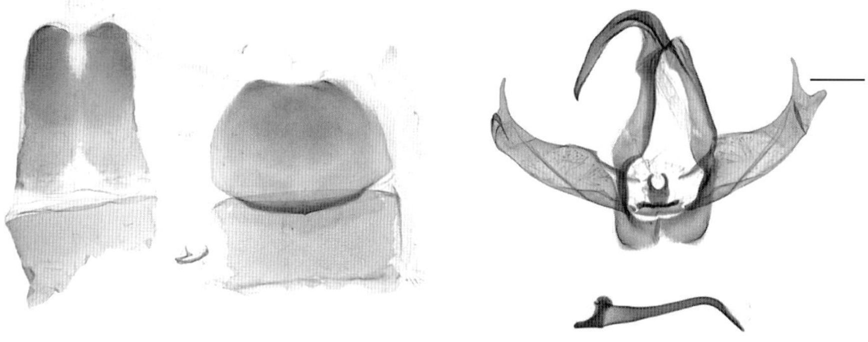

28a *Pseudogargetta orbiculata*
Obi, GU 92-87

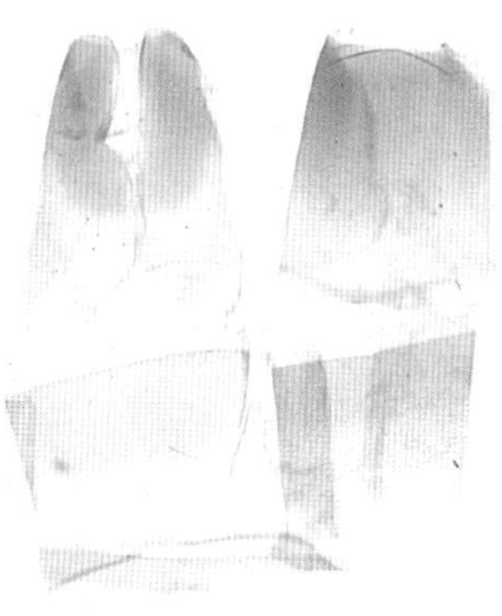

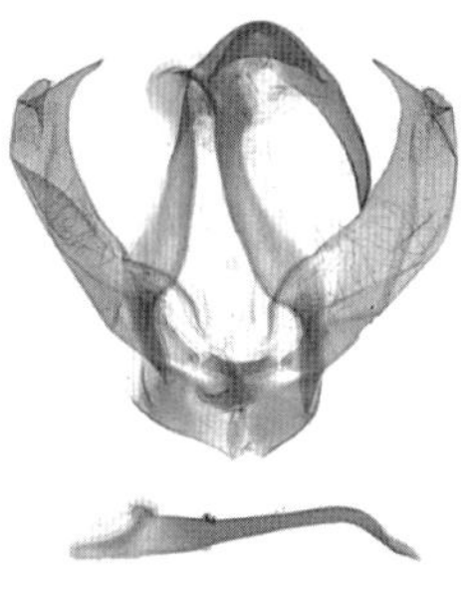

28b *Pseudogargetta orbiculata*
Ambon, GU 79-04

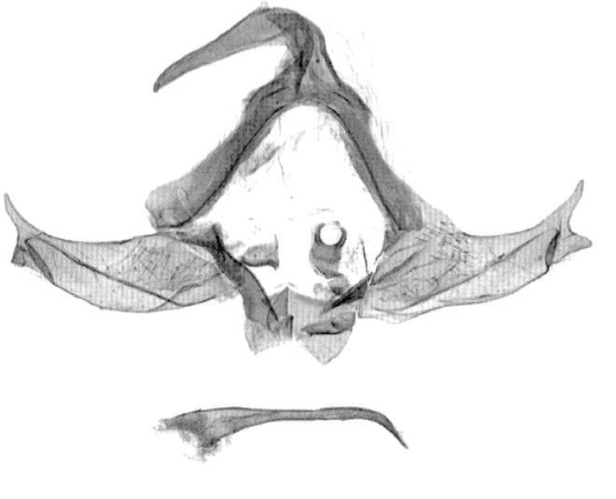

28c *Pseudogargetta orbiculata*
Halmaheira, GU 79-05

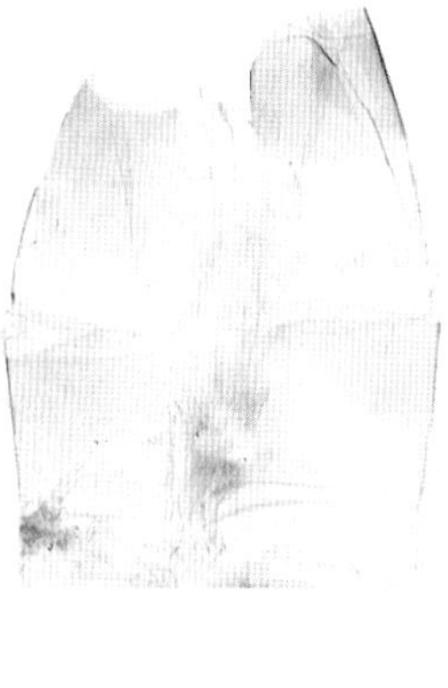

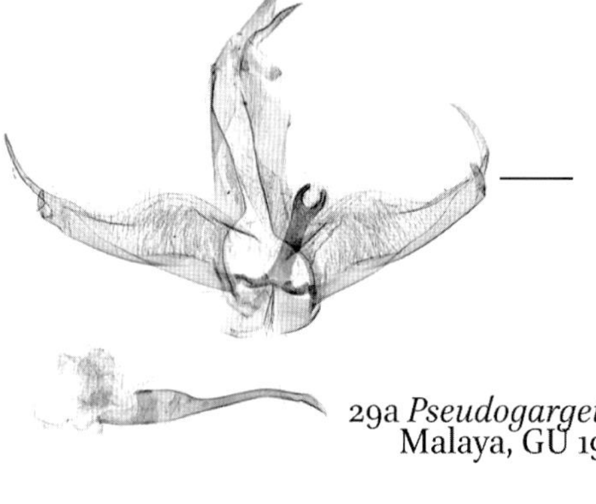

29a *Pseudogargetta marmorata*
Malaya, GU 19-47

1 mm

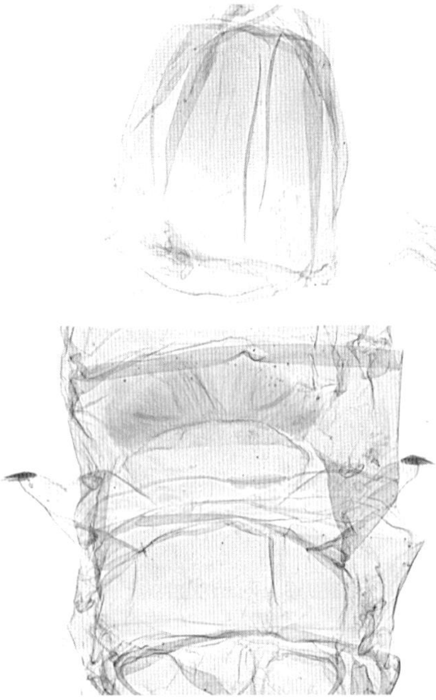

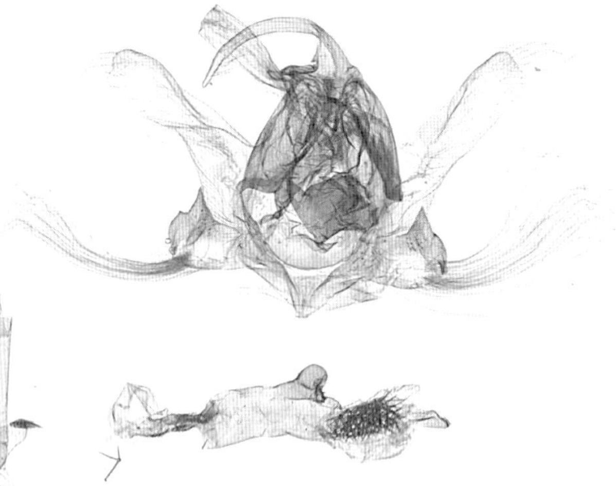

30a *Gargettiana punctatissima punctatissima*
W Papua, BM 36
(holotype of *Gargetta melanosticta*)

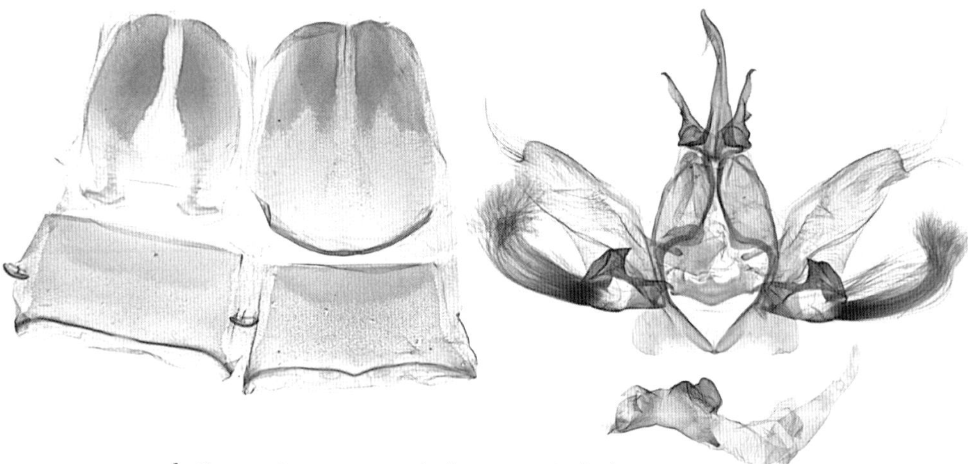

30b *Gargettiana punctatissima punctatissima*
Seram, BM 10313996

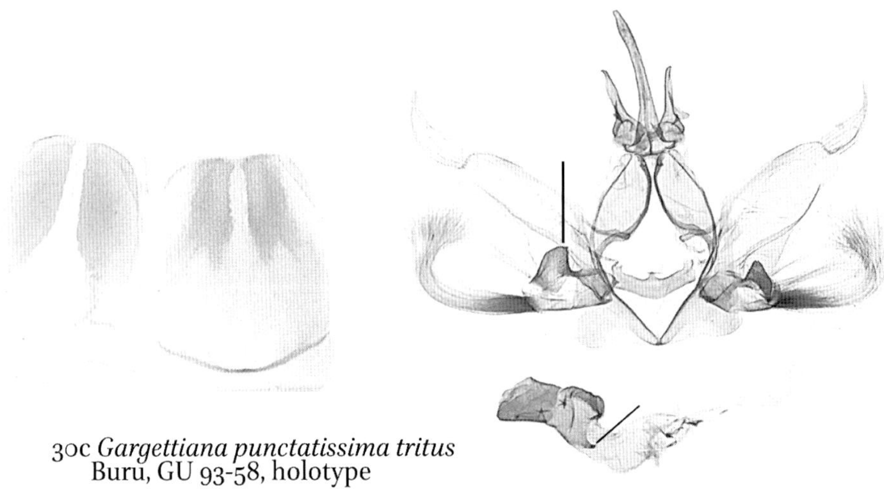

1 mm

30c *Gargettiana punctatissima tritus*
Buru, GU 93-58, holotype

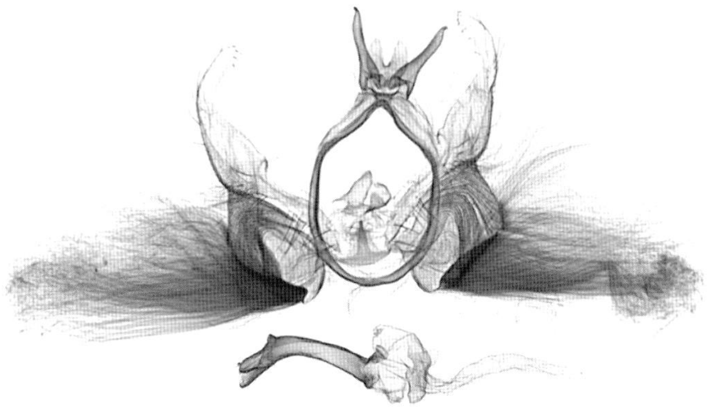

31Aa *Polychoa styphlopis*
Australia, Qsd., BM 10313985

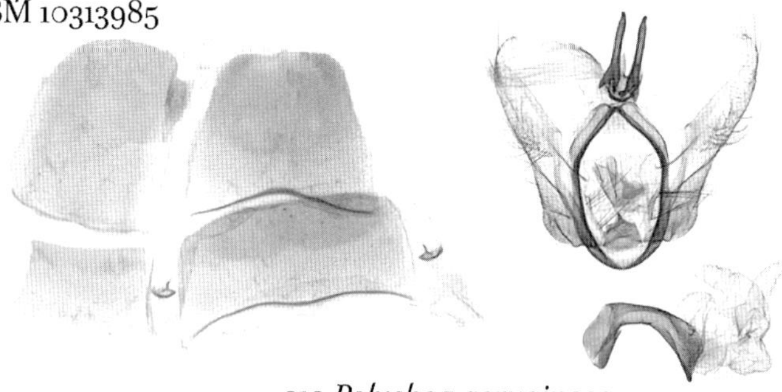

31a *Polychoa aeruginosa*
PNG, BM 10313983

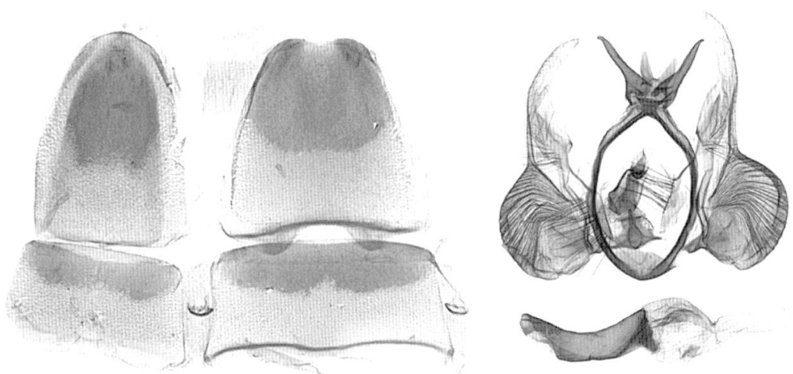

32a *Polychoa bakami*
Halmahera, GU 96-41, paratype

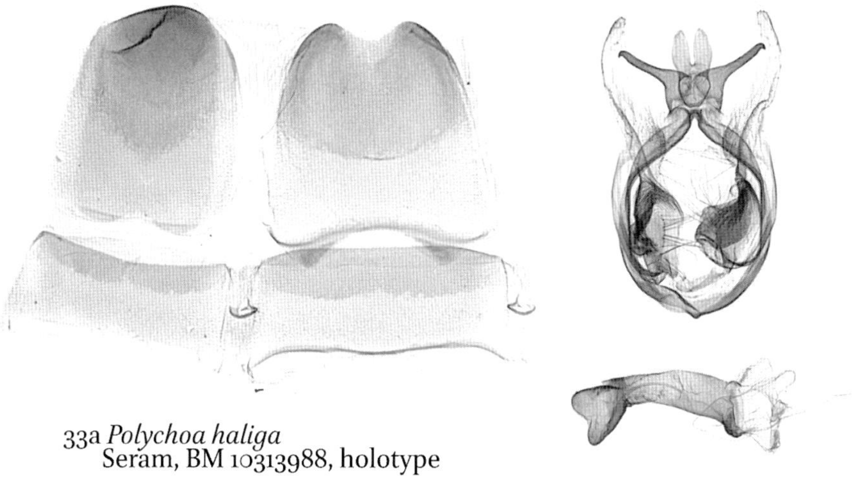

1 mm

33a *Polychoa haliga*
Seram, BM 10313988, holotype

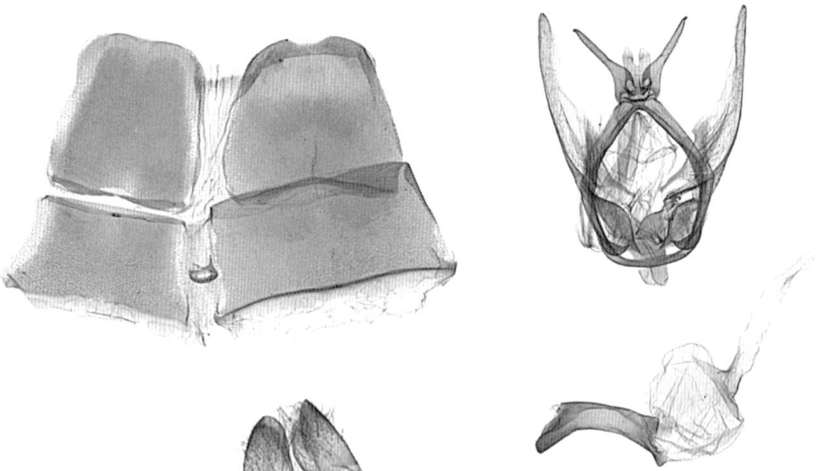

34a *Polychoa cheesmanae*
Yapen, Ihle 03-2017, holotype

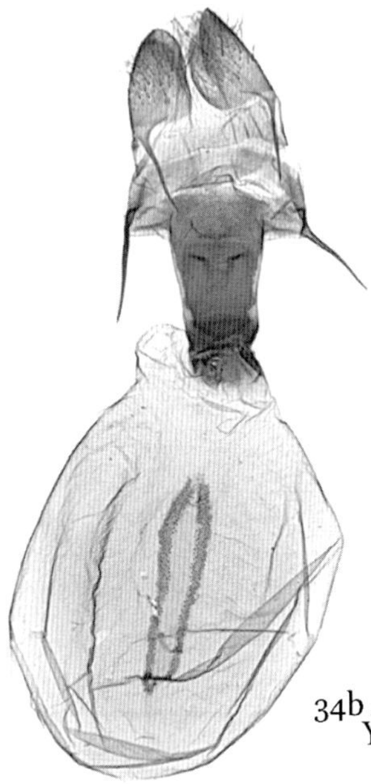

34b *Polychoa cheesmanae*
Yapen, Ihle 04-2017, paratype

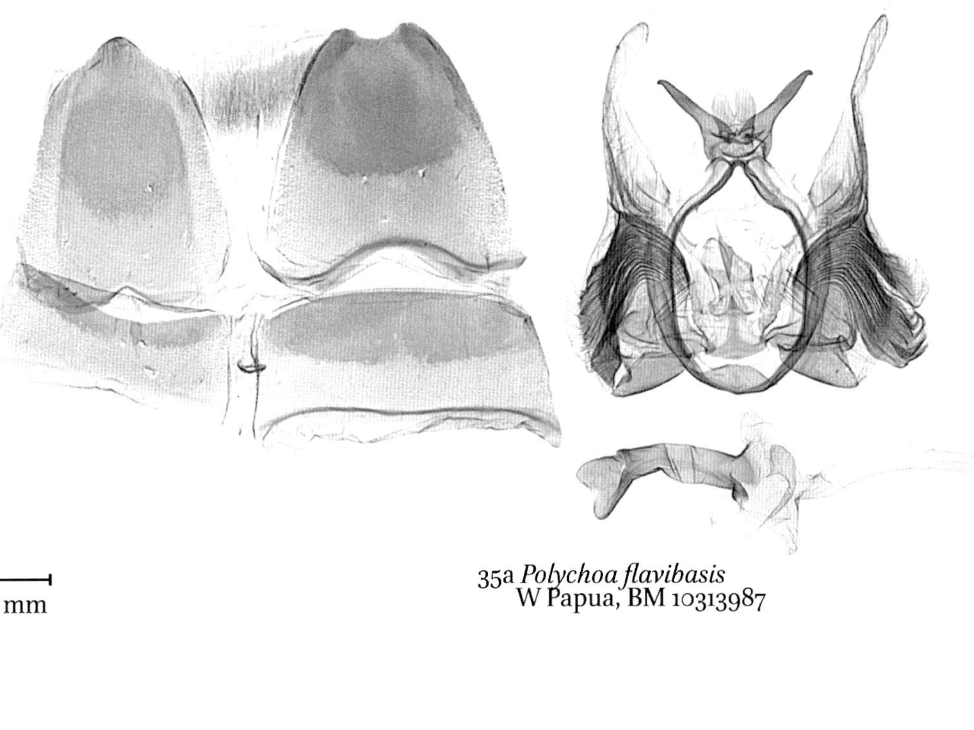

1 mm

35a *Polychoa flavibasis*
W Papua, BM 10313987

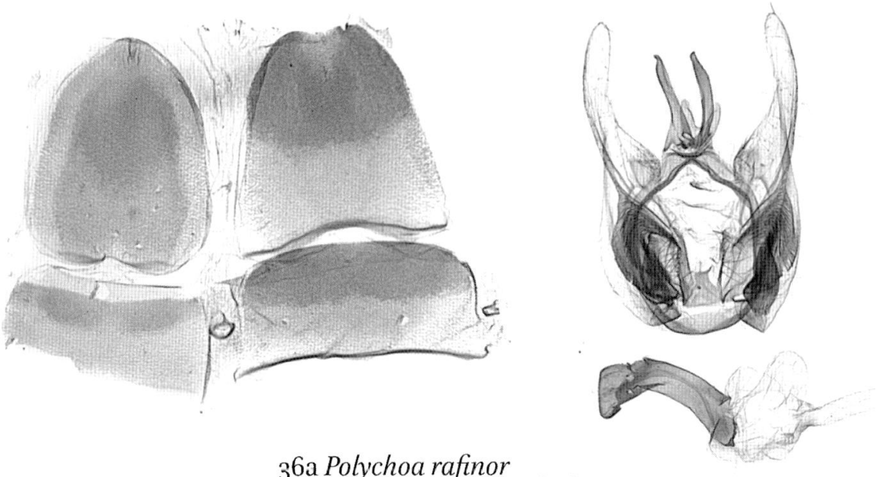

36a *Polychoa rafinor*
PNG, BM 10313986, holotype

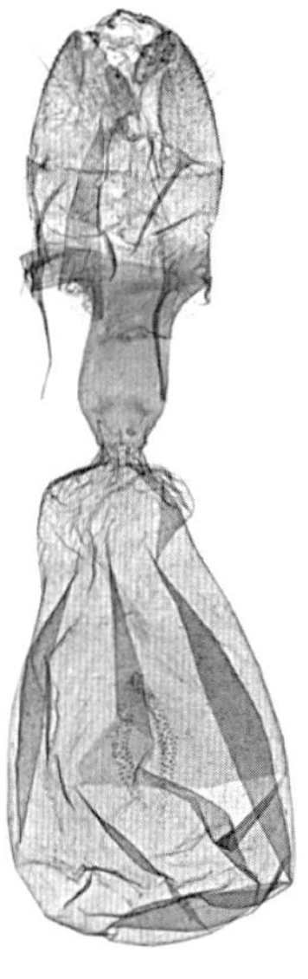

36b *Polychoa rafinor*
PNG, BM 128, paratype

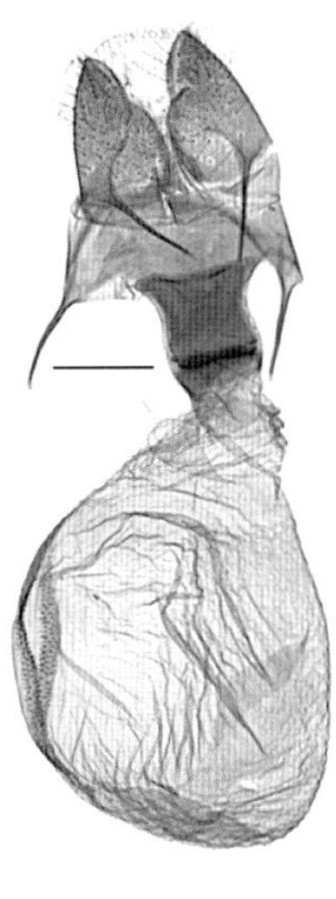

37a *Polychoa bipuncta*
PNG, BM 10313984

37b *Polychoa bipuncta*
PNG, BM 80,
holotype

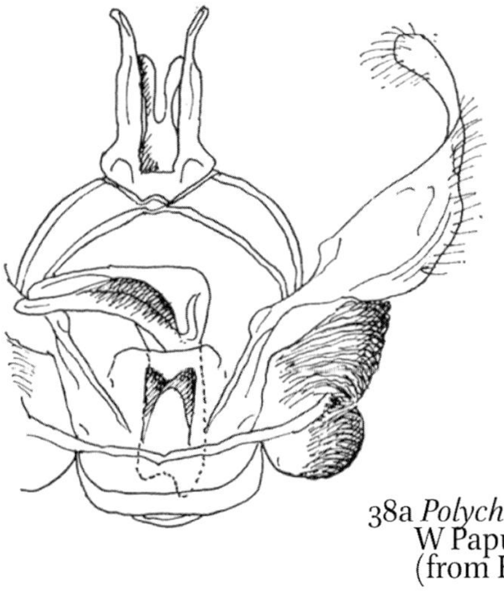

38a *Polychoa metallica*
W Papua, holotype
(from Kiriakoff, 1967, fig. 3)

1 mm

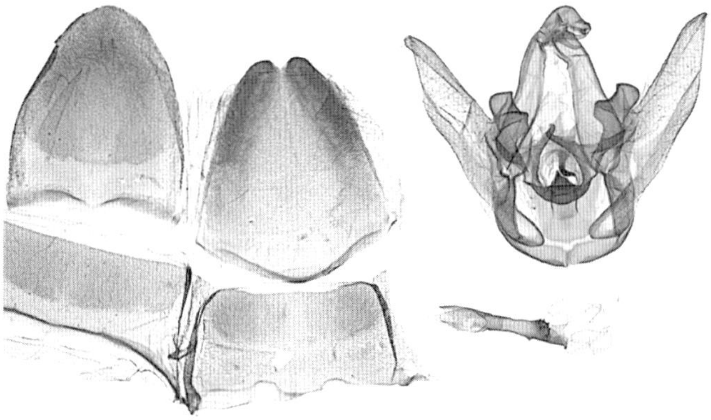

39a *Archigargetta amydra*
Australia, Qsd, USNM 127,542

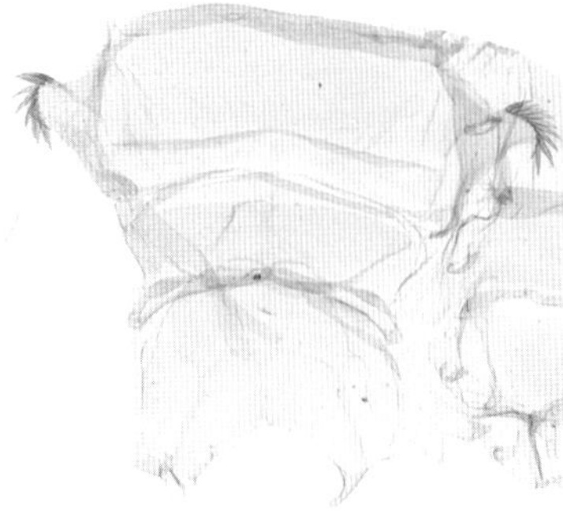

39b *Archigargetta amydra*
PNG, BM 112, Ctenophore
(Holotype *albostigmata*)

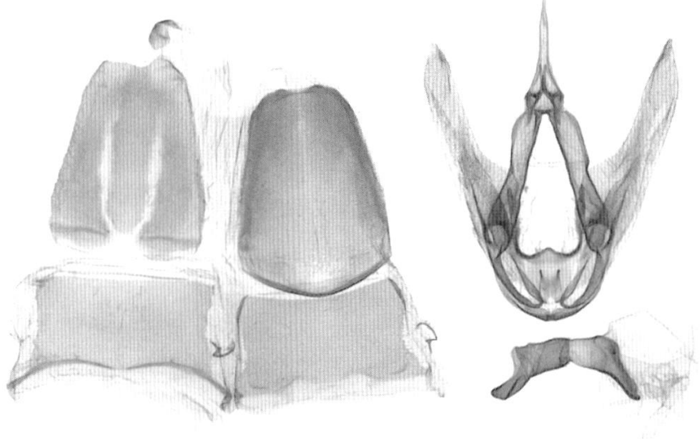

40a *Archigargetta viridigrisea*
Sumba, GU 95-49

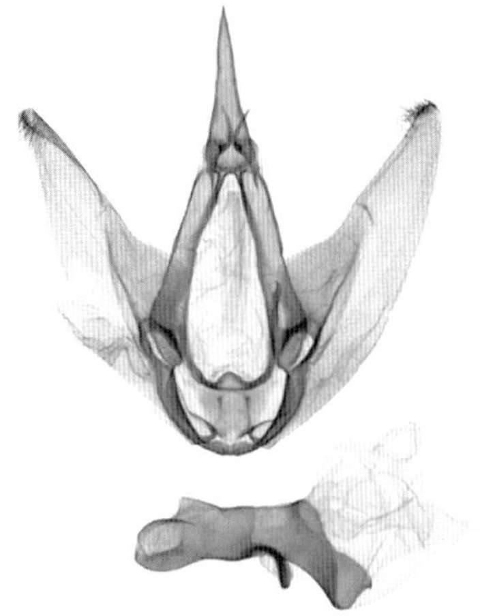

40b *Archigargetta viridigrisea*
Borneo, MWM 15.563

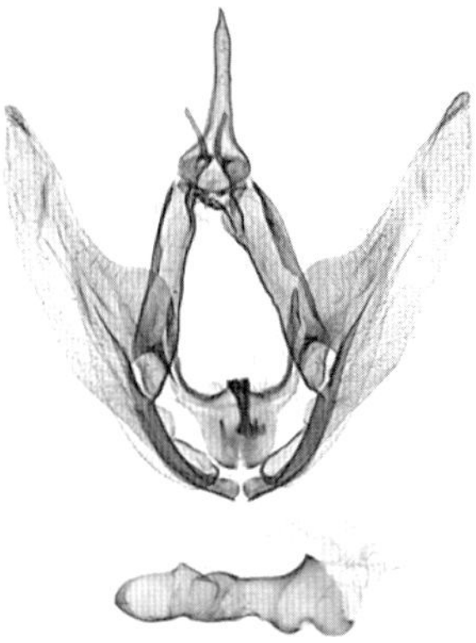

40c *Archigargetta viridigrisea*
Pantar, GU 93-13

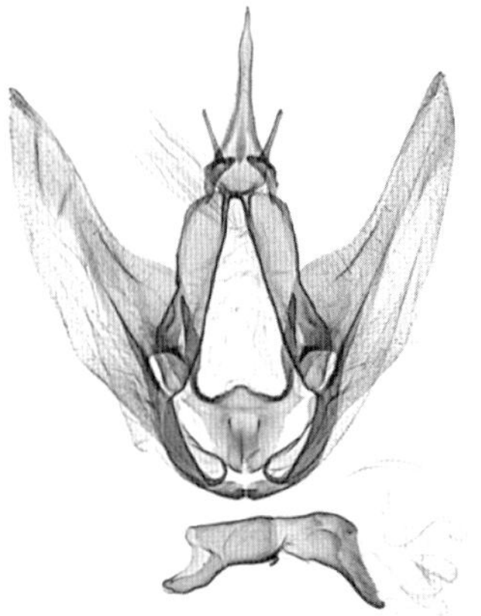

40d *Archigargetta viridigrisea*
Alor, GU 93-12

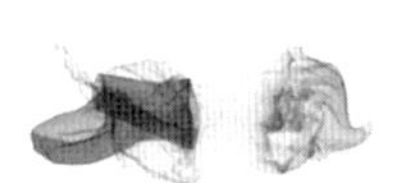

40e *A. viridigrisea*
Sumatra, GU 51-63
Phallus

1 mm

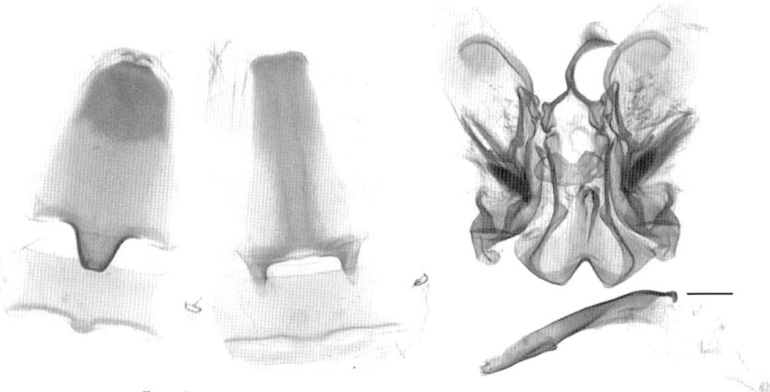

41a *Lasioceros aroa aroa*
W Papua, GU 95-61

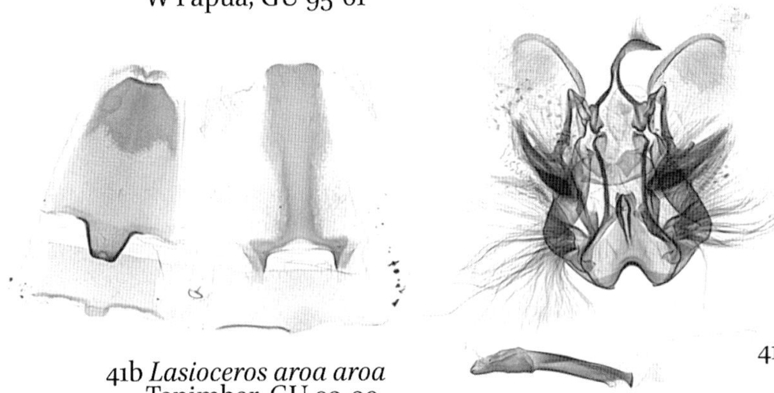

41b *Lasioceros aroa aroa*
Tanimbar, GU 93-09

41c *Lasioceros aroa aroa*
Tanimbar, GU 96-27

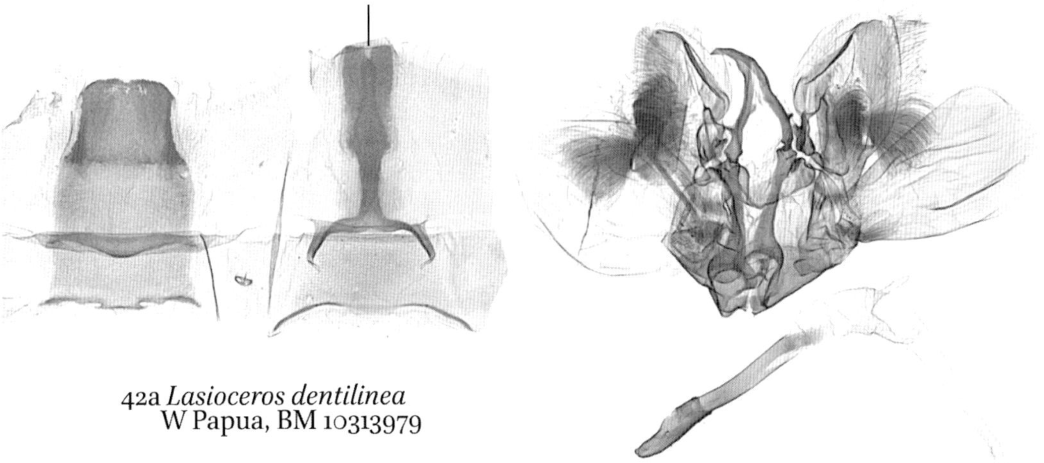

42a *Lasioceros dentilinea*
W Papua, BM 10313979

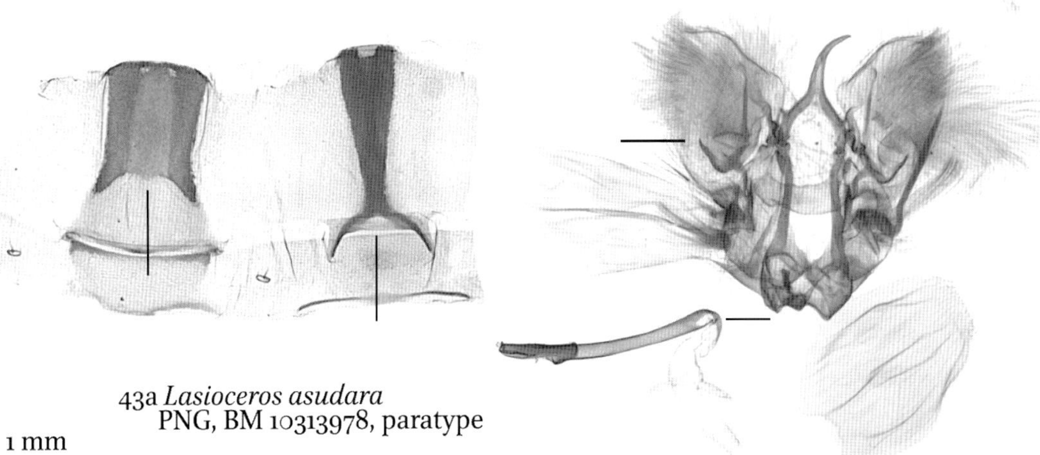

43a *Lasioceros asudara*
PNG, BM 10313978, paratype

1 mm

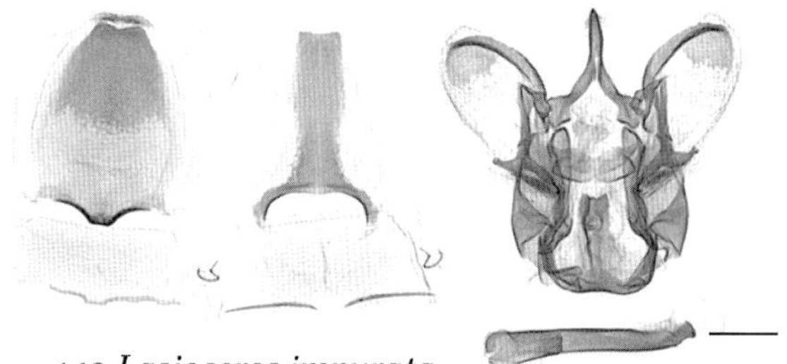

44a *Lasioceros impurata*
Timor, GU 95-61, holotype

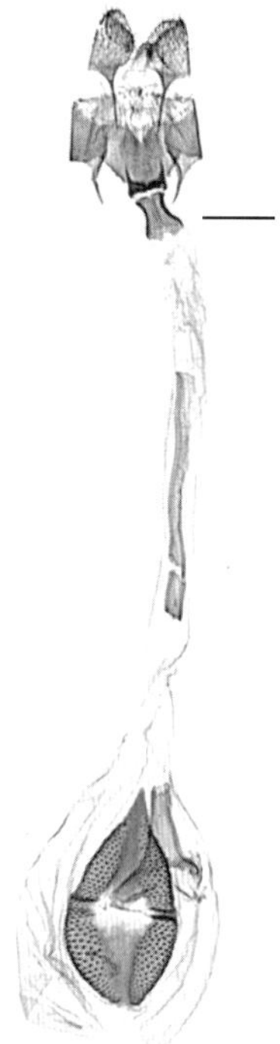

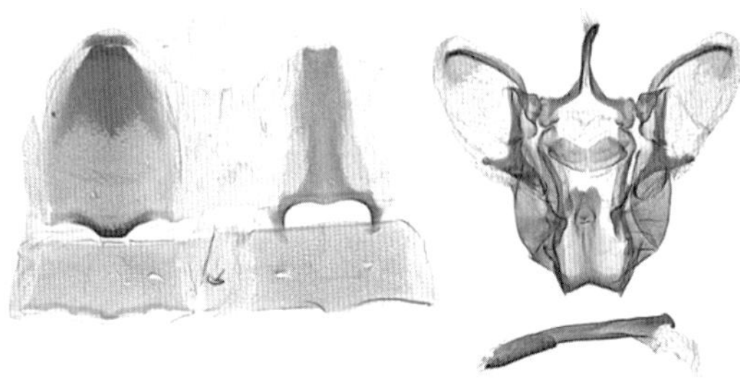

45a *Lasioceros sola*
Saleyer Isl., BM 10313976

44b *Lasioceros impurata*
Pantar, GU 93-11, PT

45Aa *Lasioceros euteles*
Palawan, MWM 14.983

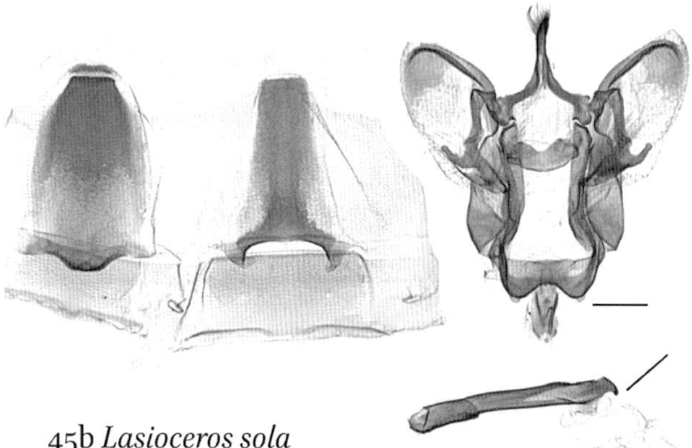

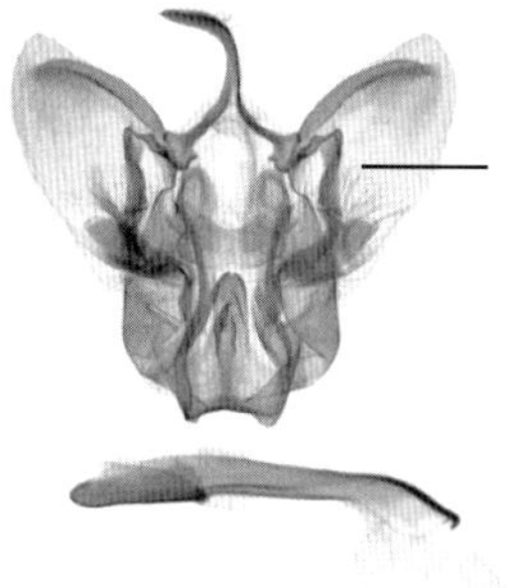

45Ab *Lasioceros euteles*
Palawan, MWM 14.984

45b *Lasioceros sola*
Sulawesi, BM 10313975

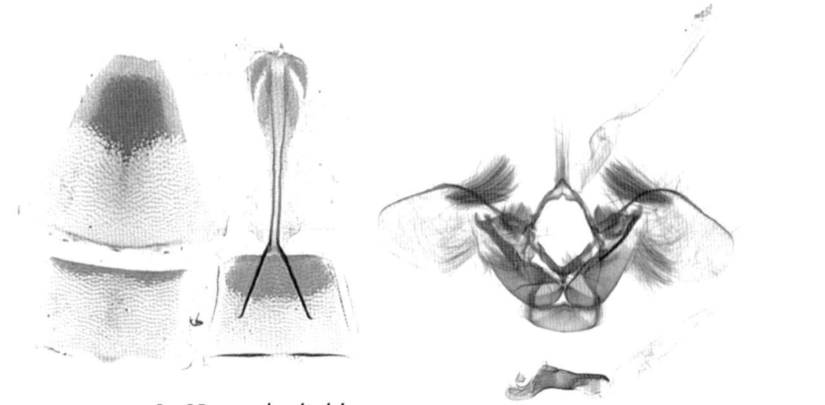

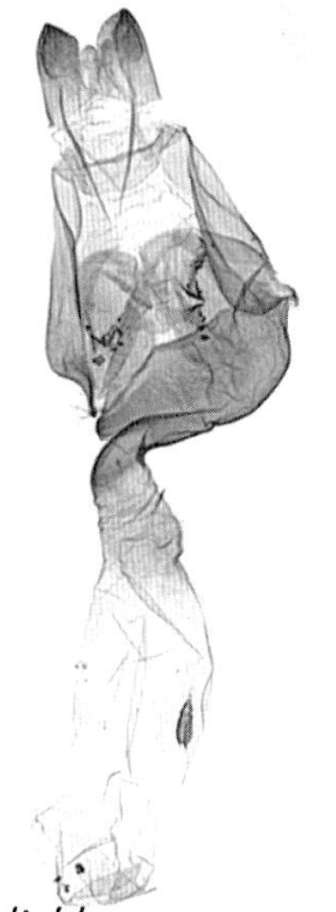

1 mm

46a *Noctuola diehlongata*
Sumatra, GU 96-54, holotype

46b *Noctuola diehlongata*
Sumatra, GU 96-53, paratype

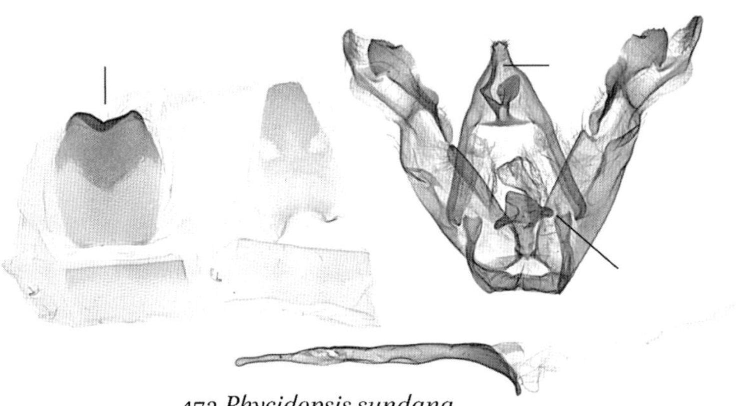

47a *Phycidopsis sundana*
Sumatra, GU 96-28, paratype

47b *Phycidopsis sundana*
Sumatra, GU 96-29

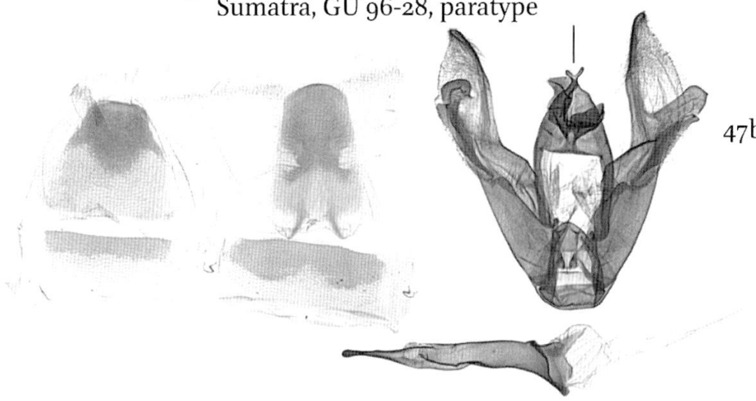

48a *Phycidopsis segregarius*
Alor, GU 96-30, holotype

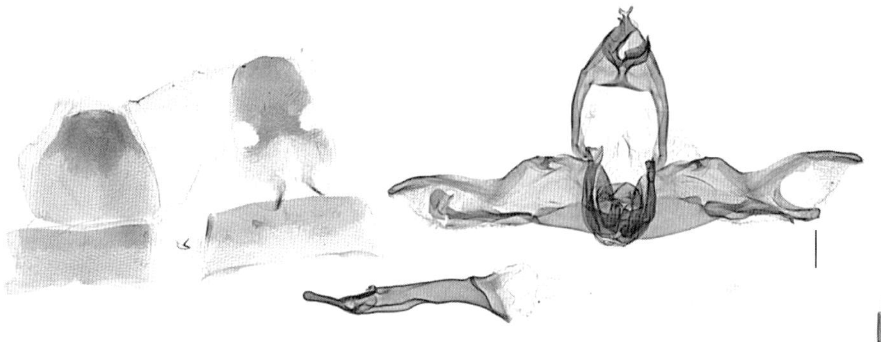

48b *Phycidopsis segregarius*
Alor, GU 93-10, paratype

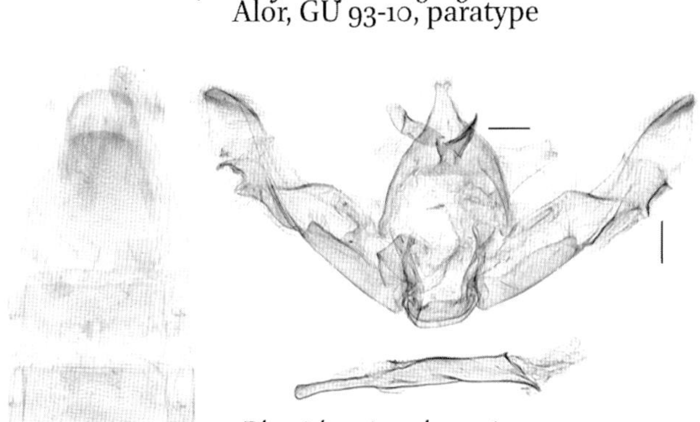

49a *Phycidopsis sulawesiana*
Sulawesi, GU 15-16, paratype

49b *Phycidopsis sulawesiana*
Sulawesi, GU 96-32

1 mm

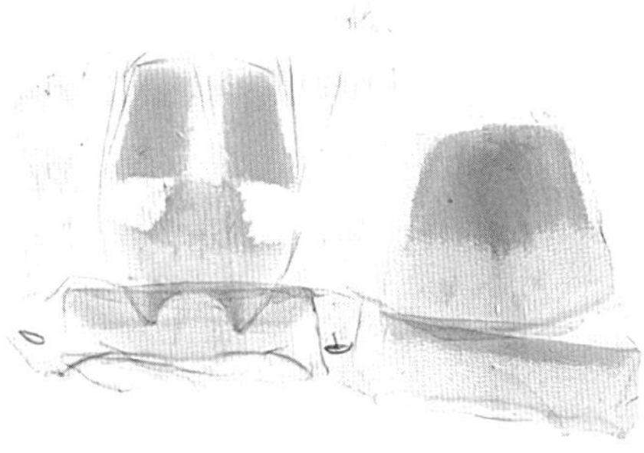

50a *Stictogargetta albimacula*
Sumatra, GU 96-33

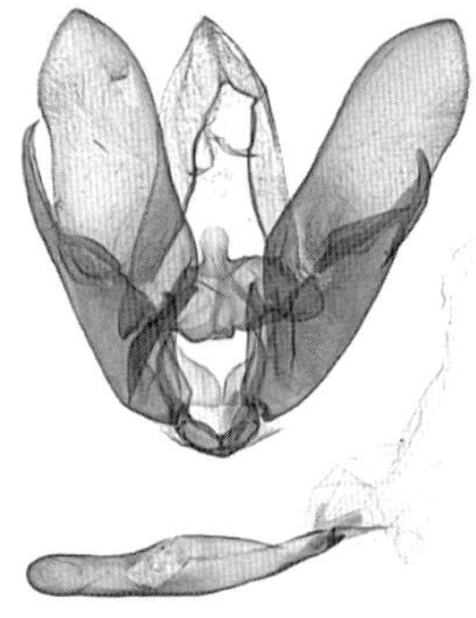

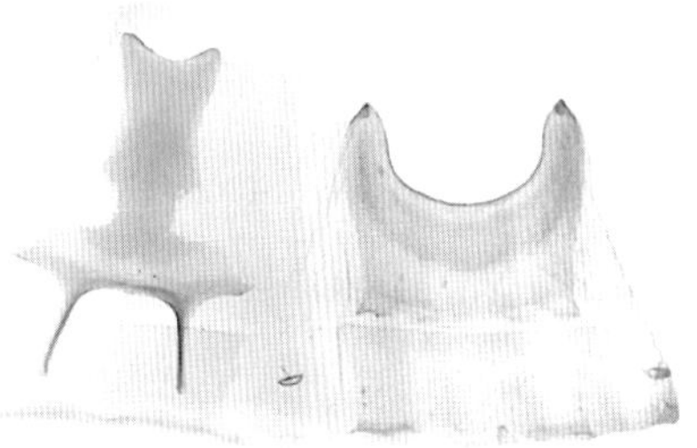

51a *Stictogargetta lithosidia*
Sumatra, GU 93-41,
(8th abdominal segments)

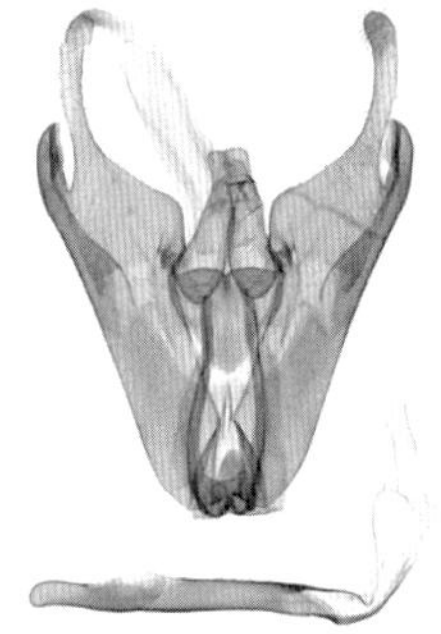

51b *Stictogargetta lithosidia*
Sumatra, MWM 14.970,

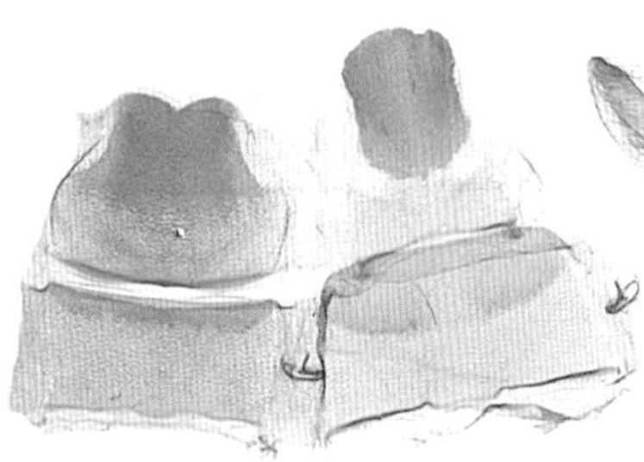

52a *Gargetta costigeroides*
Sumatra, LNK 02-2017

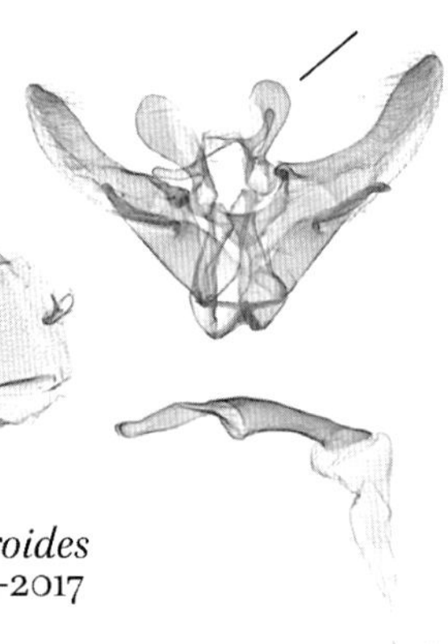

1 mm

52b *Gargetta costigeroides*
Brunei, BM 1182, holotype

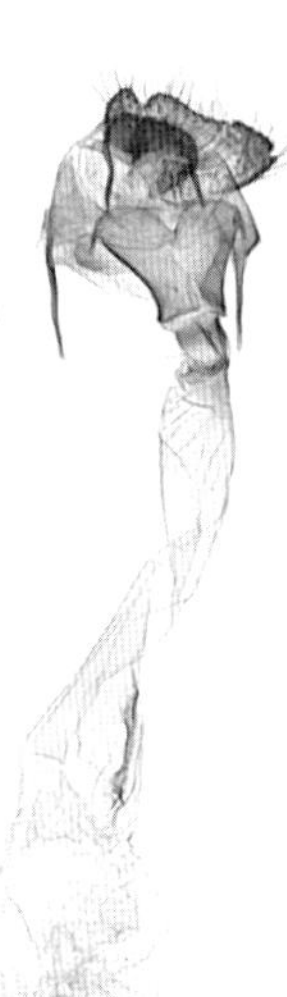

52c *Gargetta costigeroides*
Sumatra, LNK 01-2017

52d *G. costigeroides*
Sumatra, GU 96-34

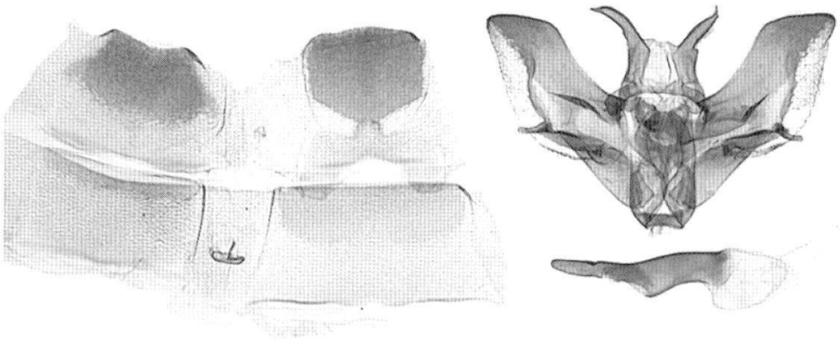

53a *Gargetta costigerana*
Sulawesi, GU 96-36, paratype

53b *Gargetta costigerana*
Sulawesi, GU 96-35, paratype

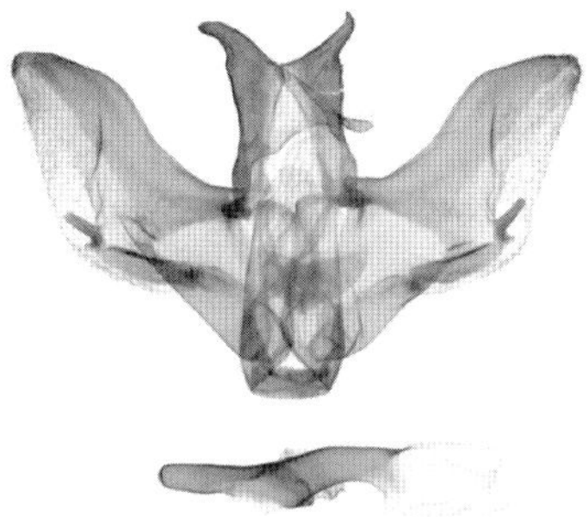

53c *Gargetta costigerana*
Sulawesi, MWM 14.971, paratype

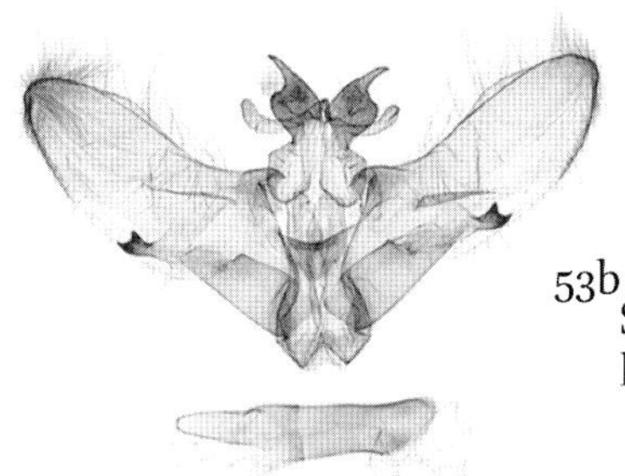

53Aa *Gargetta costigera*
India, Meghalaya, GU 74-34

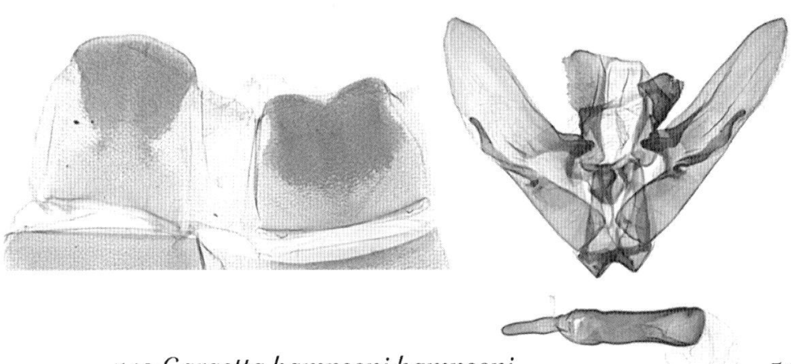

54a *Gargetta hampsoni hampsoni*
Kalimantan, GU 96-37

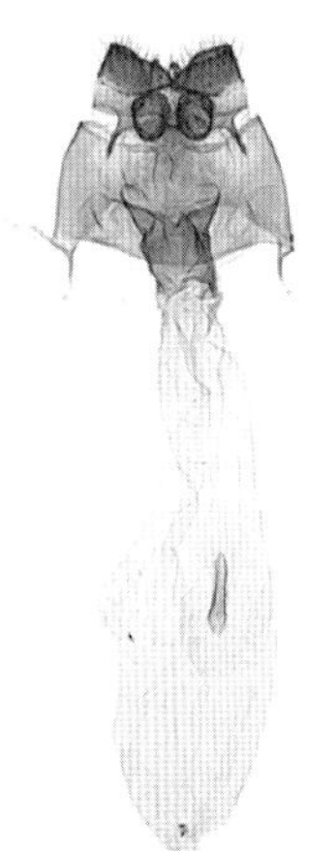

54b *G. hampsoni hampsoni*
Sumatra, GU 96-38

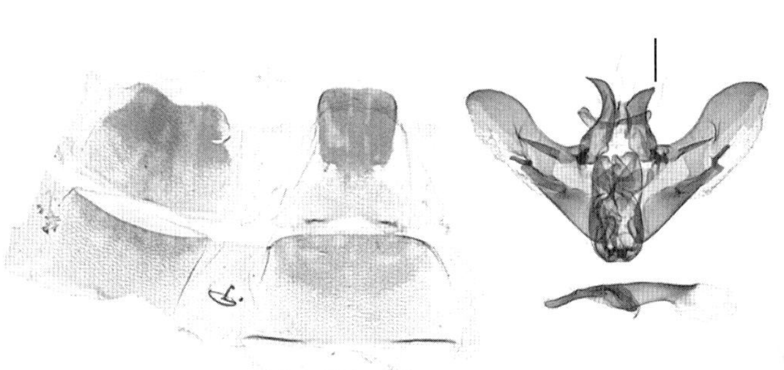

1 mm

55a *Gargetta divisa divisa*
Java, GU 96-45

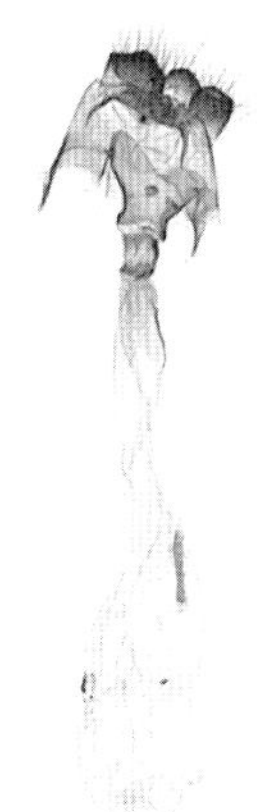

55b *Gargetta divisa divisa*
Java, GU 96-46

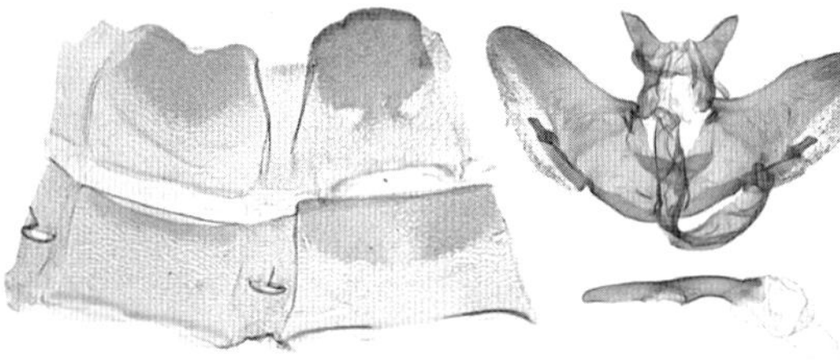

56a *Gargetta tompua*
Salayer Isl., BM 10313981

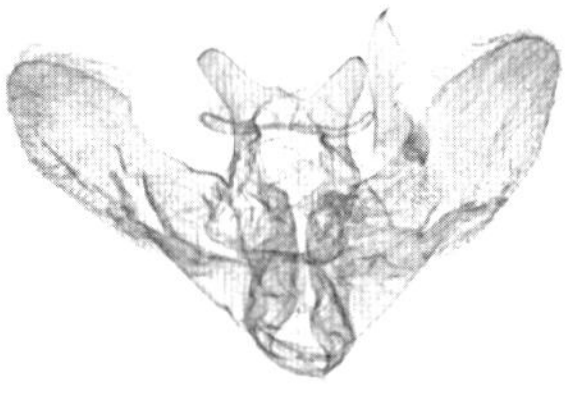

56b *Gargetta tompua*
Sulawesi, BM 150,
holotype

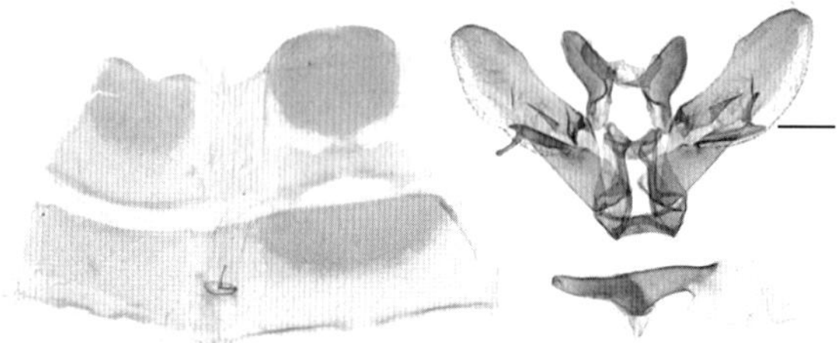

57a *Gargetta acarodes*
Tanimbar, GU 93-06

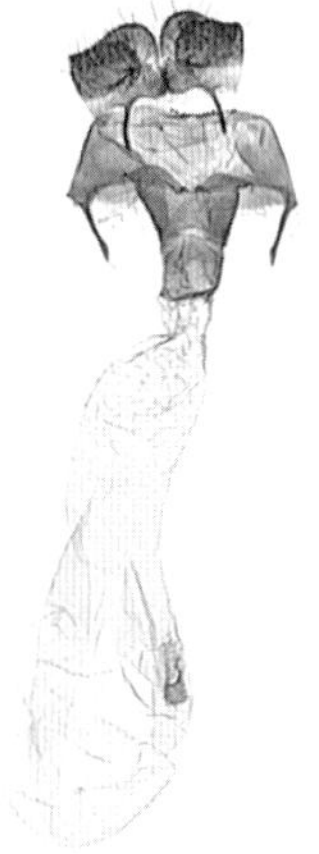

57b *Gargetta acarodes*
Australia-NT, GU 96-40

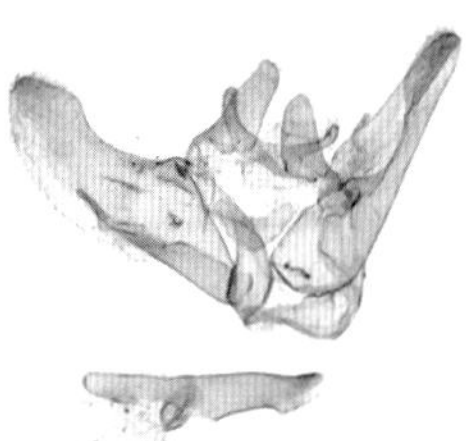

57c *Gargetta acarodes*
Australia-NT, GU 43-02a

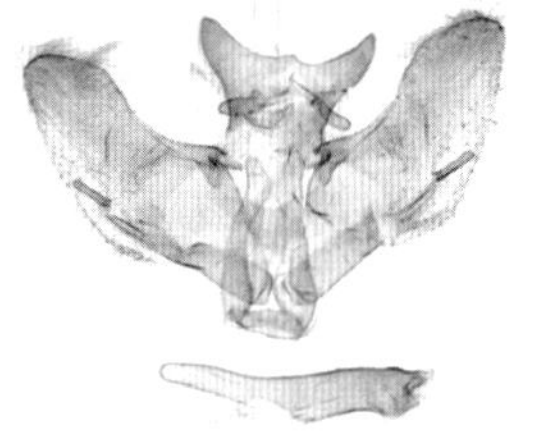

58a *Gargetta brevisocius*
Alor, GU 51-05a, paratype

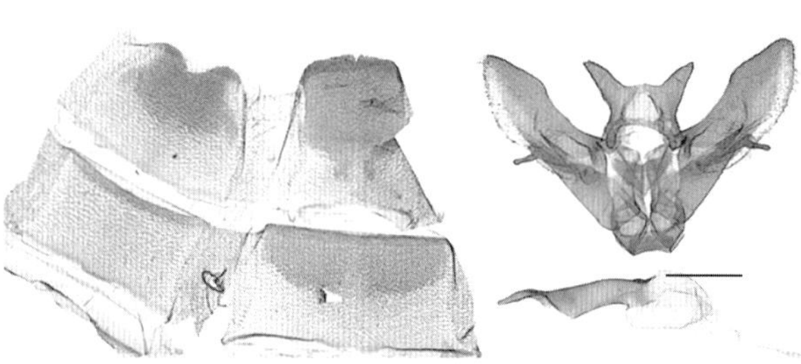

58b *Gargetta brevisocius*
Sumba, USNM 125.127,
paratype

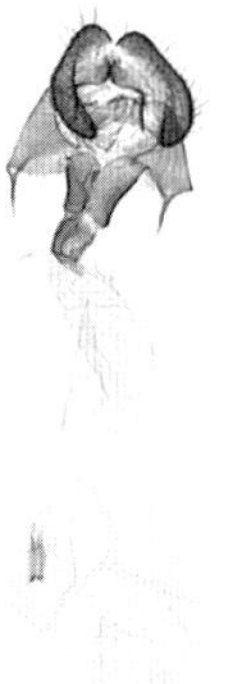

58c *Gargetta brevisocius*
Alor, GU 96-56,
paratype

58d *G. brevisocius*
Timor, GU 96-57
paratype

1 mm

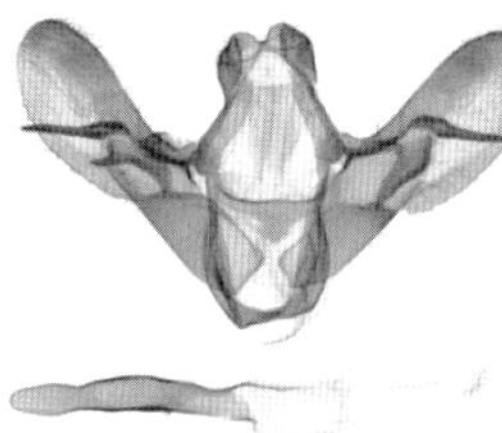

59a *Gargetta curvaria curvaria*
Java, MWM 14.980

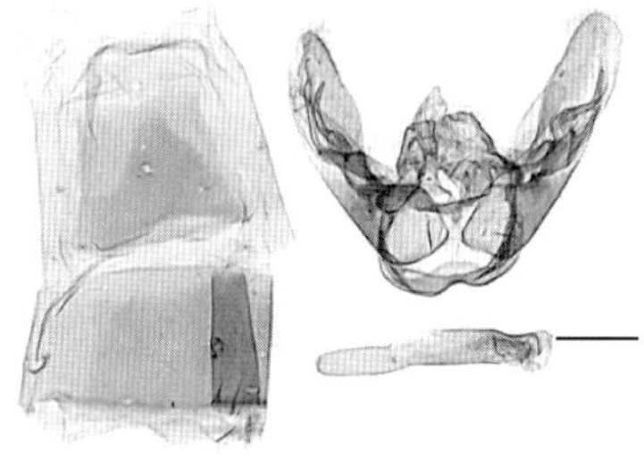

59b *Gargetta curvaria palua*
Sulawesi, GU 58-11

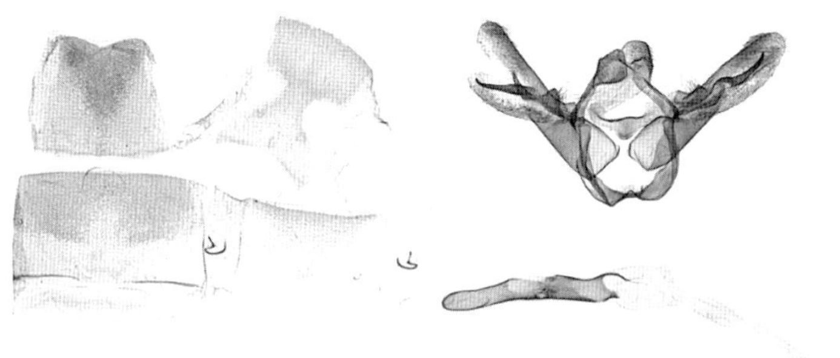

59c *Gargetta curvaria curvaria*
Sumatra, GU 96-55

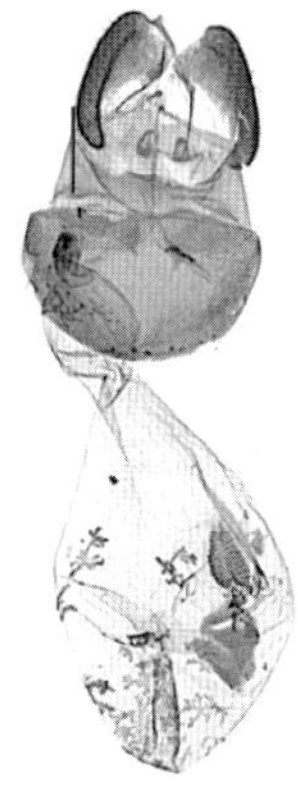

59d *Gargetta curvaria curvaria*
Borneo, SMNS 05-2017

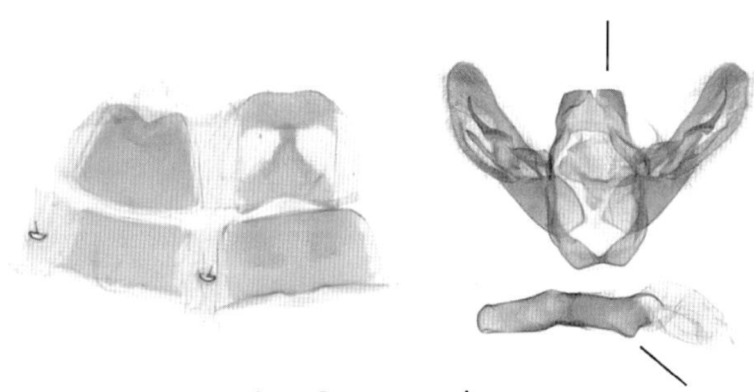

60a *Gargetta alora*
Alor, GU 95-58, paratype

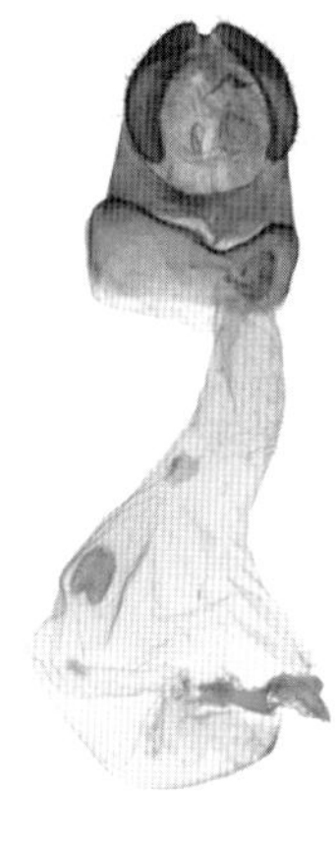

60b *Gargetta alora*
Flores, W 14979, paratype

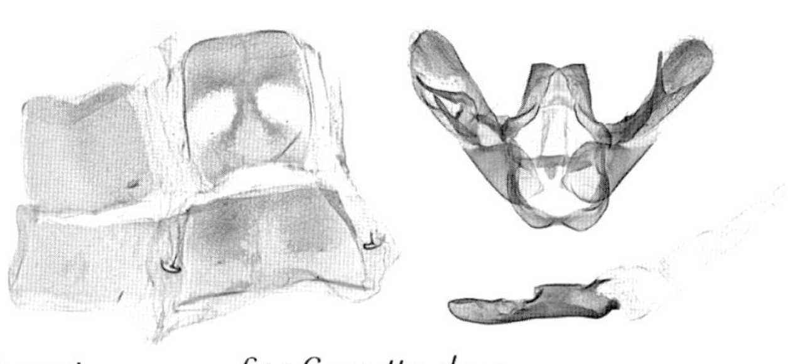

60c *Gargetta alora*
Lomblen, GU 96-39, paratype

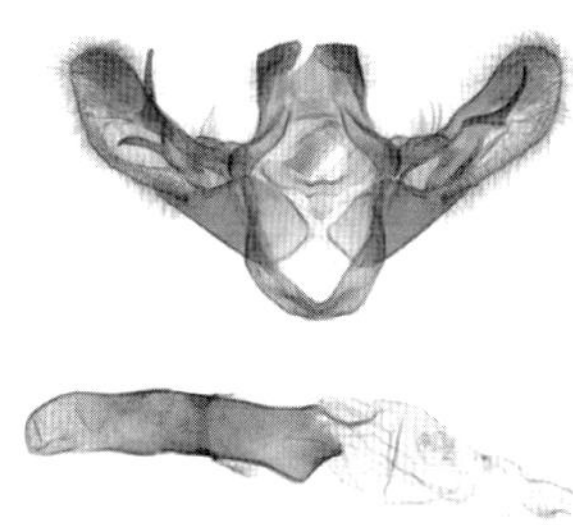

60d *Gargetta alora,* Pantar
GU 95-60, paratype

1 mm

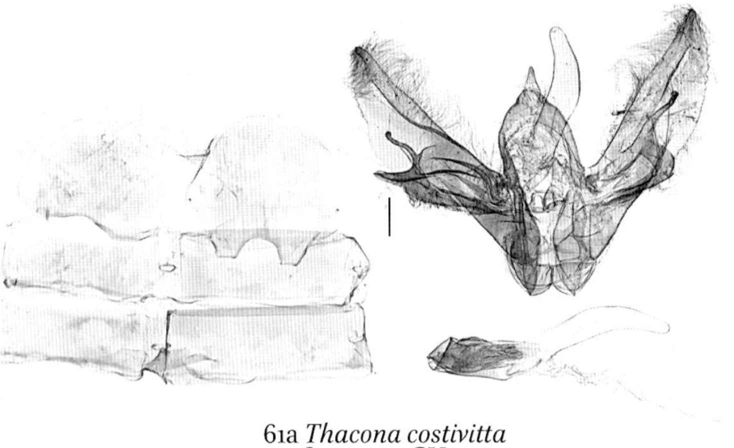

61a *Thacona costivitta*
Sumatra, GU 44-79

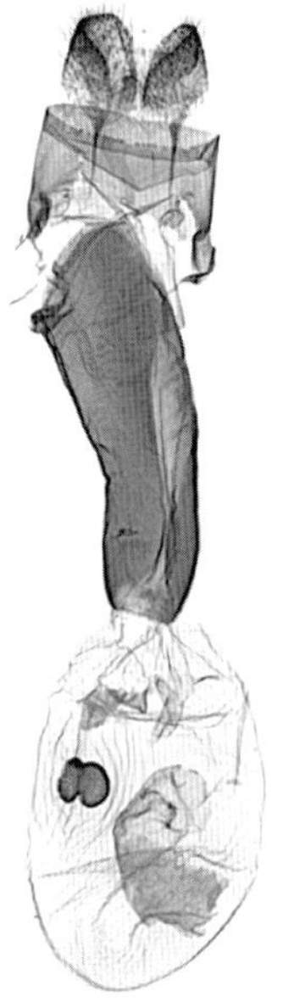

61b *Thacona costivitta*
Sumatra, GU 96-48

62a *Thacona rufocostata*
Sumatra, GU 44-100

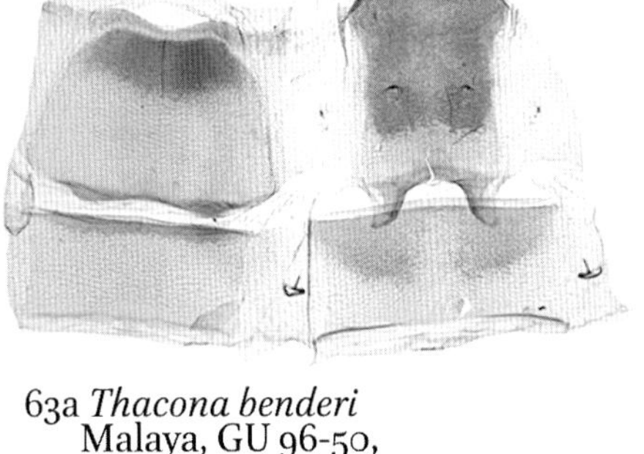

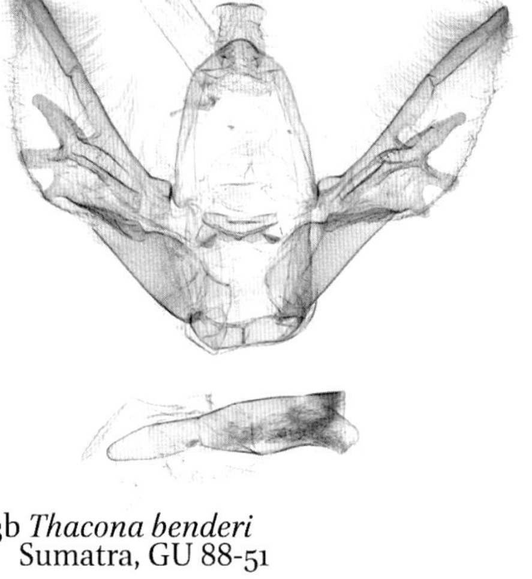

63a *Thacona benderi*
Malaya, GU 96-50,
8th abdominal segment

1 mm

63b *Thacona benderi*
Sumatra, GU 88-51

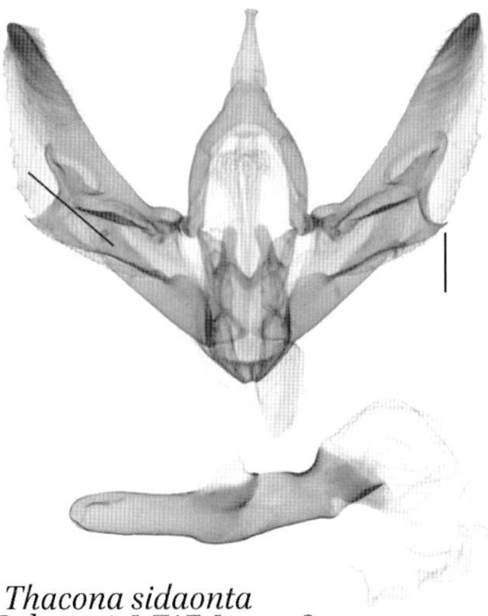

64a *Thacona sidaonta*
Sulawesi, MWM 14.976

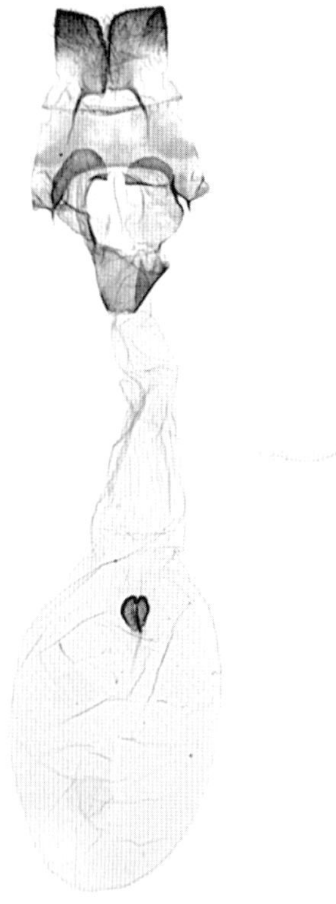

64b *Thacona sidaonta*
Sulawesi, GU 96-59

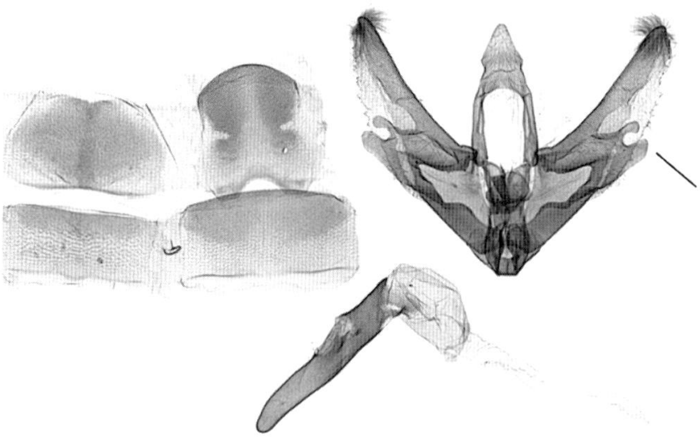

65a *Thacona ingens inopinata*
Sulmatra, GU 96-52

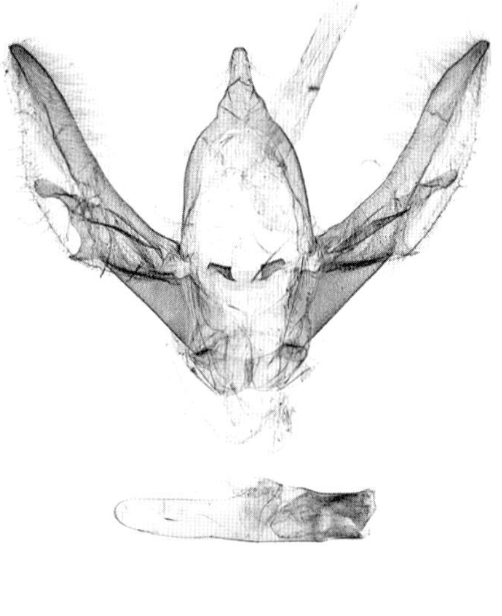

1 mm

66a *Thacona albescens*
Sumatra, GU 12-21

66b *Thacona albescens*
Sumatra, GU 96-49

68b*Thacona punctifascia*
Thailand, MWM 4.584

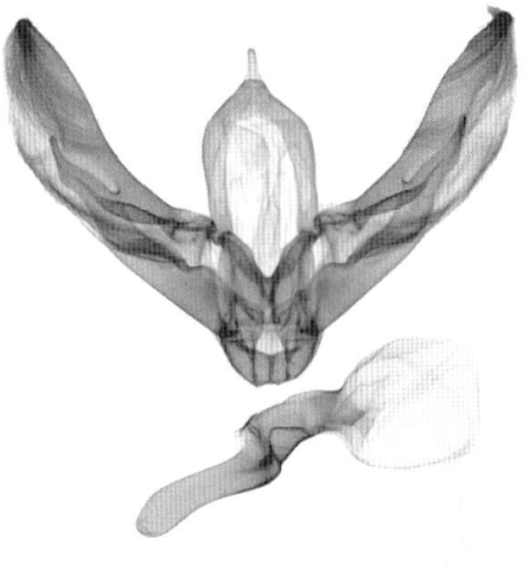

68a *Thacona punctifascia*
Sumatra, MWM 14.977

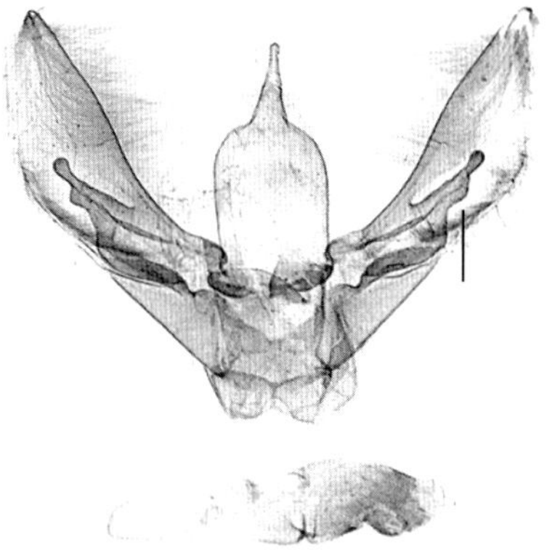

68c *Thacona punctifascia*
Sumatra, GU 88-43

69a *Thacona fallax*
Sumatra, GU 96-61

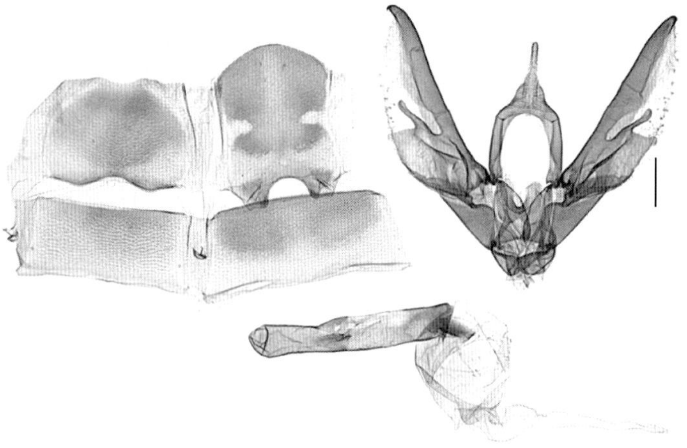

69b *Thacona fallax*
Sumatra, GU 96-58

70a *Hijracona punctifascioides*
Brunei, BM 1949 & BM 1950,
holotype

1 mm

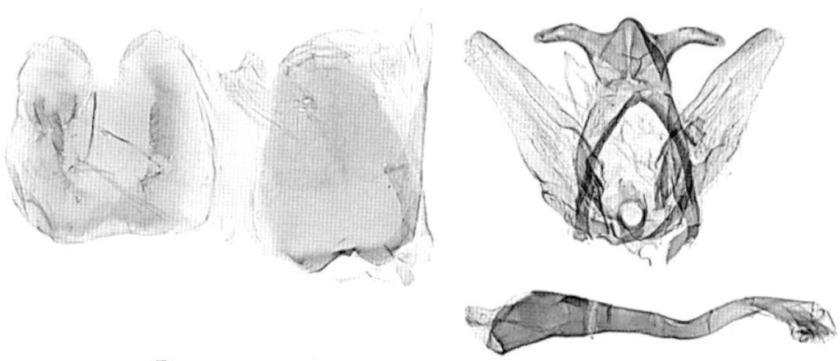

71a *Ramesa tosta*
Vietnam, GU 44-94

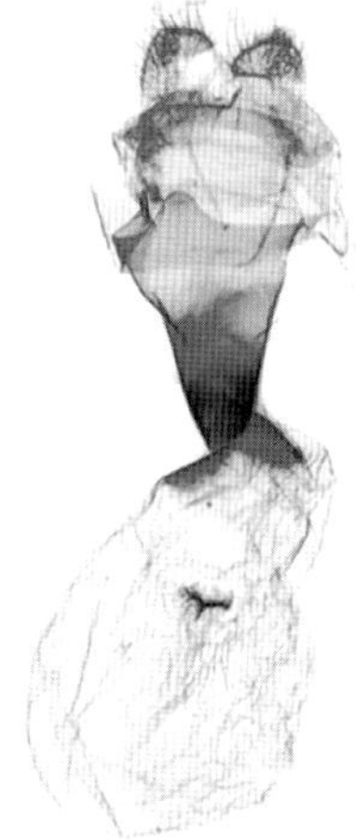

71b *Ramesa tosta*
Vietnam, GU 40-50

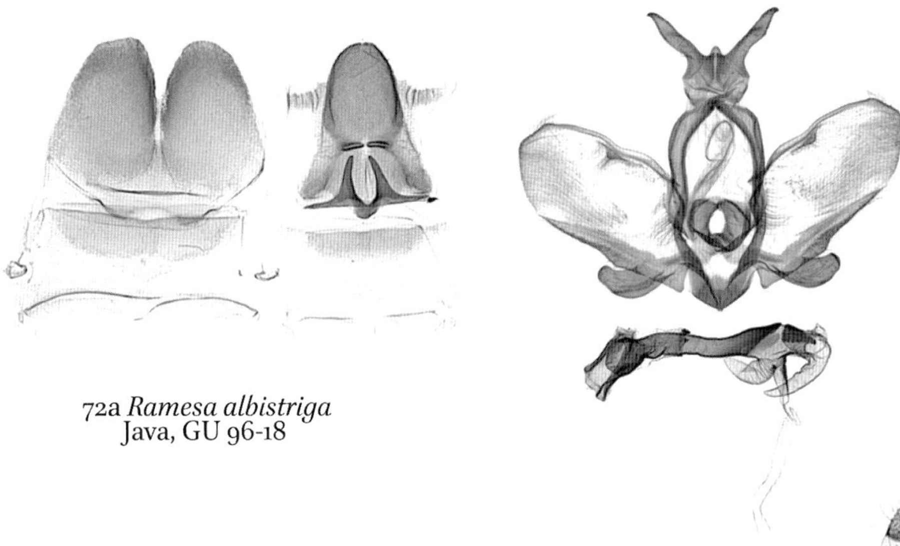

72a *Ramesa albistriga*
Java, GU 96-18

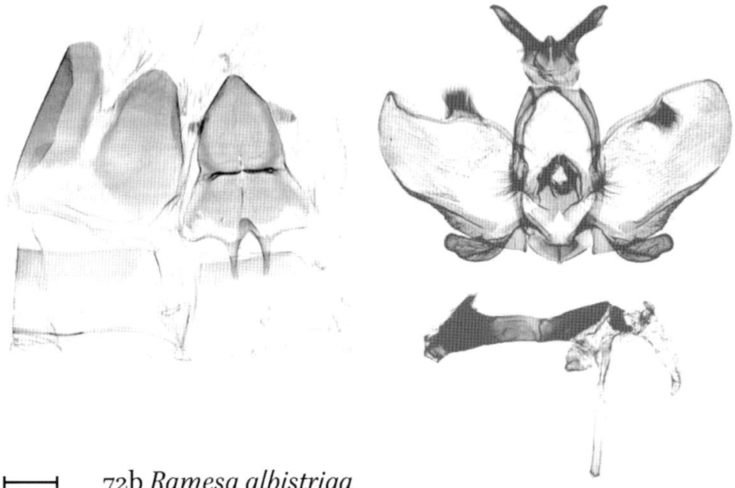

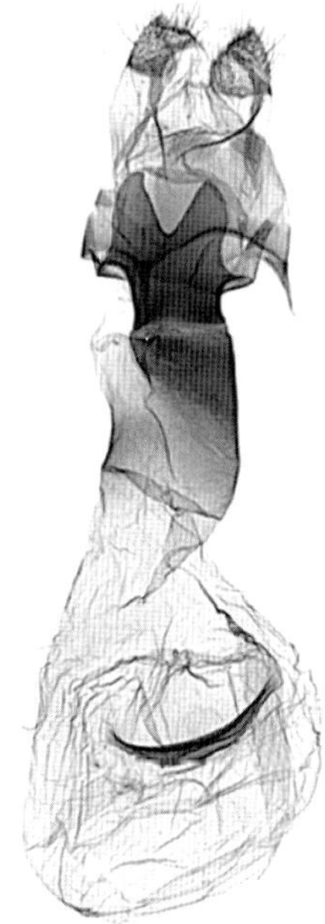

1 mm

72b *Ramesa albistriga*
Myanmar, GU 40-62

72c *Ramesa albistriga*
Sumatra, GU 96-71

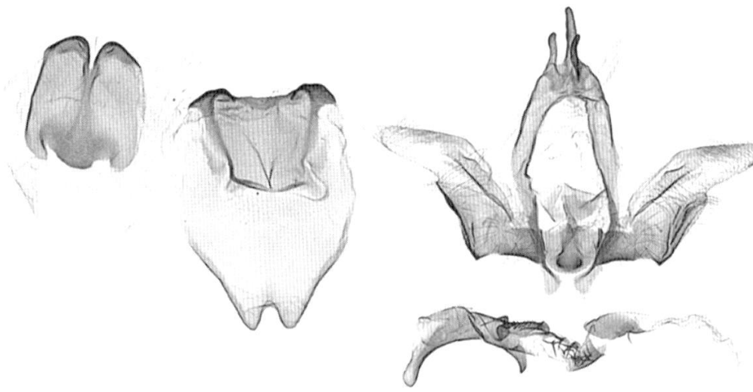

73a *Brykia horsfieldi*
Java, GU 96-24

73b *Brykia horsfieldi*
Borneo, MWM 15.564

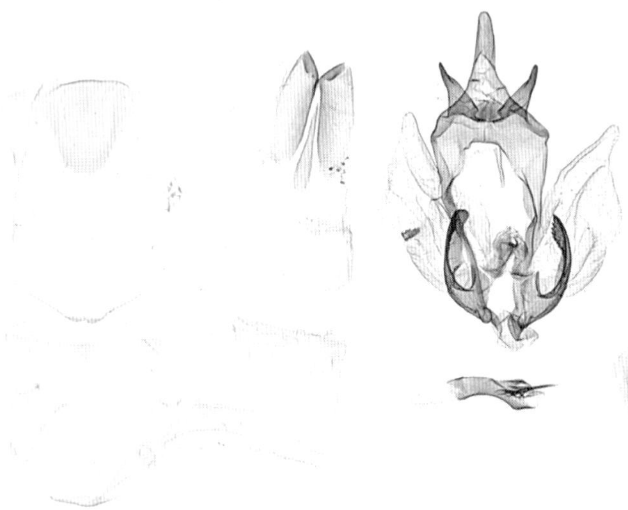

74a *Brykia samarinda*
Sumatra, GU 78-77

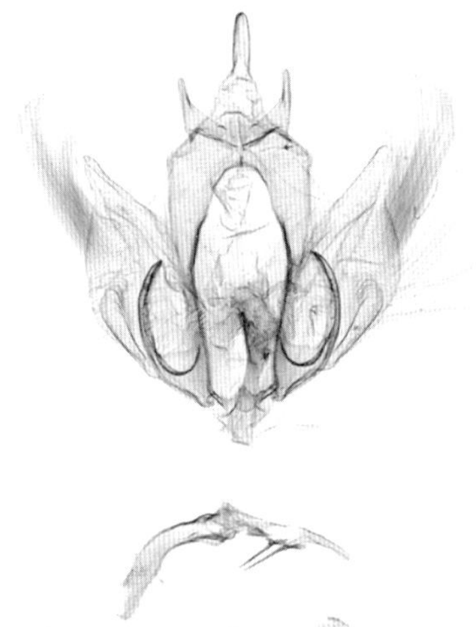

74b *Brykia samarinda*
Borneo, BM 987

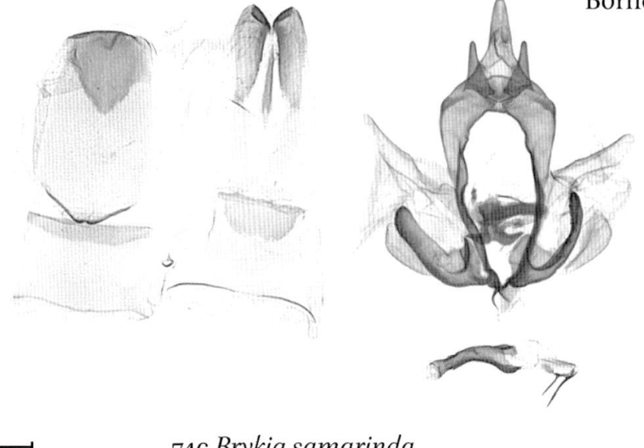

1 mm

74c *Brykia samarinda*
Borneo, GU 96-70

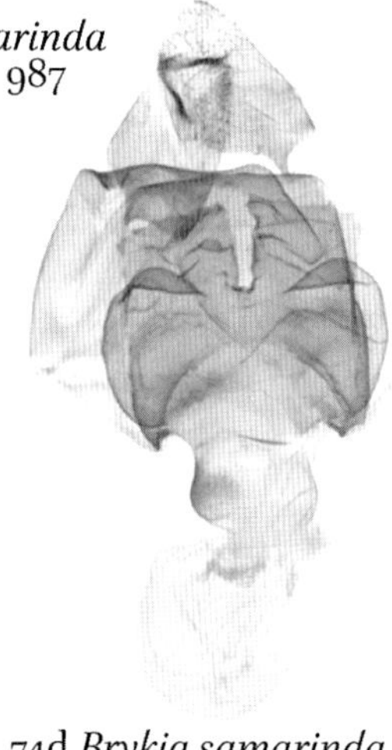

74d *Brykia samarinda*
Borneo, MWM 15.565

75a *Hyperaeschra pallida*
Sumatra, GU 96-73

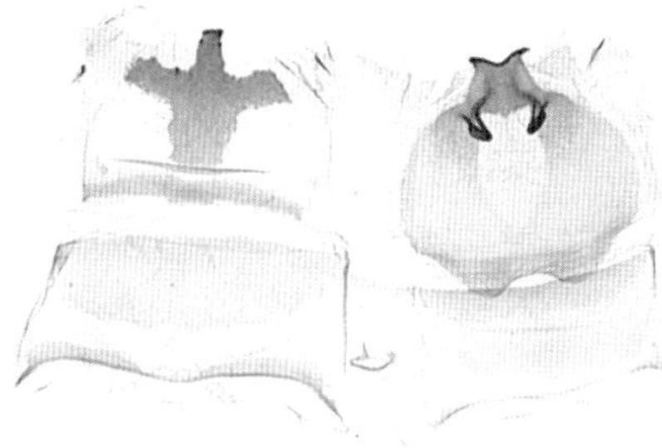

75b *Hyperaeschra pallida*
Sumatra, GU 96-75

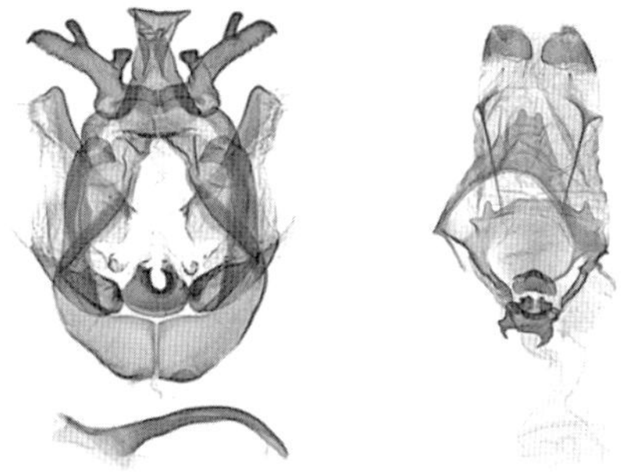

76a *Changea siamensis*
Malaya, BM 10314007

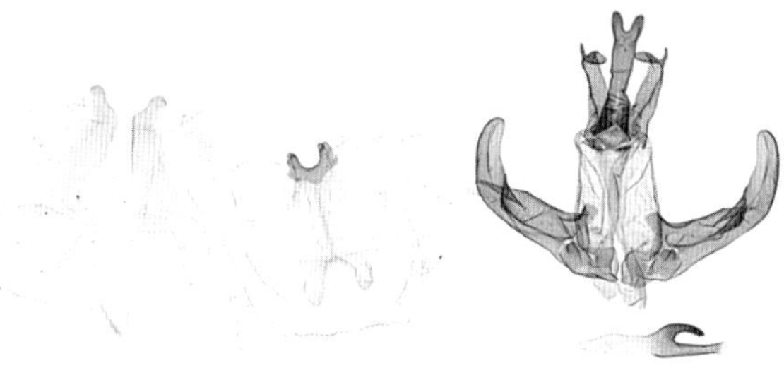

76b *Changea siamensis*
Sumatra, GU 78-82

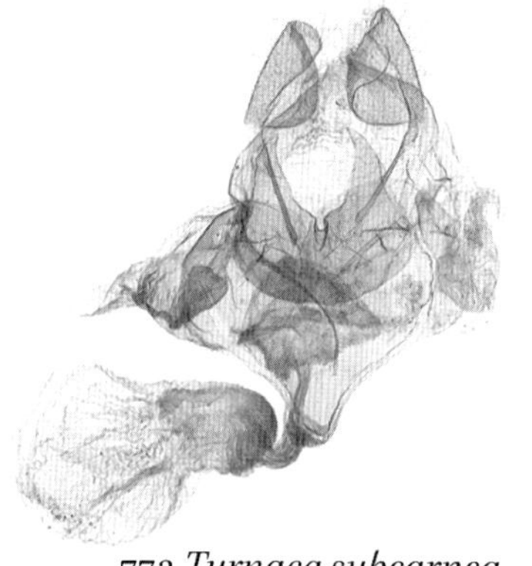

77a *Turnaca subcarnea*
Ambon, GU 52-03

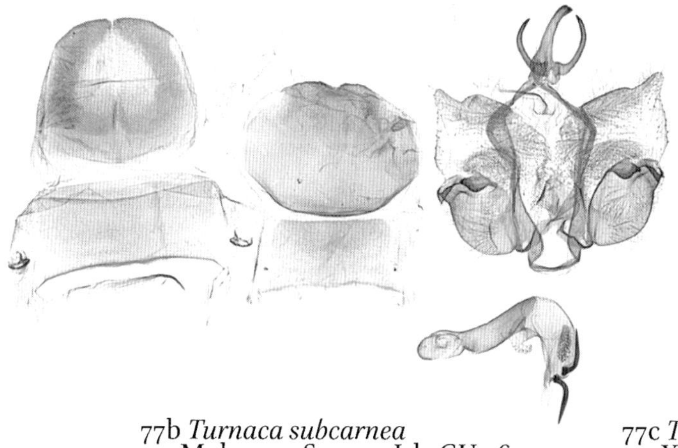

77b *Turnaca subcarnea*
Moluccas, Sanana Isl., GU 96-77

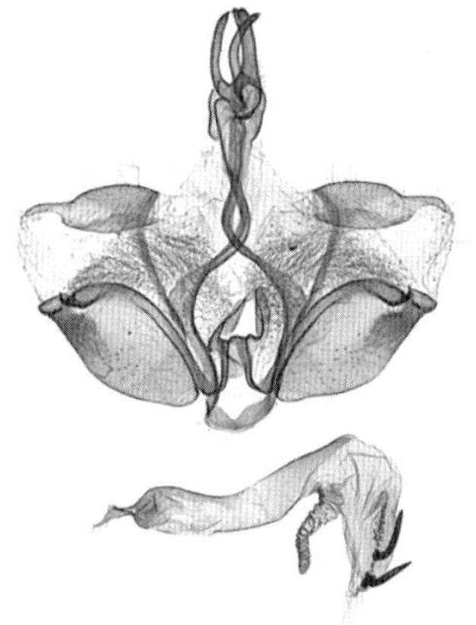

77c *Turnaca subcarnea*
Yapen Isl., Ihle 07-2017

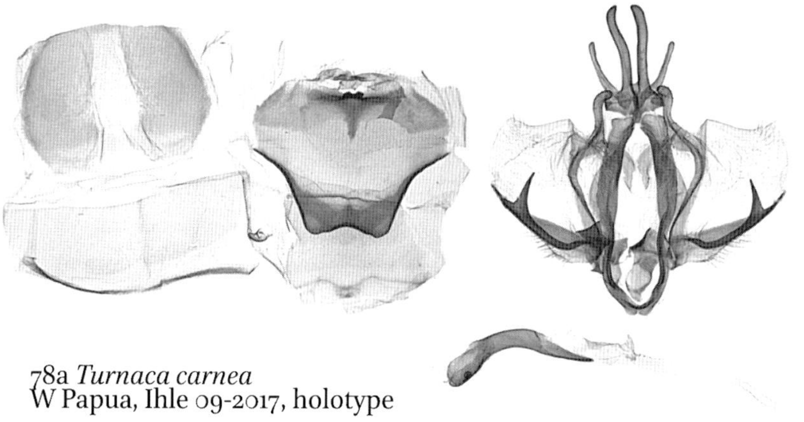

1 mm

78a *Turnaca carnea*
W Papua, Ihle 09-2017, holotype

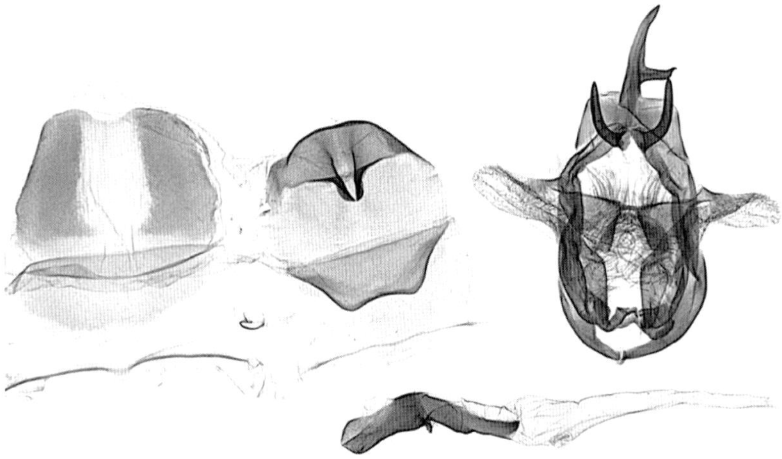

79a *Tensha delineivena*
Sumatra, GU 96-80

79b *Tensha delineivena*
Sumatra, GU 96-79

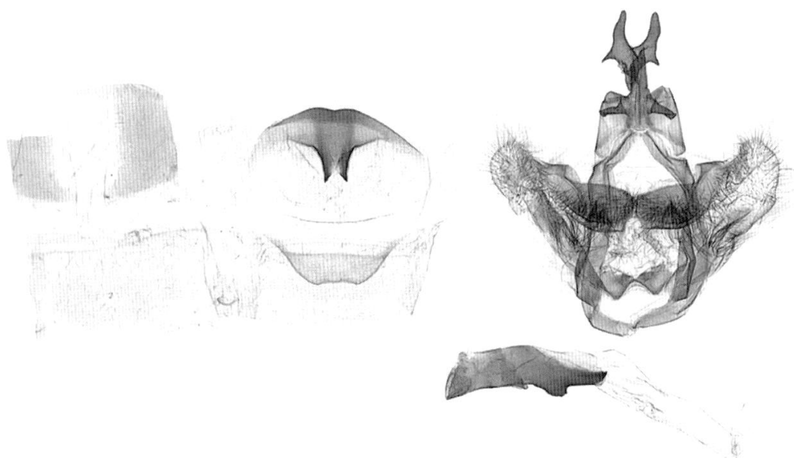

80a *Tensha postobscura*
Kalimantan, GU 77-12

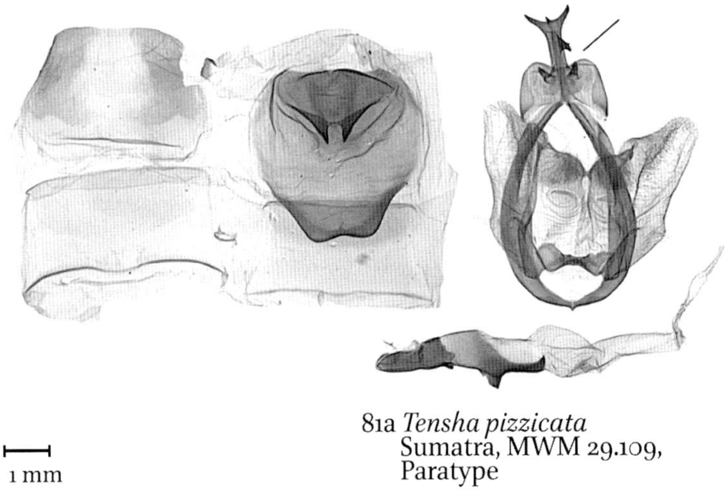

81a *Tensha pizzicata*
Sumatra, MWM 29.109,
Paratype

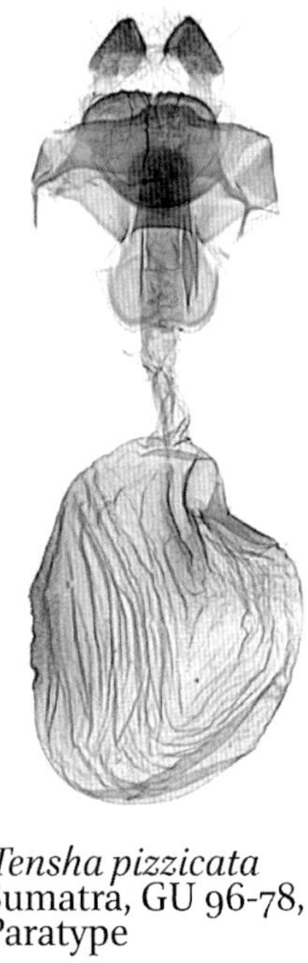

81b *Tensha pizzicata*
Sumatra, GU 96-78,
Paratype

1 mm

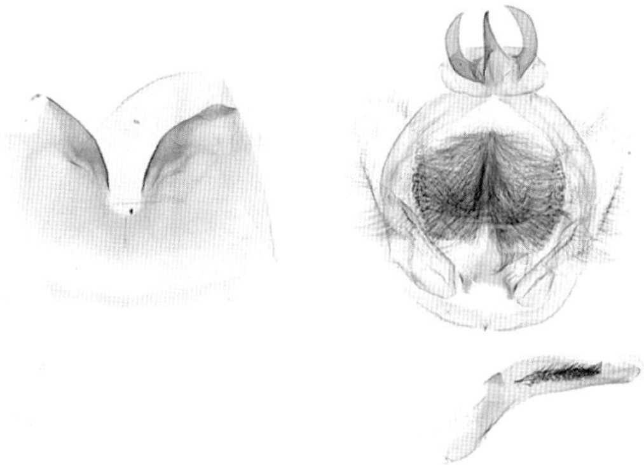

82Aa *Canatensha nigridorsalis*
Vietnam, GU 25-82

82Ab *Canatensha nigridorsalis*
Vietnam, GU 96-83, Paratype

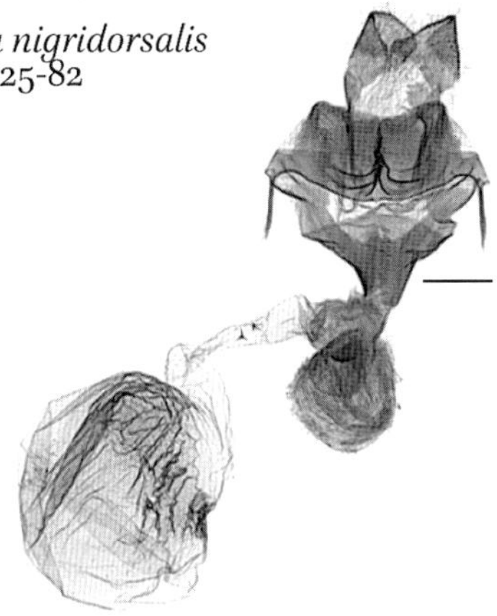

82a *Canatensha mediofascia*
Myanmar, GU 96-81

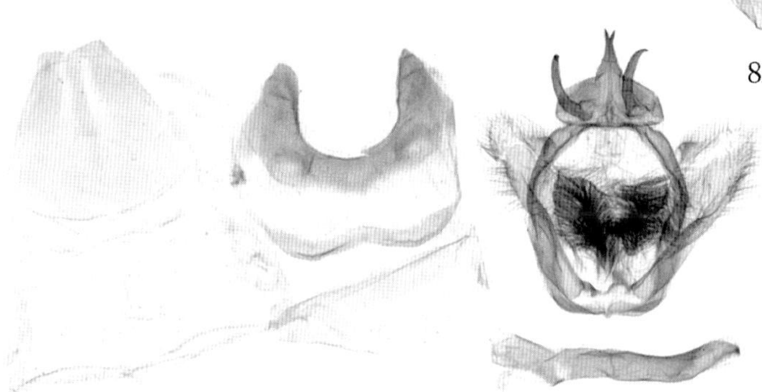

82b *Canatensha mediofascia*
Malaya, GU 17-88

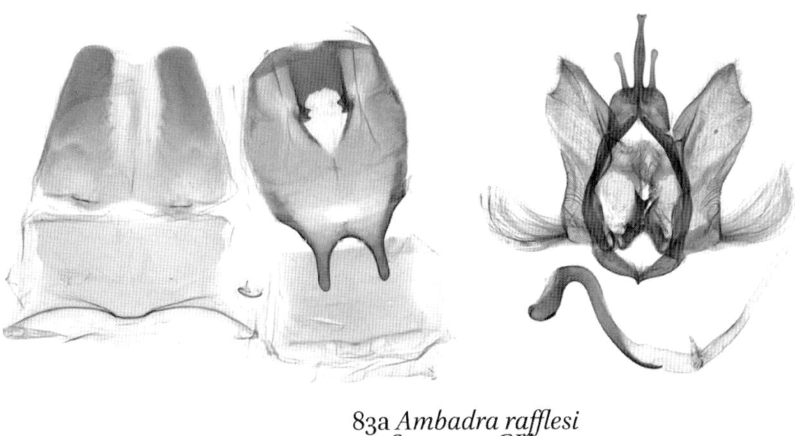

83a *Ambadra rafflesi*
Sumatra, GU 95-44

1 mm

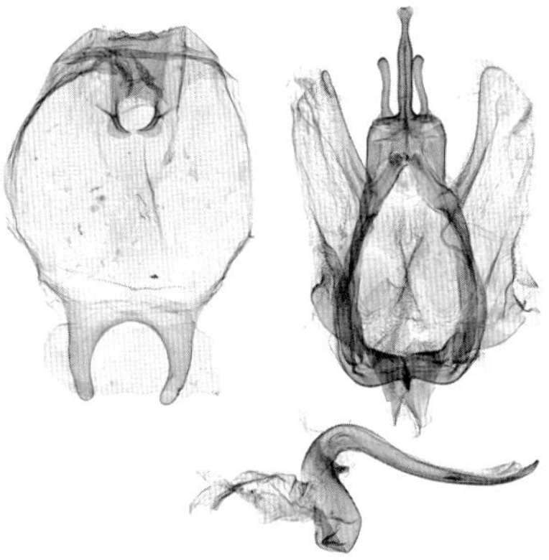

83b *Ambadra rafflesi*
Java, BM 809, holotype

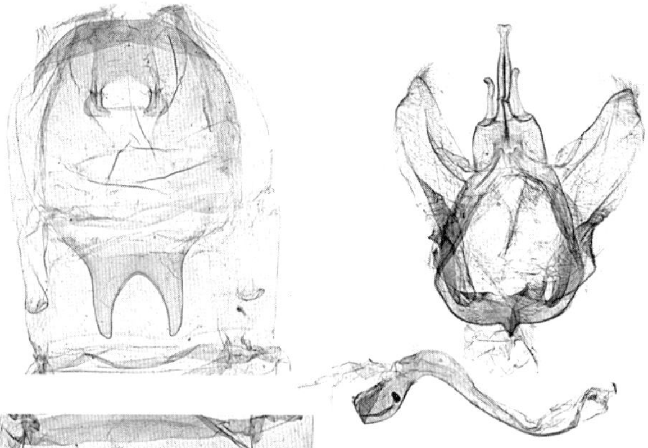

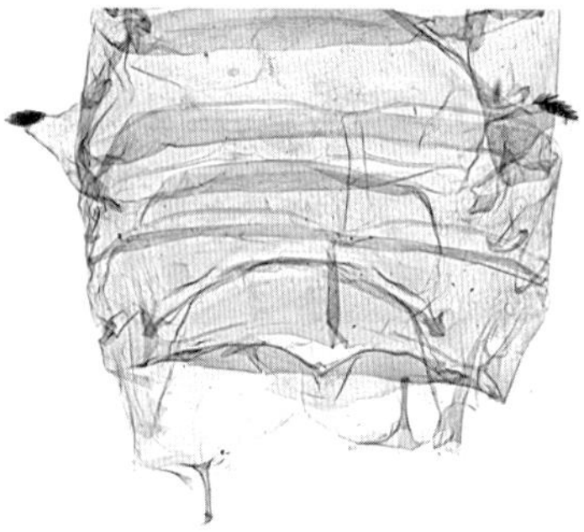

83c *Ambadra rafflesi*
Sumatra, BM 23
(holotype
Turnaca rufescens)

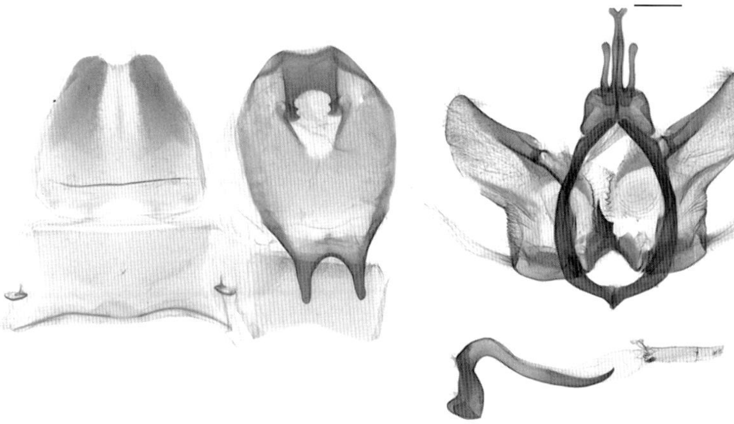

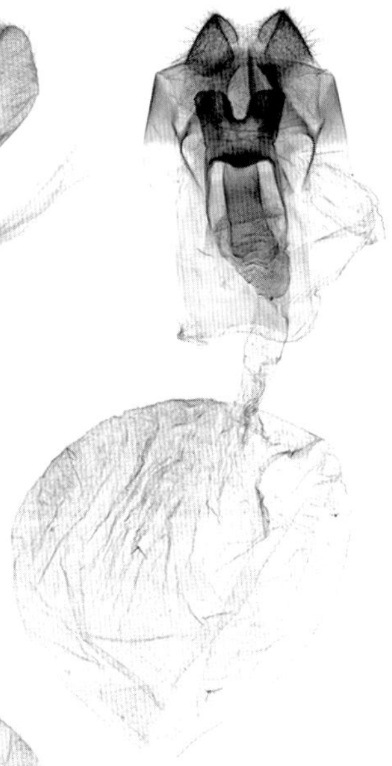

83d *Ambadra rafflesi*
Belitung, GU 95-41

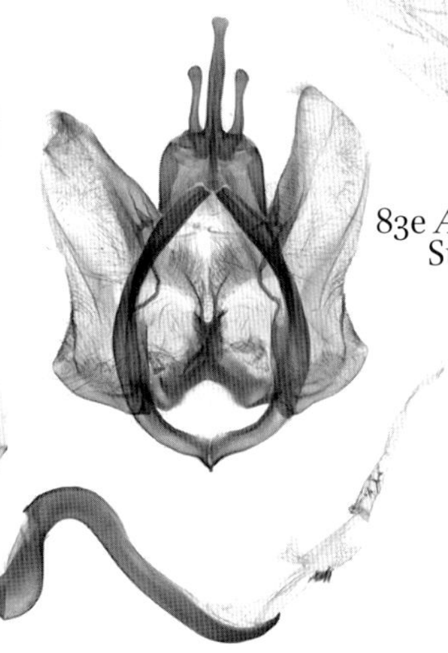

83e *Ambadra rafflesi*
Sumatra, GU 88-54

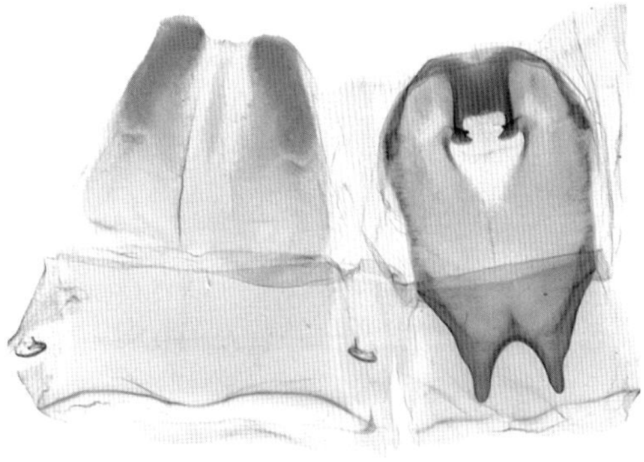

1 mm

83f *Ambadra rafflesi*
Sumba, GU 95-48

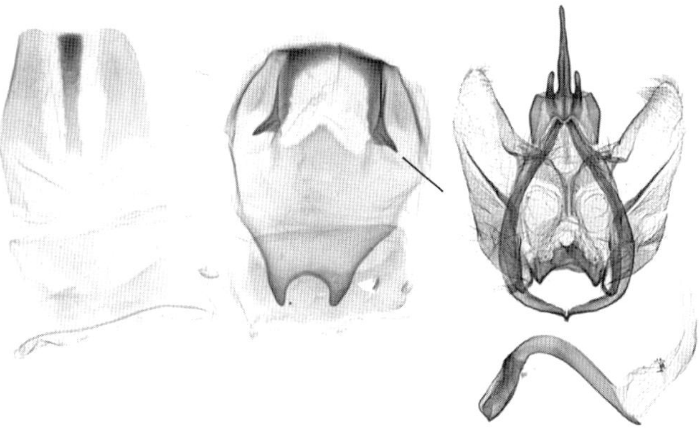

84a *Ambadra canescens*
Sulawesi, GU 96-84, paratype

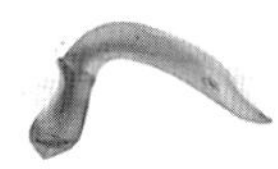

84b *Ambadra canescens*
Sulawesi, MWM 8.567, phallus

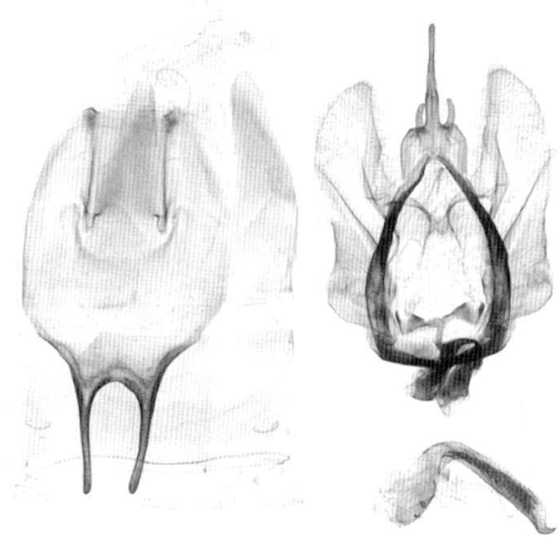

85a *Ambadra miscera*
Sulawesi, GU 88-52, paratype

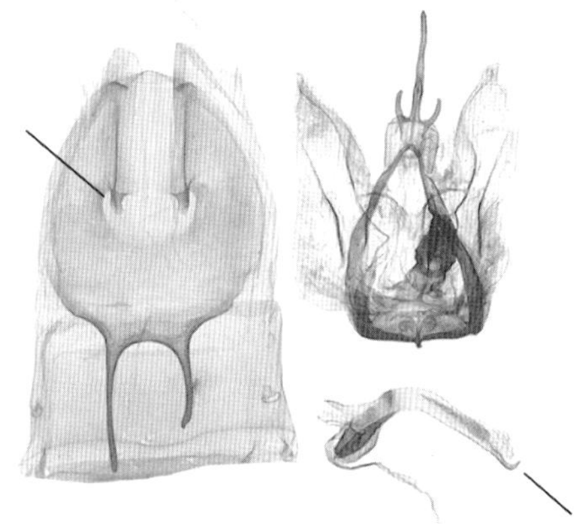

85b *Ambadra miscera*
Sulawesi, GU 54-63, paratype

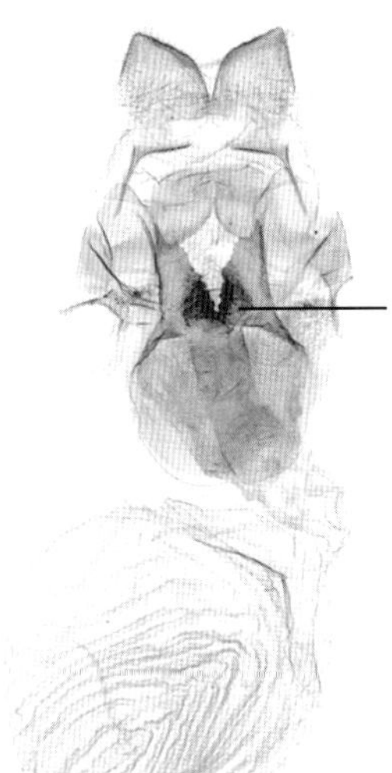

1 mm

85c *Ambadra miscera*
Sulawesi, GU 41-35, paratype

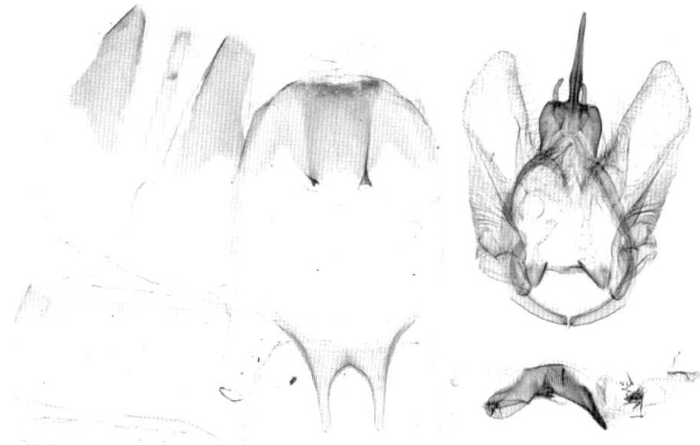

86a *Ambadra celebensis*
Sulawesi, GU41-42

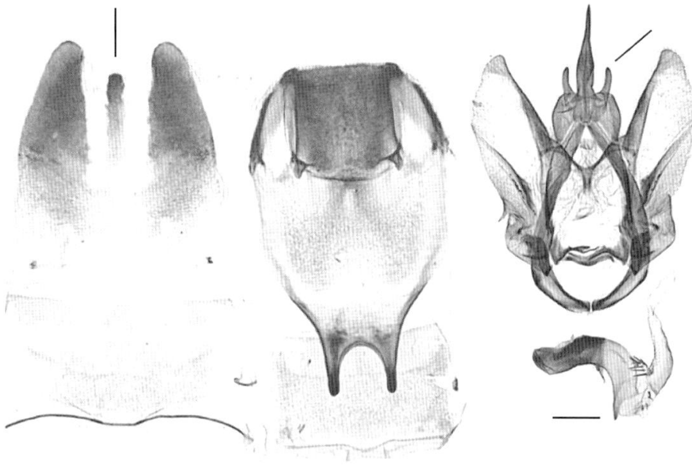

87a *Ambadra macula*
Peleng, GU 96-88

87b *Ambadra macula*
Sulawesi, GU 41-37

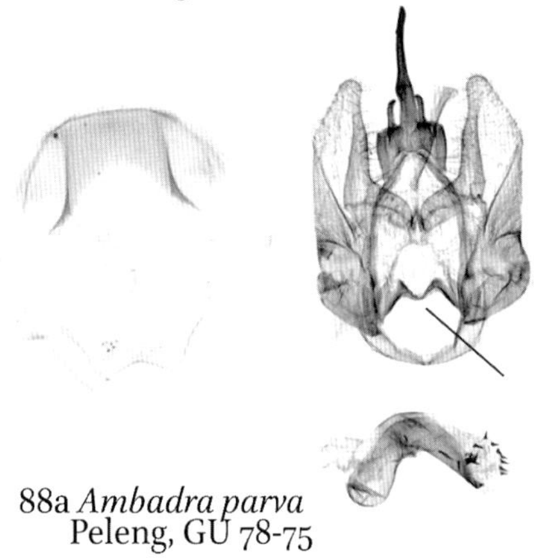

88a *Ambadra parva*
Peleng, GU 78-75

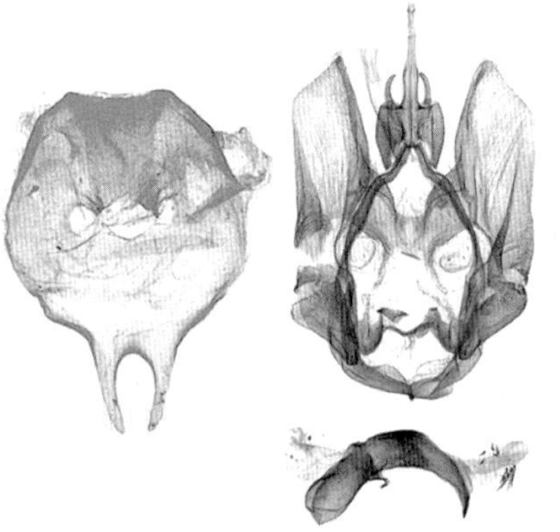

89a *Ambadra tondanoa*
Sulawesi, holotype
(from Kobayashi, 2011)

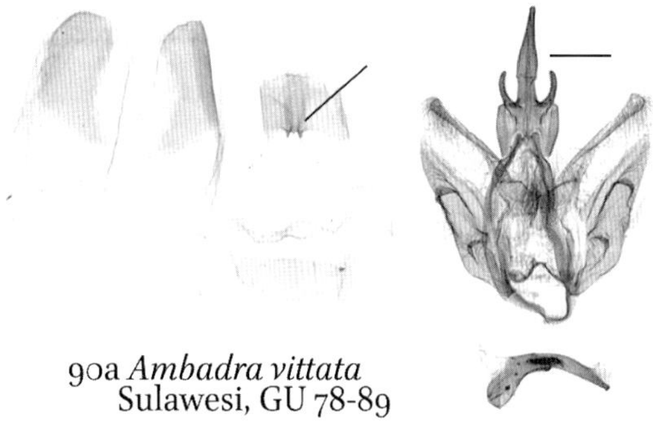

90a *Ambadra vittata*
Sulawesi, GU 78-89

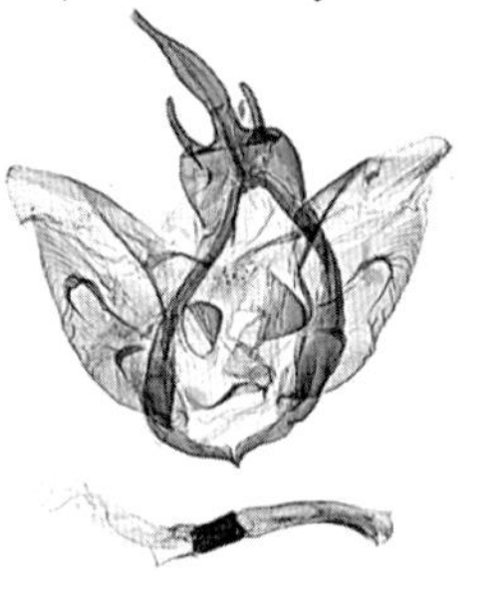

90b *Ambadra vittata*
Sulawesi, GU 78-89

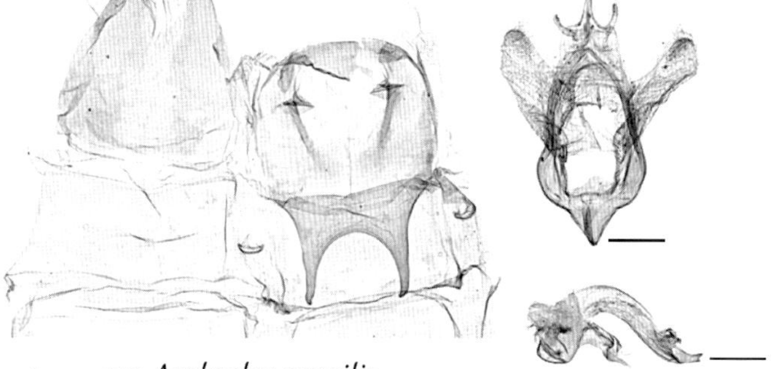

1 mm

91a *Ambadra gracilis*
Borneo, BM 1947, holotype

90c *Ambadra vittata*
Sulawesi, holotype, uncus
(from Kobayashi 2011)

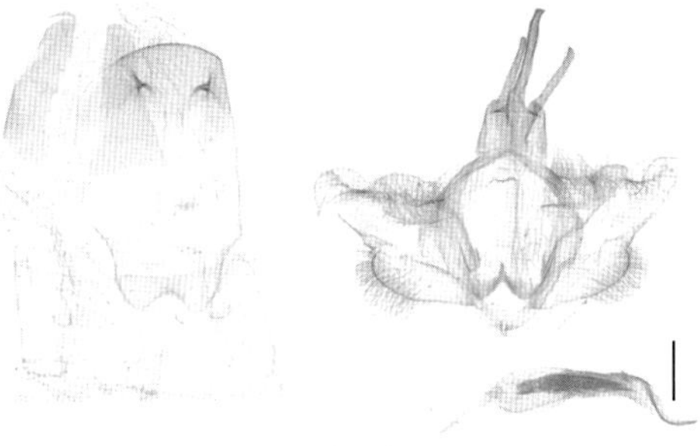

92a *Ambadra batara*
Java, GU 88-53,
holotype

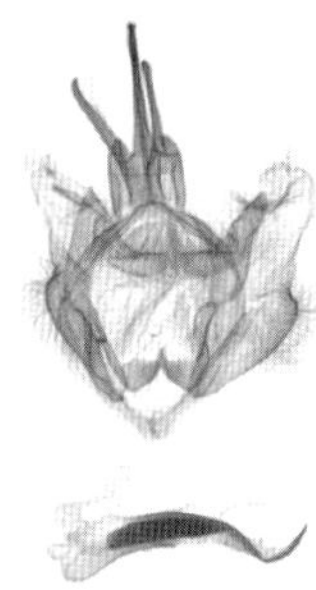

92b *Ambadra batara*
Java, GU 78-71,
paratype

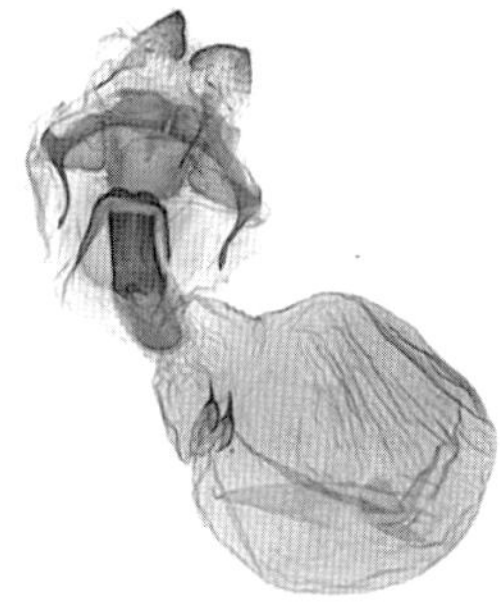

92c *Ambadra batara*
Java, GU 97-02,
paratype

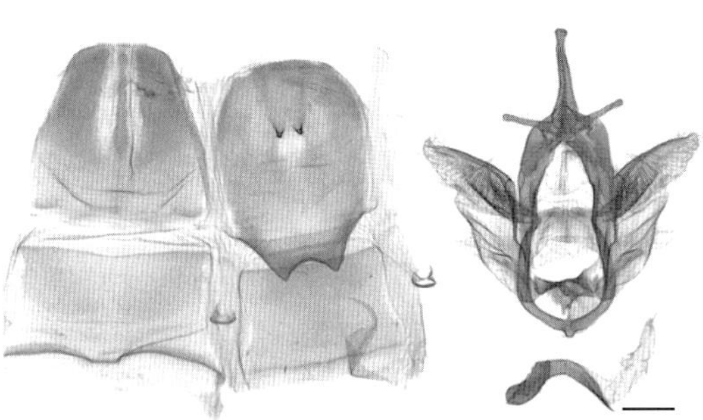

93a *Ambadra discocellularis*
Java, GU 95-43

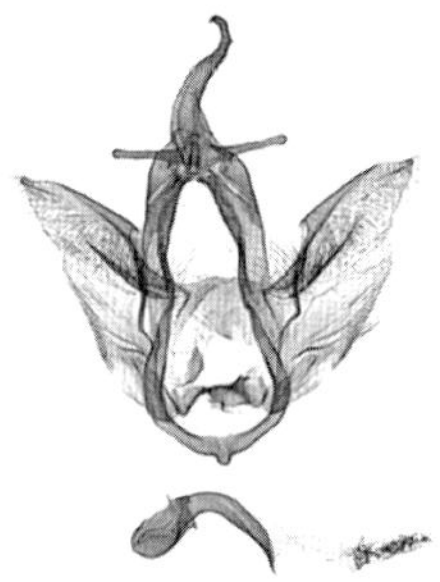

93b *A. discocellularis*
Java, GU 93-49

93c *Ambadra discocellularis*
Sumatra, GU 41-34

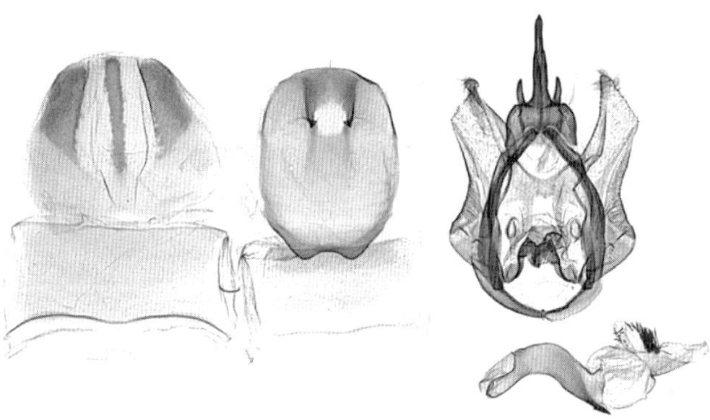

94a *Ambadra similis*
Sumatra, GU 96-86

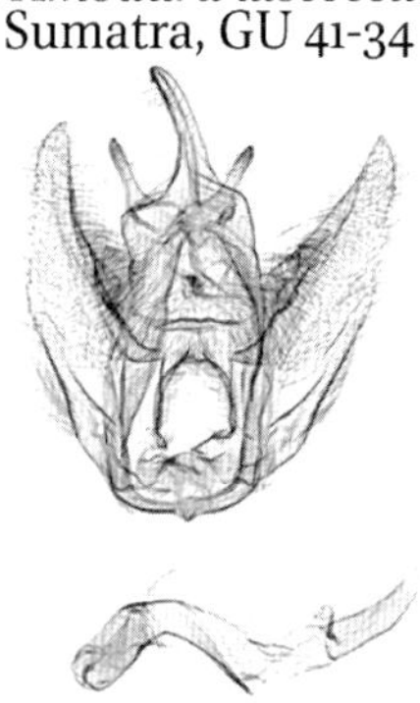

94Aa *Ambadra stigmatica*
S India, BM 755, holotype

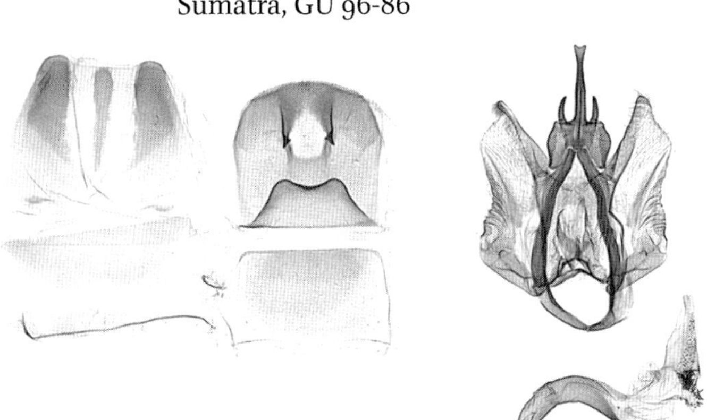

95a *Ambadra melanostriga*
Sumatra, GU 96-85

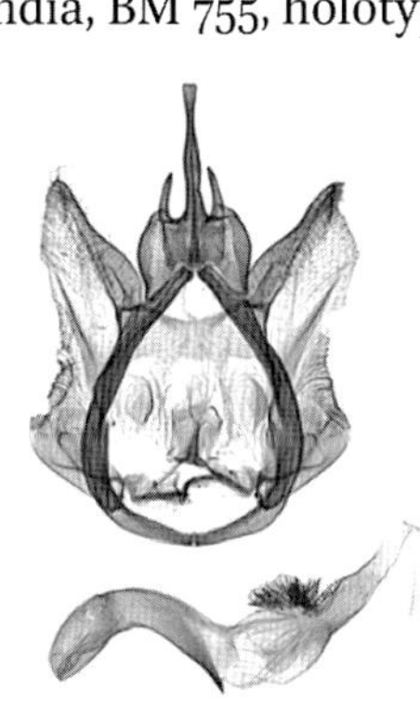

95b *Ambadra melanostriga*
Java, ZMHU 05-2017

1 mm

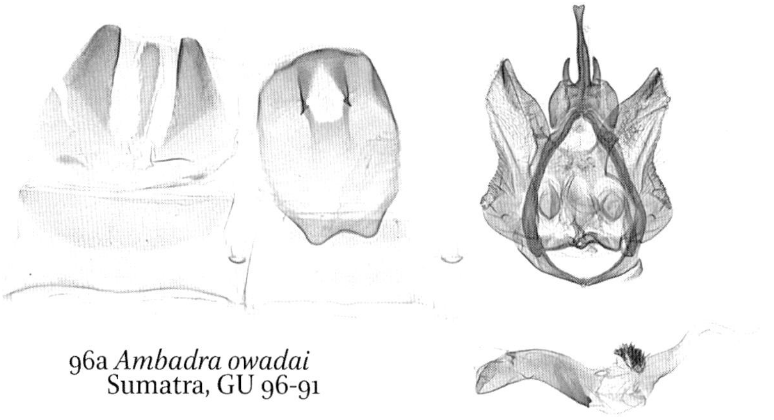

96a *Ambadra owadai*
Sumatra, GU 96-91

94b *Ambadra similis*
Sumatra, GU 96-86
phallus detail (enlarged)

95c *Ambadra melanostriga*
Sumatra, GU 96-55
phallus detail (enlarged)

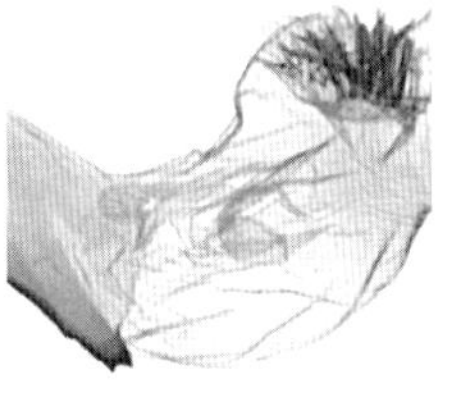

96b *Ambadra owadai*
Sumatra, GU 96-90
phallus detail (enlarged)

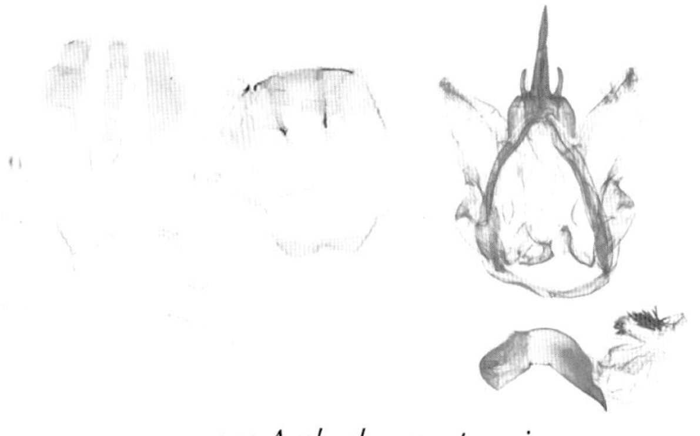

97a *Ambadra mentawai*
Mentawai, Pagai Isl., GU 41-50,
paratype

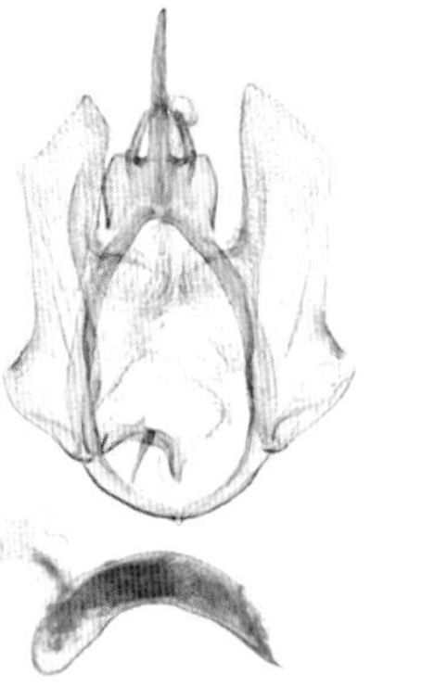

97Aa *Ambadra suriga*
Mindanao, USNM 895,
holotype

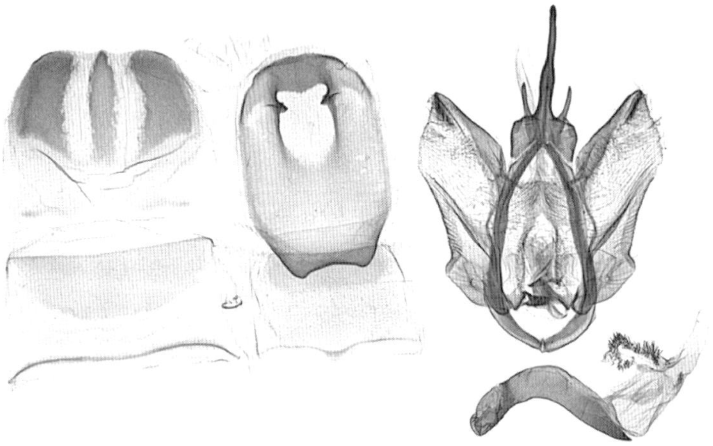

98a *Ambadra montana*
Borneo, GU 96-69

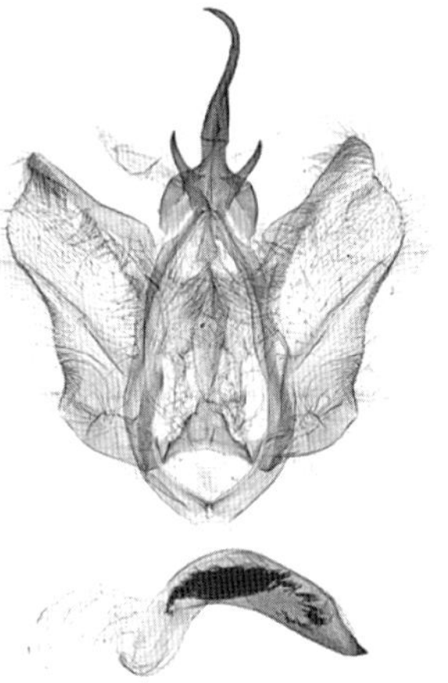

98b *Ambadra montana*
Borneo, GU 77-40

1 mm

99a *Ambadrana punctata*
Borneo, MWM 4.811

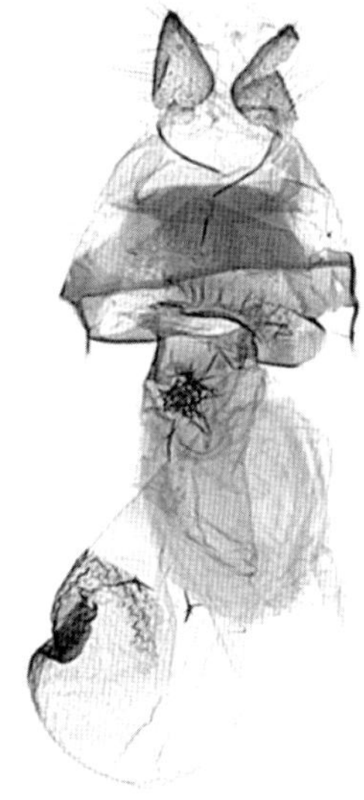

99b *Ambadrana punctata*
Kalimantan, GU 96-92

99c *Ambadrana punctata*
Sumatra, MWM 4.876

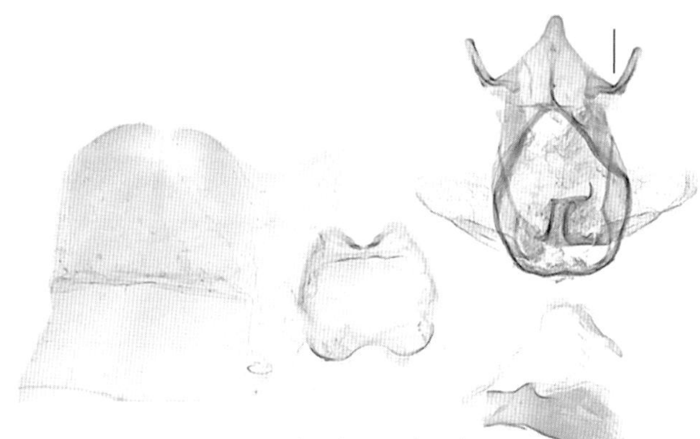

100a *Ambadrana barlowi*
Sumatra, GU 93-45

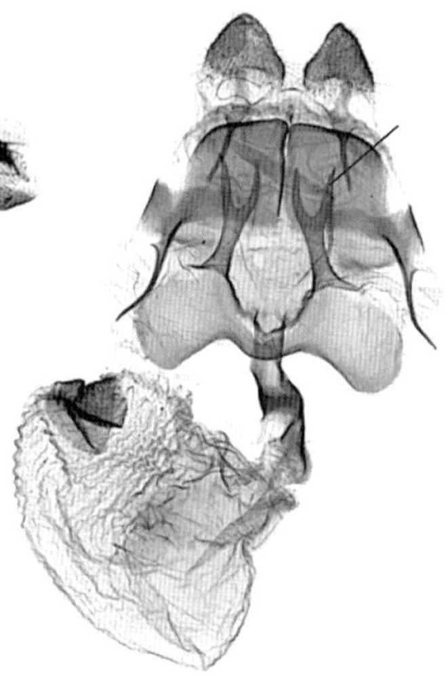

100b *Ambadrana barlowi*
S Thailand, GU 96-100

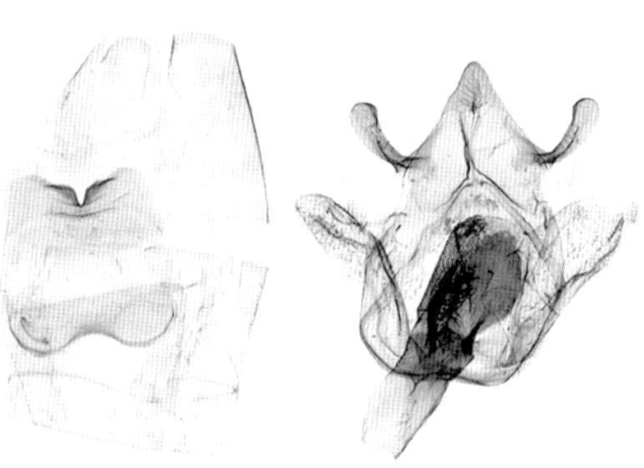

100c *Ambadrana barlowi*
Sumatra, GU 22-31, paratype

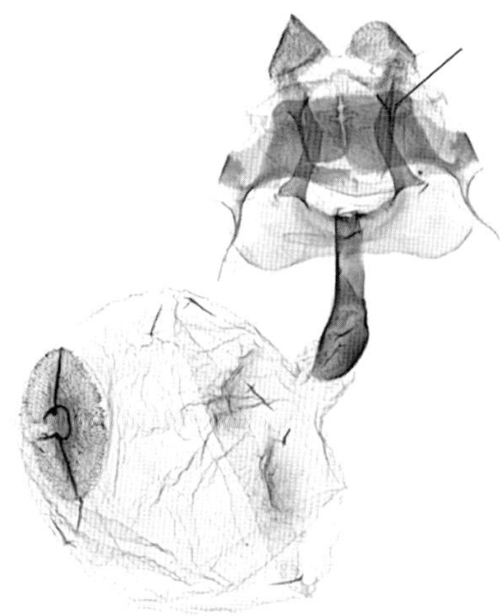

101a *Ambadrana taruma*
Sumatra, GU 96-93, paratype

1 mm

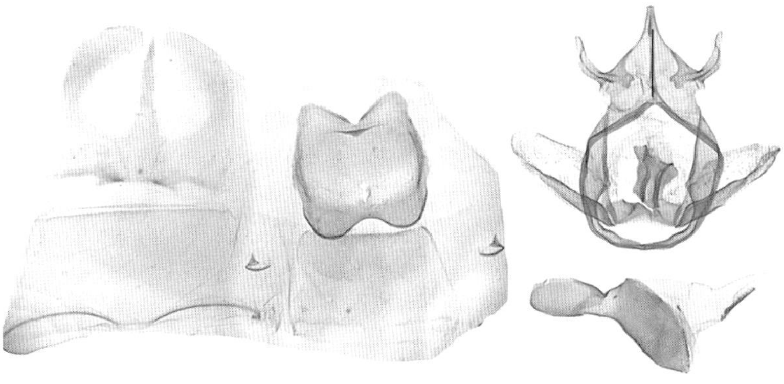

101b *Ambadrana taruma*
Java, GU 93-47, paratype

101c *Ambadrana taruma*
Java, GU 53-20, paratype

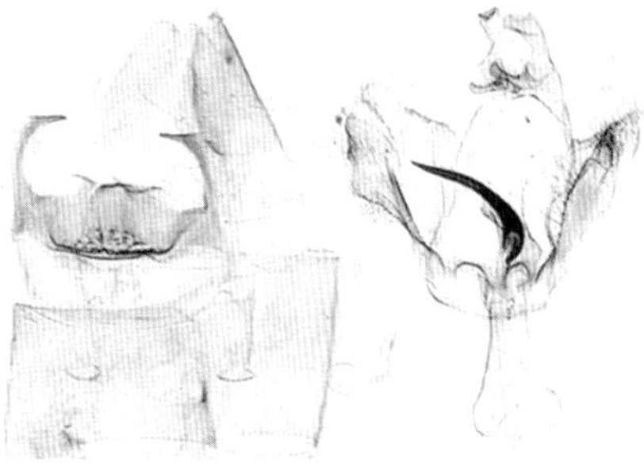

103a *Ambadrana furcata*
Sumatra, ZSM 940, holotype

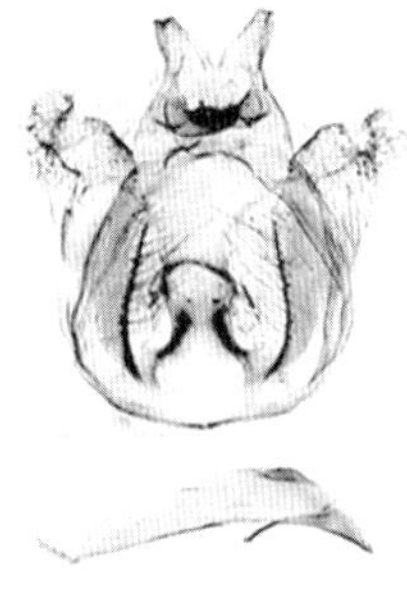

103b *Ambadrana furcata*
Sumatra, GU 19-78

102a *Ambadrana spinifera*
Kalimantan, GU 97-05

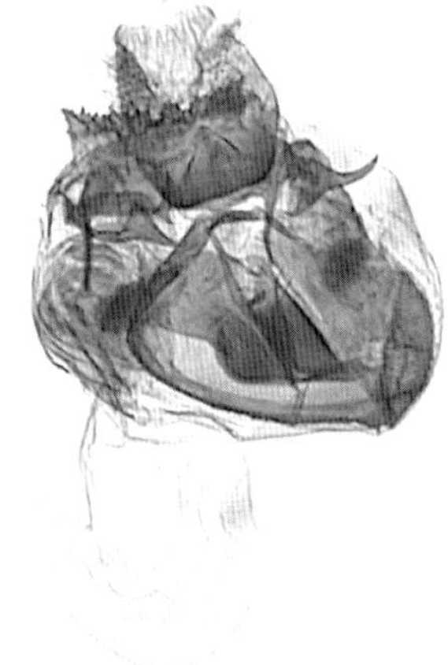

104a *Pydnella rosacea*
Thailand, MWM 8.608

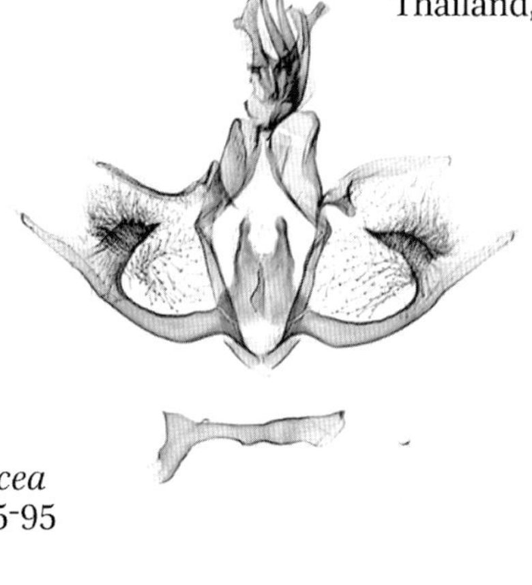

1 mm

104b *Pydnella rosacea*
Sumatra, GU 25-95

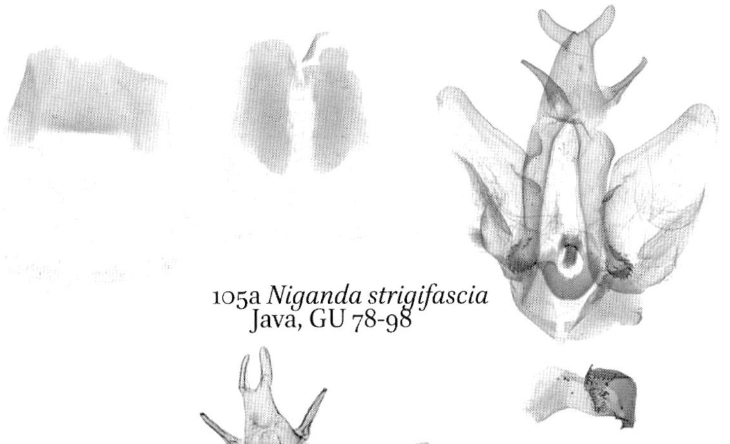

105a *Niganda strigifascia*
Java, GU 78-98

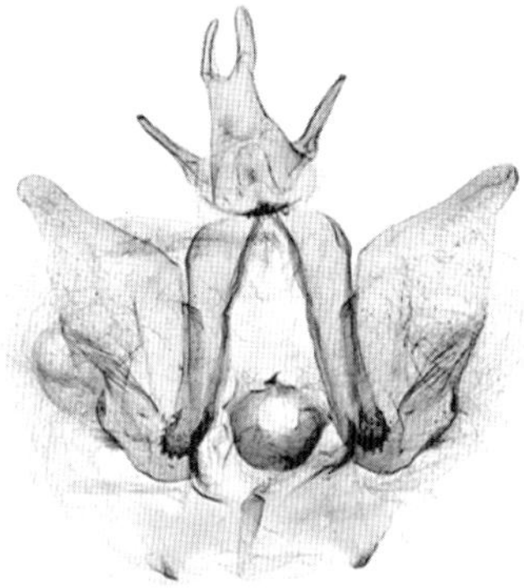

105b *Niganda strigifascia*
Java, GU 52-09

105c *Niganda strigifascia*
Thailand, MWM 8.607

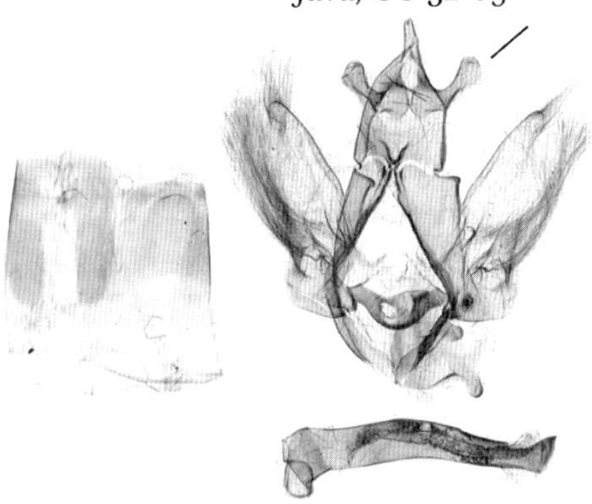

106Aa *Niganda cyttarosticta*
India, Sikim, GU 24-67

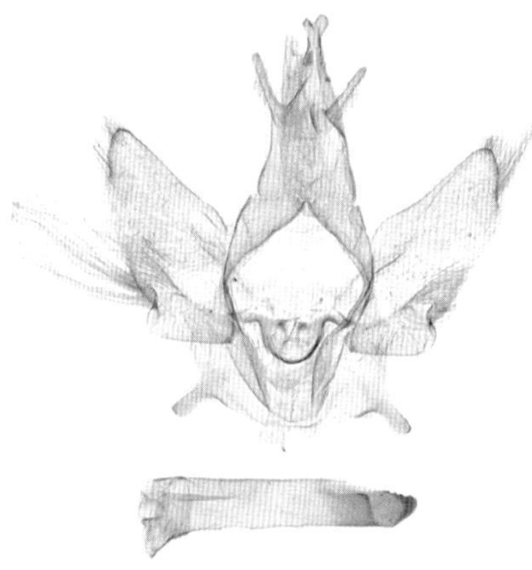

106a *Niganda pallens*
Malaya, GU 16-12, holotype

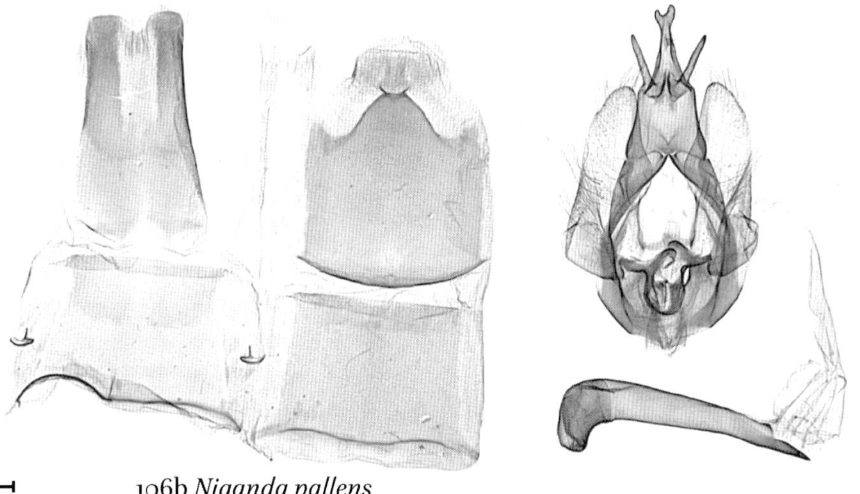

106b *Niganda pallens*
Malaya, ZMHU 08-2017, paratype

1 mm

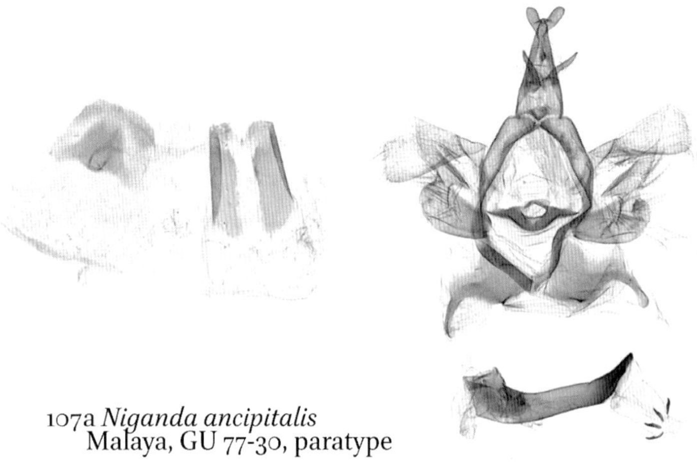

107a *Niganda ancipitalis*
Malaya, GU 77-30, paratype

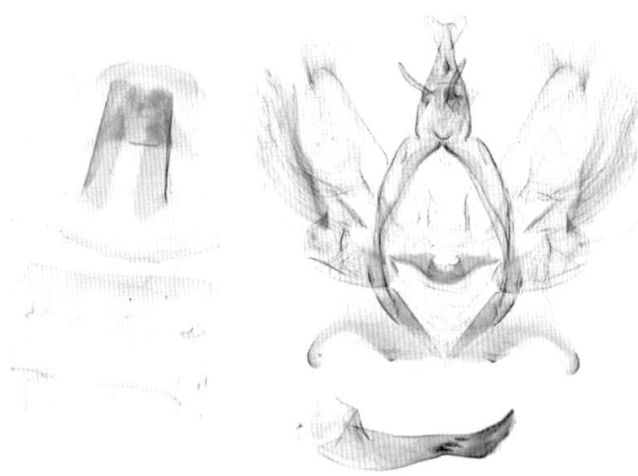

107b *Niganda ancipitalis*
Malaya, GU 15-94, paratype

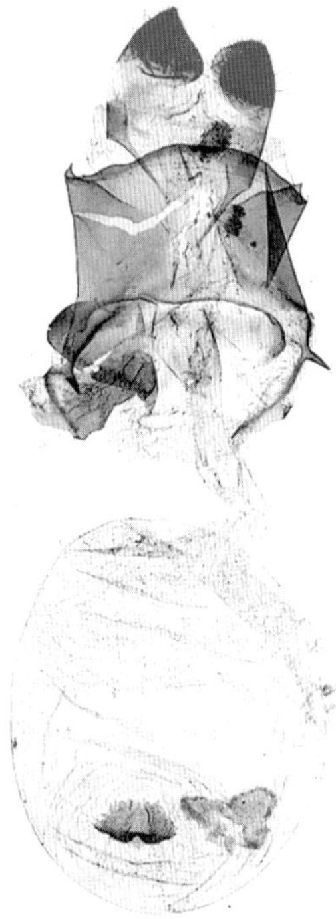

108a *Postniganda pudens*
Thailand, GU77-35, paratype

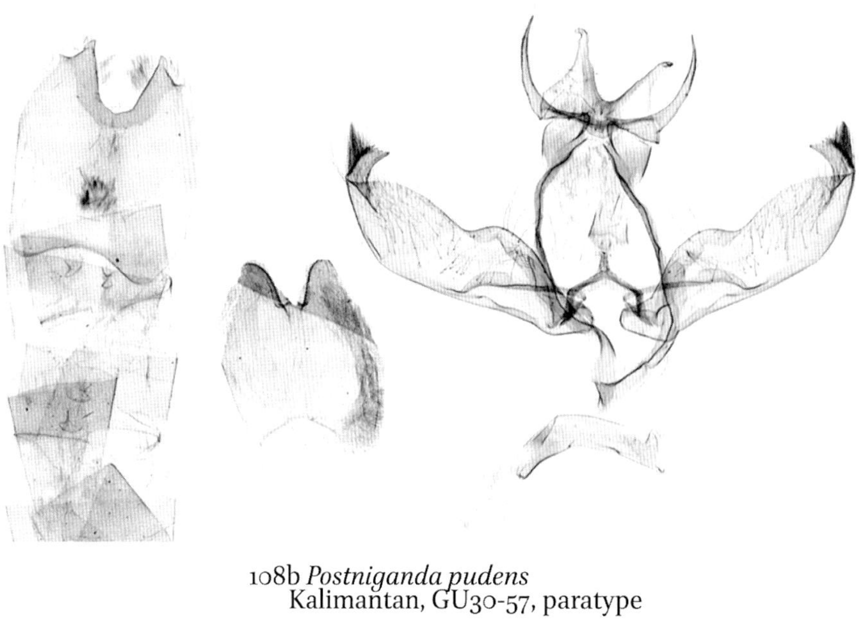

1 mm

108b *Postniganda pudens*
Kalimantan, GU30-57, paratype

109a *Norraca longipennis*
Java, GU 95-46

109b *Norraca longipennis*
Java, GU 15-44, phallus

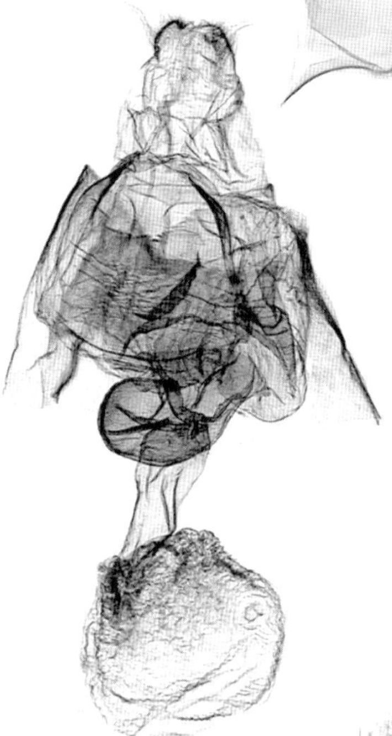

109c *Norraca longipennis*
Penang, BM 878,
holotype

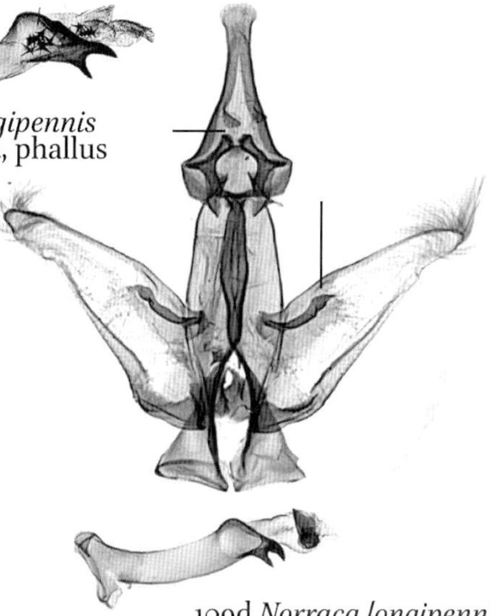

109d *Norraca longipennis*
Timor, GU 93-02

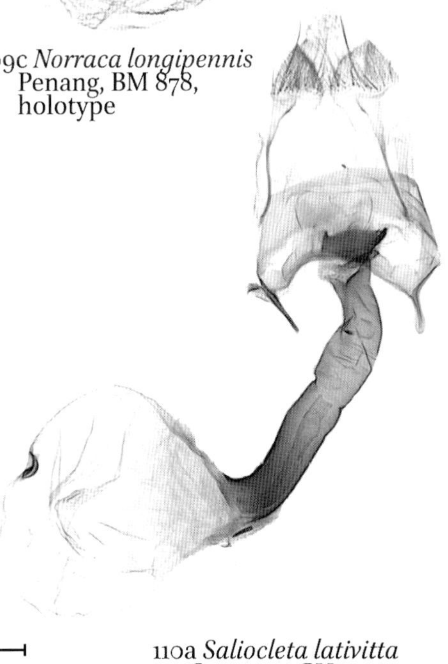

1 mm

110a *Saliocleta lativitta*
Sumatra, GU 95-17

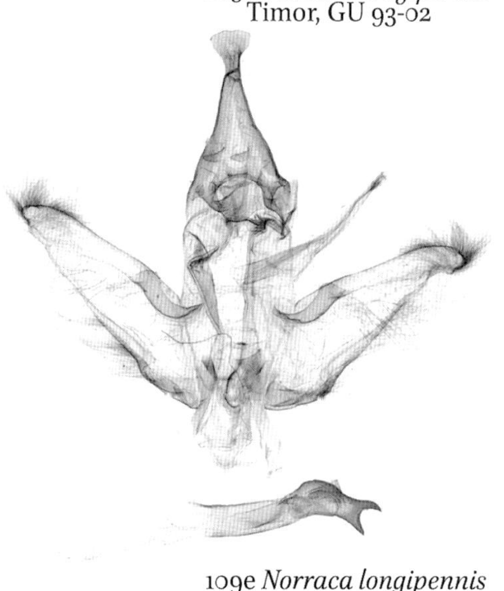

109e *Norraca longipennis*
Sumatra, GU 88-55

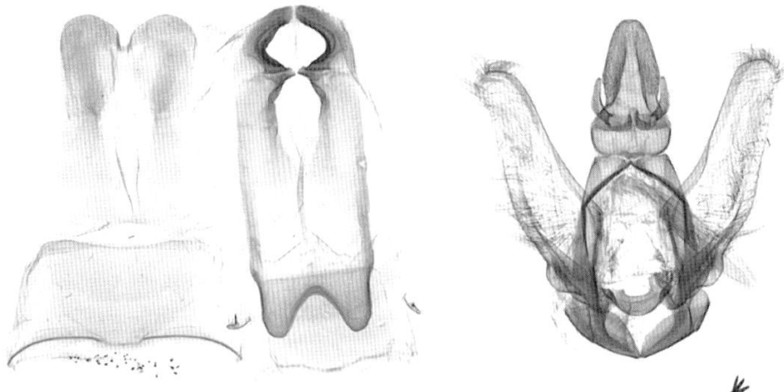

110b *Saliocleta lativitta*
Sumatra, GU 95-18

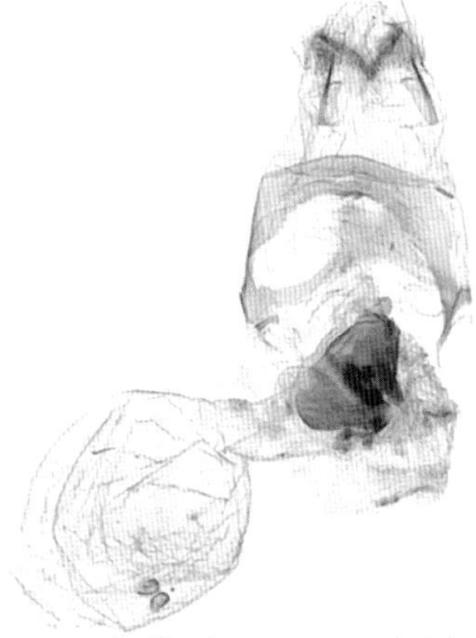

111a *Saliocleta nonagrioides*
Sumatra, GU 88-61

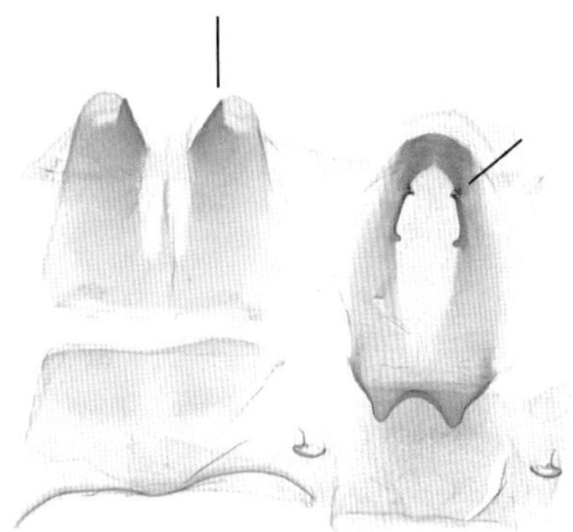

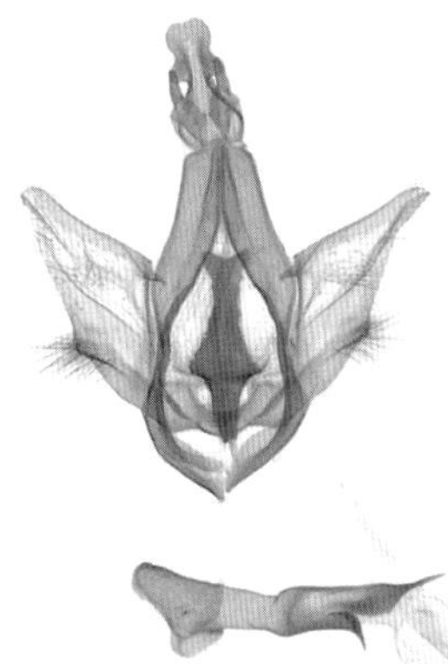

111b *Saliocleta nonagrioides*
Sumatra, GU 95-27

111c *Saliocleta nonagrioides*
Sumatra, BM 1050
(holotype *sabulosa*)

111d *Saliocleta nonagrioides*
Borneo, GU 88-60

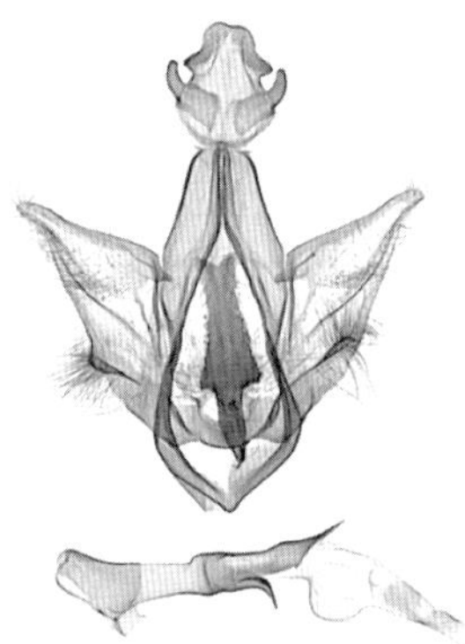

111e *Saliocleta nonagrioides*
Sumatra, GU 93-50

111f *Saliocleta nonagrioides*
Kalimantan, GU 64-73

1 mm

112a *Saliocleta sordida*
Java, GU76-91

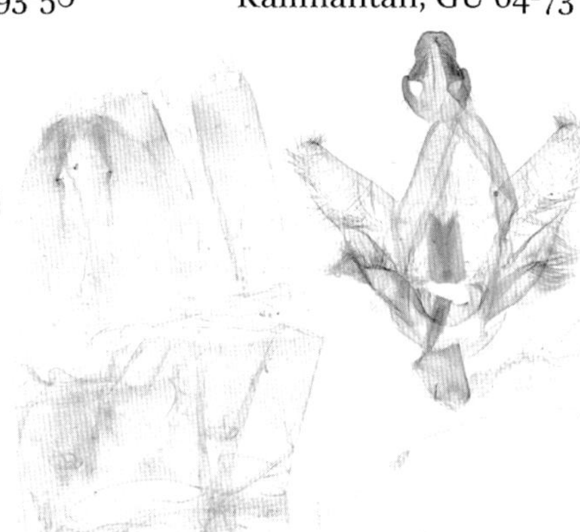

112b *Saliocleta sordida*
Java, GU 88-56

113a *Saliocleta odrana odrana*
Mindoro, MWM 15.695, phallus

113b *S. odrana odrana*
Leyte, MWM 15.742, phallus

113c *S. odrana odrana*
Negros, MWM 15.507, phallus

113d *S. odrana odrana*
Mindanao, GU 76-92, phallus

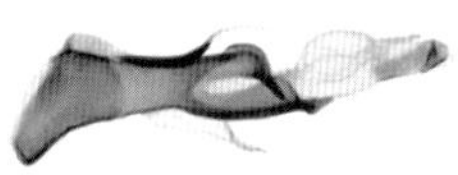

113e *S. odrana celebica*
Sulawesi, MWM 15.504, phallus

113f *Saliocleta odrana celebica*
Sulawesi, MWM 29.087

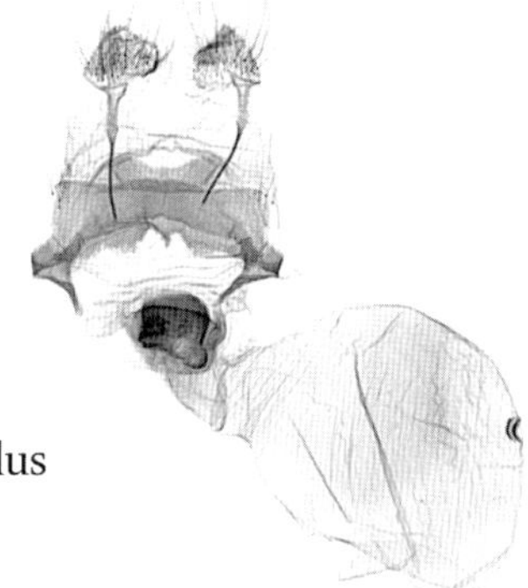

113g *S. odrana celebica*
Sulawesi,
MWM 29.088

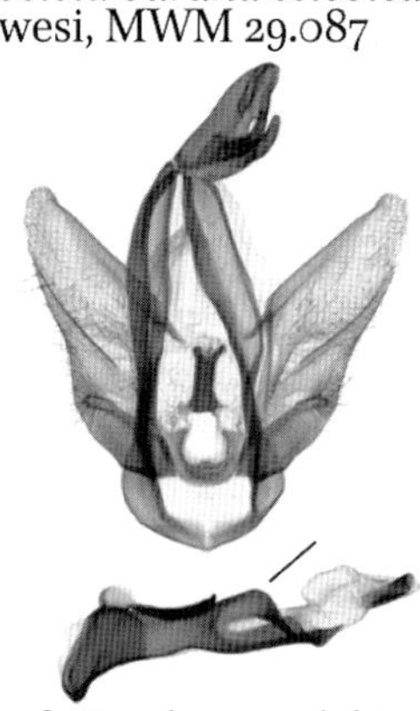

113h *S. odrana celebica*
Sulawesi,
MWM 1506

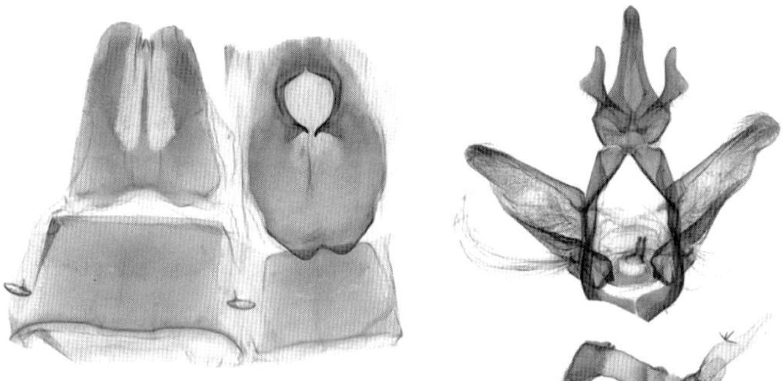

114a *Saliocleta ediehli*
Sumatra-Siberut, GU 95-32

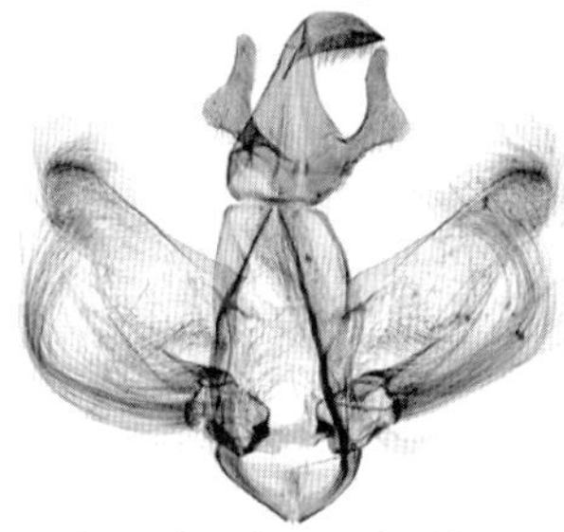

114b *Saliocleta ediehli*
Sumatra, GU 19-52, holotype

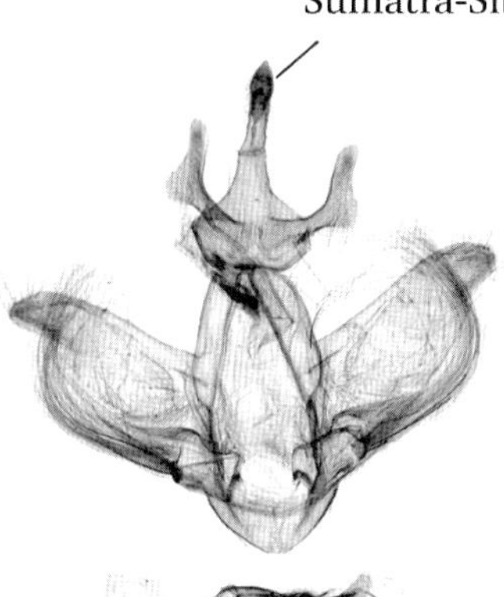

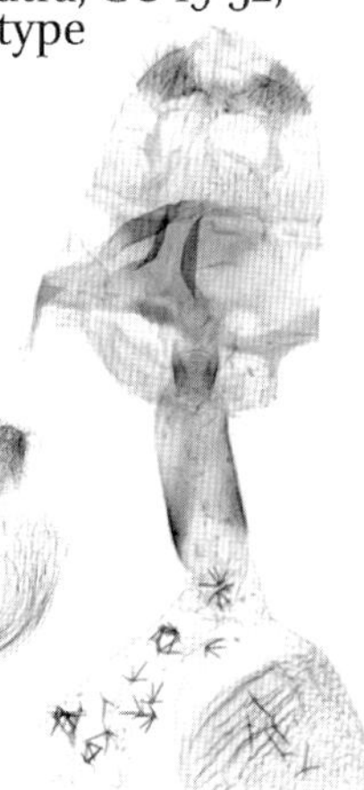

1 mm

115a *Saliocleta commutatis*
Palawan, GU 17-80, PT

115b *Saliocleta commutatis*
Mindanao, GU 17-86, PT

115c *Saliocleta commutatis*
Palawan, MV 17598, PT

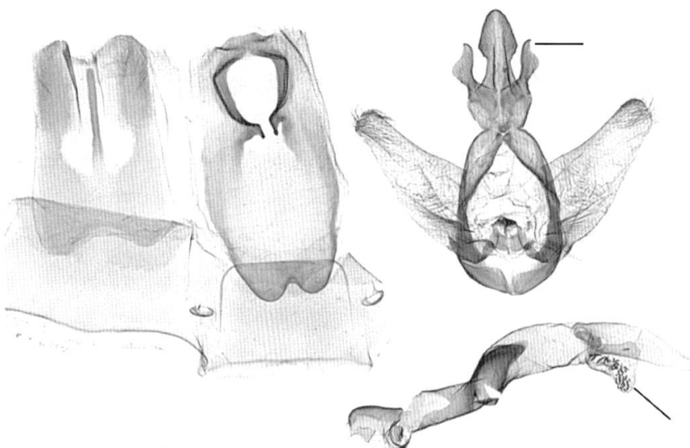

116a *Saliocleta widagdoi*
Sumatra, GU 64-92, paratype

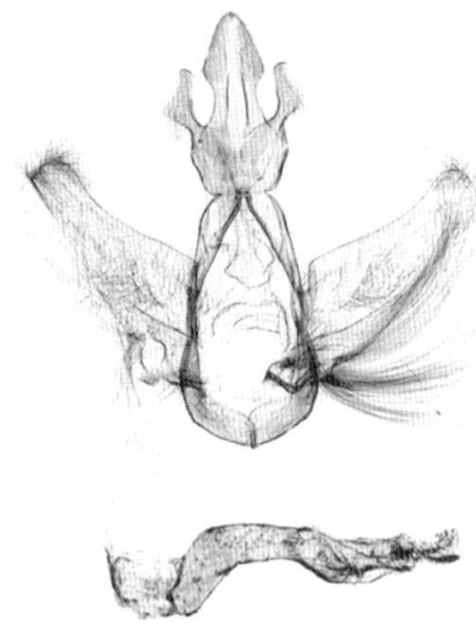

116b *Saliocleta widagdoi*
Sumatra, GU 12-85, paratype

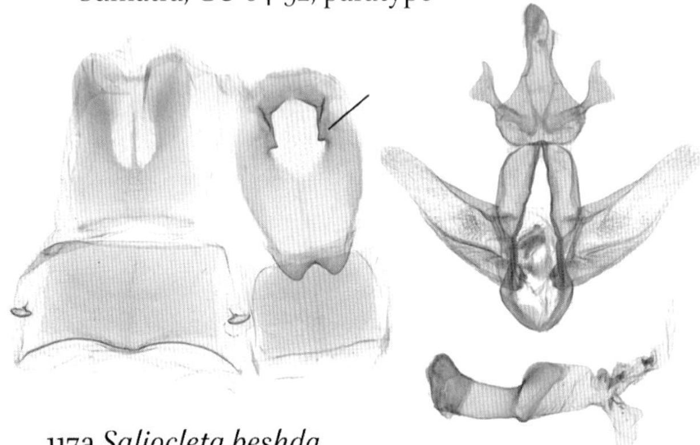

117a *Saliocleta beshda*
Java, GU 95-55, paratype

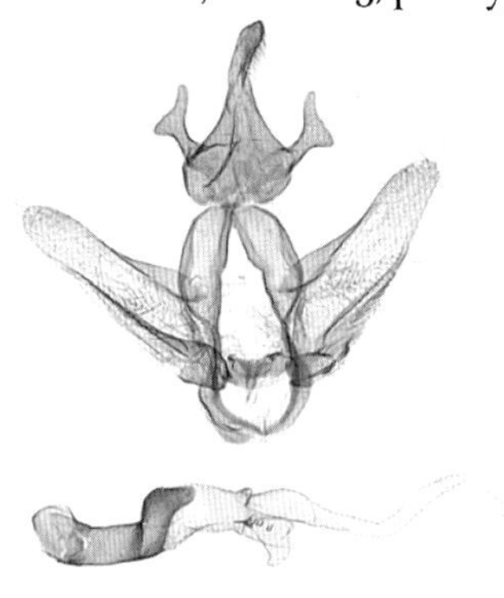

117b *Saliocleta beshda*
Java, GU 64-80, paratype

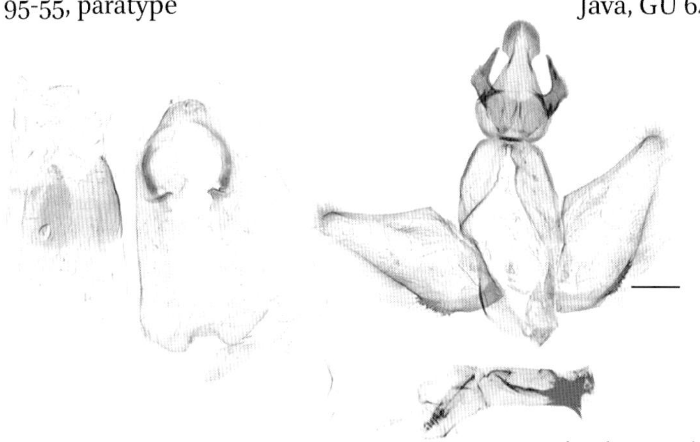

118a *Saliocleta malayana*
Malaya, GU 22-34, paratype

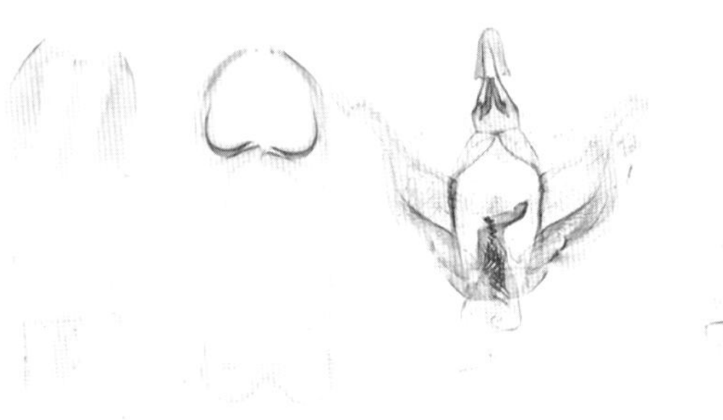

119a *Saliocleta armata*
Sumatra, ZSM 372, holotype

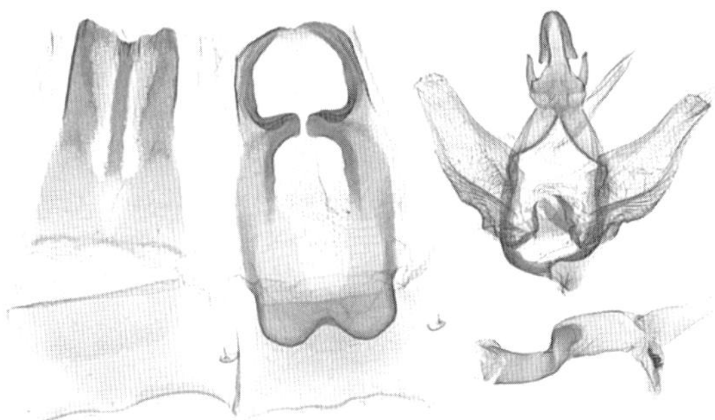

119b *Saliocleta armata*
Sumatra, GU 64-89

1 mm

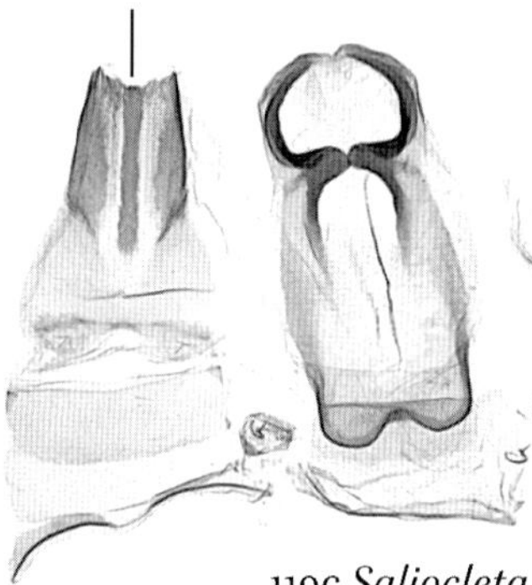

119c *Saliocleta armata*
Sumatra, LNK 04-2017

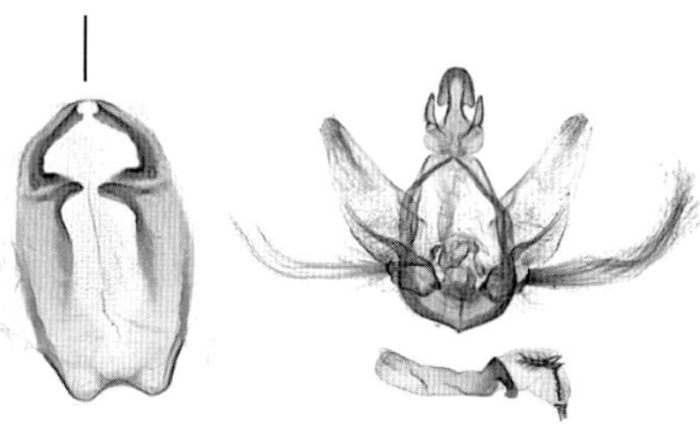

120a *Saliocleta nannion*
Kalimantan, GU 97-03

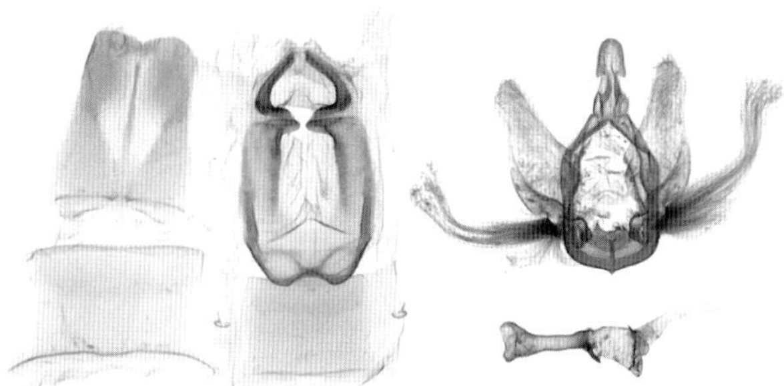

120b *Saliocleta nannion*
Sumatra, GU 95-37

120c *Saliocleta nannion*
Sumatra, ZSM 233
holotype

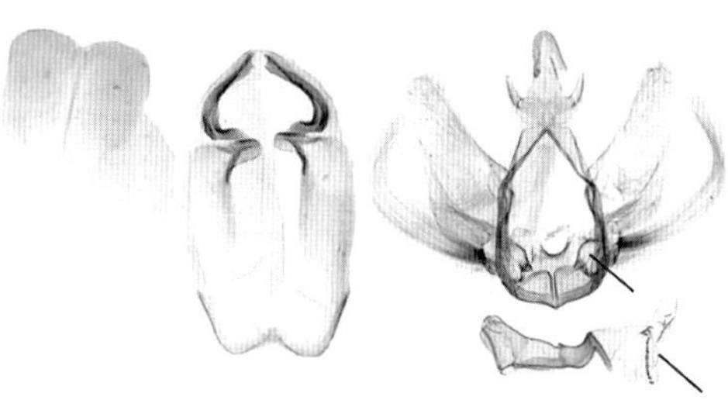

120d *Saliocleta nannion*
Sumatra, GU 93-43

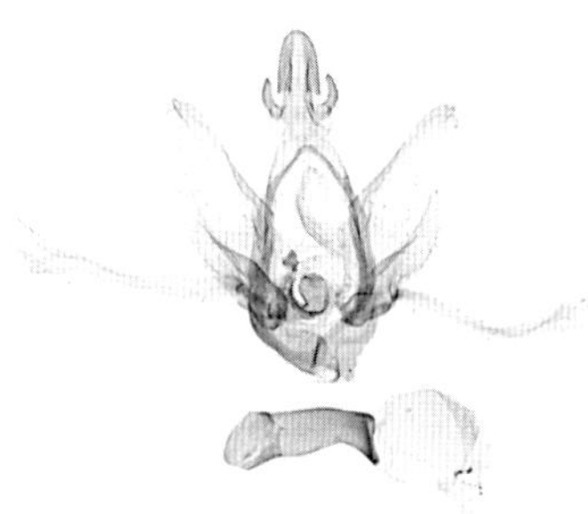

121a *Saliocleta straminea*
Lombok, GU 94-13

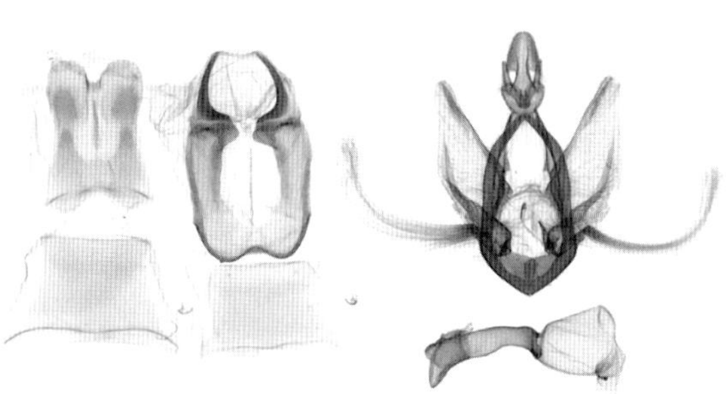

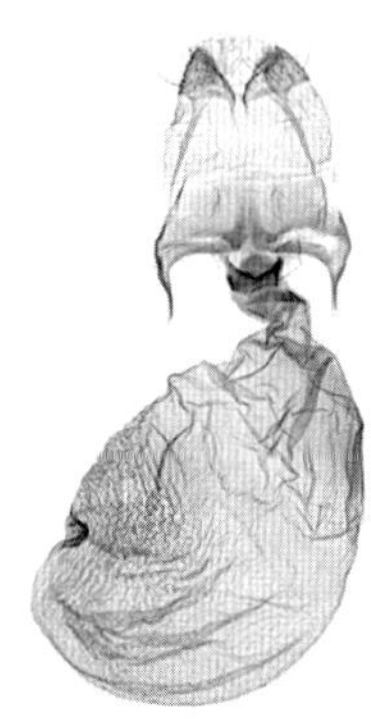

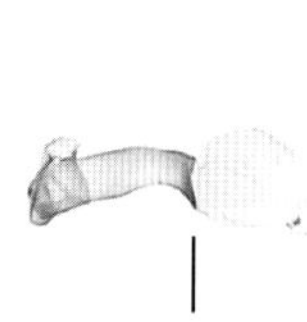

1 mm

121b *Saliocleta straminea*
Java, GU 95-34

121c *Saliocleta straminea*
Java, BM 10314009

121d *S. straminea*
Bali, GU 93-44
phallus

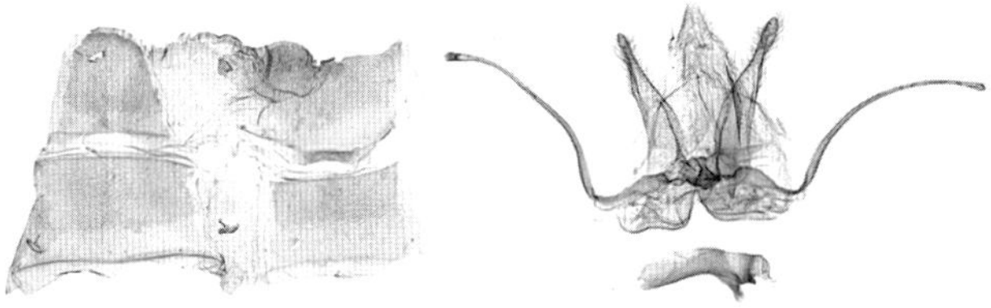

122a *Chalepa pendleburyi*
Malaya, BM 10314005
holotype

122b *Chalepa pendleburyi*
Malaya, BM 10314005
paratype

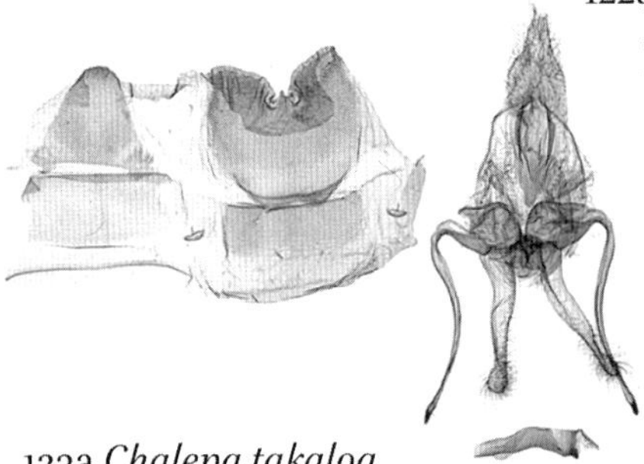

123a *Chalepa takaloa*
Sumatra, GU 64-81
holotype

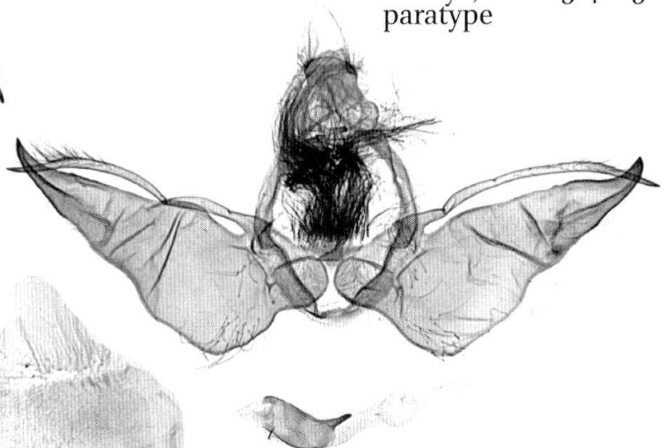

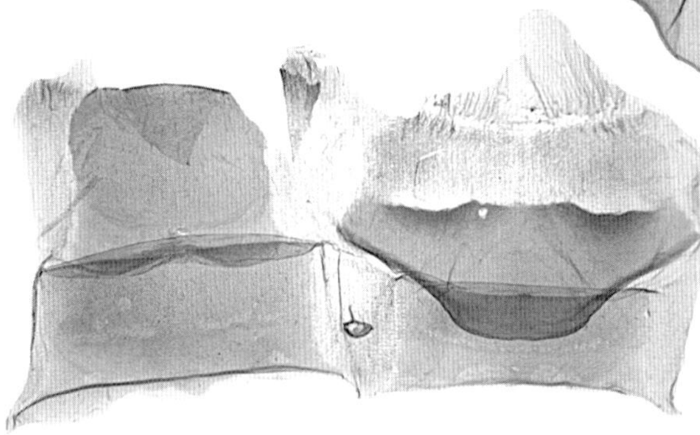

124a *Periergos postrubra*
Sumatra, GU 64-87

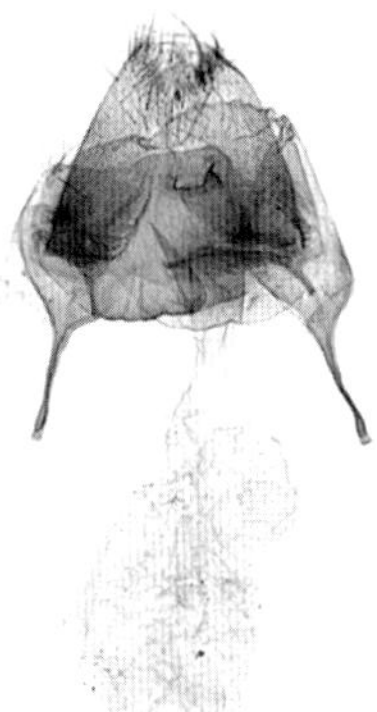

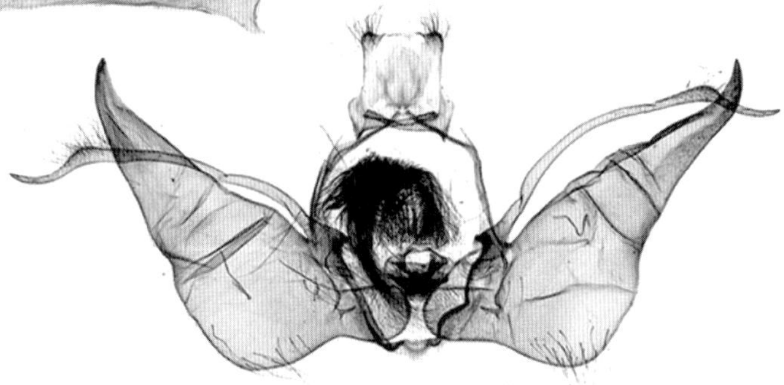

124c *Periergos postrubra*
Sumatra, GU 24-26

124b *Periergos postrubra*
Sumatra, GU 64-86

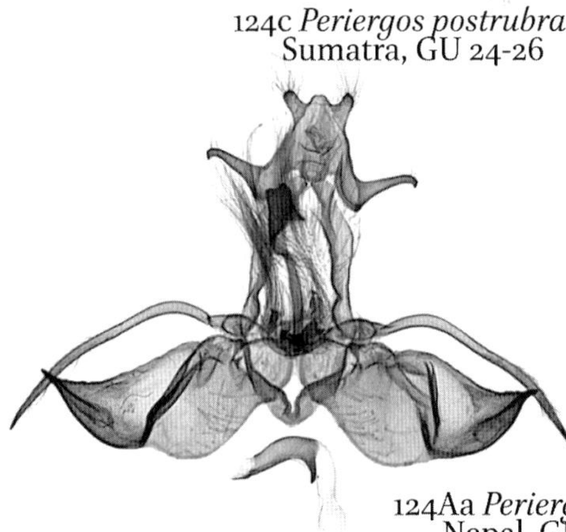

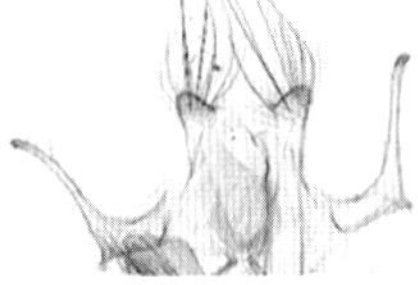

124d *Periergos postrubra*
Sumatra, ZSM 675,
uncus, socci

1 mm

124Aa *Periergos testacea*
Nepal, GU 64-84

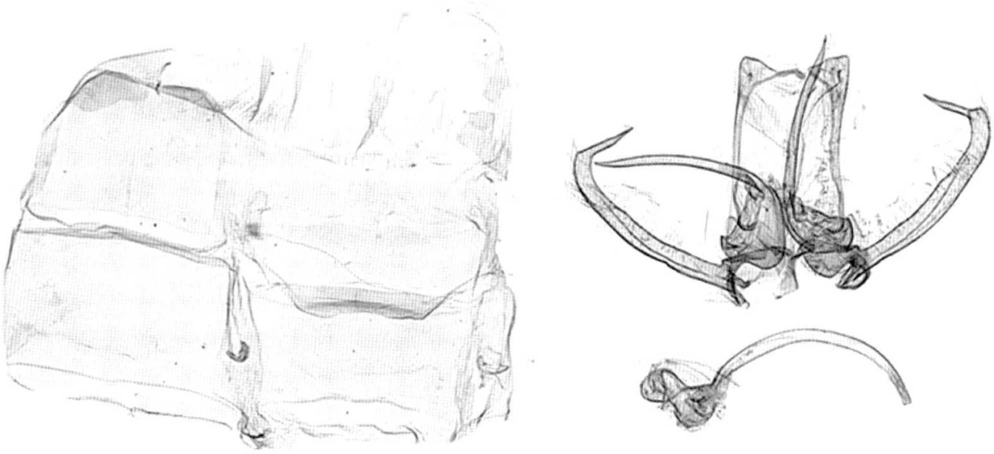

125a *Periergos hunyada*
Java, BM 800, holotype

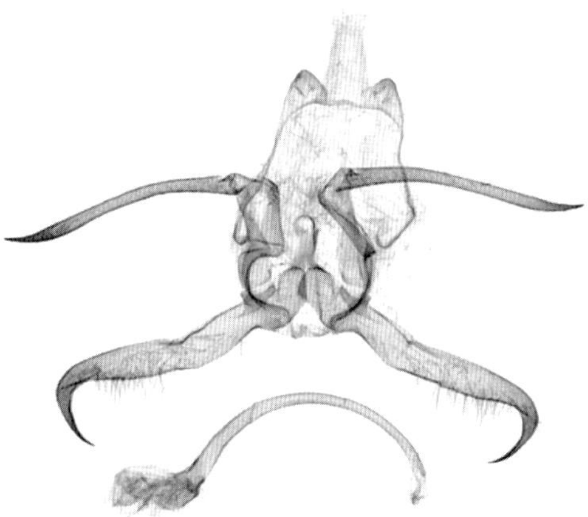

125b *Periergos hunyada*
Sumatra, Mentawai Isl., GU 88-65

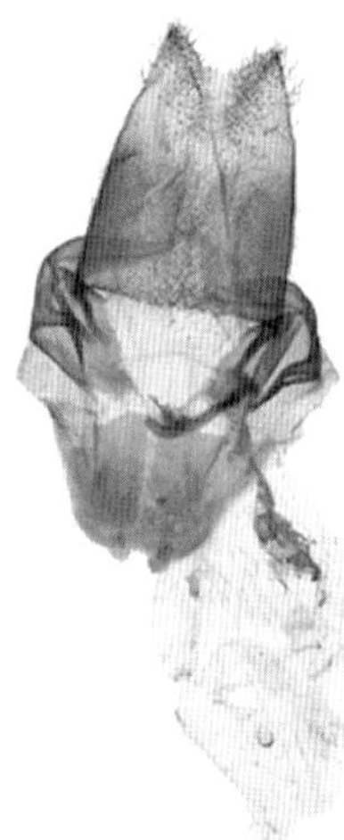

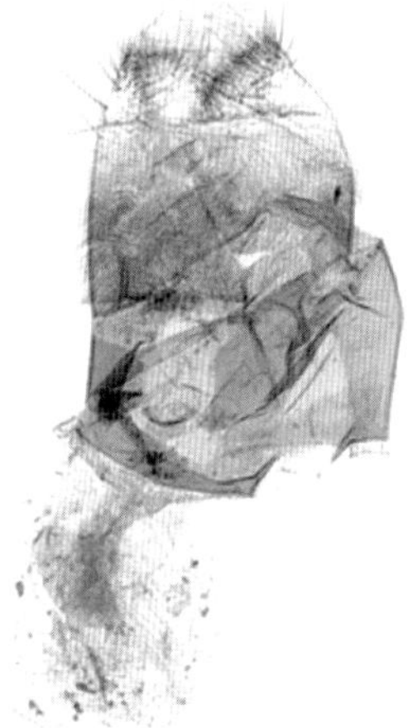

125c *Periergos hunyada*
Negros, MWM 15.748

126a *Periergos spinosa*
Palawan, GU 88-64

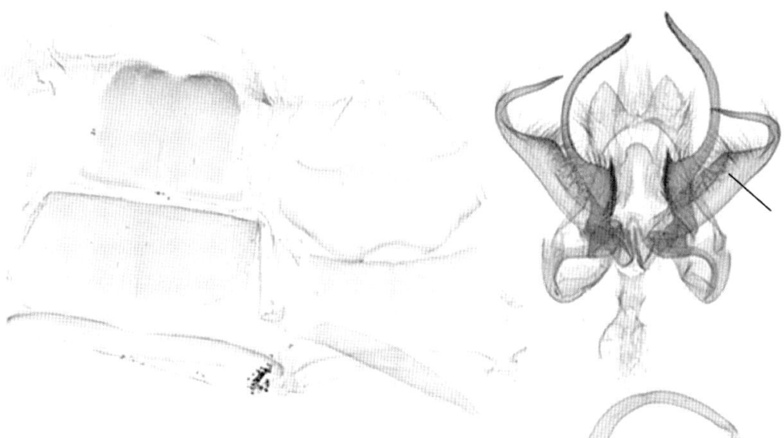

1 mm

126b *Periergos spinosa*
Borneo, MWM 8.588

127a *Neocerura liturata*
Java, MWM4.988

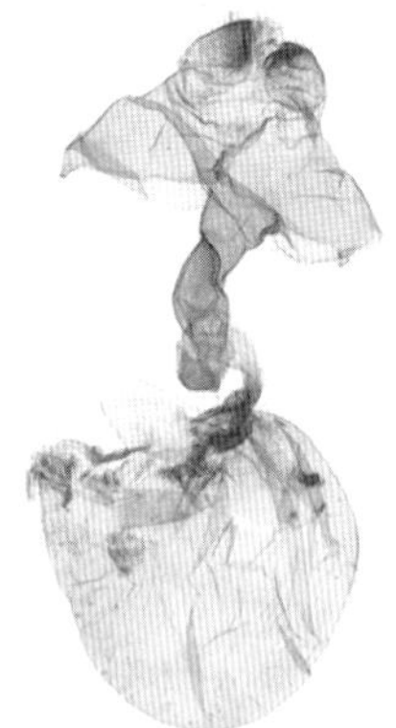

127b *Neocerura liturata*
Java, GU 51-75

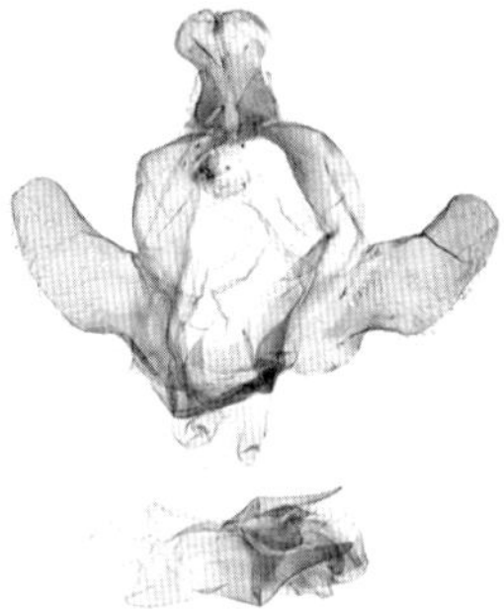

127c *Neocerura liturata*
Sumbawa, GU 51-74

128a *Neocerura longinquus*
Tanimbar, GU 92-67, holotype

129a *Kamalia malaysiana malaysiana*
Sumatra, GU 17-68

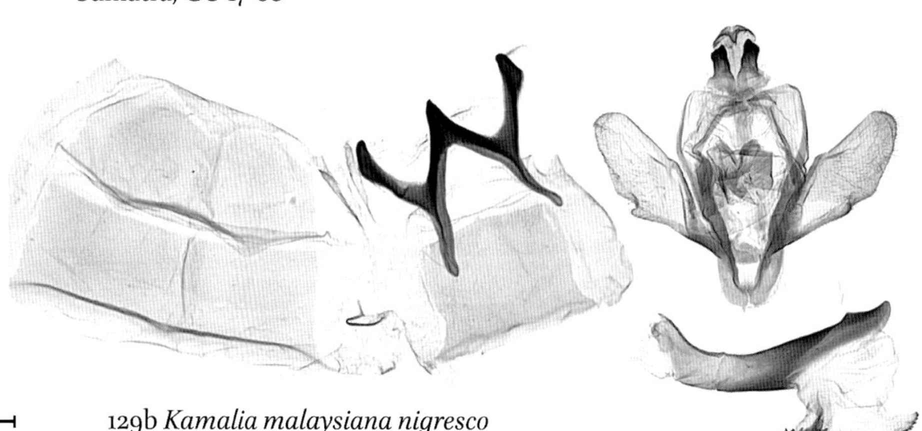

1 mm

129b *Kamalia malaysiana nigresco*
Siberut, GU 64-96, paratype

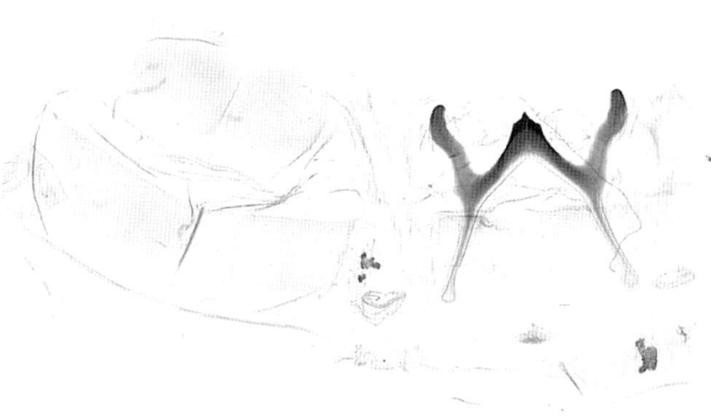

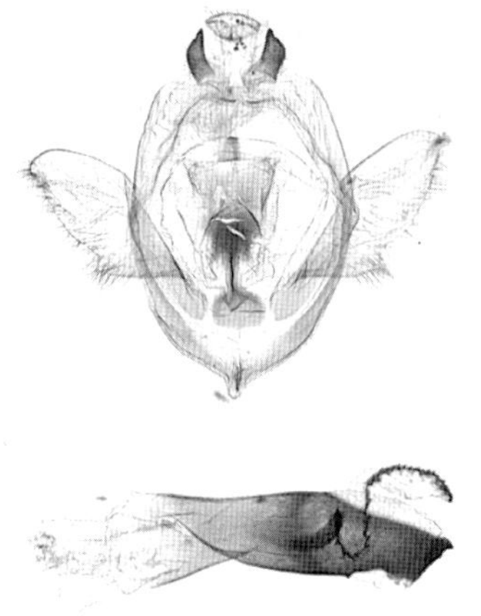

129c *Kamalia malaysiana jakli*
Sumbawa, GU 45-15, paratype

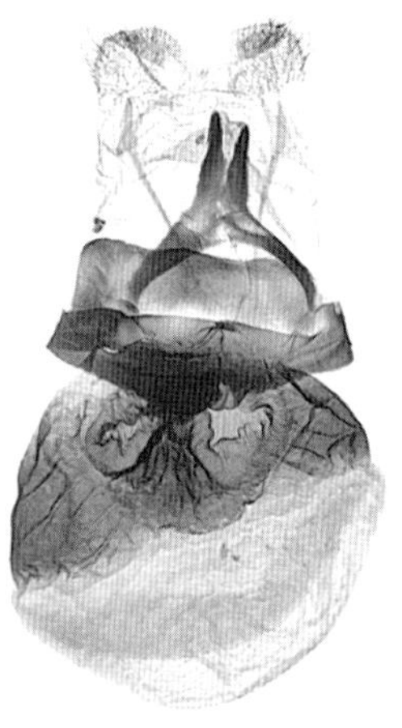

129d *Kamalia malaysiana malaysiana*
Sumatra, GU 64-98

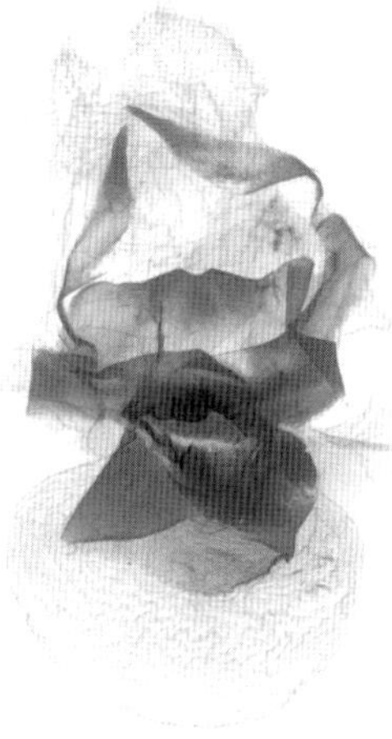

129e *Kamalia malaysiana palawana*
Palawan, MV 17.870, PT

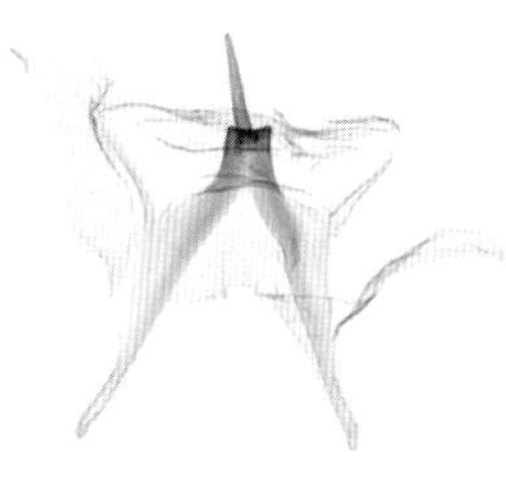

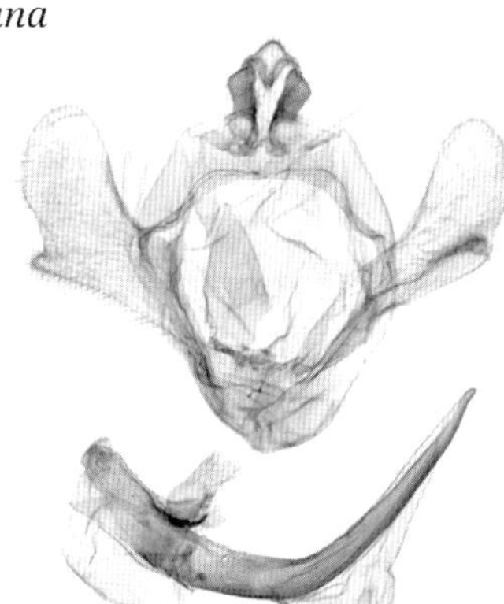

130a *Kamalia minahassae*
Sulawesi, BM 826, holotype

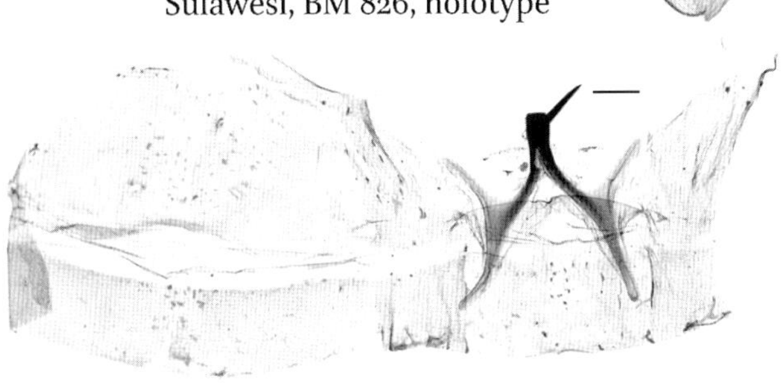

1 mm

130b *Kamalia minahassae*
Sula, GU 75-24

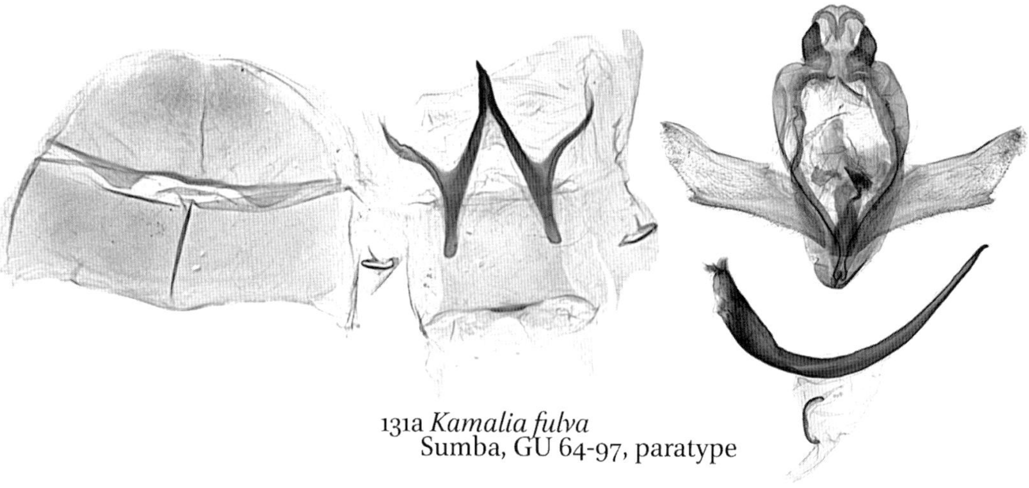

131a *Kamalia fulva*
Sumba, GU 64-97, paratype

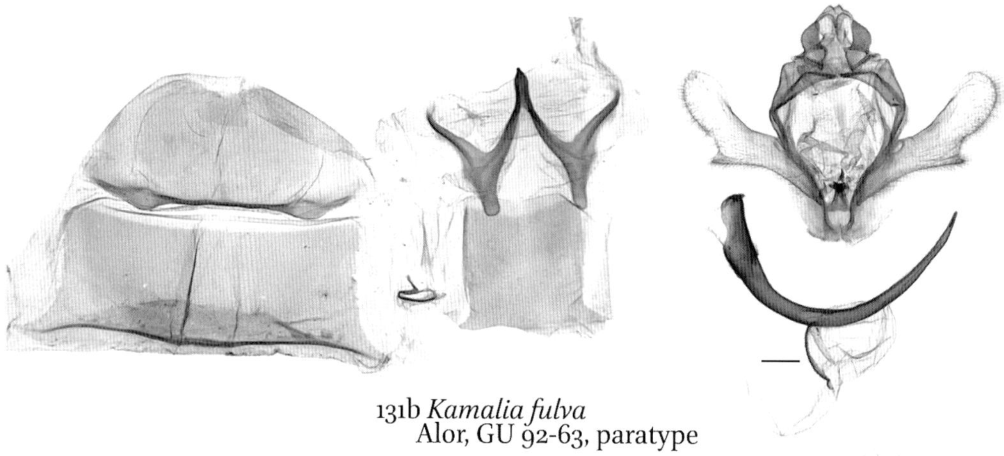

131b *Kamalia fulva*
Alor, GU 92-63, paratype

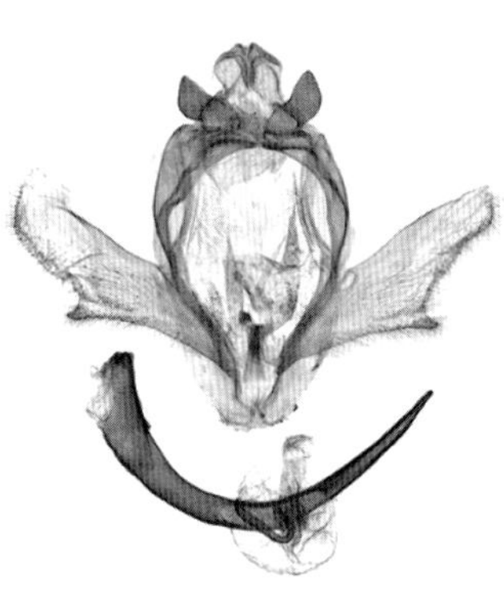

131c *Kamalia fulva*
Alor, GU 92-64, paratype

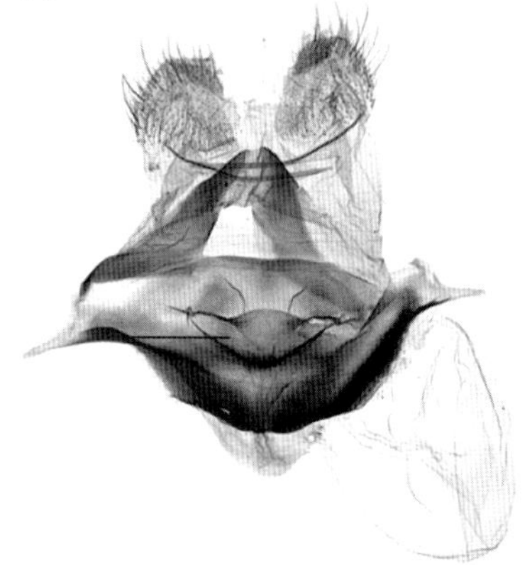

132a *Kamalia timorensis*
Timor, GU 92-66

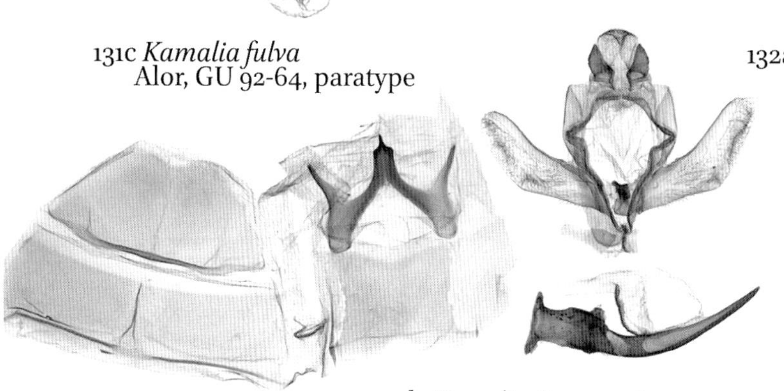

1 mm

132b *Kamalia timorensis*
Timor, GU 92-65

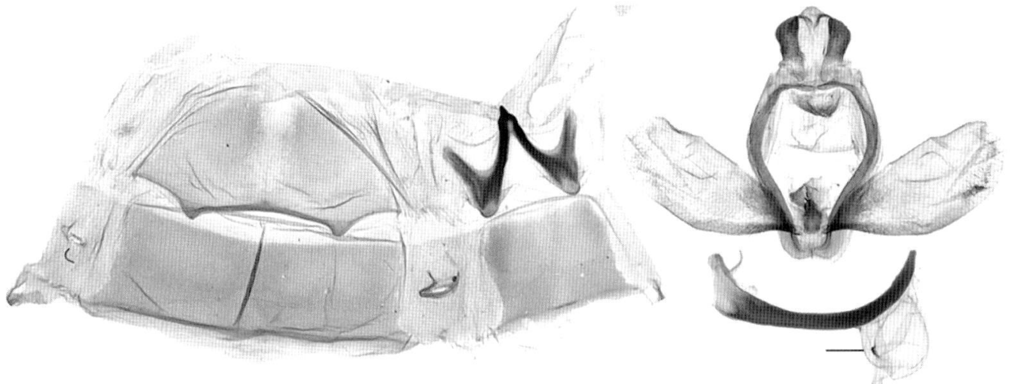

133a *Kamalia multipunctata*
W Papua, GU 92-61

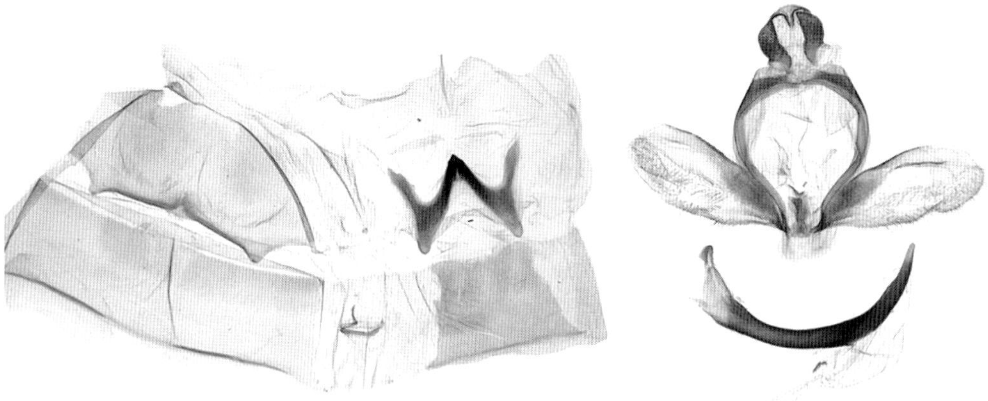

133b *Kamalia multipunctata*
Tanimbar, GU 92-57

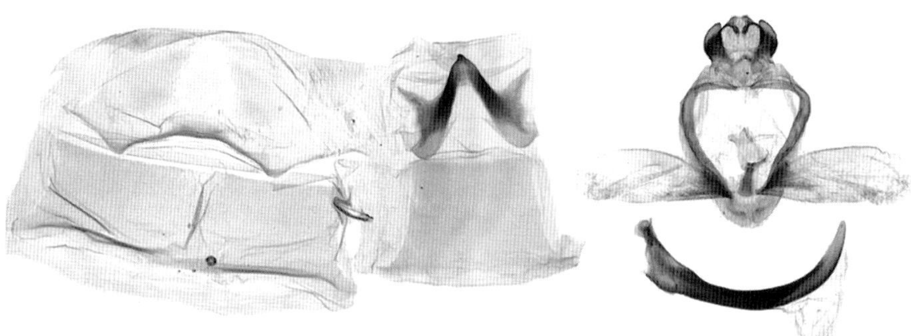

133c *Kamalia multipunctata*
Seram, GU 92-61

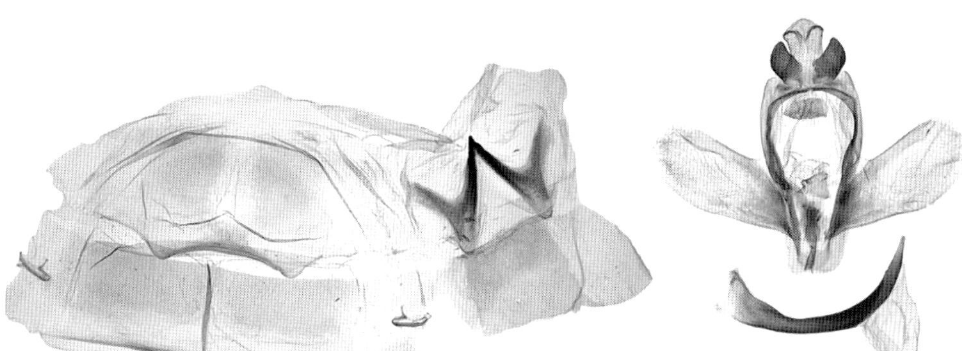

1 mm

133d *Kamalia multipunctata*
Obi, GU 92-58

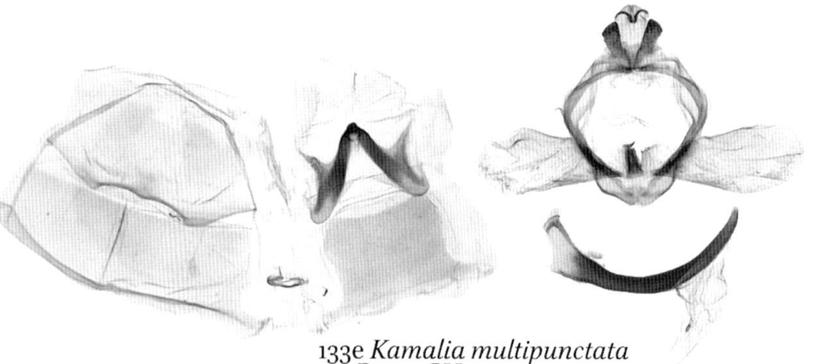

133e *Kamalia multipunctata*
Buru, GU 92-54

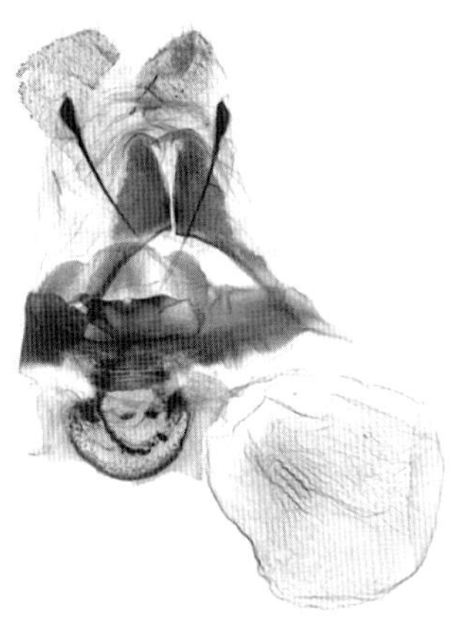

133f *Kamalia multipunctata*
Tanimbar, GU 92-56

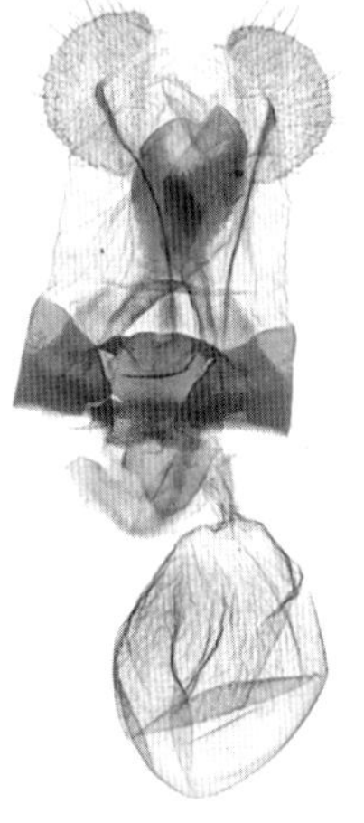

133g *Kamalia multipunctata*
New Britain, MWM 24.301

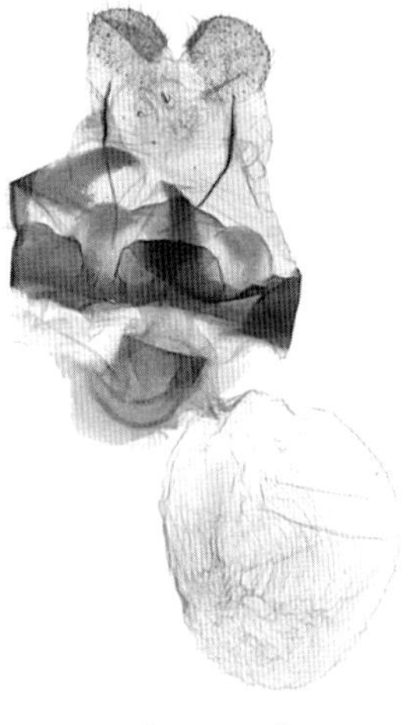

133h *K. multipunctata*
Australia, Qsd,
MWM 24.302

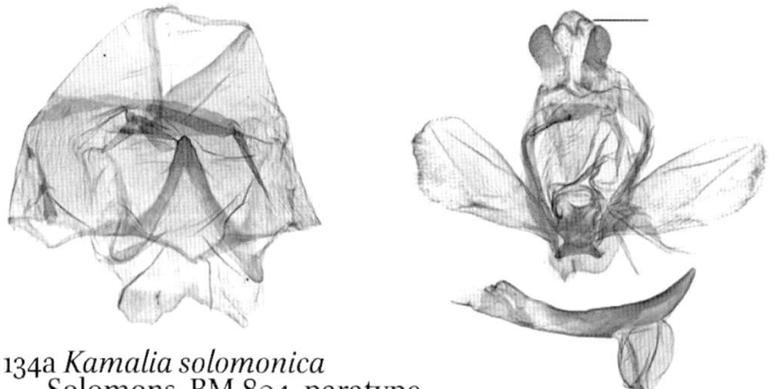

134a *Kamalia solomonica*
Solomons, BM 804, paratype

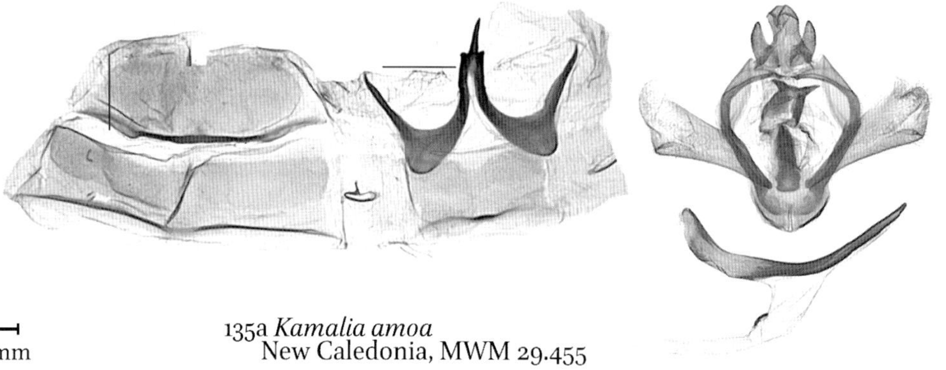

1 mm

135a *Kamalia amoa*
New Caledonia, MWM 29.455

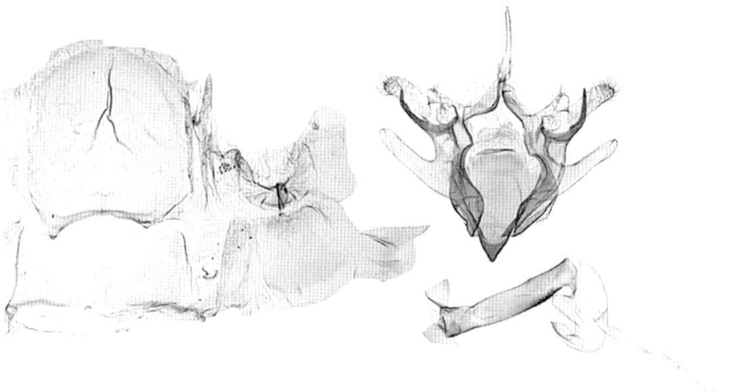

136a *Liparopsis postalbida postalbida*
NE India, Meghalaya, GU 65-35

136b *Liparopsis postalbida postalbida*
Taiwan, MWM 8.200

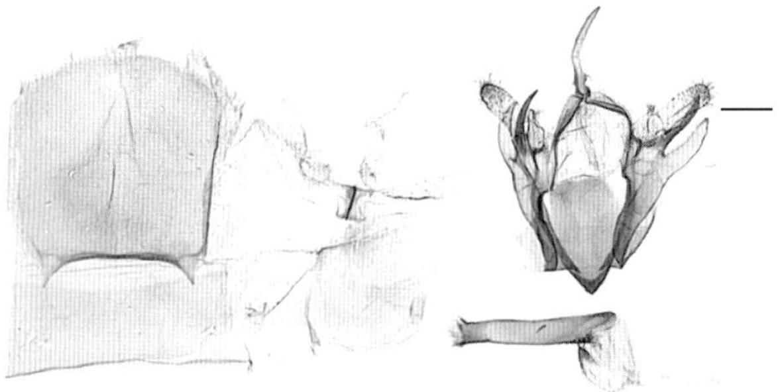

136c *Liparopsis postalbida brevispinus*
Mentawai, GU 65-32, paratype

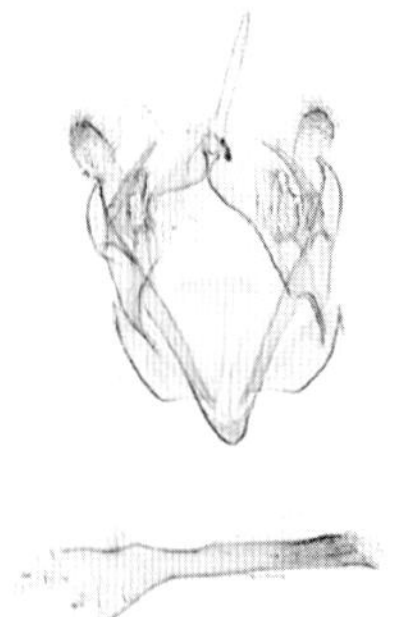

136d *Liparopsis postalbida brevispinus*, Sumatra
GU 15-44, paratype

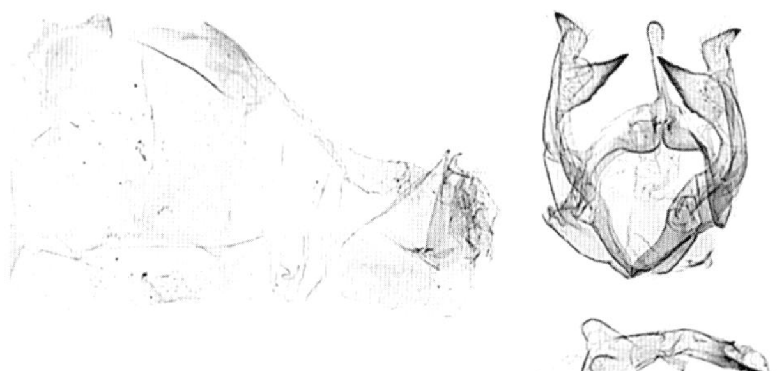

137a *Liparopsis barlowi*
Kalimantan, GU 79-84

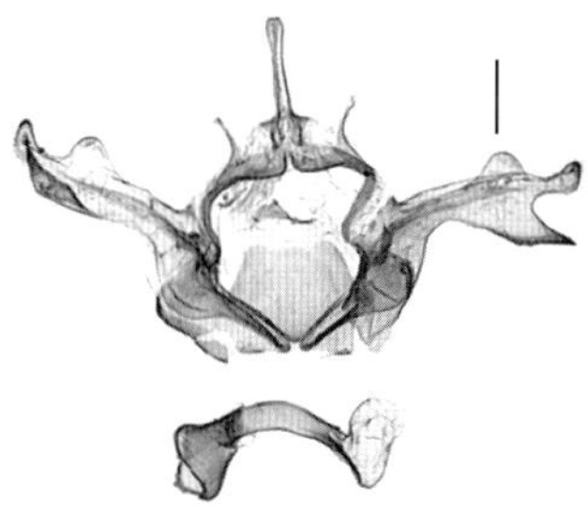

137b *Liparopsis barlowi*
Sumatra, GU 65-36

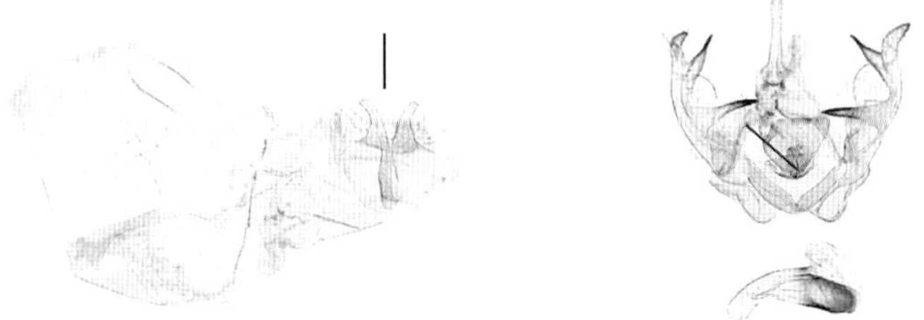

138a *Liparopsis sundana*
Kalimantan, GU 79-83

1 mm

139a *Liparopsis dierli*
Sumatra, GU 79-81

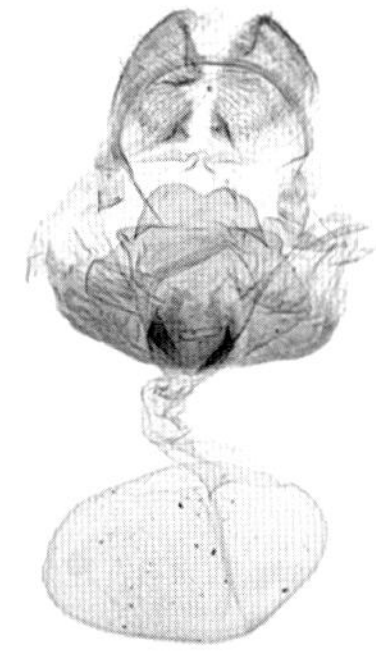

139b *Liparopsis dierli*
Kalimantan, GU 65-33

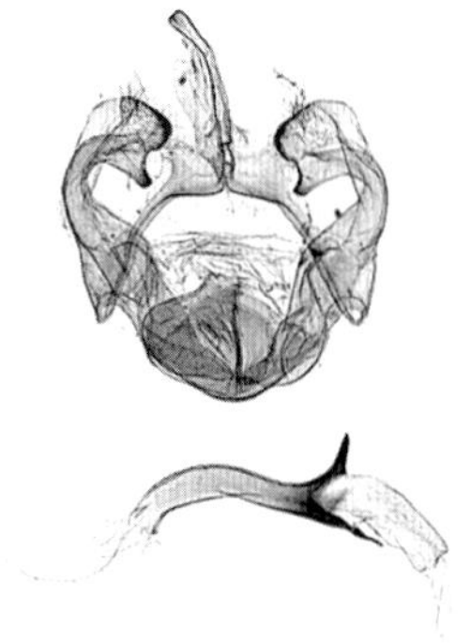

139c *Liparopsis dierli*
Kalimantan, GU 79-85

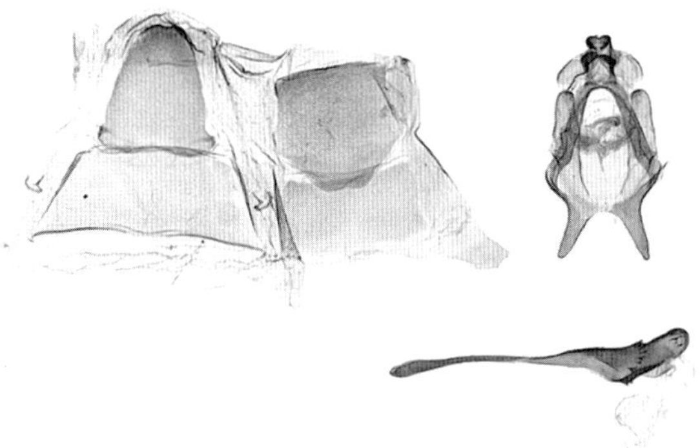

140a *Pantanopsis diehli*
Kalimantan, GU 65-38

140b *Pantanopsis diehli*
Sumatra, GU 79-74

140c *Pantanopsis diehli*
Kalimantan, W 15731

140d *Pantanopsis diehli*
Sumatra, MWM 29.120

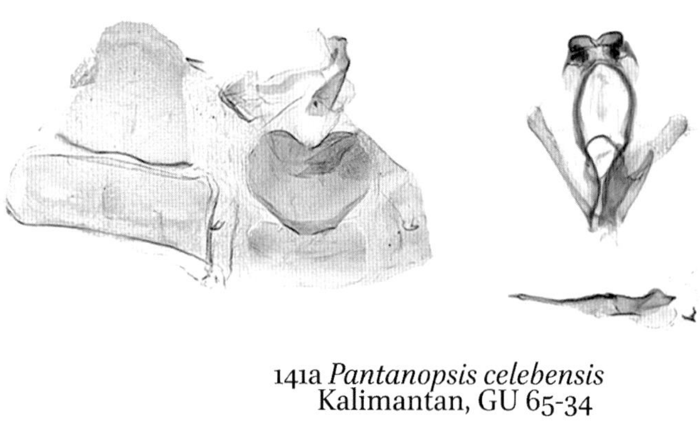

141a *Pantanopsis celebensis*
Kalimantan, GU 65-34

1 mm

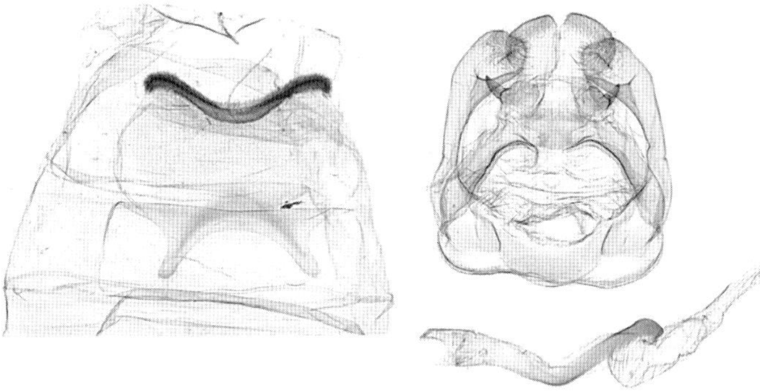

142Ba *Harpyia baibarana*
China, Hunan, MWM 15.700, phallus

142Aa *Harpyia microsticta*
NE India, BM902, holotype

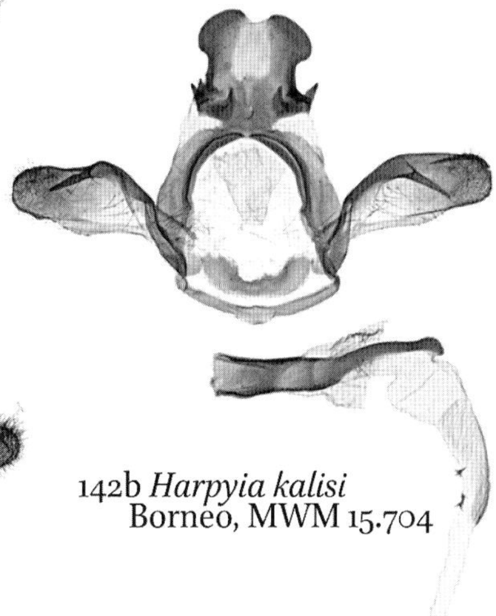

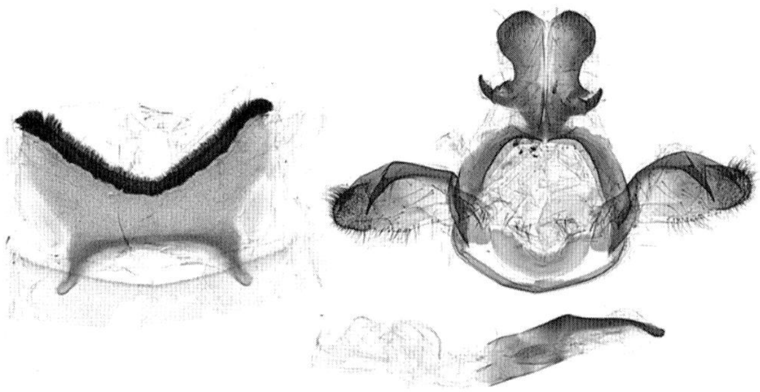

142b *Harpyia kalisi*
Borneo, MWM 15.704

142a *Harpyia kalisi*
Borneo, GU 77-38

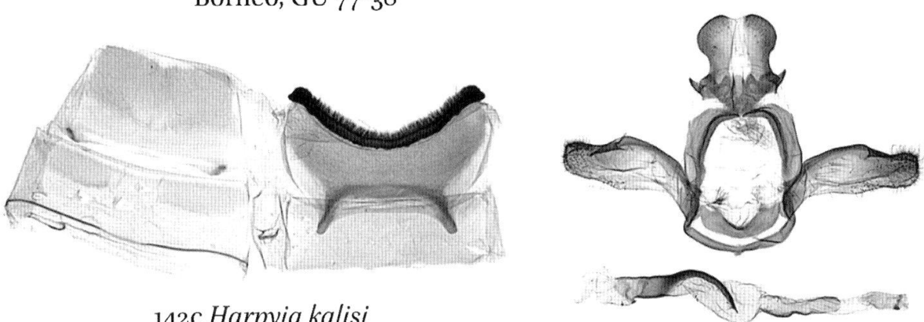

142c *Harpyia kalisi*
Borneo, GU 96-68

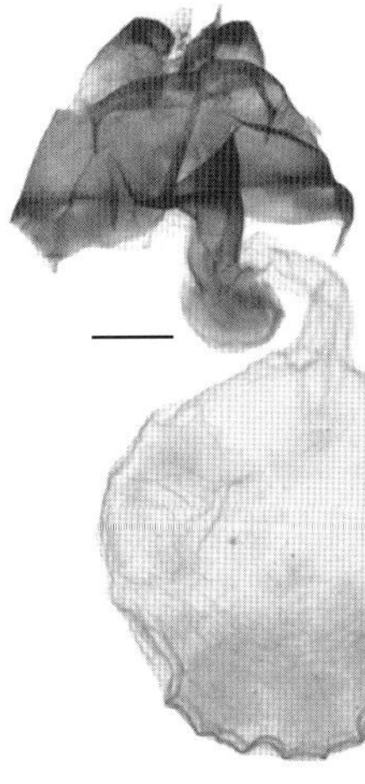

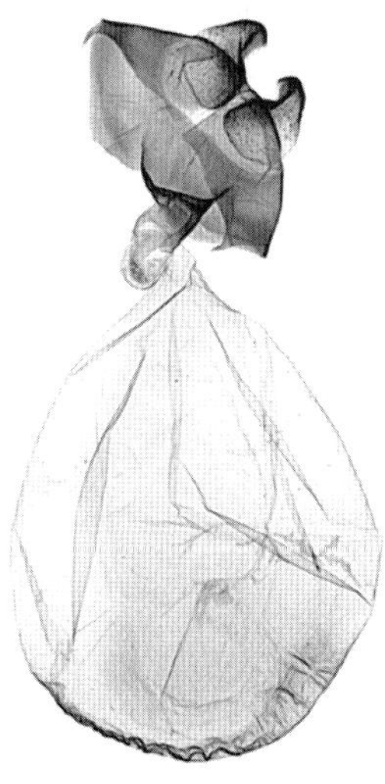

1 mm

142Bb *Harpyia baibarana*
Myanmar, MWM 15.669

142d *Harpyia kalisi*
Sumatra, GU 65-40

143a *Harpyia kobesi*
Sumatra, GU 65-39, PT

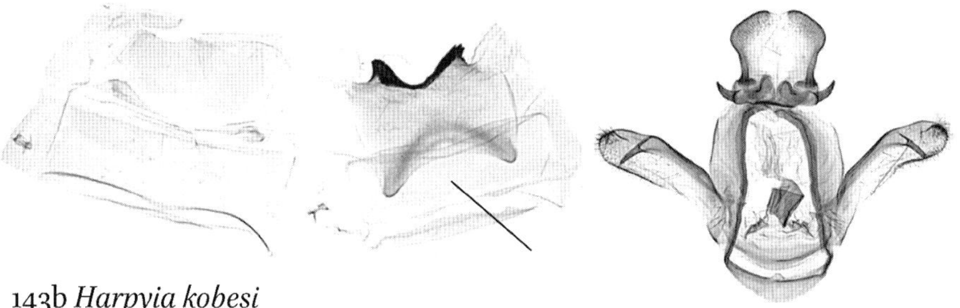

143b *Harpyia kobesi*
Sumatra, GU 65-37, paratype

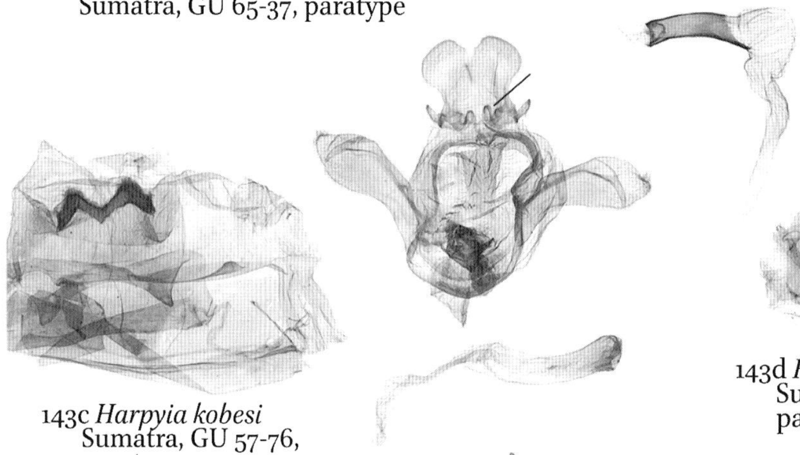

143c *Harpyia kobesi*
Sumatra, GU 57-76,
paratype

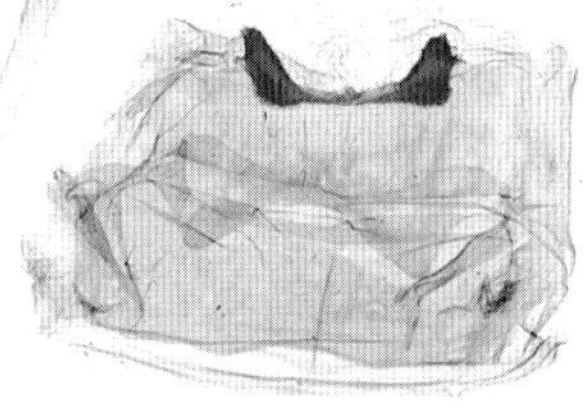

143d *Harpyia kobesi*
Sumatra, GU 57-77,
paratype, 8th sternite

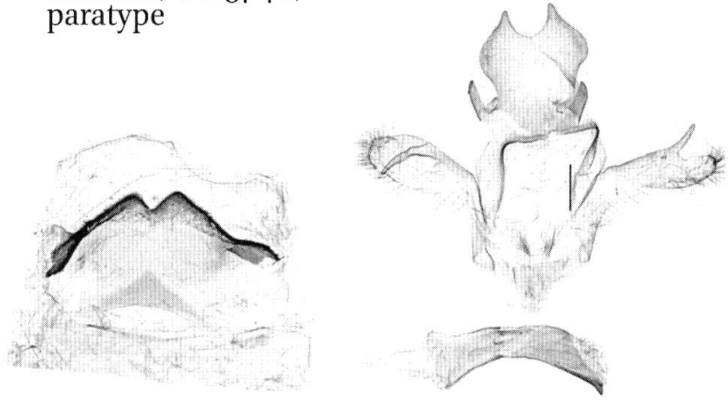

144a *Harpyia koeppeli*
Borneo, GU 17-65a, holotype

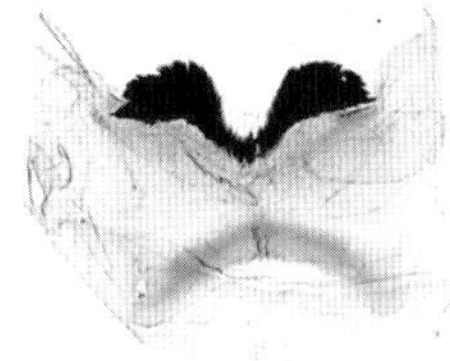

144b *Harpyia koeppeli*
Kalimantan, GU 46-34,
paratype, 8th sternite

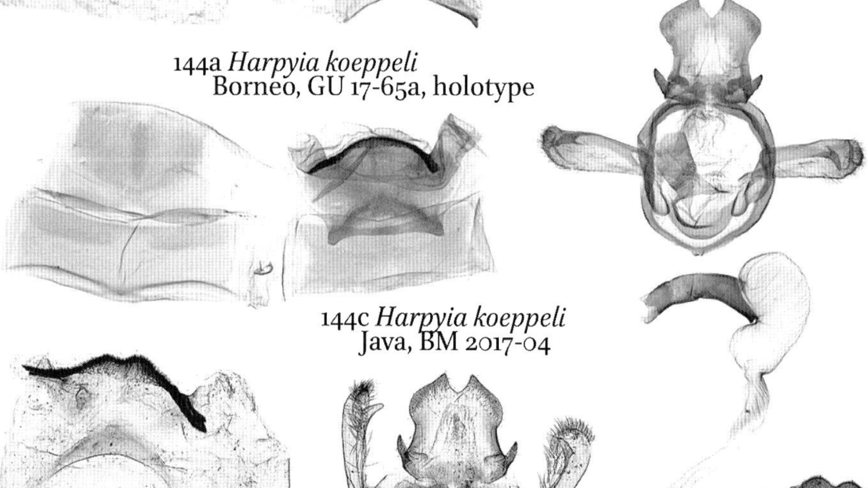

144c *Harpyia koeppeli*
Java, BM 2017-04

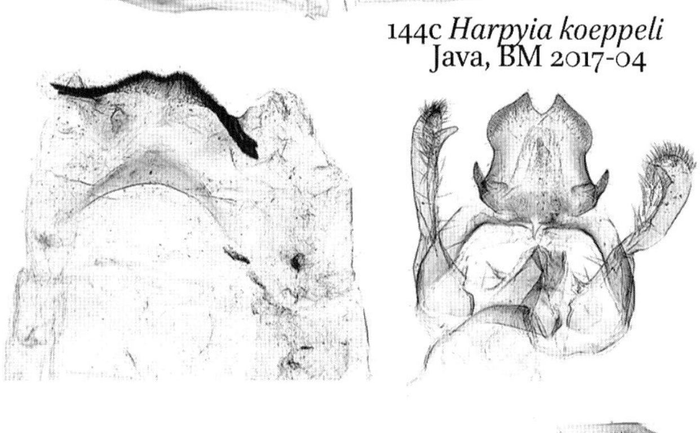

1 mm

144d *Harpyia koeppeli*
Kalimantan, GU 46-34, paratype

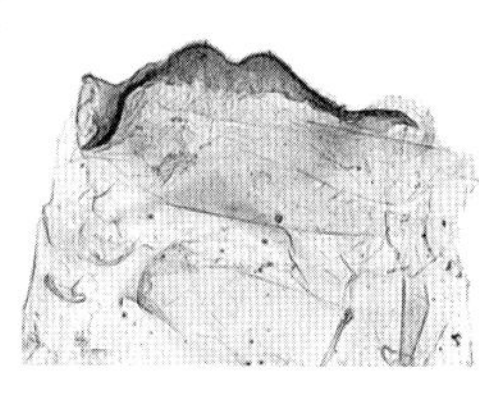

144e *Harpyia koeppeli*
Kalimantan, GU 46-34,
paratype, 8th sternite

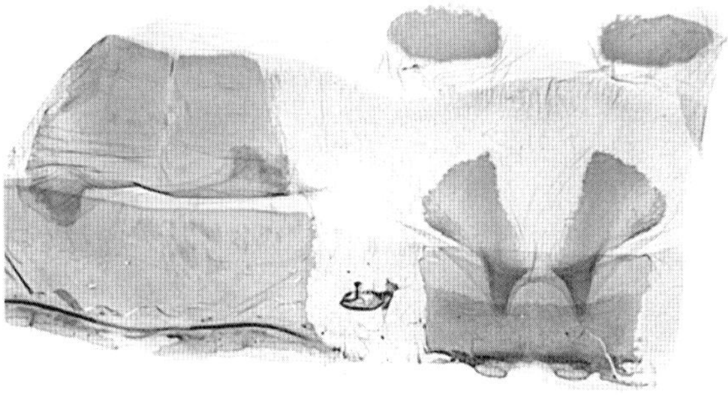
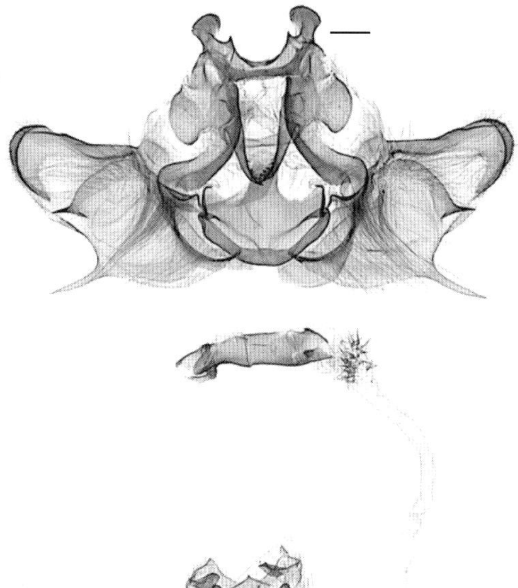

145a *Teleclita strigata cathana*
Nias, MWM 29.100

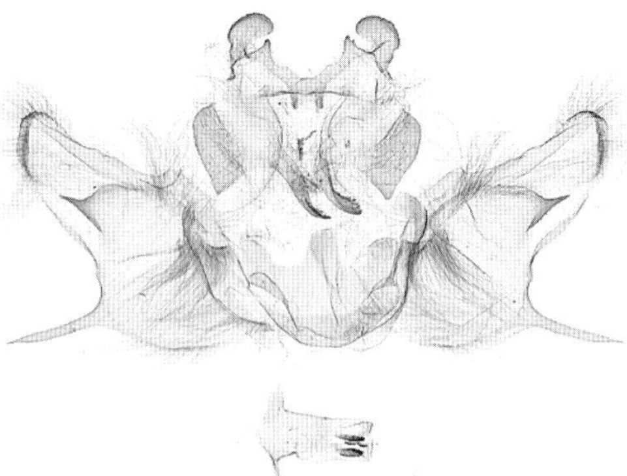

145b *Teleclita strigata cathana*
Bali, GU 46-58

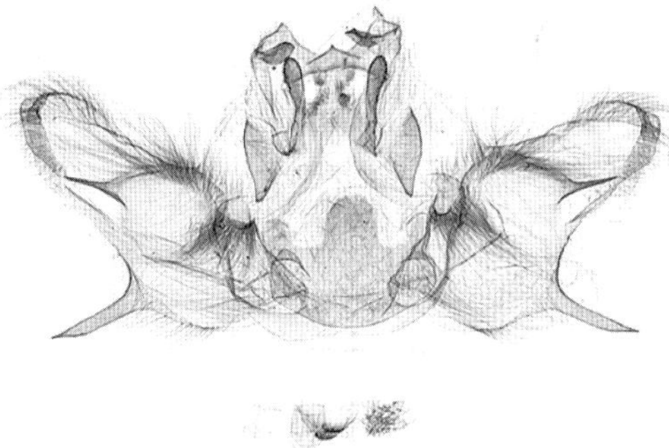

145c *Teleclita strigata strigata*
S India, GU 51-33

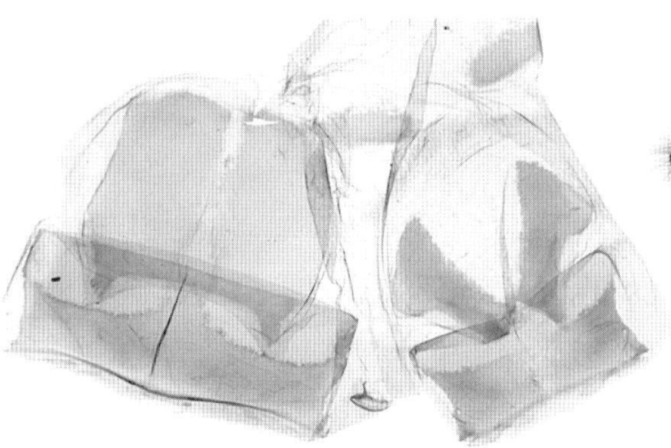
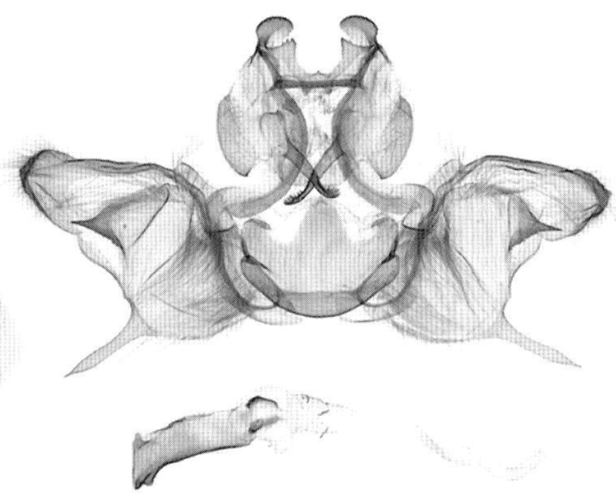

145d *Teleclita strigata mediocana*
Sumbawa, GU 94-29, paratype

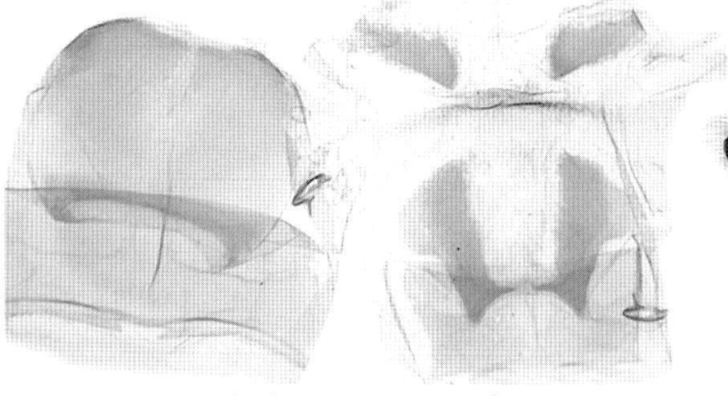
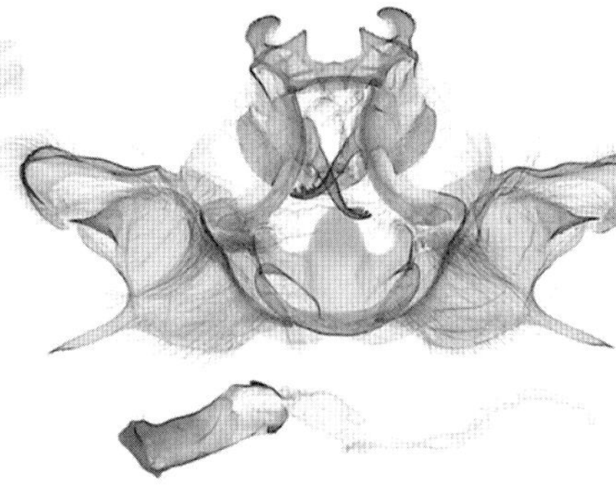

1 mm

145e *Teleclita strigata mediocana*
Alor, GU 94-37, paratype

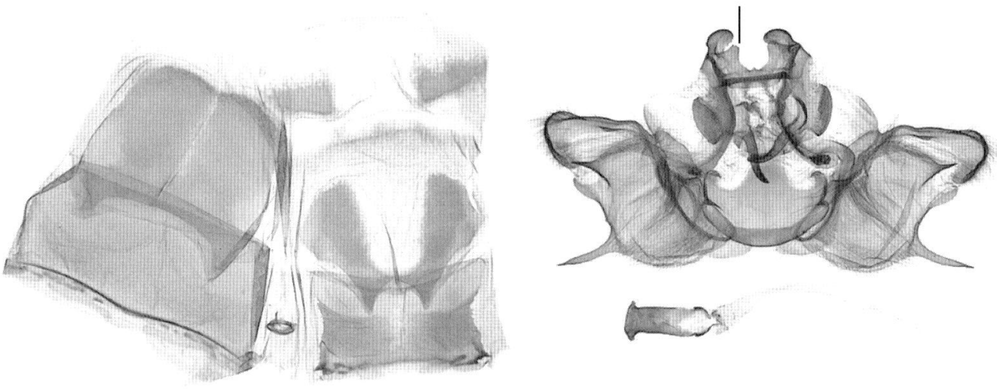

145f *Teleclita strigata multifascia*
Tanimbar, GU 94-31, paratype

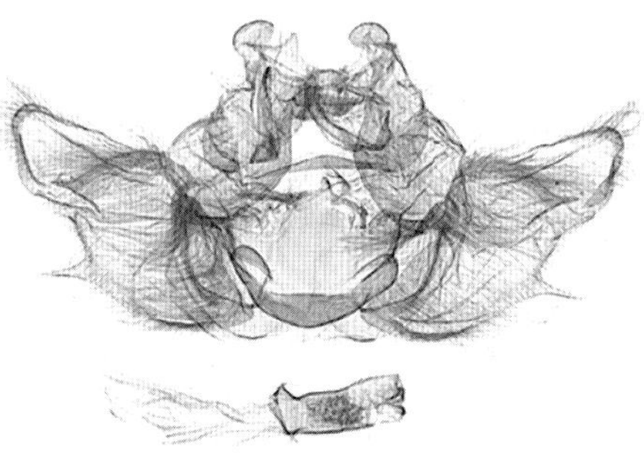

145g *Teleclita strigata insignata*
Dammar, BM 116, paralectotype

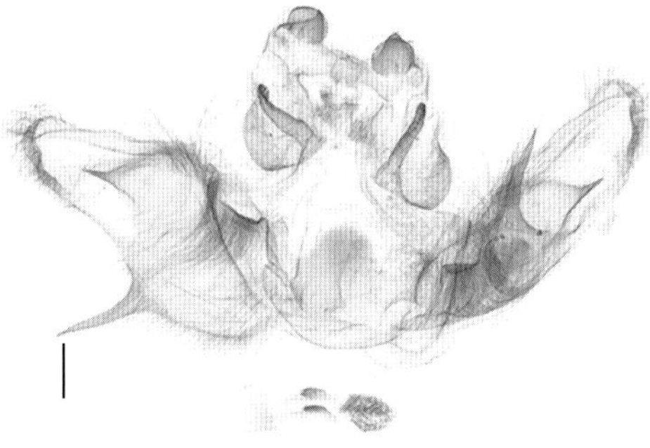

145h *Teleclita strigata insignata*
Halmahera, GU46-57

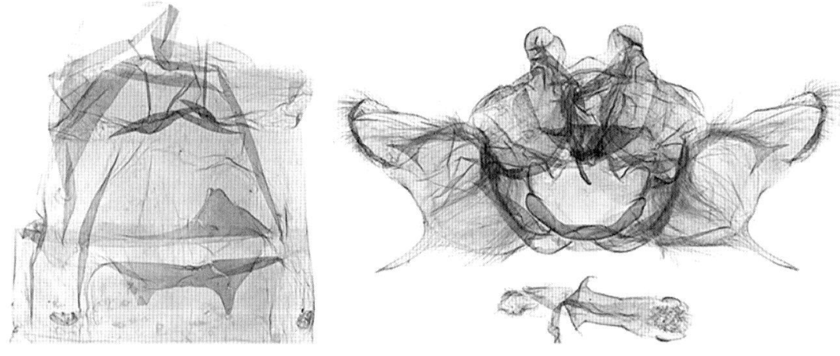

145i *Teleclita strigata cinnamomea*
W Papua, BM 71 and BM 72, holotype

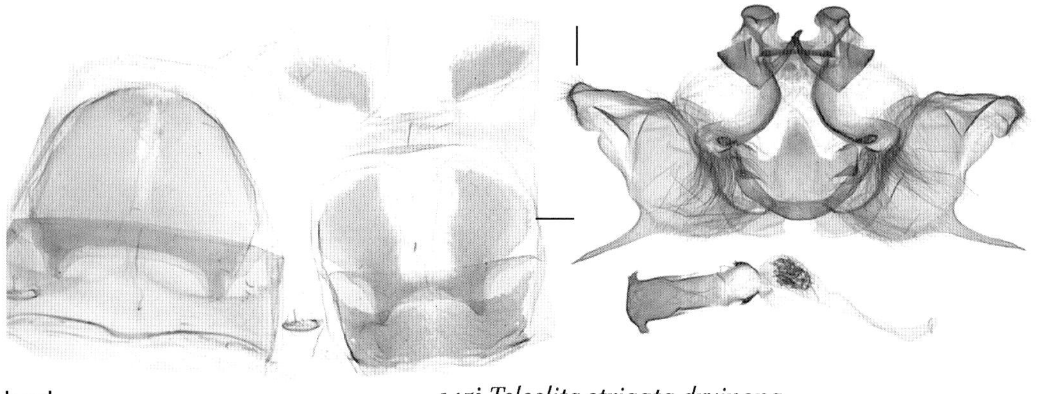

145j *Teleclita strigata dryinopa*
Australia, NT, GU 94-27

1 mm

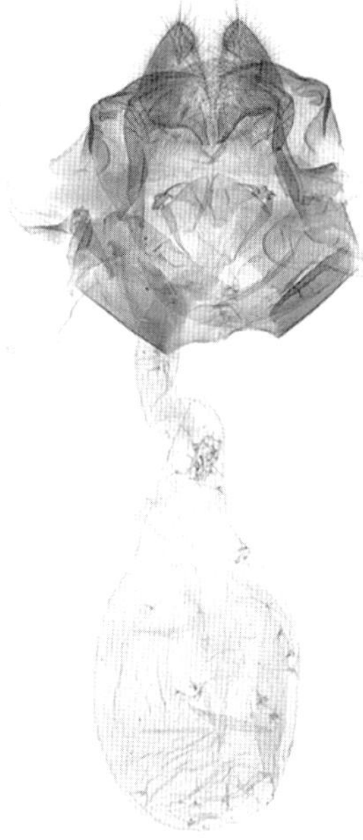

145k *Teleclita strigata cathana*
Sumatra, GU 51-47

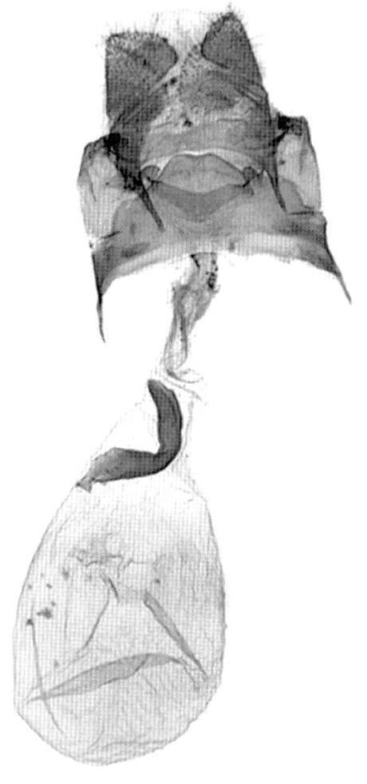

145l *Teleclita strigata mediocana*
Alor, GU 94-38, paratype

145m *Teleclita strigata multifascia*
Tanimbar, GU 94-32, paratype

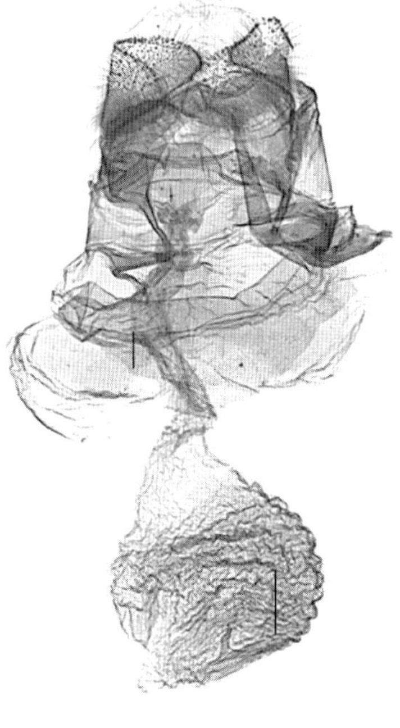

1 mm

145n *Teleclita strigata insignata*
Dammar, BM 43, lectotype

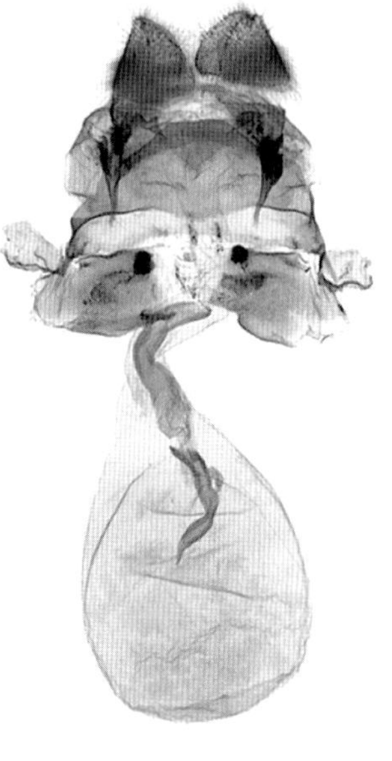

145o *Teleclita strigata dryinopa*
Australia, NT, GU 94-28

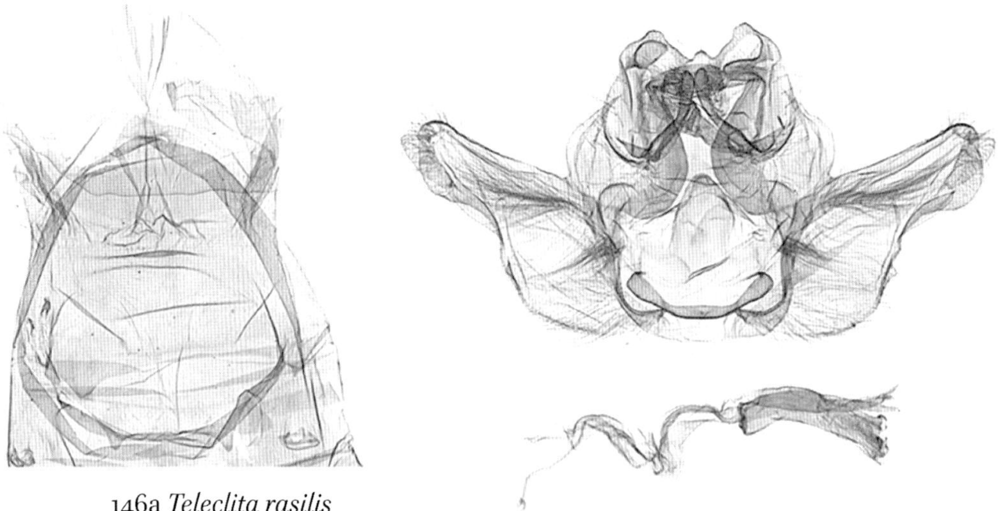

146a *Teleclita rasilis*
Sulawesi, BM 1480, paratype

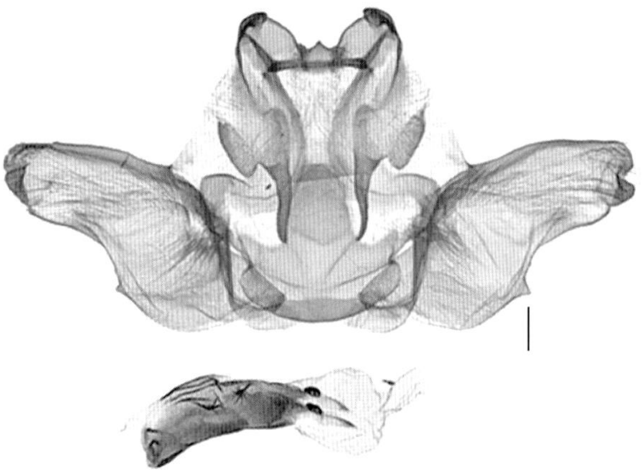

146b *Teleclita rasilis*
Peleng, MWM15.726, paratype

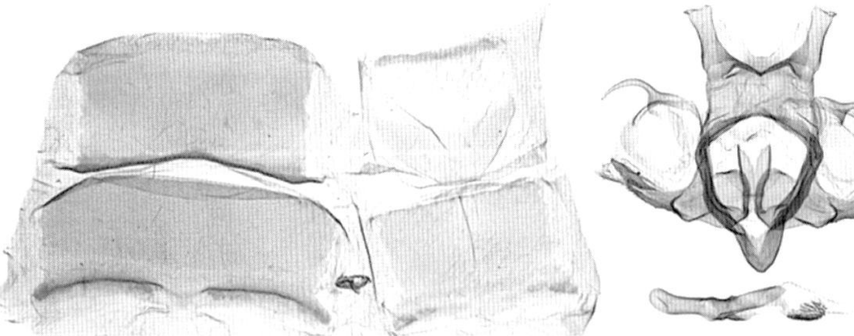

148a *Pheressaces siegfriedihlei*
W Papua, Ihle 01-2017, holotype

147a *Teleclita flavisticta*
Sumatra, MWM 29.101

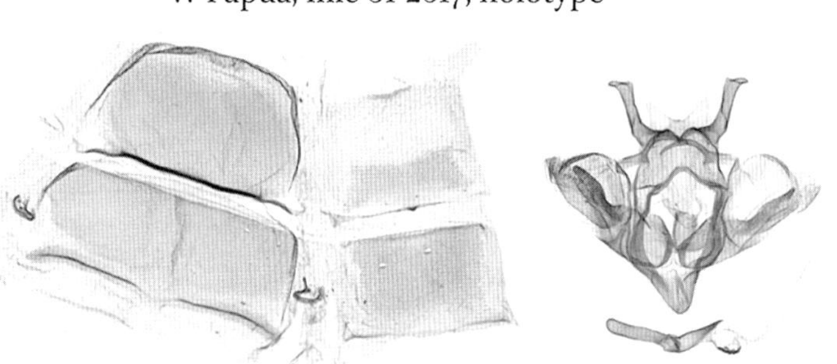

1 mm

148Aa *Pheressaces cycnoptera*
Australia, Qsd., SMS 02-2017

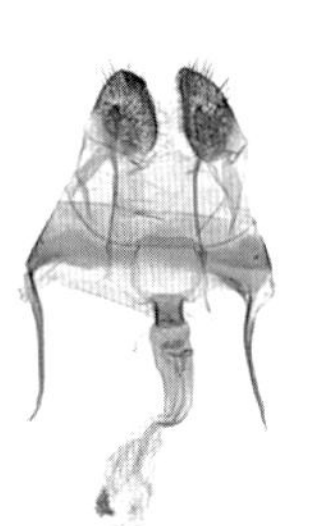

148Ab *Pheressaces cycnoptera*
Australia, Qsd., SMS 03-2017

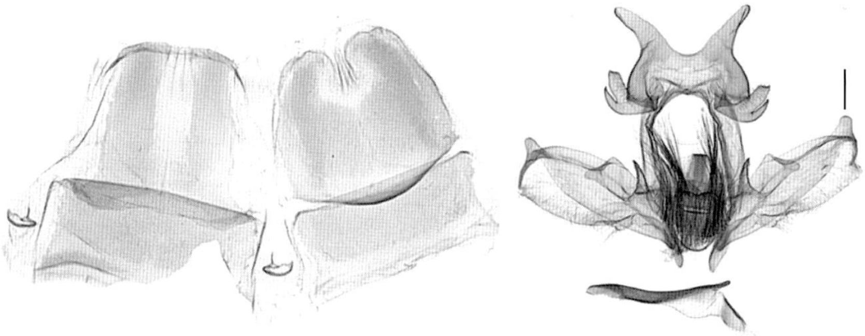

149a *Destolmia lineata*
PNG, GU 66-32

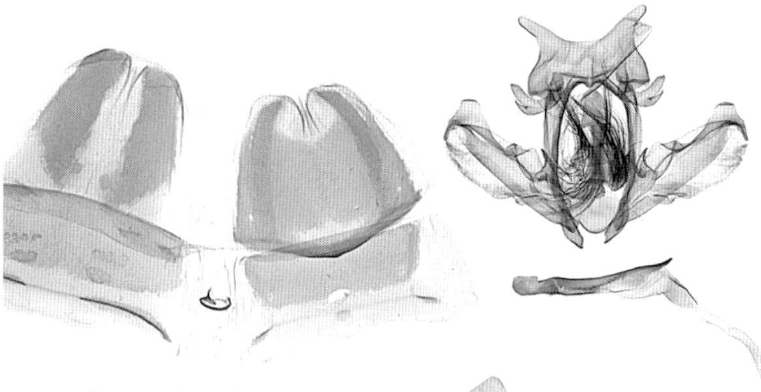

149b *Destolmia lineata*
Australia, NSW, GU 65-99

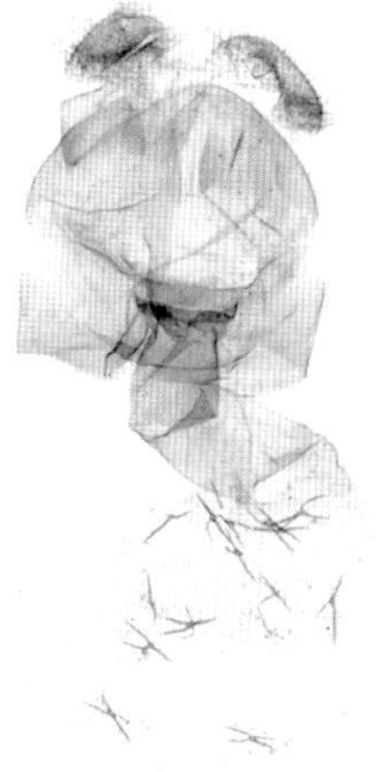

149c *Destolmia lineata*
Australia, ZSM 2011-08

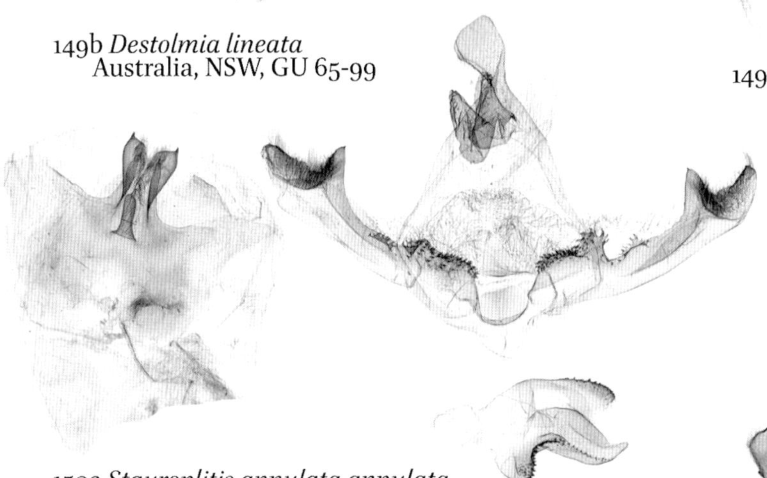

150a *Stauroplitis annulata annulata*
Sumatra, GU 46-39

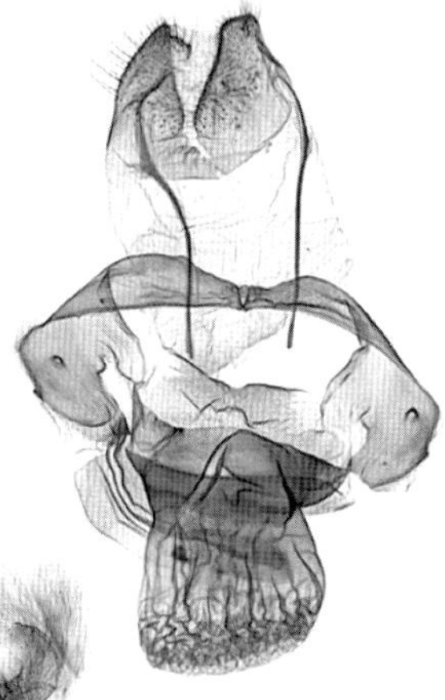

150b *S. annulata annulata*
Sumatra, GU 65-40

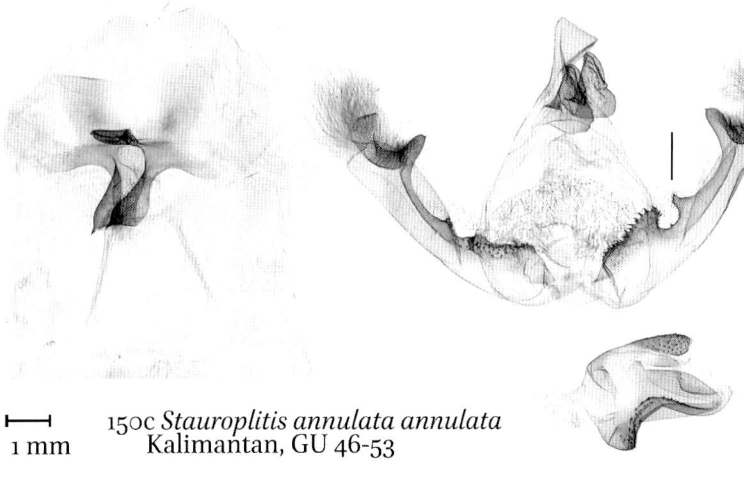

1 mm

150c *Stauroplitis annulata annulata*
Kalimantan, GU 46-53

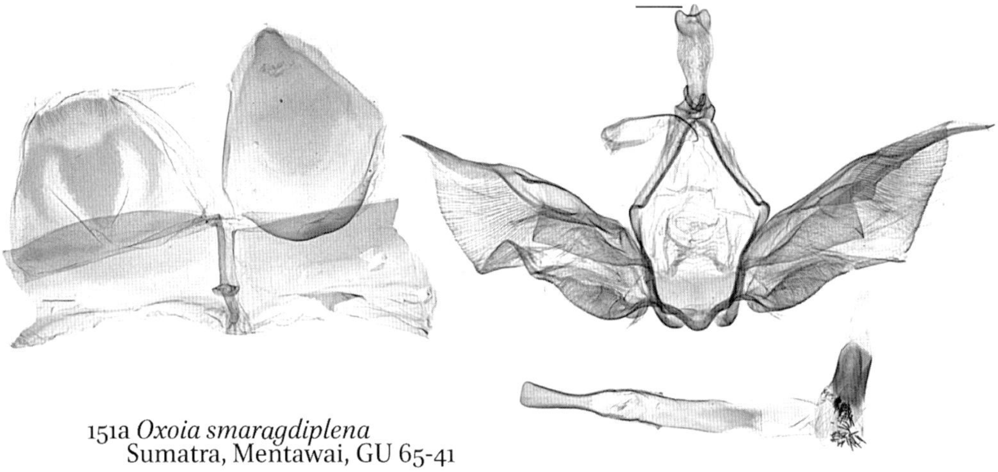

151a *Oxoia smaragdiplena*
Sumatra, Mentawai, GU 65-41

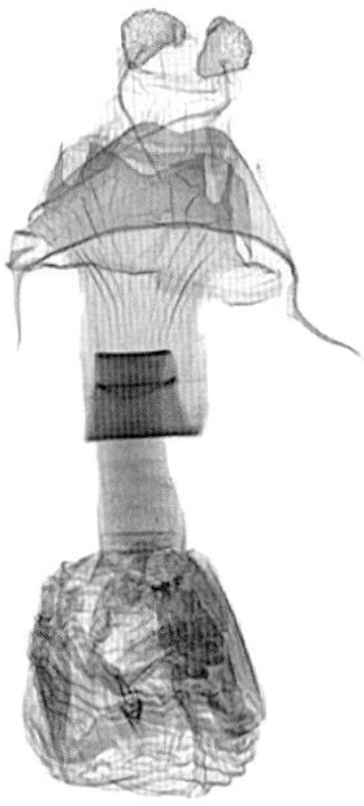

151b *Oxoia smaragdiplena*
Kalimantan, GU 45-62

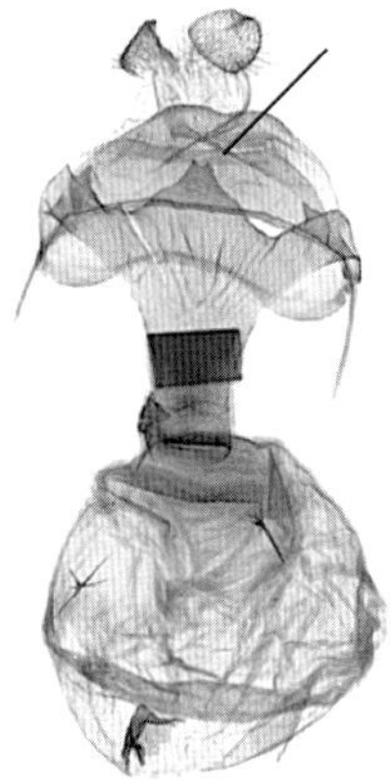

152a *Oxoia viridipicta*
Sumatra, GU 65-44

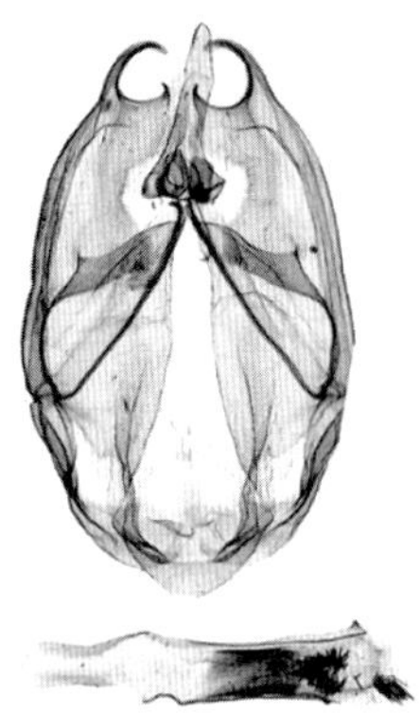

152b *Oxoia viridipicta*
Sumatra, sp398
holotype

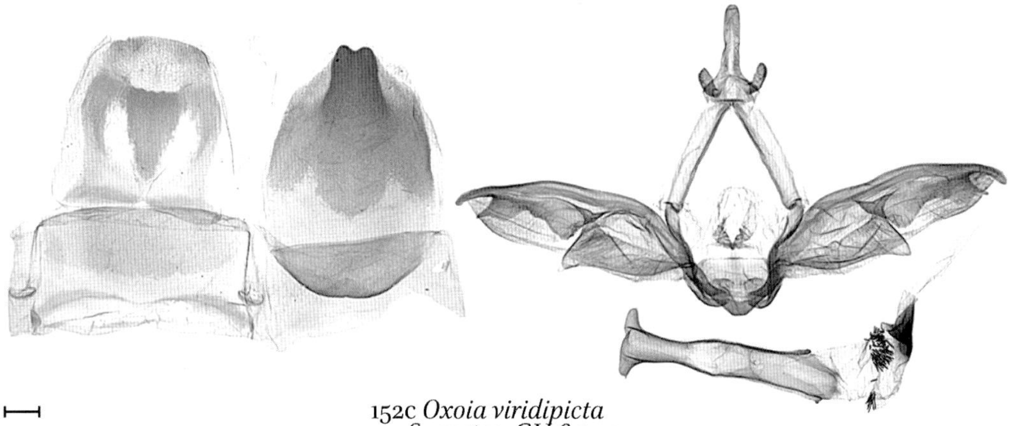

1 mm

152c *Oxoia viridipicta*
Sumatra, GU 65-43

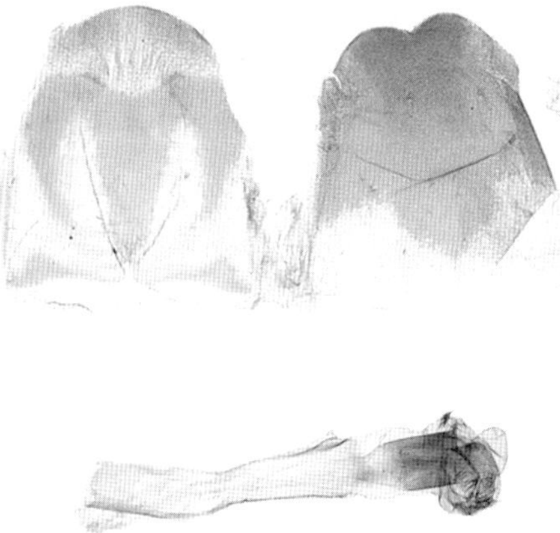

153a *Oxoia irrotativiridis irrotativiridis*
Halmahera, GU 55-46, phallus

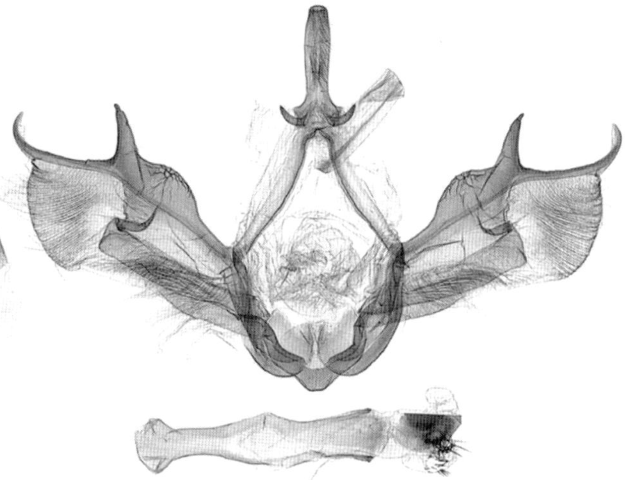

153b *Oxoia irrotativiridis irrotativiridis*
W Papua, GU 77-04

153c *Oxoia irrotativiridis irrotativiridis*
Sulawesi, GU 55-45, phallus

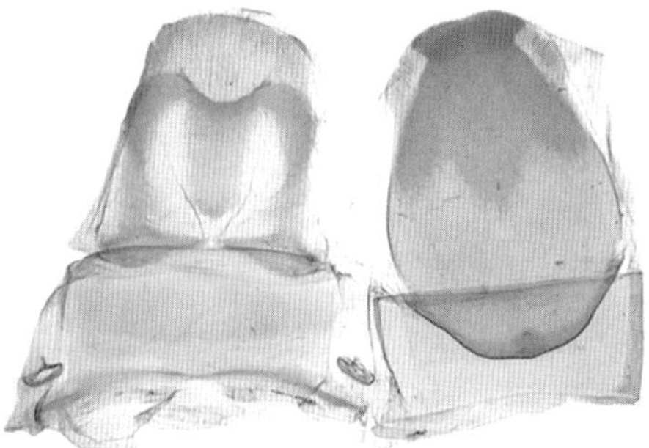

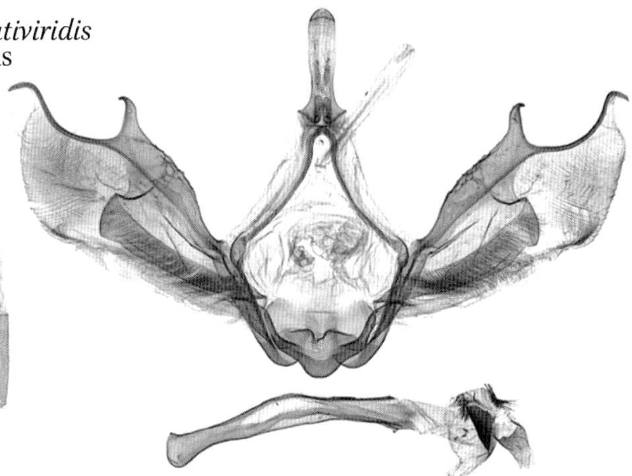

153d *Oxoia irrotativiridis triangularis*
Bacan, GU 94-41

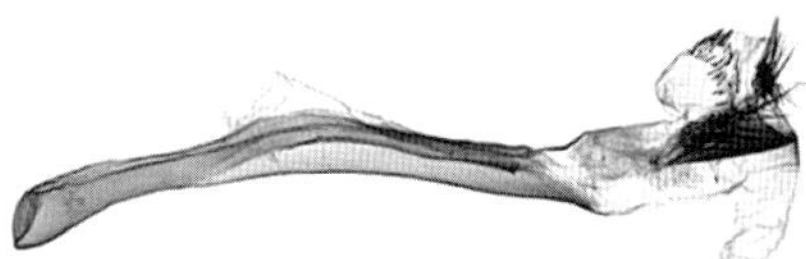

153e *Oxoia irrotativiridis triangularis*
Buru, GU 93-23, phallus

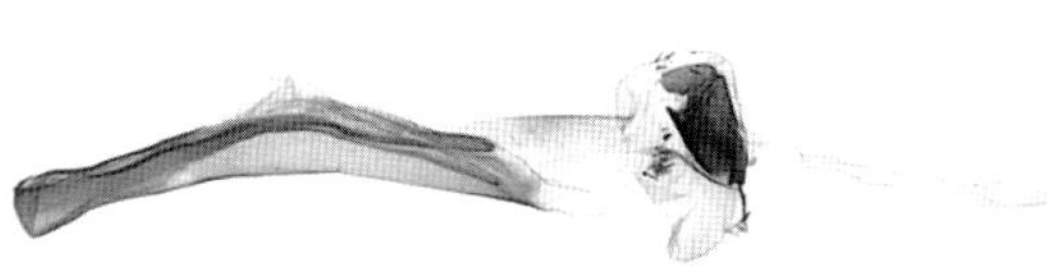

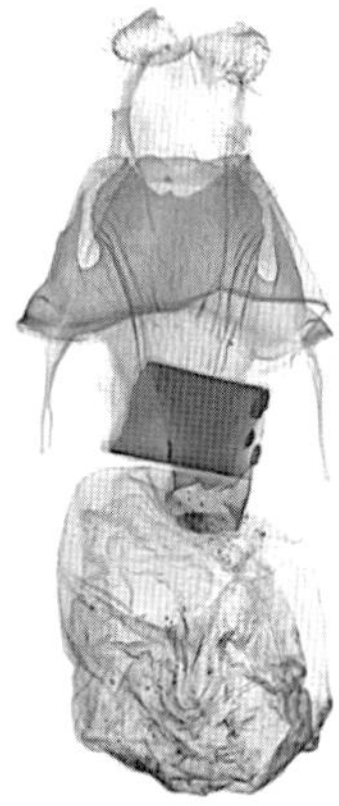

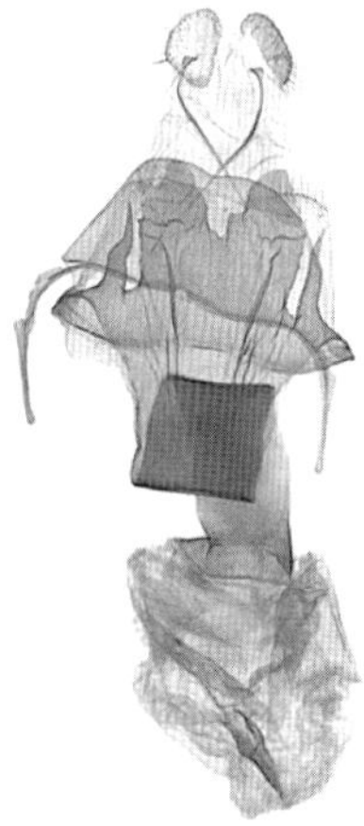

1 mm

153f *Oxoia irrotativiridis triangularis*
Obi, GU 94-42, phallus

153g *Oxoia irrotativiridis irrotativiridis*
W Papua, GU 94-02

153h *Oxoia irrotativiridis triangularis*
Buru, GU 93-24

154a *Rodneya caudata*
Sumatra, GU 26-56

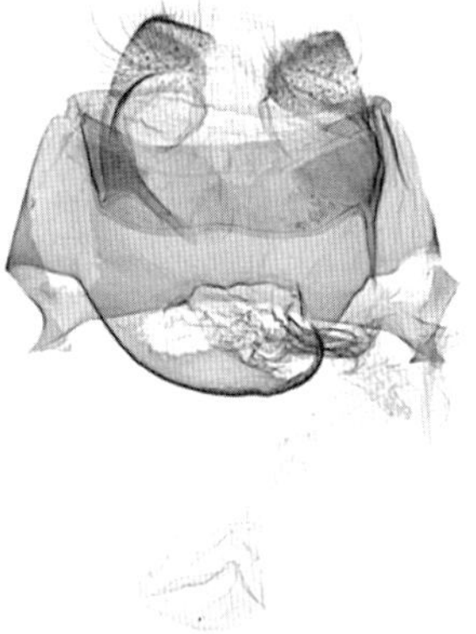

154b *Rodneya caudata*
Myanmar, GU 65-52

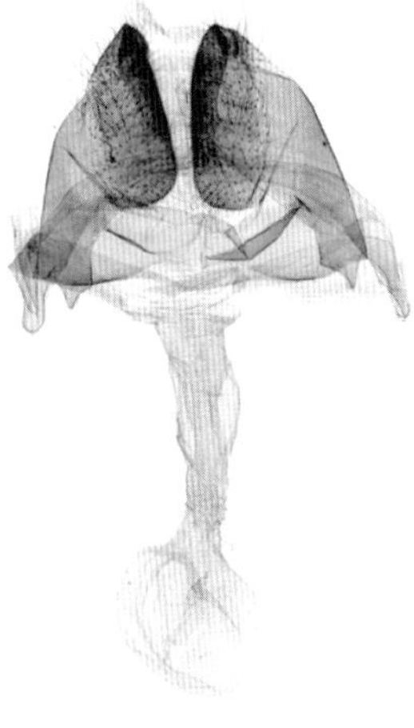

155a *Stauropus major*
Sumatra, GU 78-62

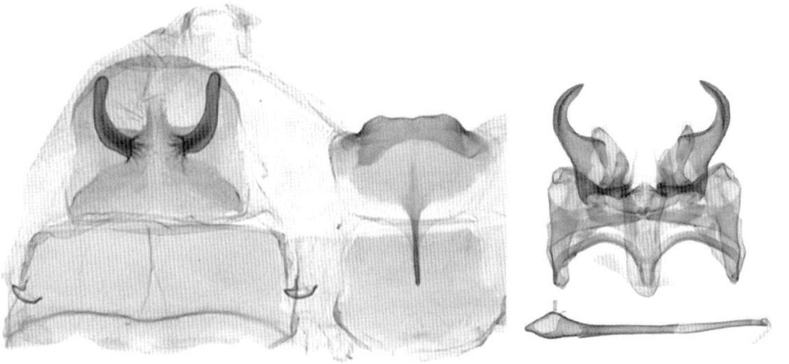

155b *Stauropus major*
Alor, GU 94-60

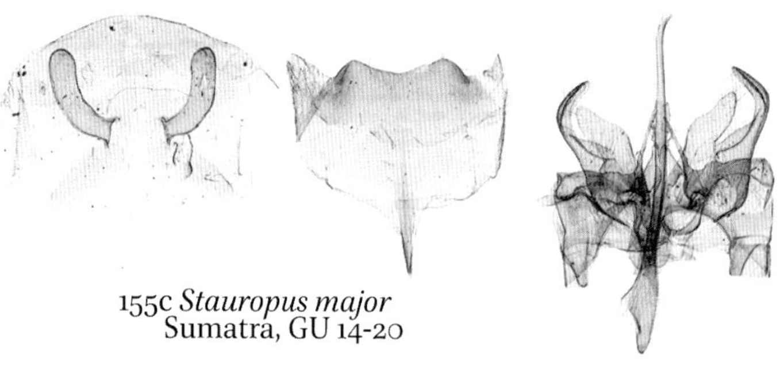

155c *Stauropus major*
Sumatra, GU 14-20

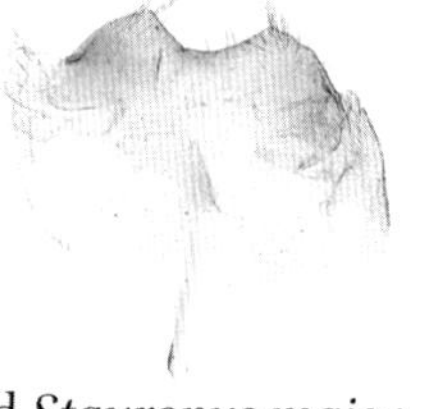

155d *Stauropus major*
Sumatra, GU 14-10
8th sternite.

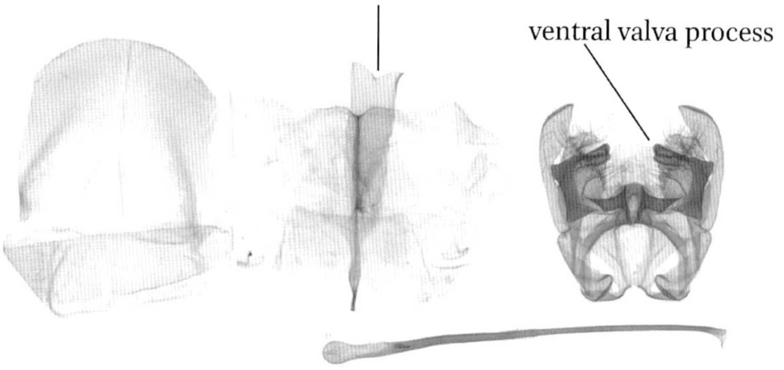

1 mm

156a *Stauropus alternus alternus*
Taiwan, MWM 8.189

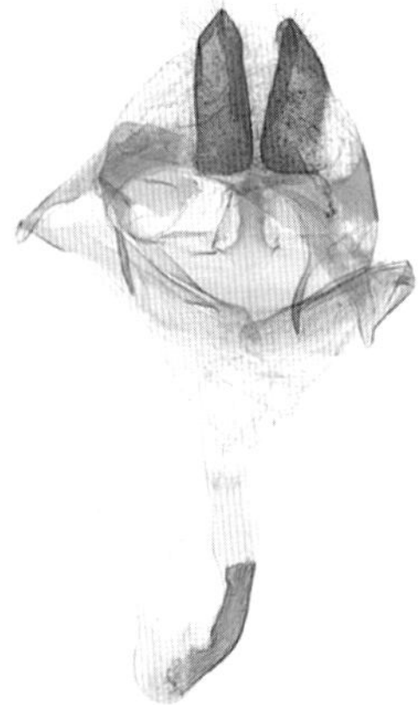

156b *Stauropus alternus alternus*
Pakistan, MWM 8.181

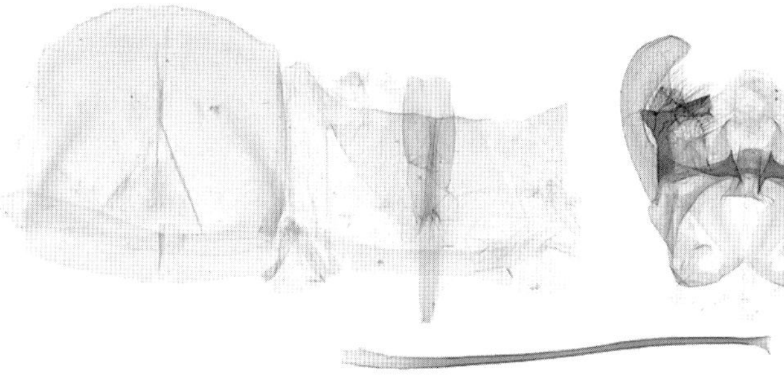

156c *Stauropus alternus brunnea*
Sumatra, Mentawei Isl., GU 78-57

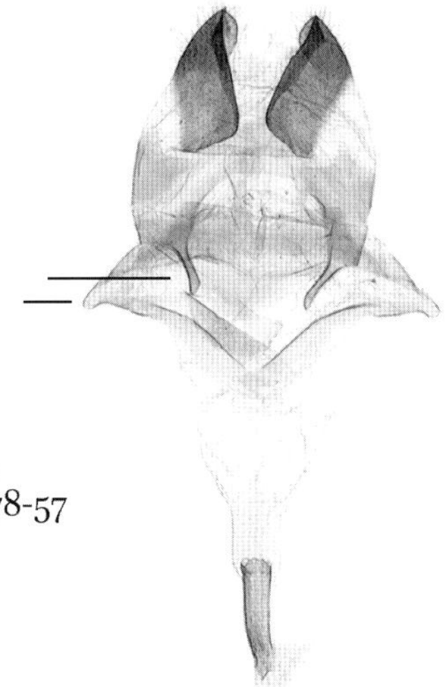

156d *Stauropus alternus brunnea,* Sumatra, GU 47-30

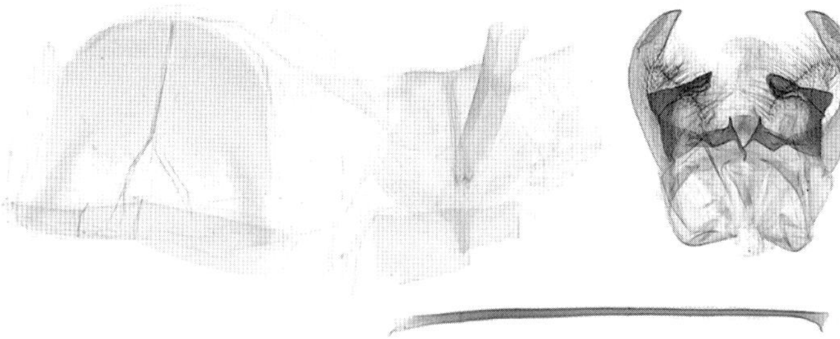

156e *Stauropus alternus hofmanae*
Java, GU 78-58, paratype

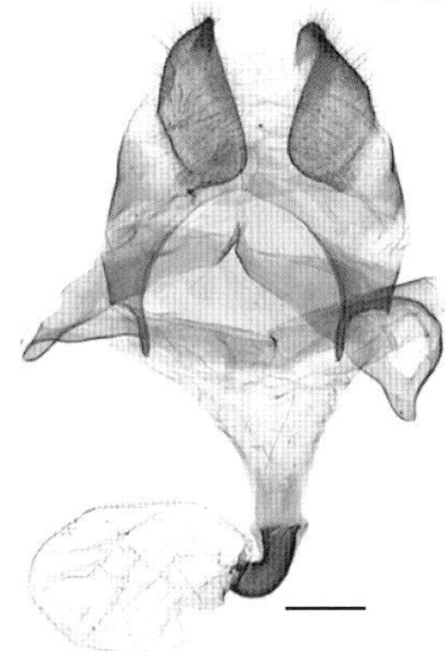

156f *Stauropus alternus hofmanae*
Sumba, GU 95-24

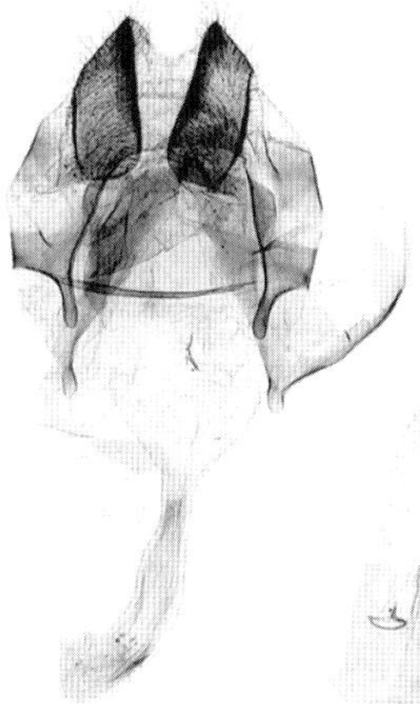

156g *Stauropus alternus sulawesianus,* Sulawesi
GU 95-24, paratype

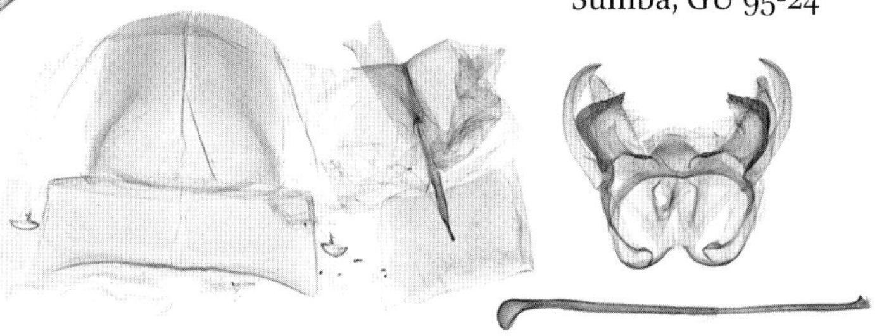

156h *Stauropus alternus hofmanae*
Pantar, GU 95-23

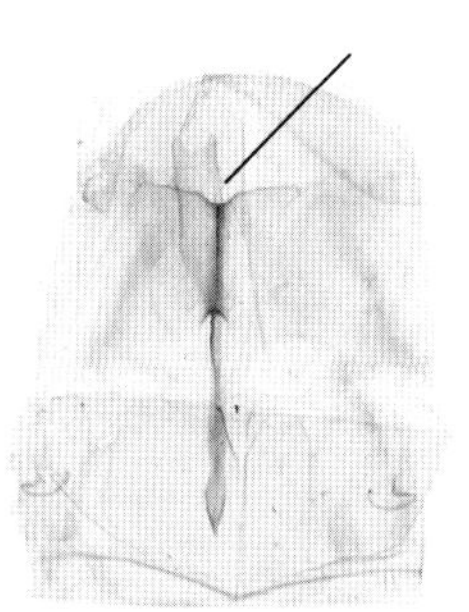

156i *S. alternus sulawesianus*
Sulawesi, GU 47-37
8th sternite
1 mm

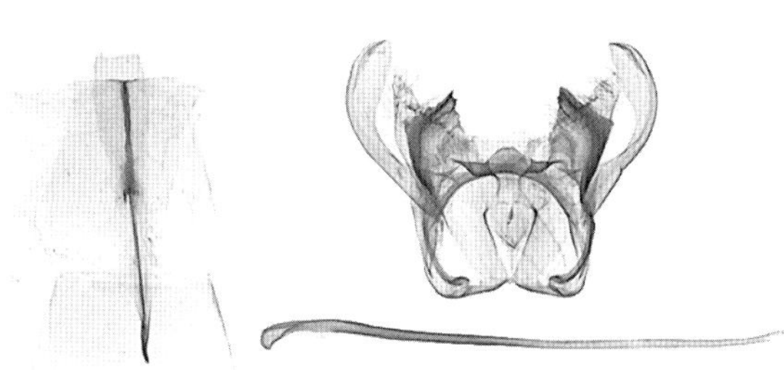

156j *Stauropus alternus sulawesianus*
Buru, GU 95-33

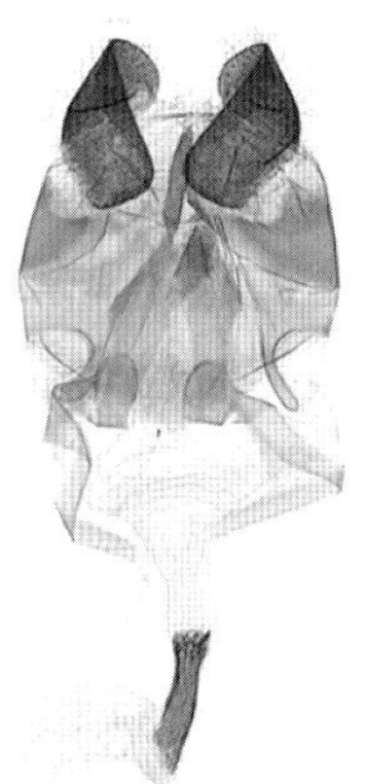

156k *Stauropus alternus melastomatis*
Seram, MWM 8.182

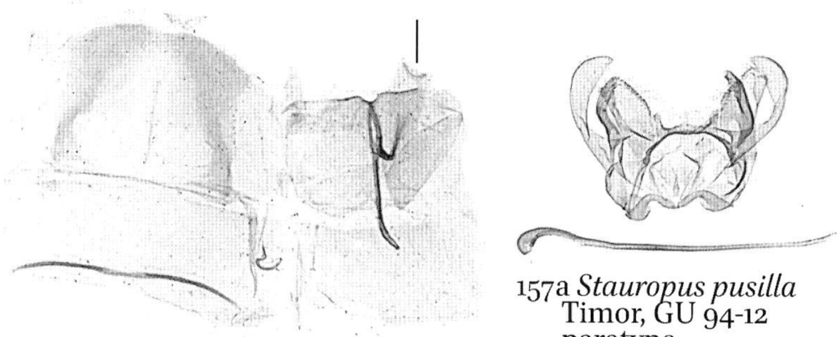

157a *Stauropus pusilla*
Timor, GU 94-12
paratype

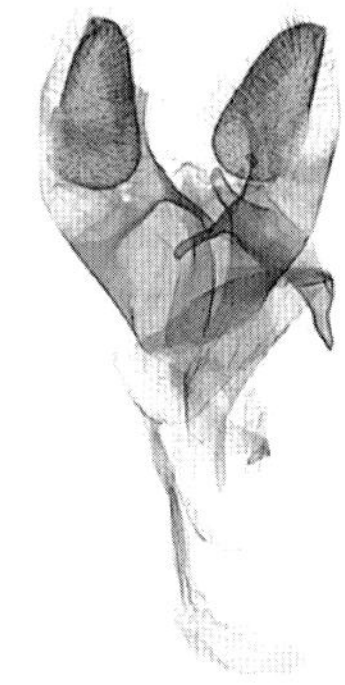

157b *Stauropus pusilla*
Timor, GU 65-46
paratype

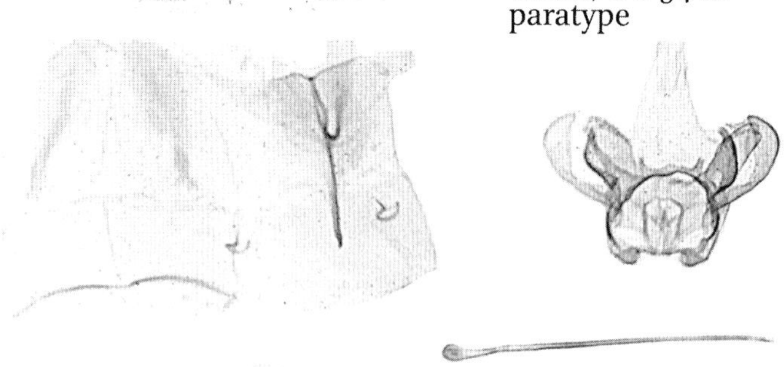

157c *Stauropus pusilla*
Timor, GU 94-11, paratype

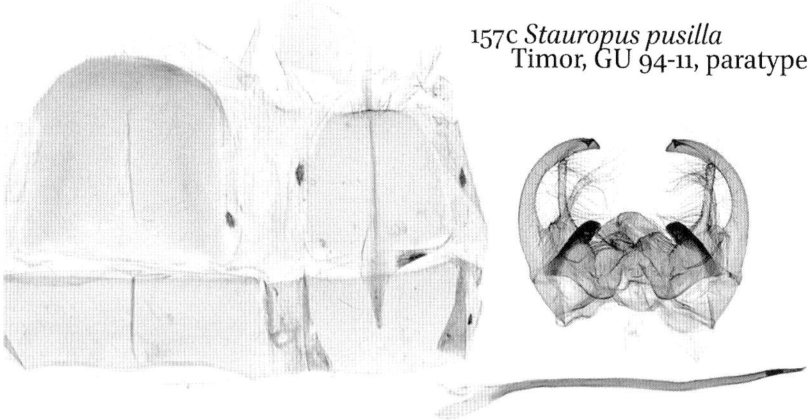

158a *Stauropus viridissimus viridissimus*
W Papua, GU 78-49

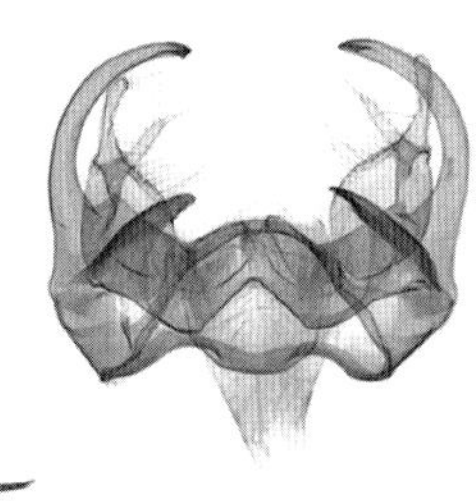

158b *S. viridissimus naumanni*
Obi, GU 92-82

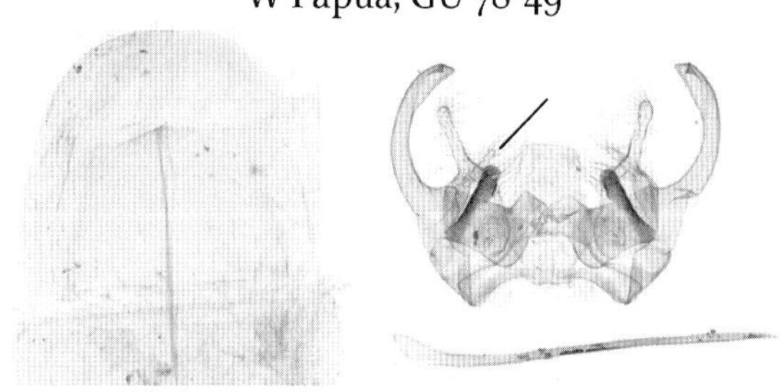

158c *Stauropus viridissimus amboynica*
Ambon, GU 47-38

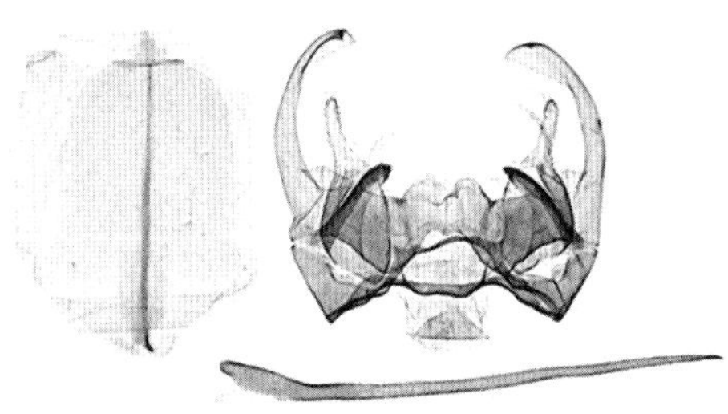

158d *Stauropus viridissimus boeroe*
Buru, GU 93-64, paratype

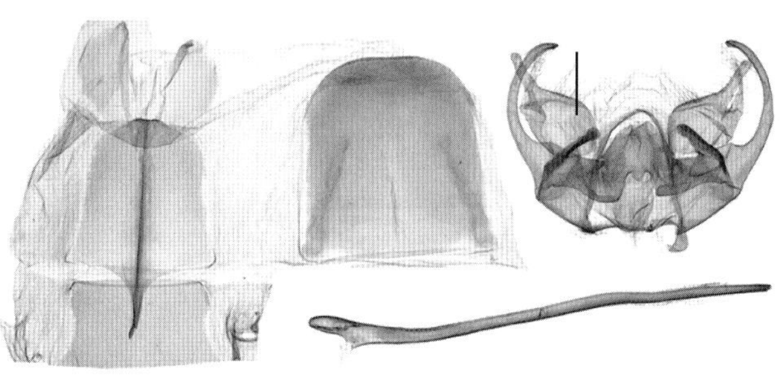

1 mm

158e *Stauropus viridissimus habrochlora*
Australia-NT, GU 94-45

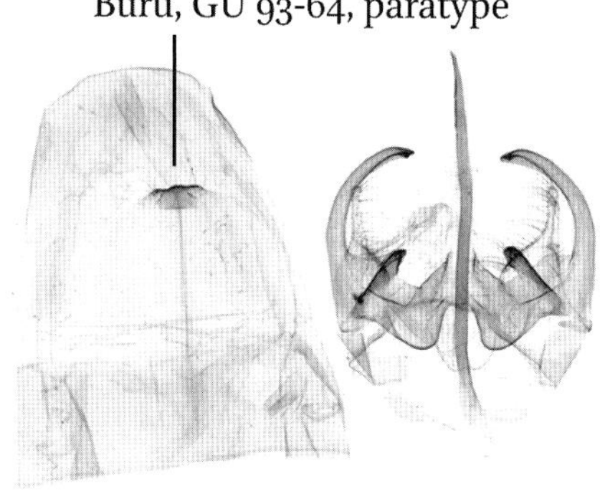

158f *S. viridissimus habrochlora*
Australia-Qsd, ZSM 2011-6

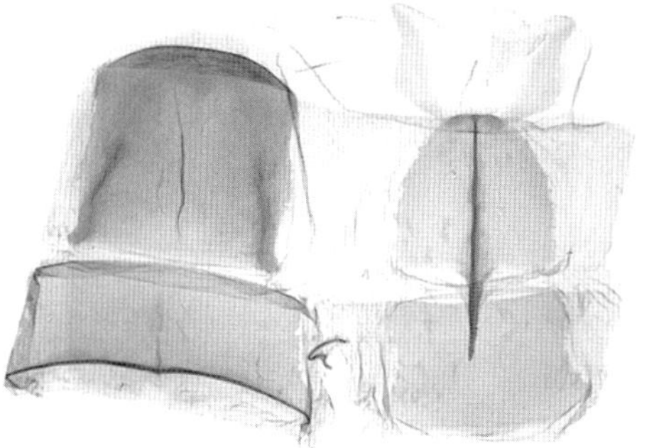
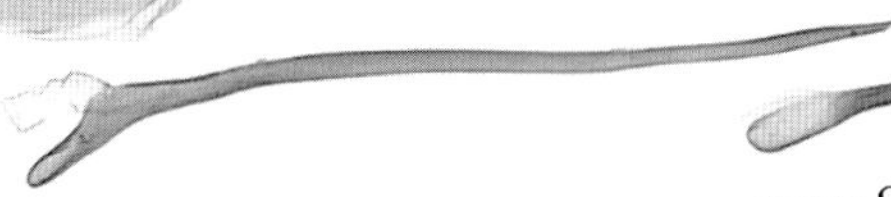

158g *Stauropus viridissimus habrochlora*
Tanimbar, GU 92-81

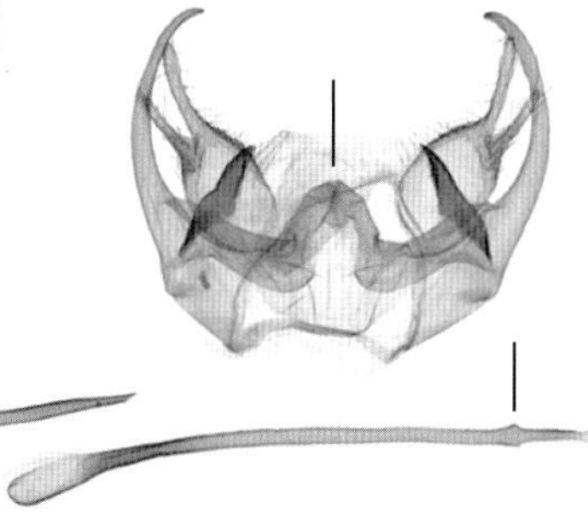

159a *Stauropus talasea*
New Britain, W 15542, holotype

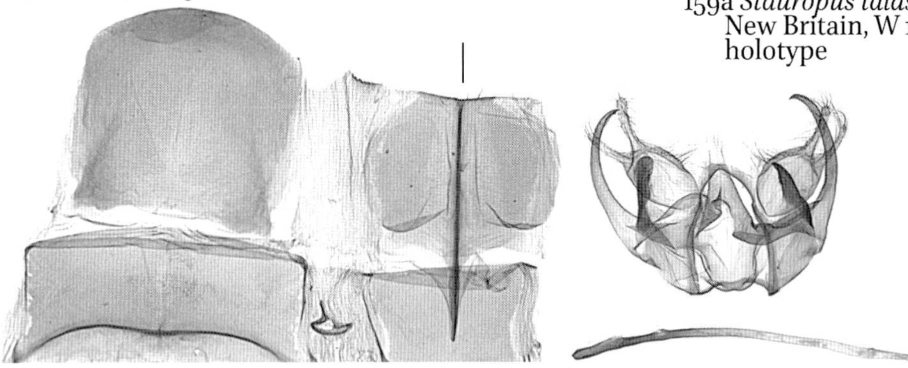

159b *Stauropus talasea*
New Ireland, BM 10314011, paratype

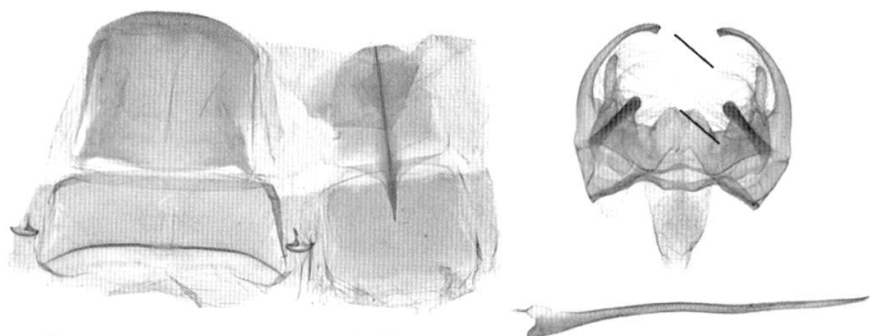

160a *Stauropus vanmastrigti*
W Papua, GU 94-24, paratype

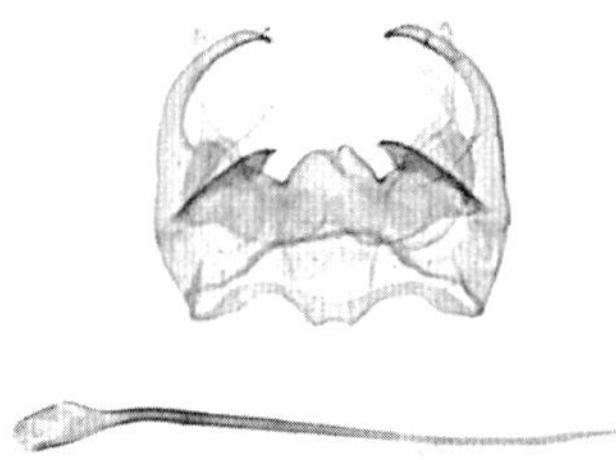

160b *Stauropus vanmastrigti*
W Papua, GU 94-08, paratype

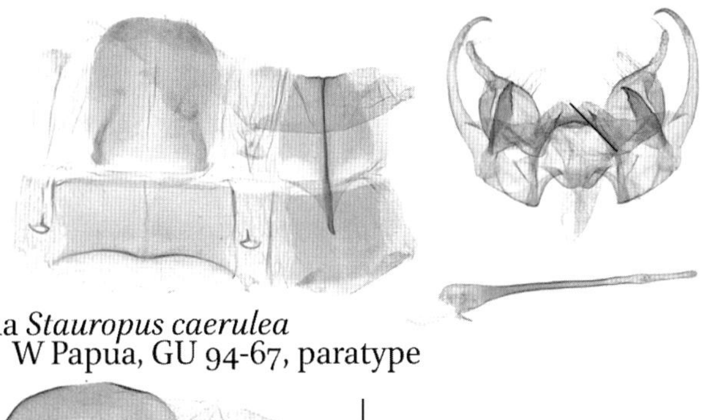

161a *Stauropus caerulea*
W Papua, GU 94-67, paratype

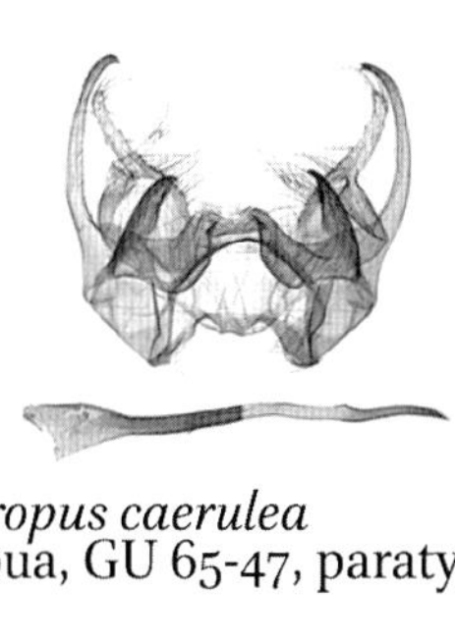

161b *Stauropus caerulea*
W Papua, GU 65-47, paratype

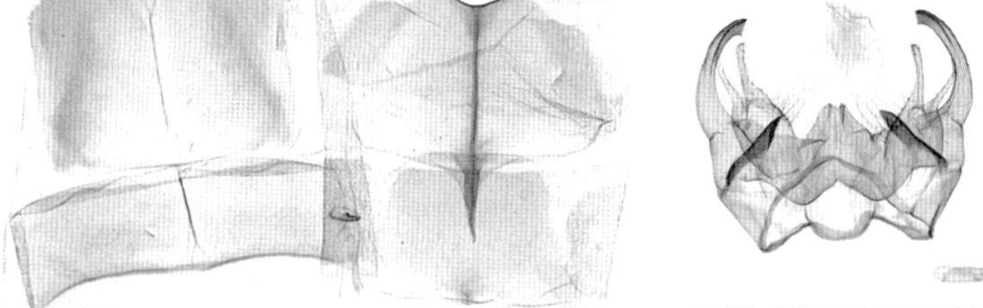
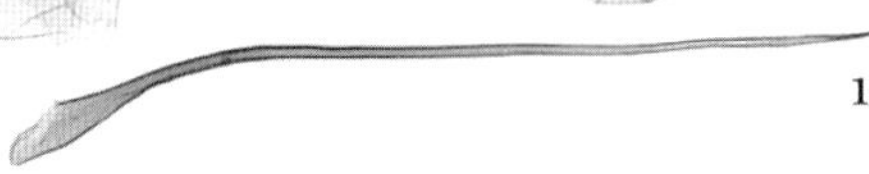

1 mm

162a *Stauropus affinis*
W Papua, GU 65-49

162b *Stauropus affinis*
PNG, GU 78-53

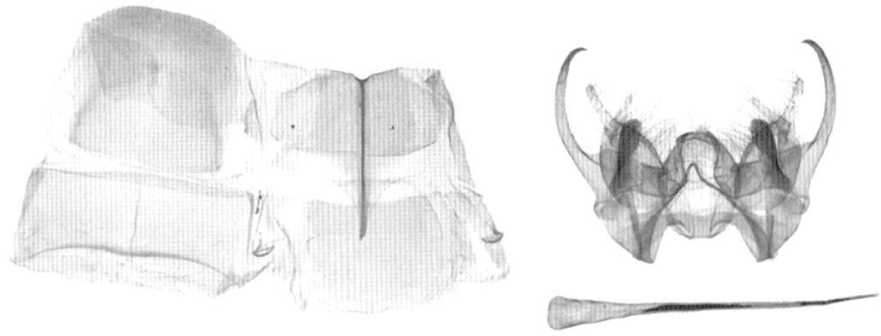

163a *Stauropus pratti*
W Papua, GU 94-68

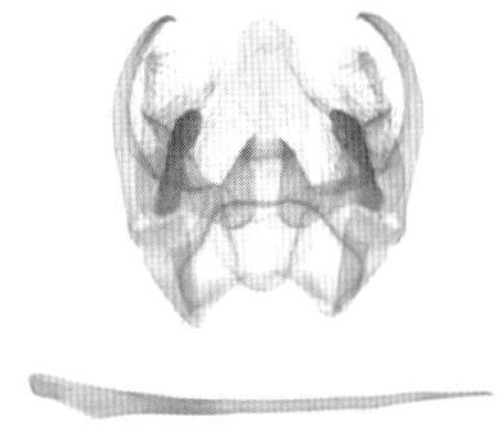

163b *Stauropus pratti*
W Papua, MWM 8.193

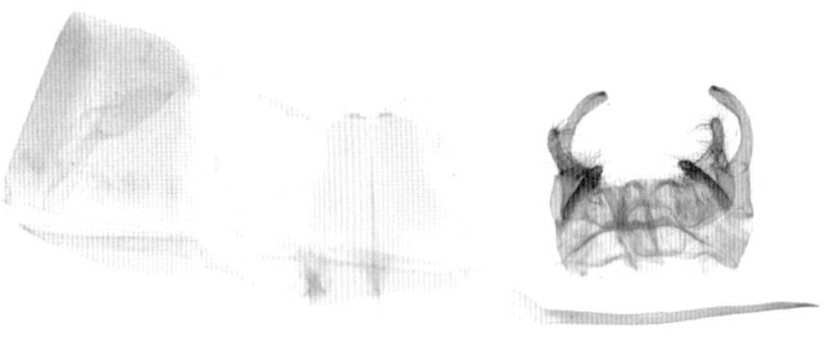

164a *Stauropus schaarschmidti*
W Papua, GU 78-21, holotype

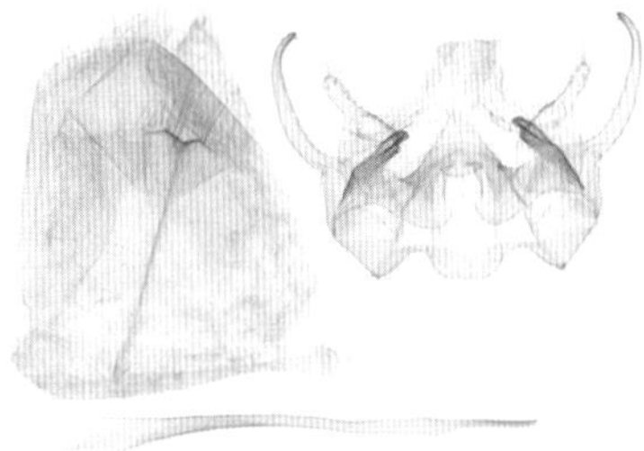

165a *Stauropus evanescens*
W Papua, GU 47-39

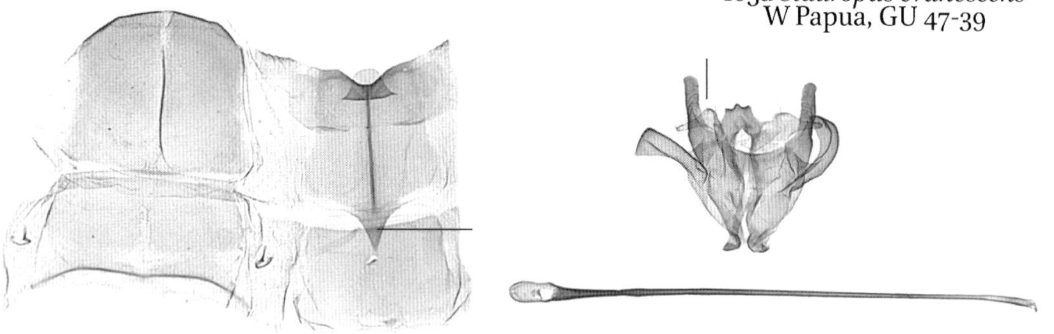

166a *Stauropus pelz*
Sumba, GU 65-45, paratype

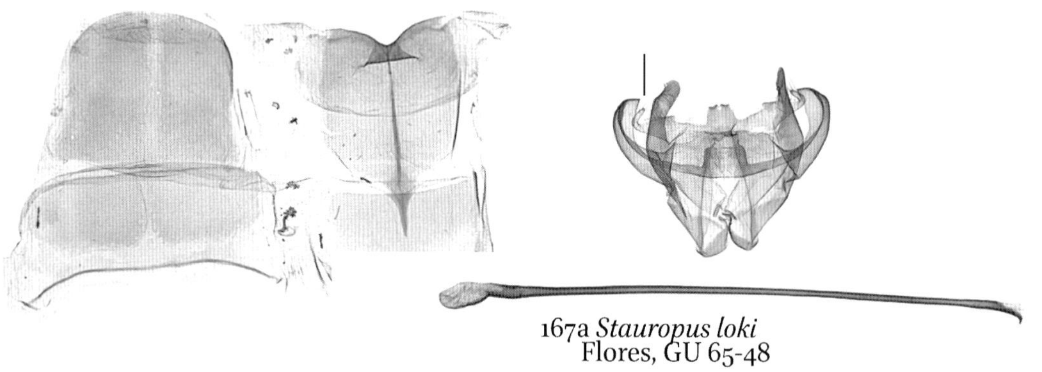

167a *Stauropus loki*
Flores, GU 65-48

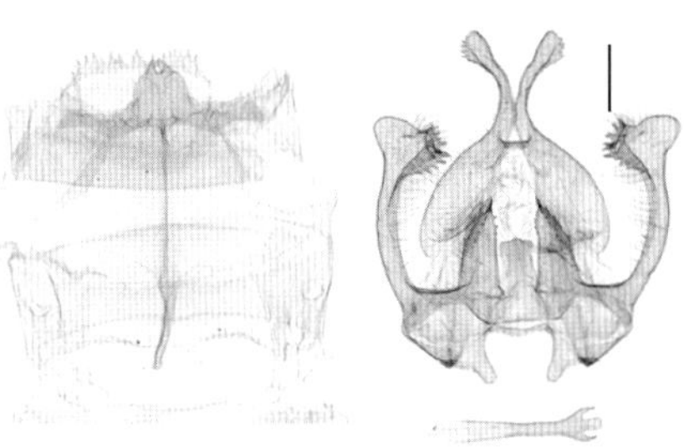

1 mm

168a *Stauropus virescens virescens*
NW India, BM 503, holotype

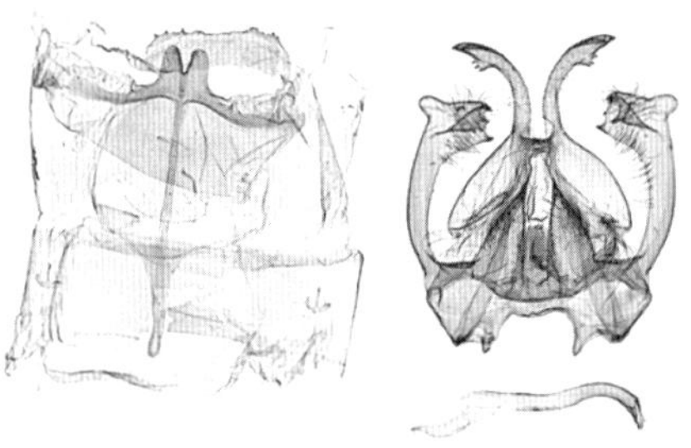

168b *Stauropus virescens obscuratus*
Sumatra, BM 502
(holotype of *Benbowia dudgeoni*)

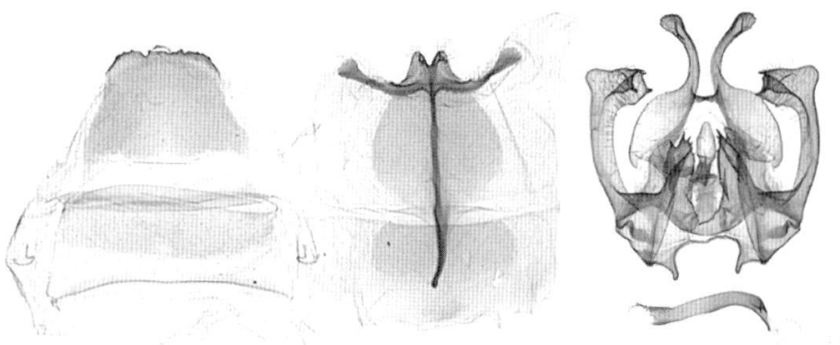

168c *Stauropus virescens obscuratus*
Java, GU 65-60

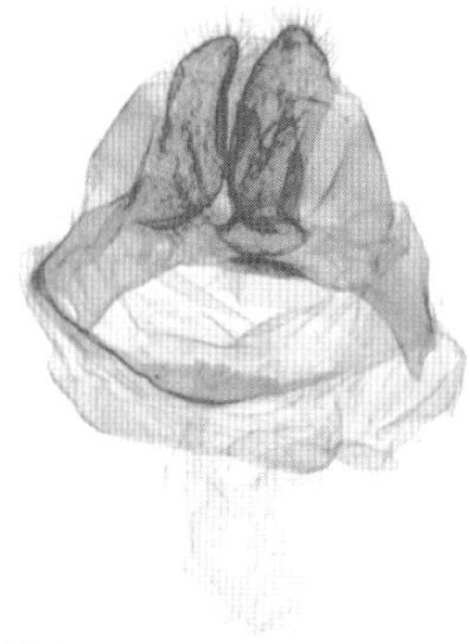

168d *Stauropus virescens obscuratus*
Java, GU 43-65

168e *Stauropus virescens obscuratus*
Sumatra, GU 65-59

168f *Stauropus virescens virescens*
NE India, Meghalaya, GU 53-42, phallus

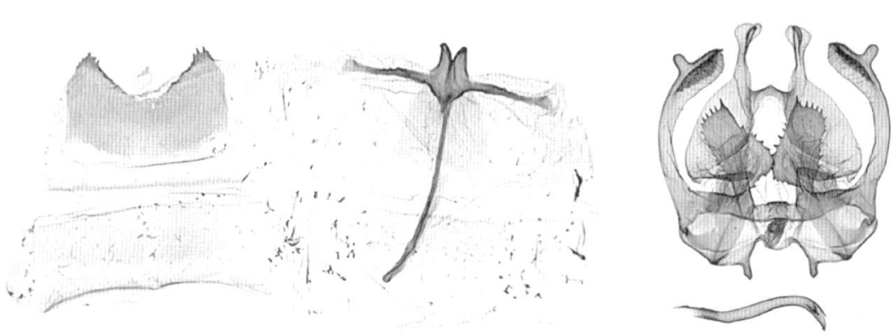

169a *Stauropus elisabethae elisabethae*
Sumatra, GU 93-38

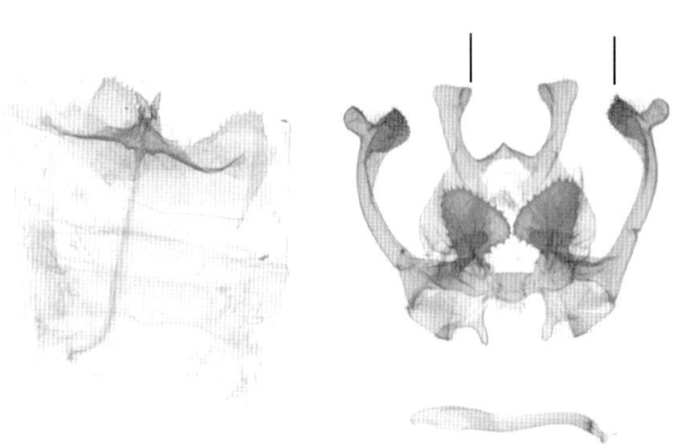

169b *Stauropus elisabethae kirikoffi*
Kalimantan, GU 76-11

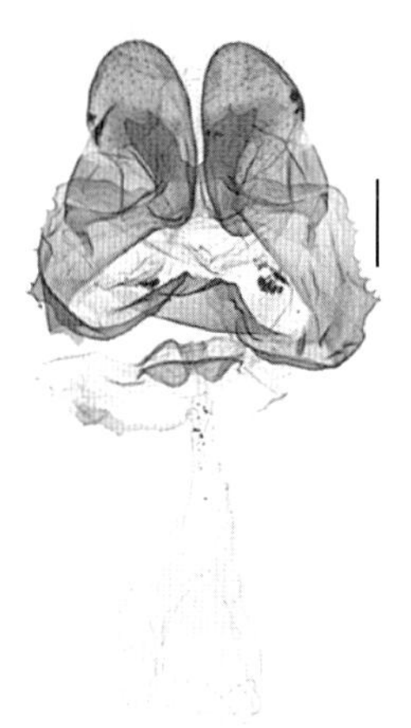

169c *Stauropus elisabethae elisabethae*
Sumatra, GU 65-53

1 mm

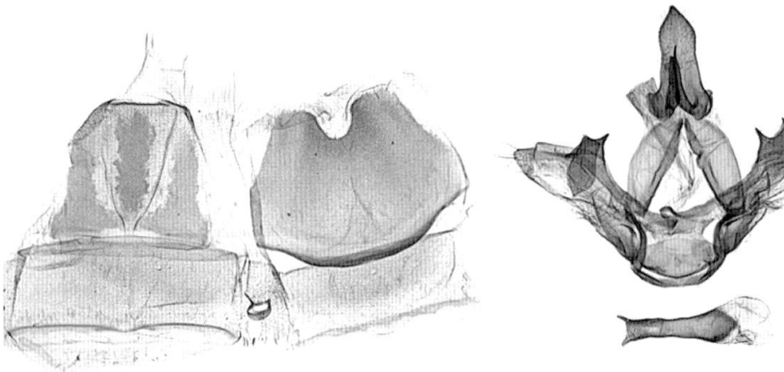

170a *Kakostauropus glaucoviridis glaucoviridis*
W Papua, GU 93-72

170b *K. glaucoviridis glaucoviridis*
W Papua, GU 93-73

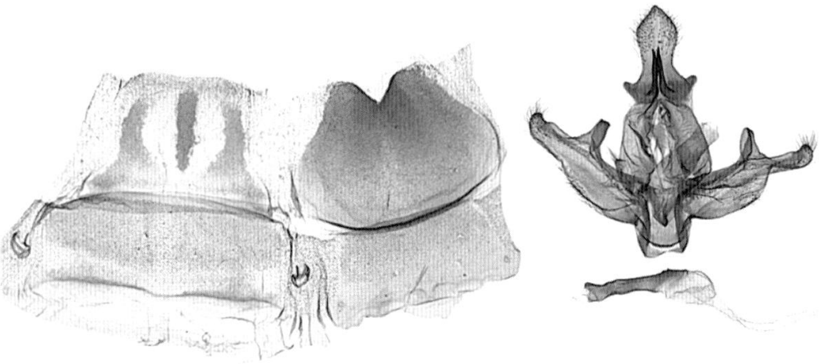

170c *Kakostauropus glaucoviridis seramica*
Seram, BM 10314132, holotype,

171a *Syntypistis kebeae*
W Papua, 93-90

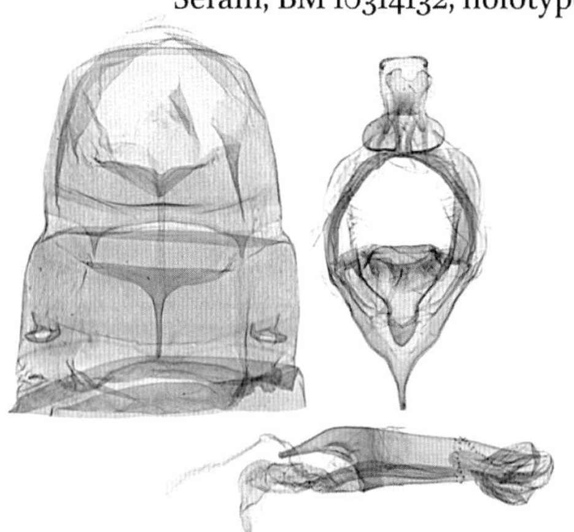

171b *Syntypistis kebeae*
PNG, BM 1459, lectotype

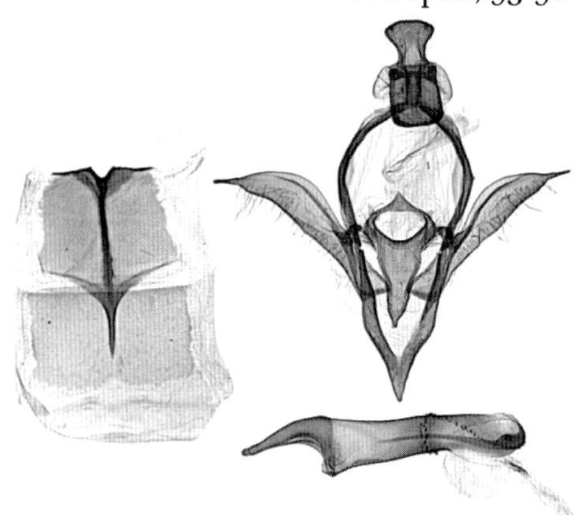

171c *Syntypistis kebeae*
W Papua, 93-86

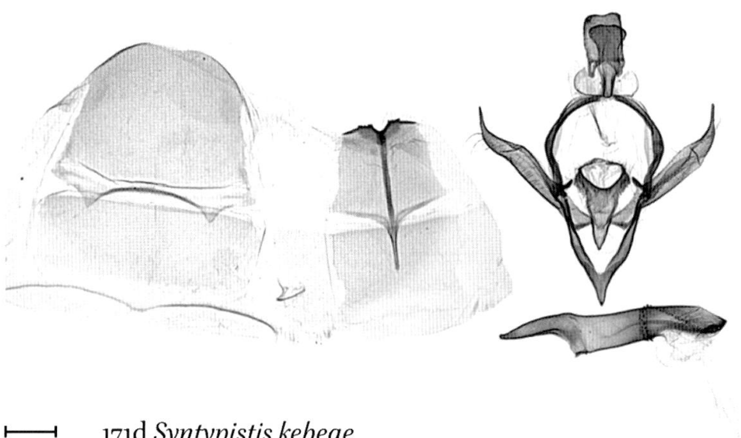

1 mm

171d *Syntypistis kebeae*
W Papua, GU 93-88

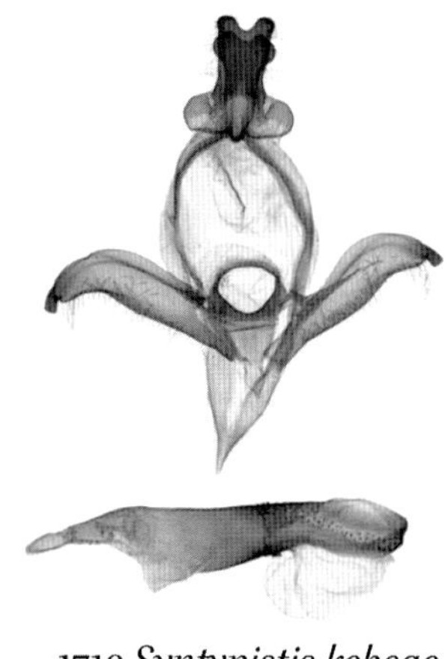

171e *Syntypistis kebeae*
PNG, MWM 15.579

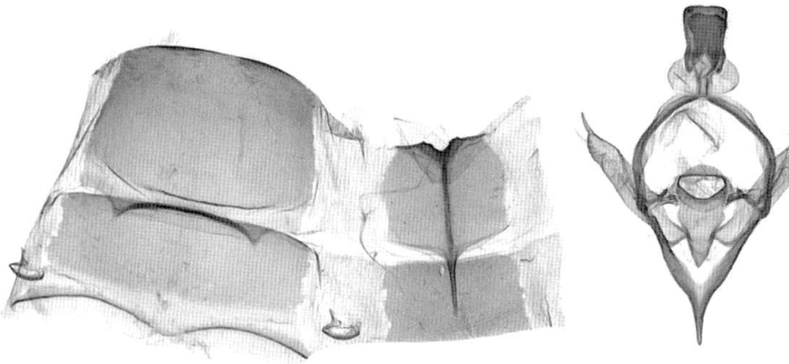
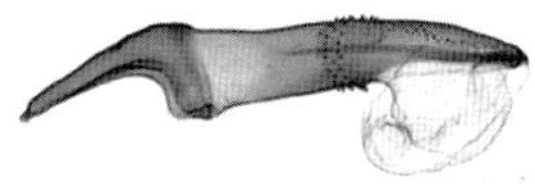

171f *Syntypistis kebeae*
Australia, Qsd., GU 94-51

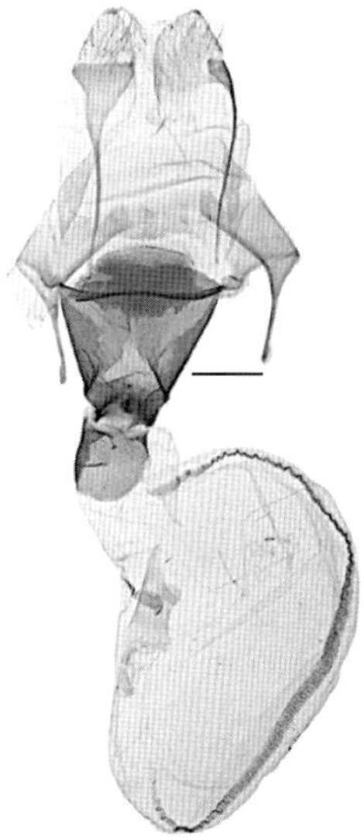

171g *Syntypistis kebeae*
W Papua, GU 93-87

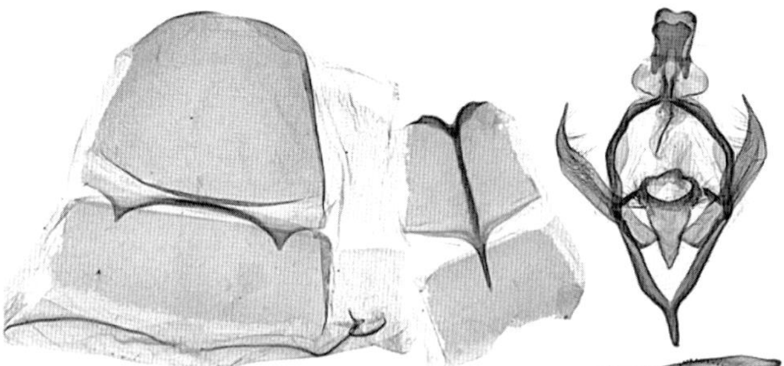
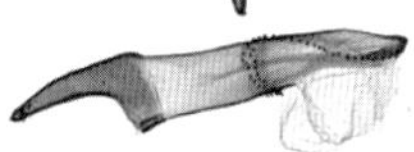

171h *Syntypistis kebeae*
Buru, GU 93-69

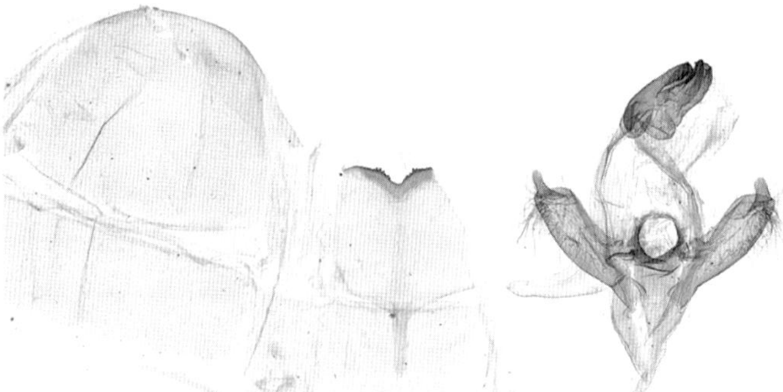

1 mm

172a *Syntypistis dasychirinus*
Sulawesi, GU 79-66

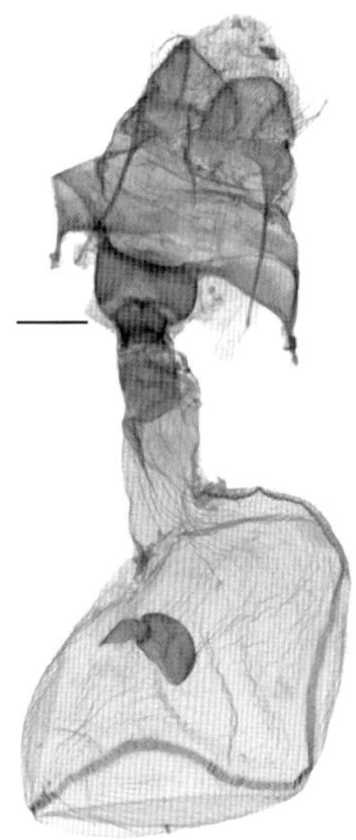

172b *Syntypistis dasychirinus*
Sulawesi, MWM 15.611

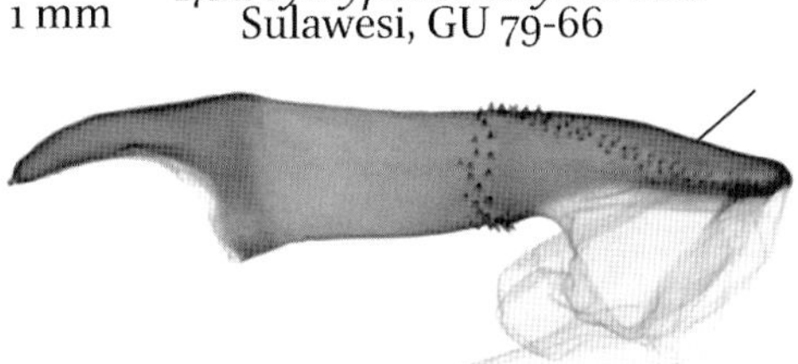

171i *Syntypistis kebeae*
PNG, MWM 15.579, phallus

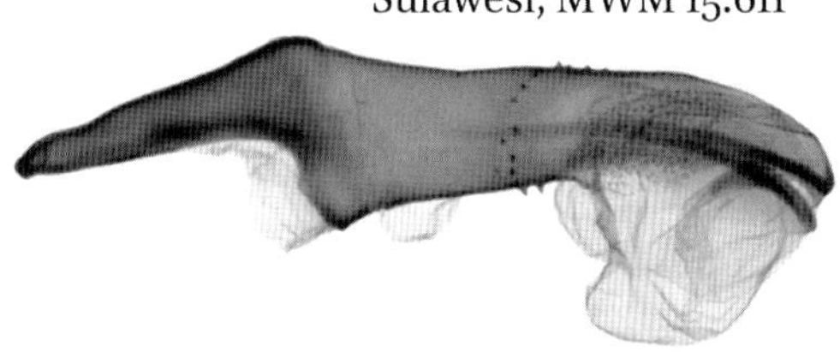

172c *Syntypistis dasychirinus*
Sulawesi, MWM 15.613, phallus

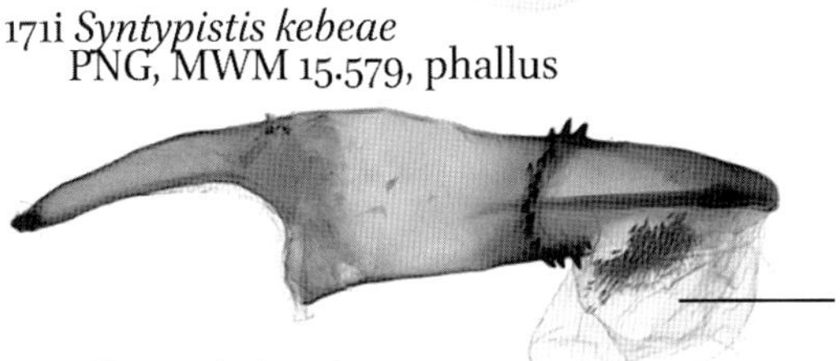

173a *Syntypistis grisescens*
Lombok, GU 94-61, phallus

1 mm

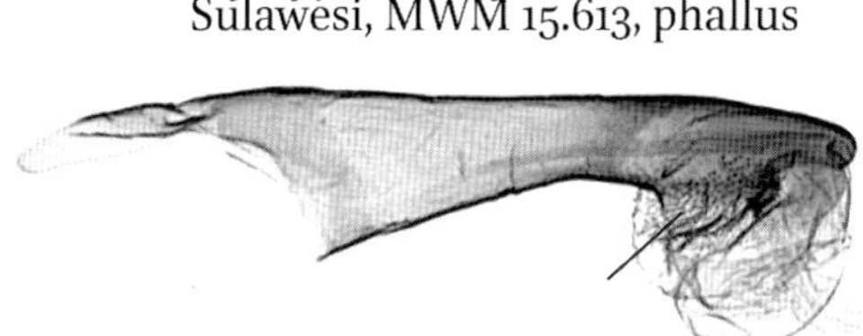

174a *Syntypistis frugilegus*
New Britain, GU 65-66, phallus

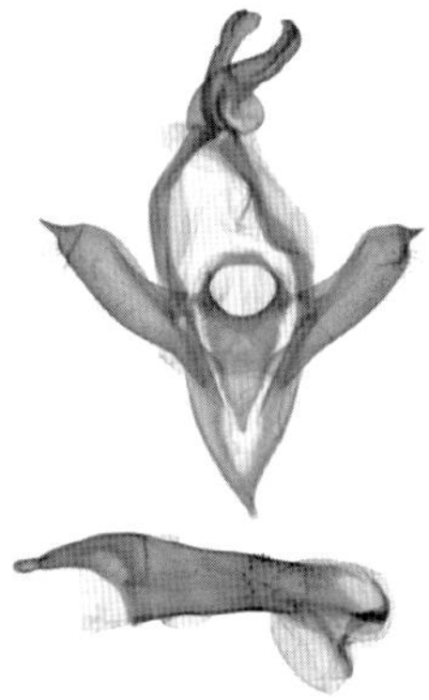

172d *Syntypistis dasychirinus*
Sulawesi, MWM 15.606

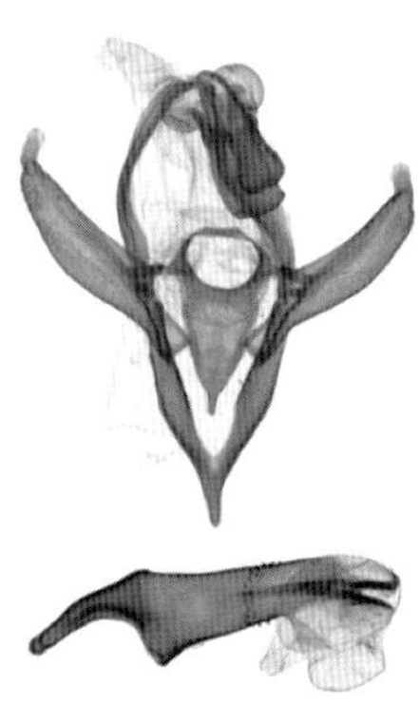

172e *Syntypistis dasychirinus*
Sulawesi, MWM 15.614

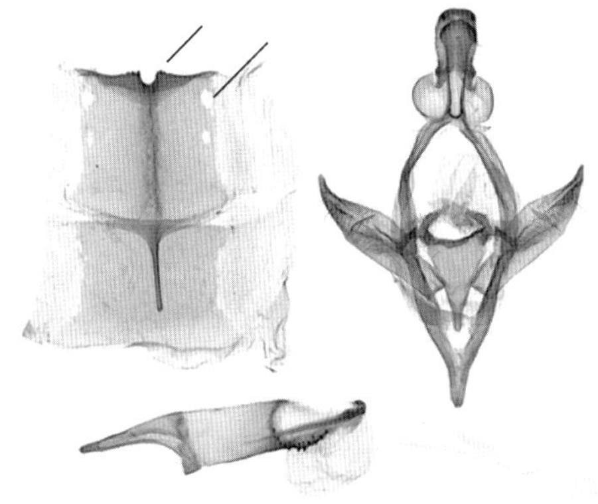

173b *Syntypistis grisescens*
Sumba, GU 94-64

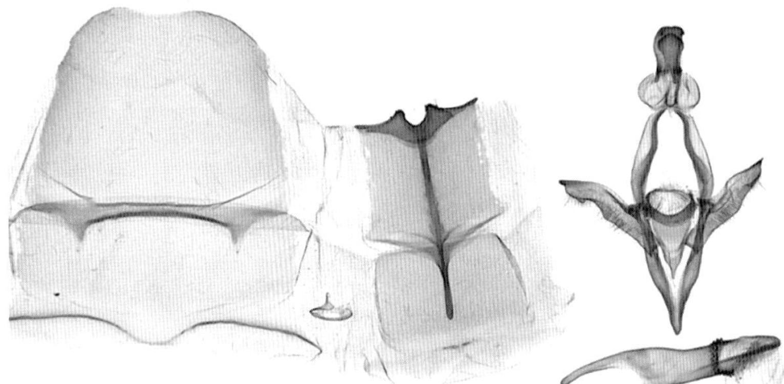

173c *Syntypistis grisescens*
Lomblen, GU 94-63

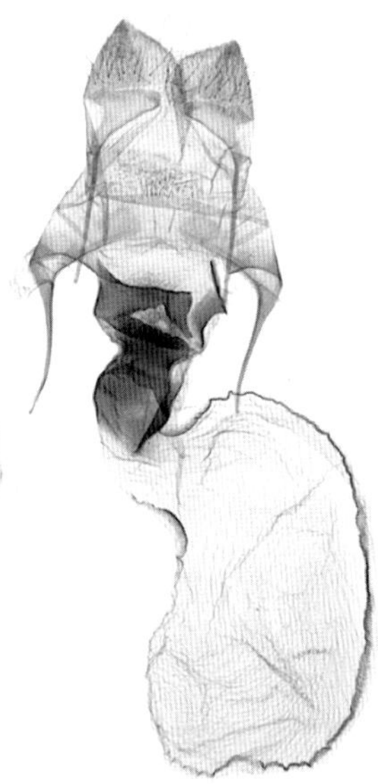

173d *Syntypistis grisescens*
Sumbawa, GU 94-55

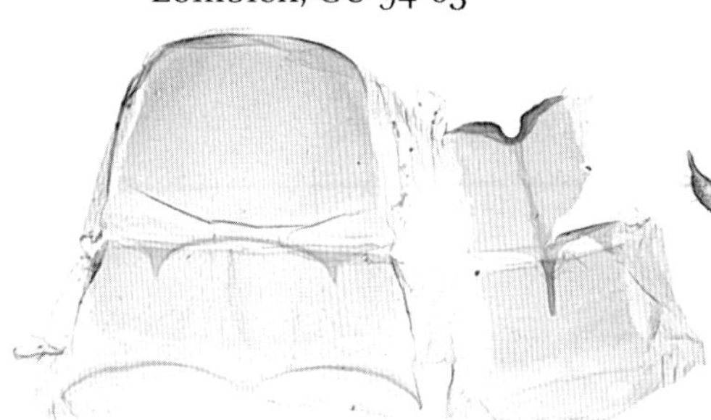

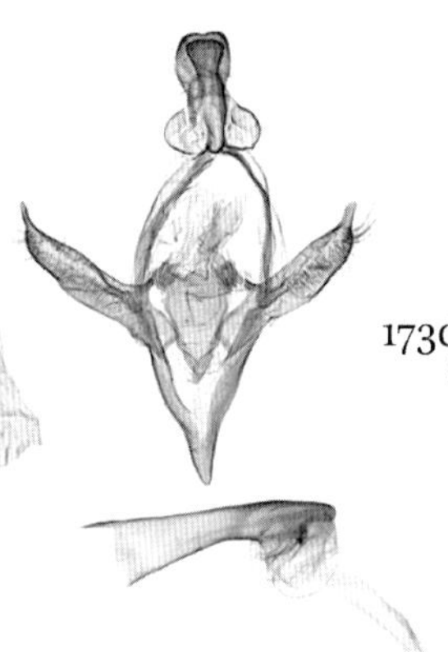

174b *Syntypistis frugilegus*
New Britain GU 95-66

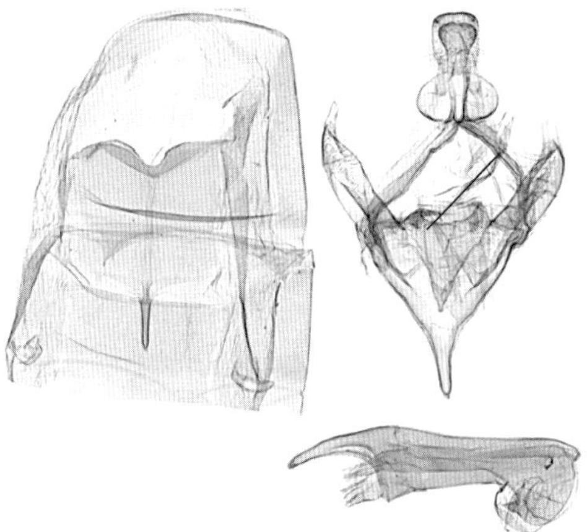

174c *Syntypistis frugilegus*
PNG, Rook Isl., BM 1534, holotype

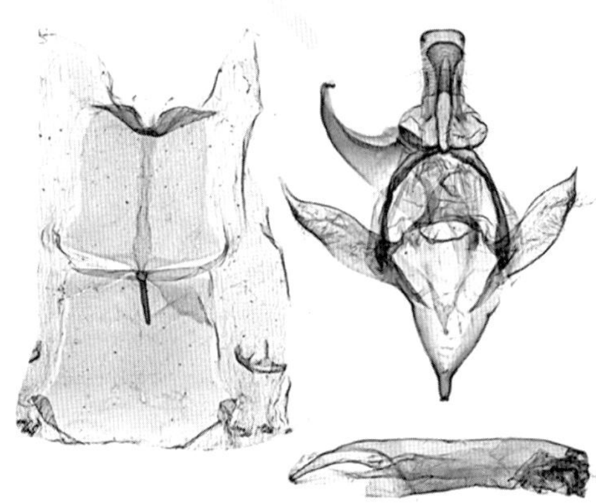

174d *Syntypistis frugilegus*, New Ireland
BM 155,(holotype *eichhorni*)

1 mm

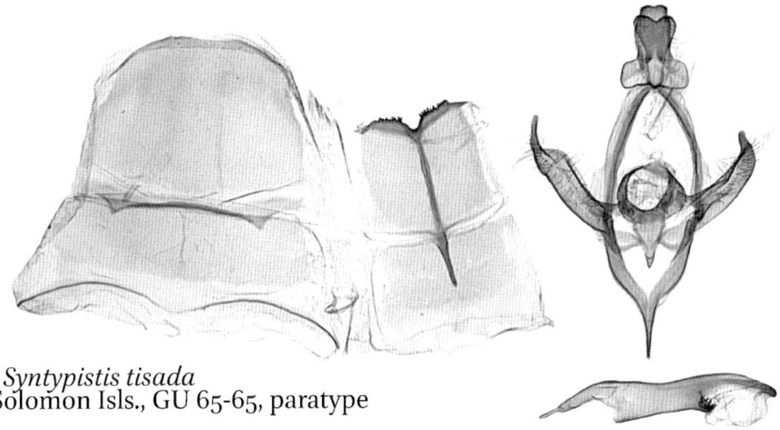

175a *Syntypistis tisada*
Solomon Isls., GU 65-65, paratype

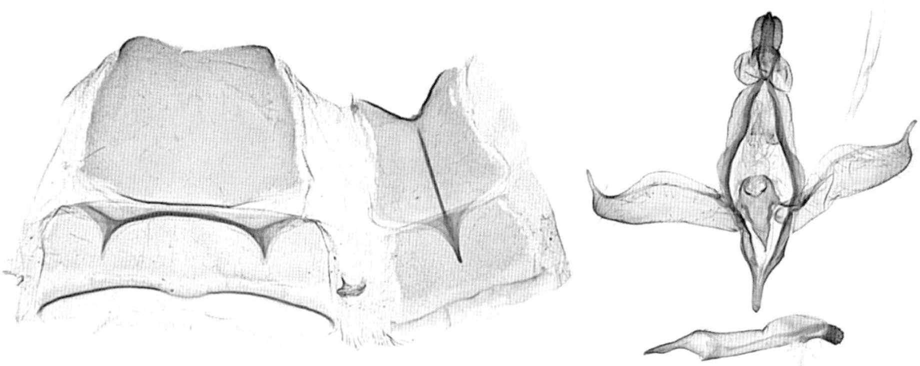

176a *Syntypistis mapia*
W Papua, GU 65-74, paratype

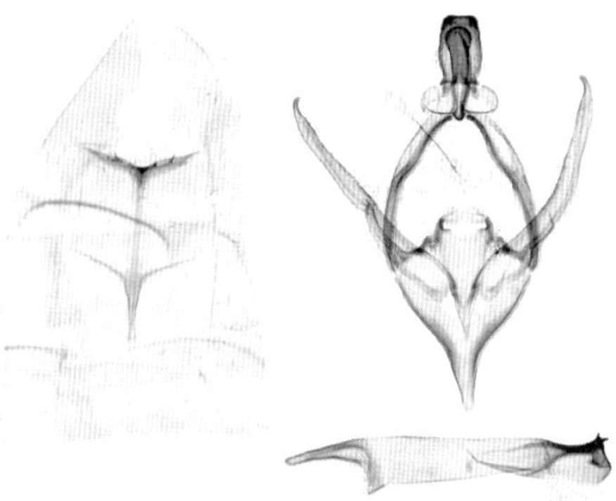

177a *Syntypistis charistera fraseriana*
Kalimantan, GU 47-60

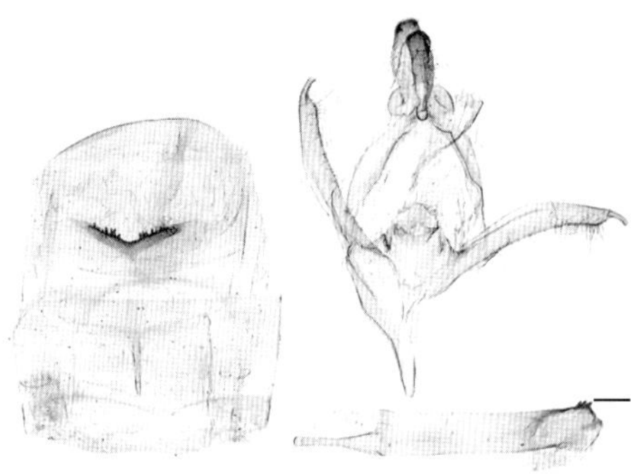

177b *Syntypistis charistera fraseriana*
Malaya, GU 18-85

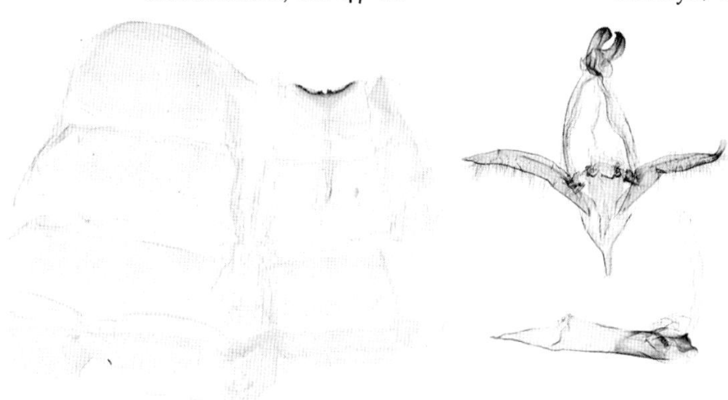

1 mm

177c *Syntypistis charistera celebensis*
Sulawesi, GU 42-76

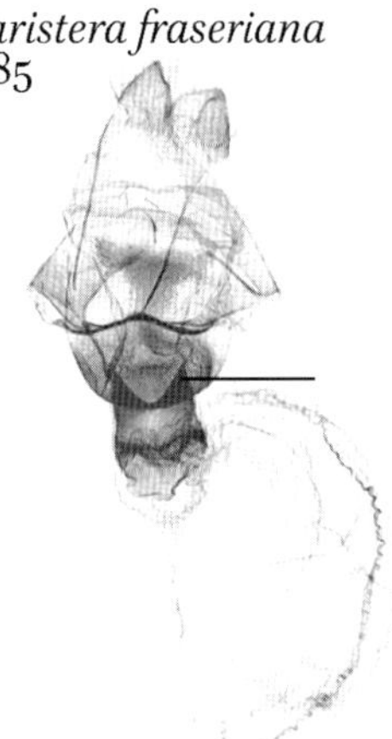

177d *Syntypistis charistera celebensis* Sulawesi, GU 42-80

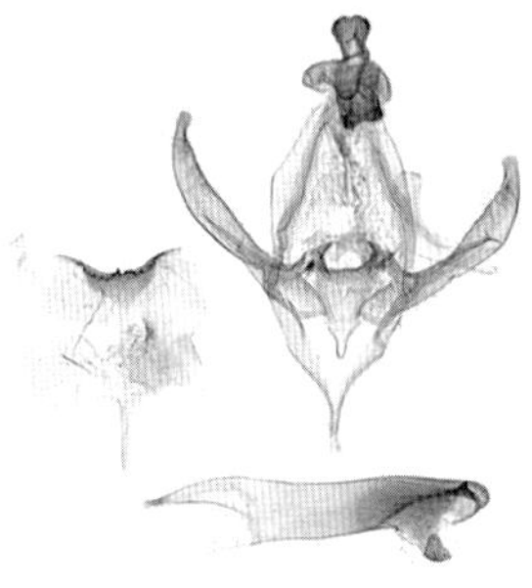

177e *Syntypistis charistera celebensis*
Peleng, GU 46-76

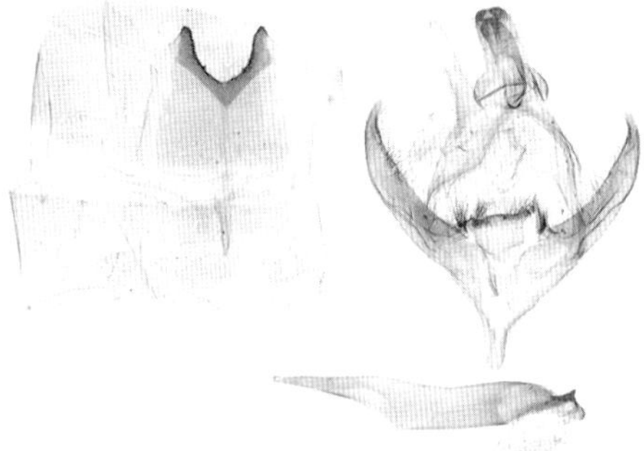

177f *S. charistera charistera*
Luzon, GU 18-88

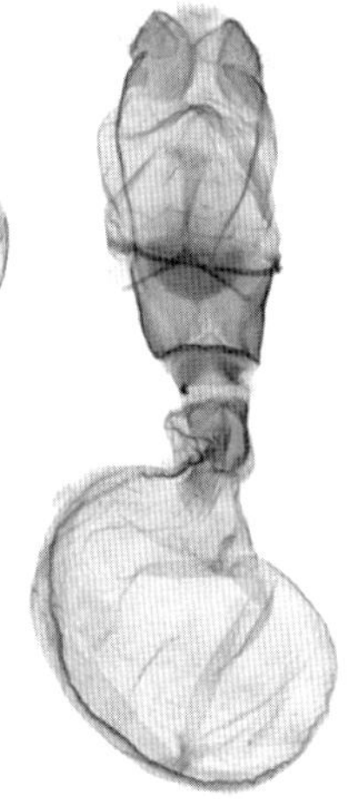

177g *Syntypistis charistera charistera*
Mindanao, MWM 15.612

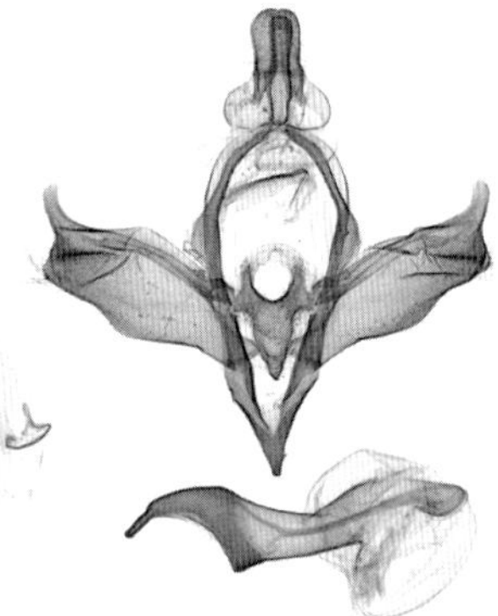

178a *Syntypistis triplagosus*
PNG, GU 94-50

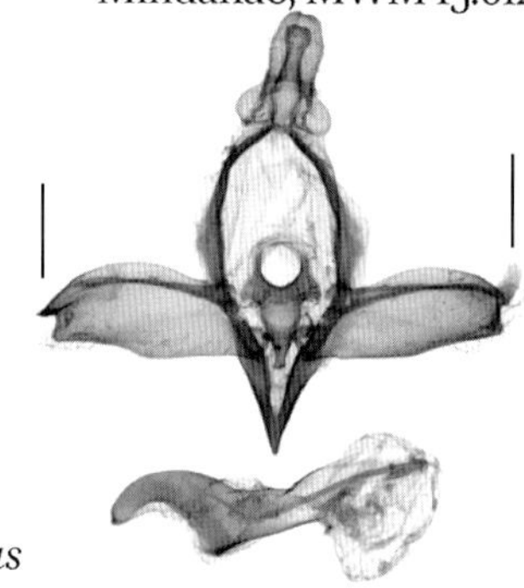

178b *S. triplagosus*
PNG, MWM 15.583

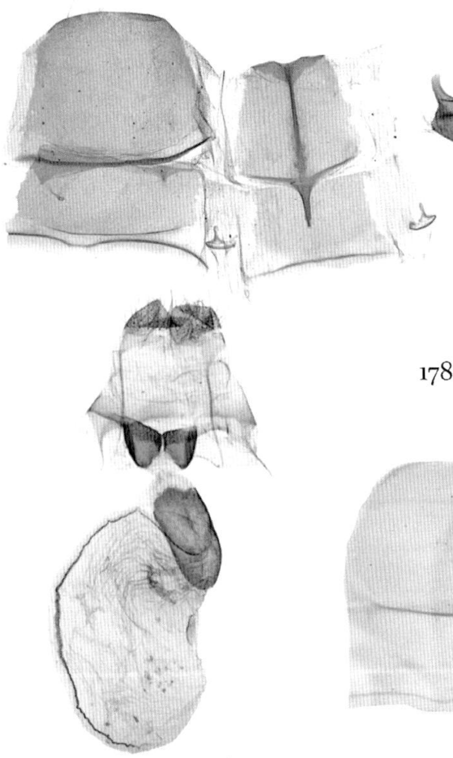

178c *Syntypistis triplagosus*
PNG,MWM 15.580

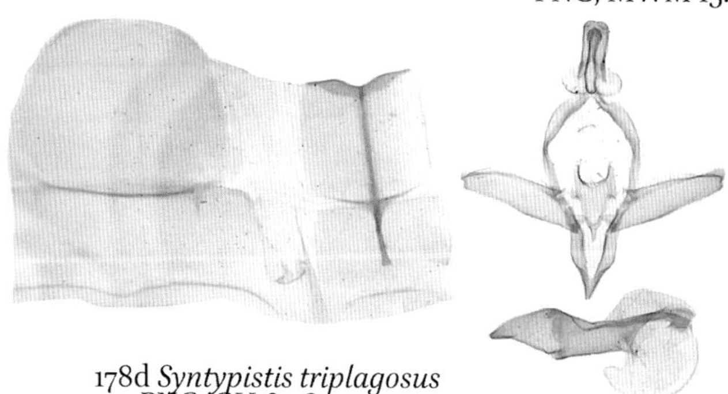

178d *Syntypistis triplagosus*
PNG, GU 65-82

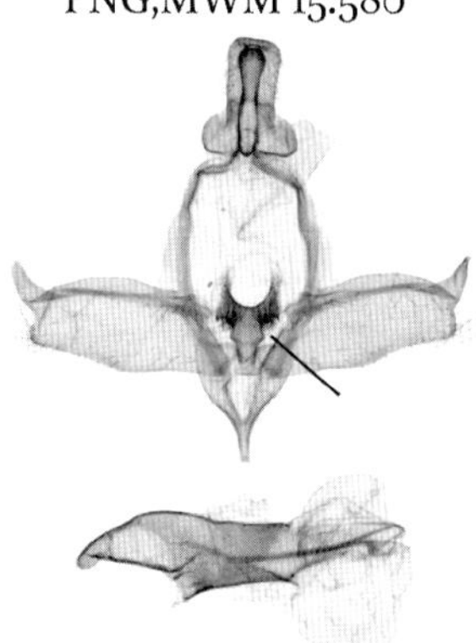

1 mm

178e *Syntypistis triplagosus*
PNG, MWM 15.881

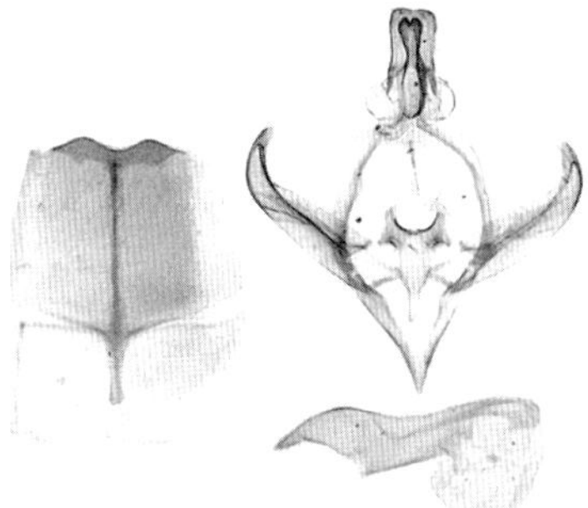

178f *Syntypistis triplagosus*
PNG, GU 65-81

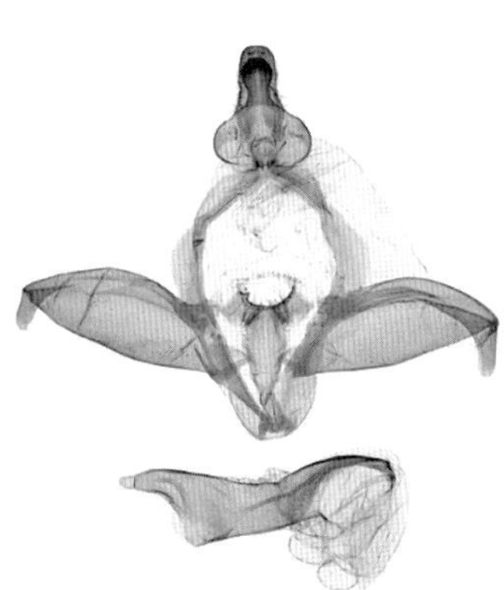

178g *Syntypistis triplagosus*
W Papua, MWM 15.732

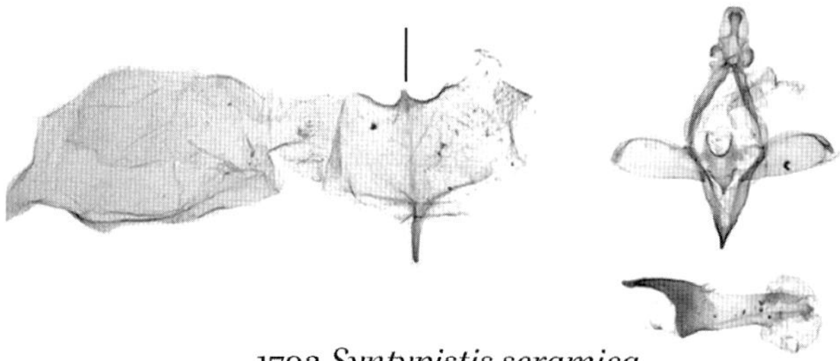

179a *Syntypistis seramica*
Seram, SS 799
(from Kobayashi 2011, fig. 42a-c, modified)

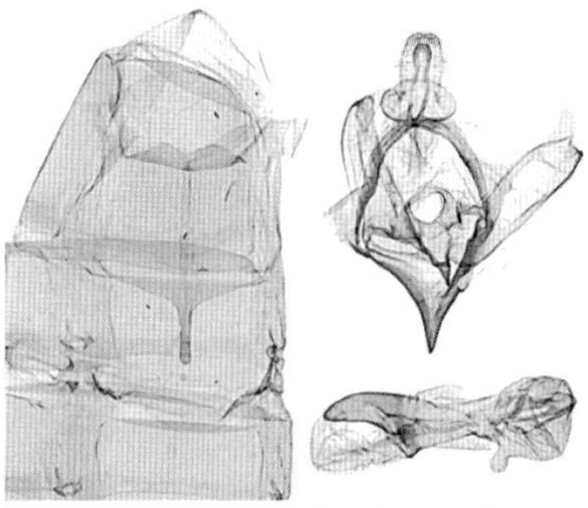

180a *Syntypistis trisospylus*
W Papua, BM 1544 (holotype *roseus*)

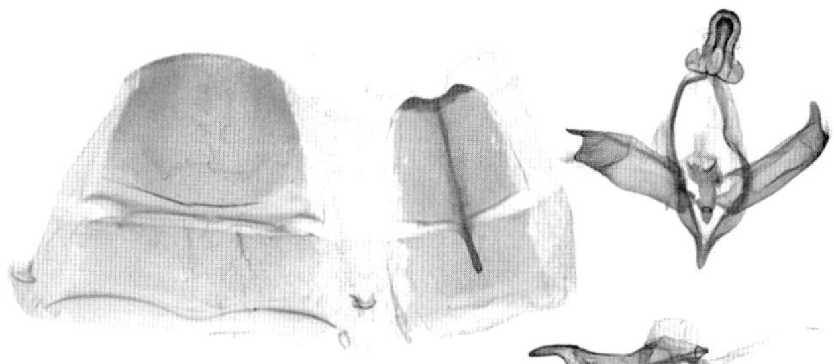

180b *Syntypistis trisospylus*
PNG, GU 65-84

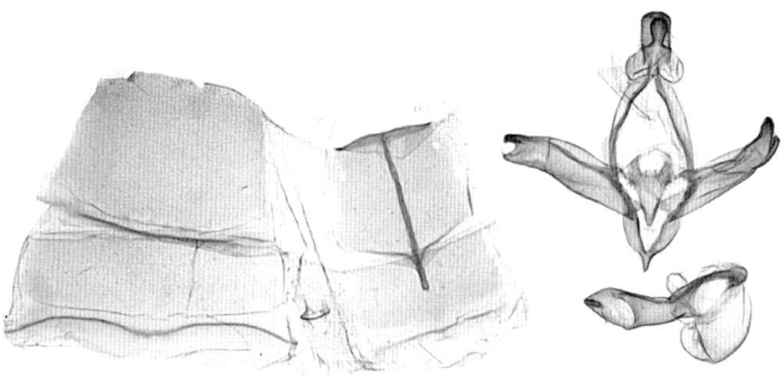

181a *Syntypistis famelica*
PNG, GU 65-71

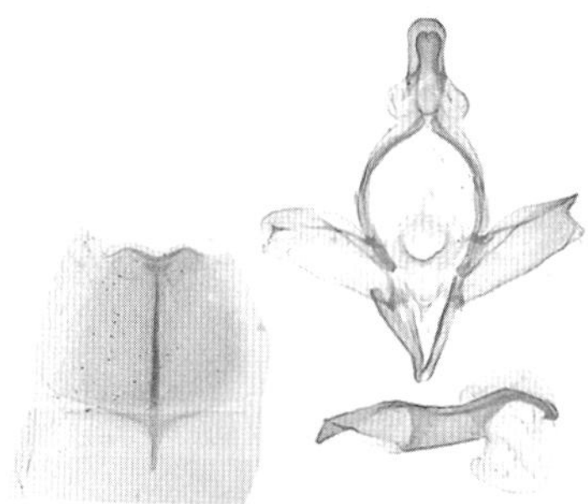

181b *Syntypistis famelica*
PNG, GU 65-80

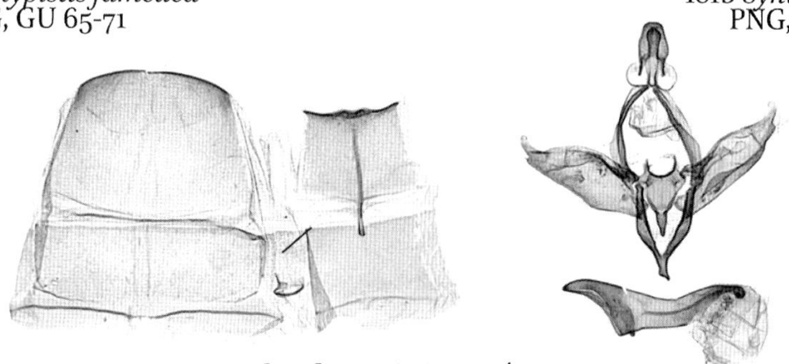

182a *Syntypistis quadrivittata*
W Papua, GU 93-91

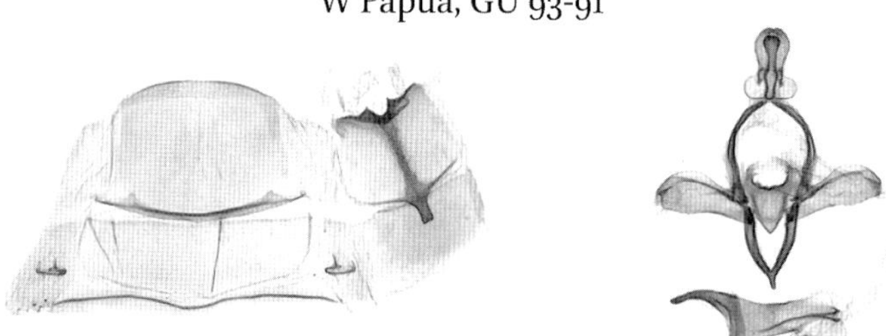

184a *Syntypistis dinawa*
W Papua, GU 94-48

1 mm

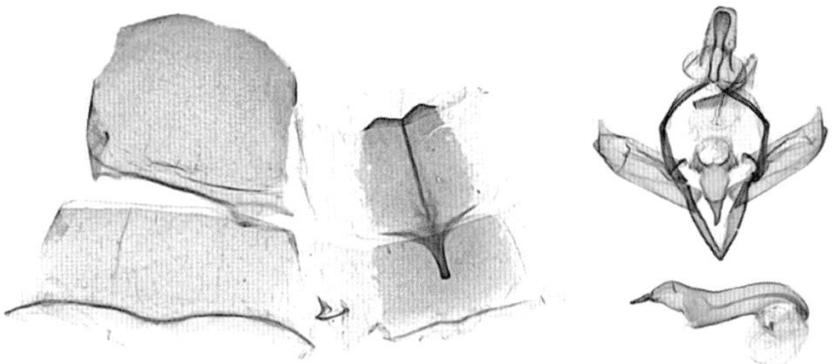

185a *Syntypistis bella*
W Papua, GU 93-82

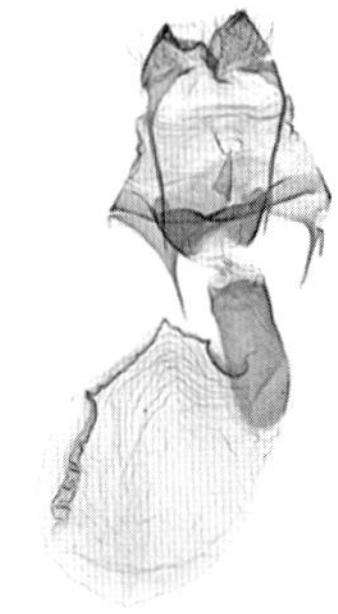

185b *Syntypistis bella*
Seram, BM 10313993

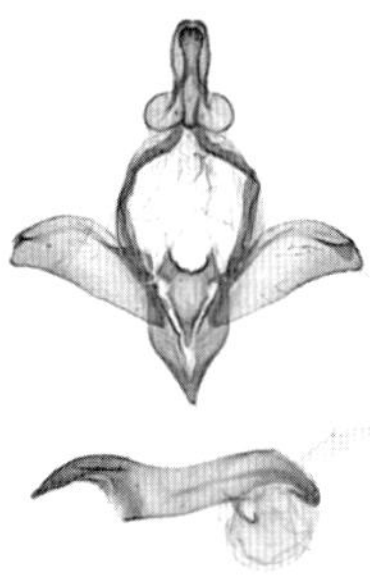

185c *Syntypistis bella*
Seram, BM 10313994

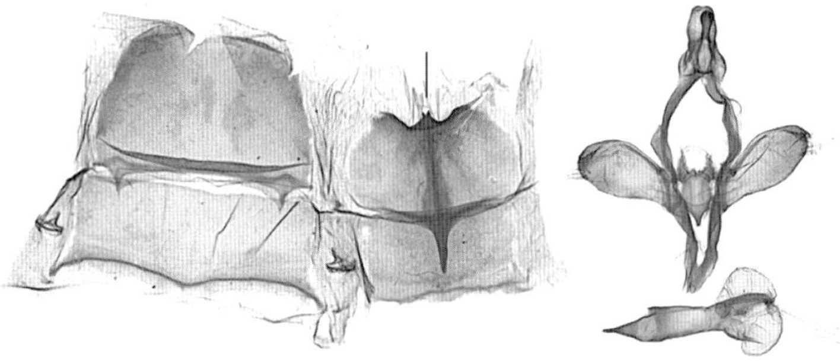

186a *Syntypistis mediogriseus binaia*
paratype Seram, BM 5-2017-

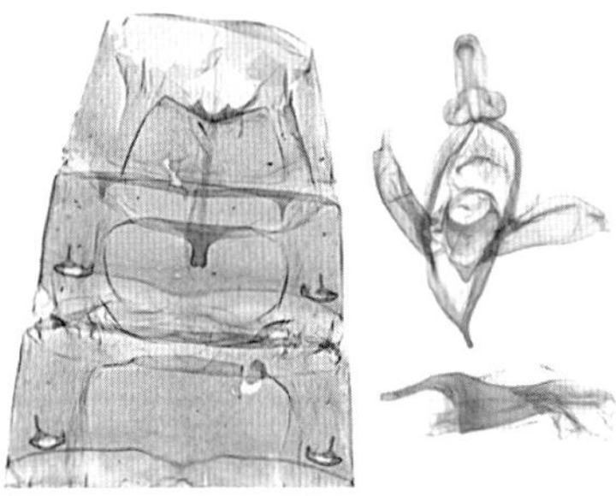

186b *Syntypistis mediogriseus mediogriseus*
W Papua, BM 1536, lectotype

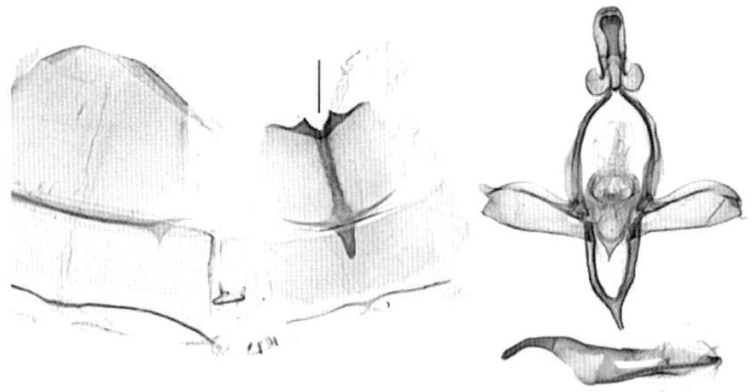

186c *Syntypistis mediogriseus mediogriseus*
W Papua, GU 93-83

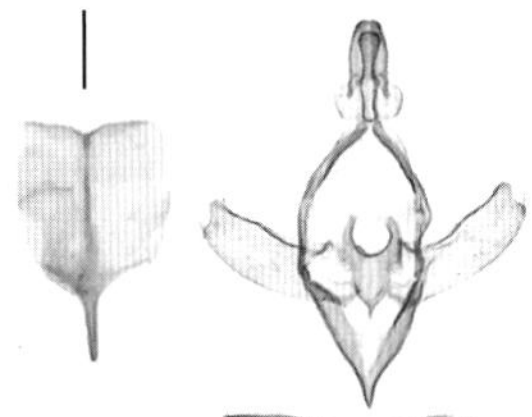

186d *Syntypistis mediogriseus inviridis*
New Britain, GU 65-78, paratype

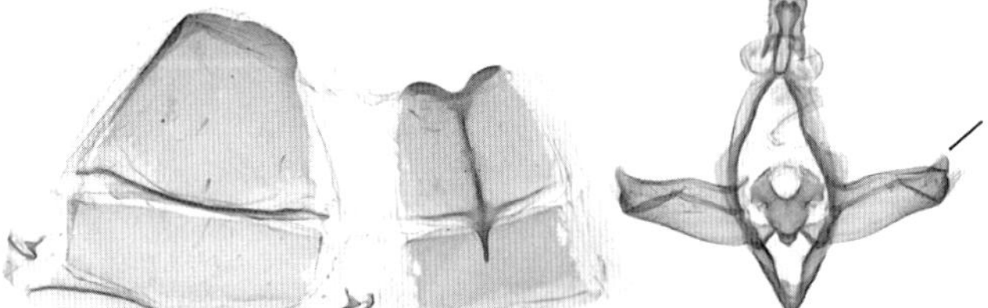

187a *Syntypistis leucaspis*
W Papua, GU 94-46, paratype

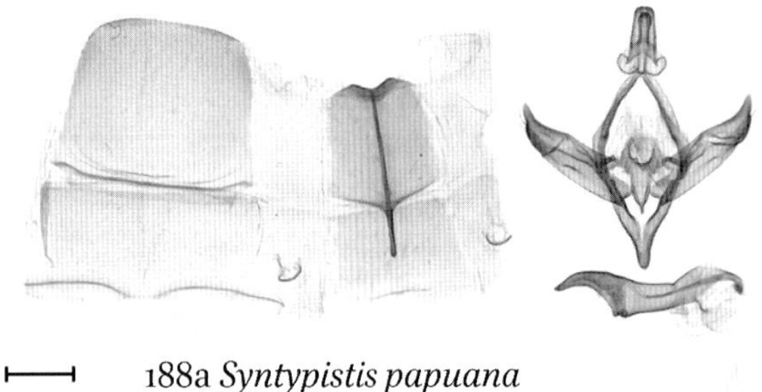

188a *Syntypistis papuana*
PNG, GU 94-54, paratype

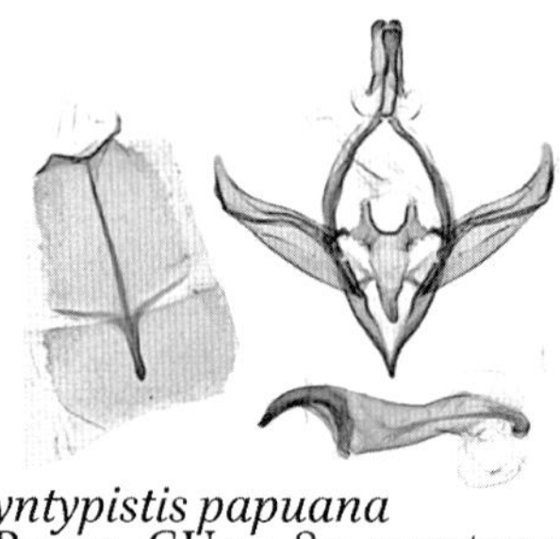

188b *Syntypistis papuana*
W Papua, GU 93-80, paratype

1 mm

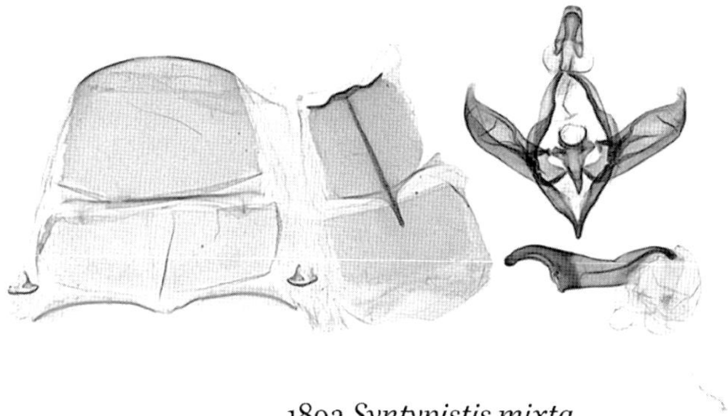

189a *Syntypistis mixta*
W Papua, GU 93-77

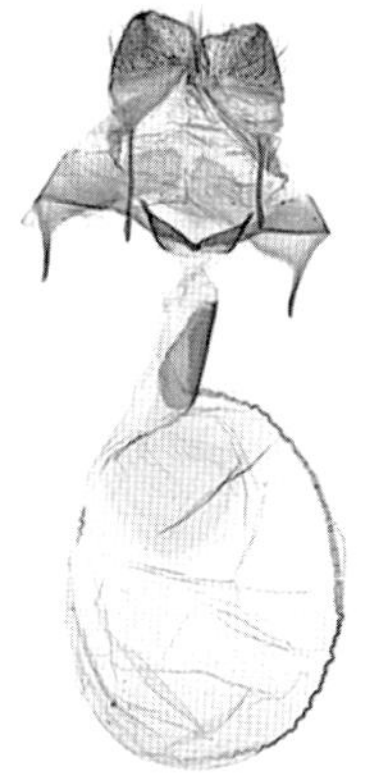

189b *Syntypistis mixta*
W Papua, GU 93-78

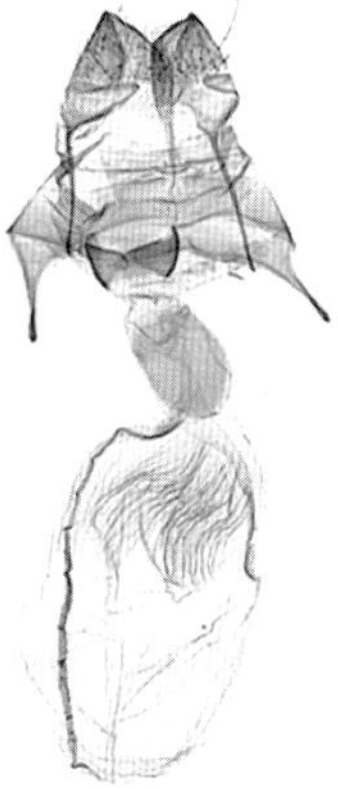

190a *Syntypistis chloriolus*
W Papua, GU 93-93

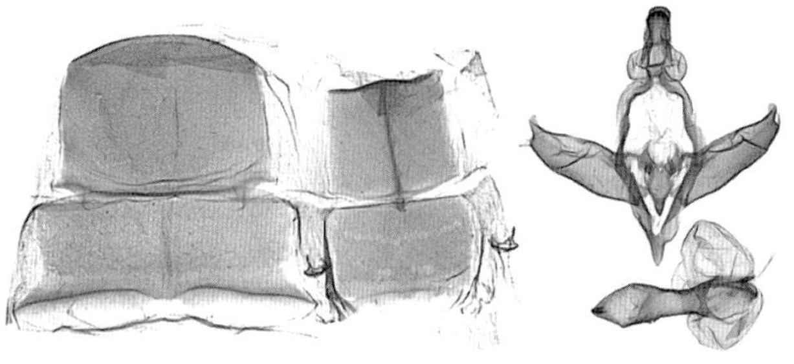

190b *Syntypistis chloriolus*
PNG, USNM 127.540

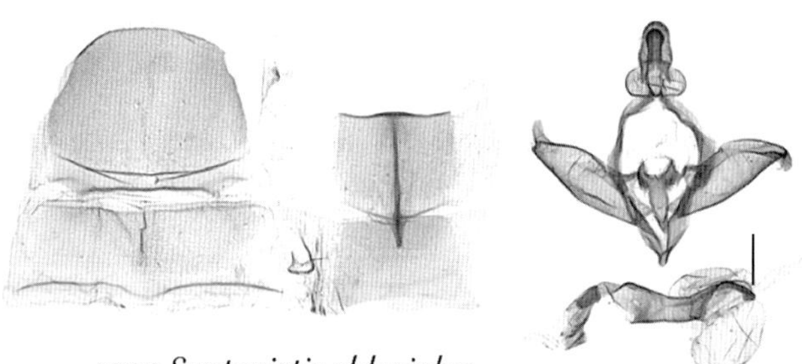

190c *Syntypistis chloriolus*
PNG, GU 65-85

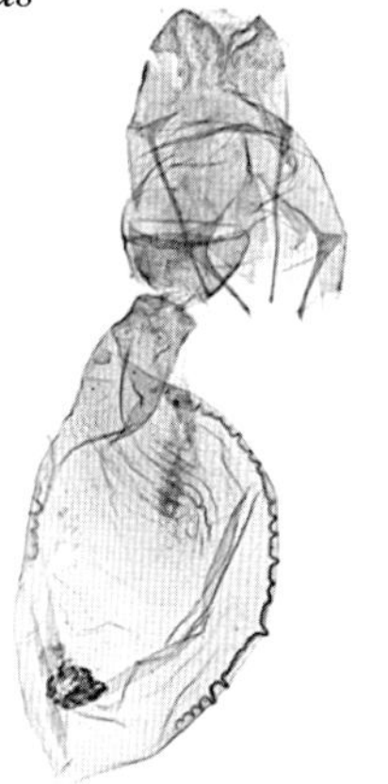

192a *Syntypistis alboviridis*
W Papua, BM 111,
paralectotype

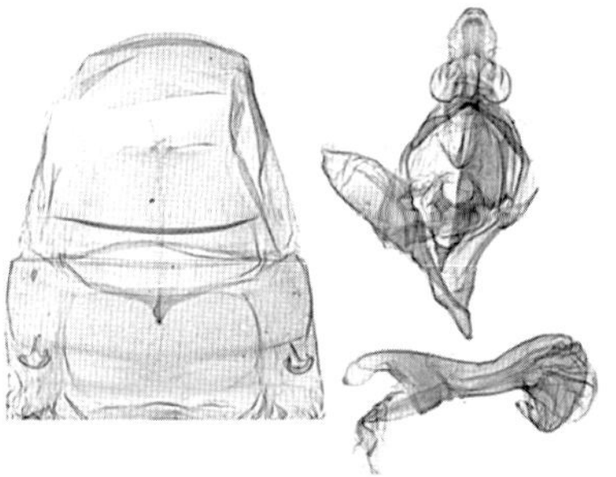

1 mm

193a *Syntypistis flavicollis*
W Papua, BM 111, holotype

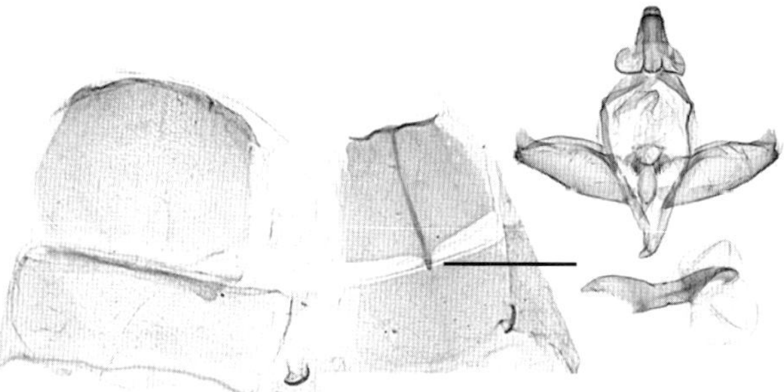

193b *Syntypistis flavicollis*
W Papua, GU 65-88

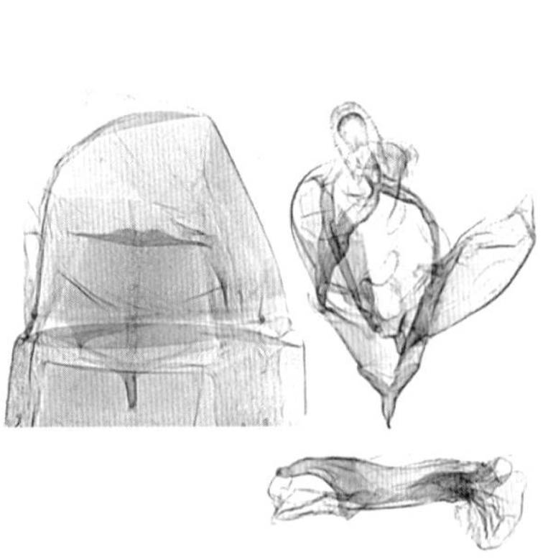

194a *Syntypistis nitidus*
PNG, BM 1467

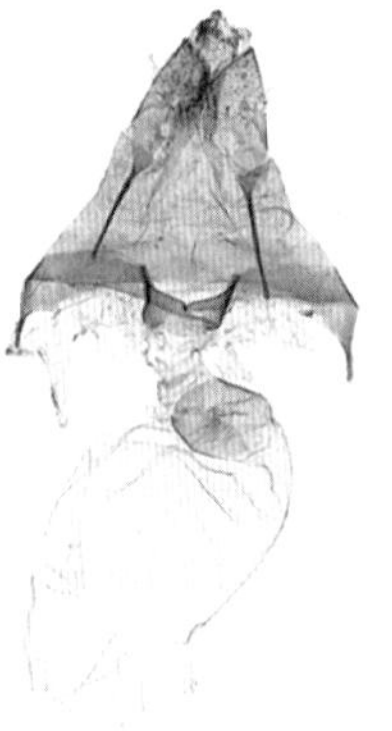

194b *Syntypistis nitidus*
PNG, MWM 15.582

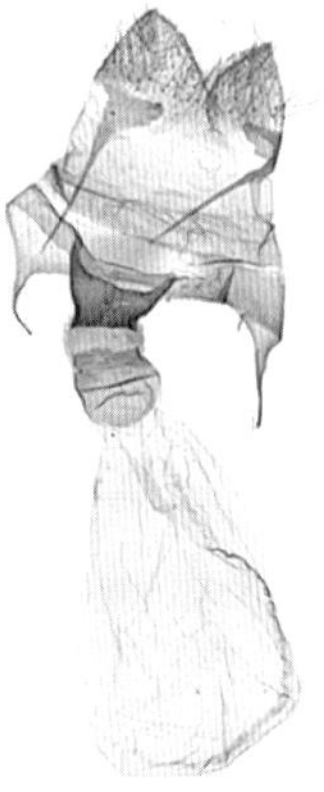

195a *Syntypistis gemmea*
Sumatra, GU 65-89
paratype

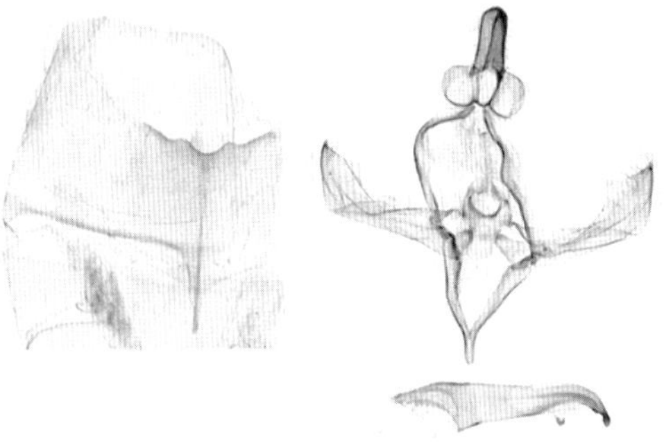

195b *Syntypistis gemmea*
Sumatra, GU 47-46, paratype

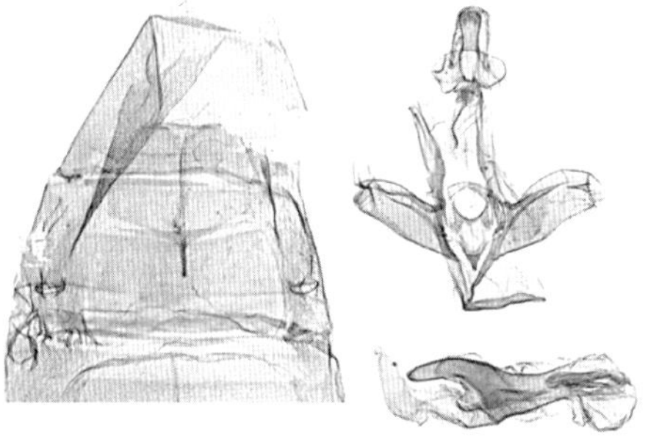

196a *Syntypistis purpurascens*
W Papua, BM 1458, paralectotype

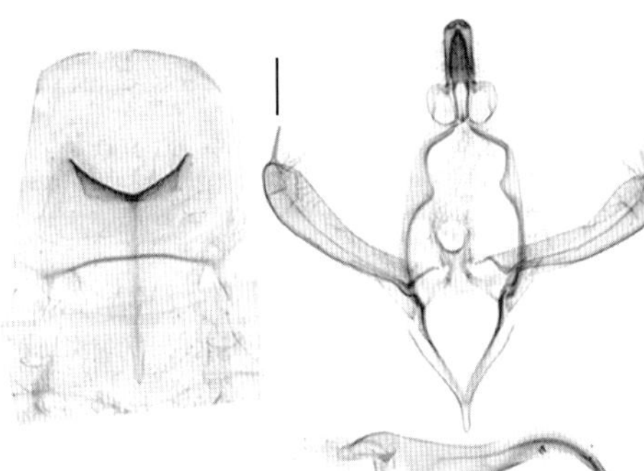

197a *Syntypistis triguttata*
Kalimantan, GU 47-48

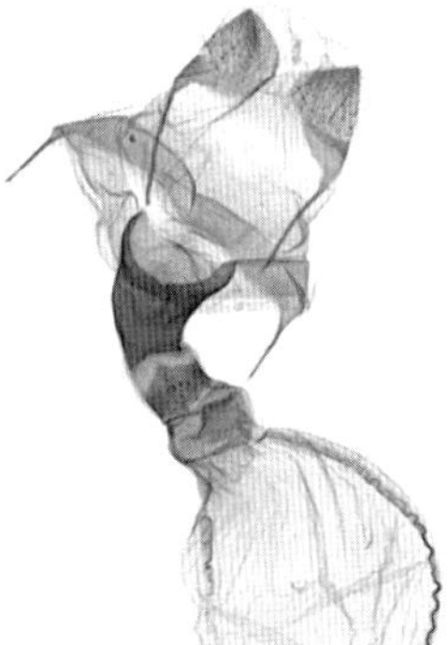

197b *Syntypistis triguttata*
Borneo, GU 96-64

198a *Syntypistis variegata*
Sumatra, GU 16-39

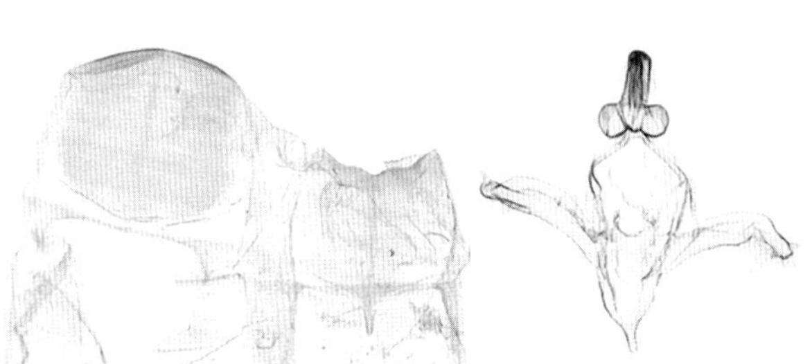

1 mm

198b *Syntypistis variegata*
Sumatra, GU 42-98

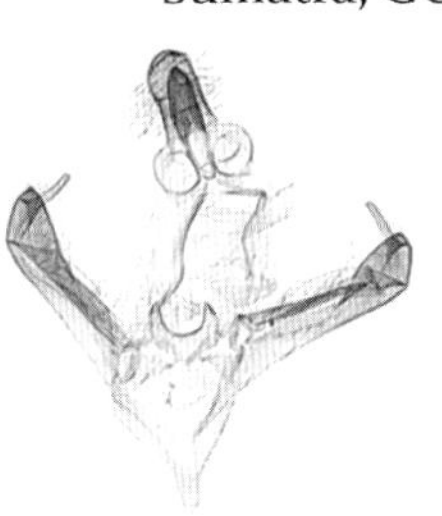

198c *Syntypistis variegata*
Sumatra, GU 02-54

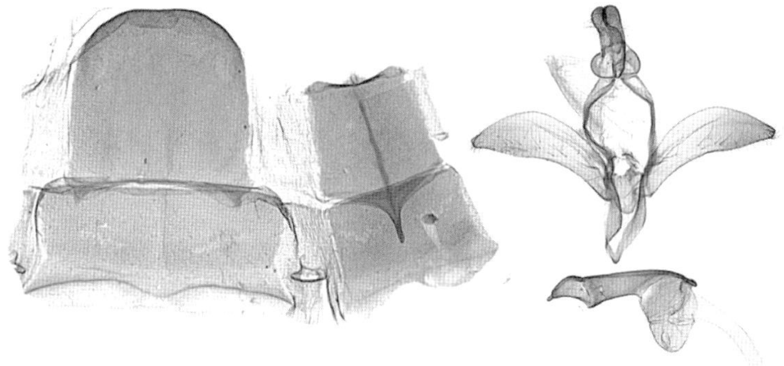

199a *Syntypistis uskwara*
PNG, GU 65-95

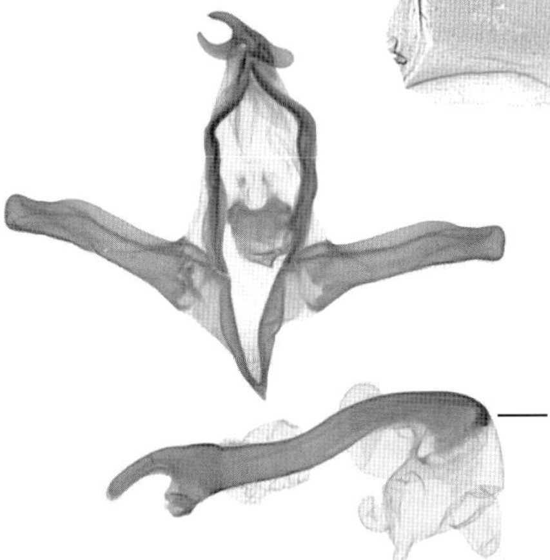

200a *Syntypistis miyanoi*
Sulawesi, MWM 15.604

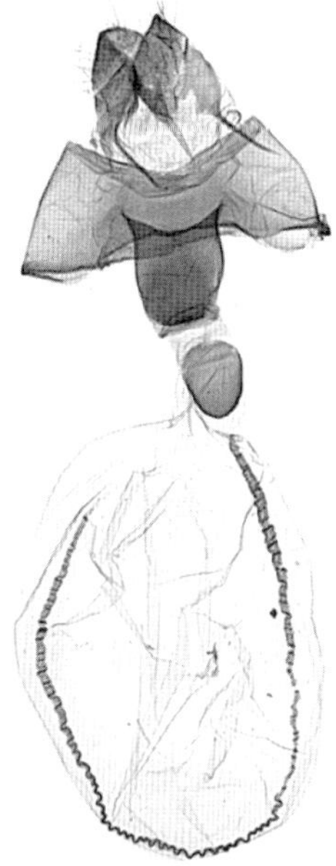

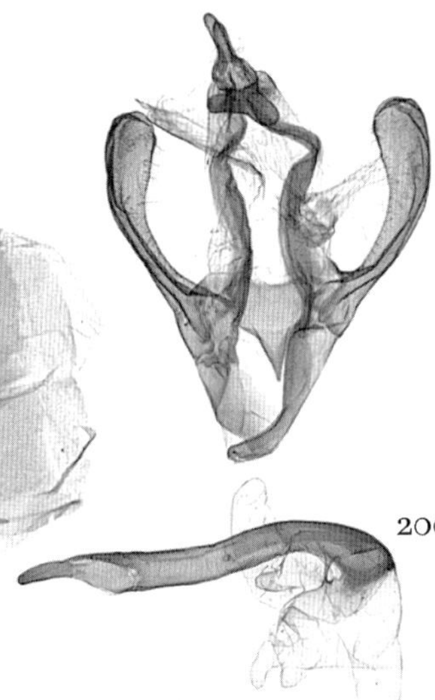

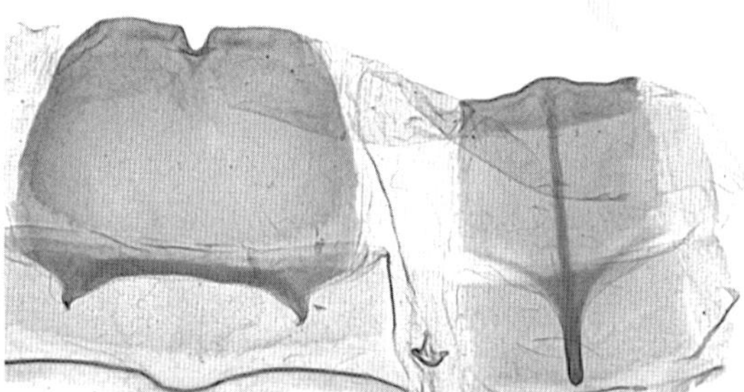

200c *Syntypistis miyanoi*
Sulawesi, GU 65-94

200b *Syntypistis miyanoi*
Peleng, GU 65-93

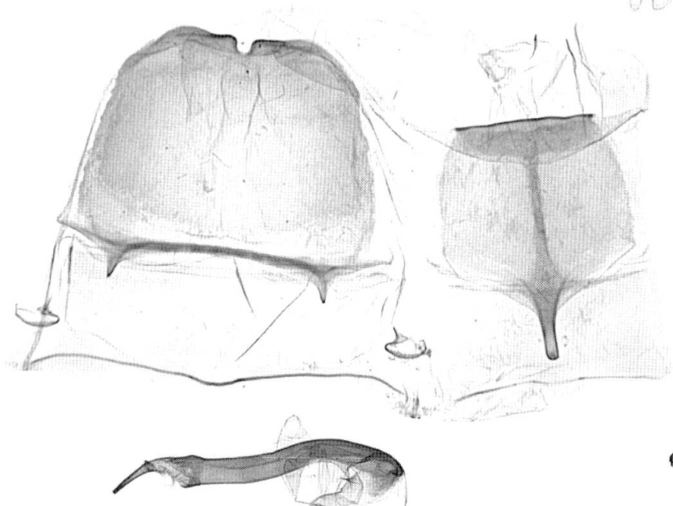

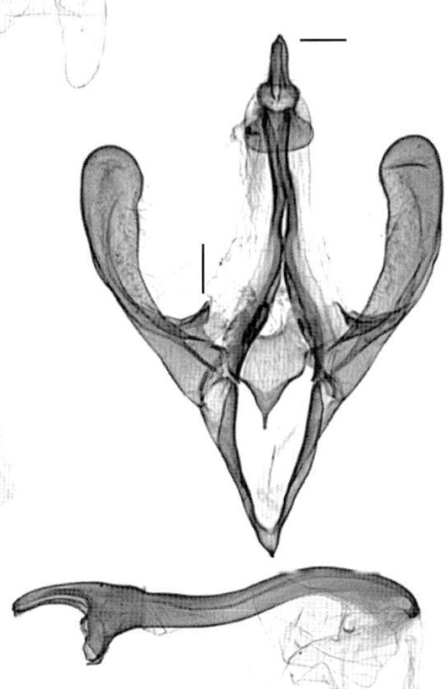

1 mm

201a *Syntypistis cognata*
Obi, GU 93-85, paratype, phallus

201b *Syntypistis cognata*
Obi, GU 93-85, paratype

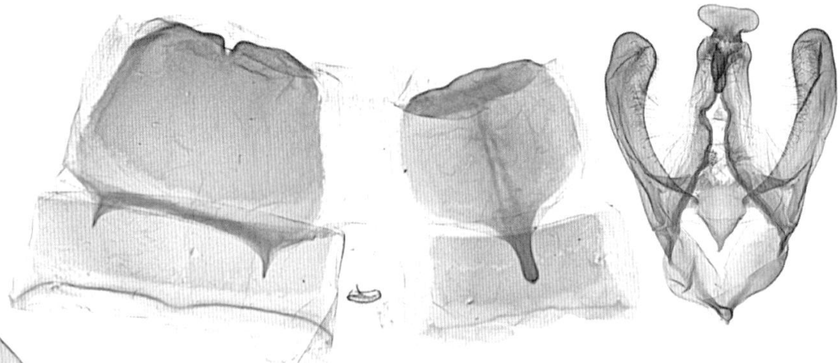

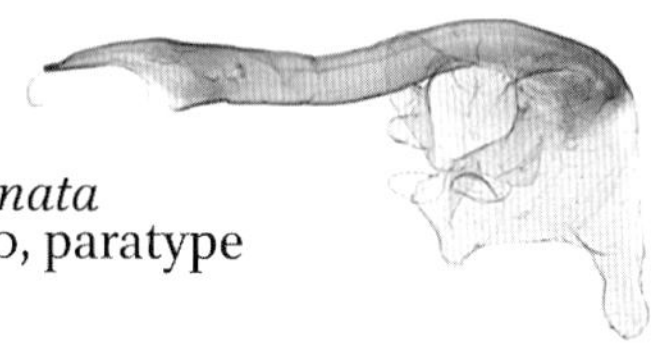

201c *Syntypistis cognata*
Bacan, GU 65-90, paratype

201d *Syntypistis cognata*
Halmahera, GU 65-91,
paratype

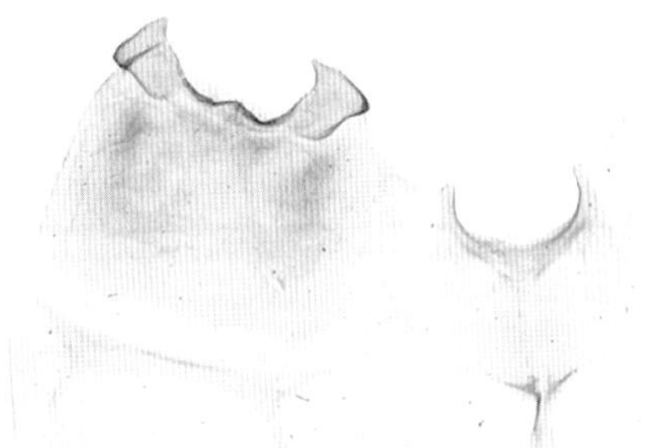

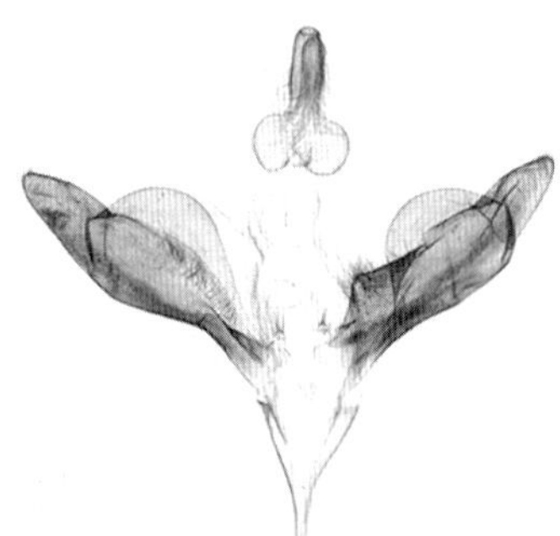

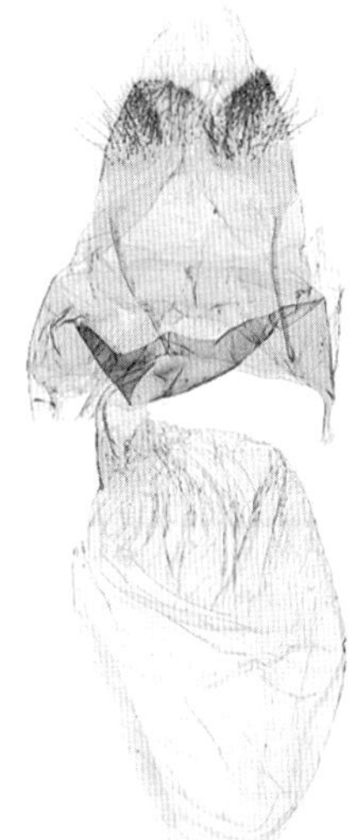

202a *Syntypistis sula*
Sulawesi, GU 42-70

203a *S. paukstadtorum*
Flores, GU42-61,
paratype

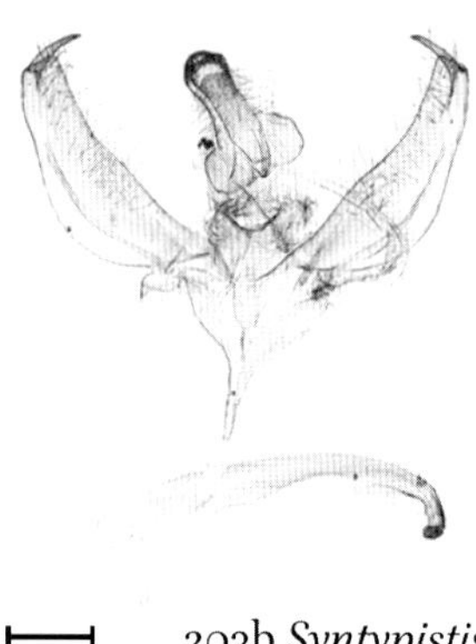

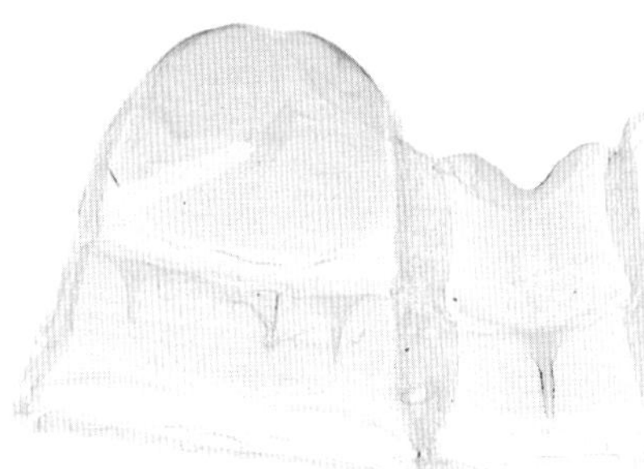

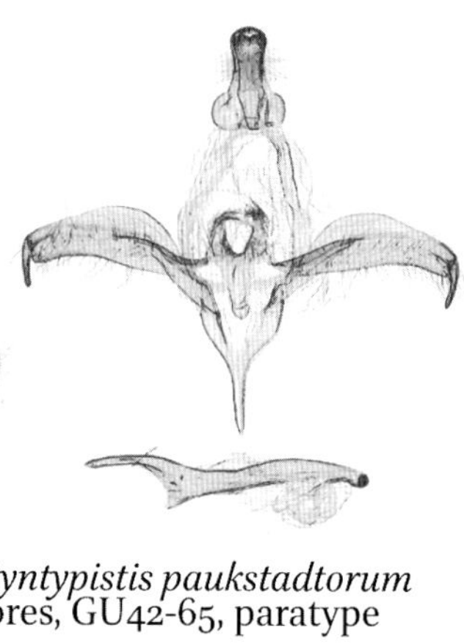

1 mm

203b *Syntypistis paukstadtorum*
Flores, GU47-87, paratype

203c *Syntypistis paukstadtorum*
Flores, GU42-65, paratype

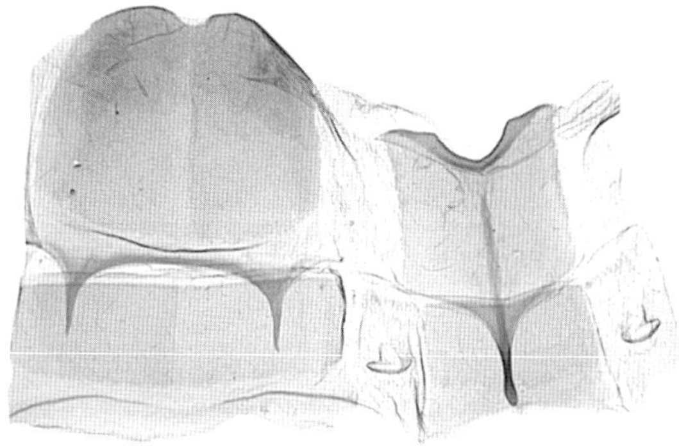

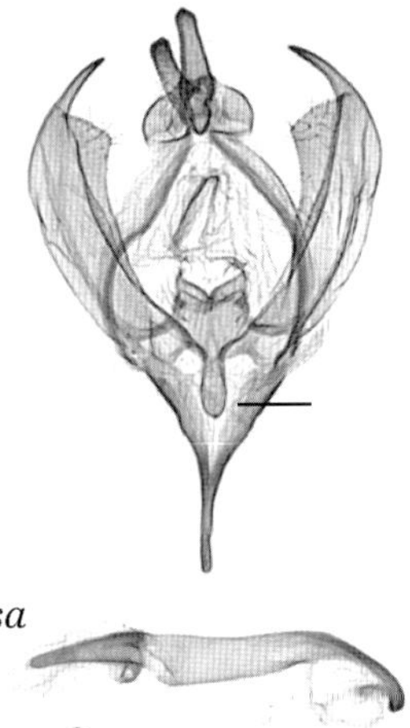

204a *Syntypistis umbrosa*
Borneo, GU 96-72

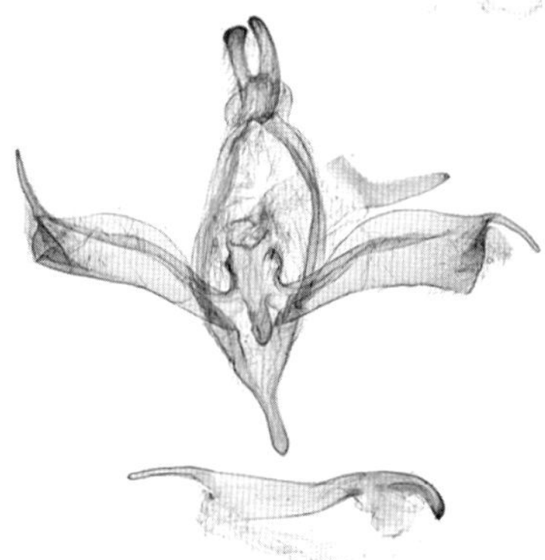

204c *Syntypistis umbrosa*
Taiwan, MWM 4.446

204b *Syntypistis umbrosa*
Borneo, GU 65-97

205a *Syntypistis garuda*
Sumatra, GU 22-26, paratype

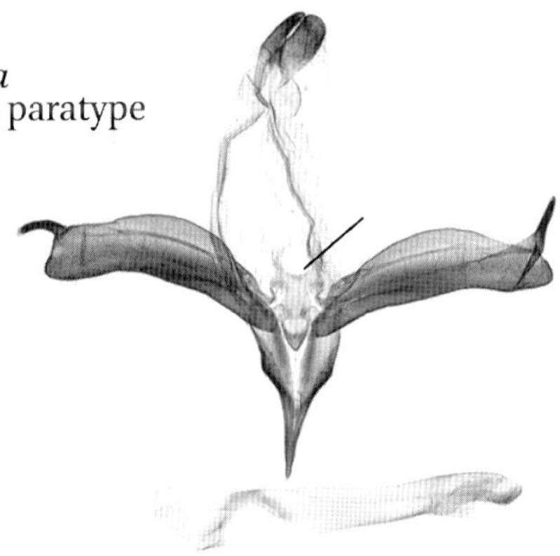

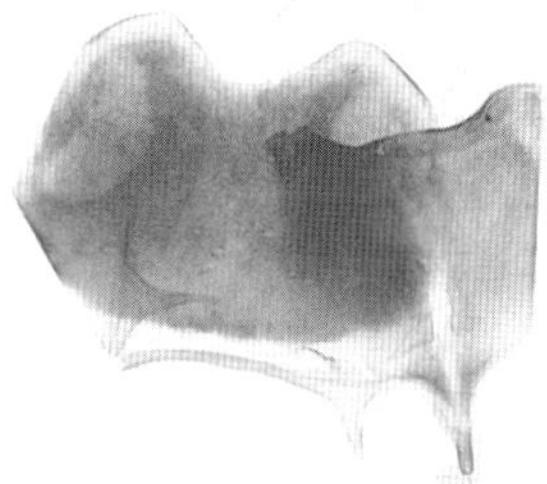

205b *Syntypistis garuda*
Sumatra, GU 65-96

1 mm

206a *Syntypistis kiriakoffi*
Sulawesi, GU 42-93

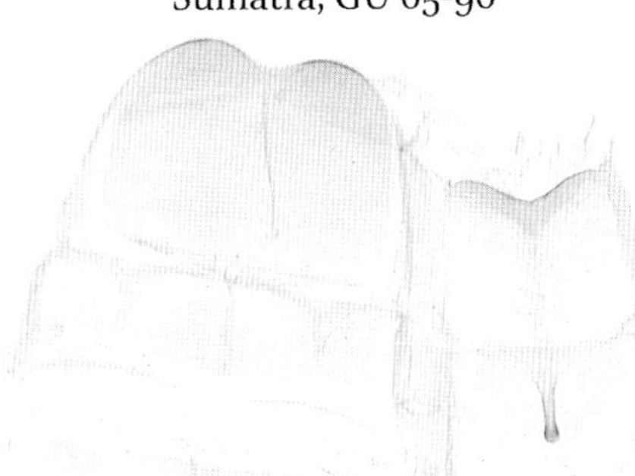

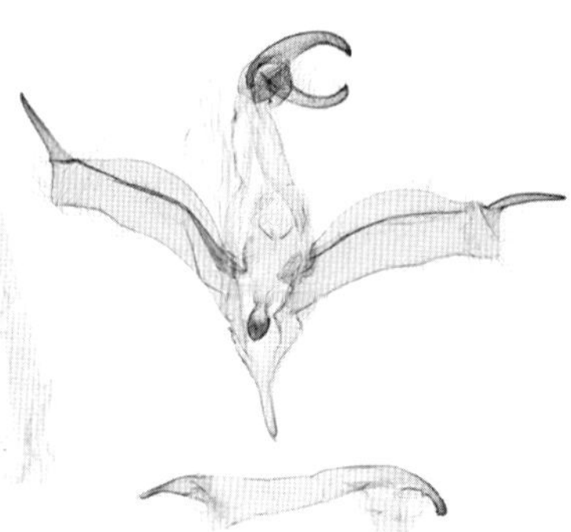

206b *Syntypistis kiriakoffi*
Sulawesi, GU 42-92

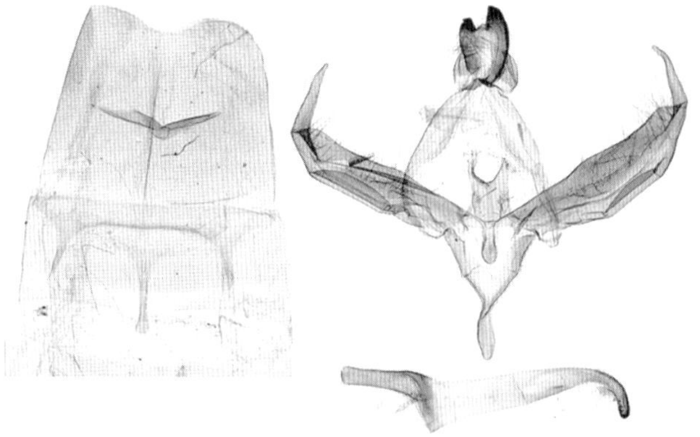

207a *Syntypistis palladina viridimargo*
Kalimantan, GU 47-91

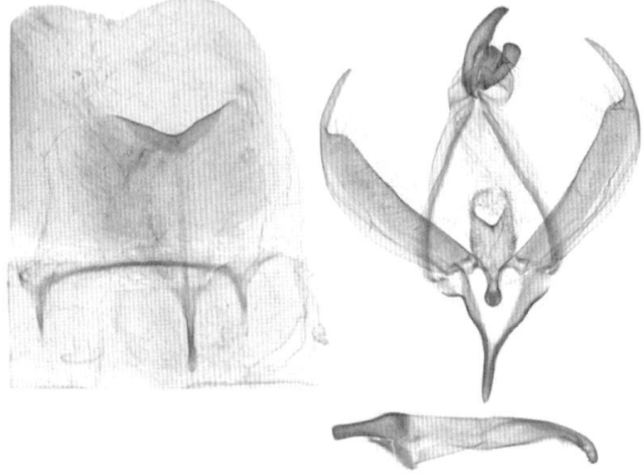

207b *Syntypistis palladina sasak*
Java, GU 47-50, paratype

207c *Syntypistis palladina sasak*
Bali, GU 94-66, paratype

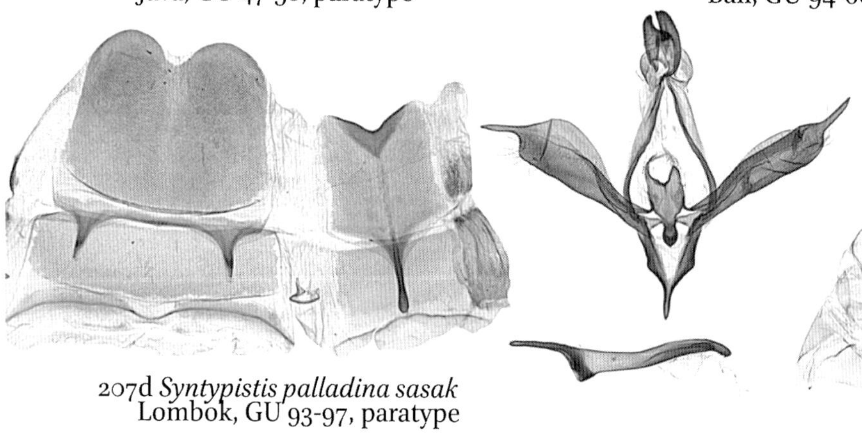

207d *Syntypistis palladina sasak*
Lombok, GU 93-97, paratype

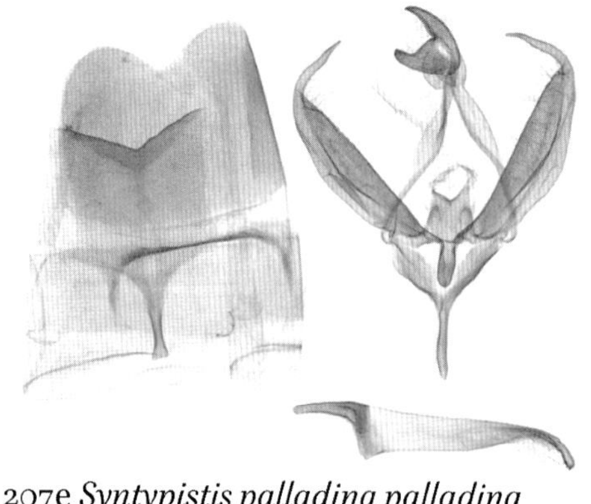

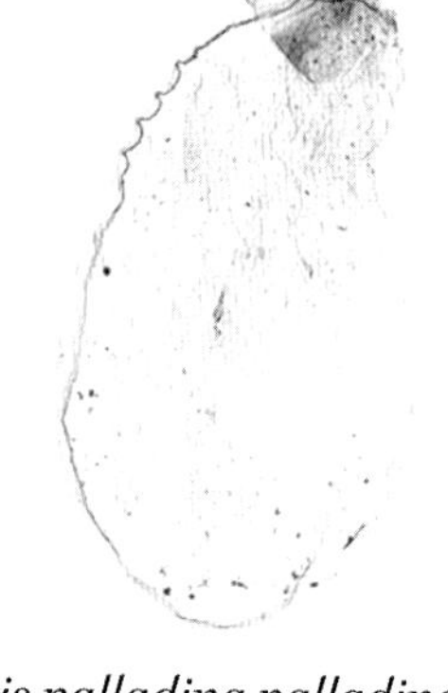

1 mm

207e *Syntypistis palladina palladina*
Mindanao, MV 17479

207f *Syntypistis palladina palladina*
Mindanao, GU 47-82

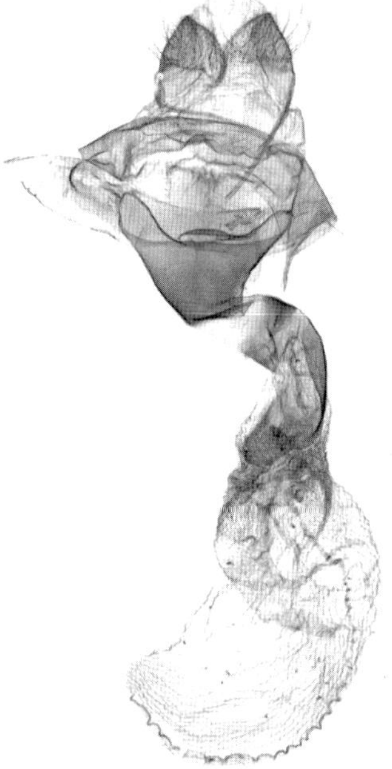

208b *Syntypistis comatus comatus*
Kalimantan, GU 54-90

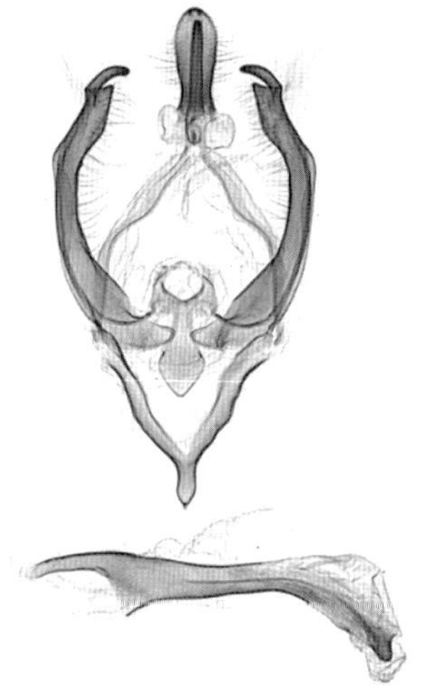

208c *Syntypistis comatus comatus*
Sumatra, BM 466
(holotype *bioculata*)

208a *Syntypistis comatus comatus*
Sumatra, GU 43-04

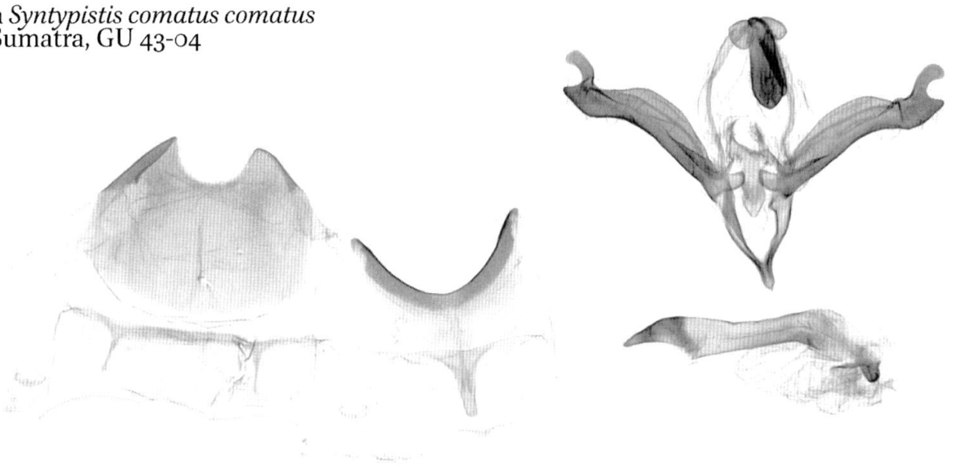

208d *Syntypistis comatus majapahit*
Java, GU 43-06, paratype

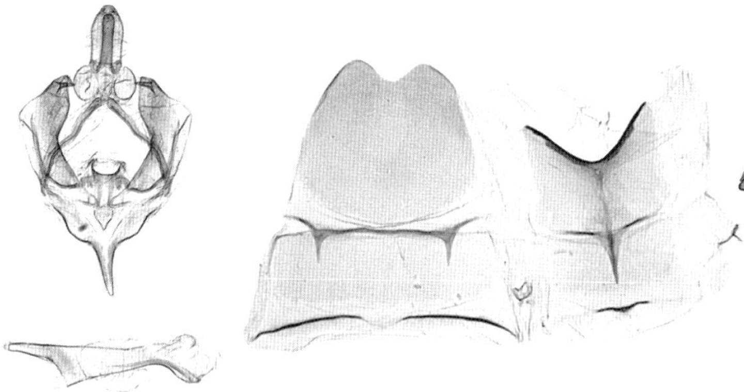

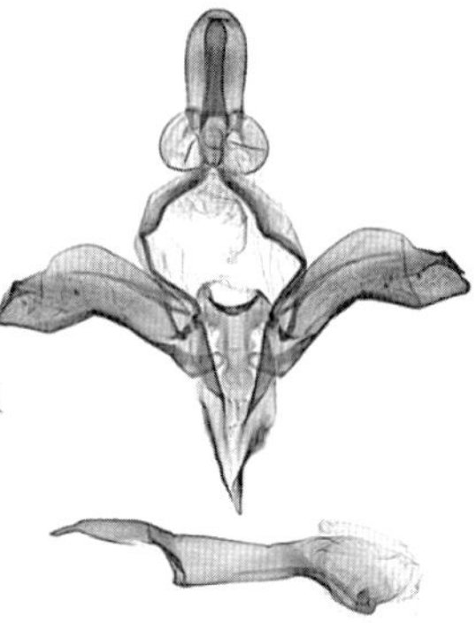

1 mm

209a *Syntypistis trioculata*
Borneo, BM 465, holotype

209b *Syntypistis trioculata*
Sumatra, GU 96-63

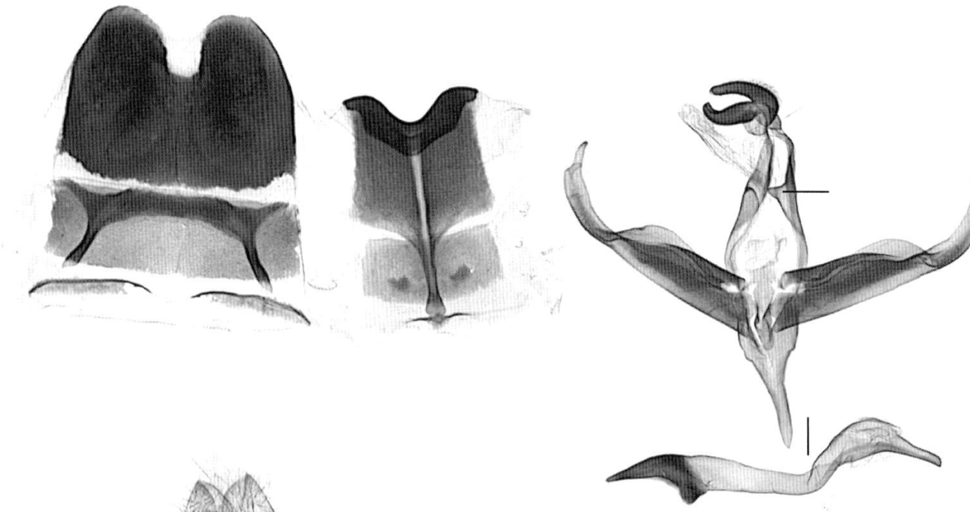

210a *Syntypistis viridipicta*
Sumatra, GU 65-98

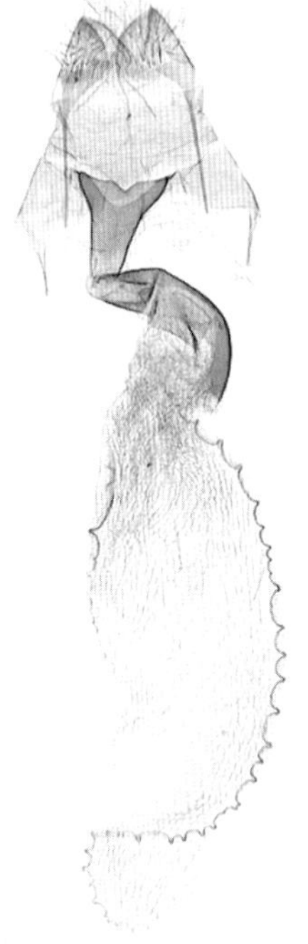

210b *Synipistis viridipicta*
Sumatra, GU47-44

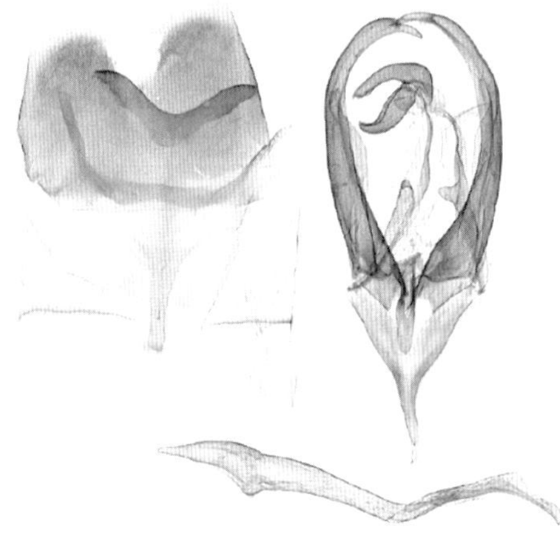

210c *Syntypistis viridipicta*
Sumatra, ZSM 360
(holotype *eusebia*)

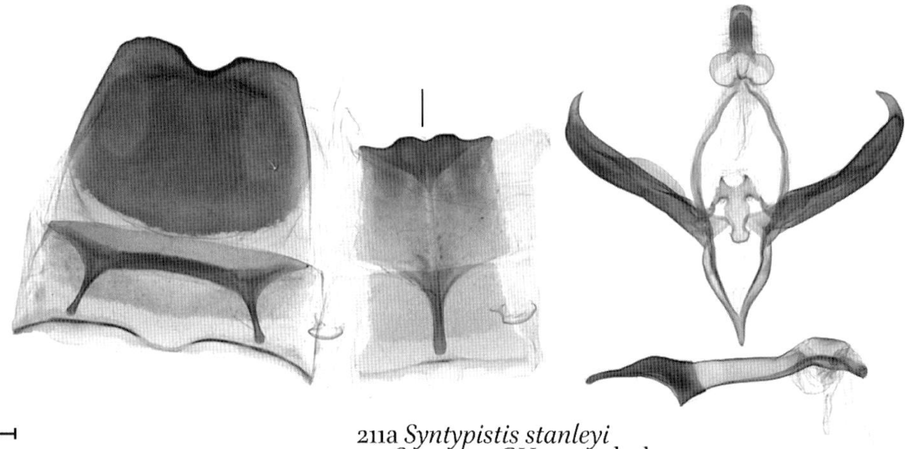

1 mm

211a *Syntypistis stanleyi*
Sumatra, GU 94-23, holotype

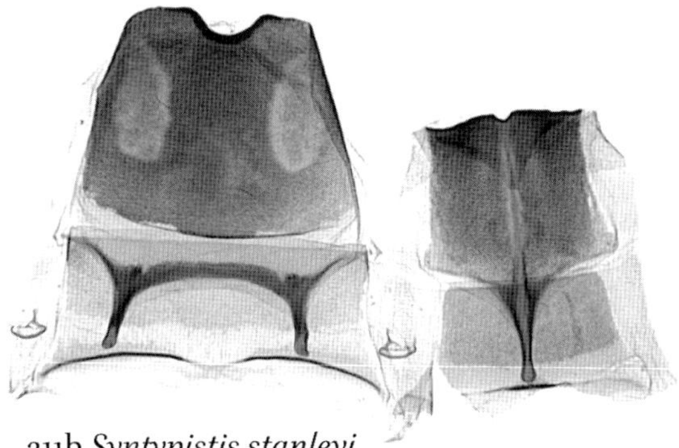

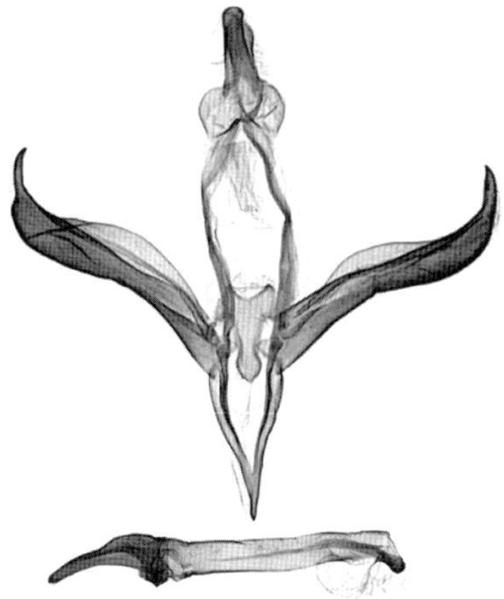

211b *Syntypistis stanleyi*
Borneo, ZMHU 02-2017, paratype

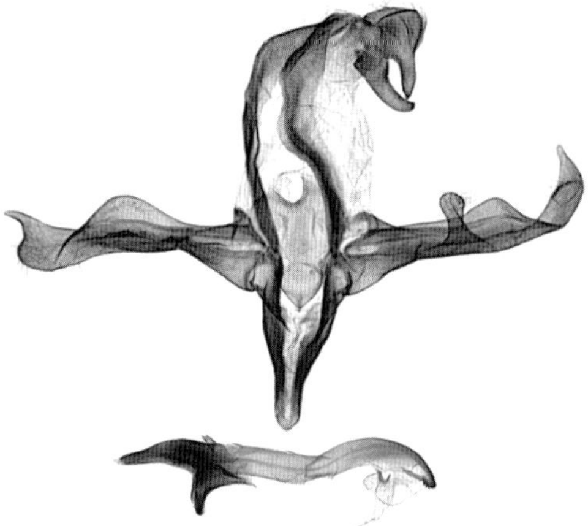

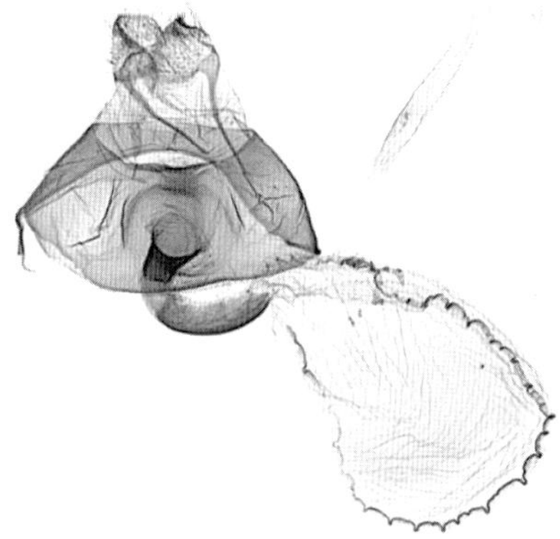

212b *Syntypistis pallidifascia pallidifascia*
Sumatra, GU 66-08

212a *Syntypistis pallidifascia pallidifascia*
Seram, BM 10313995

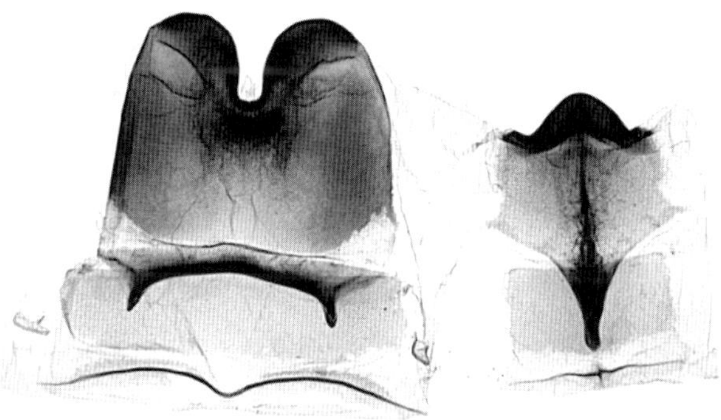

212c *Syntypistis pallidifascia pallidifascia*
Buru, GU 93-28

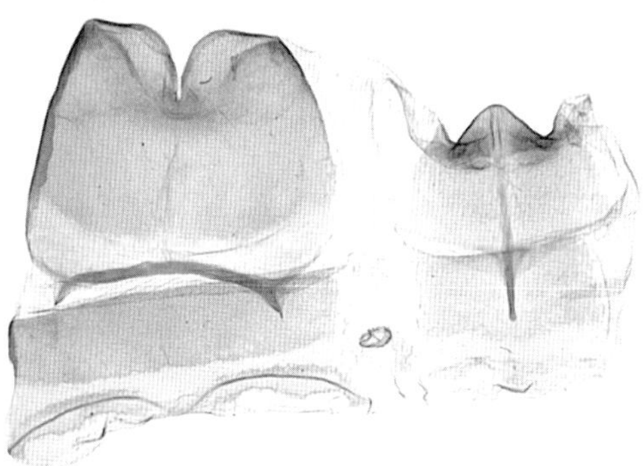

1 mm

212d *Syntypistis pallidifascia germana*
New Ireland, GU 93-28

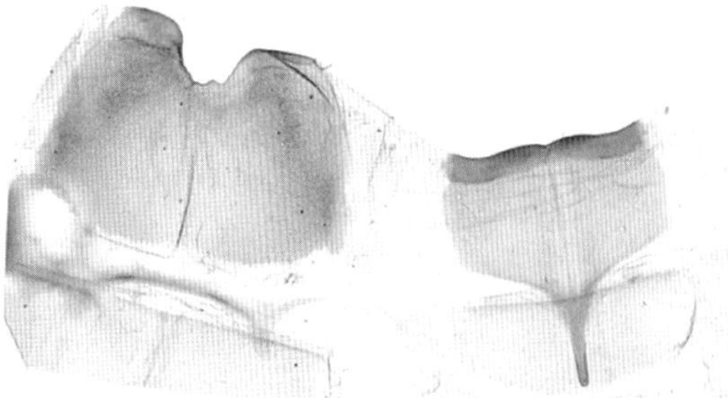

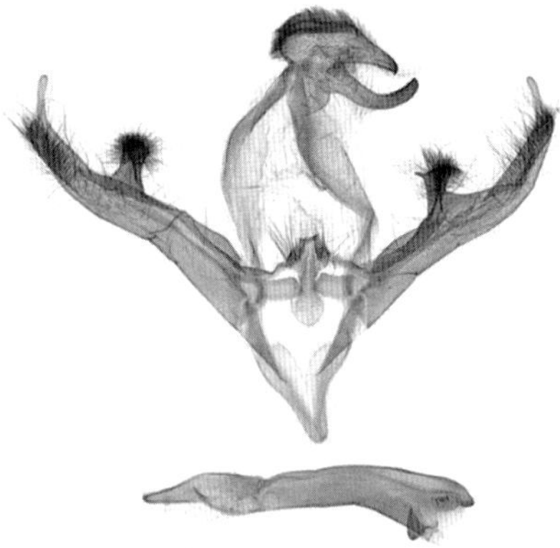

213a *Syntypistis vaneeckei*
Sulawesi, GU 79-64,
paratype

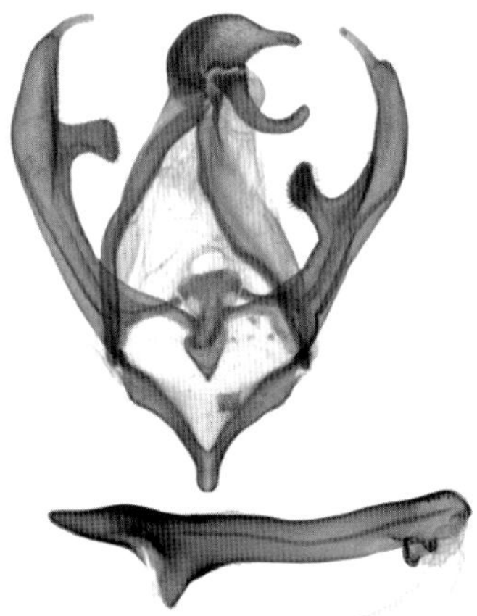

213b *Syntypistis vaneeckei*
Sulawesi, MWM 15.624,
paratype

214a *Syntypistis nigribasalis tropica*
Sulawesi, GU 66-07

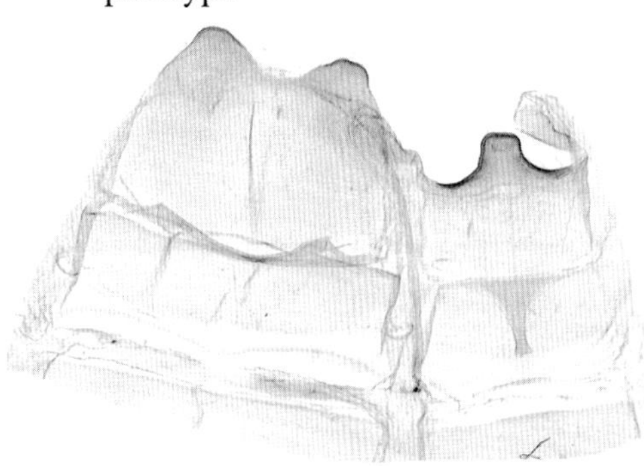

214b *Syntypistis nigribasalis tropica*
Sulawesi, GU 42-72

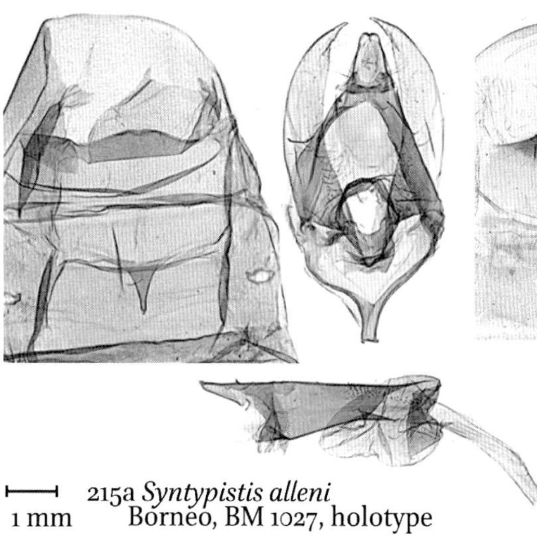

1 mm

215a *Syntypistis alleni*
Borneo, BM 1027, holotype

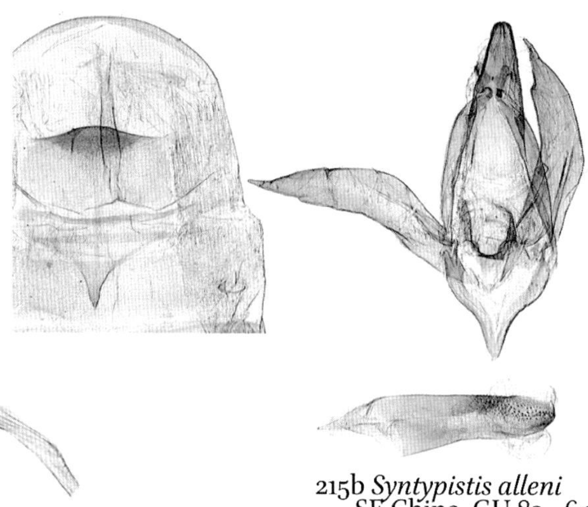

215b *Syntypistis alleni*
SE China, GU 83 - 64

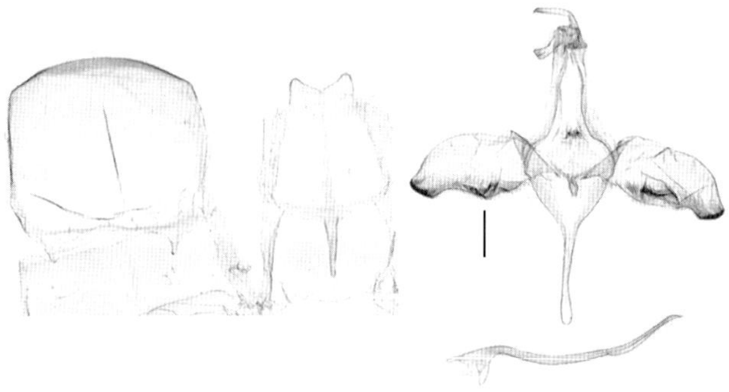

216a *Parasinga lichenina*
Sumatra, MWM 4.722

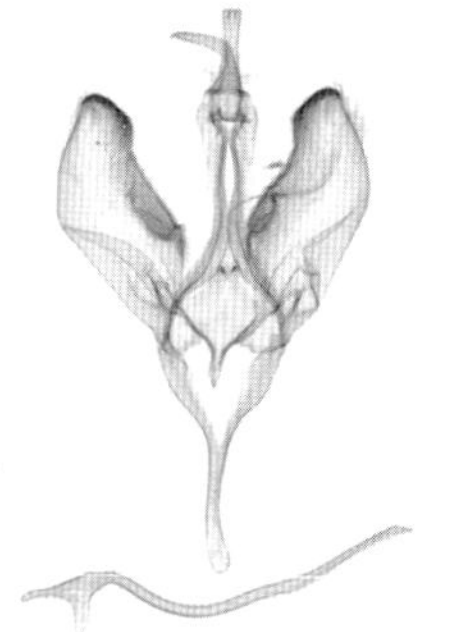

216c *Parasinga lichenina*
Sumatra, GU 88-35

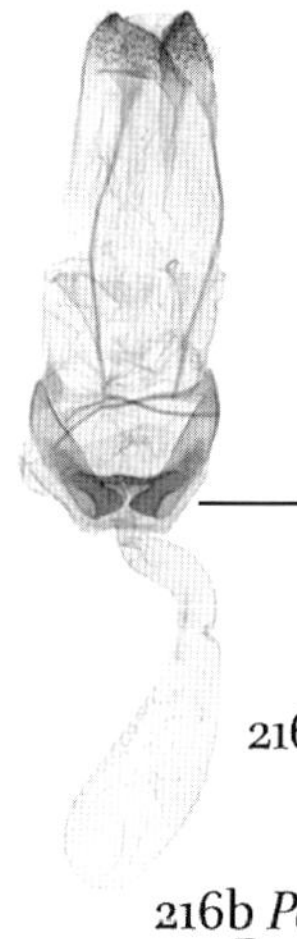

216b *Parasinga lichenina*
Borneo, MWM 15.739

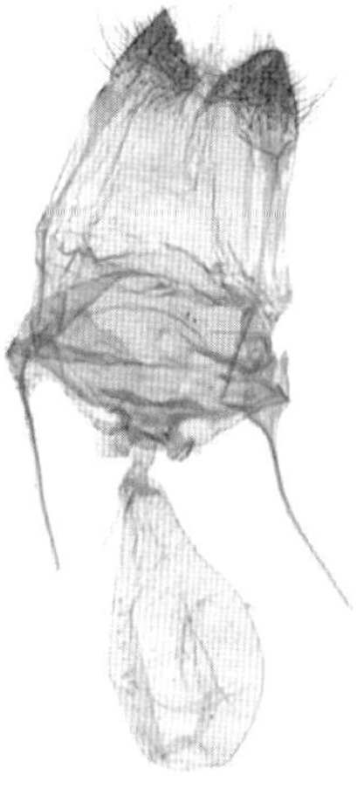

217a *Parasinga muliercula*
Sulawesi, GU 43-22, paratype

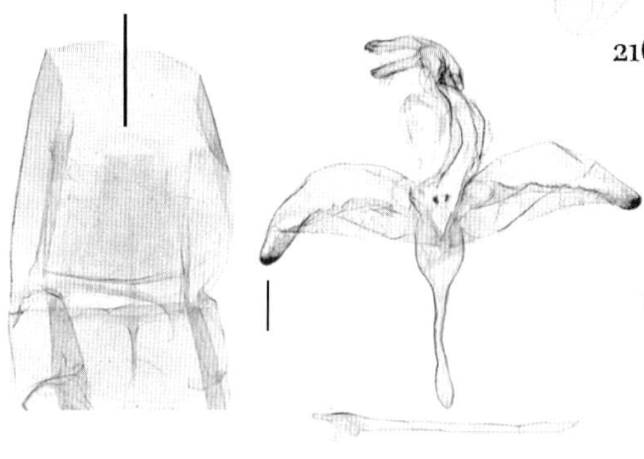

217b *Parasinga muliercula*
Sulawesi, GU 17-42, paratype

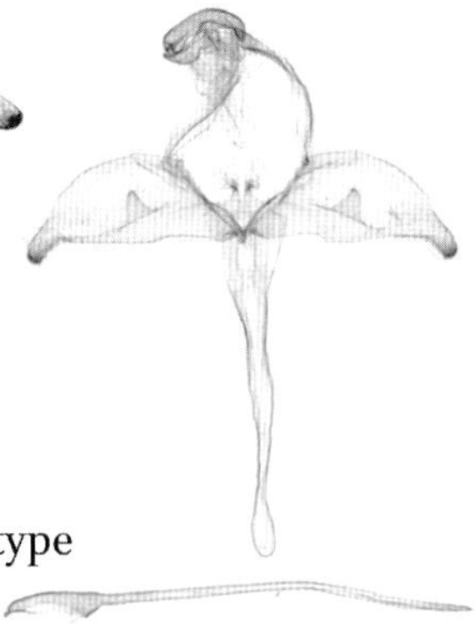

217Aa *Parasinga viridescens*
Mindanao, MV 17470, paratype

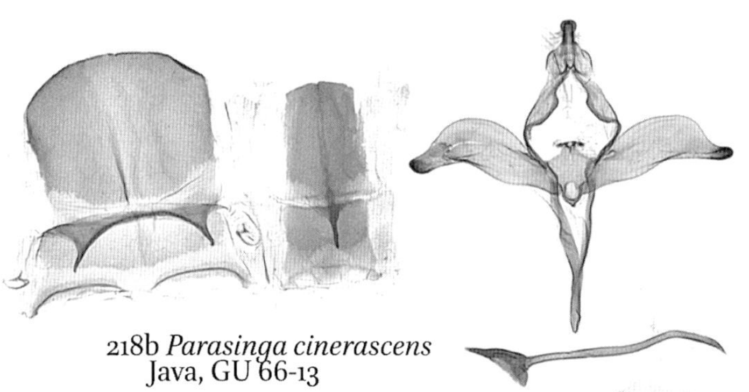

218b *Parasinga cinerascens*
Java, GU 66-13

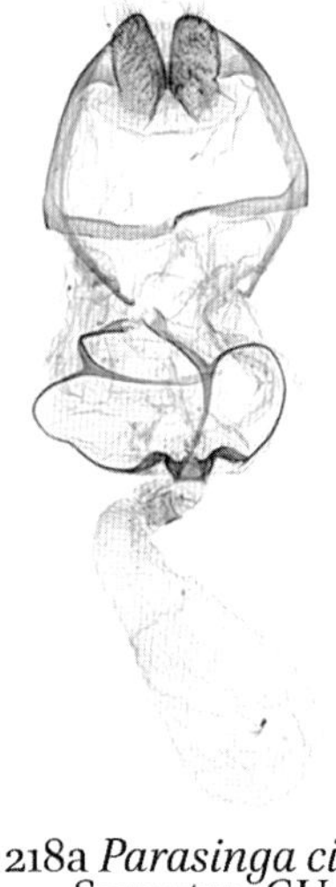

218a *Parasinga cinerascens*
Sumatra, GU 66-14

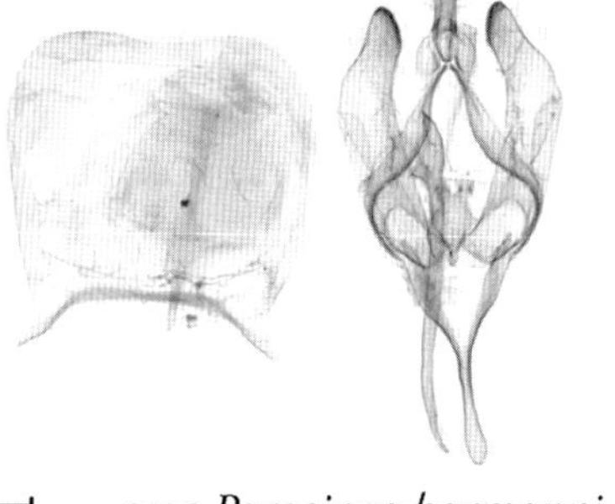

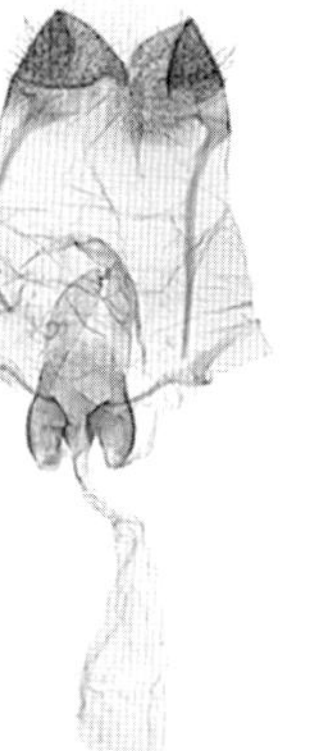

1 mm

219a *Parasinga harmanni*
Sumatra, GU 88-39

220a *Parasinga insufficiens*
Kalimantan, GU 88-38

220b *P. insufficiens*
Borneo, MWM 15.735

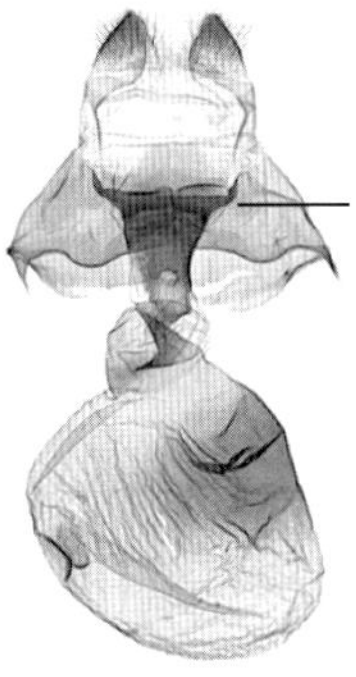

221a *Cascera marginata*
W Papua, BM 10314139

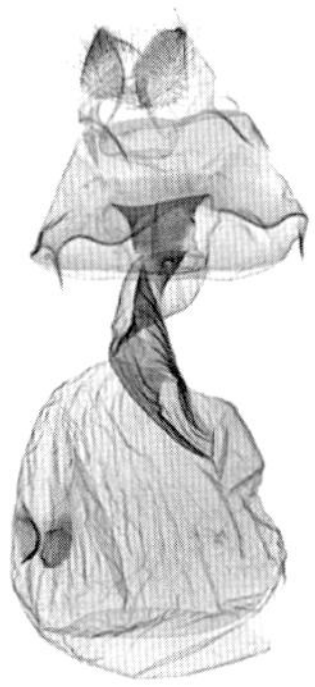

222a *Cascera inconcisa*
W Papua, BM 10314140

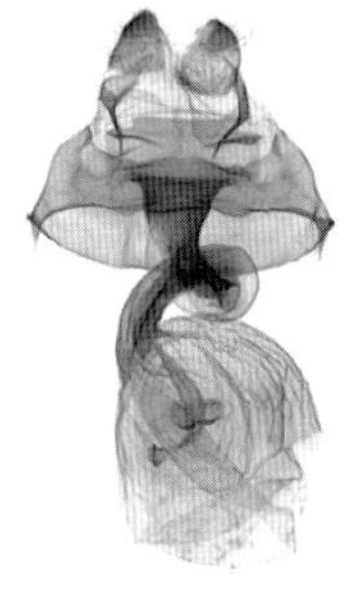

223a *Cascera bella*
W Papua, GU 94-78

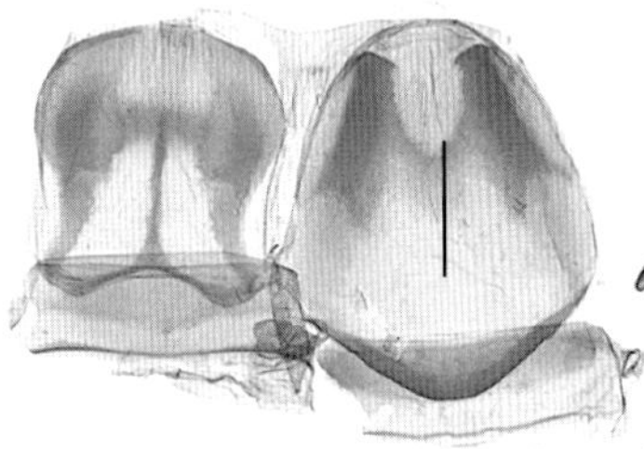

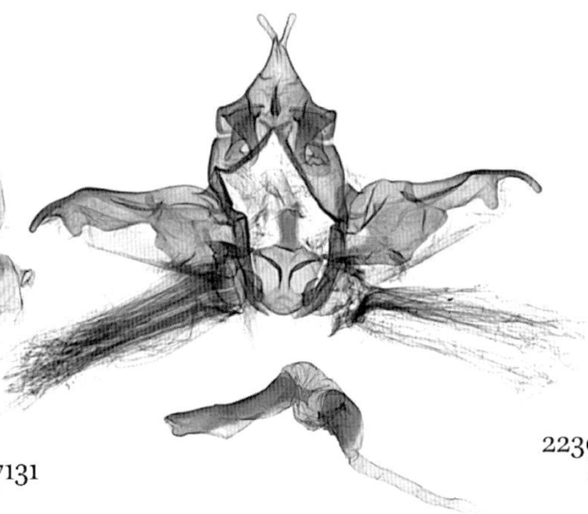

223b *Cascera bella*
PNG, USNM 127131

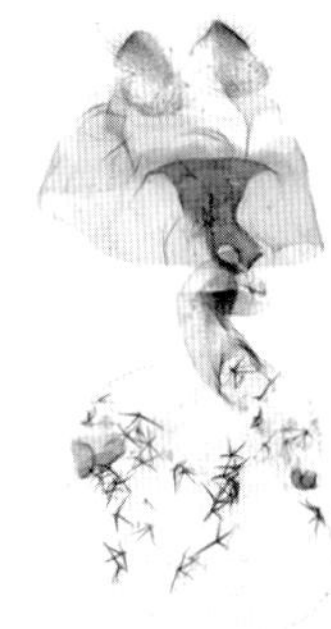

223c *Cascera bella*
PNG, ZSM 2011-10

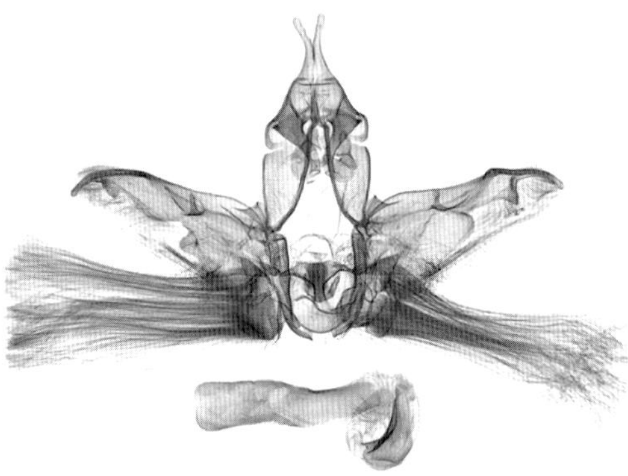

223d *Cascera bella*
PNG, GU 94-70

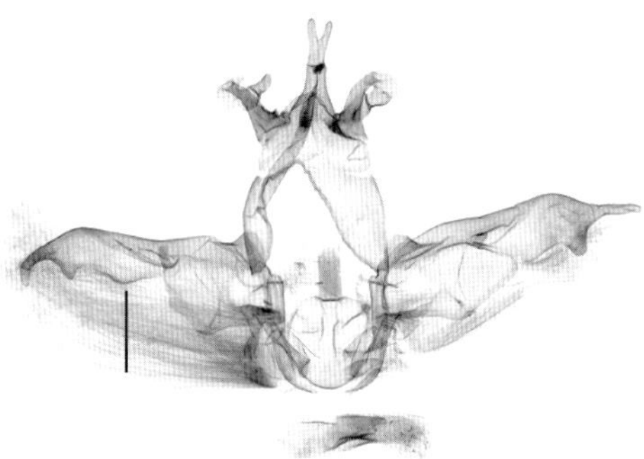

223e *Cascera bella*
Australia, Qsd., ZSM 04-2011

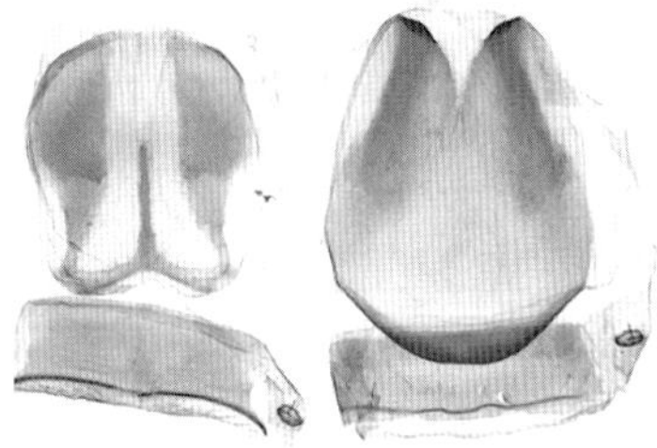

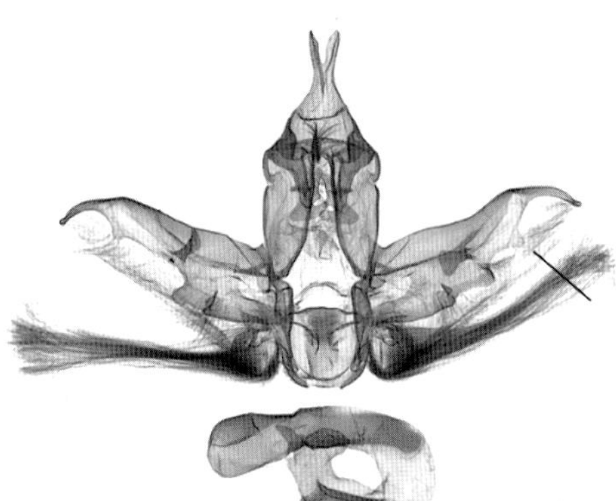

1 mm

224a *Cascera latifasciata*
W Papua, GU 94-77

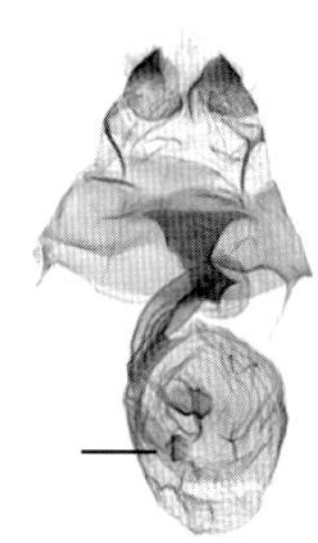

224b *Cascera latifasciata*
W Papua, BM 10314134

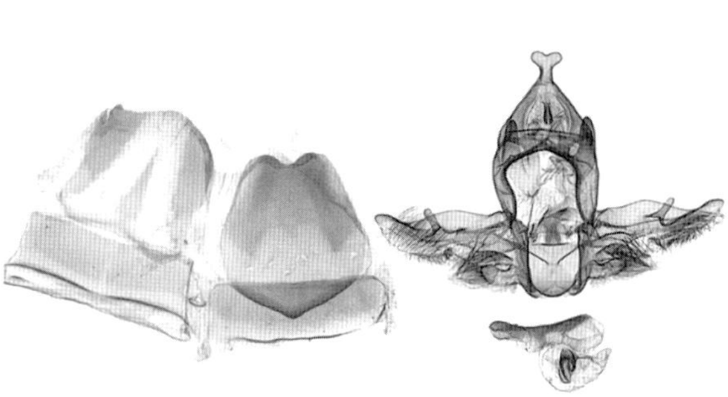

225a *Cascera albiscripta*
PNG, BM 10314131

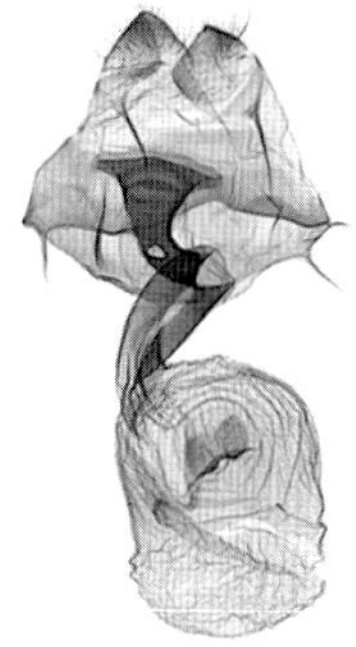

226a *Cascera olivacea*
PNG, BM 10314144

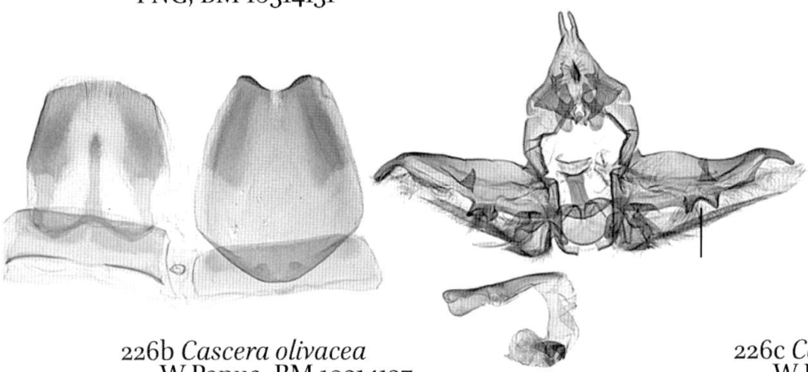

226b *Cascera olivacea*
W Papua, BM 10314137

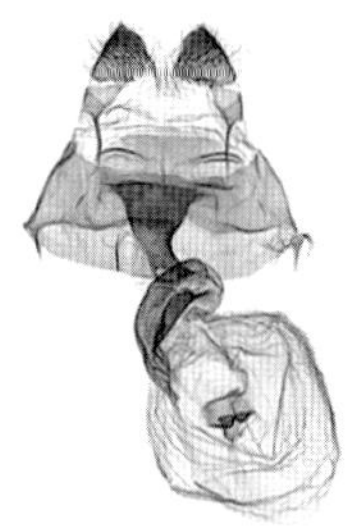

226c *Cascera olivacea*
W Papua, BM 10314137

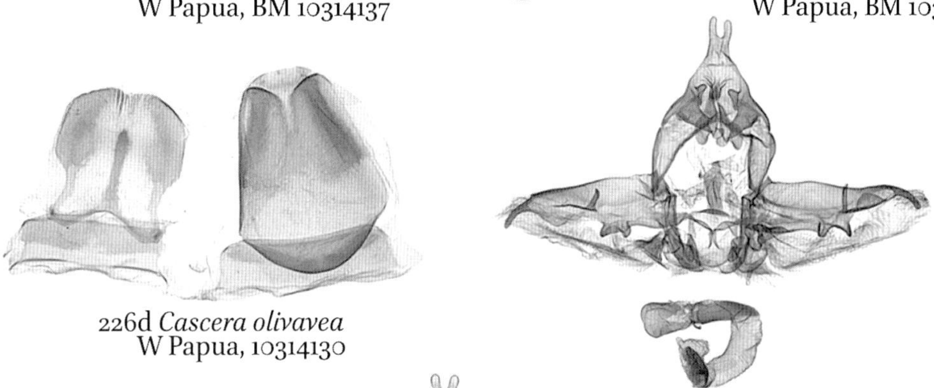

226d *Cascera olivavea*
W Papua, 10314130

227a *Cascera perscripta*
W Papua, BM 10314142

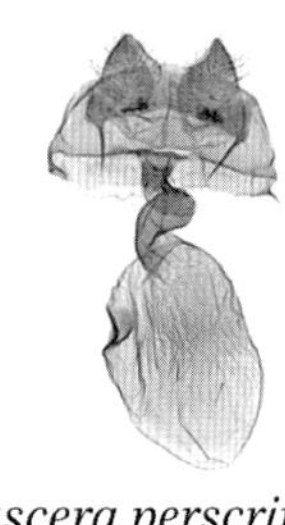

227b *Cascera perscripta*
W Papua, BM 10314141,
paralectotype

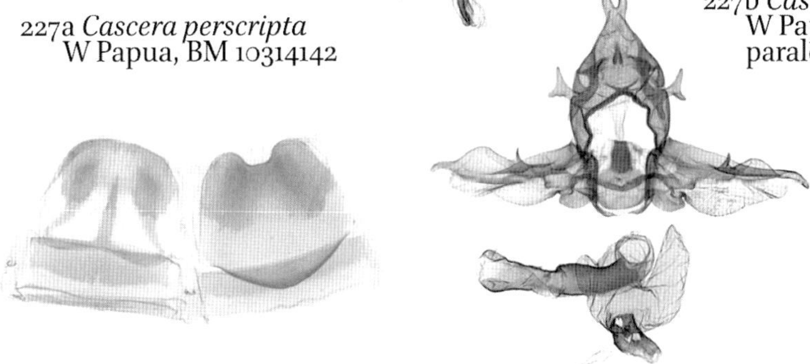

1 mm

229a *Cascera callima*
PNG, USNM 127.530

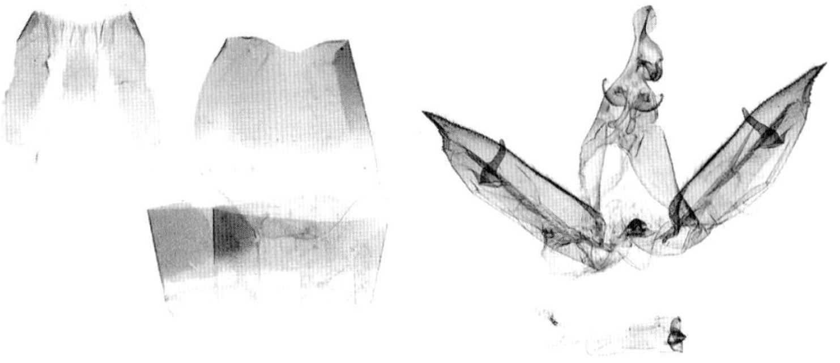

230a *Omichlis dimorpha*
Kalimantan, GU 51-54

230b *Omichlis dimorpha*
Sumatra, GU 66-15

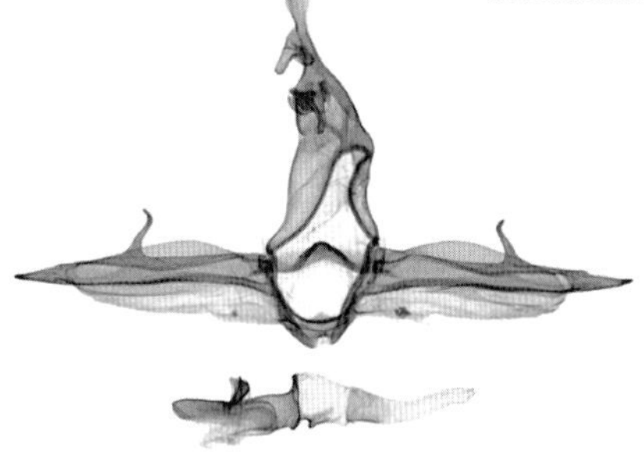

230c *Omichlis dimorpha*
Sumatra, MWM 15.595

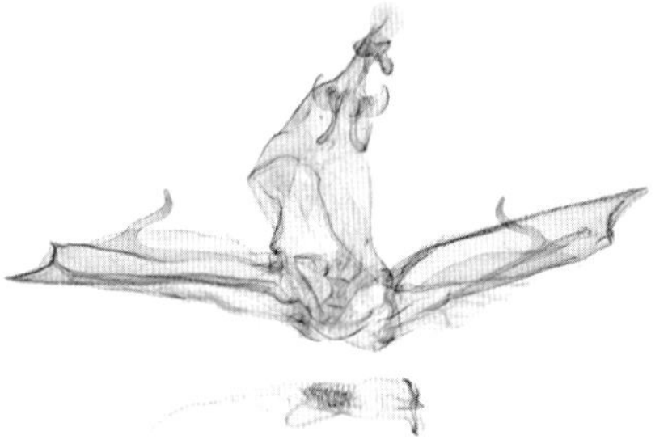

230d *Omichlis dimorpha*
Sumatra, GU 17-30

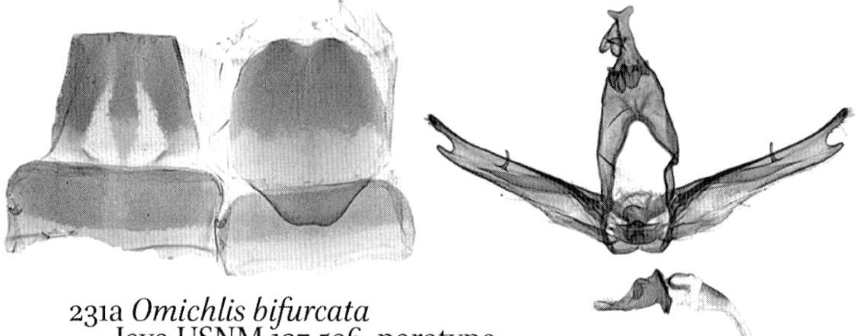

231a *Omichlis bifurcata*
Java,USNM 127.526, paratype

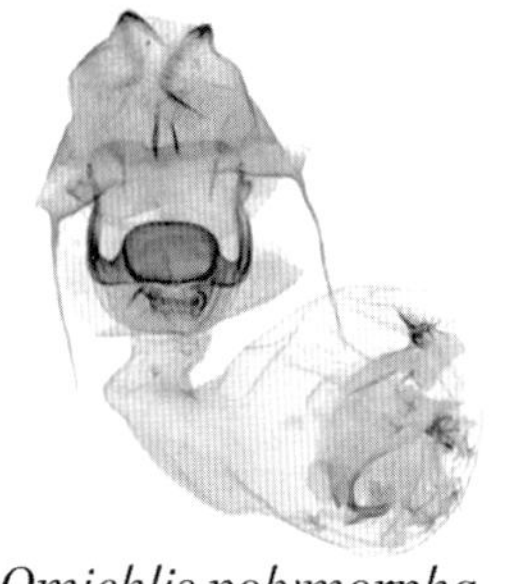

232a *Omichlis polymorpha*
Sulawesi, MWM 15.568
paratype

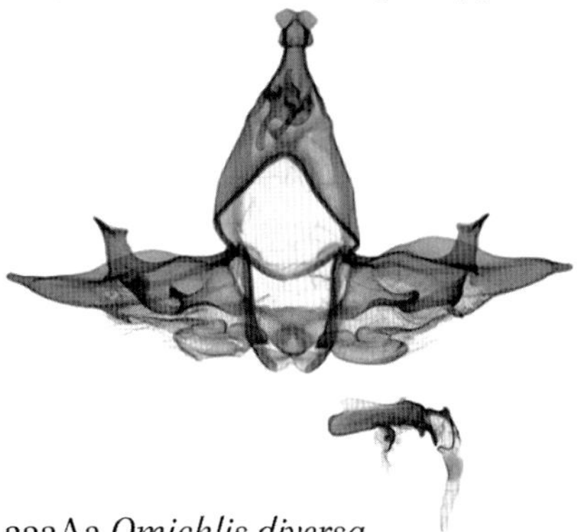

232Aa *Omichlis diversa hinumbian*
Negros, MWM 15.593

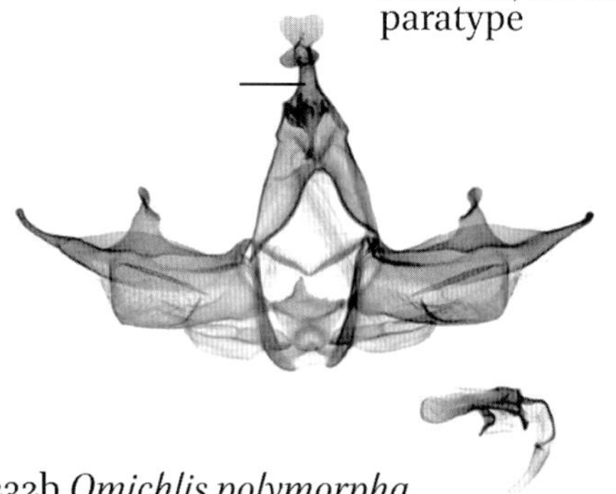

232b *Omichlis polymorpha*
Sulawesi, MWM 15.596, paratype

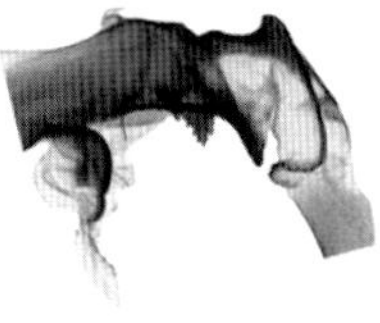

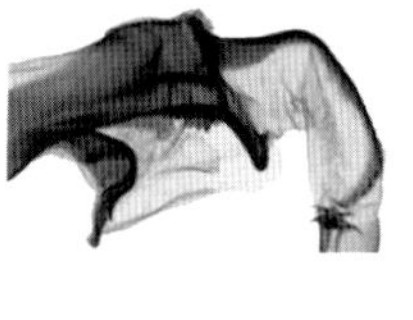

1 mm Phallus detail of fig. 232Aa, enlarged

Phallus detail of fig. 232b, enlarged

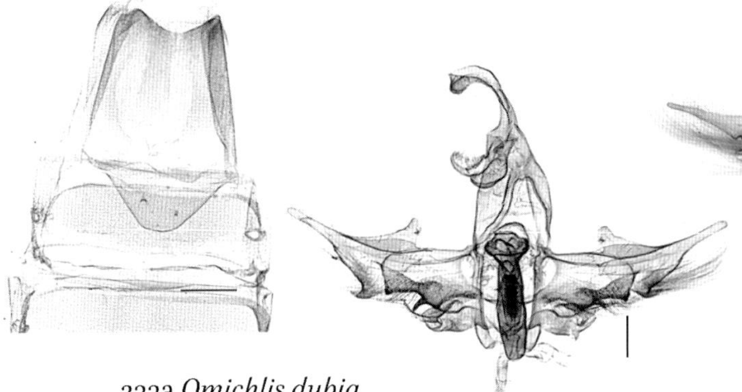

233a *Omichlis dubia*
Seram, BM 1810, paratype

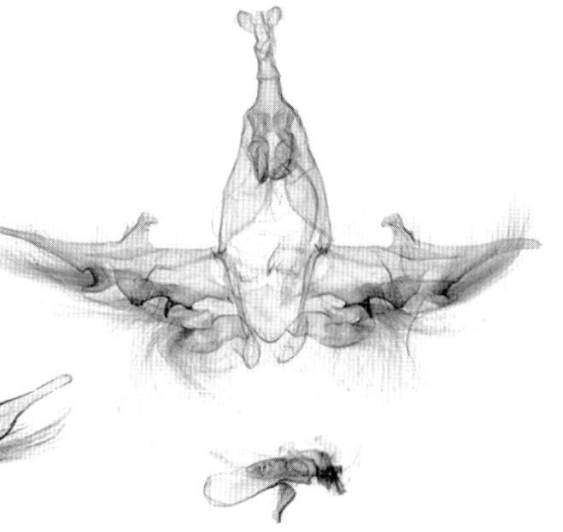

233b *Omichlis dubia*
Seram, GU 16-48, paratype

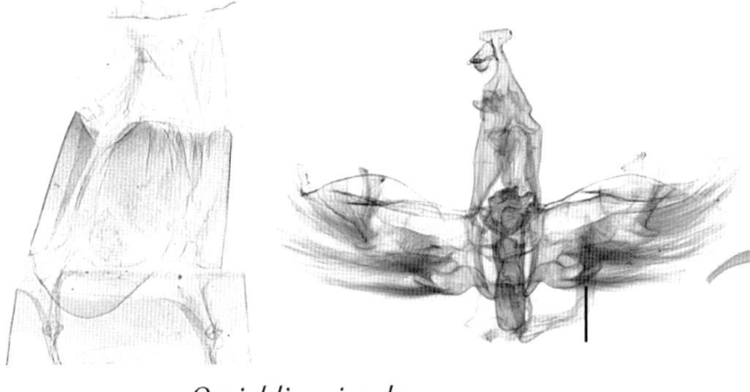

234a *Omichlis griseola*
PNG, BM 1482, holotype

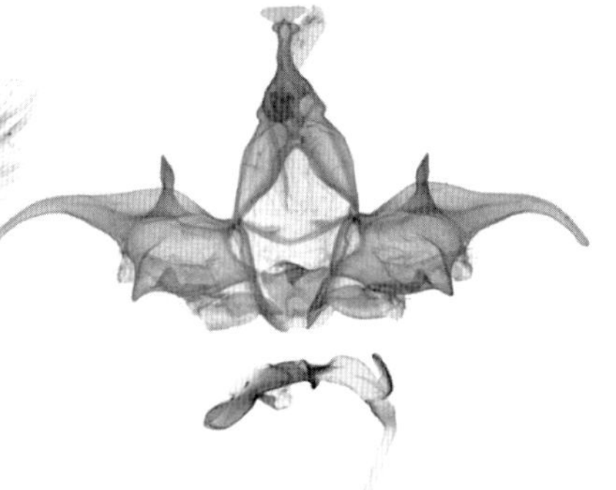

234b *Omichlis griseola*
W Papua, MWM 15.6503

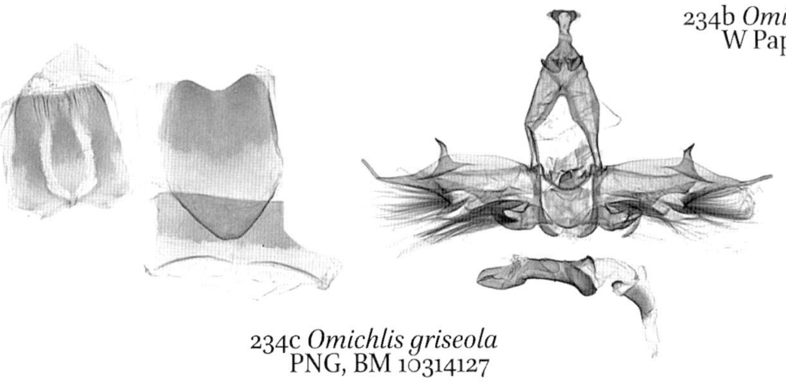

234c *Omichlis griseola*
PNG, BM 10314127

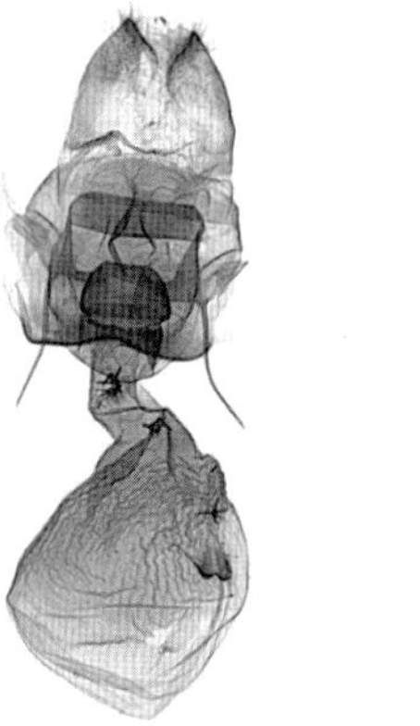

235a *Omichlis renata*
New Britain,
MWM 29.419

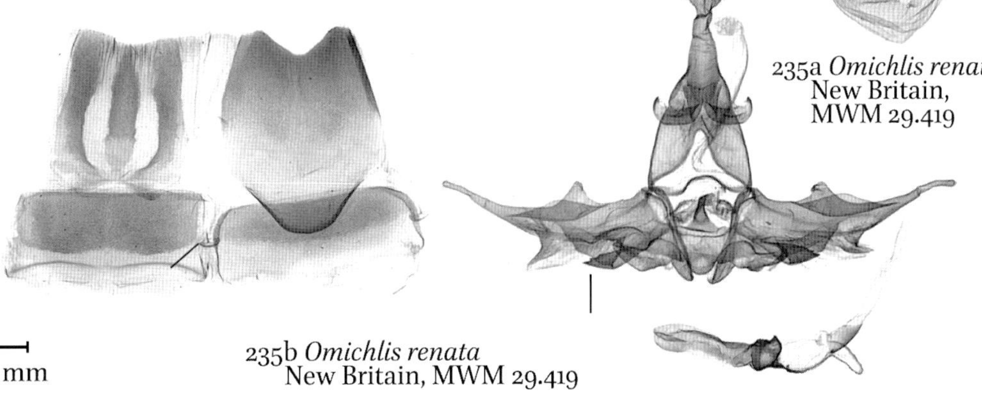

1 mm

235b *Omichlis renata*
New Britain, MWM 29.419

236a *Omichlis pratti pratti*
Rossel Isl., BM 10314129

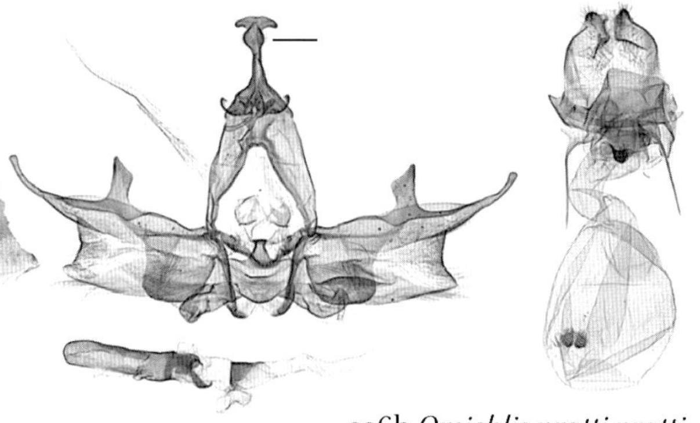

236b *Omichlis pratti pratti*
W Papua MWM 24.317

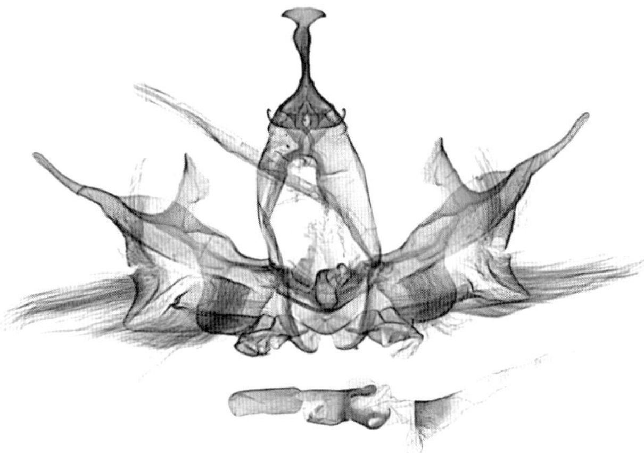

236c *Omichlis pratti pratti*
W Papua, BM 10314111

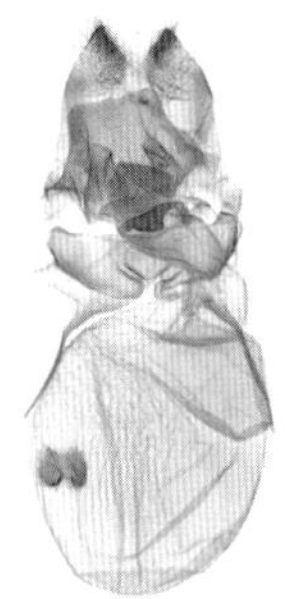

236d *Omichlis pratti pratti*
Fergusson Isl.,
BM 10314129

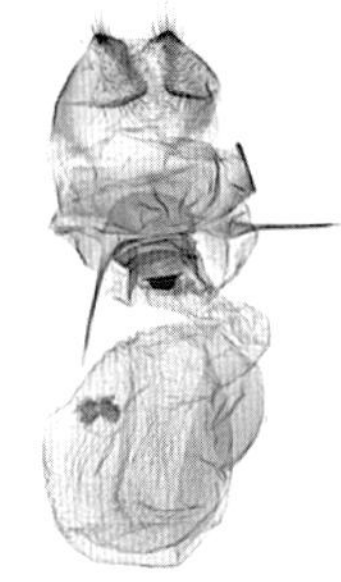

236e *O. pratti pratti*
Australia-Qsd.
BM 10314121

236f *Omichlis pratti pratti*
Fergusson Isl., BM 10314115

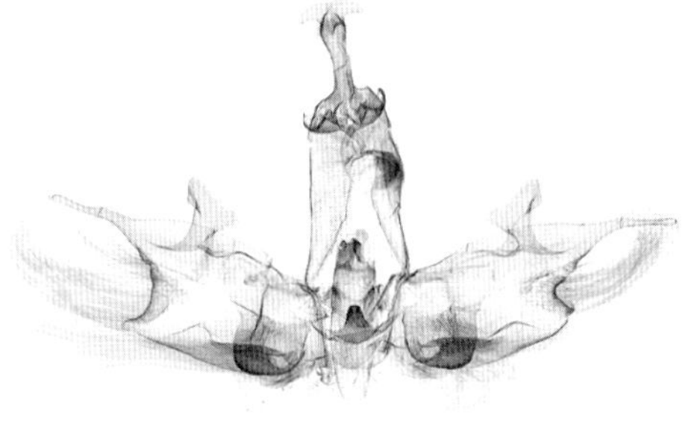

236g *Omichlis pratti falcigera*
Halmahera, GU 51-39, paratype

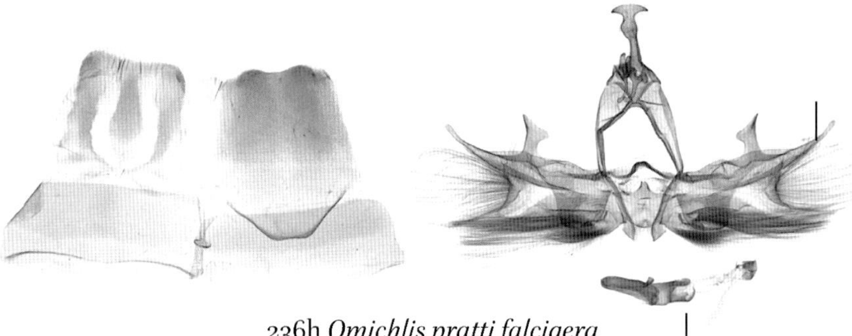

1 mm

236h *Omichlis pratti falcigera*
Obi, GU 92-90, paratype

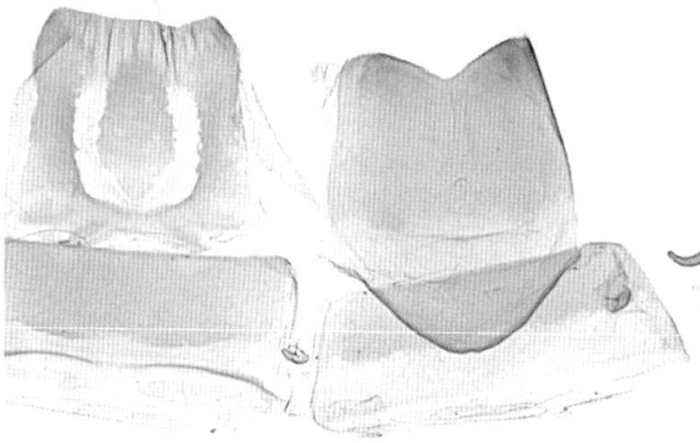

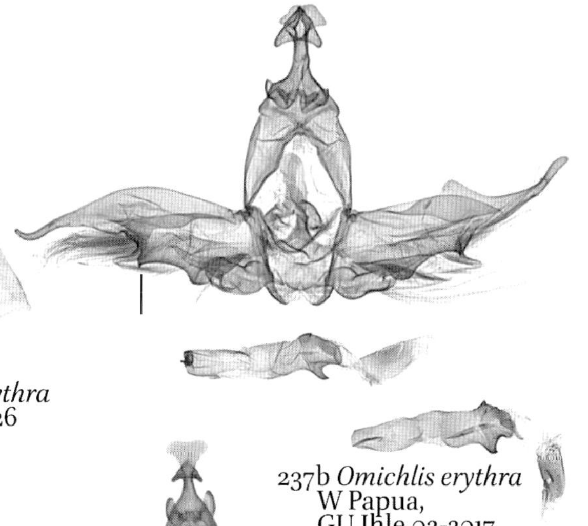

237a *Omichlis erythra*
PNG, GU 66-26

237b *Omichlis erythra*
W Papua,
GU Ihle 02-2017
phallus

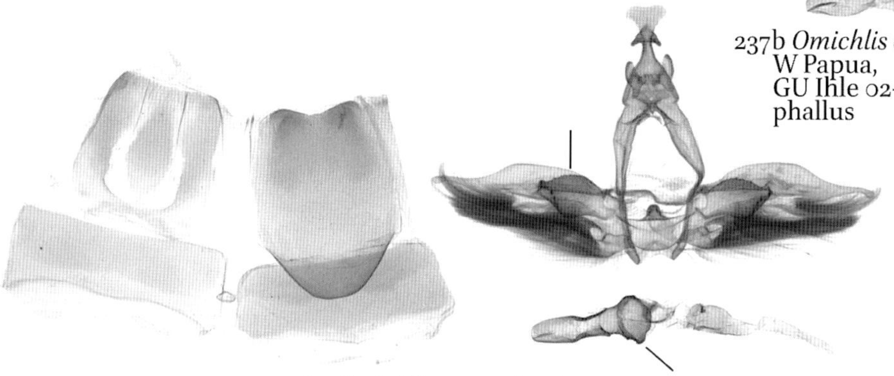

238a *Omichlis paragriseola*
W Papua, GU 94-84, holotype

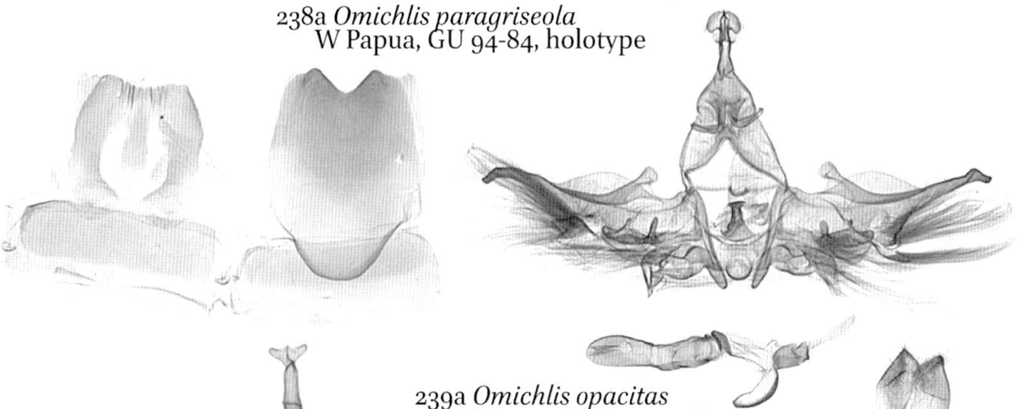

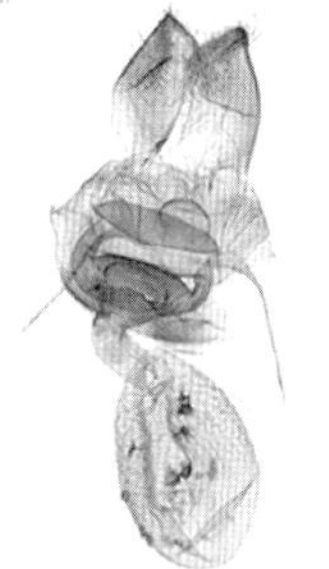

239a *Omichlis opacitas*
W Papua, BM 10314106
paratype

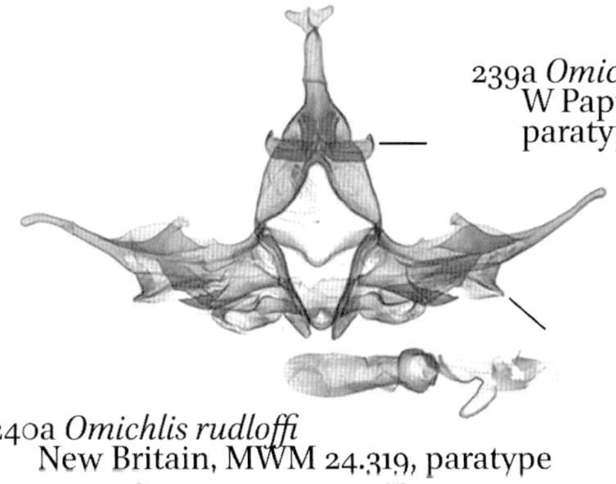

240a *Omichlis rudloffi*
New Britain, MWM 24.319, paratype

239b *Omichlis opacitas*
W Papua, BM 10314108
paratype

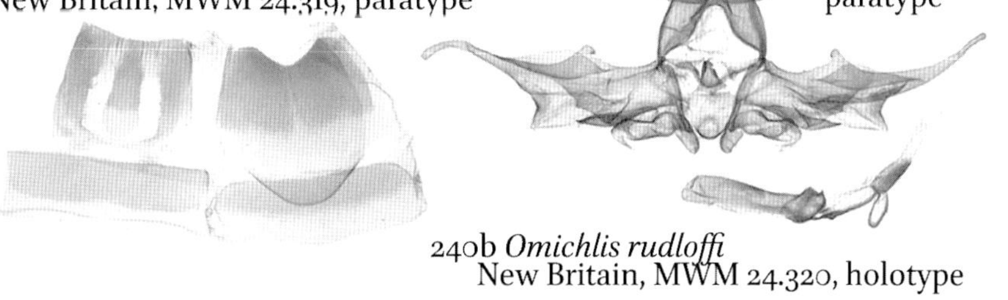

1 mm

240b *Omichlis rudloffi*
New Britain, MWM 24.320, holotype

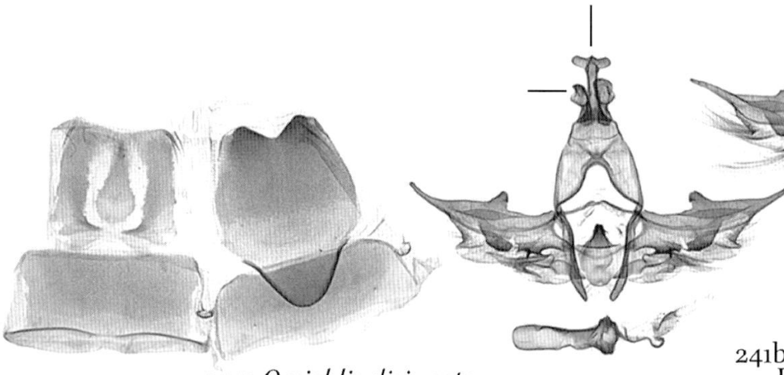

241a *Omichlis disiuncta*
Buru, GU 93-34, paratype

241b *Omichlis disiuncta*
Buru, GU 93-35, paratype

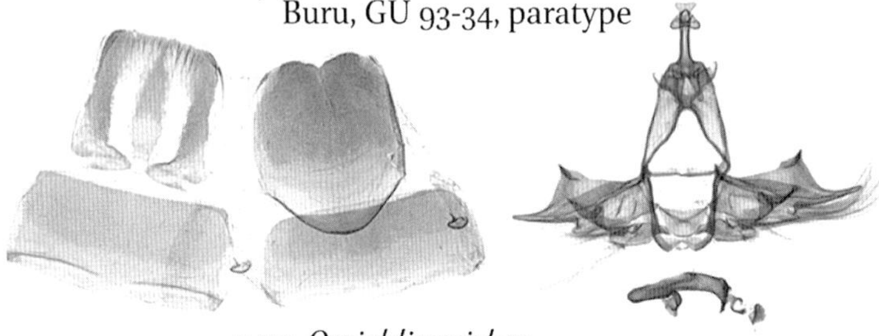

242a *Omichlis priskae*
Tanimbar GU 93-04, paratype

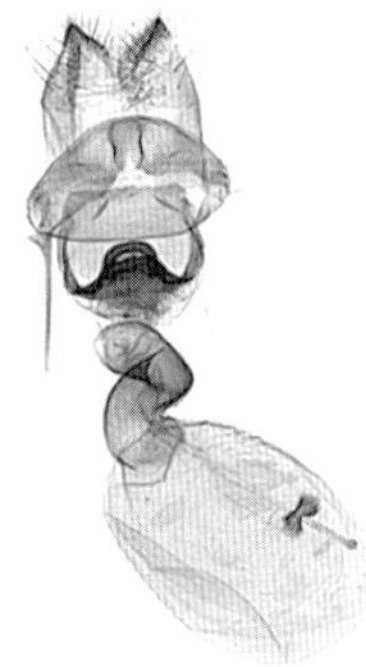

243a *Omichlis lineata*
W Papua, GU 93-76

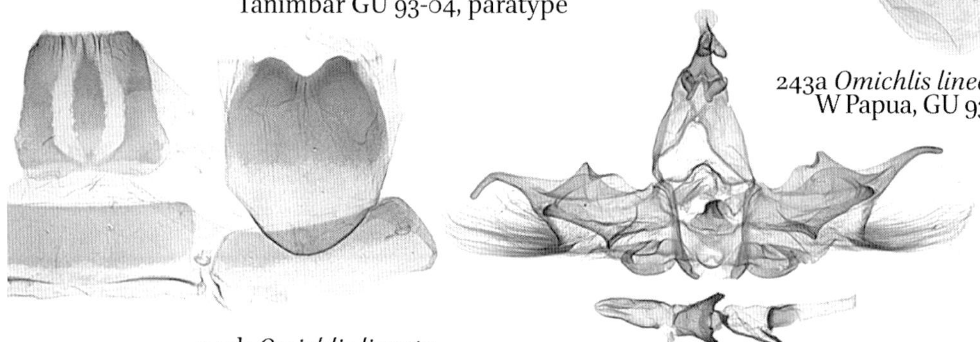

243b *Omichlis lineata*
W Papua, BM 10314101

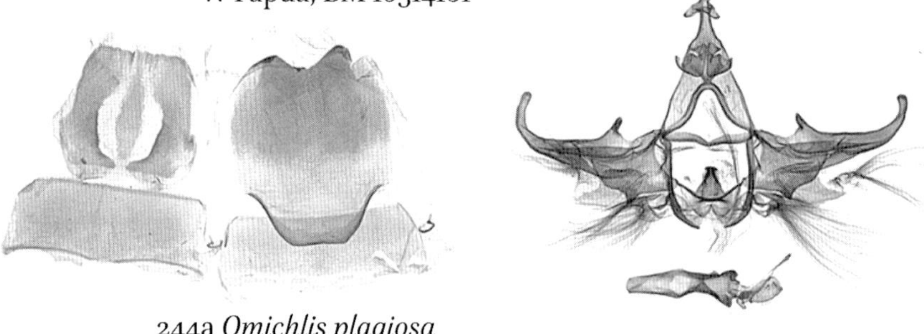

244a *Omichlis plagiosa*
W Papua, GU 93-75

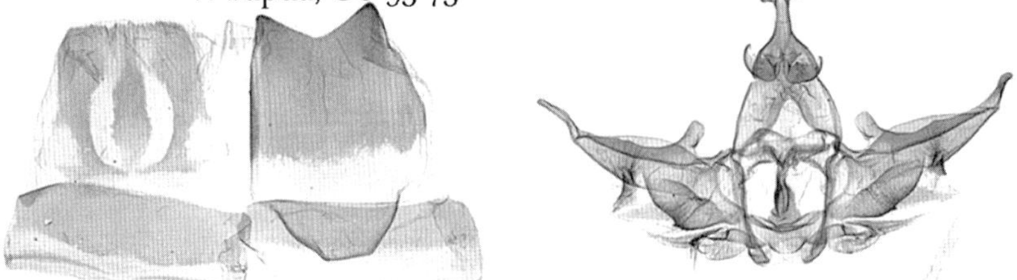

244b *Omichlis plagiosa*
PNG, BM 10314105

1 mm

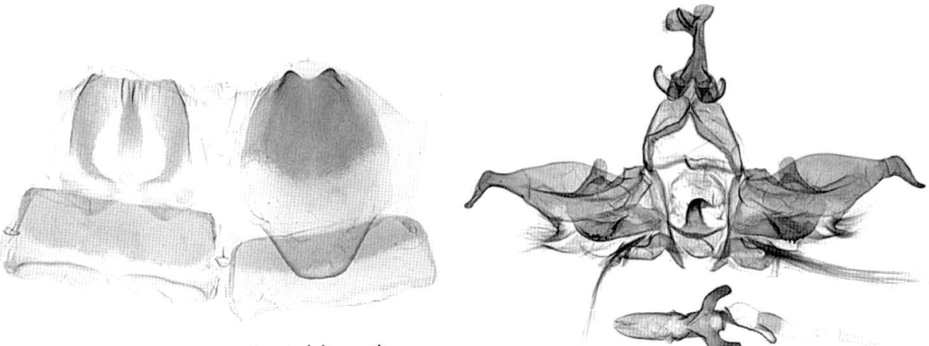

245a *Omichlis trilineata*
W Papua, BM 10314103

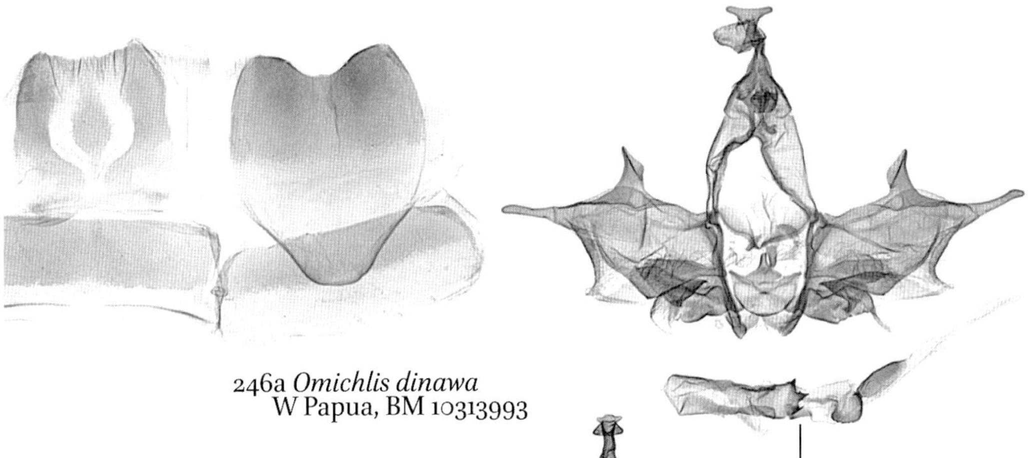

246a *Omichlis dinawa*
W Papua, BM 10313993

247a *Omichlis ochromaculata*
PNG, BM 10314126

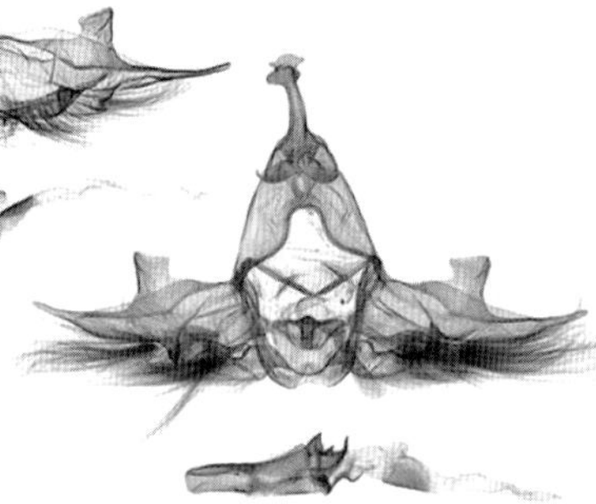

247b *Omichlis ochromaculata*
W Papua, GU 94-83

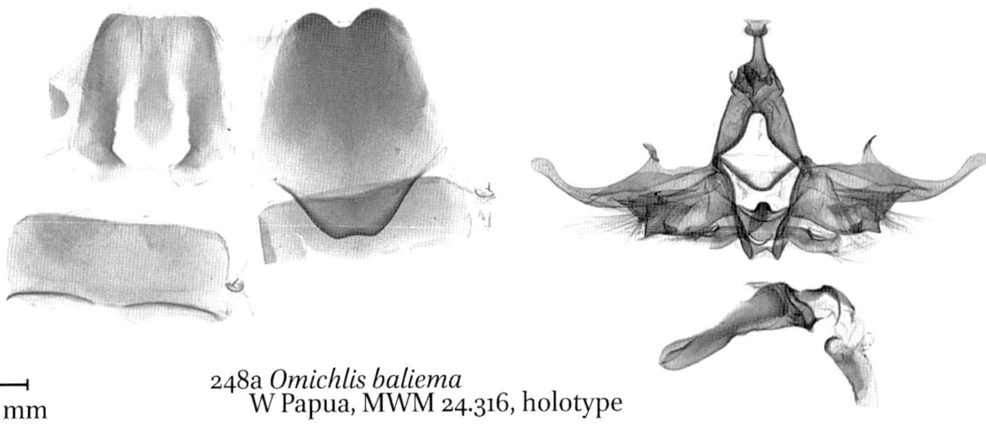

1 mm

248a *Omichlis baliema*
W Papua, MWM 24.316, holotype

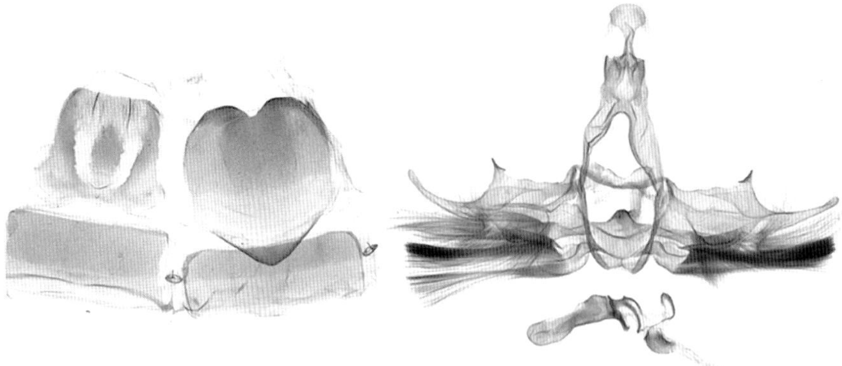

249a *Omichlis calyptis*
Ambon, GU 94-86

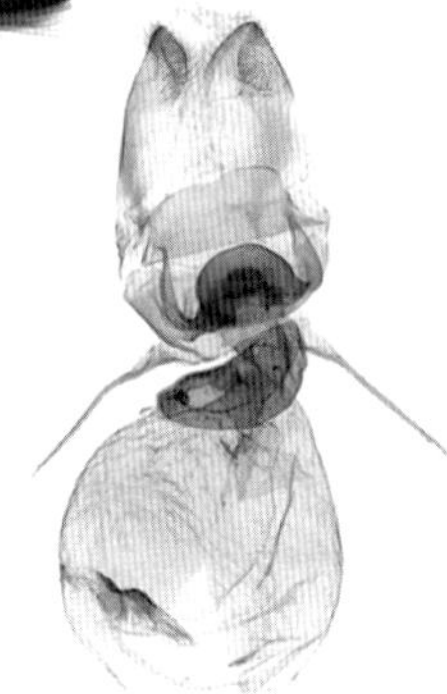

249b *Omichlis calyptis*
Seram, GU 94-85

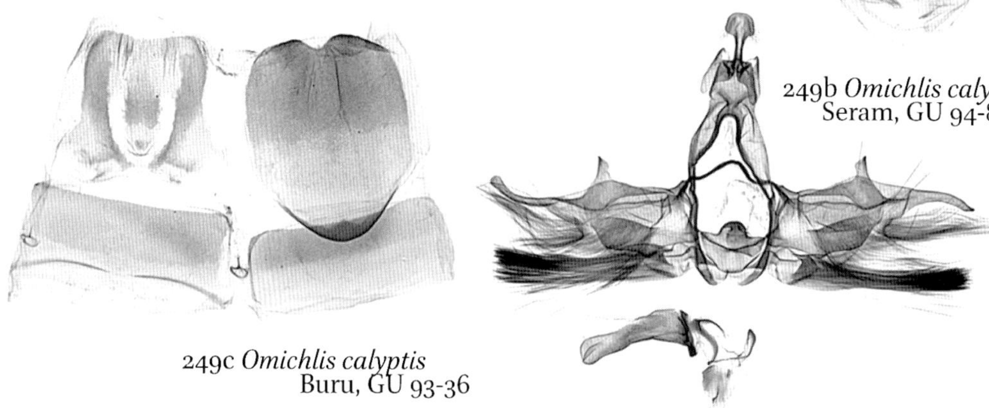

249c *Omichlis calyptis*
Buru, GU 93-36

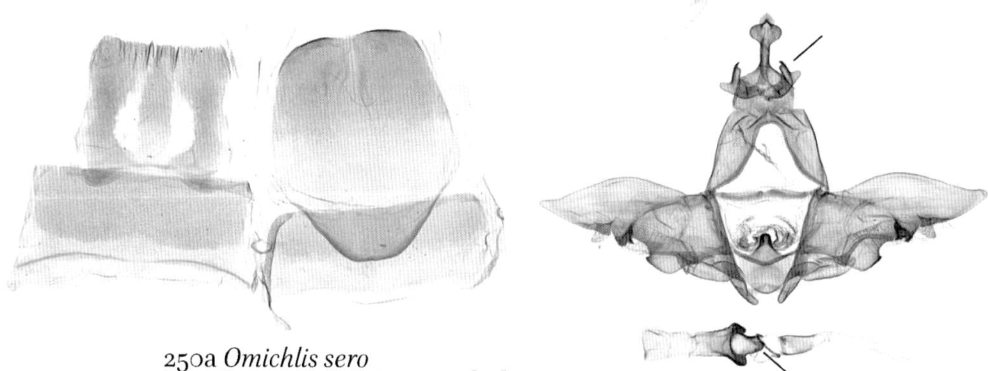

250a *Omichlis sero*
New Britain, MWM 29.420, holotype

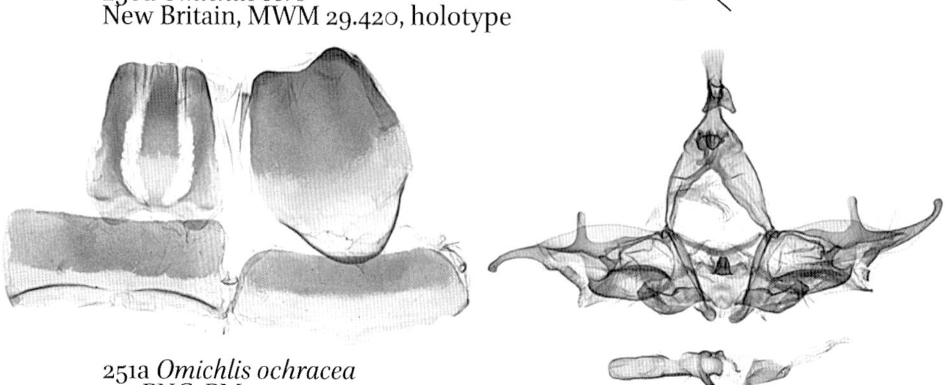

1 mm

251a *Omichlis ochracea*
PNG, BM 10314123

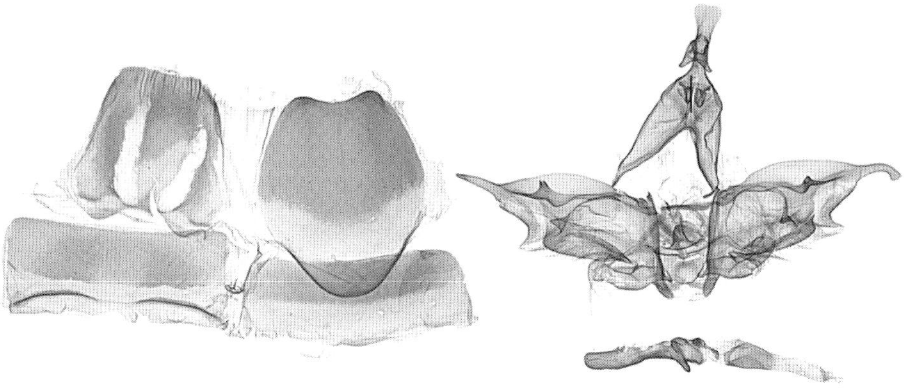

251b *Omichlis ochracea*
W Papua, BM 10314125

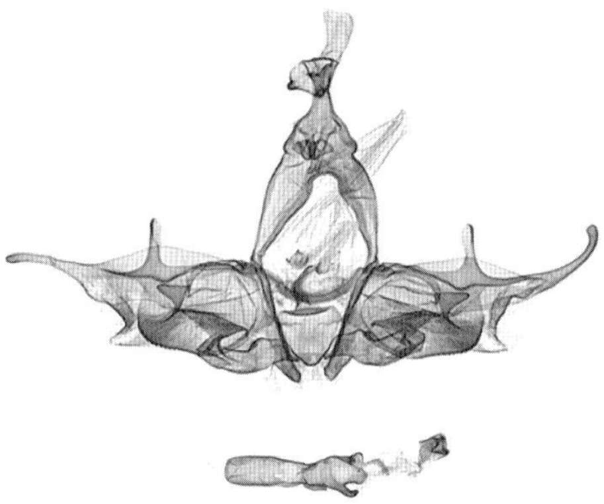

251c *Omichlis ochracea*
PNG, BM 10314124

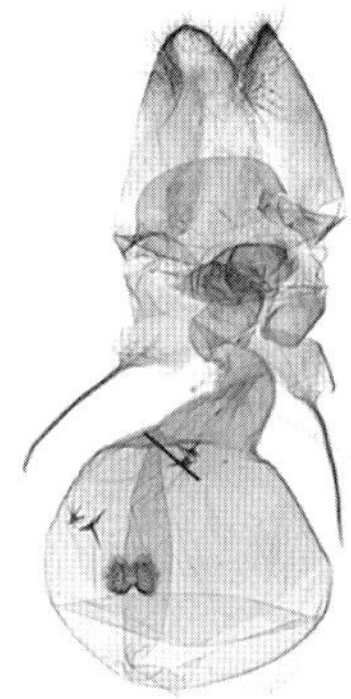

251d *Omichlis ochracea*
W Papua, BM 10314122

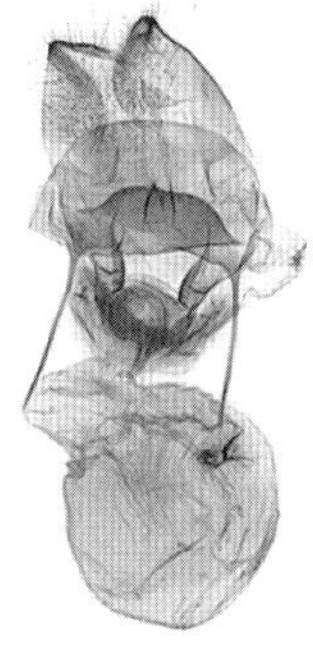

253a *Omichlis transversa*
W Papua, BM 10314129

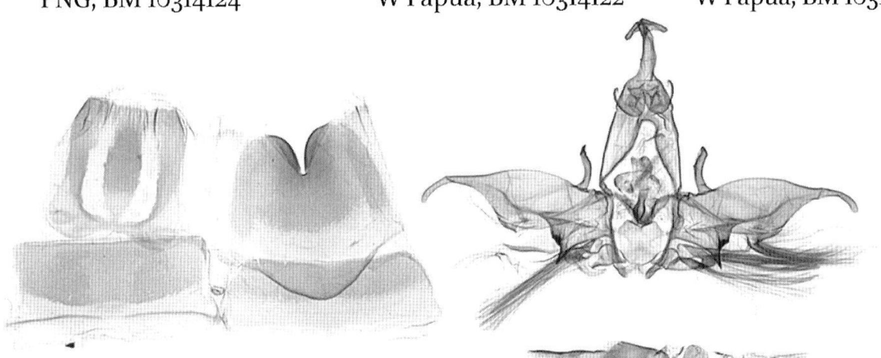

253b *Omichlis transversa*
W Papua, BM 10314130

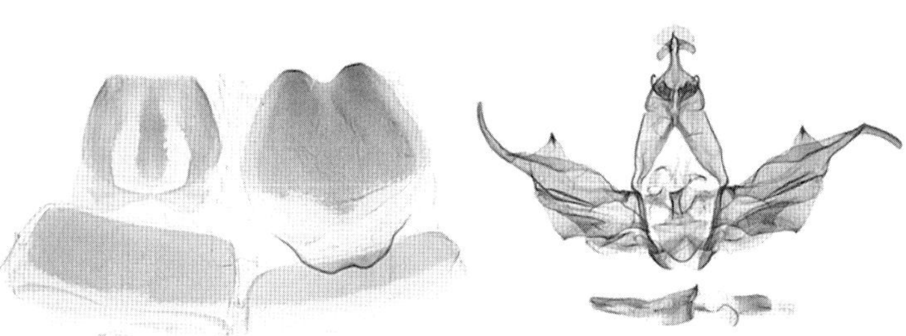

254a *Omichlis strigata*
PNG, BM 10314104

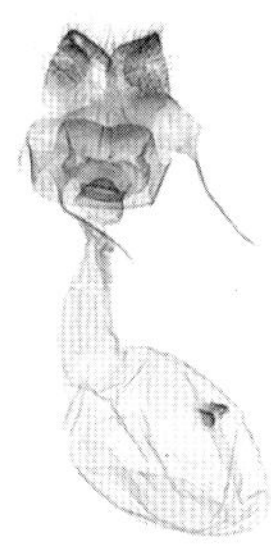

254b *Omichlis strigata*
PNG, BM 10314107

1 mm

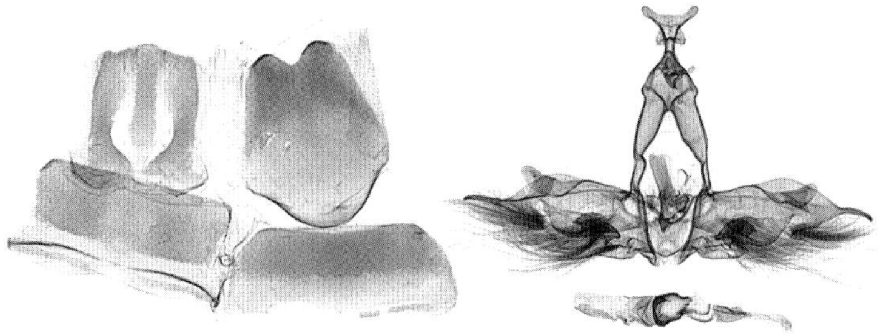

256a *Omichlis buergersi*
PNG, GU 66-25, holotype

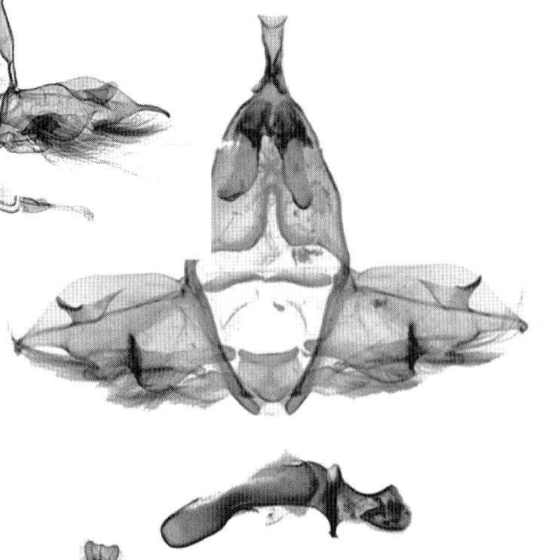

259a *Omichlis superne*
PNG, MWM 15.586,
holotype

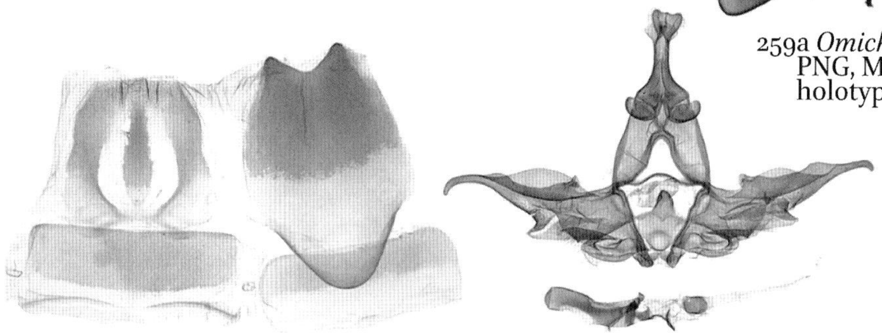

257a *Omichlis albidilinea*
W Papua, MWM 24.318

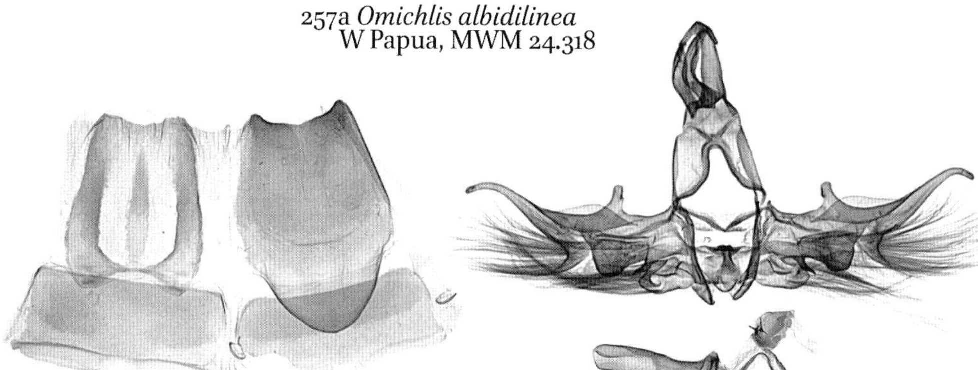

260a *Omichlis plagata*
W Papua, GU 93-74

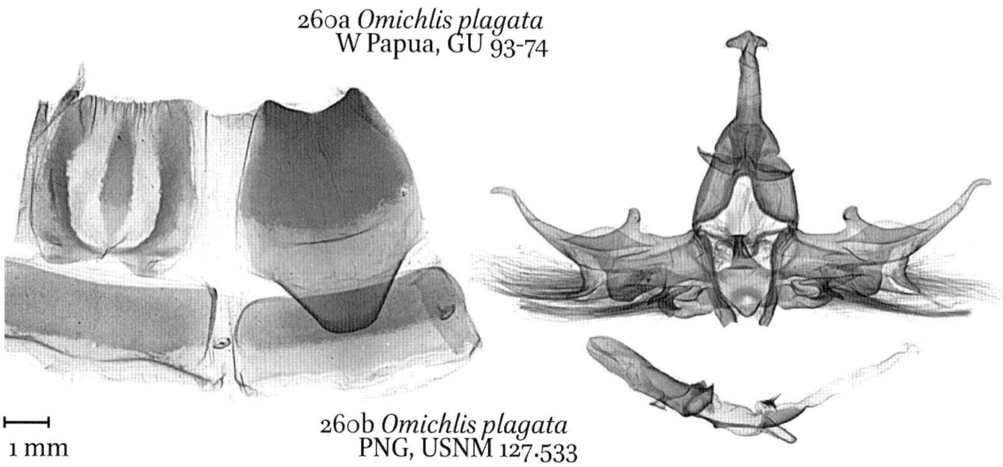

1 mm

260b *Omichlis plagata*
PNG, USNM 127.533

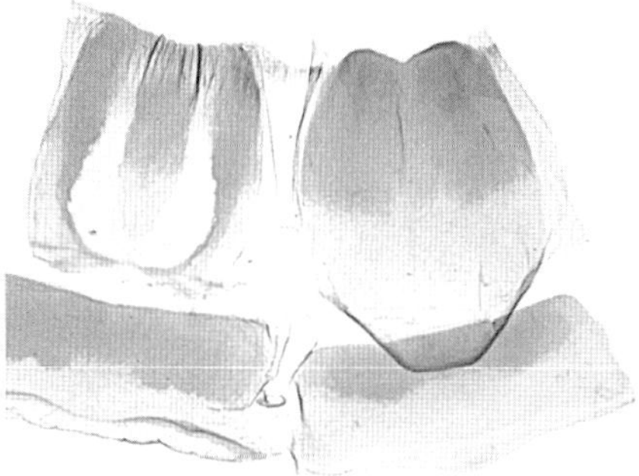

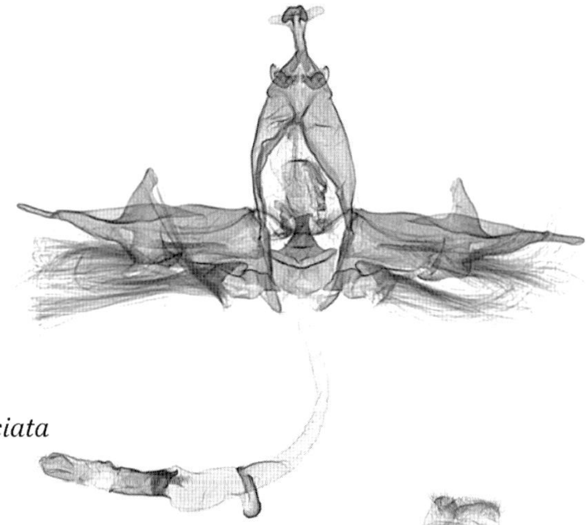

261a *Omichlis rufofasciata*
PNG, GU 66-24

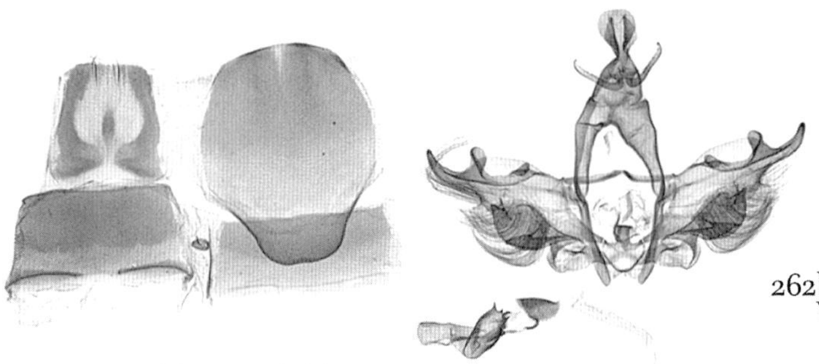

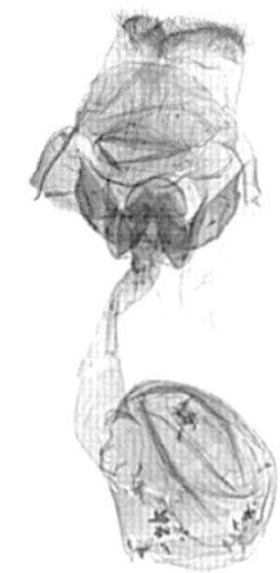

262b *Omichlis pseudolibatrix*
W Papua, BM 109, holotype

262a *Omichlis pseudolibatrix*
W Papua, Yapen Isl., Ihle 2017-06

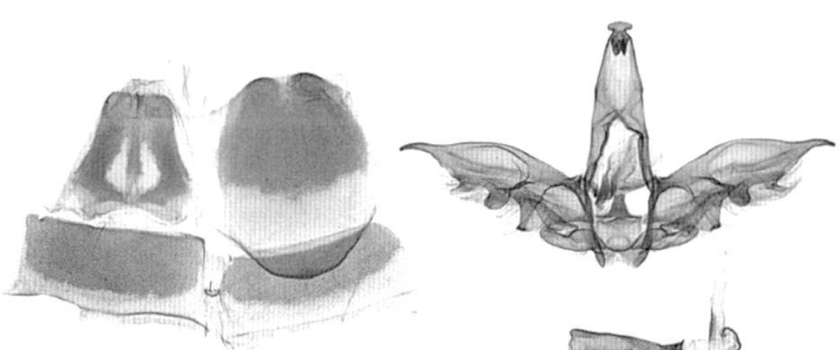

263a *Omichlis mediofasciata*
PNG, BM 10314110, paralectotype

264a *Omestia bella*
W Papua
BM10314109, paratype

261b *O. rufofasciata*
W Papua
GU 66-27

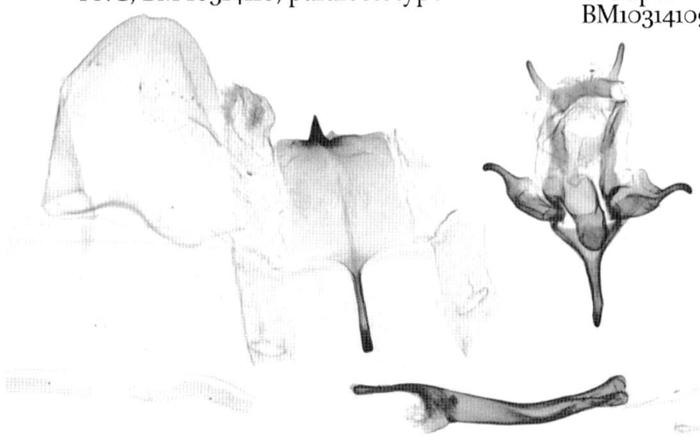

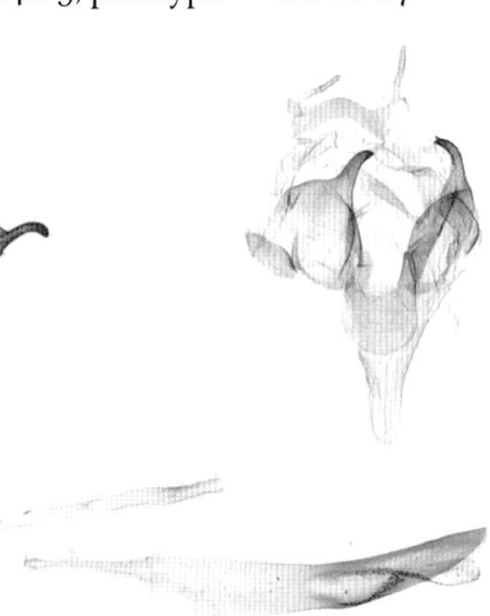

1 mm

265a *Netria viridescens viridescens*
Java, GU 43-77

265b *Netria viridescens viridescens*
Kalimantan, GU 81-23

265c *Netria viridescens viridescens*
Java, MWM 4.882

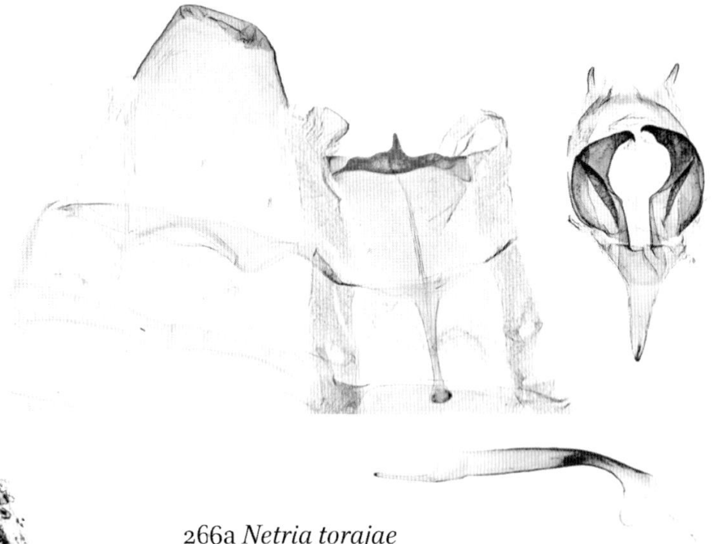

266a *Netria torajae*
Sulawesi, MWM 4.859, paratype

266b *Netria torajae*
Sulawesi, GU 43-91, paratype

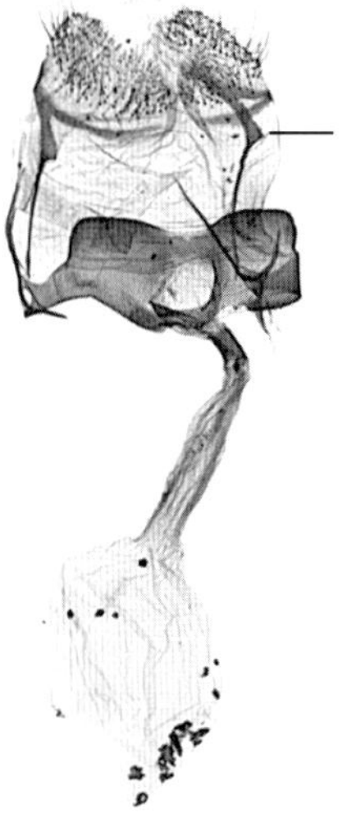

1 mm

266c *Netria torajae*
Sulawesi, GU 44-19

267a *Netria longisacci*
Sumba, GU 66-34

268a *Netria insularis*
Tanimbar, GU 66-35

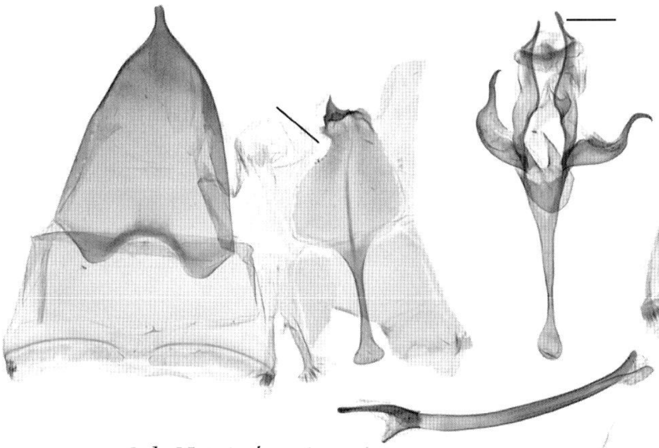

267b *Netria longisacci*
Alor, GU 92-69

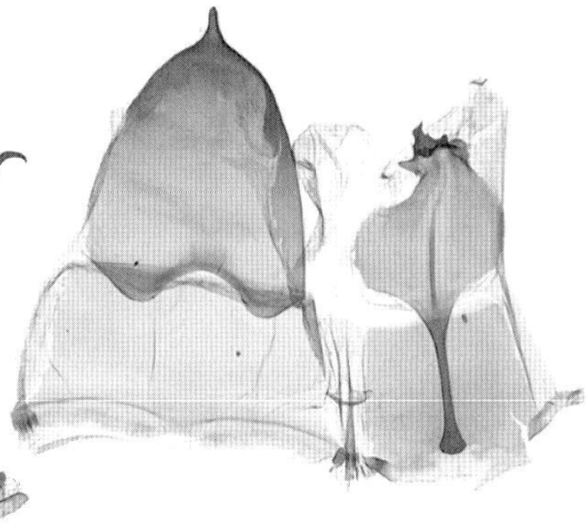

267c *Netria longisacci*
Pantar, GU 92-70
8th abdominal segments

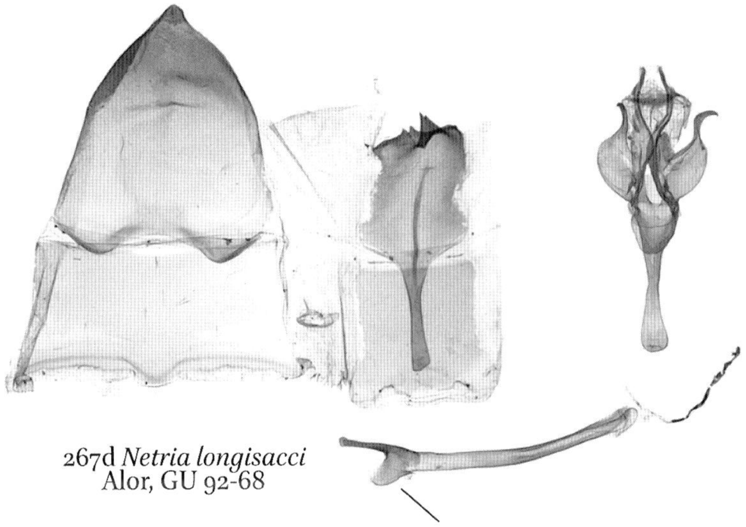

267d *Netria longisacci*
Alor, GU 92-68

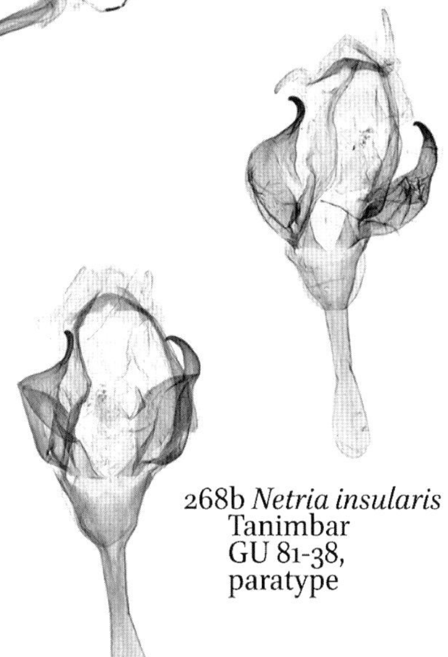

268b *Netria insularis*
Tanimbar
GU 81-38,
paratype

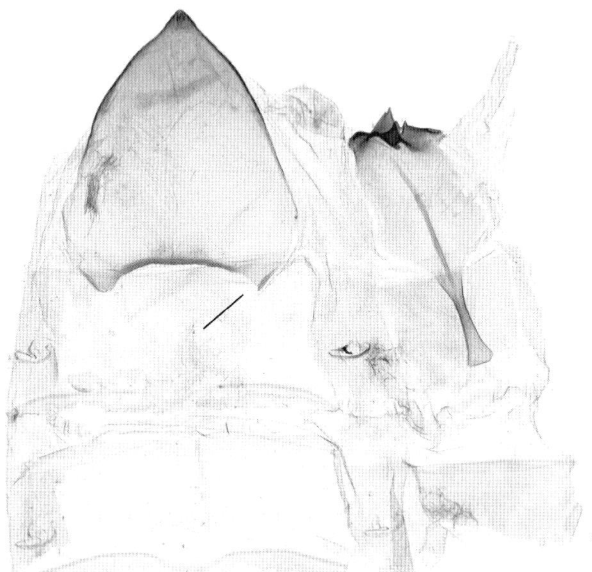

1 mm

268c *Netria insularis*
Tanimbar, GU 81-13, paratype

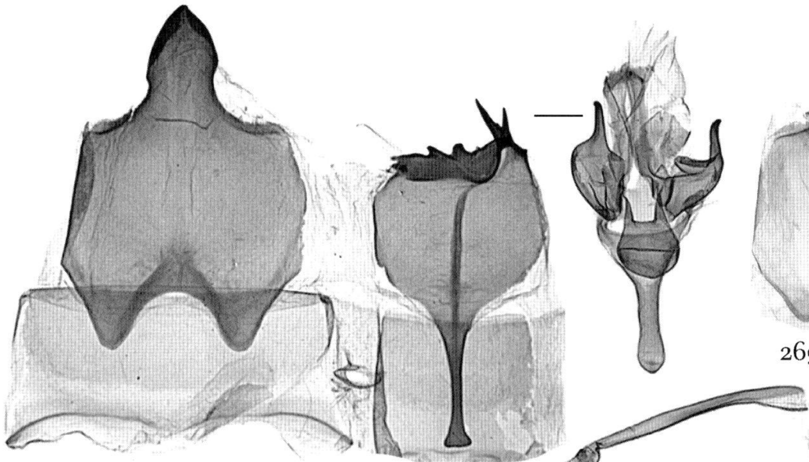

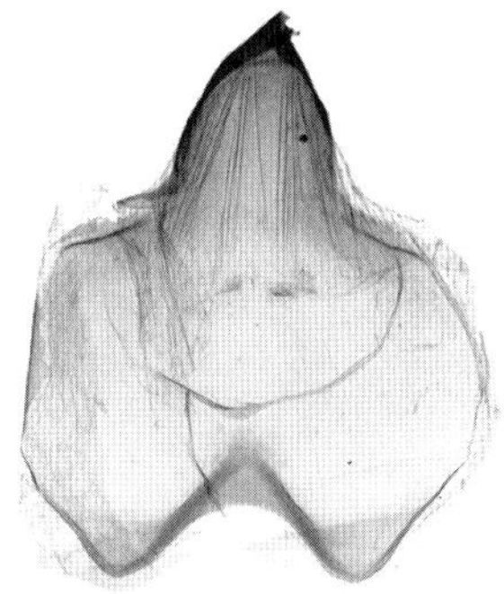

269b *Netria arisemna maiorvirens* W Papua, GU 81-32 8th tergite, paratype

269a *Netria arisemna maiorvirens* W Papua, GU 66-33, paratype

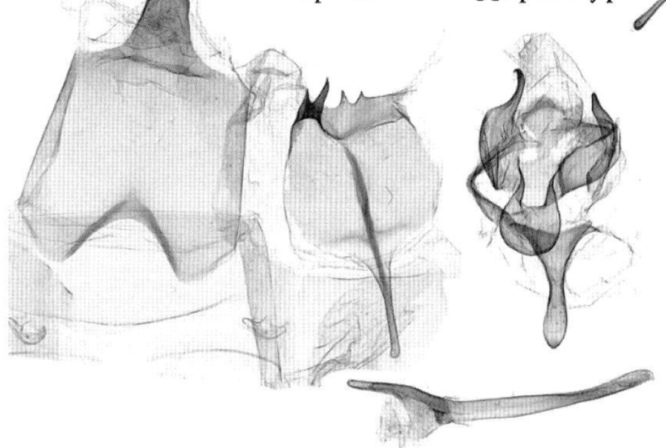

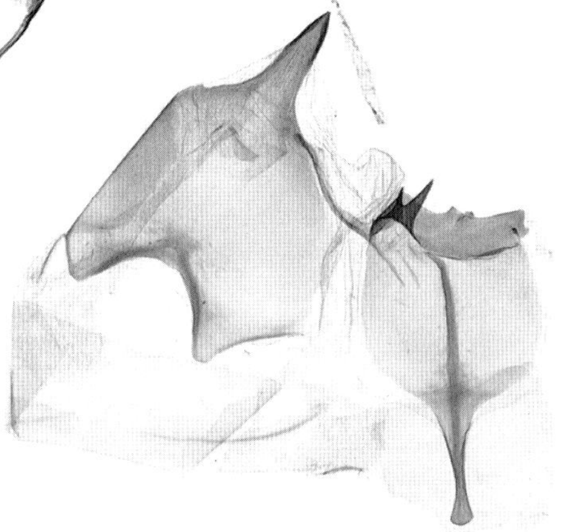

269c *Netria arisemna arisemna* Australia, Qsd., GU 43-85a

269d *Netria arisemna andreasmuelleri* Seram, GU 81-07, paratype 8th abdominal segment

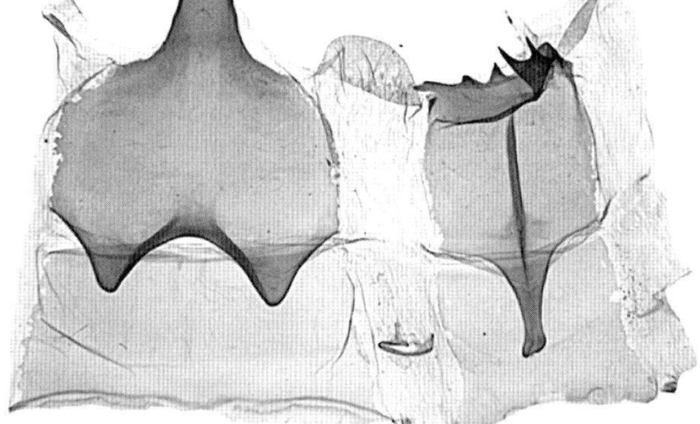

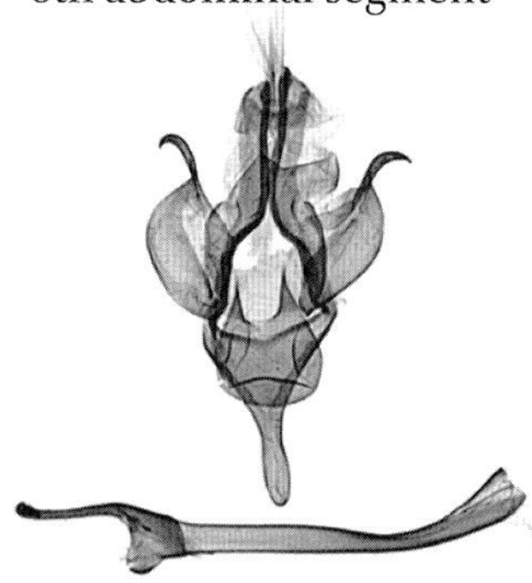

269e *Netria arisemna andreasmuelleri* Buru, GU 93-27

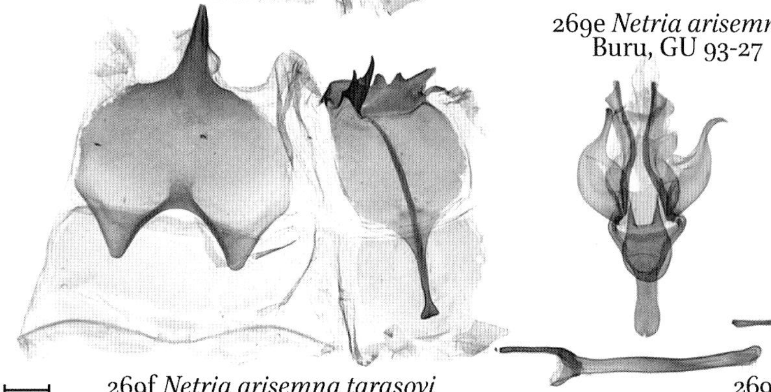

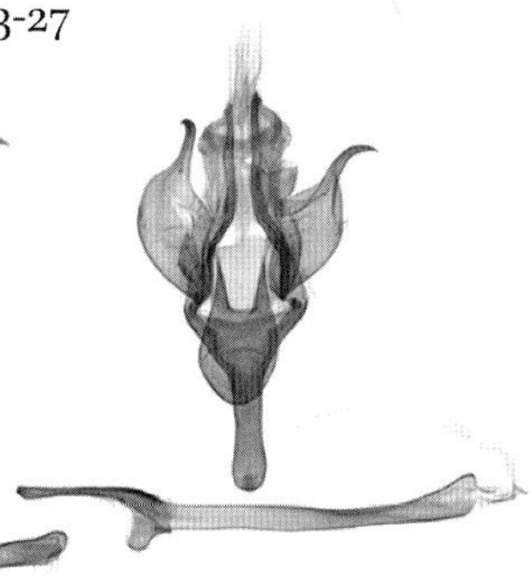

1 mm

269f *Netria arisemna tarasovi* Obi, GU 92-72

269g *Netria arisemna tarasovi,* Obi, GU 92-73

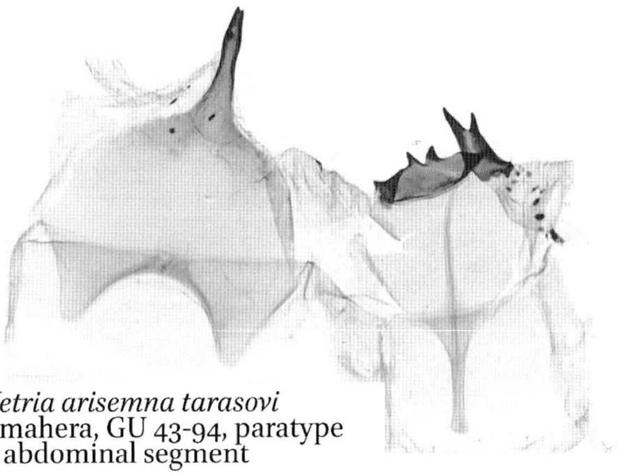

269h *Netria arisemna tarasovi*
Halmahera, GU 43-94, paratype
8th abdominal segment

269i *Netria arisemna bipartita*
Peleng, GU 66-39, paratype

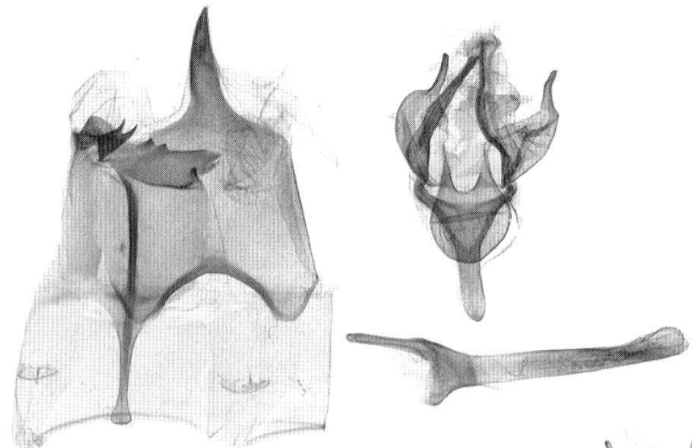

269j *N. arisemna bipartita*
Sulawesi, GU 81-39
paratype

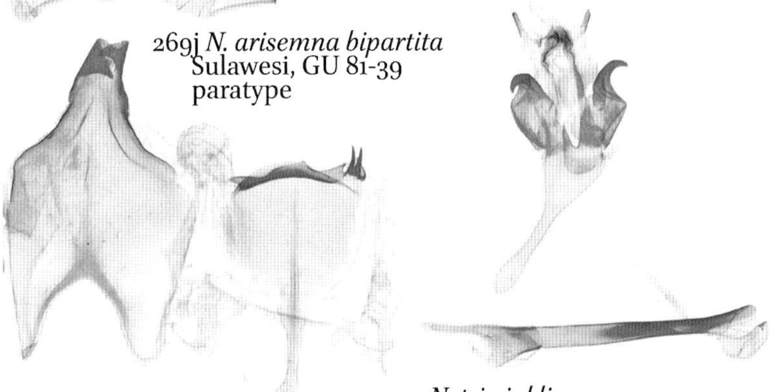

270a *Netria jakli*
Java, GU 43-74, paratype

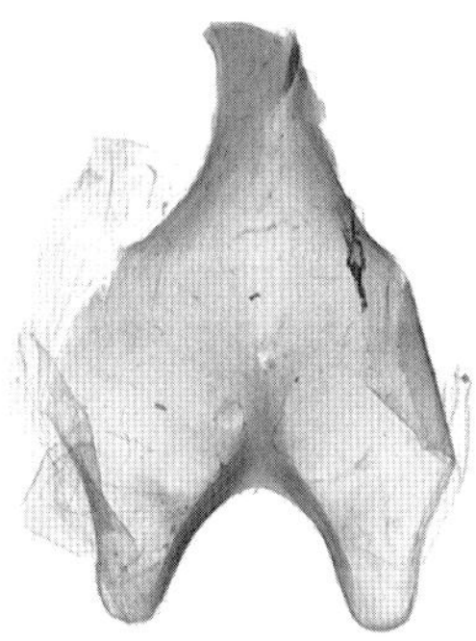

270b *Netria jakli*
Java, GU 81-11, paratype
8th tergite

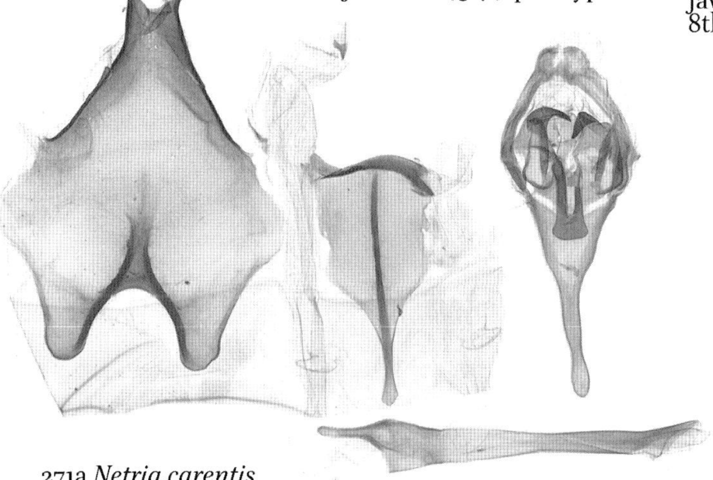

271a *Netria carentis*
Kalimantan, GU 81-21, paratype

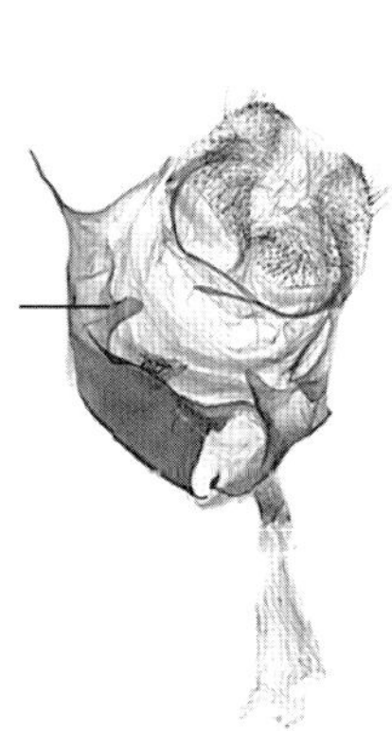

270c *Netria jakli*
Bali, GU 66-36

1 mm

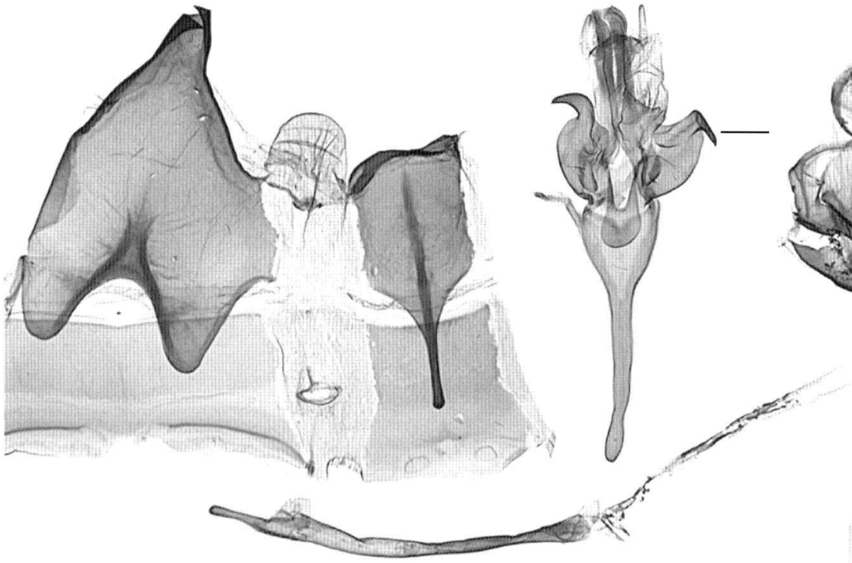

271b *Netria carentis*
Belitung, GU 66-38

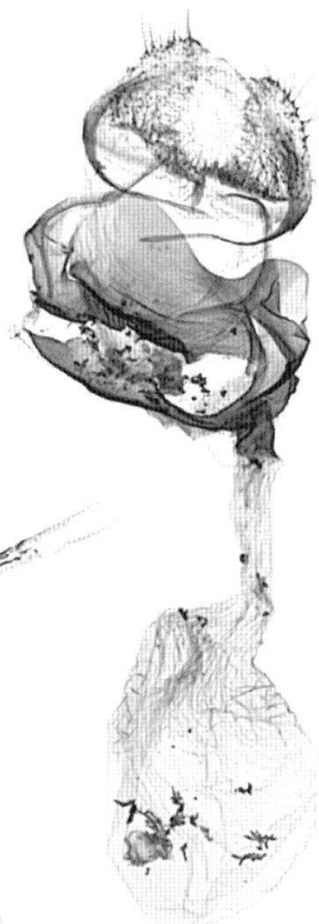

271c *Netria carentis*
Belitung
GU 66-37

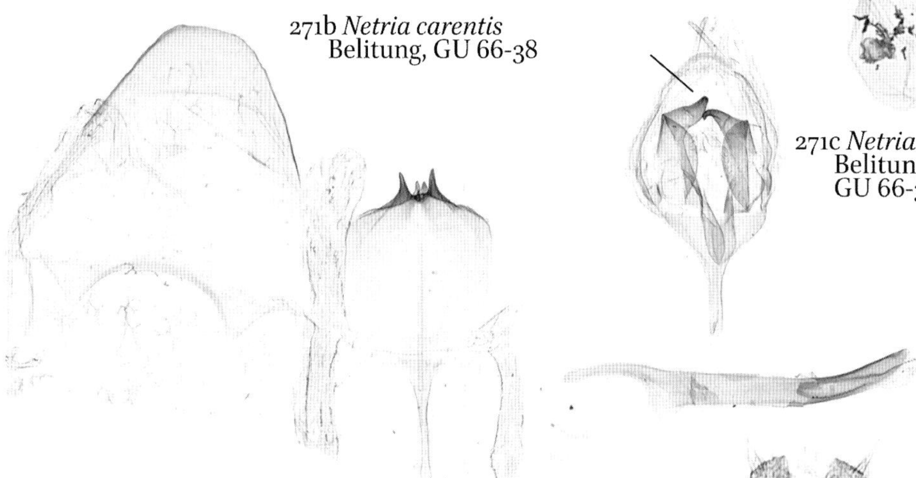

272a *Netria multispinae multispinae*
Sumatra, GU 81-24, paratype

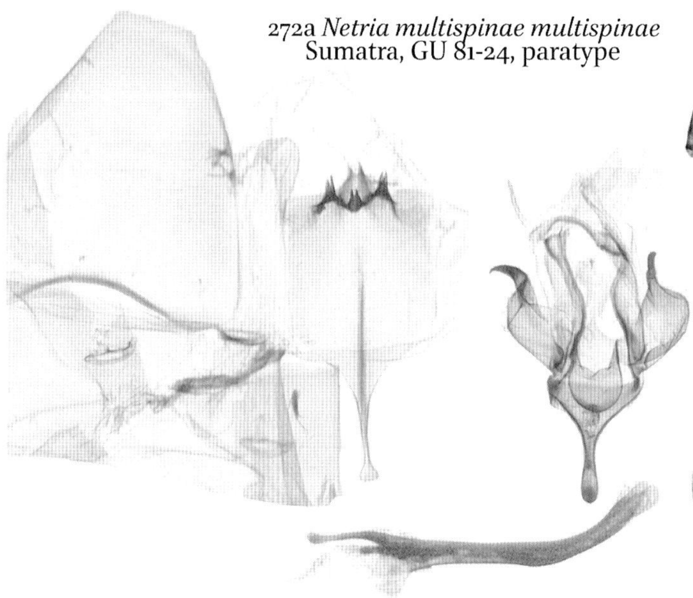

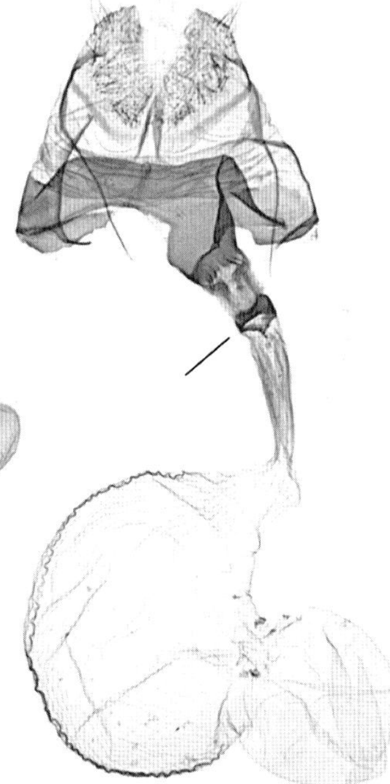

1 mm

272b *Netria multispinae multispinae*
Sumatra, GU 81-17, paratype

272c *Netria multispinae multispinae*
Sumatra, GU 66-40, paratype

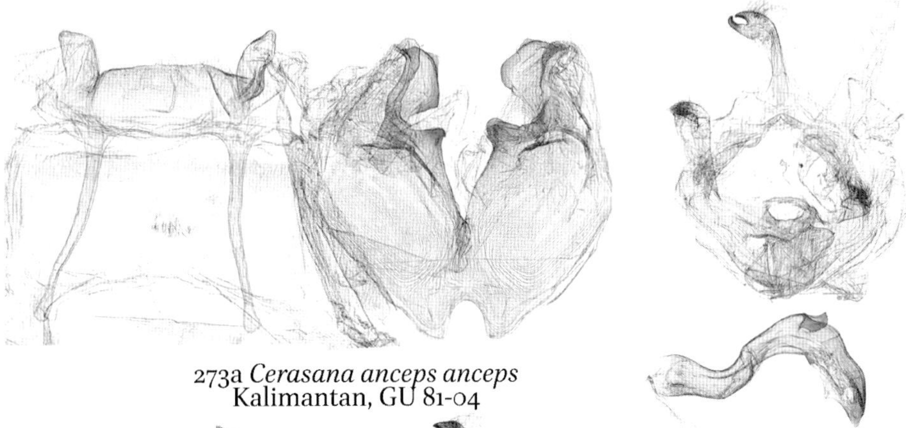

273a *Cerasana anceps anceps*
Kalimantan, GU 81-04

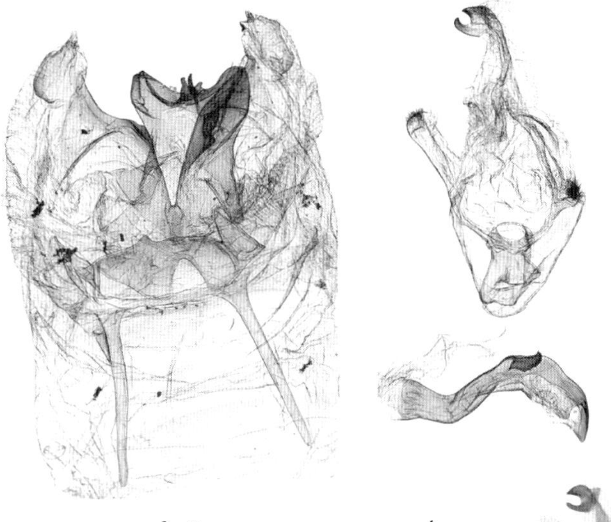

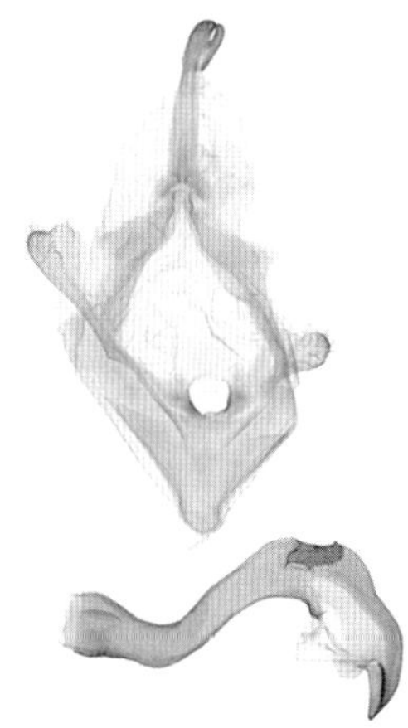

273b *Cerasana anceps ursulae*
Thailand, GU 81-03, paratype

273c *Cerasana anceps anceps*
Borneo, MWM 15.736

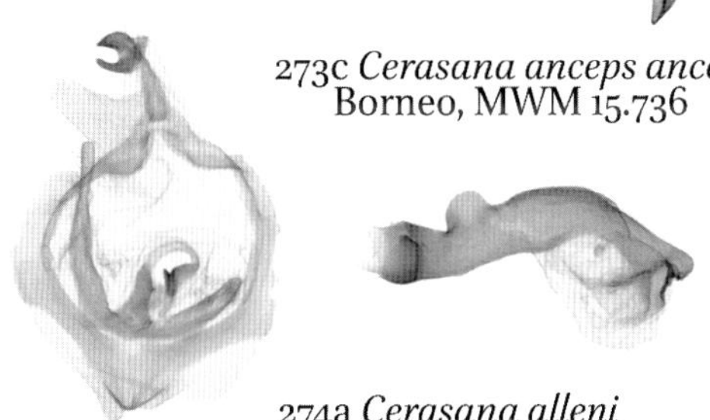

274a *Cerasana alleni*
Borneo, MWM 15.527

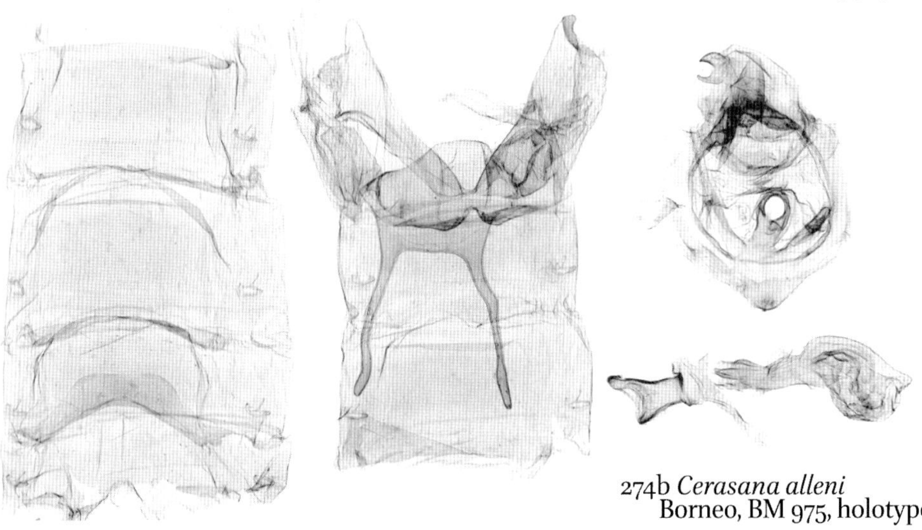

1 mm

274b *Cerasana alleni*
Borneo, BM 975, holotype

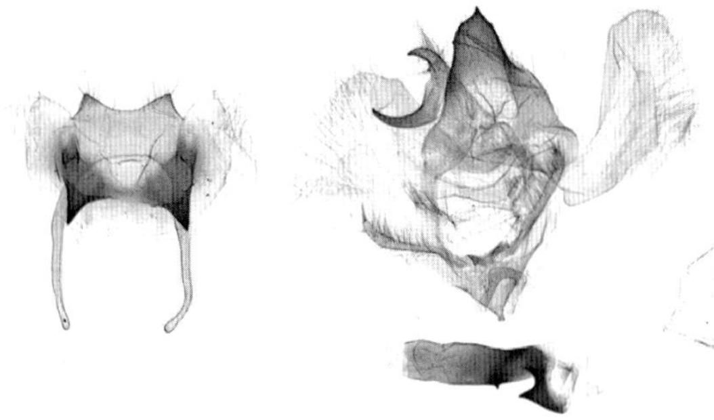

275a *Cerasana pagenstecheri pagenstecheri,* Malaya, BM 37/2003, paratype

275b *C. p. pagenstecheri* Sumatra, BM 53/2003, paratype

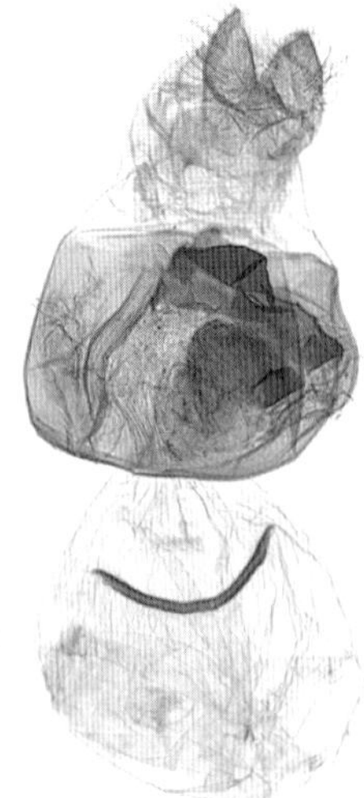

273d *C. anceps lutea,* Palawan, MWM 8.097

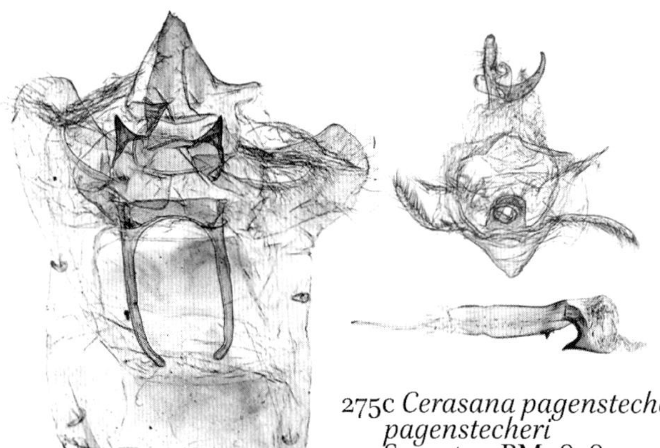

275c *Cerasana pagenstecheri pagenstecheri* Sumatra, BM 1848, paratype

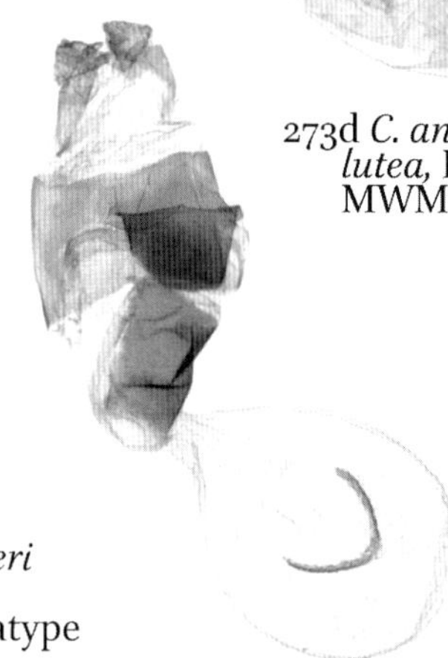

274c *Cerasana alleni* Borneo, MWM 15.528

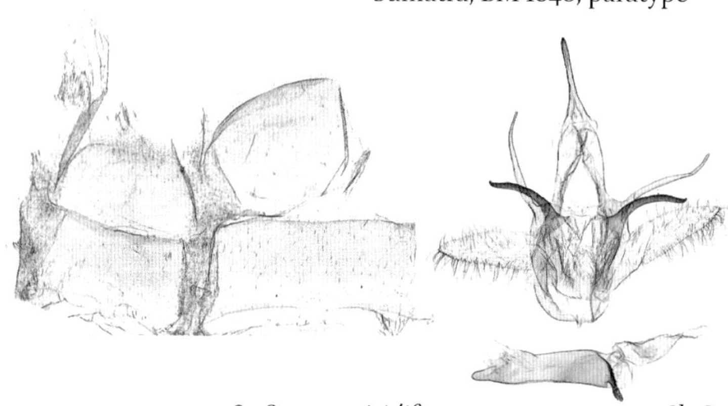

276a *Somera viridifusca sumatrana* Borneo, MWM 4.851

276b *Somera viridifusca sumatrana* Sumatra, GU 43-39, paratype

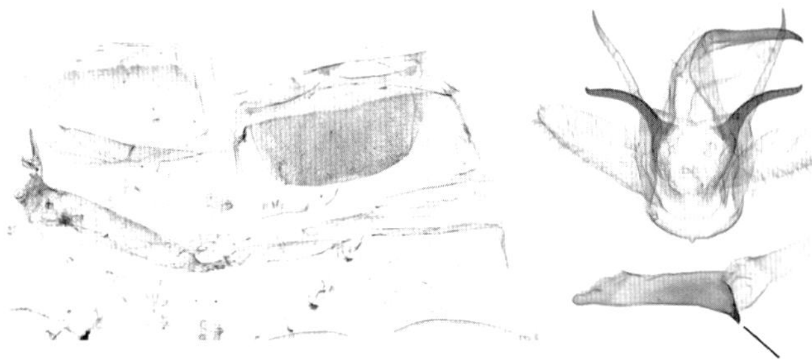

1 mm

276c *Somera viridifusca javana* Java, GU 43-45, paratype

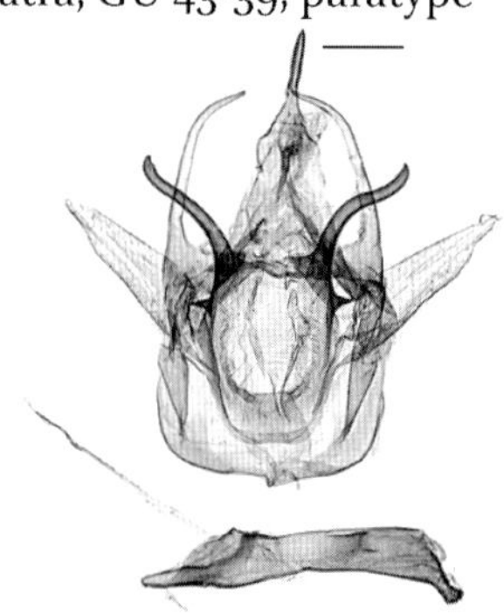

276d *Somera viridifusca celebica* Sulawesi, GU 02-42, paratype

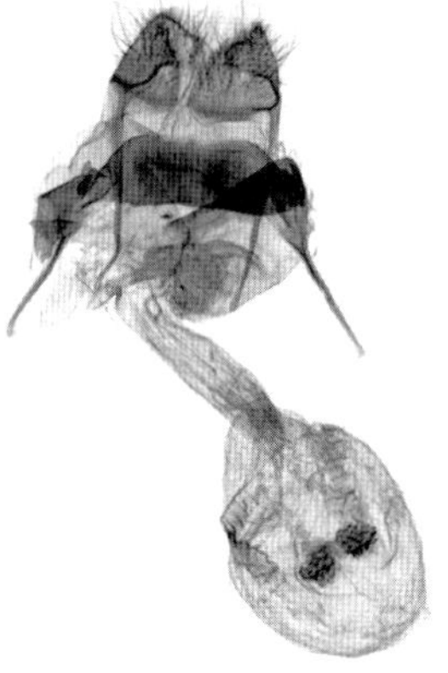

276e *Somera viridifusca viridifusca*
India, Sikkim, GU 43-37

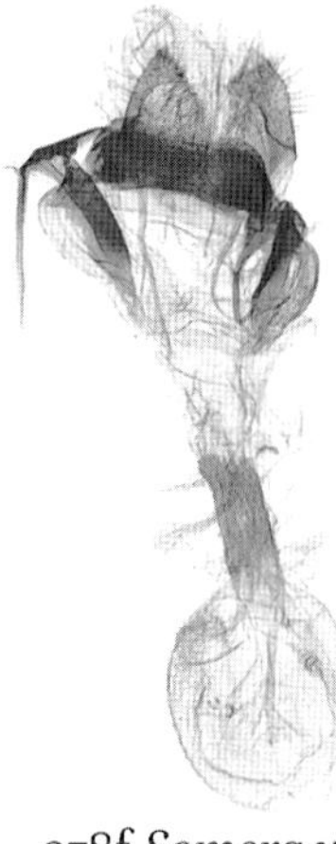

278f *Somera viridifusca luzonensis*
Luzon, GU 43-43, paratype

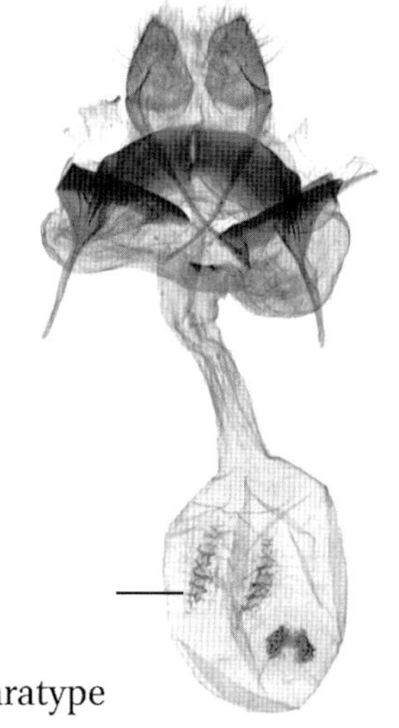

276g *Somera viridifusca celebica*
Sulawesi, GU 43-40, paratype

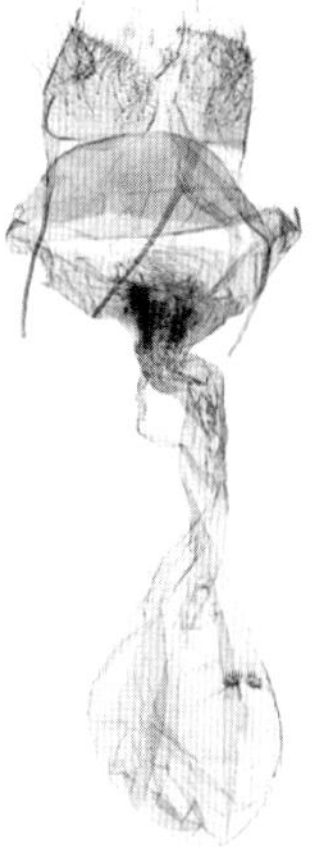

278a *Somera virens virens*
Malaya, MWM 4.855

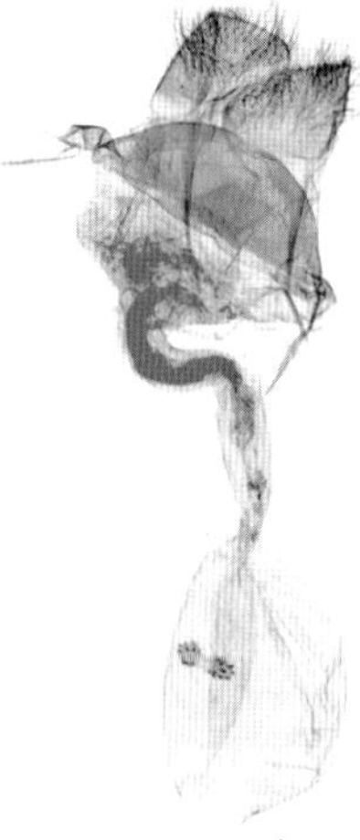

278b *Somera virens virens*
Sumatra, GU 43-48

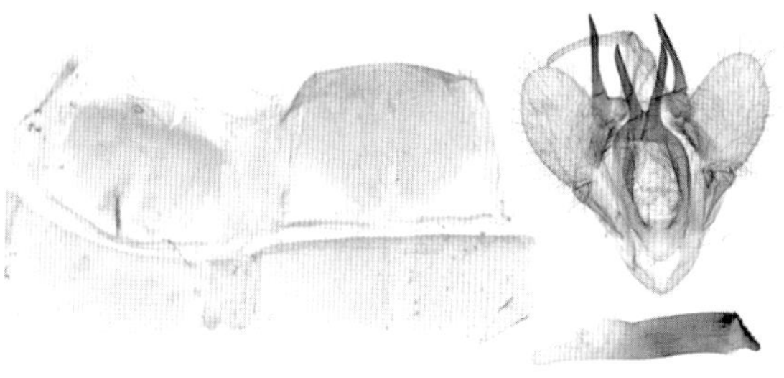

278c *Somera virens virens*
Kalimantan, GU 77-74

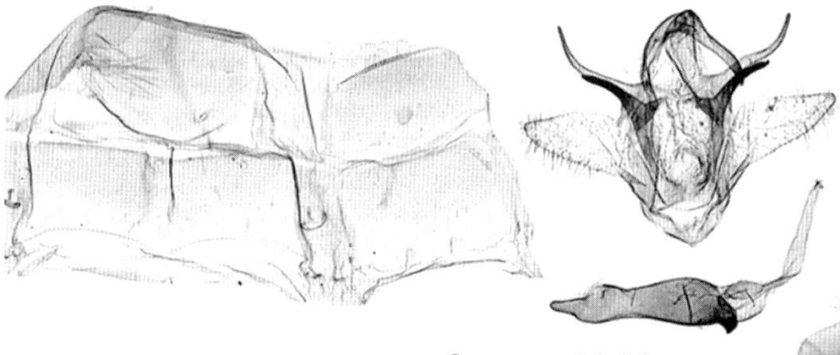

277a *Somera viriviri*
Sulawesi, MWM 1.769

277b *Somera viriviri*
Sulawesi
GU 25-55

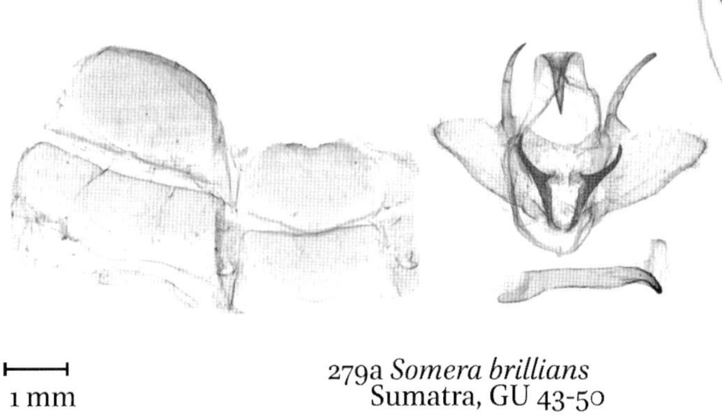

279a *Somera brillians*
Sumatra, GU 43-50

279b *Somera brillians*
Sumatra, GU 43-51

1 mm

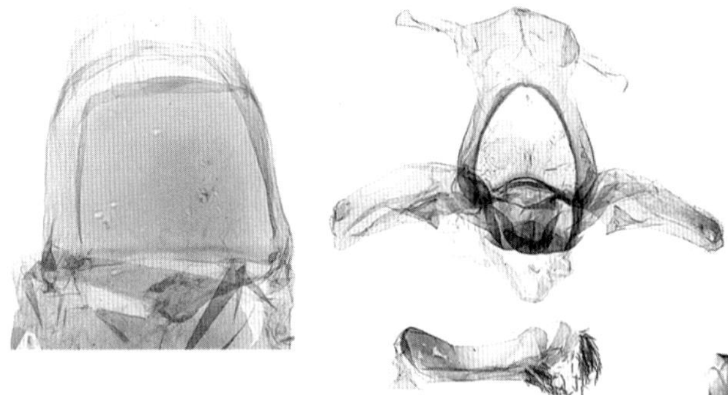

280a *Psegmaphora tripunctata*
Sumatra, GU 51-68

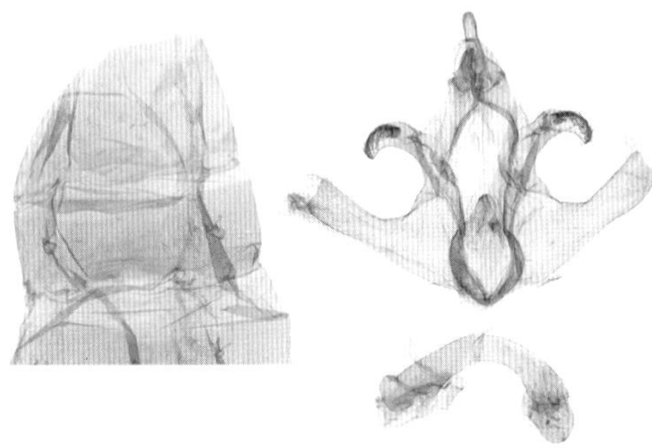

281a *Shachia vernalis infuscata*
Sumatra, GU 51-67

281b *Shachia vernalis infuscata*
Sumatra, GU 66-46

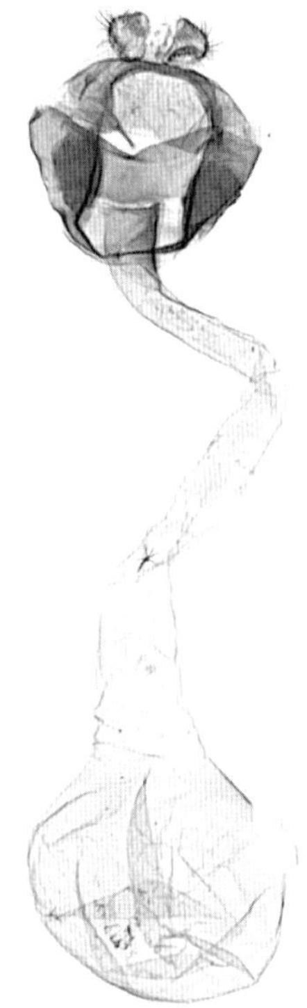

282a *Pheraspis calida*
Tanimbar, GU 66-29
paratype

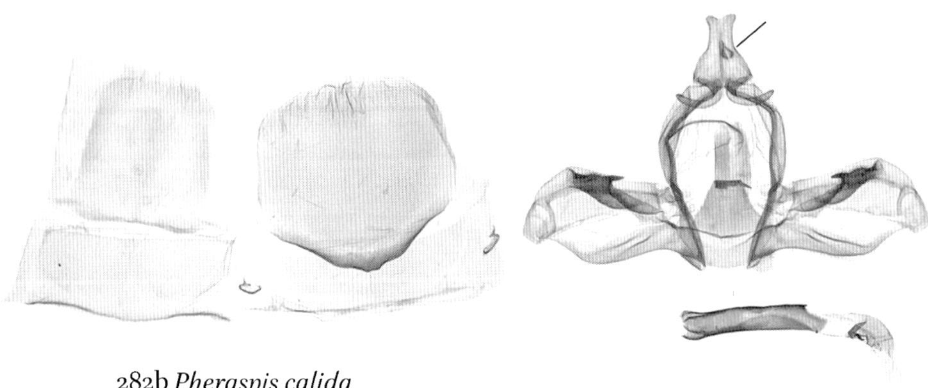

282b *Pheraspis calida*
Tanimbar, GU 92-85, paratype

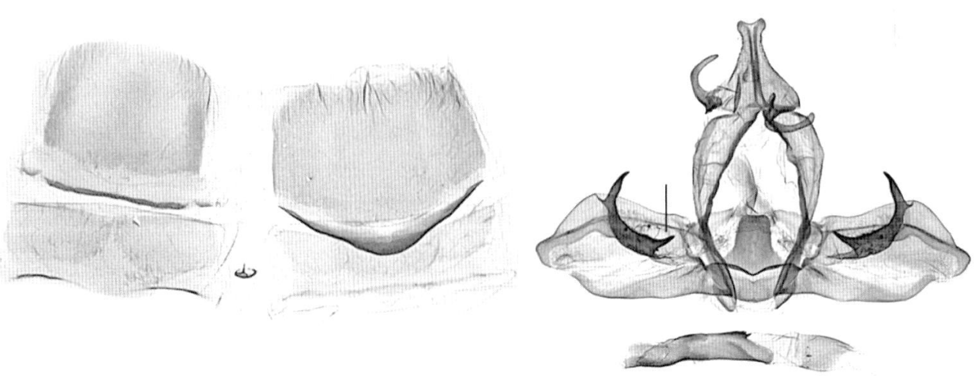

1 mm

282Aa *Pheraspis spodea*
Australia, NT, GU 66-28

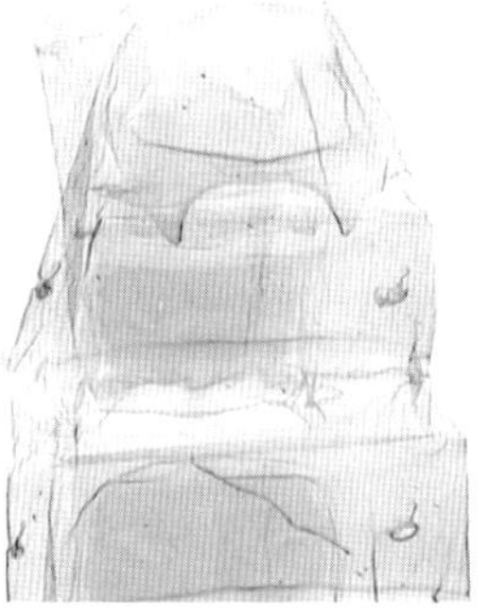
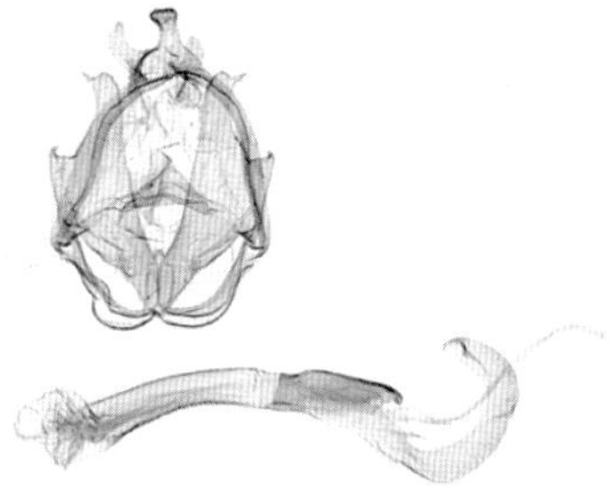

283a *Notodontella ferrifusa*
India, Sikkim, BM 1082
lectotype

283b *Notodontella ferrifusa*
Sumatra, GU 51-90

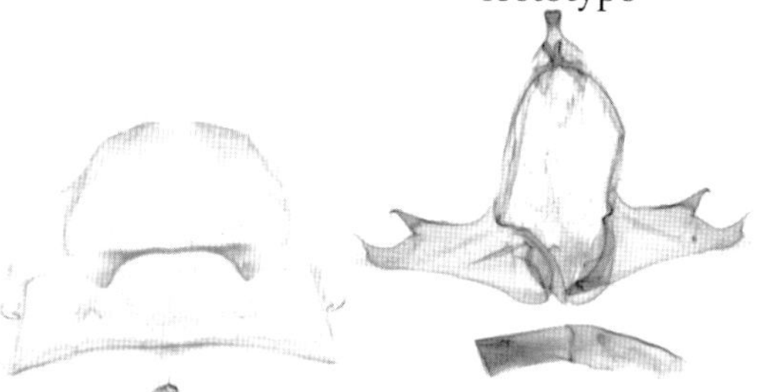

283c *Notodontella ferrifusa*
Sumatra, GU 43-60

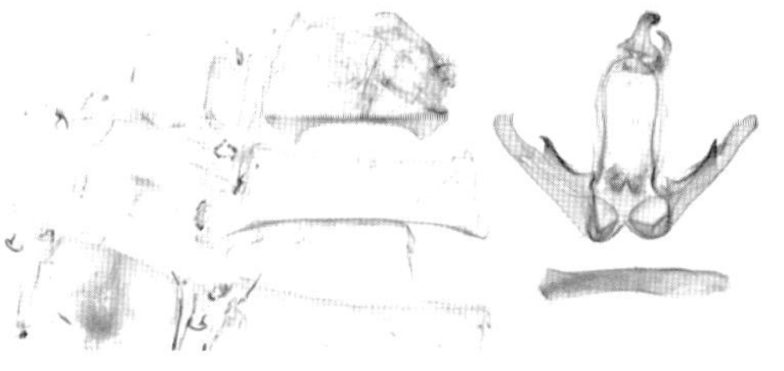

284a *Notodontella floresiensis*
Sumatra, GU 43-57, holotype

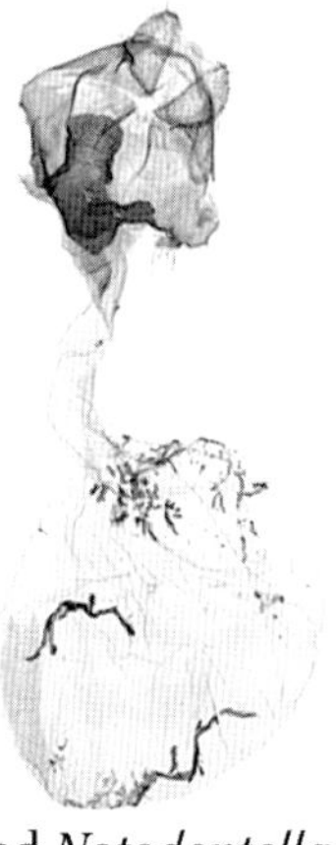

283d *Notodontella ferrifusa*
Sumatra, GU 66-47

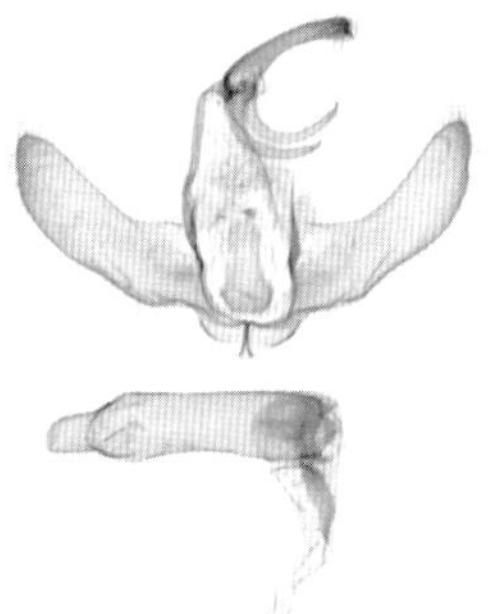

285a *Chloroceramis viridinota*
Sumatra, GU 43-59

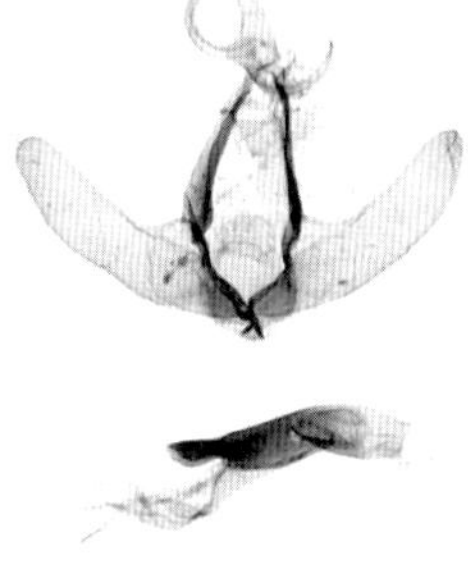

285b *Chloroceramis viridinota*
Java, GU 51-99

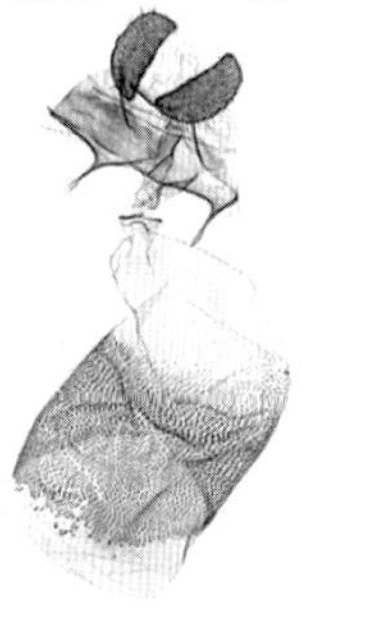

285c *Chloroceramis viridinota*
Sumatra, GU 66-48

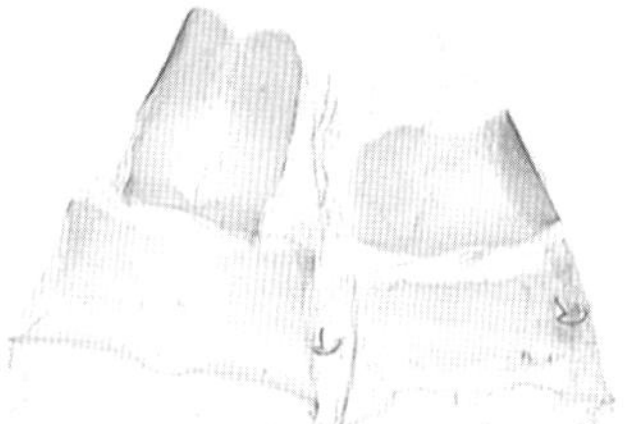

286a *Notodontella albuala*
Sulawesi, GU 43-58,
paratype

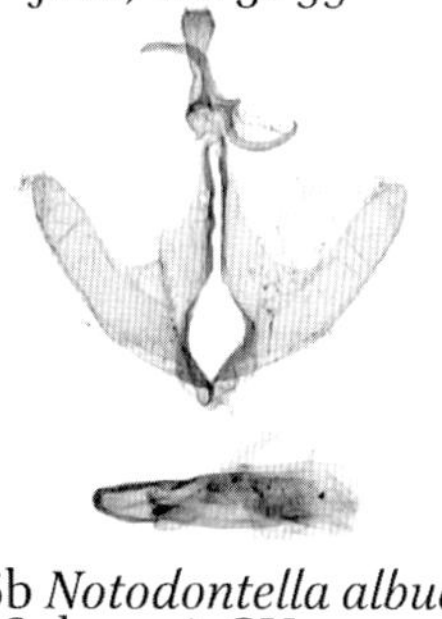

286b *Notodontella albuala*
Sulawesi, GU 51-97,
paratype

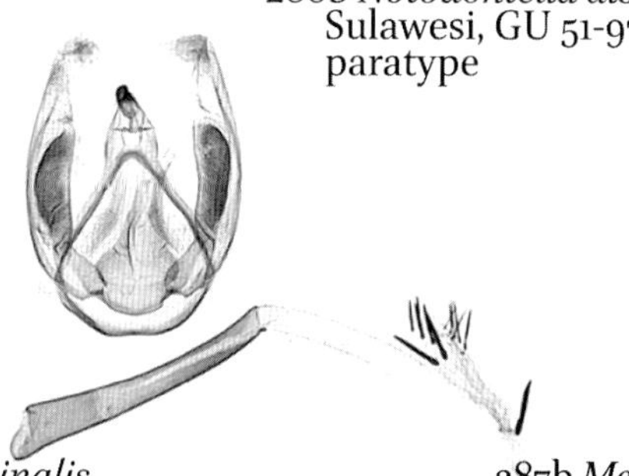

1 mm

287a *Medanella subterminalis*
Sumatra, GU 66-45

287b *Medanella subterminalis*
Sumatra, GU 66-44

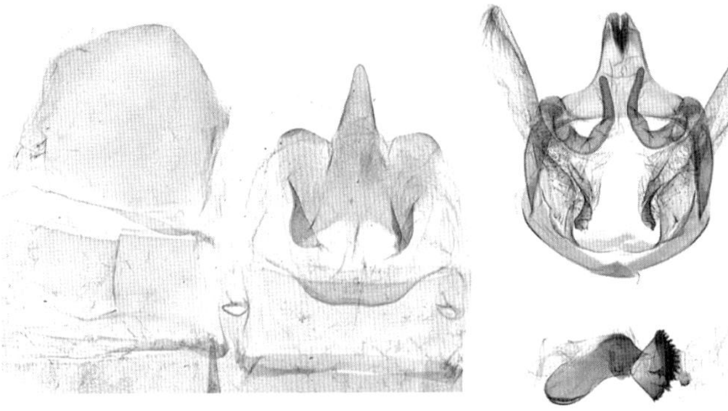

288a *Antiphalera sumatrana*
Kalimantan, GU 72-36

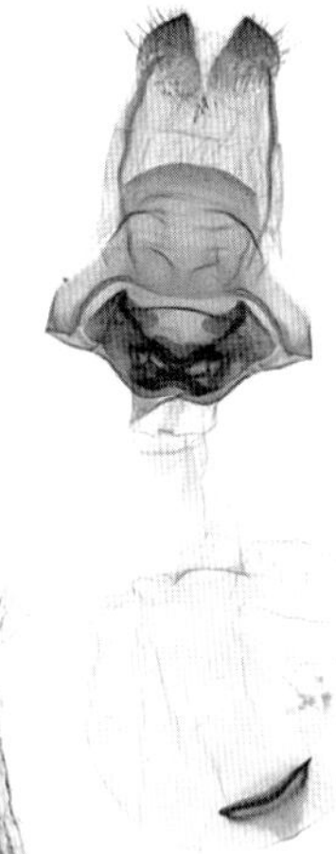

288b *Antiphalera sumatrana*
Kalimantan, GU 95-13

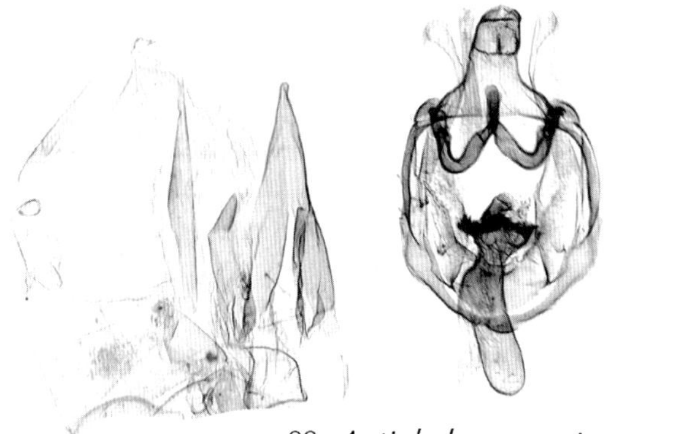

288c *Antiphalera sumatrana*
Sumatra, ZSM 383, holotype

288d *Antiphalera sumatrana*
Sumatra, W 4659

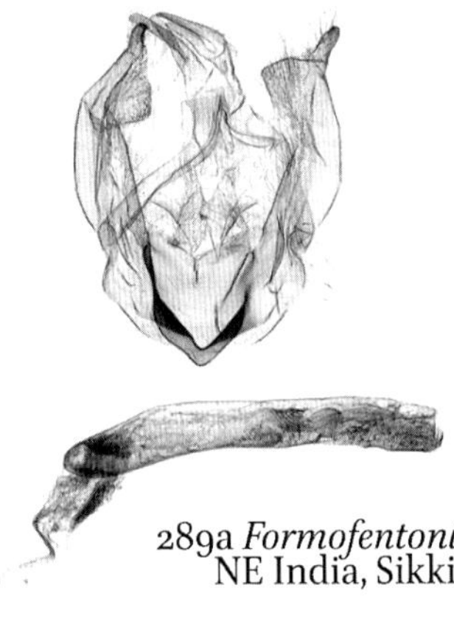

289a *Formofentonia orbifer orbifer*
NE India, Sikkim, GU 17-55

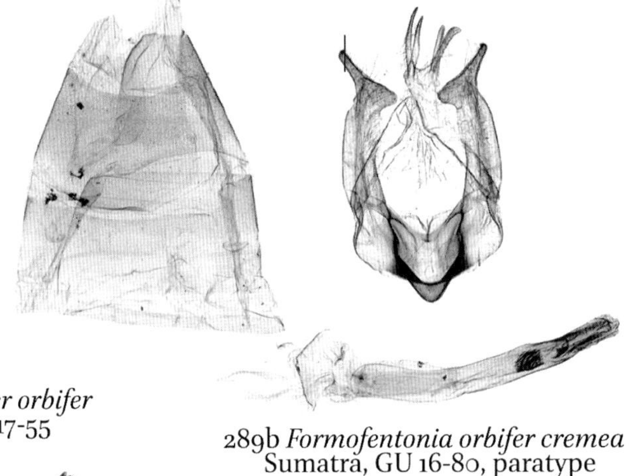

289b *Formofentonia orbifer cremea*
Sumatra, GU 16-80, paratype

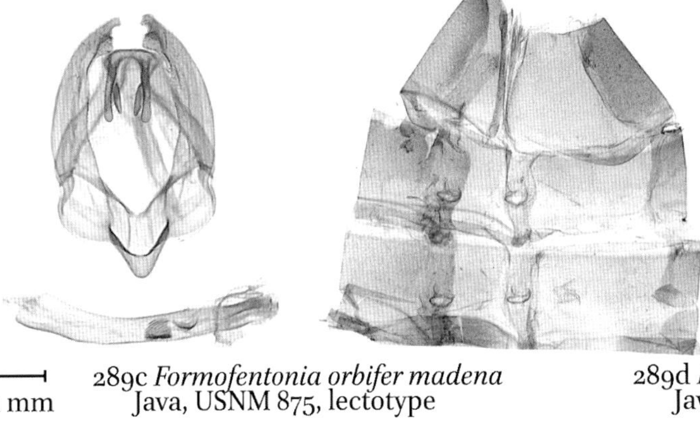

1 mm

289c *Formofentonia orbifer madena*
Java, USNM 875, lectotype

289d *Formofentonia orbifer madena*
Java, GU 72-37

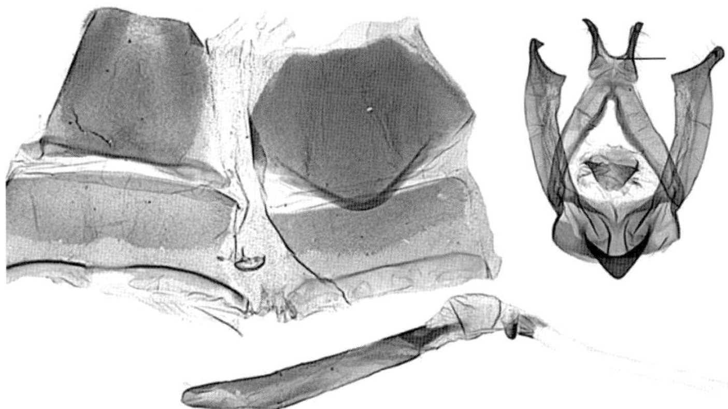

289e *Formofentonia orbifer sampuraga* Sulawesi, GU 66-51, paratype

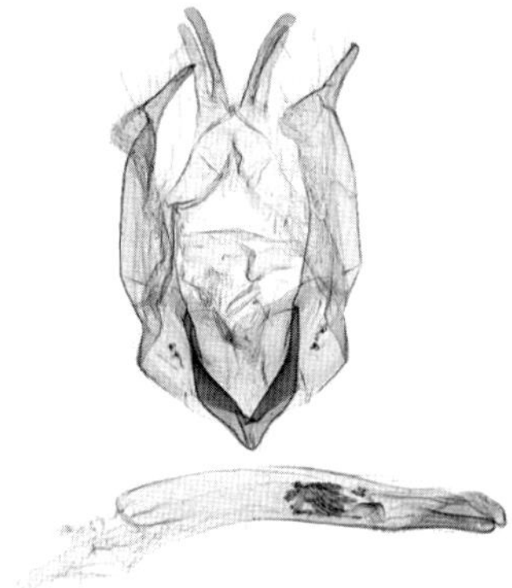

289f *Formofentonia orbifer binansilang,* Mindanao GU 17-57, paratype

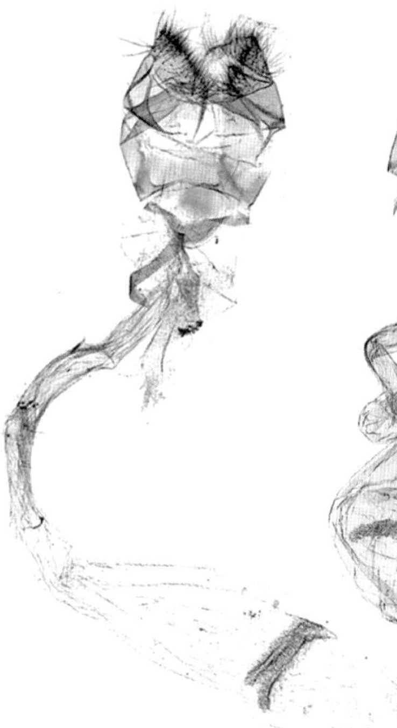

289h *Formofentonia orbifer rotundata* Taiwan, W 4696

289i *F. orbifer cremea* Malaya, GU 66-50, paratype

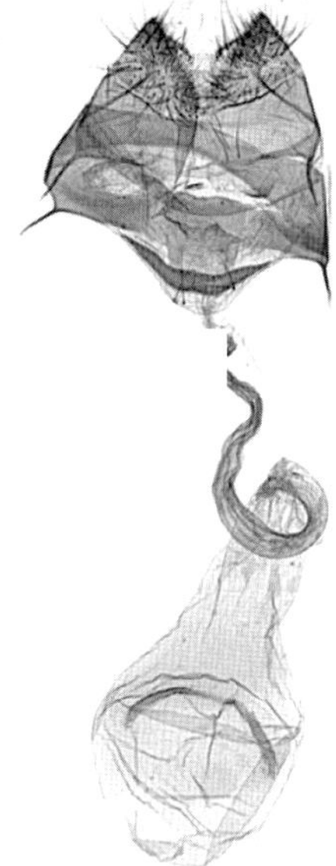

289j *Formofentonia orbifer sampuraga* Sulawesi, GU 66-59 paratype

289g *Formofentonia orbifer orbifer* NE India, GU 07-79a

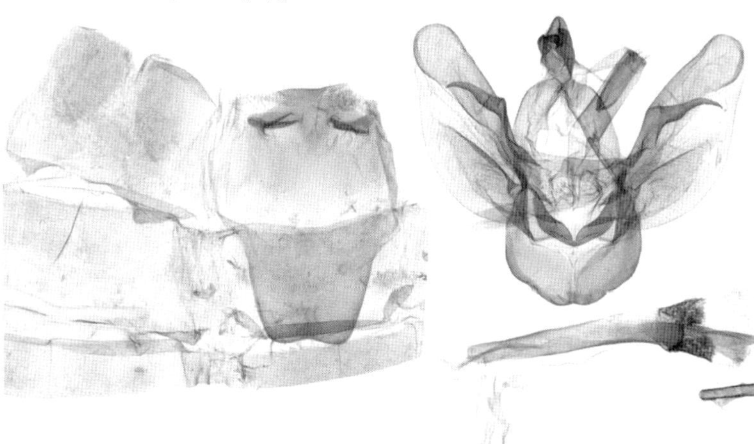

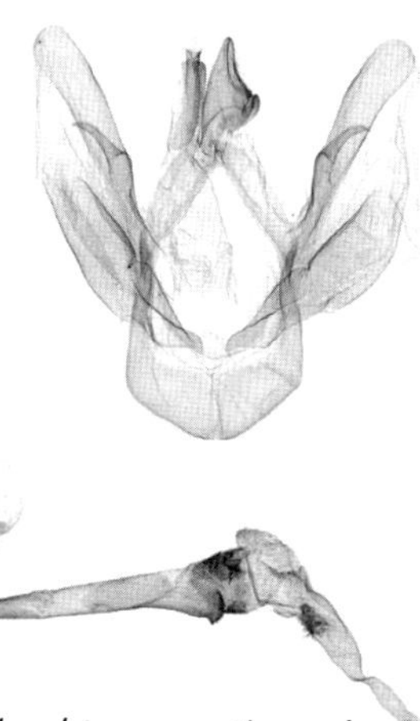

1 mm

290a *Betashachia angustipennis tropica,* Sumatra, GU 74-79

290b *Betashachia angustipennis angustipennis* Thailand, W 1737

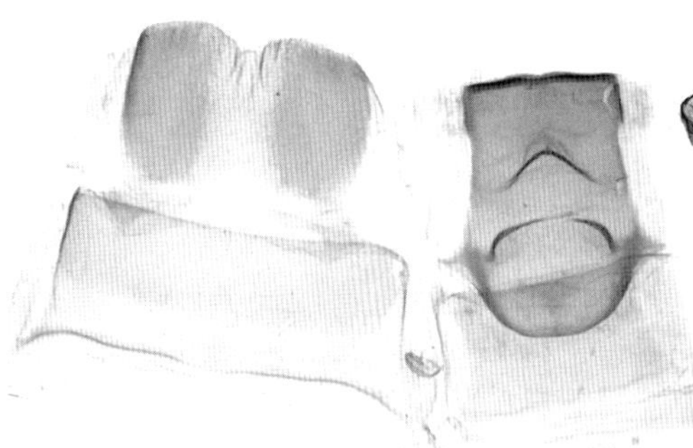

291a *Fentonia sumatrana*
Belitung, GU 66-52

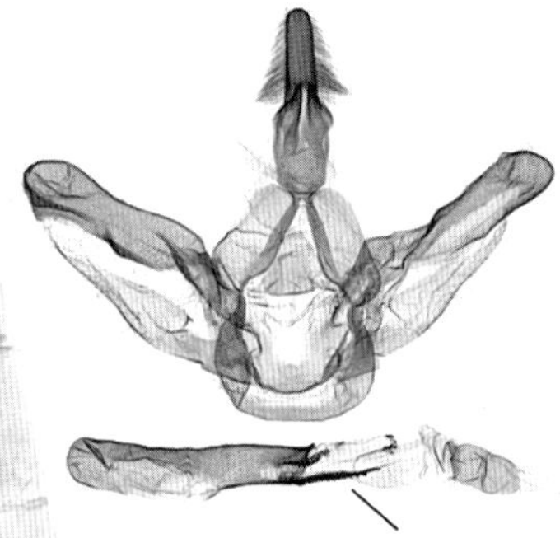

290c *Betashachia a. angustipennis*
Vietnam, GU 74-51

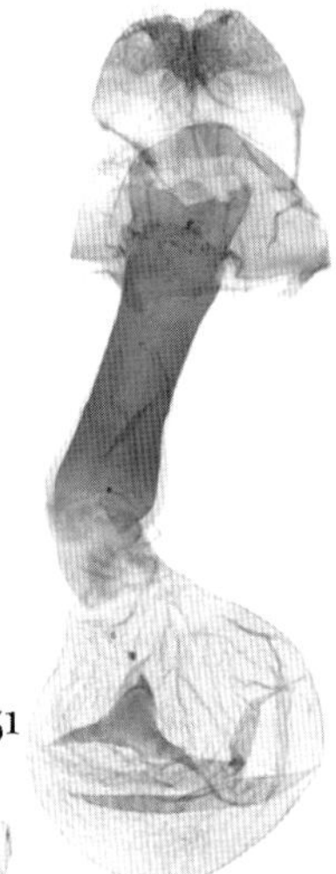

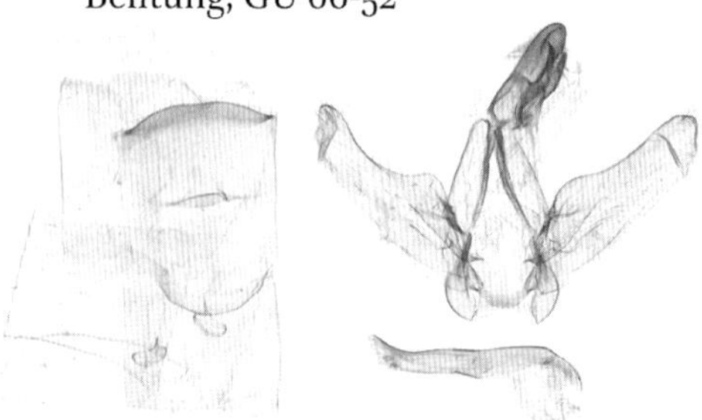

292a *Fentonia helena*
Sumatra, GU 79-89

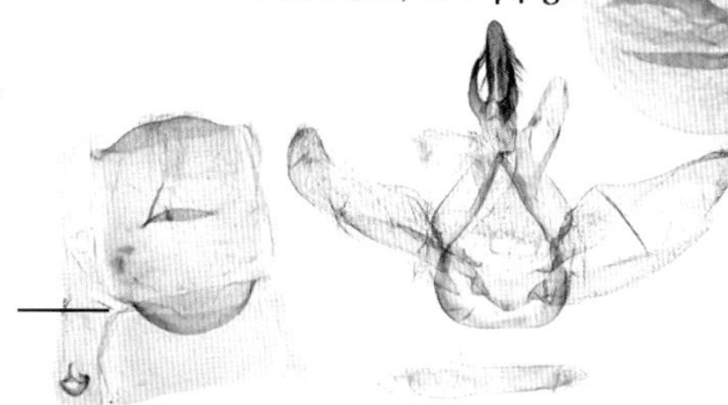

292b *Fentonia helena*
Java, GU 72-22

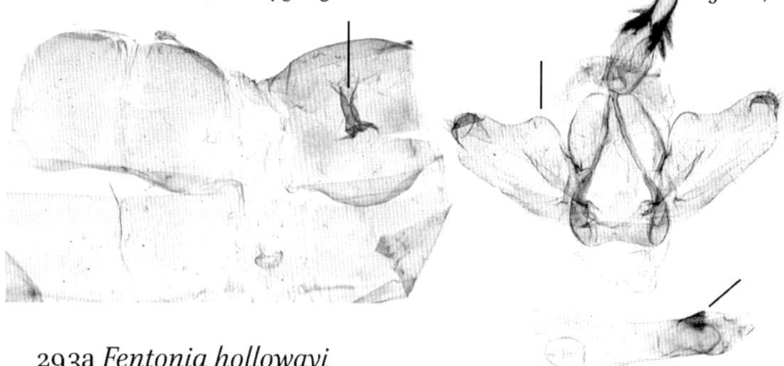

293a *Fentonia hollowayi*
Kalimantan, GU 72-23, paratype

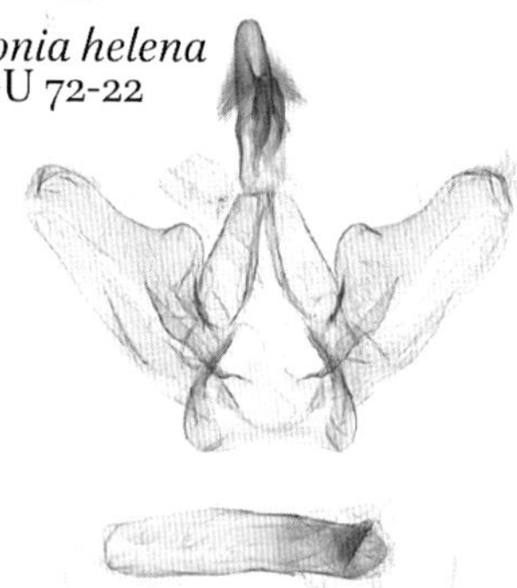

293b *Fentonia hollowayi*
Sumatra, GU 79-90, PT

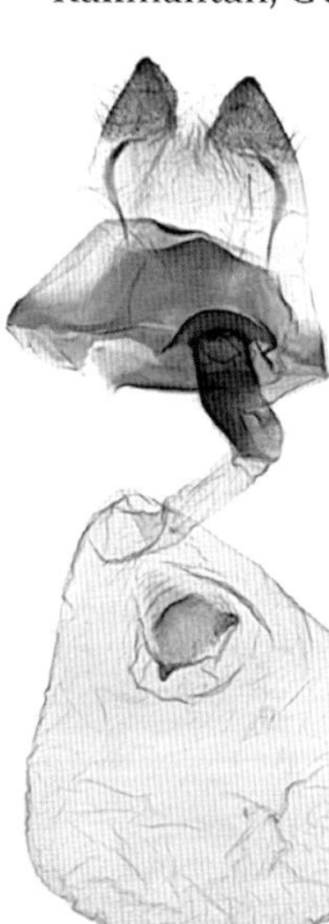

1 mm

291b *Fentonia sumatrana*
Sumatra, GU 66-53

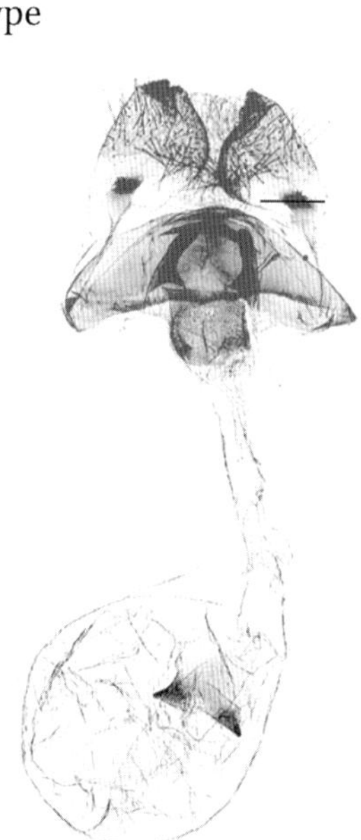

293c *Fentonia hollowayi*
Kalimantan, GU 72-23, PT

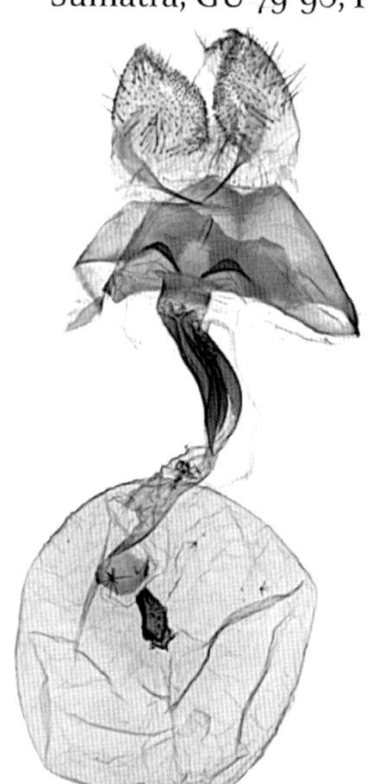

294a *F. talboti ingridae*
Sumatra, GU 66-54

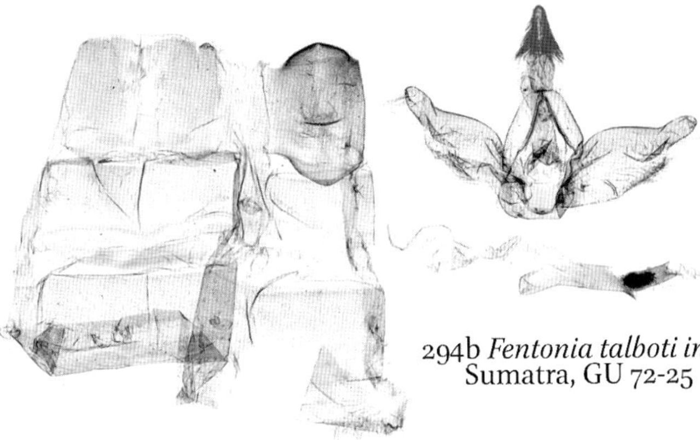

294b *Fentonia talboti ingridae*
Sumatra, GU 72-25

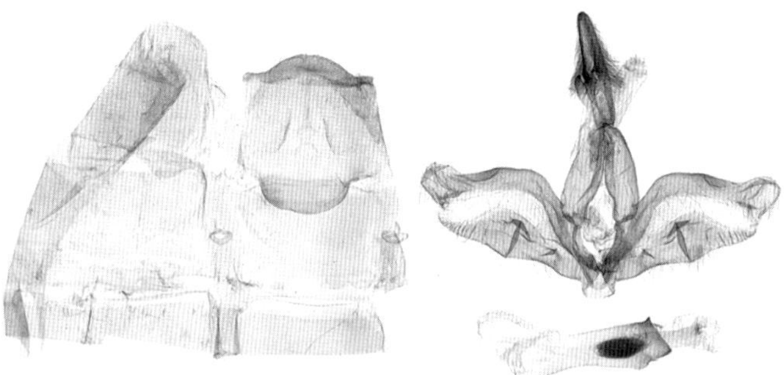

295a *Fentonia bipunctus*
Kalimantan, GU 72-28

295b *Fentonia bipunctus*
Borneo, W 15714

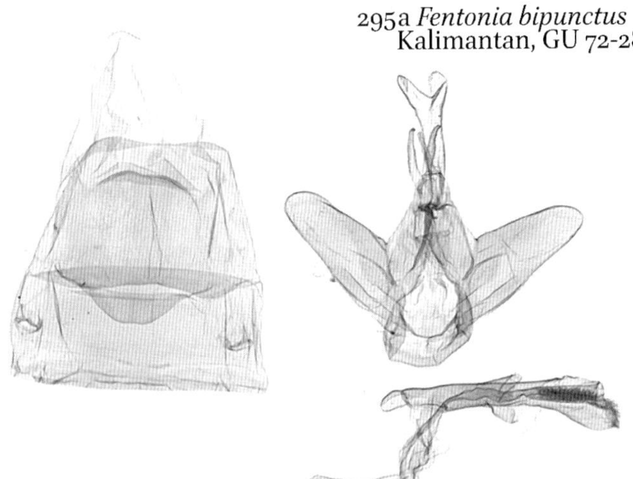

296a *Neopheosia fasciata fasciata*
India, Himachal Pradesh
BM1072, lectotype

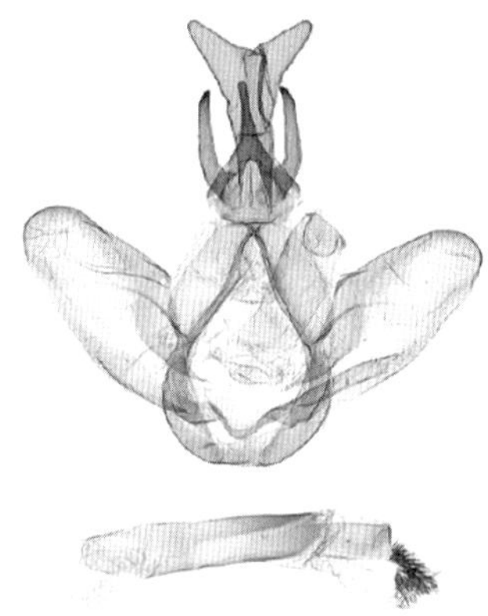

296b *Neopheosia fasciata fasciata*
India, Meghalaya, GU 72-30

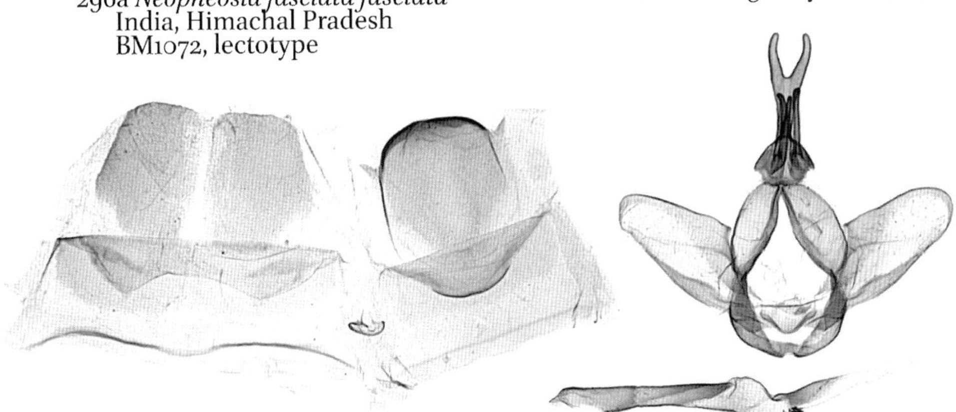

1 mm

296c *Neopheosia fasciata fasciata*
Sulawesi, GU 66-58

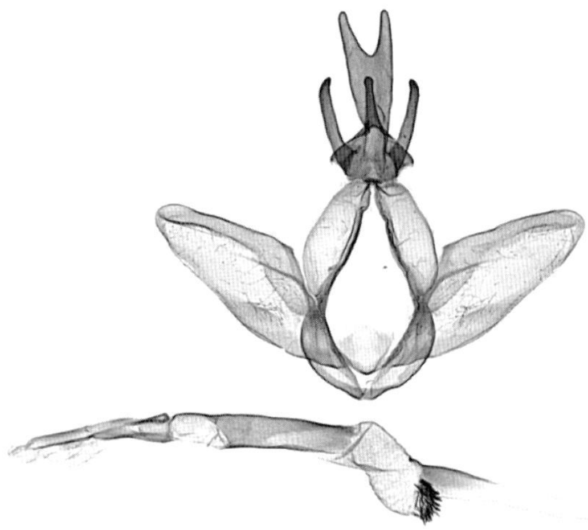

296d *Neopheosia fasciata fasciata*
Lombok, GU 72-33

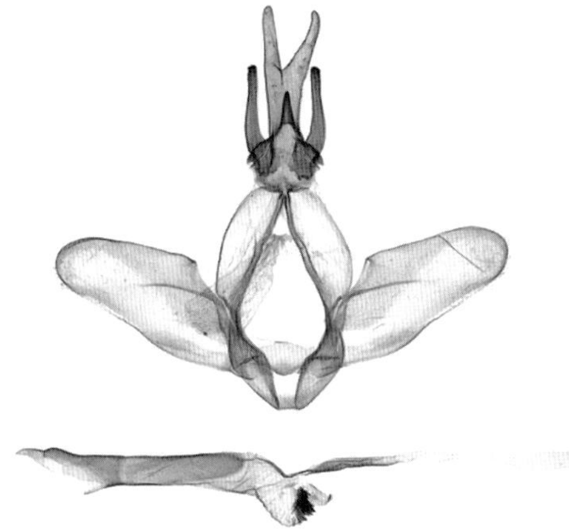

296e *Neopheosia fasciata buruana*
Obi, GU92-79, paratype

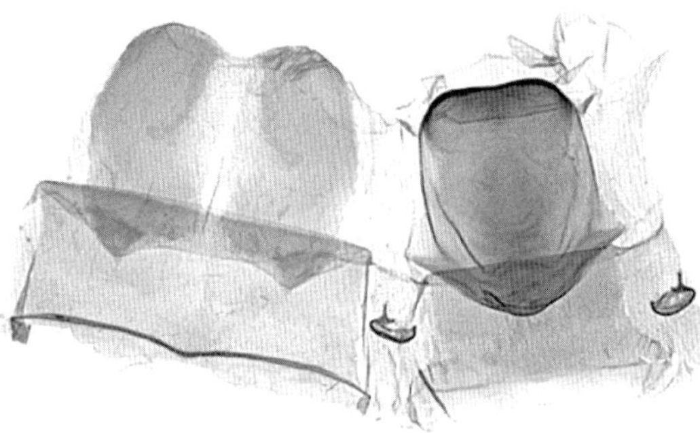

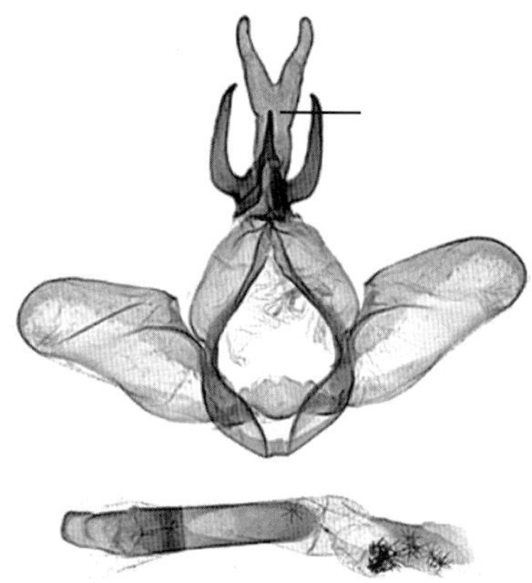

296f *Neopheosia fasciata buruana*
Buru, GU 93-30, paratype

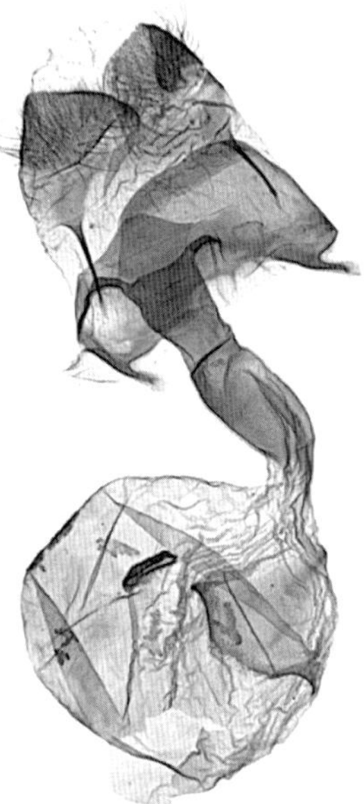

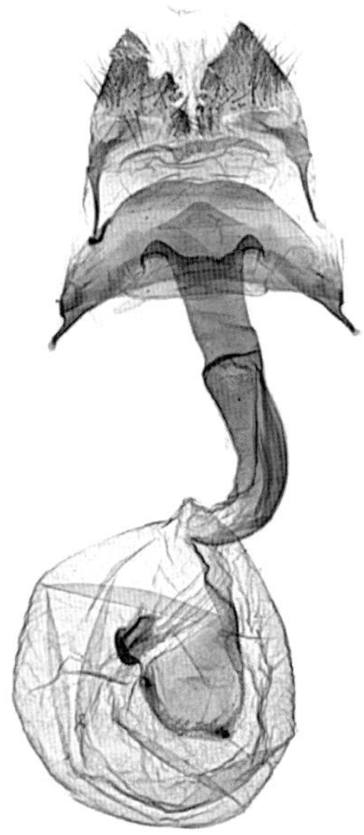

1 mm

296g *N. fasciata fasciata*
Sumatra, GU 66-55

296h *Neopheosia fasciata fasciata,* Java, GU 66-56

296i *N. fasciata fasciata*
Sulawesi, GU 66-57

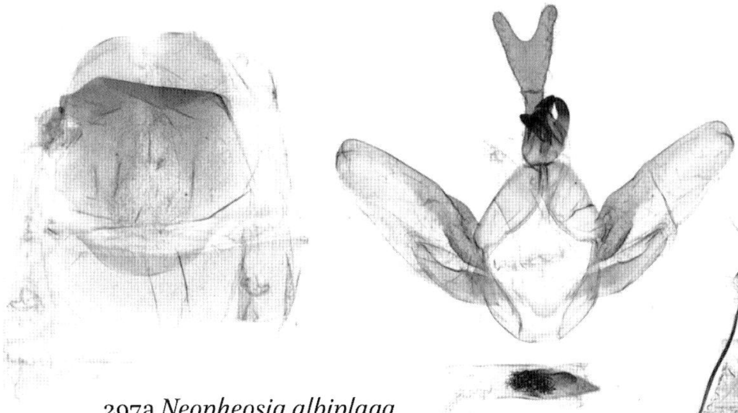

297a *Neopheosia albiplaga*
Sumatra, GU 80-72

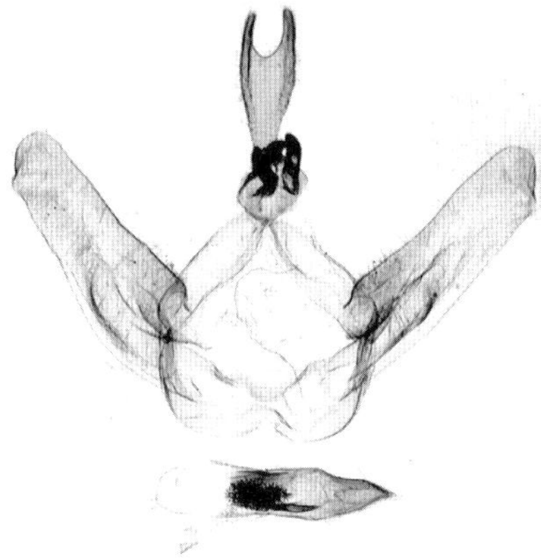

297b *Neopheosia albiplaga*
Sumatra, ZSM 1240

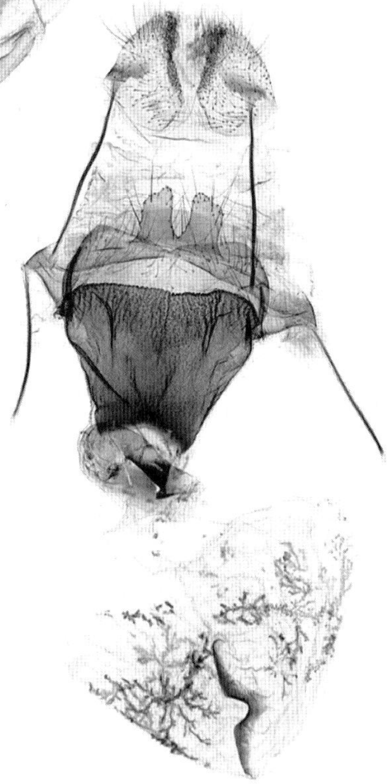

298a *Fusadonta albipuncta*
Sumatra, GU 45-25

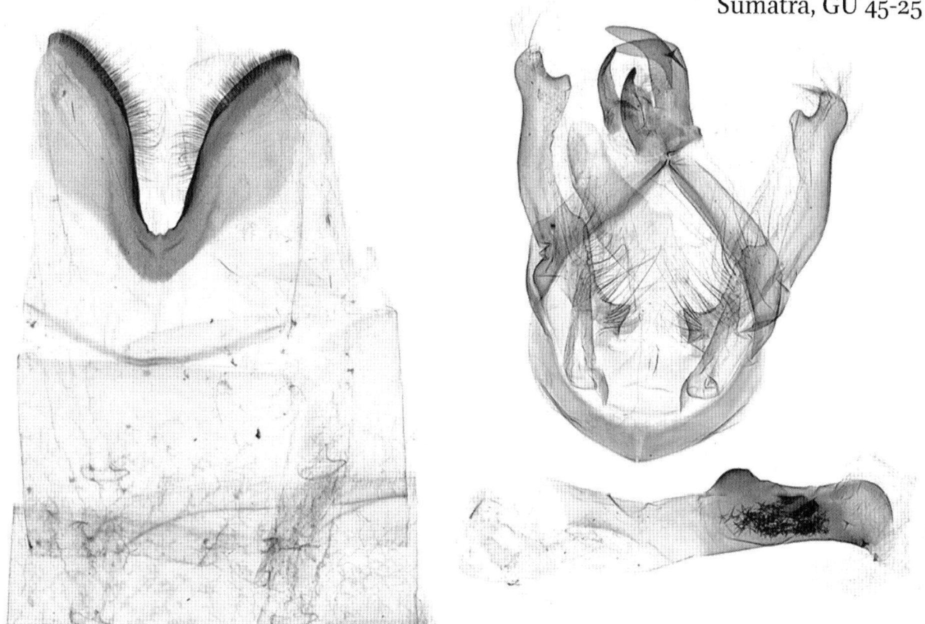

1 mm

298b *Fusadonta albipuncta*
Sumatra, GU 45-26

Color Plates

1c *Dudusa synopla*
♂, Sumatra

1d *Dudusa synopla*
♀, SE Myanmar

2b *Dudusa roepkei*
♂, Java, holotype

2c *Dudusa roepkei*
♂, Java, paratype

3b *Dudusa celebensis*
♂, Sulawesi

3c *Dudusa celebensis*
♀, Sulawesi, syntype

 | DOI:10.1163/9789004354258_009

4g *Dudusa vethi vethi*
♂, Nias

4h *Dudusa vethi vethi*
♀, Sumatra

4i *Dudusa vethi ferruginea*
♀, Bali, paratype

4j *Dudusa vethi ferruginea*
♂, Bali, holotype

4k *Dudusa vethi javana*
♂, Java, lectotype

4l *Dudusa vethi javana*
♀, Java,paralectotype

4m *Dudusa vethi borneensis* ♂, Kalimantan

4n *Dudusa vethi borneensis* ♀, Kalimantan

5e *Tarsolepis remicauda remicauda* ♀, Java

5f *Tarsolepis r. remicauda* ♂, Lombok

5g *Tarsolepis r. remicauda* ♂, Lombok, underside

6c *Tarsolepis punicea* ♂, Sulawesi

6d *Tarsolepis punicea* ♀, Sulawesi

6e *Tarsolepis punicea* ♂, Sulawesi underside

7b *Tarsolepis adultus*
♂, Halmahera, holotype

7c *Tarsolepis adultus*
♂, Halmahera, holotype, underside

8d *Tarsolepis sommeri*
♂, Java

8e *Tarsolepis sommeri*
♀, Bali

8f *Tarsolepis sommeri*
♂, Borneo

9f *Tarsolepis dinawensis dinawensis,* ♂, W Papua

9g *Tarsolepis dinawensis dinawensis,* ♀, W Papua

9h *Tarsolepis dinawensis extera*
♂, Halmahera, holotype

9i *Tarsolepis dinawensis extera*
♀, Halmahera, paratype

10d *Tarsolepis malayana*
♂, Sumatra

10e *Tarsolepis malayana*
♀, Sumatra

11b *Tarsolepis sericeus*
♂, Sumatra

11Ab *Tarsolepis hammamelis*
♀, Thailand

12e *Gangarides rosea*
♂, Sumatra

12f *Gangarides rosea*
♂, Java

12g *Gangarides rosea*
♂, Sumatra

12h *Gangarides rosea*
♀, S Myanmar

13d *Gangarides splendidus*
♂, Sumatra, holotype

13e *Gangarides splendidus*
♀, Sumatra

13f *Gangarides splendidus*
♂, Siberut

14e *Gangarides wallacei*
♂, Halmahera

14f *Gangarides wallacei*
♀, Seram

14g *Gangarides wallacei*
♀, Sulawesi

15c *Gangarides brunneus*
♀, Sumatra

15d *Gangarides brunneus*
♂, Sumatra, paratype

16d *Gangarides vittipalpis*
♂, Sumatra

16e *Gangarides vittipalpis*
♀, Sumatra

17c *Gangarides vardena*
♂, Kalimantan

17d *Gangarides vardena*
♂, Sumatra

17e *Gangarides vardena*
♂, Kalimantan

17f *Gangarides vardena*
♀, Sumatra

18c *Diehleana gigantea*
♂, W Papua

18d *Diehleana gigantea*
♀, W Papua

19c *Diehleana gilesi*
♂, Kalimantan

20i *Euhampsonia roepkei sumatrana*
♂, Sumatra, holotype

20j *Euhampsonia roepkei roepkei*
♀, Borneo

20k *Euhampsonia roepkei javana*
♂, Java, holotype

21c *Ortholomia moluccana*
♂, W Papua

21d *Ortholomia moluccana*
♂, W Papua

21e *Ortholomia moluccana*
♂, W Papua

21f *Ortholomia moluccana*
♀, Ambon

21g *Ortholomia moluccana*
♀, W Papua

21h *Ortholomia moluccana*
♀, W Papua

22b *Ortholomia gilolo*
♂, Halmahera, holotype

22c *Ortholomia gilolo*
♀, Halmahera, paratype

23f *Ortholomia subtilis*
♂, Alor, holotype

23g *Ortholomia subtilis*
♂, Timor, paratype

23h *Ortholomia subtilis*
♀, Alor, paratype

23i *Ortholomia subtilis*
♀, Tanimbar, paratype

24d *Ortholomia xylinata*
♂, Kalimantan

24e *Ortholomia xylinata*
♀, Java

25c *Ortholomia besida*
♂, Sumatra

25d *Ortholomia besida*
♀, Sumatra

26e *Ortholomia osica sulawesiensis*, ♂, Sulawesi

26f *Ortholomia osica osica* ♂, Mindanao, paratype

26g *Ortholomia o. osica* ♂, Samar, paratype

26h *Ortholomia o. osica* ♀, Samar, paratype

27c *Pseudogargetta diversa* ♂, W Papua, Yapen Isl.

27d *Pseudogargetta diversa* ♀, W Papua

27e *Pseudogargetta diversa* ♀, W Papua

28d *Pseudogargetta orbiculata* ♂, Obi

28e *Pseudogargetta orbiculata* ♀, Sulawesi

28f *Pseudogargetta orbiculata*, ♂, Sulawesi

29b *Pseudogargetta marmorata*, ♂, Malaya

29c *Pseudogargetta marmorata* ♀, Borneo

30d *Gargettiana punctatissima punctatissima*, ♀, W Papua

30e *Gargettiana. p. punctatissima*, ♂, W Papua

30f *G. p. punctatissima* ♀, Halmahera

30g *G. punctatissima* ♂, Bacan

30h *G. punctatissima tritus* ♂, Buru, holotype

31Ab *Polychoa styphlopis* ♀, Australia, Qsd.

31b *Polychoa aeruginosa* ♂, PNG

31c *Polychoa aeruginosa* ♂, PNG

31d *Polychoa aeruginosa* ♀, PNG

32b *Polychoa bakami* ♂, Halmahera, holotype

33b *Polychoa haliga* ♂, Seram, holotype

33c *Polychoa haliga* ♀, Seram, paratype

34c *Polychoa cheesmanae* ♀, W Papua, paratype

34d *Polychoa cheesmanae* ♂, Yapen Isl., holotype

34e *Polychoa cheesmanae* ♀, Yapen Isl., paratype

35b *Polychoa. flavibasis* ♂, W Papua

35c *Polychoa flavibasis* ♀, W Papua

36c *Polychoa rafinor* ♂, PNG, holotype

36d *Polychoa rafinor* ♂, PNG, paratype

36e *Polychoa rafinor* ♀, PNG, paratype

37c *Polychoa bipuncta* ♂, W Papua

37d *Polychoa bipuncta* ♀, PNG

37e *Polychoa bipuncta* ♀, PNG

38b *Polychoa metallica* ♂, W Papua, holotype

38c *Polychoa metallica* ♀, PNG

39c *Archigargetta amydra* ♂, PNG, Kar Kar Isl.

39d *Archigargetta amydra* ♂, W Papua, Batanta Isl.

39e *Archigargetta amydra*, ♀, PNG

39f *Archigargetta amydra* ♂, W Papua, Biak Isl. (holotype *diakonoffi*)

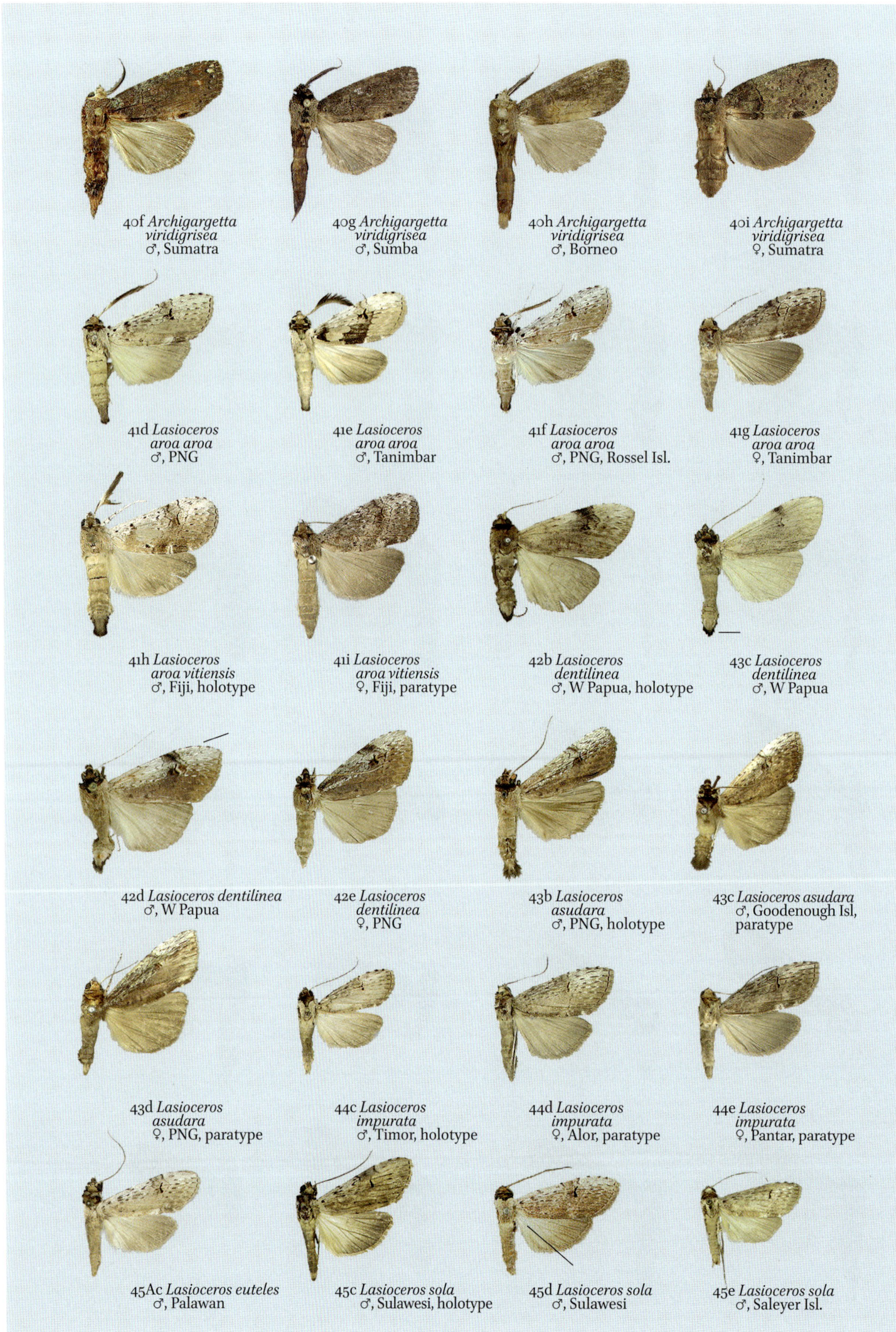

40f *Archigargetta viridigrisea* ♂, Sumatra
40g *Archigargetta viridigrisea* ♂, Sumba
40h *Archigargetta viridigrisea* ♂, Borneo
40i *Archigargetta viridigrisea* ♀, Sumatra

41d *Lasioceros aroa aroa* ♂, PNG
41e *Lasioceros aroa aroa* ♂, Tanimbar
41f *Lasioceros aroa aroa* ♂, PNG, Rossel Isl.
41g *Lasioceros aroa aroa* ♀, Tanimbar

41h *Lasioceros aroa vitiensis* ♂, Fiji, holotype
41i *Lasioceros aroa vitiensis* ♀, Fiji, paratype
42b *Lasioceros dentilinea* ♂, W Papua, holotype
43c *Lasioceros dentilinea* ♂, W Papua

42d *Lasioceros dentilinea* ♂, W Papua
42e *Lasioceros dentilinea* ♀, PNG
43b *Lasioceros asudara* ♂, PNG, holotype
43c *Lasioceros asudara* ♂, Goodenough Isl, paratype

43d *Lasioceros asudara* ♀, PNG, paratype
44c *Lasioceros impurata* ♂, Timor, holotype
44d *Lasioceros impurata* ♀, Alor, paratype
44e *Lasioceros impurata* ♀, Pantar, paratype

45Ac *Lasioceros euteles* ♂, Palawan
45c *Lasioceros sola* ♂, Sulawesi, holotype
45d *Lasioceros sola* ♂, Sulawesi
45e *Lasioceros sola* ♂, Saleyer Isl.

46c *Noctuola diehlongata* ♂, Sumatra, holotype

46d *Noctuola diehlongata* ♀, Sumatra, paratype

47c *Phycidopsis sundana* ♂, Sumatra, paratype

47d *Phycidopsis sundana* ♂, Java

47e *Phycidopsis sundana* ♀, Kalimantan

48c *Phycidopsis segregarius* ♂, Alor, holotype

48d *Phycidopsis segregarius* ♂, Alor, paratype

49c *Phycidopsis sulawesiana* ♂, Sulawesi, holotype

49d *Phycidopsis sulawesiana* ♀, Sulawesi

50b *Stictogargetta albimacula* ♂, Sumatra

50c *S. albimacula* ♀, Thailand

51c *Stictogargetta lithosidia* ♂, Sumatra

51d *Stictogargetta lithosidia* ♀, Sumatra

52e *Gargetta costigeroides* ♂, Brunei, holotype

52f *Gargetta costigeroides* ♂, Sumatra

52g *Gargetta costigeroides* ♀, Sumatra

53d *Gargetta costigerana* ♂, Sulawesi, holotype

53e *Gargetta costigerana* ♂, Sulawesi, paratype

53f *Gargetta costigerana* ♀, Sulawesi, paratype

53Ab *Gargetta costigera* ♂, NE India

54c *Gargetta hampsoni* ♂, Sumatra

54d *Gargetta hampsoni* ♀, Kalimantan

55c *Gargetta divisa divisa* ♂, Java, holotype

55d *Gargetta divisa divisa* ♀, Java

55e *Gargetta divisa divisa* ♂, Sumatra

55f *Gargetta dives divisa* ♂, Sumatra

55g *Gargetta divisa divisa* ♀, Sumatra

55h *Gargetta divis divisa* ♀, Sumatra

55i *Gargetta divisa luzonica* ♂, Luzon, lectotype

55j *Gargetta divisa luzonica,* ♀, Luzon paralectotype

56c *Gargetta tompua* ♂, Sulawesi

56d *Gargetta tompua* ♀, Saleyer Isl.

57d *Gargetta acarodes* ♂, Australia, NT

57e *Gargetta acarodes* ♀, Australia, NT

58e *Gargetta brevisocius* ♂, Alor, holotype

58f *Gargetta brevisocius* ♀, Alor, paratype

58g *Gargetta brevisocius* ♂, Timor, paratype

58h *Gargetta brevisocius* ♂, Pantar, paratype

59e *Gargetta curvaria curvaria* ♂, Sumatra

59f *Gargetta curvaria palua* ♂, Sulawesi,

59g *Gargetta curvaria palua* ♀, Sulawesi

60e *Gargetta alora* ♂, Alor, holotype

60f *Gargetta alora* ♂, Pantar, Paratype

60g *Gargetta alora* ♂, Lomblen, PT

61c *Thacona costivitta* ♂, Sumatra

61d *Thacona costivitta* ♀, Sumatra

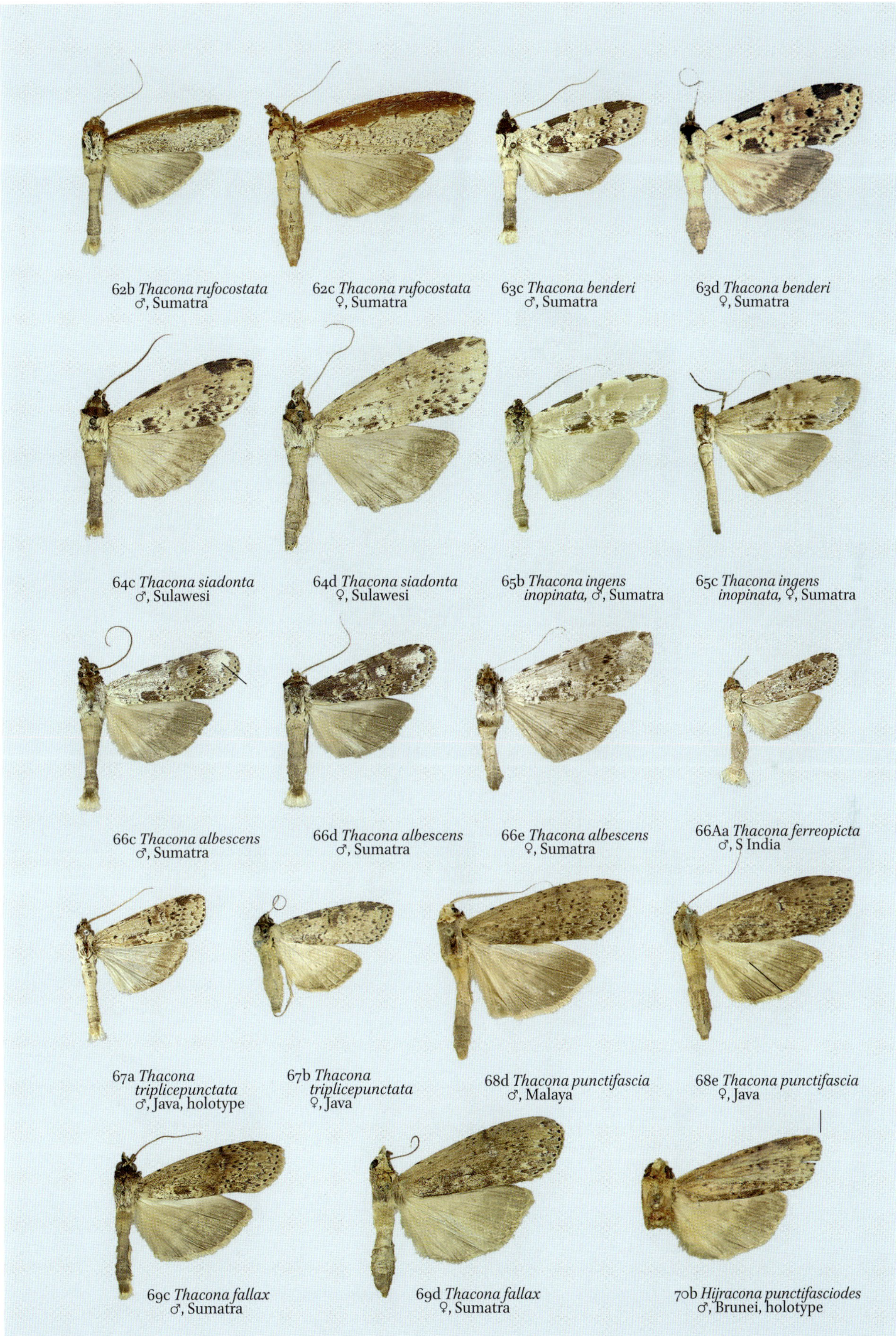

62b *Thacona rufocostata* ♂, Sumatra

62c *Thacona rufocostata* ♀, Sumatra

63c *Thacona benderi* ♂, Sumatra

63d *Thacona benderi* ♀, Sumatra

64c *Thacona siadonta* ♂, Sulawesi

64d *Thacona siadonta* ♀, Sulawesi

65b *Thacona ingens inopinata*, ♂, Sumatra

65c *Thacona ingens inopinata*, ♀, Sumatra

66c *Thacona albescens* ♂, Sumatra

66d *Thacona albescens* ♂, Sumatra

66e *Thacona albescens* ♀, Sumatra

66Aa *Thacona ferreopicta* ♂, S India

67a *Thacona triplicepunctata* ♂, Java, holotype

67b *Thacona triplicepunctata* ♀, Java

68d *Thacona punctifascia* ♂, Malaya

68e *Thacona punctifascia* ♀, Java

69c *Thacona fallax* ♂, Sumatra

69d *Thacona fallax* ♀, Sumatra

70b *Hijracona punctifasciodes* ♂, Brunei, holotype

71c *Ramesa tosta* ♂, Java

71d *Ramesa tosta* ♀, Java

72d *Ramesa albistriga* ♂, Java

72e *Ramesa albistriga* ♀, Sumatra

73b *Brykia horsfieldi* ♂, Java

73c *Brykia horsfieldi* ♂, Borneo

73d *Brykia horsfieldi* ♀, Sumatra

74e *Brykia samarinda* ♂, Sumatra

74f *Brykia samarinda* ♀, Sumatra

75c *Hyperaeschra pallida* ♂, Sumatra

75d *Hyperaeschra pallida* ♀, Sumatra

76c *Changea siamensis* ♂, Thailand, Paratype

76d *Changea siamensis* ♂, Sumatra

76e *Changea siamensis* ♀, Malaya

77d *Turnaca subcarnea* ♂, Seram

77e *Turnaca subcarnea* ♀, Buru

77f *Turnaca subcarnea* ♂, Yapen isl.

77g *Turnaca subcarnea* ♂, W Papua

77h *Turnaca subcarnea* ♀, Ambon

78b *Turnaca carnea* ♂, W Papua, holotype

79c *Tensha delineivena*
♂, Sumatra

79d *Tensha delineivena*
♀, Sumatra

80b *Tensha postobscura*
♀, Borneo, paratype

80c *Tensha postobscura*
♂, Kalimantan

80d *Tensha postobscura*
♀, Borneo, paratype

81c *Tensha pizzicata*
♂, Sumatra, holotype

81d *Tensha pizzicata*
♀, Sumatra, paratype

82c *Canatensha mediofascia*
♂, Sumatra

82d *Canatensha mediofascia*
♀, Sumatra

82Ac *Canatensha nigridorsalis*
♂, Vietnam, holotype

82Ad *Canatensha nigridorsalis*
♀, Vietnam, paratype

83g *Ambadra rafflesi*
♂, Java
(holotype *pallida*)

83h *Ambadra rafflesi*
♂, Sumatra

83i *Ambadra rafflesi*
f. *rufescens*
♂, Sumatra

83j *Ambadra rafflesi*
♂, Sumatra

83k *Ambadra rafflesi*
♂, Flores

83l *Ambadra rafflesi*
♂, Sulawesi

83m *Ambadra rafflesi*
♀, Sumatra

83n *Ambadra rafflesi*
♀, Sumatra

84c *Ambadra canescens* ♂, Sulawesi, holotype

84d *Ambadra canescens* ♂, Sulawesi,paratype

84e *Ambadra canescens* ♂, Sulawesi, paratype

84f *Ambadra canescens* ♂, Sulawesi, paratype

85d *Ambadra miscera* ♂, Sulawesi, holotype

85e *Ambadra miscera* ♂, Sulawesi, paratype

85f *Ambadra miscera* ♂, Sulawesi, paratype

85g *Ambadra miscera* ♀, Sulawesi, paratype

86b *Ambadra celebensis* ♂, Sulawesi, holotype

86c *Ambadra celebensis* ♂, Peleng

86d *Ambadra celebensis* ♂, Sulawesi

86e *Ambadra celebensis* ♀, Sulawesi

87c *Ambadra macula* ♂, Sulawesi

87d *Ambadra macula* ♂, Sulawesi

87e *Ambadra macula* ♂, Sulawesi

87f *Ambadra macula* ♂, Sulawesi

88b *Ambadra parva* ♂, Sulawesi

89b *Ambadra tondanoa* ♂, Sulawesi

89c *Ambadra tondanoa* ♂, Sulawesi

89d *Ambadra tondanoa* ♀, Sulawesi

90d *Ambadra vittata* ♂, Sulawesi

90e *Ambadra vittata* ♂, Sulawesi

90f *Ambadra vittata* ♀, Sulawesi, (from Kobyashi, 2011)

91b *Ambadra gracilis* ♂, Borneo, holotype

92d *Ambadra batara* ♂, Java, holotype

92e *Ambadra batara* ♂, Java, paratype

92f *Ambadra batara* ♀, Java, paratype

93Aa *Ambadra modesta modesta* ♂, Thailand, paratype

93d *Ambadra discocellularis* ♂, Sumatra

93e *Ambadra discocellularis* ♂, Java

93f *Ambadra discocellularis* ♀, Sumatra

94c *Ambadra similis* ♂, Sumatra

94d *Ambadra similis* ♂, Sumatra

95d *Ambadra melanostriga* ♂, Sumatra

95e *Ambadra melanostriga* ♂, Java

96c *Ambadra owadai* ♂, Malaya

96d *Ambadra owadai* ♂, Borneo

96e *Ambadra owadai* ♂, Java

97b *Ambadra mentawai* ♂, Mentawai, holotype

98c *Ambadra montana* ♂, Borneo

98d *Ambadra montana* ♂, Borneo

99d *Ambadrana punctata* ♂, Kalimantan

99e *Ambadrana punctata* ♂, Kalimantan

99f *Ambadrana punctata* ♂, Borneo

99g *Ambadrana punctata* ♀, Borneo

100d *Ambadrana barlowi* ♀, Belitung

100e *Ambadrana barlowi* ♂, Malaya

100f *Ambadrana barlowi* ♂, Sumatra

100g *Ambadrana barlowi* ♀, S Thailand

101d *Ambadrana taruma* ♂, Java, holotype

101e *Ambadrana taruma* ♀, Java, paratype

102b *Ambadrana spinifera* ♂, Borneo
103c *Ambadrana furcata* ♂, Sumatra, paratype
103d *Ambadrana furcata* ♂, Sumatra
104c *Pydnella rosacea* ♂, Sumatra
104d *Pydnella rosacea* ♀, Thailand
105d *Niganda strigifascia* ♂, Sumatra
105e *Niganda strigifascia* ♀, Java
105f *Niganda strigifascia* ♀, Sumatra
106c *Niganda pallens* ♂, Malaya, holotype
106d *Niganda pallens* ♂, Malaya, paratype
106e *Niganda pallens* ♀, Malaya, paratype
106Ab *Niganda cyttarosticta* ♂, India, West Bengal
107c *Niganda ancipitalis* ♂, Malaya, holotype
107d *Niganda ancipitalis* ♀, Malaya
108c *Postniganda pudens* ♂, Kalimantan, holotype
108d *Postniganda pudens* ♂, Borneo
108e *Postniganda pudens* ♀, Borneo
109f *Norraca longipennis* ♂, Sumatra
109g *Norraca longipennis* ♂, Timor
109h *Norraca longipennis* ♀, Bali
110c *Saliocleta lativitta* ♂, Java
110d *Saliocleta lativitta* ♀, Sumatra
111g *Saliocleta nonagrioides* ♀, Sumatra

111h *Saliocleta nonagrioides* ♂, Sumatra

111i *S. nonagrioides* ♂, Sumatra

111j *Saliocleta nonagrioides* ♂, Sumatra

111k *Saliocleta nonagrioides* ♂, Sumatra

111l *Saliocleta nonagrioides* ♂, Sumatra

111m *S. nonagrioides* ♂, Kalimantan

111n *Saliocleta nonagrioides* ♂, Kalimantan

111o *Saliocleta nonagrioides* ♀, Sumatra

112c *Saliocleta sordida* ♂, Java

112d *Saliocleta sordida,* ♂, Java

112e *Saliocleta sordida* ♂, Bali

112f *Saliocleta sordida* ♀, Java

113i *Saliocleta odrana celebica,* ♂, Sulawesi

113j *Saliocleta odrana celebica,* ♂, Sulawesi

113k *Saliocleta odrana celebica,* ♂, Sulawesi

113l *Saliocleta odrana celebica* ♂, Sulawesi, Peleng

113m *Saliocleta odrana celebica* ♂, Sulawesi

113n *Saliocleta odrana celebica* ♀, Sulawesi

113o *Saliocleta odrana celebica* ♀, Sulawesi, Peleng

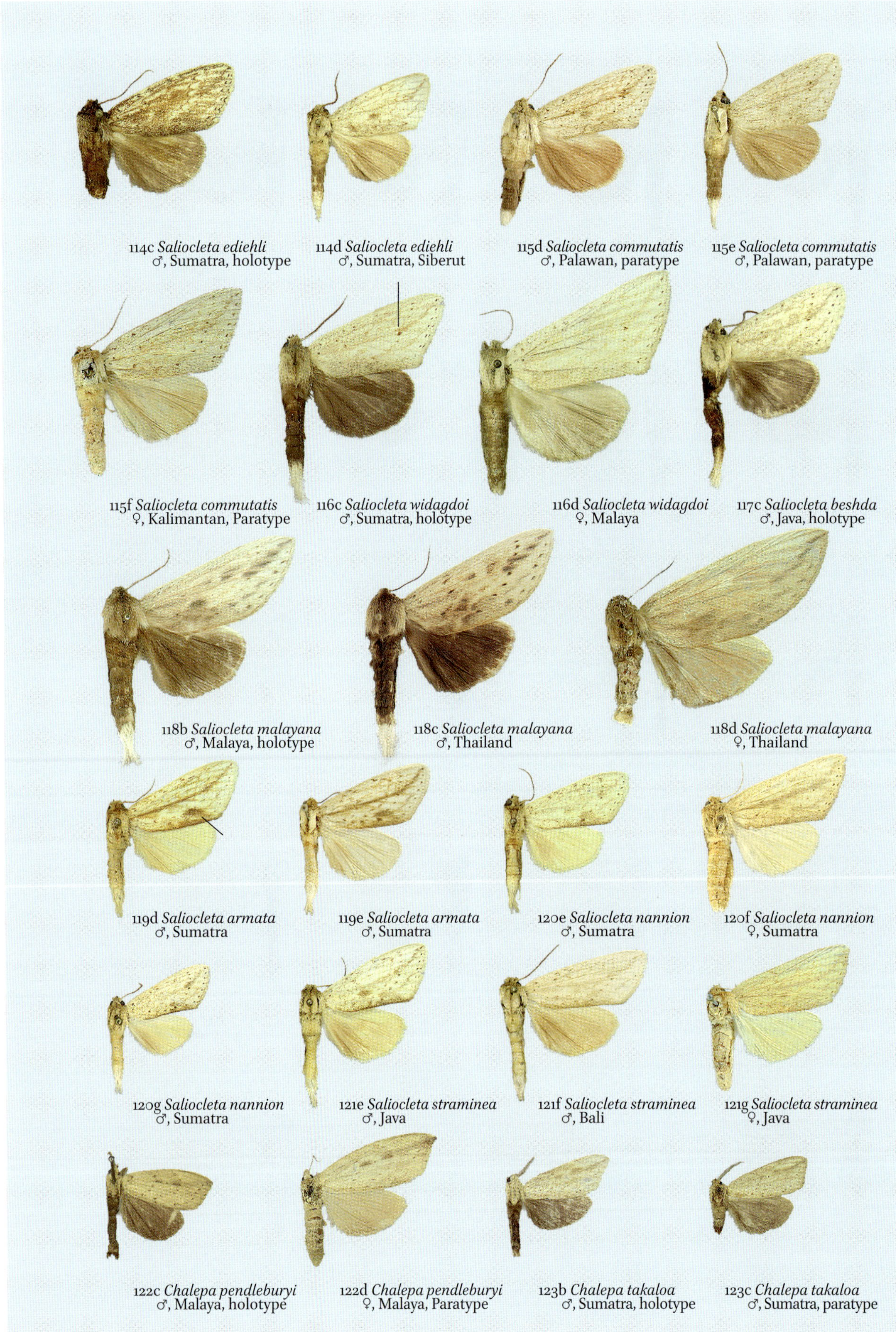

114c *Saliocleta ediehli* ♂, Sumatra, holotype

114d *Saliocleta ediehli* ♂, Sumatra, Siberut

115d *Saliocleta commutatis* ♂, Palawan, paratype

115e *Saliocleta commutatis* ♂, Palawan, paratype

115f *Saliocleta commutatis* ♀, Kalimantan, Paratype

116c *Saliocleta widagdoi* ♂, Sumatra, holotype

116d *Saliocleta widagdoi* ♀, Malaya

117c *Saliocleta beshda* ♂, Java, holotype

118b *Saliocleta malayana* ♂, Malaya, holotype

118c *Saliocleta malayana* ♂, Thailand

118d *Saliocleta malayana* ♀, Thailand

119d *Saliocleta armata* ♂, Sumatra

119e *Saliocleta armata* ♂, Sumatra

120e *Saliocleta nannion* ♂, Sumatra

120f *Saliocleta nannion* ♀, Sumatra

120g *Saliocleta nannion* ♂, Sumatra

121e *Saliocleta straminea* ♂, Java

121f *Saliocleta straminea* ♂, Bali

121g *Saliocleta straminea* ♀, Java

122c *Chalepa pendleburyi* ♂, Malaya, holotype

122d *Chalepa pendleburyi* ♀, Malaya, Paratype

123b *Chalepa takaloa* ♂, Sumatra, holotype

123c *Chalepa takaloa* ♂, Sumatra, paratype

124e *Periergos postrubra*
♂, Sumatra

124f *Periergos postrubra*
♂, Malaya

124g *Periergos postrubra*
♀, Sumatra
(paratype *indonesiae*)

125d *Periergos hunyada*
♂, Java

125e *Periergos hunyada*
♂, Sumatra

125f *Periergos hunyada*
♀, Sumatra

126c *Periergos spinosa*
♂, Borneo

126d *Periergos spinosa*
♂, Palawan

126e *Periergos spinosa*
♀, Palawan

127d *Neocerura liturata*
♂, Java

127e *Neocerura liturata*
♂, Java

127f *Neocerura liturata*
♂, Selayar Isl.

127g *Neocerura liturata*
♂, NW India, lectotype

127h *Neocerura liturata*
♀, Sumatra

127i *Neocerura liturata*
♀, Java

128b *Neocerura longinquus*
♂, Tanimbar, holotype

128c *Neocerura longinquus*
♀, Tanimbar, paratype

129f *Kamalia malaysiana malaysiana,* ♂, Borneo

129g *Kamalia malaysiana malaysiana,* ♀, Sumatra

129h *Kamalia malaysiana malaysiana,* ♀, Sumatra

129i *Kamalia malaysiana niasica,* ♂, Nias, holotype

129j *Kamalia malaysiana nigresco,* ♂, Siberut, holotype

129k *Kamalia malaysiana nigresco,* ♀, Siberut, paratype

129l *Kamalia malaysiana jakli,* ♂, Java

129m *Kamalia malaysiana jakli,* ♂, Java, paratype

129n *Kamalia malaysiana jakli,* ♂, Sumba, paratype

129o *Kamalia malaysiana jakli* ♀, Bali

130c *Kamalia minahassae* ♂, Peleng Isl.

130d *Kamalia minahassae* ♂, Sulawesi

130e *Kamalia minahassae* ♀, Sulawesi

131d *Kamalia fulva* ♂, Alor, holotype

131e *Kamalia fulva* ♂, Sumba, paratype

131f *Kamalia fulva* ♀, Flores paratype

132c *Kamalia timorensis* ♀, Timor

132d *Kamalia timorensis* ♂, Timor

133i *Kamalia multipunctata* ♀, New Britain

133j *Kamalia multipunctata* ♂, W Papua

133k *Kamalia multipunctata* ♂, W Papua

133l *Kamalia multipunctata* ♂, Seram

133m *Kamalia multipunctata* ♀, W Papua

133n *Kamalia multipunctata* ♂, Tanimbar

133o *Kamalia multipunctata* ♀, Tanimbar

134b *Kamalia solomonica* ♂, Solomons, holotype

134c *Kamalia solomonica* ♂, Solomons, paratype

135b *Kamalia amoa* New Caledonia paratype

136e *Liparopsis postalbida brevispinus* ♂, Sumatra, holotype

136f *Liparopsis postalbida brevispinus* ♂, Mentawai paratype

136g *Liparopsis postalbida postalbida*, ♀, Vietnam

137c *Liparopsis barlowi* ♂, Kalimantan

137d *Liparopsis barlowi* ♂, Sumatra

138b *Liparopsis sundana* ♂, Sumatra

138c *Liparopsis sundana* ♂, Sumatra

138d *Liparopsis sundana* ♂, Malaya

138e *Liparopsis sundana* ♀, Kalimantan

138f *Liparopsis sundana* ♀, Sumatra

139d *Liparopsis dierli* ♂, Sumatra

139e *Liparopsis dierli* ♀, Sumatra

140e *Pantanopsis diehli* ♂, Sumatra

140f *Pantanopsis diehli* ♀, Sumatra

141b *Pantanopsis celebensis* ♂, Sulawesi

141c *Pantanopsis celebensis* ♀, Sulawesi

142e *Harpyia kalisi* ♀, Java, holotype

142f *Harpyia kalisi* ♂, Java

142g *Harpyia kalisi* ♂, Sumatra

143e *Harpyia kobesi* ♀, Sumatra, paratype

143f *Harpyia kobesi* ♂, Sumatra, holotype

144f *Harpyia koeppeli* ♂, Borneo, holotype

144g *Harpyia koeppeli* ♂, Sumatra, paratype

145p *Teleclita strigata strigata* ♂, South India, lectotype

145q *Teleclita strigata strigata* ♂, India, Kerala

145r *Teleclita strigata strigata* ♀, India, Sikkim

145s *Teleclita strigata cathana* ♂, Sumatra

145t *Teleclita strigata cathana* ♀, Bali

145u *Teleclita strigata mediocana* ♀, Alor, paratype

145v *Teleclita strigata mediocana* ♂, Pantar, holotype

145w *Teleclita strigata mediocana* ♀, Alor, paratype

145x *Teleclita strigata mediocana* ♂, Kanguan, paratype

145y *Teleclita strigata multifascia* ♂, Tanimbar, holotype

42z *Teleclita strigata multifascia* ♂, Tanimbar, paratype

145aa *Teleclita strigata multifascia* ♀, Tanimbar, paratype

145ab *Teleclita strigata insignifica* ♂, Halmahera

145ac *Teleclita strigata insignifica,* ♂, Dammer, lectotype

145ad *Teleclita strigata dryinopa* ♂, Australia, NT

145ae *Teleclita strigata cinnamomea* ♂, W Papua

145af *Teleclita strigata cinnamomea* ♀, W Papua

146c *Teleclita rasilis* ♂, Sulawesi holotype

146d *Teleclita rasilis* ♀, Sulawesi, paratype

147b *Teleclita flavisticta* ♂, Sumatra

147c *Teleclita flavisticta* ♀, Sumatra

148b *Pheressaces siegfriedihlei* ♂, W Papua, holotype

148c *Pheressaces siegfriedihlei* ♂, W Papua, paratype

148Ac *Pheressaces cycnoptera* ♂, Australia, Qlsd.

148Ad *Pheressaces cycnoptera* ♀, Australia, Qlsd.

149d *Destolmia lineata* ♂, PNG

149e *Destolmia lineata* ♂, Australia, NT

149f *Destolmia lineata* ♀, Australia, NT

149g *Destolmia lineata* ♀, Australia, Qlsd.

150d *Stauroplitis annulata annulata,* ♀, Sumatra

150e *Stauroplitis annulata annulata,* ♂, Sumatra

151c *O. smaragdiplena* ♂, Malaya

151d *Oxoia smaragdiplena* ♀, Belitung

150f *Stauroplitis annulata annulata,* ♂, Borneo

152d *Oxoia viridipicta* ♂, Sumatra

152e *Oxoia viridipicta* ♀, Malaya

153i *Oxoia irrorativiridis irrorativiridis* ♂, W Papua

153j *Oxoia irrotativiridis irrorativiridis* ♀, W Papua

153k *Oxoia irrotativiridis irrorativiridis* ♂, Sulawesi

153l *Oxoia irrotativiridis triangularis*, ♂, Buru

153m *Oxoia irrotativiridis triangularis*, ♂, Bacan

153n *Oxoia irrotativiridis triangularis*, ♀, Buru

154c *Rodneya caudata* ♂, Sumatra

154d *Rodneya caudata* ♀, Myanmar

155e *Stauropus major* ♀, Sumatra

155f *Stauropus major* ♂, Lombok

155g *Stauropus major* ♂, Sumba

156l *Stauropus alternus brunnea*, ♂, Sumatra

156m *Stauropus alternus brunnea*, ♀, Sumatra

156n *Stauropus alternus niasicus* ♀, Nias, paratype

156o *Stauropus alternus niasicus* ♂, Nias, holotype

156p *Stauropus alternus hofmanae* ♀, Java, paratype

156q *Stauropus alternus sulawesianus* ♂, Sulawesi, holotype

156r *Stauropus alternus sulawesianus* ♀, Sulawesi, paratype

156s *Stauropus alternus hofmanae* ♂, Java, paratype

156t *Stauropus alternus hofmanae* ♂, Komodo, paratype

156u *Stauropus alternus melastomatis* ♂, Ambon

156v *Stauropus alternus melastomatis* ♀, Seram

157d *Stauropus pusilla* ♂, Timor, holotype

157e *S. pusilla* ♂, Timor paratype

157f *Stauropus pusilla* ♀, Timor, paratype

158h *Stauropus viridissimus viridissimus,* ♂, W Papua

158i *Stauropus viridissimus viridissimus,* ♂, Biak

158j *Stauropus viridissimus viridissimus* ♀, PNG, paralectotype

158k *S. viridissimus habrochlora* ♂, Australia, Qld.

158l *Stauropus viridissimus habrochlora* ♂, Tanimbar

158m *Stauropus viridissimus habrochlora* ♂, Australia, NT

158n *Stauropus viridissimus naumanni,* ♀, Obi

158o *Stauropus viridissimus naumanni,* ♂, Halmahera

158p *Stauropus viridissimus amboynica,* ♀, Seram

158q *S. viridissimus amboynica* ♂, Ambon

158r *Stauropus viridissimus boeroe* ♂, Buru, holotype

158s *Stauropus viridissimus boeroe* ♂, Buru, paratype

159c *Stauropus talasea* ♂, New Britain holotype

159d *Stauropus talasea* ♂, New Britain paratype

160c *Stauropus vanmastrigti* ♂, W Papua, holotype

160d *S. vanmastrigti* ♂, W Papua paratype

161c *Stauropus caerulea* ♂, W Papua holotype

161d *Stauropus caerulea* ♂, W Papua paratype

161e *Stauropus caerulea* ♀, W Papua paratype

162c *Stauropus affinis* ♀, PNG

162d *Stauropus affinis* ♂, PNG

163c *Stauropus pratti* ♂, W Papua

163d *Stauropus pratti* ♂, W Papua

164b *Stauropus schaarschmidti* ♂, W Papua

165b *Stauropus evanescens* ♂, W Papua

166b *Stauropus pelz*, ♂, Sumba holotype

167b *Stauropus loki* ♂, Flores paratype

168g *Stauropus virescens virescens* ♂, Bhutan

168h *S. virescens obscuratus* ♂, Sumatra holotype

168i *Stauropus virescens obscuratus* ♂, Belitung

168j *Stauropus virescens obscuratus*, ♀, Java

168k *Stauropus virescens obscuratus*, ♀, Sumatra

169d *Stauropus elisabethae elisabethae* ♀, Sumatra

169d *Stauropus elisabethae kiriakoffi*, ♀, Borneo

169e *Stauropus elisabethae elisabethae* ♂, Sumatra, holotype

169f *Stauropus elisabethae kiriakoffi* ♂, Kalimantan

170d *Kakostauropus glaucoviridis glaucoviridis* ♂, W Papua

170e *Kakostauropus g. glaucoviridis* ♀, W Papua

170f *Kakostauropus glaucoviridis seramica*, ♂, Seram holotype

171j *Syntypistis kebeae* ♂, W Papua

171k *Syntypistis kebeae* ♂, Buru

171l *Syntypistis kebeae* ♂, Buru

171m *Syntypistis kebeae* ♂, W Papua

171n *Syntypistis kebeae* ♂, W Papua

171o *Syntypistis kebeae* ♂, Biak

171p *Syntypistis kebeae* ♂, Halmahera

171q *Syntypistis kebeae* ♀, W Papua

171r *Syntypistis kebeae* ♀, W Papua

171s *Syntypistis kebeae* ♀, Obi

171t *Syntypistis. kebeae* ♀, Ambon

172f *Syntypistis dasychirinus* ♂, Sulawesi

172g *Syntypistis dasychirinus* ♂, Sulawesi

172h *Syntypistis dasychirinus* ♂, Sulawesi

172i *S Syntypistis dasychirinus* ♀, Peleng

173e *Syntypistis grisescens* ♂, Java

173f *Syntypistis grisescens* ♂, Sumba

173g *Syntypistis grisescens* ♂, Lombok

173h *Syntypistis grisescens* ♂, Lombok

173i *Syntypistis grisescens* ♂, Alor

173j *Syntypistis grisescens* ♀, Bali

173k *Syntypistis grisescens* ♀, Sumbawa

173l *Syntypistis grisescens* ♀, Tanimbar

173m *Syntypistis grisescens* ♂, Lombok

174e *Syntypistis frugilegus* ♂, PNG, Umboi Isl. lectotype

174f *Syntypistis frugilegus* ♂, New Britain

174g *Syntypistis frugilegus* ♂, New Ireland

174h *Syntypistis frugilegus* ♀, New Britain

175b *Syntypistis tisada* ♂, Solomon, holotype

175c *Syntypistis tisada* ♂, Solomon, paratype

176b *Syntypistis mapia* ♂, W Papua, holotype

176c *Syntypistis mapia* ♀, W Papua, paratype

177h *Syntypistis charistera fraseriana* ♂, Malaya

177i *Syntypistis charistera fraseriana* ♀, Sumatra

177j *S. charistera fraseriana* ♂, Sumatra

177k *Syntypistis charistera fraseriana* ♂, Sumatra

177l *Syntypistis charistera fraseriana* ♀, Sumatra

177m *Syntypistis charistera celebensis* ♀, Sulawesi

177n *Syntypistis charistera celebensis* ♀, Sulawesi

177o *S. charistera celebensis* ♀, Peleng

177p *Syntypistis charistera celebensis* ♂, Sulawesi

177q *Syntypistis charistera celebensis* ♂, Sulawesi

177r *Syntypistis charistera celebensis* ♂, Sulawesi

177s *Syntypistis charistera charistera* ♂, Luzon

178h *Syntipistis triplagosus* ♂, W Papua, (holotype *nitidula*)

178i *Syntypistis triplagosus* ♀, W Papua, holotype

178j *Syntipistis triplagosus* ♂, PNG

178k *Syntypistis triplagosus* ♂, W Papua

178l *Syntipistis triplagosus* ♂, W Papua (holotype *testacea*)

178m *Syntypistis triplagosus* ♂, W Papua (holotype *didyma*)

178n *Syntypistis triplagosus* ♂, W Papua (holotype *trivia*)

178o *Syntypistis triplagosus* ♂, PNG

178p *Syntypistis triplagosus* ♂, PNG

178q *Syntypistis triplagosus* ♂, PNG

178r *Syntypistis triplagosus* ♀, W Papua

178s *Syntypistis triplagosus* ♀, PNG

180c *Syntypistis trisospylus* ♂, W Papua, holotype

180d *Syntypistis trisospylus* ♂, W Papua,

180e *Syntypistis trisospylus* ♀, W Papua

180f *Syntypistis trisospylus* ♀, W Papua

181c *Syntypistis famelica*, ♂, W Papua, holotype

181d *Syntypistis famelica* ♂, PNG

181e *Syntypistis famelica* ♂, W Papua

181f *Syntypistis famelica* ♀, W Papua

181g *Syntypistis famelica* ♀, W Papua

182b *Syntypistis quadrivittata*, ♂ W Papua, holotype

182c *Syntypistis quadrivittata*, ♂ W Papua, paratype

182d *S. quadrivittata* ♂, W Papua (holotype *nana*)

182e *Syntypistis quadrivittata* ♂, W Papua

183a *Syntypistis histrionica* ♂, W Papua

184b *Syntipistis dinawa* ♂, W Papua

184c *Syntipistis dinawa* ♀, W Papua

185d *Syntipistis bella* ♂, PNG

185e *Syntypistis bella* ♀, Seram

186e *S.mediogriseus binaia*, ♀, Seram holotype

186f *S. mediogriseus mediogriseus* ♂, W Papua

186g *S. mediogriseus mediogriseus* ♀, W Papua

186h *S. mediogriseus inviridis* ♂, New Britain, holotype

187b *S. leucaspis* ♂, W Papua holotype

187c *S. leucaspis* ♀, W Papua paratype

188c *Syntypistis papuana*, ♂, W Papua holotype

188d *S. papuana*, ♂, W Papua paratype

188e *S.papuana* ♂, W Papua paratype

188f *S.papuana* ♂, W Papua paratype

188g *S. papuana* ♂, W Papua paratype

189c *Syntypistis mixta* ♀, W Papua (HT *leucophaeus*)

189d *Syntypistis mixta* ♀, W Papua

189e *Syntypistis mixta* ♂, W Papua

189f *Syntypistis mixta* ♀, W Papua

190d *Syntypistis chloriolus* ♂, PNG

190e *Syntypistis chloriolus* ♂, PNG

190f *Syntypistis chloriolus* ♀, W Papua lectotype

190g *Syntypistis chloriolus* ♀, W Papua

190h *Syntypistis chloriolus* ♀, PNG (holotype *viridinitens*)

191a *Syntypistis viridigriseus*, ♂, PNG holotype

191b *Syntypistis viridigriseus*, ♀, PNG paratype

192b *Syntypistis alboviridis*, ♀, PNG lectotype

193c *Syntypistis flavicollis* ♂, W Papua, paratype

193d *Syntypistis flavicollis* ♂, W Papua, (holotype *poecilochroa*)

193e *Syntypistis flavicollis* ♂, W Papua

193f *Syntypistis flavicollis* ♀, W Papua, holotype

194c *Syntypistis nitidus* ♂, W Papua, lectotype

194d *Synypistis nitidus* ♂, W Papua

194e *Syntypistis nitidus* ♂, W Papua

194f *Syntypistis nitidus* ♀, W Papua

195c *Syntypistis gemmea* ♂, Sumatra, holotype

195d *Syntypistis gemmea* ♂, Sumatra, paratype

195e *Syntypistis gemmea* ♀, Sumatra, paratype

196b *S. purpurascens* ♂, W Papua

196c *Syntypistis purpurascens* ♂, W Papua, lectotype

196d *Syntypistis purpurascens* ♀, W Papua

197c *Syntypistis triguttata* ♂, Kalimantan

197d *Syntypistis triguttata* ♂, Borneo

197e *Syntypistis triguttata* ♀, Borneo

198d *Syntypistis variegata* ♀, Sumatra

198e *Syntypistis variegata* ♂, Sumatra

198f *Syntypistis variegata* ♀, Sumatra

199b *Syntypistis uskwara* ♂, W Papua

199c *Syntypistis uskwara* ♂, W Papua

199d *Syntypistis uskwara* ♂, W Papua

199e *Syntypistis uskwara* ♀, W Papua

200d *Syntypistis miyanoi* ♂, Sulawesi

200e *Syntypistis miyanoi* ♂, Sulawesi

200f *Syntypistit miyanoi* ♀, Sulawesi

200g *Syntypistis miyanoi* ♂, Solayar Isl.

201e *Syntypistis cognata* ♂, Obi, holotype

201f *Syntypistis cognata* ♂, Bacan, paratype

201g *Syntypistis cognata* ♀, Halmahera, paratype

202b *Syntypistis sula* ♂, Sulawesi

203d *Syntypistis paukstadtorum* ♂, Flores, holotype

203e *Syntypistis paukstadtorum* ♂, Flores, paratype

203f *Syntypistis paukstadtorum* ♀, Flores, paratype

204d *Syntypistis umbrosa umbrosa*, ♂, Kalimantan

204e *Syntypistis umbrosa umbrosa* ♀, Myanmar

204f *Syntypistis umbrosa umbrosa*, ♂, Borneo

205c *Syntypistis garuda* ♂, Sumatra, holotype

205d *Syntypistis garuda* ♂, Sumatra

205e *Syntypistis garuda* ♀, Sumatra

206c *Syntypistis kiriakoffi* ♂, Sulawesi

206d *Syntypistis kiriakoffi* ♂, Sulawesi

206e *Syntypistis kiriakoffi* ♀, Sulawesi

207g *Synypistis palladina viridimargo* ♂, Borneo

207h *Syntypistis palladina viridimargo* ♀, Borneo

207i *Syntypistis palladina sasak* ♂, Lombok, holotype

207j *Syntypistis palladina sasak* ♂, Bali, paratype

207k *Syntypistis palladina sasak* ♀, Bali, paratype

207l *Syntypistis palladina sasak* ♀, Lombok, paratype

207m *Syntypistis palladina sasak*, ♀, Java paratype

207n *Syntypistis palladina palladina*, ♂, Mindanao lectotype

208e *Syntyistis comatus comatus* ♂, Sumatra

208f *Syntyistis comatus comatus* ♂, Kalimantan

208g *Syntyistis comatus comatus* ♀, Kalimantan

208h *Syntypistis comatus majapahit*, ♀, Java paratype

208i *Syntypistis comatus majapahit* ♂, Java, holotype

208j *Syntypistis comatus majapahit* ♂, Java, paratype

209c *Syntypistis trioculata* ♀, Borneo

209d *Syntypistis trioculata* ♀, Borneo

210d *Syntypistis viridipicta* ♂, Sumatra

210e *Syntypistis viridipicta* ♂, Sumatra

210f *Syntypistis viridipicta* ♂, Sumatra

210g *Syntypistis viridipicta* ♀, Sumatra

213h *Syntypistis viridipicta* ♀, Sumatra

211c *Syntypistis stanleyi* ♂, Sumatra, holotype

211d *Syntypistis stanleyi* ♂, Borneo, paratype

212d *Syntypistis pallidifascia pallidifascia,* ♀, Sumatra

212e *Syntypistis pallidifascia pallidifascia,* ♂, Java

212f *Syntypistis pallidifascia pallidifascia,* ♀, Belitung

212g *Syntypistis pallidifascia germana,* ♂, W Papua

212h *S. pallidifascia germana,* ♀, W Papua

213c *Syntypistis vaneeckei* ♂, Sulawesi, holotype

213d *Syntypistis vaneeckei* ♂, Sulawesi, paratype

213e *Syntypistis vaneeckei* ♂, Sulawesi, paratype

214c *Syntypistis nigribasalis tropica,* ♂, Sumatra paratype

214d *Synypistis nigribasalis tropica,* ♀, Sumatra

215c *Syntypistis alleni* ♂, China, Guangxi

215d *Syntypistis alleni* ♂, Borneo, paratype

215e *Syntypistis alleni* ♀, Borneo, paratype

216d *Parasinga lichenina lichenina,* ♂, Belitung

216e *Parasinga lichenina lichenina,* ♂, Belitung

216f *Parasinga lichenina lichenina,* ♂, Belitung

216g *Parasinga lichenina lichenina,* ♂, Belitung

216h *Parasinga lichenina lichenina* ♀, Sumatra

216i *Parasinga lichenina lichenina* ♀, Sumatra

217c *Parasinga muliercula* ♂, Sulawesi, holotype

217d *Parasinga muliercula* ♀, Sulawesi, paratype

218c *Parasinga cinerascens* ♂, Sumatra

218d *Parasinga cinerascens* ♂, Borneo

218e *Parasinga cinerascens* ♂, Sumatra

218f *Parasinga cinerascens* ♂, Nias

218g *Parasinga cinerascens* ♂, Java

218h *Parasinga cinerascens* ♀, Sumatra

218i *Parasinga cinerascens* ♀, Borneo

218j *Parasinga cinerascens* ♀, Brunei

220c *Parasinga insufficiens* ♂, Kalimantan

220d *Parasinga insufficiens* ♀, Borneo

219b *Parasinga harmani* ♂, Brunei, holotype

219c *Parasinga harmani* ♂, Sumatra

219d *Parasinga harmani* ♀, Sumatra

221A *Cascera muscosa* ♀, Australia, Qld.

221b *Cascera marginata* ♂, W Papua, holotype

221c *Cascera marginata* ♀, W Papua

221d *Cascera marginata* ♀, W Papua

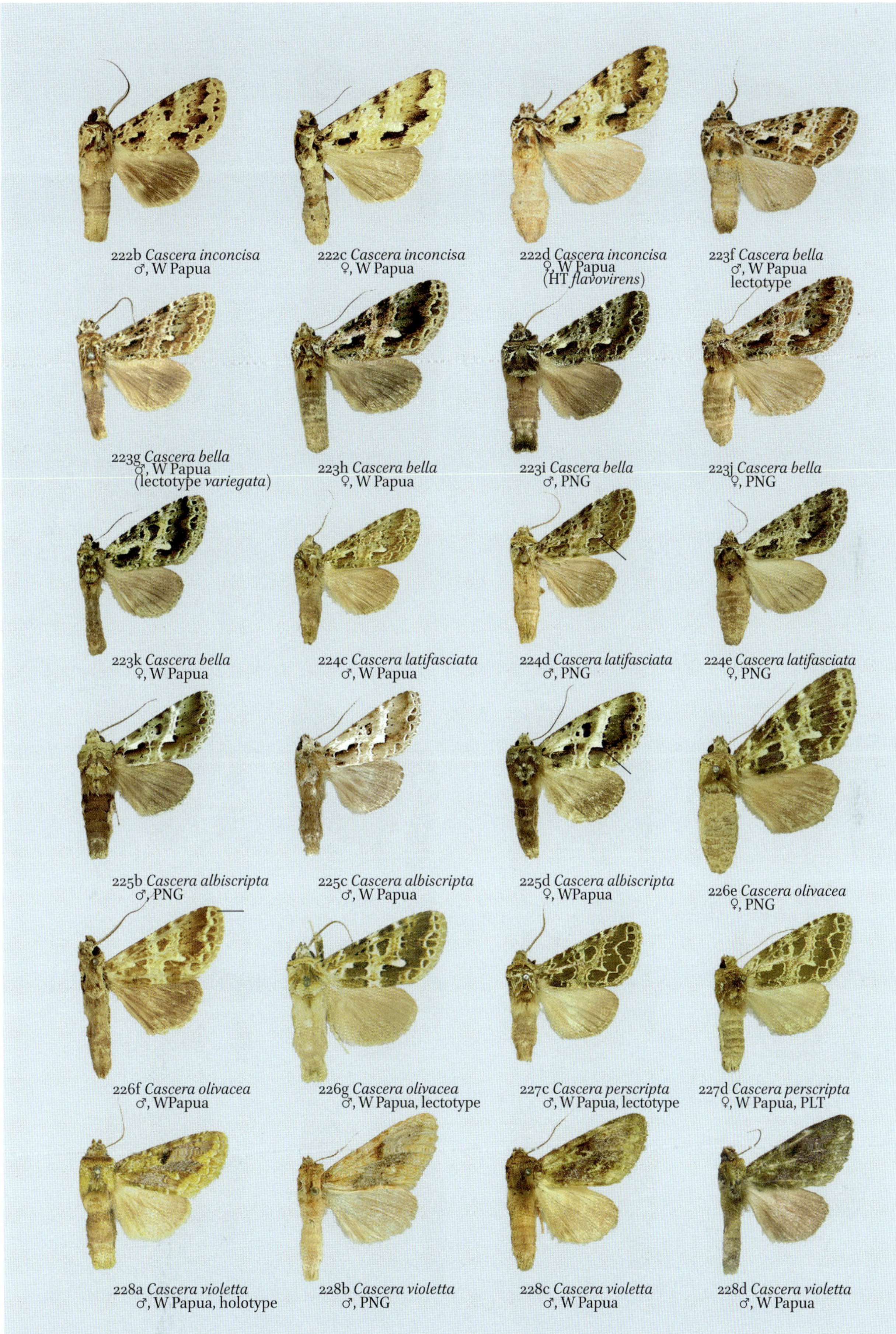

222b *Cascera inconcisa*
♂, W Papua

222c *Cascera inconcisa*
♀, W Papua

222d *Cascera inconcisa*
♀, W Papua
(HT *flavovirens*)

223f *Cascera bella*
♂, W Papua
lectotype

223g *Cascera bella*
♂, W Papua
(lectotype *variegata*)

223h *Cascera bella*
♀, W Papua

223i *Cascera bella*
♂, PNG

223j *Cascera bella*
♀, PNG

223k *Cascera bella*
♀, W Papua

224c *Cascera latifasciata*
♂, W Papua

224d *Cascera latifasciata*
♂, PNG

224e *Cascera latifasciata*
♀, PNG

225b *Cascera albiscripta*
♂, PNG

225c *Cascera albiscripta*
♂, W Papua

225d *Cascera albiscripta*
♀, WPapua

226e *Cascera olivacea*
♀, PNG

226f *Cascera olivacea*
♂, WPapua

226g *Cascera olivacea*
♂, W Papua, lectotype

227c *Cascera perscripta*
♂, W Papua, lectotype

227d *Cascera perscripta*
♀, W Papua, PLT

228a *Cascera violetta*
♂, W Papua, holotype

228b *Cascera violetta*
♂, PNG

228c *Cascera violetta*
♂, W Papua

228d *Cascera violetta*
♂, W Papua

229b *Cascera callima* ♂, PNG

229c *Cascera callima* ♂, W Papua

229d *Cascera callima* ♀, W Papua

230e *Omichlis dimorpha* ♂, Sumatra, paratype

230f *Omichlis dimorpha* ♂, Sumatra, paratype

230g *Omichlis dimorpha* ♂, Sumatra

230h *Omichlis dimorpha* ♀, Kalimantan

230i *Omichlis dimorpha* ♀, Sumatra

231b *Omichlis bifurcata* ♂, Java, holotype

231c *Omichlis bifurcata* ♂, Java, paratype

232c *Omichlis polymorpha* ♀, Sulawesi, holotype

232d *Omichlis polymorpha* ♀, Sulawesi, paratype

232e *Omichlis polymorpha* ♂, Sulawesi, paratype

232f *Omichlis polymorpha* ♂, Sulawesi, paratype

232g *Omichlis polymorpha* ♀, Sulawesi, paratype

232h *Omichlis polymorpha* ♂, Sulawesi, paratype

233c *Omichlis dubia* ♂, Seram, holotype

233d *Omichlis dubia* ♂, Seram, paratype

233e *Omichlis dubia* ♂, Seram, Paratype

234d *Omichlis griseola* ♂, PNG

234e *Omichlis griseola* ♂, PNG, holotype

234f *Omichlis griseola* ♂, PNG

234g *Omichlis griseola* ♀, PNG

234h *Omichlis griseola* ♂, W Papua

235c *Omichlis renata* ♂, New Britain

235d *Omichlis renata* ♂, New Britain

235e *Omichlis renata* ♀, New Britain, holotype

235f *Omichlis renata* ♀, New Britain

236i *Omichlis pratti pratti* ♂, W Papua

236j *Omichlis pratti pratti* ♂, W Papua

236k *Omichlis pratti pratti* ♂, W Papua (holotype *parallelistriga*)

236l *Omichlis pratti pratti* ♀, PNG, Rossel I.

236m *O. pratti pratti* ♂, PNG, Fergusson I.

236n *Omichlis pratti pratti* ♂, W Papua

236o *O. pratti pratti* ♂, Australia-Qld.

236p *Omichlis pratti pratti* ♀, Australia-Qld.

236q *Omichlis pratti falcigera* ♂, Halmahera, holotype

236r *O. pratti falcigera* ♂, Halmahera, PT

236s *O. pratti falcigera* ♀, Halmahera, paratype

237c, *Omichlis erythra* ♂, PNG

237d, *Omichlis erythra* ♂, W Papua

237e, *Omichlis erythra* ♀, W Papua

238b *O. paragriseola* ♂, W Papua, holotype

238c *Omichlis paragriseola* ♀, W Papua, paratype

239c *Omichlis rudloffi* ♂, New Britain, Holotype

239d *Omichlis rudloffi* ♂, New Britain, paratype

240c *Omichlis opacitas* ♂, PNG, holotype

240d *Omichlis opacitas* ♀, W Papua, paratype

241c *Omichlis disiuncta*
♂, Buru, holotype

241d *Omichlis disiuncta*
♀, Buru, paratype

242b *Omichlis priskae*
♂, Tanimbar, holotype

242c *Omichlis priskae*
♂, Tanimbar, paratype

243c, *Omichlis lineata*
♂, W Papua

243d, *Omichlis lineata*
♀, W Papua

244c *Omichlis plagiosa*
♂, W Papua

244d *Omichlis plagiosa*
♂, PNG

244e *Omichlis plagiosa*
♀, W Papua

244b *Omichlis trilineata*
♂, PNG, holotype

245c *Omichlis trilineata*
♂, W Papua

245d *Omichlis trilineata*
♀, W Papua

246b *Omichlis dinawa*
♂, PNG, lectotype

246c *Omichlis dinawa*
♀, PNG

247c *O. ochromacula*
♂, W Papua, holotype

247d *Omichlis ochromacula*
♂, W Papua

248b, *Omichlis baliema*
♂, W Papua, holotype

249d *Omichlis calyptis*
♂, W Papua
(lectotype *hampsoni*)

249e *Omichlis calyptis*
♂, Buru

249f *Omichlis calyptis*
♀, Seram

250b *Omichlis sero*
♂, New Britain, holotype

251e *Omichlis ochracea*
♂, PNG, lectotype

251f *Omichlis ochracea*
♂, W Papua

251g *Omichlis ochracea*
♀, W Papua

252a *Omichlis leucosticta* ♂, PNG, lectotype

252b *Omichlis leucosticta* ♂, W Papua

253c *Omichlis transversa* ♂, W Papua

253d *Omichlis transversa* ♀, PNG, holotype

254c *Omichlis strigata* ♂, PNG

254d *Omichlis strigata* ♂, W Papua

255a *Omichlis geminata* ♂, New Britain, holotype

255a *Omichlis geminata* ♀, New Britain,

256b *Omichlis buergersi* ♂, PNG, holotype

257b *Omichlis albidilinea* ♂, W Papua

258a *Omichlis stueberi* ♂, W Papua, holotype

258b *Omichlis stueberi* ♂, W Papua, paratype

259b *Omichlis superne* ♂, PNG, holotype

260c *Omichlis plagata* ♂, W Papua

260d *Omichlis plagata* ♂, PNG

260e *Omichlis plagata* ♀, PNG, holotype

261c *Omichlis rufofasciata* ♂, PNG, lectotype

261d *O. rufofasciata* ♀, W Papua

262c *Omichlis pseudolibatrix* ♂, W Papua, Yapen

262d *Omichlis pseudolibatrix* ♂, W Papua

262e *Omichlis pseudolibatrix* ♀, W Papua, lectotype

263b *O. mediofascata* ♂, W Papua

263c *Omichlis mediofascata* ♂, PNG, lectotype

263d *Omichlis mediofascata* ♀, W Papua

264b *Omestia bella*
♂, W Papua

264c *Omestia bella*
♂, PNG, holotype

264d *Omestia bella*
♀, PNG
(holotype *plusiotis*)

264e *Omestia bella*
♀, W Papua,
(paratype *plusiotis*)

265d *Netria viridescens viridescens,* ♂, Java

265e *Netria viridescens viridescens,* ♂, Kalimantan

265f *Netria viridescens viridescens,* ♀, Sumatra

265g *Netria viridescens viridescens,* ♂, Belitung

266d *Netria torajae*
♂, Sulawesi, holotype

266e *Netria torajae*
♀, Peleng, paratype

267e *Netria longisacci*
♂, Alor

267f *Netria longisacci*
♂, Alor

267g *Netria longisacci*
♀, Sumbawa, paratype

268d *Netria insularis*
♂, Tanimbar

268e *Netria insularis*
♀, Tanimbar

269k *Netria arisemna arisemna*
♀, Australia, Qsd.

269l *Netria arisemna arisemna*
♂, Australia, Qsd.

269m *Netria arisemna maiorvirens*
♂, W Papua, holotype

269n *Netria arisemna maiorvirens*
♂, W Papua, paratype

269o *Netria arisemna maiorvirens*
♀, W Papua, paratype

269p *Netria arisemna tarasovi*
♀, Halmahera,

269q *Netria arisemna andreasmuelleri*, ♂, Buru

269r *Netria arisemna tarasovi*
♂, Halmahera, holotype

269s *Netria arisemna tarasovi*
♂, Obi

269t *Netria arisemna bipartita*
♀, Sulawesi, paratype

269u *Netria arisemna bipartita*
♂, Sulawesi, holotype

270d *Netria jakli*
♂, Java, holotype

270e *Netria jakli*
♀, Java, paratype

270f *Netria jakli*
♀, Bali

271d *Netria carentis*
♂, Sumatra, holotype

271e *Netria carentis*
♂, Belitung

272d *Netria multispinae multispinae*
♂, Sumatra, paratype

271f *Netria carentis*
♀, Belitung

272e *Netria multispinae multispinae*
♀, Sumatra, paratype

273e *Cerasana anceps anceps* ♂, Sumatra

273f *Cerasana anceps anceps* ♂, Borneo

273g *Cerasana anceps anceps* ♀, Sumatra

273h *Cerasana anceps anceps* ♂, Kalimantan

273i *Cerasana anceps ursulae* ♀, Malaya, paratype

274d *Cerasana alleni* ♂, Borneo

274e *Cerasana alleni* ♂, Borneo

274f *Cerasana alleni* ♀, Borneo

275d *Cerasana pagenstecheri pagenstecheri* ♂, Belitung, paratype

275e *Cerasana pagenstecheri pagenstecheri,* ♀, Borneo

275f *C. pagenstecheri pagenstecheri* ♀, Java

276h *Somera viridifusca viridifusca* ♀, India, Meghalaya

276i *Somera viridifusca sumatrana,* ♂, Sumatra, paratype

276j *Somera viridifusca sumatrana* ♀, Borneo

276k *Somera viridifusca javana,* ♀, Java, holotype

276l *Somera viridifusca javana,* ♂, Bali, paratype

276m *Somera viridifusca javana,* ♀, Java, paratype

276n *Somera viridifusca celebica,* ♂, Sulawesi

276o *Somera viridifusca celebica,* ♀, Sulawesi

277c *Somera viriviri* ♂, Sulawesi

278d *Somera virens virens,* ♂, Sumatra

278e *Somera virens virens,* ♂, Kalimantan

278f *Somera virens virens,* ♀, Sumatra

278g *Somera virens watsoni,* ♂, Thailand

279c *Somera brillians* ♂, Sumatra

279d *Somera brillians* ♂, Sumatra

279e *Somera brillians* ♀, Sumatra

280b *Psegmaphora tripunctata,* ♂, Sumatra

280c *Psegmaphora tripunctata,* ♀, Sumatra

281c *Shachia vernalis vernalis,* ♂, Sulawesi

281d *Shachia vernalis infuscata,* ♂, Sumatra

281e *Shachia vernalis infuscata* ♀, Kalimantan

282c *Pheraspis calida* ♂, Tanimbar holotype

282d *Pheraspis calida* ♀, Tanimbar paratype

282Ab *Pheraspis spodea* ♂, Australia, NT

283e *Notodontella ferrifusa* ♂, Thailand

283f *Notodontella ferrifusa* ♀, Sumatra

283g *Notodontella ferrifusa* ♀, Sumatra

284b *Notodontella floresiensis* ♂, Flores, holotype

284c *Notodontella floresiensis* ♀, Flores, paratype

285d *Chloroceramis viridinota* ♂, Sumatra

285e *Chloroceramis viridinota* ♀, Sumatra

286c *Chloroceramis albuala,* ♀, Sulawesi holotype

287c *Medanella subterminalis* ♂, Sumatra

287d *Medanella subterminalis* ♂, Sumatra

288e *Antiphalera sumatrana*
♂, Sumatra, paratype

288f *Antiphalera sumatrana*
♀, Sumatra

289k *Formofentonia orbifer orbifer*
♂, India, Sikkim

289l *Formofentonia orbifer cremea*
♂, Sumatra, holotype

289m *Formofentonia orbifer cremea*
♀, Sumatra, paratype

289n *Formofentonia orbifer sampuraga*
♂, Sulawesi, holotype

289o *Formofentonia orbifer sampuraga*
♀, Sulawesi, paratype

289p *Formofentonia orbifer binansilang*
♂, Mindanao, holotype

289q *Formofentonia orbifer madena*
♂, Java

289r *Formofentonia orbifer madena*
♀, Java

290d *Betashachia angustipennis tropica*
♂, Sumatra

290e *Betashachia angustipennis tropica*
♀, Sumatra

291c *Fentonia sumatrana*
♂, Sumatra

291d *Fentonia sumatrana*
♂, Belitung

291e *Fentonia sumatrana*
♀, Sumatra

292c *Fentonia helena*
♂, Sumatra

292d *Fentonia helena*
♂, Java

292e *Fentonia helena*
♀, Java

293d *Fentonia hollowayi*
♂, Borneo, holotype

293e *Fentonia hollowayi*
♀, Borneo, paratype

294c *Fentonia talboti talboti*, ♀, Borneo

294d *Fentonia talboti ingridae*, ♀, Sumatra

294e *Fentonia talboti ingridae*, ♀, Sumatra

294f *Fentonia talboti ingridae*, ♂, Sumatra

295c *Fentonia bipunctus* ♂, Belitung

295d *Fentonia bipunctus* ♀, Borneo

296j *Neopheosia fasciata fasciata* ♀, NW India, lectotype

296k *Neopheosia fasciata fasciata* ♂, Java

296l *Neopheosia fasciata fasciata* ♂, Sumatra

296m *Neopheosia fasciata fasciata* ♀, Sumatra

296n *Neopheosia fasciata fasciata* ♂, Lombok

296o *Neopheosia fasciata buruana* ♂, Buru, holotype

296p *Neopheosia fasciata buruana* ♂, Buru, paratype

296q *Neopheosia fasciata fasciata* ♀, Sulawesi

297c *Neopheosia albiplaga* ♂, Sumatra

297d *Neopheosia albiplaga* ♂, Sumatra

297e *Neopheosia albiplaga* ♀, Sumatra

298c *Fusadonta albipuncta* ♂, Sumatra

298d *Fusadonta albipuncta* ♀, Sumatra

Synopsis of Taxonomic Changes

Newly Described Taxa

Diehleana **Gen. nov.**
Noctuola **Gen. nov.**
Canatensha **Gen. nov.**
Ambadrana **Gen. nov.**
Postniganda **Gen. nov.**

Dudusa roepkei **spec. nov.**
Tarsolepis adultus **spec. nov.**
Ortholomia gilolo **spec. nov.**
Ortholomia subtilis **spec. nov.**
Polychoa bakami **spec. nov.**
Polychoa haliga **spec. nov.**
Polychoa cheesmanae **spec. nov.**
Polychoa rafinor **spec. nov.**
Lasioceros asudara **spec. nov.**
Lasioceros impurata **spec. nov.**
Phycidopsis segregarius **spec. nov.**
Gargetta costigerana **spec. nov.**
Gargetta brevisocius **spec. nov.**
Gargetta alora **spec. nov.**
Turnaca carnea **spec. nov.**
Tensha pizzicata **spec. nov.**
Ambadra miscera **spec. nov.**
Ambadra batara **spec. nov.**
Ambadra mentawai **spec. nov.**
Ambadrana taruma **spec. nov.**
Niganda pallens **spec. nov.**
Saliocleta beshda **spec. nov.**
Chalepa pendleburyi **spec. nov.**
Chalepa takaloa **spec. nov.**
Neocerura longiquus **spec. nov.**
Kamalia fulva **spec. nov.**
Kamalia solomonica **spec. nov.**
Harpyia kobesi **spec. nov.**
Harpyia koeppeli **spec. nov.**
Teleclita rasilis **spec. nov.**
Pheressaces siegfriedihlei spec. nov.
Stauropus pusilla **spec. nov.**
Stauropus talasea **spec. nov.**
Stauropus vanmastrigti **spec. nov.**
Stauropus caerulea **spec. nov.**
Stauropus pelz **spec. nov.**
Pheressaces siegfriedihlei **spec. nov.**
Syntypistis tisada **spec. nov.**
Syntypistis mapia **spec. nov.**
Syntypistis leucaspis **spec. nov.**
Syntypistis papuana **spec. nov.**
Syntypistis gemmea **spec. nov.**
Syntypistis cognata **spec. nov.**
Syntypistis paukstadtorum **spec. nov.**
Syntypistis stanleyi **spec. nov.**
Syntypistis vaneeckei **spec. nov.**
Parasinga muliercula **spec. nov.**
Omichlis bifurcata **spec. nov.**
Omichlis polymorpha **spec. nov.**
Omichlis dubia **spec. nov.**
Omichlis paragriseola **spec. nov.**
Omichlis opacitas **spec. nov.**
Omichlis rudloffi **spec. nov.**
Omichlis disiuncta **spec. nov.**
Omichlis priskae **spec. nov.**
Omichlis ochromacula **spec. nov.**
Omichlis baliema **spec. nov.**
Omichlis sero **spec. nov.**
Omichlis buergersi **spec. nov.**
Omichlis stueberi **spec. nov.**
Omichlis superne **spec. nov.**
Pheraspis calida **spec. nov.**
Notodontella floresiensis **spec. nov.**
Chloroceramis albuala **spec. nov.**
Dudusa vethi ferruginea **ssp. nov.**
Tarsolepis dinawensis extera **ssp. nov.**
Euhampsonia roepkei sumatrana **ssp. nov.**
Euhampsonia roepkei javana **ssp. nov.**
Euhampsonia roepkei palawana **ssp. nov.**
Gargetta punctatissima tritus **ssp. nov.**
Kamalia malaysiana nigresco **ssp. nov.**
Liparopsis postalbida brevispinus **ssp. nov.**
Teleclita strigata mediocana **ssp. nov.**
Teleclita strigata multifascia **ssp. nov.**
Stauropus viridissimus boeroe **ssp. nov.**
Kakostauropus glaucoviridis seramica **ssp. nov.**
Syntypistis mediogriseus inviridis **ssp. nov.**
Syntypistis mediogriseus binaia **ssp. nov.**
Syntypistis palladina sasak **ssp. nov.**
Syntypistis comatus majapahit **ssp. nov.**
Omichlis pratti falcigera **ssp. nov.**
Netria arisemna maiorvirens **ssp. nov.**
Formofentonia orbifer cremea **ssp. nov.**
Formofentonia orbifer sampuraga **ssp. nov.**
Formofentonia orbifer binansilang **ssp. nov.**
Neopheosia fasciata buruana **ssp. nov.**

 | DOI:10.1163/9789004354258_010

New Synonymies

(The original genus of the description is given)

Pathemerastis turneri melanistis Bethune-Baker, 1916 **syn. nov.** of *Ortholomia moluccana* Felder, 1861
Stictogargetta umbrina Kiriakoff, 1974 **syn. nov.** of *Gargetta curvaria curvaria* Hampson, 1893
Hybambadra speculigera Kiriakoff, 1974 **syn. nov.** of *Turnaca delineivena* Swinhoe, 1894
Ceira pallida Heylaerts, 1891 **syn. nov.** of *Eumeta rafflesi* Moore, 1883
Turnaca rafflesi borneensis Roepke, 1943 **syn. nov.** of *Eumeta rafflesi* Moore, 1883
Turnaca rufescens Gaede, 1930 **syn. nov.** of *Eumeta rafflesi* Moore, 1883
Ambadra arta Kobayashi, 2012 **syn. nov.** of *Ambadra sibena canescens* Schintlmeister & Lourens, 2010
Ambadra angustifolia Schintlmeister, 2013 **syn. nov.** of *Ambadra sibena canescens* Schintlmeister & Lourens, 2010
Niganda strigifascia insularis Kiriakoff, 1960, **syn. nov.** of *Niganda strigifascia* Moore, 1879
Bireta (*Niganda*) *strigifascia coelestis* Kiriakoff, 1962, **syn. nov.** of *Niganda strigifascia* Moore, 1879
Stauropus dubiosus Bethune-Baker, 1904, **syn. nov.** of *Stauropus kebeae* Bethune-Baker, 1904
Syntypistis chloropasta Turner, 1907, **syn. nov.** of *Stauropus kebeae* Bethune-Baker, 1904
Desmeocraera frugilegus form *aeruginosus* Gaede, 1930, **syn. nov.** of *Stauropus kebeae* Bethune-Baker, 1904
Quadricalcarifera ceramensis Kiriakoff, 1967, **syn. nov.** of *Stauropus kebeae* Bethune-Baker, 1904
Quadricalcarifera eichhorni Kiriakoff, 1970, **syn. nov.** of *Stauropus frugilegus* Rothschild, 1917
Quadricalcarifera nitidula Kiriakoff, 1967 **syn. nov.** of *Stauropus triplagosus* Rothschild, 1917
Quadricalcarifera testacea Kiriakoff, 1967 **syn. nov.** of *Stauropus triplagosus* Rothschild, 1917
Quadricalcarifera didyma Kiriakoff, 1967 **syn. nov.** of *Stauropus triplagosus* Rothschild, 1917
Quadricalcarifera trivia Kiriakoff, 1967 **syn. nov.** of *Stauropus triplagosus* Rothschild, 1917
Quadricalcarifera nana (Kiriakoff, 1967) **syn. nov.** of *Desmeocraera quadrivittata* Kiriakoff, 1967
Stauropus leucophaeus Rothschild, 1917 **syn. nov.** of *Stauropus mixta* Bethune-Baker, 1916
Stauropus viridinitens Rothschild, 1917 **syn. nov.** of *Stauropus chloriolus* Joicey & Talbot, 1917
Quadricalcarifera poecilochroa Kiriakoff, 1967 **syn. nov.** of *Stauropus flavicollis* Rothschild, 1917
Stauropus germana Rothschild, 1917 **syn. nov.** of *Stauropus pallidifascia pallidifascia* Hampson, 1893
Syntypistis aspera Kobayashi & Kishida, 2004 **syn. nov.** of *Desmeocraera alleni* Holloway, 1983
Cascera flavovirens Rothschild, 1917 **syn. nov.** of *Cascera inconcisa* Swinhoe, 1892
Cascera irrorata irrorata Rothschild, 1917 **syn. nov.** of *Cascera bella* Bethune-Baker, 1904
Cascera irrorata pallida Rothschild, 1917 **syn. nov.** of *Cascera bella* Bethune-Baker, 1904
Cascera variegata Rothschild, 1917 **syn. nov.** of *Cascera bella* Bethune-Baker, 1904
Cascera albiscripta ***virens*** Rothschild, 1917 **syn. nov.** of *Cascera albiscripta* Rothschild, 1917
Cascera olivacea flavolavata Rothschild, 1917 **syn. nov.** of *Cascera flavolavata* Rothschild, 1917
Cascera archboldiana Kiriakoff, 1967 **syn. nov.** of *Cascera callima* Bethune-Baker, 1916
Omichlis parallelistriga Rothschild, 1915 **syn. nov.** of *Omichlis pratti* Bethune-Baker, 1904
Omichlis hampsoni Bethune-Baker, 1904 **syn. nov.** of *Pydna calyptis* Swinhoe, 1892
Omichlis macrospila Kiriakoff, 1970 **syn. nov.** of *Omichlis ochracea* Bethune-Baker, 1904
Omichlis plusiotis Rothschild, 1917 **syn. nov.** of *Omestia bella* Bethune-Baker, 1908

New Combinations

Diehleana gigantea (Druce, 1909) **comb. nov.** (*Nadata gigantea*)
Diehleana gilesi (Holloway, 1983) **comb. et stat. nov.** (*Euhampsonia gigantea gilesi*)
Canatensha mediofascia (Rothschild, 1917) **comb. nov.** (*Turnaca mediofascia*)
Canatensha nigrodorsalis (Schintlmeister, 1997) **comb. nov.** (*Tensha nigridorsalis*)
Ambadrana punctata (Holloway, 1983) **comb. nov.** (*Turnaca punctata*)
Ambadrana barlowi (Schintlmeister, 1994) **comb. nov.** (*Turnaca barlowi*)
Ambadrana spinifera (Holloway, 1983) **comb. nov.** (*Turnaca spinifera*)
Ambadrana furcata (Dierl, 1984) **comb. nov.** (*Turnaca furcata*)
Postniganda pudens (Schintlmeister, 2007) **comb. nov.** (*Niganda pudens*)

Status Changes

Stigmatophorina hammamelis Mell, 1922 **stat. rev.**
Osica glauca Walker, 1865 **stat. rev.**
Polychoa aeruginosa Schintlmeister, 2013 **stat. nov.**
Phycidopsis sundana **stat. nov.**
Ambadra melanostriga (Schintlmeister, 1994) **spec. rev.**
Harpyia kalisi (Roepke, 1943) **stat. nov.**
Harpyia baibarana (Matsumura, 1927) **spec. rev.**
Teleclita strigata cathana (Schaus, 1928) **stat. nov.**
Teleclita strigata insignifica (Rothschild, 1917) **stat. nov.**
Teleclita strigata cinnamomea (Rothschild, 1917) **stat. nov.**
Teleclita strigata dryinopa (Dodd, 1902) **stat. nov.**
Pheressaces Turner, 1903 **stat. rev.**
Oxoia irrotativiridis triangularis (Gaede, 1930) **stat. nov.**
Stauropus viridissimus habrochlora Turner, 1922 **stat. nov.**
Stauropus virescens obscuratus Van Eecke, 1929 **stat. nov.**
Stauropus elisabethae kiriakoffi Holloway, 1983 **stat. nov.**
Netria arisemna andreasmuelleri Schintlmeister, 2006 **stat.nov.**
Netria arisemna tarasovi Schintlmeister, 2006 **stat. nov.**
Netria arisemna bipartita Schintlmeister, 2006 **stat. nov.**
Chloroceramis Kiriakoff, 1968 **stat. rev.**

Lectotypes Designations

(The original combination and spelling of the description is given)

Dudusa synopla Swinhoe, 1907
Tarsolepis dinawensis Bethune-Baker, 1904
Pseudogargetta diversa Bethune-Baker, 1904
Gargettiana punctatissima Bethune-Baker, 1916
Hoplitis strigata Moore, 1879
Hoplitis insignifica Rothschild, 1917
Stauropus chloriolus Joicey & Talbot, 1917
Stauropus alboviridis Rothschild, 1917
Stauropus nitidus Rothschild, 1917
Cascera bella Bethune-Baker, 1904
Cascera olivacea Rothschild, 1917
Cascera perscripta Rothschild, 1917
Omichlis mediofasciata Rothschild, 1917
Omichlis rufofasciata Bethune-Baker, 1904
Omichlis leucosticta Joicey & Talbot, 1917
Omichlis ochracea Bethune-Baker, 1904
Pheosia fasciata Moore, 1888

References

Bänziger, H. (1972): Biologie der lacriphagen Lepidopteren in Thailand und Malaya – Revue Suisses Zoologie **79**: 1381-1469.

– (1988): The heaviest tear drinkers: Ecology and systematics of new and unusual Notodontid moths – Nat. Hist. Bulletin Siam Society **36**: 17-53.

– (1992): Remarkable new cases of Moths dringking human tears in Thailand (Lepidopterera: Thyatiridae, Sphingidae, Notodontidae) – Natural History Bulletin of the Siam Society **40**: 91-102.

Barlow, H. (1982): An introduction to the moths of South East Asia – The author, Kuala Lumpur, 305 pp., 50 pls.

Bender, R. (1985): Notodontidae von Sumatra – Heterocera Sumatrana **5**, 122 pp., 14 pls., Heterocera Sumatrana Society, Keltern.

– (1988): A new species of *Hyperaeschrella* Strand (Lep., Notodontidae) – Heterocera Sumatrana, Keltern **2** (6): 111-114.

Bender, R. & Bender, E. (1984): 2. Beitrag zur Kenntnis der Notodontidae von Sumatra: Eine neue Art der Gattung *Liparopsis* Kempson (Lep. Notodontidae) – Heterocera Sumatrana, Keltern **2**: 27-28.

Bender, R. and W. Dierl (1977): Kommentiertes Verzeichnis der Notodontidae Sumatras (Lepidoptera) – Zeitschrift Arbeitsgemeinschaft Österreichischer Entomologen **28**: 117-131.

Bender, R. & H. Steiniger (1984): 1. Beitrag zur Kenntnis der Notodontidae von Sumatra: Eine neue Gattung und eine neue Art (Lep., Notodontidae) – Heterocera Sumatrana, Keltern **2**: 26-27.

Bethune-Baker, G. T. (1904): New Lepidoptera from British New Guinea – Novitates Zoologicae **11**: 367-381; pls. 4 – 6.

– (1908): New Heterocera from British New Guinea – Novitates Zoologicae **15**: 175-243.

– (1916): Descriptions of new Species of Lepidoptera – The Annals and Magazine of Natural History including Zoology, Botany, and Geology (8th series) **17**: 378-385.

Bryk, F. (1949): Entomological Results from the Swedish Expedition 1934 to Burma and British India. Lepidoptera: Notodontidae Stephens, Cossidae Newman und Hepialidae Stephens – Arkiv för Zoologi **42** A: 1-51, pl. 1-4.

– (1950): Eine kleine Ausbeute von nordindischen Spinnern – Entomologisk Tidskrift **71** (1): 55-59.

Butler, A. G. (1872): Description of a new Genus and Species of Heterocerous Lepidoptera – The Annales and Magazine of natural History including Zoology, Botany, and Geology (4th series) **10**: 125-126.

– (1880): Descriptions of new Species of Asiatic Lepidoptera Heterocera – The Annales and Magazine of natural History including Zoology, Botany, and Geology (5th series) **6**: 61-69.

– (1881): Descriptions of new Genera and Species of Heterocous Lepidoptera from Japan – Transactions of the Entomological Society of London **29** (1): 1-23; **29** (2): 171-200; **29** (3): 401-426; **29** (4): 579-600.

– (1892): On a collection of Lepidoptera from Sandakan, N. E. Borneo – Proceedings of the Zoological Society of London **1892**: 120–133, pl. 6.

Dierl, W. (1976): Neue Notodontidae (Lep.) aus Asien – Entomologische Zeitschrift, Frankfurt/M. **86**: 83-85.

– (1976): Notizen zur Kenntnis der Gattungsgruppe *Allata* Walker (Lepidoptera, Notodontidae) – Entomologische Zeitschrift, Frankfurt/M. **86**: 209-214.

– (1978): Zwei Neue Notodontidae (Lepidoptera) aus dem Himalaya – Nachrichtenblatt der Bayerischen Entomologen **27**: 71-73.

– (1981): Eine neue *Benbowia*-Art aus Sumatra: *Benbowia elisabethae* sp.n. (Lepidoptera, Notodontidae) – Entomofauna **2**(15): 169-174.

– (1984): Neue Arten von Notodontidae aus Sumatra (Lepidoptera) – Heterocera Sumatrana **2**: 18-19.

Dodd, F. P. (1902): Notes and observations – The Entomologist **35**: 42-43; 193-194.

Draudt, M (1931-1933): Notodontidae. pp. 905-1070. In A. Seitz (1913-1940): Die Großschmetterlinge der Erde. Vol. **6**. Die Amerikanischen Spinner und Schwärmer. Alfred Kernen, Stuttgart, 1327 pp., 198 pls.

Druce, H. (1888): List of the Lepidoptra Heterocera, with Descriptions of the new Species, collected by Mr. C. M. Woodford at Aola, Guadalcanar Island, Salomon Islands – Proceedings of the Zoological Society of London **1888**: 570-580; pl. 29.

– (1901): Descriptions of some new Species of Heterocera – The Annals and Magazine of Natural History including Zoology, Botany, and Geology(7th series) **7**: 74-79.

– (1909): Descriptions of Three new Species of Heterocera from Dutch New Guinea – The Annales and Magazine of natural History including Zoology, Botany, and Geology (8th series) **3**: 74-79.

Dudgeon, G. C. (1898): A Catalogue of the Heterocera of Sikhim and Bhutan Part III – Journal of the Bombay Natural History Society **11**: 624-634, pl. 1.

– (1898): A Catalogue of the Heterocera of Sikhim and Bhutan IV – Journal of the Bombay Natural History Society **12**: 32-42; pl. 1.

Eckweiler, W. (2001): Entomologische Praxis: Schattenfreie Fotografie von Insekten mit Hilfe einer Ringleuchte – Nachrichten entomologischer Verein Apollo, Frankfurt/ Main, Neue Folge, **22** (3): 136.

Eecke, R. van (1929): De Heterocera van Sumatra VII – Zoologische Mededeelingen **12**: 28-175.

– (1930): De Heterocera van Sumatra, eerste deel – E. J. Brill, Leiden 456 pp., 13 pls.

Felder, C. & R. Felder (1874): Heft 4, Heterocera. Bombyces & Sphinges, pls. 75-107, in Felder, R. & A. F. Rogenhofer [1865-1875]: Reise der österreichischen Fregatte Novara um die Erde in den Jahren 1857, 1858, 1859 unter den Beihilfen des Commodore B. von Wüllerstorf-Urbair. Zoologischer Theil, Zweiter Band, Zweiter Abtheilung: Lepidoptera. Atlas – K.-k. Hof- und Staatsdruckerei pp. 1-9, 1-10, 1-20; pls. 1-140.

Gaede, M. (1930): Notodontidae. In A. Seitz (ed.): Die Großschmetterlinge der Erde, Band **10**. Die Indoaustralische Spinner und Schwärmer – A. Kernen Verlag, Stuttgart, 9 pp., 909 pp., vii pp., 109 pls.

– (1933): Notodontidae. In A. Seitz (ed.): Die Großschmetterlinge der Erde, Band **2**. Supplement. Die Palaearktische Spinner und Schwärmer – A. Kernen, Stuttgart, vii pp., 315 pp, iii pp., 16 pls.

– (1934): Einige Drepanidae und Notodontidae (Lepid.) aus Celebes – Bulletin du Musée royal d'Histoire naturelle de Belgique **10** (26): 1-6.

– (1934): In E. Strand (ed.): Lepidopterorum Catalogus pars **59** Notodontidae – 351 pp., Junk, Berlin.

Günther, A. C. L. G. (1906): The history of the collections contained in the Natural History departments of the British Museum **2** Separate historical accounts of the several collections included in the Department of zoology – Trustees of the British Museum, London, 782 pp.

Hämäläinen, M. (2016): Wilhelm Stüber (1877-1942) collector extraordinaire of New Guinean dragonflies, discoverer of the fabulous Sepik blue orchid, tragic victim of war – Agrion 20(2): 68 – 88.

Hampson, G. F. (1893): The Fauna of British India including Ceylon and Burma. Moths **1** – Taylor & Francis, London, 527 pp.

– (1893): Illustrations of typical specimens of Lepidoptera Heterocera in the collection of the British Museum. **IX**. The Macrolepidoptera Heterocera of Ceylon – British Museum, London i–v., 1-182 pp., pls. 157-186.

– (1895): Descriptions of new Heterocera from India – Transactions of the Entomological Society of London **1895**: 277-315.

– (1896): The Fauna of British India including Ceylon and Burma. Moths **4** – Taylor & Francis, London, 494 pp.

– (1897): The Moths of India. Supplementary paper to the volumes in "The fauna of British India." Part 1 – Journal of the Bombay Natural History Society **11**: 277-297.

– (1910): The moths of India. Supplementary paper to the volumes in "The fauna of British India." Series iv. Part 1 – Journal of the Bombay Natural History Society **20**: 83-125.

– (1912): The moths of India. Supplementary paper to the volumes in "The fauna of British India." Series iv. Part 5 – Journal of the Bombay Natural History Society **21**(4): 1222-1272.

Herrich-Schäffer, G. A. W. (1850-1858): Sammlung neuer und wenig bekannter aussereuropäischer Schmetterlinge **1** (1) – G. J. Manz, Regensburg 84 pp., 4, pp.,120 pls.

Holloway, J. D. (1976): Moths of Borneo with special Reference to Mount Kinabalu – The Malayan Nature Society, Kuala Lumpur, 264 pp.

– (1979): A Survey of the Lepidoptera, Biogeography and Ecology of New Caledonia. Series Entomologica **15** – W. Junk, The Hague, xii pp., 588 pp.

– (1982): Appendix. In Barlow, H. (1982): An introduction to the moths of South East Asia – H. Barlow, Kuala Lumpur, 305 pp., 50 pls.

– (1983): The Moths of Borneo **4**. Notodontidae – The Malayan Nature Journal **37**: 1-107; pls. 1-9.

– (1987): Lepidoptera patterns involving Sulawesi: what do they indicate of past geography? – In: Whitmore, T. C. (ed.) Biogeographic Evolution of the Malay Archipelago, pp. 103-118, Clarendon press, Oxford.

– (1989): The Moths of Borneo **12**. Noctuidae, trifine subfamilies: Noctuinae, Heliothinae, Hadeninae, Acronictinae, Amphipyrinae, Agaristinae – The Malayan Nature Journal **42** (2-3): 1-226; pls. 1-8.

– (2011): The Moths of Borneo **2**. Phaudidae, Himantopteridae, Zygaenidae, Complete Checklist, Checklist Notes, Historical Appendix, Index – The Malayan Nature Journal **63** (1-2): 1-548; pls. 1-12.

Holloway, J. D. & R. Bender (1985): Further notes on the Notodontidae of Sumatra, with description of seven new species – Heterocera Sumatrana, Göttingen **5**: 102-112.

Horsfield, T. & F. Moore [1858]: A catalogue of the lepidopterous Insects in the Museum at the East-India House – Vol. **1**: Allan & Co., London, v pp., 278 pp., 11 pp., 12 pls.

Hübner, J. (1819-[1827]): Sammlung exotischer Schmetterlinge – Band **2** – Im Verlag des Hübnerschen Werkes bey C. Geyer, Augsburg, Selbstverlag, 225 pls.

Joannis, J. de (1929): Lépidoptères Hétérocères du Tonkin 2e partie – Annales de la Société d'entomologique de France **98**: 361-552.

Joicey, J. J., A. Noakes & G. Talbot [1916]: New Lepidoptera from Dutch New Guinea – Transactions of the Entomological Society of London **63** (3-4): 361- 386; pls. 55-62 (1915).

Joicey J. J. & G. Talbot (1915): New Heterocera from Dutch New Guinea – The Annals and Magazine of Natural History including Zoology, Botany, and Geology (8th series) **15**: 295-301; pl. 12.

– (1917): New Heterocera from Dutch New Guinea – The Annals and Magazine of Natural History including Zoology, Botany, and Geology (8th series) **20**: 50-87; pls. 1-4.

Jordan, K. (1898): Contributions to the morphology of Lepidoptera – Novitates Zoologicae **5**: 374–415.

Kiriakoff, S. G. (1959): Entomological results of the Swedish expedition 1934 to Burma and British India. Lepidoptera: Family Notodontidae – Arkiv för Zoologi Ser. 2 Band **12** (20): 313-333+1 pl.

– (1960): Notes sur les Notodontidae (Lepidoptera) *Pydna* Walker et genres voisins – Bulletin & Annales de la Société Royale d'Entomologie de Belgique **97**: 99-117

– (1960): Les Notodontidae (Lepidoptera) de la collection van Delden – Bulletin Institut royal des Sciences naturelles de Belgique **36** (17): 1-12.

– (1962): Notes sur les Notodontidae (Lepidoptera) *Pydna* Walker et genres voisins (Deuxieme partie) – Bulletin et Annales de la Société d'Entomologie de Belgique **98**: 149-214, 6 pls.

– (1963): The systematic position of the so-called Subfamily Platychasmatinae Nakamura, 1956 (Notodontidae) – Journal of the Lepidopterists' Society **17**: 33-34.

– (1967): New genera and species of Oriental Notodontidae (Lepidoptera) – Tijdschrift voor Entomologie **110**: 37-64.

– (1967): Lepidoptera Familia Notodontidae Pars secunda Genera Palaearctica. In: P. Wytsman (ed.) Genera Insectorum fasc. **217 B**. – Kraainem, 238 pp., 7 pls.

– (1967): On the Notodontidae (Lepidoptera) from New Guinea in the Leiden Museum – Zoologische Mededelingen **42**: 189-214.

– (1968): Lepidoptera Familia Notodontidae Pars tertia Genera Indo-Australica In P. Wytsman (ed.): Genera Insectorum fasc. **217 C** – Kraainem 269 pp., 11 pls.

– (1970): New or less known Indo-Australian Notodontidae (Lepidoptera) – Tijdschrift voor Entomologie **113**: 105-123.

– (1974): Neue und wenig bekannte asiatische Notodontidae (Lepidoptera) – Veröffentlichungen der Zoologischen Staatssammung München **17**: 371-421, 5 pls.

– (1977): Neue asiatische Notodontidae (Lepidoptera) nebst Beschreibung zweier Neallotypen – Mitteilungen der Münchner Entomologischen Gesellschaft **66**: 29-35.

Kitching, I. J. & Rawlins, J. E. (1999): The Noctuoidea. pp. 355-401. In Kristensen, N. P. (ed.): Lepidoptera, Moths and Butterflies. 1. Evolution, Systematics, and Biogeography. In Fischer, M (Ed.): Handbuch der Zoologie 4. Aerthropoda: Insecta (35) – de Gruyter, Berlin, New York. 491 pp.

Kobayashi, H. (2008): A new species of the genus *Tarsolepis* from Sulawesi (Lepidoptera, Notodontidae) – Tinea **21** (4): 171-176.

– (2011): Some new Notodontidae species in Southeast Asia (Lepidoptera, Notodontidae) – Tinea **21**(5): 288-312.

Kobayashi, H. & Y. Kishida (2004): New species of the genus *Saliocleta* (Lepidoptera, Notodontidae) from Myanmar and Vietnam – Transactions of the lepidopterological Society of Japan **55** (2): 83-87.

Kobayashi, H., Y. Kishida & M. Wang (2004): Three new species of *Syntypistis* (Notodontidae, Lepidoptera) from Guangdong, China, and on *Syntypistis synechochlora* (Kiriakoff, 1963), stat. rev. – Tinea **18**: 199-208.

– (2007): Several new species of the genus *Phalera* (Notodontidae, Lepidoptera) – Tinea **19**: 274-292.

Kobayashi, H. & Y. Kishida (2004): New species of the genus *Saliocleta* (Lepidoptera, Notodontidae) from Myanmar and Vietnam – Transactions of the lepidopterological Society of Japan **55**: 83-87.

Kobayashi, H. & M. Nonaka (2016): Molecular phylogeny of the Notodontidae: Subfamilies inferred from 258rRnA sequences (Lepidoptera, Noctuoidea, Notodontidae) – Tinea **23** (Supplement 1): pp. 1-83.

Leech, J. H. (1989): On the Lepidoptera of Japan and Corea. Part ii Heterocera Sect. I – Proceedings of the Zoological Society London **1889**: 474-571, pls 50-53.

Matsumura, S. (1919): [New species of the Notodontidae from Japan] – Zoological Magazine, Tokyo **31**: 74-80 (in Japanese).

– (1920): [New genera and new species of the Notodontidae from Japan] – Zoological Magazine Tokyo **32**: 139-151 (in Japanese).

– (1922): [A critical review to Marumo's paper on the Notodontidae, with descriptions of new species] – Zoological Magazine Tokyo **34**: 517-523 (in Japanese).

– (1925): [On the Formosan Notodontidae] – Zoological Magazine Tokyo **37**: 391-409, pls. 6-7 (in Japanese).

– (1927): New species and subspecies of moths from the Japan Empire – Journal of the College of Hokkaido Imperial University **19**: 1-91, pls. 1-5.

– (1929): Generic revision of the Palaearctic Notodontidae – Insecta matsumurana **4**: 78-93, 1 pl.

Mell, R. (1922): Neue südchinesische Lepidoptera – Deutsche Entomologische Zeitschrift **1922**: pp. 113-129.

Moore, F. (1860): In Horsfield, T. & F. Moore: A catalogue of the lepidopterous Insects in the Museum of Natural History at the East-India House. Vol. **2**: 279-440, 1-6, pls. 13-23, 7A-13A – Allan & Co, London.

– (1879): Heterocera. In Hewitson, W.C. & F. Moore: Descriptions of new Indian lepidopterous insects from the collection of the late Mr. W. S. Atkinson, M.A., F.I.S. &c. Part **1**

– Asiatic Society of Bengal, Calcutta, Taylor & Francis, London, xi pp., 198 pp., 5 pls.

– (1879): Descriptions of new genera and species of Asiatic Lepidoptera Heterocera – Proceedings of the Zoological Society of London **1879**: 387-417; pls. 32-34.

– (1881): Descriptions of new Genera and Species of Asiatic Lepidoptera Heterocera – Proceedings of the Zoological Society of London **1881**: 326-380; pls. 37-38.

– (1882-1883): The Lepidoptera of Ceylon, Vol. 2. – L. Reeve & Co., London, 162 pp., pls. 72-143.

– (1883): Descriptions of new Genera and Species od Asiatic Lepidoptera Heterocera – Proceedings of the Zoological Society of London **1883**: 15-29; pls. 5-6.

Nakamura, M. – (1960): Third note on nomenclature of some Notodontid-species (Lep.) – Tyo to Ga **11**: 34-38.

– (1974): Notodontidae of Eastern Nepal based on the collection of the Lepidopterological research expedition to Nepal Himalaya by the Lepidopterological Society of Japan in 1963 (Lepidoptera) – Tyo to Ga **25**: 115-129.

– (1976): Some Notodontidae from Malaysia and Burnei (Lepidoptera) – Tyo to Ga **27**: 35-42.

– (1976): Philippine Notodontidae (Lepidoptera) taken by the Entomological survey of the Japan Entomological Academy in Cooperation with Nanzan University and San Carlos University in 1970 – Tyo to Ga **27**: 138-142.

Nakatomi, K. (1980): Note on the Genus *Vaneeickia* Kiriakoff, with Description of a New Species (Lepidoptera: Notodontidae) – Tyo to Ga **31**: 74-80.

Orhant, G. E. R. J. (1995): Description d'un nouveau *Somera* de Malaisie (Lepidoptera Notodontidae) – Bulletin de la Societe Entomologique de Mulhouse **51**: 71-73.

– (1996): A contribution to knowledge of the moth fauna of Malaysia (Lepidoptera Notodontidae) – Tinea **14**: 254-255.

Pagenstecher, A. F. (1894): Beiträge zur Lepidopteren-Fauna des malayischen Archipels. IX. Über javanische Schmetterlinge und einige von der Insel Sumba – Jahrbücher des Nassauischen Vereins für Naturkunde **47**: 25-58; pl. 1.

– (1900): Die Lepidopterenfauna des Bismarck-Archipels. Zweiter Theil: Die Nachtfalter – Zoologica **29**: 1-XX, 2 pls.

Pellinen, M. J. (2017): Description of a new species of *Saliocleta* Walker, 1862 (Lepidoptera, Notodontidae, Spatalinae, Ceirini) and description of *S. notia* Schintlmeister, 1997 female from Thailand – Zootaxa 4353 (3): 584-590.

Pellinen, M. J. & N. Wahlberg (2015): A new species of *Niganda* Moore, 1879 from Thailand, with descriptions of variation in male genitalia and female facies of *N. radialis* Moore (Lepidoptera: Notodontidae, Ceirinae) – Zootaxa **4033** (1): 94–102.

Robinson, G. S. (1975): Macrolepidoptera of Fiji and Rotuma. A taxonomic and biogeographic Study – E. W. Classey, Faringdon, 362 pp., 15 maps, 17 pls.

Roepke, W. (1943): On the genera *Dudusa* Walk. and *Tarsolepis* Butl. in the Dutch East Indies (Lepidopt. Het., Fam. Notodontidae) – Tijdschrift voor Entomologie **86**: 77-83.

– (1943): Remarks on new or little known Indomalayan Moths (Lepid. Heteroc.) – Natuurhistorisch Maandblad **32**: 78-80, 93-94, 102-103.

– (1944): Remarks on new or little known Indomalayan Moths (Lepid. Heteroc.) – Natuurhistorisch Maandblad **33**: 1-5, 19-22, 39-40.

– (1950): Aantekeningen over Synonymy – Entomologische Berichten **13**: 25-28.

– (1951): Lepidoptera Heterocera, gesammelt von Dr. F. Schneider in Sumatra – Mitteilungen der schweizerischen Entomologischen Gesellschaft **24** (4): 377-385.

Rothschild, W. L. (1910): Catalogue of the Arctianae in the Tring Museum, with Notes and Descriptions of New Species – Novitates Zoologicae **17**: 113-188.

Rothschild, W. L. (1912): New Lithosiinae – Novitates Zoologicae **19**: 212–246.

– (1915): Macrolepidoptera of the British Ornithologists' Union and Wollaston Expeditions in the Snow Mountains, Southern Dutch New Guinea, during 1912-1913, including descriptions of some new species from laocalities others than Dutch New Guinea. In: Rothschild & Durrant, Macrolepidoptera of the British Ornithologists' Union and Wollaston Expeditions in the Snow Mountains, Southern Dutch New Guinea – Hazell, Watson & Viney, London & Aylessbury, 182 pp., pls 1-2.

– (1917): On some apparently new Notodontidae – Novitates Zoologicae **24**: 231-264; pls. 3-8.

Schaus, W. (1928): New Moths oft the Family Ceruridae (Notodontidae) in the United States National Museum – Proceedings of the United States National Museum **73** (19): 1-90.

Schintlmeister, A. (1981): Fünf neue Notodontidae aus Sumatra – Atalanta **12**: 285-291.

– (1991): Die Gattung *Stauropus* Germar, 1812 in den Philippinen (Lepidoptera, Notodontidae) – Deutsche Entomologische Zeitschrift, N.F. **38**: 109-118.

– (1994): Check-list of the Notodontidae of Sundaland (excluding Java) with description of new species (Lepidoptera, Notodontidae) – Heterocera Sumatrana, Göttingen **7**: 207-252.

– (1997): Moths of Vietnam with special reference to Mt. Fan-si-pan. Family: Notodontidae – Entomofauna Supplement **9**: 33-248.

– (2002): Das Genus *Paracerura* gen. nov. und seine Arten in der orientalischen Region (Lepidoptera: Notodontidae) – Nachrichten des entomologischer Vereins Apollo, Neue Folge **23** (3): 105-117.

– (2003): Die Gattung *Stauropus* Germar, 1812 (Lepidoptera: Notodontidae) – Nachrichten des entomologischen Vereins Apollo, Neue Folge **24** (3): 97-118.

– (2004): Die Gattung *Cerasana* Walker, 1862 (Lepidoptera: Notodontidae) – Nachrichten entomologischer Verein Apollo, Frankfurt/Main, Neue Folge, **25** (4): 195-204.

– (2006): Die Gattung *Netria* Walker, 1855 (Lepidoptera, Notodontidae) – Nachrichten entomologischer Verein Apollo, Frankfurt/Main, Neue Folge, **27** (1/2): 65-94.

– (2008): Palaearctic Macrolepidoptera vol. **1**. Notodontidae – Apollo Books, Stenstrup, 482 pp., 40 pls.

– (2013): Notodontidae & Oenosandridae (Lepidoptera). World Catalogue of Insects **11** – Brill, Leiden & Boston, 605 pp.

– (2016): An Illustrated Type Catalogue of the Notodontidae in the National Museum of Natural History Washington D.C. (Lepidoptera: Notodontidae) – Proceedings of the Museum Witt Munich **4**: 1-606, Munich -Vilnius.

Schintlmeister, A. & J. H. Lourens (2010): The Philippine Notodontidae (Lepidoptera) – Quadrifina **8**: 1-349.

Schintlmeister, A. & A. Pinratana (2007): Moths of Thailand **5**. Notodontidae – Brothers of St. Gabriel in Thailand, Bangkok, 320 pp. 45 pls.

Snellen, P. C. T. (1892): Achtste Afdeeling. Lepidoptera. In Veth, P. J.: Midden Sumatra. Reizen en onderzoekingen der Sumatra-Expeditie, uitgerust door het Aardrijkskundig Genootschap, 1877-1879. **4** (1) (Laatste Helft): 12-92; pls. 1-5 – E. J. Brill, Leiden.

– (1895): Verzeichnis der Lepidoptera Heterocera von Dr. B. Hagen gesammelt in Deli (Ost-Sumatra) – Deutsche Entomologische Zeitschrift Iris **8**: 121-151.

Sugi, S. – (1987): The Thailand Notodontidae 1 – Tinea **12** (Supplement): 303-307.

– (1994): *Paracyphanta* new genus and *Cyphanta* Walker, two genera in the 'quadrifid' Notodontidae (Lepidoptera) – Tinea **14**: 53-60.

Swinhoe, Ch. (1892): Catalogue of eastern and Australian Lepidoptera Heterocera in the Collection of the Oxford University Museum Part **1** Sphinges and Bombyces – Claredon Press, Oxford, viii pp., 324 pp., 8 pls.

– (1894): A list of the Lepidoptera of the Khasia Hills. Pt. II. –Transactions of the Entomological Society London **1894**: 145-223; pl. 2.

– (1899): New Species of Oriental Lepidoptera – The Annals and Magazine of Natural History including Zoology, Botany, and Geology (7th series) **3**: 102- 116.

– (1903): Descriptions of new Eastern Moths – The Annals and Magazine of Natural History including Zoology, Botany, and Geology (7th series) **12**: 193- 200.

– (1904): New Species of Indo-Australian and African Heterocera – The Annals and Magazine of Natural History including Zoology, Botany, and Geology (7th series) **14**: 131- 134.

– (1904): New species of Eastern and Australian and African Heterocera in the National Collection – Transactions of the Entomological Society of London **52** (1): 139-158.

– (1907): New species of Eastern and African Heterocera – The Annals and Magazine of Natural History including Zoology, Botany, and Geology (7th series) **19**: 201- 208.

Turner, A. J. (1903): Revision of Australian Lepidoptera – Proceedings of the Linnean Society New South Wales **28**: 42-92.

– (1905): Revision of Australian Lepidoptera II – Proceedings of the Linnean Society New South Wales **29**: 832-859.

– (1907): Revision of Australian Lepidoptera. III – Proceedings of the Linnean Society New South Wales **31**: 678-710.

– (1922): Revision of Australian Lepidoptera. Saturniidae, Bombycidae, Eupterotidae, Notodontidae. Proceedings of the Linnean Society of New South Wales **47**: 348-390.

– (1936): New Australian Lepidoptera – Proceedings of the Royal Society of Queensland **47**(4): 25-50.

– (1944): Studies in Australian Lepidoptera – Transactions of the Royal Society of South Australia **68**: 3-17.

Walker, F. (1855): List of specimens of Lepidopterous Insects in the Collection of the British Museum **4**: 776-976 – Edward Newman, London.

– (1855): List of specimens of Lepidopterous Insects in the Collection of the British Museum **5**: 977-1257 – Edward Newman, London.

– (1855): List of specimens of Lepidopterous Insects in the Collection of the British Museum **6**: 1258-1507 – Edward Newman, London.

– (1856): List of the specimens of Lepidopterous Insects in the Collection of the British Museum. Lepidoptera Heterocera. **7**: 1509–1786, Edward Newman, London.

– (1859): Catalogue of the Heterocerous Lepidoptera collected at Singapore by Mr. W. A. Wallace, with Descriptions of New Species – Journal of the Proceedings of the Linnean Society of London. Zoology **3**: 183-196.

– (1862): Catalogue of the Heterocerous Lepidopterous Insects collected at Sarawak, in Borneo, by Mr. A. R. Wallace, with Descriptions of New Species. Journal of the Proceedings of the Linnean Society, Zoology, 6: 82 – 145.

– (1862): List of the specimens of Lepidopterous Insects in the Collection of the British Museum **26**: 708-1120 – Edward Newman, London.

– (1864): Catalogue of the Heterocerous Lepidopterous Insects collected at Sarawak, in Borneo, by Mr. A. R. Wallace, with Descriptions of New Species –Journal and Proceedings of the Linnean Society (Zoology) **7**: 160-198.

– [1865]: List of the specimens of Lepidopterous Insects in the Collection of the British Museum. Supplement: **31**: 1–321 – Edward Newman, London, (1864).
– (1866): List of specimens of Lepidopterous Insects in the Collection of the British Museum **35**: 1535-2040 – Edward Newman, London.
– (1869): Characters of undescribed species of Heterocerous Lepidoptera – E. W. Janson, London, 112 pp.
Warren, W. (1903): New Moths from British New Guinea – Novitates Zoologicae 10: 120-126.
Wileman, A. E. (1910): Some new Lepidoptera-Heterocera from Formosa – The Entomologist **43**: 136-139, 285-291, 285-291, 309-313, 344-349.
– (1914): Some new species of Lepidoptera from Formosa – The Entomologist **47**: 266-268.
– (1914): New species of Heterocera from Formosa – The Entomologist **47**: 318-323.

Index of Scientific Names